Lecture Notes in Computer Science 10803

Advanced Research in Computing and Software Science

Subline of Lecture Notes in Computer Science

Christel Baier · Ugo Dal Lago (Eds.)

Foundations of Software Science and Computation Structures

21st International Conference, FOSSACS 2018 Held as Part of the European Joint Conferences on Theory and Practice of Software, ETAPS 2018 Thessaloniki, Greece, April 14–20, 2018 Proceedings

Springer Open

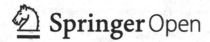

Editors
Christel Baier
TU Dresden
Dresden
Germany

Ugo Dal Lago
Università di Bologna
Bologna
Italy

ISSN 0302-9743 ISSN 1611-3349 (electronic)
Lecture Notes in Computer Science
ISBN 978-3-319-89365-5 ISBN 978-3-319-89366-2 (eBook)
https://doi.org/10.1007/978-3-319-89366-2

Library of Congress Control Number: 2018937398

LNCS Sublibrary: SL1 – Theoretical Computer Science and General Issues

Printed on acid-free paper

This Springer imprint is published by the registered company Springer International Publishing AG part of Springer Nature
The registered company address is: Gewerbestrasse 11, 6330 Cham, Switzerland

ETAPS Foreword

Welcome to the proceedings of ETAPS 2018! After a somewhat coldish ETAPS 2017 in Uppsala in the north, ETAPS this year took place in Thessaloniki, Greece. I am happy to announce that this is the first ETAPS with gold open access proceedings. This means that all papers are accessible by anyone for free.

ETAPS 2018 was the 21st instance of the European Joint Conferences on Theory and Practice of Software. ETAPS is an annual federated conference established in 1998, and consists of five conferences: ESOP, FASE, FoSSaCS, TACAS, and POST. Each conference has its own Program Committee (PC) and its own Steering Committee. The conferences cover various aspects of software systems, ranging from theoretical computer science to foundations to programming language developments, analysis tools, formal approaches to software engineering, and security. Organizing these conferences in a coherent, highly synchronized conference program facilitates participation in an exciting event, offering attendees the possibility to meet many researchers working in different directions in the field, and to easily attend talks of different conferences. Before and after the main conference, numerous satellite workshops take place and attract many researchers from all over the globe.

ETAPS 2018 received 479 submissions in total, 144 of which were accepted, yielding an overall acceptance rate of 30%. I thank all the authors for their interest in ETAPS, all the reviewers for their peer reviewing efforts, the PC members for their contributions, and in particular the PC (co-)chairs for their hard work in running this entire intensive process. Last but not least, my congratulations to all authors of the accepted papers!

ETAPS 2018 was enriched by the unifying invited speaker Martin Abadi (Google Brain, USA) and the conference-specific invited speakers (FASE) Pamela Zave (AT & T Labs, USA), (POST) Benjamin C. Pierce (University of Pennsylvania, USA), and (ESOP) Derek Dreyer (Max Planck Institute for Software Systems, Germany). Invited tutorials were provided by Armin Biere (Johannes Kepler University, Linz, Austria) on modern SAT solving and Fabio Somenzi (University of Colorado, Boulder, USA) on hardware verification. My sincere thanks to all these speakers for their inspiring and interesting talks!

ETAPS 2018 took place in Thessaloniki, Greece, and was organised by the Department of Informatics of the Aristotle University of Thessaloniki. The university was founded in 1925 and currently has around 75,000 students; it is the largest university in Greece. ETAPS 2018 was further supported by the following associations and societies: ETAPS e.V., EATCS (European Association for Theoretical Computer Science), EAPLS (European Association for Programming Languages and Systems), and EASST (European Association of Software Science and Technology). The local organization team consisted of Panagiotis Katsaros (general chair), Ioannis Stamelos,

Lefteris Angelis, George Rahonis, Nick Bassiliades, Alexander Chatzigeorgiou, Ezio Bartocci, Simon Bliudze, Emmanouela Stachtiari, Kyriakos Georgiadis, and Petros Stratis (EasyConferences).

The overall planning for ETAPS is the main responsibility of the Steering Committee, and in particular of its Executive Board. The ETAPS Steering Committee consists of an Executive Board and representatives of the individual ETAPS conferences, as well as representatives of EATCS, EAPLS, and EASST. The Executive Board consists of Gilles Barthe (Madrid), Holger Hermanns (Saarbrücken), Joost-Pieter Katoen (chair, Aachen and Twente), Gerald Lüttgen (Bamberg), Vladimiro Sassone (Southampton), Tarmo Uustalu (Tallinn), and Lenore Zuck (Chicago). Other members of the Steering Committee are: Wil van der Aalst (Aachen), Parosh Abdulla (Uppsala), Amal Ahmed (Boston), Christel Baier (Dresden), Lujo Bauer (Pittsburgh), Dirk Beyer (Munich), Mikolaj Bojanczyk (Warsaw), Luis Caires (Lisbon), Jurriaan Hage (Utrecht), Rainer Hähnle (Darmstadt), Reiko Heckel (Leicester), Marieke Huisman (Twente), Panagiotis Katsaros (Thessaloniki), Ralf Küsters (Stuttgart), Ugo Dal Lago (Bologna), Kim G. Larsen (Aalborg), Matteo Maffei (Vienna), Tiziana Margaria (Limerick), Flemming Nielson (Copenhagen), Catuscia Palamidessi (Palaiseau), Andrew M. Pitts (Cambridge), Alessandra Russo (London), Dave Sands (Göteborg), Don Sannella (Edinburgh), Andy Schürr (Darmstadt), Alex Simpson (Ljubljana), Gabriele Taentzer (Marburg), Peter Thiemann (Freiburg), Jan Vitek (Prague), Tomas Vojnar (Brno), and Lijun Zhang (Beijing).

I would like to take this opportunity to thank all speakers, attendees, organizers of the satellite workshops, and Springer for their support. I hope you all enjoy the proceedings of ETAPS 2018. Finally, a big thanks to Panagiotis and his local organization team for all their enormous efforts that led to a fantastic ETAPS in Thessaloniki!

February 2018 Joost-Pieter Katoen

Preface

This volume contains the papers presented at the 21st International Conference on Foundations of Software Science and Computation Structures (FoSSaCS 2018), which was held April 16–19, 2018, in Thessaloniki, Greece. The conference is dedicated to foundational research with a clear significance for software science and brings together research on theories and methods to support the analysis, integration, synthesis, transformation, and verification of programs and software systems.

The program consisted of 31 contributed papers, selected from among 103 submissions. Each submission was reviewed by at least three Program Committee members, with the help of external experts. After a three-day rebuttal phase, the selection was made based on discussions via the EasyChair conference management system, which was also used to assist with the compilation of the proceedings.

We wish to thank all authors who submitted to FoSSaCS 2018, all the Program Committee members for their excellent work, and the external reviewers for their thorough evaluation of the submissions. In addition, we would like to thank the ETAPS organization for providing an excellent environment for FoSSaCS and other conferences and workshops.

March 2018

Christel Baier
Ugo Dal Lago

Preface

Organization

Program Committee

Andreas Abel	Gothenburg University, Sweden
Christel Baier	TU Dresden, Germany
Nathalie Bertrand	Inria, France
Mikolaj Bojanczyk	Warsaw University, Poland
Udi Boker	Interdisciplinary Center (IDC) Herzliya, Israel
Luis Caires	Universidade NOVA de Lisboa, Portugal
Ugo Dal Lago	University of Bologna, Italy
Yuxin Deng	East China Normal University, China
Mariangiola Dezani-Ciancaglini	Università di Torino, Italy
Ichiro Hasuo	National Institute of Informatics, Japan
Radha Jagadeesan	DePaul University, UK
Stefan Kiefer	University of Oxford, UK
Barbara König	Universität Duisburg-Essen, Germany
David Monniaux	CNRS, VERIMAG, France
Andrzej Murawski	The University of Warwick, UK
Joel Ouaknine	Max Planck Institute for Software Systems, Germany
Catuscia Palamidessi	Inria, France
Kirstin Peters	TU Berlin, Germany
Damien Pous	CNRS, ENS Lyon, France
Jean-Francois Raskin	Université Libre de Bruxelles, Belgium
Helmut Seidl	Technical University of Munich, Germany
Alexandra Silva	University College London, UK
Alex Simpson	University of Ljubljana, Slovenia
Jiri Srba	Aalborg University, Denmark
Jean-Marc Talbot	Aix-Marseille Université, France
Christine Tasson	Université Denis Diderot, France
Kazushige Terui	Kyoto University, Japan

Additional Reviewers

Aler Tubella, Andrea
Almagor, Shaull
Asada, Kazuyuki
Atkey, Robert
Bacci, Giorgio
Bacci, Giovanni
Bagnol, Marc
Baldan, Paolo
Basold, Henning
Bavera, Francisco
Beffara, Emmanuel
Benveniste, Albert
Beohar, Harsh
Berardi, Stefano
Bertolissi, Clara
Berwanger, Dietmar
Blondin, Michael
Bocchi, Laura

Boreale, Michele
Boulmé, Sylvain
Bouyer, Patricia
Brazdil, Tomas
Brotherston, James
Brunet, Paul
Bruni, Roberto
Bucchiarone, Antonio
Busatto-Gaston, Damien
Bønneland, Frederik M.
Cabrera, Benjamin
Cadilhac, Michaël
Carayol, Arnaud
Castellan, Simon
Chen, Tzu-Chun
Clouston, Ranald
Cockx, Jesper
Coppo, Mario
Corbineau, Pierre
Cristescu, Ioana
Doumane, Amina
Dubut, Jérémy
Eberhart, Clovis
Emmi, Michael
Enea, Constantin
Enevoldsen, Søren
Enqvist, Sebastian
Exibard, Léo
Falcone, Ylies
Feng, Yuan
Figueira, Diego
Fijalkow, Nathanaël
Fournier, Paulin
Fujii, Soichiro
Galmiche, Didier
Geeraerts, Gilles
Genest, Blaise
Gorogiannis, Nikos
Graham-Lengrand, Stéphane
Grellois, Charles
Haar, Stefan
Haase, Christoph
Halfon, Simon
Hartmann, Nico
Hautem, Quentin
Hirschkoff, Daniel
Hirschowitz, Tom
Hsu, Justin
Huang, Mingzhang
Jacobs, Bart
Jacquemard, Florent
Jansen, Nils
Jaskelioff, Mauro
Jecker, Ismaël
Junges, Sebastian
Kakutani, Yoshihiko
Kanovich, Max
Kaufmann, Isabella
Kerjean, Marie
King, Andy
Klein, Felix
Klin, Bartek
Kołodziejczyk, Leszek
Kretinsky, Jan
Krivine, Jean
Kupke, Clemens
Kutsia, Temur
Küpper, Sebastian
Laarman, Alfons
Laird, Jim
Lanese, Ivan
Lang, Frederic
Lazic, Ranko
Lefaucheux, Engel
Leifer, Matthew
Lepigre, Rodolphe
Letouzey, Pierre
Levy, Paul Blain
Li, Xin
Liang, Hongjin
Licata, Daniel R.
Litak, Tadeusz
Lohrey, Markus
Lombardy, Sylvain
Long, Huan
Luttik, Bas
López, Hugo A.
Mackie, Ian
Madnani, Khushraj
Maggi, Fabrizio Maria
Mallet, Frederic
Maranget, Luc
Markey, Nicolas
Martens, Wim
Mayr, Richard
Mazowiecki, Filip
Mikučionis, Marius
Milius, Stefan
Mio, Matteo
Moggi, Eugenio
Monmege, Benjamin
Muniz, Marco
Nestmann, Uwe
New, Max
Nielsen, Mogens
Nolte, Dennis
Nordvall Forsberg, Fredrik
Nyman, Ulrik
Okudono, Takamasa
Orchard, Dominic
Oualhadj, Youssouf
Padovani, Luca
Panangaden, Prakash
Pang, Jun
Pavlovic, Dusko
Perez, Guillermo
Pitts, Andrew
Plump, Detlef
Pouly, Amaury
Power, John
Pruekprasert, Sasinee
Ramsay, Steven
Regnier, Laurent
Rehak, Vojtech
Roggenbach, Markus
Rot, Jurriaan
Sacerdoti Coen, Claudio
Sammartino, Matteo
Sankur, Ocan
Saurin, Alexis
Schalk, Andrea
Scherer, Gabriel
Schmidt-Schauß, Manfred
Selinger, Peter
Shirmohammadi, Mahsa
Sickert, Salomon

Sighireanu, Mihaela
Sistla, A. Prasad
Sojakova, Kristina
Soloviev, Sergei
Sozeau, Matthieu
Sprunger, David
Strassburger, Lutz
Tang, Qiyi
Torres Vieira, Hugo
Tsuiki, Hideki
Tsukada, Takeshi
Turrini, Andrea
Tzevelekos, Nikos
Valencia, Frank
Valiron, Benoît
van Ditmarsch, Hans
Varacca, Daniele
Vial, Pierre
Vicary, Jamie
Vijayaraghavan, Muralidaran
Villevalois, Didier
Waga, Masaki
Wagner, Christoph
Wojtczak, Dominik
Wolff, Sebastian
Worrell, James
Yamada, Akihisa
Yang, Pengfei
Yoshimizu, Akira
Yu, Tingting
Zimmermann, Martin

Contents

Semantics

Non-angelic Concurrent Game Semantics 3
Simon Castellan, Pierre Clairambault, Jonathan Hayman, and Glynn Winskel

A Trace Semantics for System F Parametric Polymorphism 20
Guilhem Jaber and Nikos Tzevelekos

Categorical Combinatorics for Non Deterministic Strategies on Simple Games 39
Clément Jacq and Paul-André Melliès

A Syntactic View of Computational Adequacy 71
Marco Devesas Campos and Paul Blain Levy

Linearity

A New Linear Logic for Deadlock-Free Session-Typed Processes 91
Ornela Dardha and Simon J. Gay

A Double Category Theoretic Analysis of Graded Linear Exponential Comonads 110
Shin-ya Katsumata

Depending on Session-Typed Processes 128
Bernardo Toninho and Nobuko Yoshida

FabULous Interoperability for ML and a Linear Language 146
Gabriel Scherer, Max New, Nick Rioux, and Amal Ahmed

Concurrency

Automata for True Concurrency Properties 165
Paolo Baldan and Tommaso Padoan

A Theory of Encodings and Expressiveness (Extended Abstract) 183
Rob van Glabbeek

A Framework for Parameterized Monitorability 203
Luca Aceto, Antonis Achilleos, Adrian Francalanza, and Anna Ingólfsdóttir

Logics for Bisimulation and Divergence 221
Xinxin Liu, Tingting Yu, and Wenhui Zhang

Lambda-Calculi and Types

Call-by-Need, Neededness and All That 241
Delia Kesner, Alejandro Ríos, and Andrés Viso

Fitch-Style Modal Lambda Calculi 258
Ranald Clouston

Realizability Interpretation and Normalization of Typed Call-by-Need λ-calculus with Control 276
Étienne Miquey and Hugo Herbelin

Quotient Inductive-Inductive Types 293
Thorsten Altenkirch, Paolo Capriotti, Gabe Dijkstra, Nicolai Kraus, and Fredrik Nordvall Forsberg

Category Theory and Quantum Control

Guarded Traced Categories 313
Sergey Goncharov and Lutz Schröder

Proper Semirings and Proper Convex Functors 331
Ana Sokolova and Harald Woracek

From Symmetric Pattern-Matching to Quantum Control 348
Amr Sabry, Benoît Valiron, and Juliana Kaizer Vizzotto

Quantitative Models

The Complexity of Graph-Based Reductions for Reachability in Markov Decision Processes 367
Stéphane Le Roux and Guillermo A. Pérez

A Hierarchy of Scheduler Classes for Stochastic Automata 384
Pedro R. D'Argenio, Marcus Gerhold, Arnd Hartmanns, and Sean Sedwards

Symbolically Quantifying Response Time in Stochastic Models Using Moments and Semirings 403
Hugo Bazille, Eric Fabre, and Blaise Genest

Comparator Automata in Quantitative Verification 420
Suguman Bansal, Swarat Chaudhuri, and Moshe Y. Vardi

Logics and Equational Theories

Modular Tableaux Calculi for Separation Theories 441
Simon Docherty and David Pym

Differential Calculus with Imprecise Input and Its Logical Framework. 459
Abbas Edalat and Mehrdad Maleki

The Effects of Adding Reachability Predicates in Propositional Separation Logic. 476
Stéphane Demri, Étienne Lozes, and Alessio Mansutti

The Equational Theory of the Natural Join and Inner Union is Decidable. . . . 494
Luigi Santocanale

Graphs and Automata

Minimization of Graph Weighted Models over Circular Strings 513
Guillaume Rabusseau

Games on Graphs with a Public Signal Monitoring 530
Patricia Bouyer

WQO Dichotomy for 3-Graphs. 548
Sławomir Lasota and Radosław Piórkowski

Verifying Higher-Order Functions with Tree Automata 565
Thomas Genet, Timothée Haudebourg, and Thomas Jensen

Author Index 583

Semantics

Non-angelic Concurrent Game Semantics

Simon Castellan[1(✉)], Pierre Clairambault[2], Jonathan Hayman[3], and Glynn Winskel[3]

[1] Imperial College London, London, UK
simon@phis.me
[2] Univ Lyon, CNRS, ENS de Lyon, UCB Lyon 1, LIP, Lyon, France
[3] Computer Laboratory, University of Cambridge, Cambridge, UK

Abstract. The *hiding* operation, crucial in the compositional aspect of game semantics, removes computation paths not leading to observable results. Accordingly, games models are usually biased towards *angelic* non-determinism: diverging branches are forgotten.

We present here new categories of games, not suffering from this bias. In our first category, we achieve this by avoiding hiding altogether; instead morphisms are *uncovered* strategies (with neutral events) up to *weak bisimulation*. Then, we show that by hiding only certain events dubbed *inessential* we can consider strategies up to *isomorphism*, and still get a category – this partial hiding remains sound up to weak bisimulation, so we get a concrete representations of programs (as in standard concurrent games) while avoiding the angelic bias. These techniques are illustrated with an interpretation of affine nondeterministic PCF which is adequate for weak bisimulation; and may, must and fair convergences.

1 Introduction

Game semantics represents programs as strategies for two player games determined by the types. Traditionally, a strategy is simply a collection of execution traces, each presented as a play (a structured sequence of events) on the corresponding game. Beyond giving a compositional framework for the formal semantics of programming languages, game semantics proved exceptionally versatile, providing very precise (often fully abstract) models of a variety of languages and programming features. One of its rightly celebrated achievements is the realisation that combinations of certain effects, such as various notions of state or control, could be characterised via corresponding conditions on strategies (innocence, well bracketing, ...) in a single unifying framework. This led Abramsky to propose the *semantic cube* programme [1], aiming to extend this success to further programming features: concurrency, non-determinism, probabilities, etc...

However, this elegant picture soon showed some limitations. While indeed the basic category of games was successfully extended to deal with concurrency [10,13], non-determinism [11], and probabilities [9] among others, these extensions (although fully abstract) are often incompatible with each other, and really, incompatible as well with the central condition of innocence. Hence a semantic

C. Baier and U. Dal Lago (Eds.): FOSSACS 2018, LNCS 10803, pp. 3–19, 2018.
https://doi.org/10.1007/978-3-319-89366-2_1

hypercube encompassing all these effects remained out of reach. It is only recently that some new progress has been made with the discovery that some of these effects could be reconciled in a more refined, more intensional games framework. For instance, in [6,16] innocence is reconciled with non-determinism, and in [15] with probabilities. In [7], innocence is reconciled with concurrency.

But something is still missing: the works above dealing with non-deterministic innocence consider only *may-convergence*; they ignore execution branches leading to divergence. To some extent this seems to be a fundamental limitation of the game semantics methodology: at the heart of the composition of strategies lies the *hiding* operation that removes unobservable events. Diverging paths, by nature non-observable, are forgotten by hiding. Some models of must-testing do exist for particular languages, notably McCusker and Harmer's model for non-deterministic Idealized Algol [11]; the model works by annotating strategies with *stopping traces*, recording where the program may diverge. But this approach again mixes poorly with other constructions (notably innocence), and more importantly, is tied to may and must equivalences. It is not clear how it could be extended to support richer notions of convergence, such as *fair-testing* [2].

Our aim is to present a basis for non-deterministic game semantics which, besides being compatible with innocence, concurrency, *etc.*, is not biased towards may-testing; it is *non-angelic*. It should not be biased towards must-testing either; it should in fact be *agnostic* with respect to the testing equivalence, and support them all. Clearly, for this purpose it is paramount to remember the non-deterministic branching information; indeed in the absence of that information, notions such as *fair-testing* are lost. In fact, there has been a lot of activity in the past five years or so around games model that *do* observe the branching information. It is a feature of Hirschowitz's work presenting strategies as presheaves or sheaves on certain categories of cospans [12]; of Tsukada and Ong's work on nondeterministic innocence via sheaves [16]; and of our own line of work presenting strategies as certain event structures [5,7,14].

But observing branching information is not sufficient. Of the works mentioned above, those of Tsukada and Ong and our own previous work are still angelic, because they rely on hiding for composition. On the other hand, Hirschowitz's work gets close to achieving our goals; by refraining from hiding altogether, his model constructs an agnostic and precise representation of the operational behaviour of programs, on which he then considers fair-testing. But by not considering hiding he departs from the previous work and methods of game semantics, and from the methodology of denotational semantics. In contrast, we would like an agnostic games model that still has the categorical structure of traditional semantics. A games model with partial hiding was also recently introduced by Yamada [18], albeit for a different purpose: he uses partial hiding to represent normalization steps, whereas we use it to represent fine-grained nondeterminism.

Contributions. In this paper, we present the first category of games and strategies equipped to handle non-determinism, but agnostic with respect to the notion of convergence (including fair convergence). We showcase our model by interpreting $\mathbf{APCF}_+$, an affine variant of non-deterministic PCF: it is the simplest language featuring the phenomena of interest. We show adequacy with respect to may, must and fair convergences. The reader will find in the first author's PhD thesis [3] corresponding results for full non-deterministic PCF (with detailed proofs), and an interpretation of a higher-order language with shared memory concurrency. In [3], the model is proved compatible with our earlier notions of innocence, by establishing a result of full abstraction for may equivalence, for nondeterministic PCF. We have yet to prove full abstraction in the fair and must cases; finite definability does not suffice anymore.

Outline. We begin Sect. 2 by introducing $\mathbf{APCF}_+$. To set the stage, we describe an angelic interpretation of $\mathbf{APCF}_+$ in the category **CG** built in [14] with strategies up to isomorphism, and hint at our two new interpretations. In Sect. 3, starting from the observation that the cause of "angelism" is hiding, we omit it altogether, constructing an *uncovered* variant of our concurrent games, similar to that of Hirschowitz. Despite not hiding, when restricting the location of non-deterministic choices to internal events, we can still obtain a category up to *weak bisimulation.* But weak bisimulation is not perfect: it does not preserve must-testing, and is not easily computed. So in Sect. 4, we reinstate some hiding: we show that by hiding all synchronised events except some dubbed *essential*, we arrive at the best of both worlds. We get an agnostic category of games and strategies *up to isomorphism*, and we prove our adequacy results.

2 Three Interpretations of Affine Nondeterministic PCF

2.1 Syntax of $\mathbf{APCF}_+$

The language $\mathbf{APCF}_+$ extends affine PCF with a nondeterministic boolean choice, `choice`. Its types are $A, B ::= \mathbb{B} \mid A \multimap B$, where $A \multimap B$ represents affine functions from A to B. The following grammar describes terms of $\mathbf{APCF}_+$:

$$M, N ::= x \mid M\,N \mid \lambda x.\,M \mid \mathtt{tt} \mid \mathtt{ff} \mid \mathtt{if}\ M\,N_1\,N_2 \mid \mathtt{choice} \mid \bot$$

Typing rules are standard, we show application and conditionals. As usual, a conditional eliminating to arbitrary types can be defined as syntactic sugar.

$$\frac{\Gamma \vdash M : A \multimap B \qquad \Delta \vdash N : A}{\Gamma, \Delta \vdash M\,N : B} \qquad \frac{\Gamma \vdash M : \mathbb{B} \qquad \Delta \vdash N_1 : \mathbb{B} \qquad \Delta \vdash N_2 : \mathbb{B}}{\Gamma, \Delta \vdash \mathtt{if}\ M\,N_1\,N_2 : \mathbb{B}}$$

The first rule is *multiplicative*: Γ and Δ are disjoint. The operational semantics is that of PCF extended with the (only) two nondeterministic rules $\mathtt{choice} \to \mathtt{tt}$ and $\mathtt{choice} \to \mathtt{ff}$.

2.2 Game Semantics and Event Structures

Game semantics interprets an open program by a strategy, recording the behaviour of the program (Player) against the context (Opponent) in a 2-player game. Usually, the executions recorded are represented as *plays*, *i.e.* linear sequences of computational events called *moves*; a strategy being then a set of such plays. For instance, the nondeterministic boolean would be represented as the (even-prefix closure of the) set of plays $\{\mathrm{q}^- \cdot \mathtt{tt}^+, \mathrm{q}^- \cdot \mathtt{ff}^+\}$ on the game for booleans. In the play $\mathrm{q}^- \cdot \mathtt{tt}^+$, the context starts the computation by asking the value of the program (q^-) and the program replies ($\mathtt{tt}^+$). Polarity indicates the origin (Program (+) or Opponent/Environment (−)) of the event.

Being based on sequences of moves, traditional game semantics handles concurrency via interleavings [10]. In contrast, in concurrent games [14], plays are generalised to partial orders which can express concurrency as a primitive. For instance, the execution of a parallel implementation of **and** against the context ($\mathtt{tt}, \mathtt{tt}$) gives the following partial order:

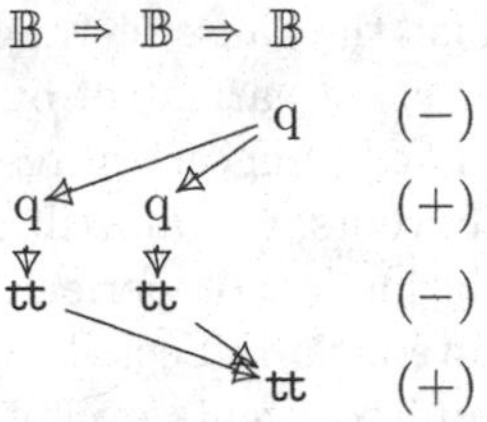

In this picture, the usual chronological linear order is replaced by an explicit partial order representing **causality**. Moves are concurrent when they are incomparable (as the two Player questions here). Following the longstanding convention in game semantics, we show which component of the type a computational event corresponds to by displaying it under the corresponding occurrence of a ground type. For instance in this diagram, Opponent first triggers the computation by asking the output value, and then **and** concurrently evaluates his two arguments. The arguments having evaluated to $\mathtt{tt}$, **and** can finally answer Opponent's initial question and provide the output value.

In [7], we have shown how deterministic pure functional parallel programs can be interpreted (in a *fully abstract* way) using such representations.

Partial-Orders and Non-determinism. To represent nondeterminism in this partial order setting, one possibility is to use sets of partial orders [4]. This representation suffers however from two drawbacks: firstly it forgets the point of non-deterministic branching; secondly, one cannot talk of an *occurrence* of a move independently of an execution. Those issues are solved by moving to *event structures* [17], where the nondeterministic boolean can be represented as:

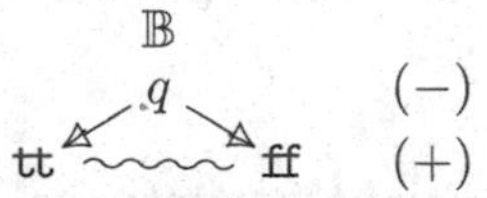

The wiggly line (~~) indicates *conflict*: the boolean values cannot coexist in an execution. Together this forms an *event structure*, defined formally later.

2.3 Interpretations of APCF$_+$ with Event Structures

Let us introduce informally our interpretations by showing which event structures they associate to certain terms of **APCF$_+$**.

Angelic Covered Interpretation. Traditional game semantics interpretations of nondeterminism are angelic (with exceptions, see *e.g.* [11]); they only describe what terms may do, and forget where they might get stuck. The interpretation of $M = (\lambda b.\ \texttt{if}\ b\,\texttt{tt}\,\bot)\ \texttt{choice}$ for instance, in usual game semantics is the same as that of $\texttt{tt}$. This is due to the nature of composition which tends to forget paths that do not lead to a value. Consider the strategy for the function $\lambda b.\ \texttt{if}\ b\,\texttt{tt}\,\bot$:

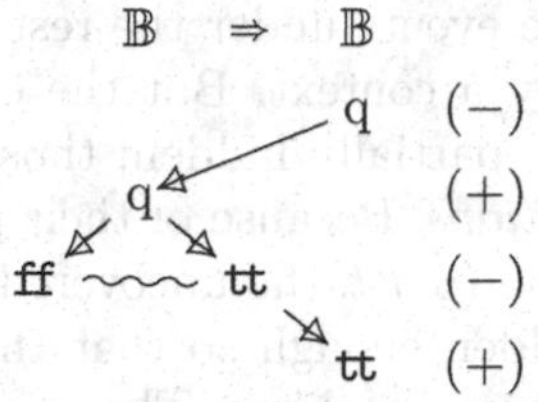

The interpretation of M arises as the *composition* of this strategy with the nondeterministic boolean. Composition is defined in two steps: interaction (Fig. 1a) and then hiding (Fig. 1b). Hiding removes intermediate behaviour which does not correspond to visible actions in the output type of the composition.

Hiding is crucial in order for composition to satisfy basic categorical properties (without it, the identity candidate, copycat, is not even idempotent). Strategies on event structures are usually considered *up to isomorphism*, which is the strongest equivalence relation that makes sense. Without hiding, there is no hope to recover categorical laws up to isomorphism. However, it turns out that, treating events in the middle as τ-transitions ($*$ in Fig. 1a), weak bisimulation equates enough strategies to get a category. Following these ideas, a category of *uncovered* strategies up to *weak bisimilarity* is built in Sect. 3.

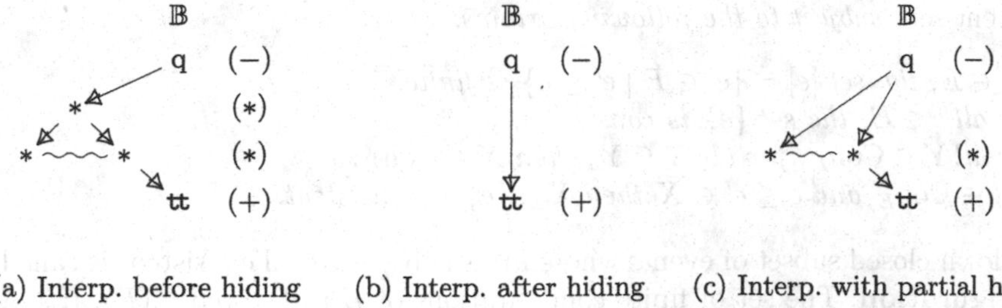

(a) Interp. before hiding (b) Interp. after hiding (c) Interp. with partial hiding

Fig. 1. Three interpretations of $(\lambda b.\ \texttt{if}\ b\,\texttt{tt}\,\bot)\ \texttt{choice}$

Interpretation with Partial Hiding. However, considering uncovered strategies up to weak bisimulation blurs their concrete nature; *causal information* is lost, for instance. Moreover checking for weak bisimilarity is computationally expensive, and because of the absence of hiding, a term evaluating to **skip** may yield a very large representative. However, there is a way to cut down the strategies to reach a compromise between hiding *no* internal events, or hiding *all* of them and collapsing to an angelic interpretation.

In our games based on event structures, having a non-ambiguous notion of an occurrence of event allows us to give a simple definition of the internal events we need to retain (Definition 9). Hiding other internal events yields a strategy still weakly bisimilar to the original (uncovered) strategy, while allowing us to get a category *up to isomorphism*. The interpretation of M in this setting appears in Fig. 1c. As before, only the events under the result type (not labelled $*$) are now *visible*, *i.e.* observable by a context. But the events corresponding to the argument evaluation are only partially hidden; those remaining are considered *internal*, treated like τ*-transitions*. Because of their presence, the partial hiding performed loses no information (*w.r.t.* the uncovered interpretation) up to weak bisimilarity. But we have hidden enough so that the required categorical laws between strategies hold *w.r.t.* isomorphism. The model is more precise and concrete than that of weak bisimilarity, preserves causal information and preserves must-convergence (unlike weak bisimilarity).

Following these ideas, a category of partially covered strategies up to iso (the target of our adequacy results) is constructed in Sect. 4.

3 Uncovered Strategies up to Weak Bisimulation

We now construct a category of "uncovered strategies", up to weak bisimulation. Uncovered strategies are very close to the *partial strategies* of [8], but [8] focused on connections with operational semantics rather than categorical structure.

3.1 Preliminaries on Event Structures

Definition 1. *An* ***event structure*** *is a triple* $(E, \leq_E, \mathrm{Con}_E)$ *where* $(E, \leq_E)$ *is a partial-order and* Con_E *is a non-empty collection of finite subsets of* E *called* consistent sets *subject to the following axioms:*

- *If* $e \in E$*, the set* $[e] = \{e' \in E \mid e' \leq e\}$ *is finite,*
- *For all* $e \in E$*, the set* $\{e\}$ *is consistent,*
- *For all* $Y \in \mathrm{Con}_E$*, for all* $X \subseteq Y$*, then* $X \in \mathrm{Con}_E$*.*
- *If* $X \in \mathrm{Con}_E$ *and* $e \leq e' \in X$ *then* $X \cup \{e\}$ *is consistent.*

A down-closed subset of events whose finite subsets are all consistent is called a **configuration**. The set of finite configurations of E is denoted $\mathscr{C}(E)$. If $x \in \mathscr{C}(E)$ and $e \notin x$, we write $x \overset{e}{-\!\!\subset} x'$ when $x' = x \cup \{e\} \in \mathscr{C}(E)$; this is the **covering relation** between configurations, and we say that e gives an **extension** of x.

Two extensions e and e' of x are **compatible** when $x \cup \{e, e'\} \in \mathscr{C}(E)$, **incompatible** otherwise. In the latter case, we have a **minimal conflict** between e and e' **in context** x (written $e \rightsquigarrow_x e'$).

These event structures are based on *consistent sets* rather than the more commonly-encountered binary *conflict* relation. Consistent sets are more general, and more handy mathematically, but throughout this paper, event structures concretely represented in diagrams will only use *binary conflict*, *i.e.* the relation $e \rightsquigarrow_x e'$ does not depend on x, meaning $e \rightsquigarrow_y e'$ whenever y extends with e, and with e' – in which case we only write $e \rightsquigarrow e'$. Then consistent sets can be recovered as those finite $X \subseteq E$ such that $\neg(e \rightsquigarrow e')$ for all $e, e' \in X$. Our diagrams display the relation $\rightsquigarrow$, along with the *Hasse diagram* of $\leq_E$, called **immediate causality** and denoted by $\rightarrowtriangle_E$. All the diagrams above denote event structures. The missing ingredient in making the diagrams formal is the *names* accompanying the events ($\mathbf{q}, \mathtt{tt}, \mathtt{ff}, \dots$). These will arise as annotations by events from *games*, themselves event structures, representing the types.

The **parallel composition** $E_0 \parallel E_1$ of event structures E_0 and E_1 has for *events* $(\{0\} \times E_0) \cup (\{1\} \times E_1)$. The *causal order* is given by $(i, e) \leq_{E_0 \parallel E_1} (j, e')$ when $i = j$ and $e \leq_{E_i} e'$, and *consistent sets* by those finite subsets of $E_0 \parallel E_1$ that project to consistent sets in both E_0 and E_1.

A **(partial) map of event structures** $f : A \rightharpoonup B$ is a (partial) function on events which *(1)* maps any finite configuration of A to a configuration of B, and *(2)* is locally injective: for $a, a' \in x \in \mathscr{C}(A)$ and $fa = fa'$ (both defined) then $a = a'$. We write $\mathscr{E}$ for the category of event structures and total maps and $\mathscr{E}_\perp$ for the category of event structures and partial maps.

An **event structure with partial polarities** is an event structure A with a map $pol : A \to \{-, +, *\}$ (where events are labelled "negative", "positive", or "internal" respectively). It is a **game** when no events are internal. The dual $A^\perp$ of a game A is obtained by reversing polarities. Parallel composition naturally extends to games. If x and y are configurations of an event structure with partial polarities we use $x \subseteq^p y$ where $p \in \{-, +, *\}$ for $x \subseteq y$ & $pol(y \setminus x) \subseteq \{p\}$.

Given an event structure E and a subset $V \subseteq E$ of events, there is an event structure $E \downarrow V$ whose events are V and causality and consistency are inherited from E. This construction is called the **projection** of E to V and is used in [14] to perform hiding during composition.

3.2 Definition of Uncovered Pre-strategies

As in [14], we first introduce *pre-strategies* and their composition, and then consider *strategies*, those pre-strategies well-behaved with respect to copycat.

Uncovered Pre-strategies. An **uncovered pre-strategy** on a game A is a partial map of event structures $\sigma : S \rightharpoonup A$. Events in the domain of σ are called **visible** or **external**, and events outside **invisible** or **internal**. Via σ, visible events inherit polarities from A.

Uncovered pre-strategies are drawn just like the usual strategies of [14]: the event structure S has its events drawn as their labelling in A if defined or $*$ if undefined. The drawing of Fig. 1a is an example of an uncovered pre-strategy. From an (uncovered) pre-strategy, one can get a pre-strategy in the sense of [14]: for $\sigma : S \rightharpoonup A$, define $S_\downarrow = S \downarrow \mathrm{dom}(\sigma)$ where $\mathrm{dom}(\sigma)$ is the domain of σ. By restriction σ yields $\sigma_\downarrow : S_\downarrow \to A$, called a **covered pre-strategy**. A configuration x of S can be decomposed as the disjoint union $x_\downarrow \cup x_*$ where $x_\downarrow$ is a configuration of $S_\downarrow$ and x_* a set of internal events of S.

A pre-strategy **from a game** A **to a game** B is a (uncovered) pre-strategy on $A^\perp \parallel B$. An important pre-strategy from a game A to itself is the **copycat pre-strategy**. In $A^\perp \parallel A$, each move of A appears twice with dual polarity. The copycat pre-strategy Ϲ_A simply waits for the negative occurrence of a move a before playing the positive occurrence. See [5] for a formal definition.

Isomorphism of strategies [14] can be extended to uncovered pre-strategies:

Definition 2. *Pre-strategies* $\sigma : S \rightharpoonup A, \tau : T \rightharpoonup A$ *are* ***isomorphic*** *(written* $\sigma \cong \tau$*) if there is an iso* $\varphi : S \cong T$ s.t. $\tau \circ \varphi = \sigma$ *(equality of partial maps).*

Interaction of Pre-strategies. Recall that in the covered case, composition is performed first by interaction, then hiding; where interaction of pre-strategies is described as their pullback in the category of *total maps* [14]. Even though $\mathscr{E}_\perp$ has pullbacks, those pullbacks are inadequate to describe interaction. In [8], uncovered strategies are seen as total maps $\sigma : S \to A \parallel N$, and their interaction as a pullback involving these. This method has its awkwardness so, instead, here we give a direct universal construction of interaction, replacing pullbacks.

We start with the simpler case of a **closed** interaction of a pre-strategy $\sigma : S \rightharpoonup A$ against a counter pre-strategy $\tau : T \rightharpoonup A^\perp$. As in [5] we first describe the expected *states* of the closed interaction in terms of *secured bijections*, from which we construct an event structure; before characterising the whole construction via a universal property.

Definition 3 (Secured bijection). *Let* $\mathbf{q}, \mathbf{q}'$ *be partial orders and* $\varphi : \mathbf{q} \simeq \mathbf{q}'$ *be a bijection between the carrier sets (non necessarily order-preserving). It is* ***secured*** *when the following relation* $\lhd_\varphi$ *on the graph of* φ *is acyclic:*

$$(s, \varphi(s)) \lhd_\varphi (s', \varphi(s')) \text{ iff } s \rightarrowtriangle_{\mathbf{q}} s' \vee \varphi(s) \rightarrowtriangle_{\mathbf{q}'} \varphi(s') \tag{1}$$

If so, the resulting partial order $(\lhd_\varphi)^*$ *is written* $\leq_\varphi$.

Let $\sigma : S \rightharpoonup A$ and $\tau : T \rightharpoonup A$ be partial maps of event structures (we dropped polarities, as the construction is completely independent of them). A pair $(x, y) \in \mathscr{C}(S) \times \mathscr{C}(T)$ such that $\sigma_\downarrow x = \tau_\downarrow y \in \mathscr{C}(A)$, induces a bijection $\varphi_{x,y} : x \parallel y_* \simeq x_* \parallel y$ defined by local injectivity of σ and τ:

$$\begin{aligned}
\varphi_{x,y}(0, s) &= (0, s) && (s \in x_*)\\
\varphi_{x,y}(0, s) &= (1, \tau^{-1}(\sigma s)) && (s \in x_\downarrow)\\
\varphi_{x,y}(1, t) &= (1, t)
\end{aligned}$$

The configurations x and y have a partial order inherited from S and T. Viewing y_* and x_* as discrete orders (the ordering relation is the equality), $\varphi_{x,y}$

is a bijection between carrier sets of partial orders. An **interaction state** of σ and τ is $(x, y) \in \mathscr{C}(S) \times \mathscr{C}(T)$ with $\sigma_{\downarrow} x = \tau_{\downarrow} y$ for which $\varphi_{x,y}$ is secured. As a result (the graph of) $\varphi_{x,y}$ is naturally partial ordered. Write $\mathscr{S}_{\sigma,\tau}$ for the set of interaction states of σ and τ. As usual [5], we can recover an event structure:

Definition 4 (Closed interaction of uncovered pre-strategies). *Let A be an event structure, and $\sigma : S \rightharpoonup A$ and $\tau : T \rightharpoonup A$ be partial maps of event structures. The following data defines an event structure $S \wedge T$:*

- events: *those interaction states (x, y) such that $\varphi_{x,y}$ has a top element,*
- causality: $(x, y) \leq_{S \wedge T} (x', y')$ *iff $x \subseteq x'$ and $y \subseteq y'$,*
- consistency: *a finite set of interaction states $X \subseteq S \wedge T$ is consistent iff its union $\bigcup X$ is an interaction state in $\mathscr{S}_{\sigma,\tau}$.*

This event structure comes with partial maps $\Pi_1 : S \wedge T \rightharpoonup S$ and $\Pi_2 : S \wedge T \rightharpoonup T$, analogous to the usual projections of a pullback: for $(x, y) \in S \wedge T$, $\Pi_1(x, y)$ is defined to $s \in S$ whenever the top-element of $\varphi_{x,y}$ is $((0, s), w_2)$ for some $w_2 \in x \,_* \| \, y$. The map Π_1 is undefined only on events of $S \wedge T$ corresponding to internal events of T (*i.e.* (x, y) with top element of $\varphi_{x,y}$ of the form $((1, t), (1, t))$). The map Π_2 is defined symmetrically, and undefined on events corresponding to internal events of S. We write $\sigma \wedge \tau$ for $\sigma \circ \Pi_1 = \tau \circ \Pi_2 : S \wedge T \rightharpoonup A$.

Lemma 1. *Let $\sigma : S \rightharpoonup A$ and $\tau : T \rightharpoonup A$ be partial maps. Let $(X, f : X \rightharpoonup S, g : X \rightharpoonup T)$ be a triple such that the following outer square commutes:*

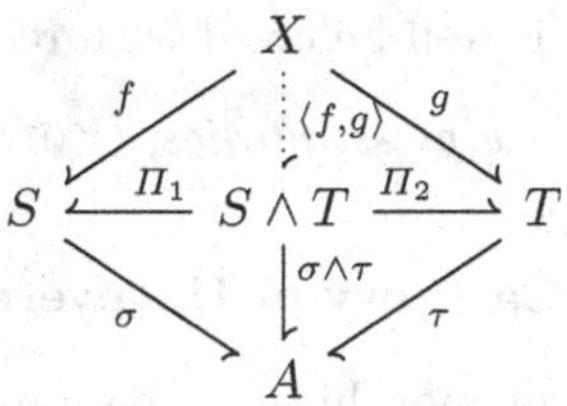

If for all $p \in X$ with $f\,p$ and $g\,p$ defined, $\sigma(f\,p) = \tau(g\,p)$ is defined, then there exists a unique $\langle f, g\rangle : X \rightharpoonup S \wedge T$ making the two upper triangles commute.

From this closed interaction, we define the open interaction as in [14]. Given two pre-strategies $\sigma : S \to A^{\perp} \parallel B$ and $\tau : T \to B^{\perp} \parallel C$, their interaction

$$\tau \circledast \sigma : (S \parallel C) \wedge (A \parallel T) \rightharpoonup A^{\perp} \parallel C$$

is defined as the composite partial map $(S \parallel C) \wedge (A \parallel T) \rightharpoonup A \parallel B \parallel C \rightharpoonup A \parallel C$, where the "pullback" is first computed ignoring polarities – the codomain of the resulting partial map is $A^{\perp} \parallel C$, once we reinstate polarities.

Weak Bisimulation. To compare uncovered pre-strategies, we cannot use isomorphisms as in [14], since as hinted earlier, $\text{æ}_A \circledast \sigma$ comprises synchronised events not corresponding to those in σ. To solve this, we introduce weak bisimulation between uncovered strategies:

Definition 5. *Let $\sigma : S \rightharpoonup A$ and $\tau : T \rightharpoonup A$ be uncovered pre-strategies. A weak bisimulation between σ and τ is a relation $\mathscr{R} \subseteq \mathscr{C}(S) \times \mathscr{C}(T)$ containing $(\emptyset, \emptyset)$, such that for all $x \mathrel{\mathscr{R}} y$, we have:*

- *If $x \overset{s}{\longrightarrow\!\!\!\subset} x'$ such that s is visible, then there exists $y \subseteq^* y' \overset{t}{\longrightarrow\!\!\!\subset} y''$ with $\sigma s = \tau t$ and $x' \mathrel{\mathscr{R}} y''$ (and the symmetric condition for τ)*
- *If $x \overset{s}{\longrightarrow\!\!\!\subset} x'$ such that s is internal, then there exists $y \subseteq^* y'$ such that $x' \mathrel{\mathscr{R}} y'$ (and the symmetric condition for τ)*

Two uncovered pre-strategies σ, τ are weakly bisimilar (written $\sigma \simeq \tau$) when there is a weak bisimulation between them.

Associativity of interaction (up to isomorphism, hence up to weak bisimulation) follows directly from Lemma 1. Moreover, it is straightforward to check that weak bisimulation is a congruence (*i.e.* compatible with composition).

Composition of Covered Strategies. From interaction, we can easily define the composition of covered strategies. If $\sigma : S \to A^\perp \parallel B$ and $\tau : T \to B^\perp \parallel C$ are covered pre-strategies, their composition (in the sense of [14]) $\tau \odot \sigma$ is defined as $(\tau \circledast \sigma)_\downarrow$. The operation $\downarrow$ is well-behaved with respect to interaction:

Lemma 2. *For σ, τ composable pre-strategies, $(\tau \circledast \sigma)_\downarrow \cong \tau_\downarrow \odot \sigma_\downarrow$.*

3.3 A Compact-Closed Category of Uncovered Strategies

Although we have a notion of morphism (pre-strategies) between games and an associative composition, we do not have a category up to weak bisimulation yet. Unlike in [14], races in a game may cause copycat on this game to not be idempotent (see [3] for a counterexample), which is necessary for it to be an identity. To ensure that, we restrict ourselves to **race-free** games: those such that whenever a configuration x can be extended by a_1, a_2 of distinct polarities, the union $x \cup \{a_1, a_2\}$ is consistent. From now on, games are assumed race-free.

Lemma 3. *For a race-free game A, $\text{æ}_A \circledast \text{æ}_A \simeq \text{æ}_A$.*

Proof. It will follow from the forthcoming Lemma 4.

Uncovered Strategies. Finally, we characterise the pre-strategies invariant under composition with copycat. The two ingredients of [5,14], receptivity and courtesy (called *innocence* in [14]) are needed, but this is not enough: we need another condition as witnessed by the following example.

Consider the strategy $\sigma : \oplus_1 \sim\sim \oplus_2$ on the game $A = \oplus_1 \oplus_2$ playing non-deterministically one of the two moves. Then the interaction $\mathrm{æ}_A \circledast \sigma$ is:

$$\begin{array}{ccc} A^* & & A \\ *_1 & \longrightarrow & \oplus_1 \\ \wr & & \\ *_2 & \longrightarrow & \oplus_2 \end{array}$$

It is not weakly bisimilar to σ: $\mathrm{æ}_A \circledast \sigma$ can do $*_1$, an internal transition, to which σ can only respond by not doing anything. Then σ can still do $\oplus_1$ and $\oplus_2$ whereas $\mathrm{æ}_A \circledast \sigma$ cannot: it is committed to doing $\oplus_1$. To solve this problem, we need to force strategies to decide their nondeterministic choices *secretly*, by means of internal events – so σ will not be a valid uncovered strategy, but $\mathrm{æ}_A \circledast \sigma$ will. Indeed, $\mathrm{æ}_A \circledast (\mathrm{æ}_A \circledast \sigma)$ below is indeed weakly bisimilar to $\mathrm{æ}_A \circledast \sigma$.

$$\begin{array}{ccccc} A^* & & A^* & & A \\ *_1 & \longrightarrow & *_1 & \longrightarrow & \oplus_1 \\ \wr & & & & \\ *_2 & \longrightarrow & *_2 & \longrightarrow & \oplus_2 \end{array}$$

Definition 6. *An (uncovered) strategy is a pre-strategy $\sigma : S \rightharpoonup A$ satisfying:*

- receptivity: *if $x \in \mathscr{C}(S)$ is such that $\sigma x \overset{a}{\longrightarrow\!\!\subset}$ with $a \in A$ negative, then there exists a unique $x \overset{s}{\longrightarrow\!\!\subset}$ with $\sigma s = a$.*
- courtesy: *if $s \rightarrowtriangle s'$ and s is positive or s' is negative, then $\sigma s \rightarrowtriangle \sigma s'$.*
- secrecy: *if $x \in \mathscr{C}(S)$ extends with s_1, s_2 but $x \cup \{s_1, s_2\} \notin \mathscr{C}(S)$, then s_1 and s_2 are either both negative, or both internal.*

Receptivity and courtesy are stated exactly as in [14]. As a result, hiding the internal events of an uncovered strategy yields a strategy $\sigma_\downarrow$ in the sense of [14].

For any game A, $\mathrm{æ}_A$ is an uncovered strategy: it satisfies secrecy as its only minimal conflicts are inherited from the game and are between negative events.

The Category $\mathbf{CG}_\circledast$. Our definition of uncovered strategy does imply that copycat is neutral for composition.

Lemma 4. *Let $\sigma : S \rightharpoonup A$ be an uncovered strategy. Then $\mathrm{æ}_A \circledast \sigma \simeq \sigma$.*

The result follows immediately:

Theorem 1. *Race-free games and uncovered strategies up to weak bisimulation form a compact-closed category $\mathbf{CG}_\circledast$.*

3.4 Interpretation of Affine Nondeterministic PCF

From now on, strategies are by default considered uncovered. We sketch the interpretation of $\mathbf{APCF}_+$ inside $\mathbf{CG}_\circledast$. As a compact-closed category, $\mathbf{CG}_\circledast$ supports an interpretation of the linear λ-calculus. However, the empty game 1 is not terminal, as there are no natural transformation $\epsilon_A : A \to 1$ in $\mathbf{CG}_\circledast$.

The negative category $\mathbf{CG}_\circledast^-$. We solve this issue as in [4], by looking at negative strategies and negative games.

Definition 7. *An event structure with partial polarities is* **negative** *when all its minimal events are negative.*

A strategy $\sigma : S \rightharpoonup A$ is negative when S is. Copycat on a negative game is negative, and negative strategies are stable under composition:

Lemma 5. *There is a subcategory* $\boldsymbol{CG}_{\circledast}^{-}$ *of* $\boldsymbol{CG}_{\circledast}$ *consisting in negative race-free games and negative strategies. It inherits a monoidal structure from* $\boldsymbol{CG}$ *in which the unit (the empty game) is terminal.*

Moreover, $\boldsymbol{CG}_{\circledast}^{-}$ has products. The **product** $A \,\&\, B$ of two games A and B, has events, causality, polarities as for $A \parallel B$, but consistent sets restricted to those of the form $\{0\} \times X$ or $\{1\} \times X$ with X consistent in A or B. The **projections** are $\varpi_A : \mathbb{C}\mathbb{C}_A \to (A \,\&\, B)^{\perp} \parallel A$, and $\varpi_B : \mathbb{C}\mathbb{C}_B \to (A \,\&\, B)^{\perp} \parallel B$.

Finally, the **pairing** of negative strategies $\sigma : S \rightharpoonup A^{\perp} \parallel B$ and $\tau : T \to A^{\perp} \parallel C$ is the obvious map $\langle \sigma, \tau \rangle : S \,\&\, T \rightharpoonup A^{\perp} \parallel B \,\&\, C$, and the laws for the cartesian product are direct verifications.

We also need a construction to interpret the function space. However, for A and B negative, $A^{\perp} \parallel B$ is not usually negative. To circumvent this, we introduce a negative variant $A \multimap B$, the linear arrow. To simplify the presentation, we only define it in a special case. A game is **well-opened** when it has at most one initial event. When B is well-opened, we define $A \multimap B$ to be 1 if $B = 1$; and otherwise $A^{\perp} \parallel B$ with the exception that every move in A depends on the single minimal move in B. As a result $\multimap$ preserves negativity. We get:

Lemma 6. *If* B *is well-opened,* $A \multimap B$ *is well-opened and is an exponential object of* A *and* B.

In other words, well-opened games are an exponential ideal in $\boldsymbol{CG}_{\circledast}^{-}$. We interpret types of **APCF**$_+$ inside well-opened games of $\boldsymbol{CG}_{\circledast}^{-}$:

$$\llbracket \mathbf{com} \rrbracket = \begin{array}{c} \mathbf{run}^{-} \\ \downarrow \\ \mathbf{done}^{+} \end{array} \qquad \llbracket \mathbb{B} \rrbracket = \begin{array}{c} q^{-} \\ \swarrow \quad \searrow \\ \mathbf{tt}^{+} \; \sim \; \mathbf{ff}^{+} \end{array} \qquad \llbracket A \multimap B \rrbracket = \llbracket A \rrbracket \multimap \llbracket B \rrbracket$$

Interpretation of Terms. Interpretation of the affine λ-calculus in $\boldsymbol{CG}_{\circledast}^{\circledast}$ follows standard methods. First, the primitives $\mathbf{tt}, \mathbf{ff}, \bot, \mathbf{if}$ are interpreted as:

$$\llbracket \mathbf{tt} \rrbracket : \mathbb{B} \qquad \mathbf{ff} : \mathbb{B} \qquad \llbracket \bot \rrbracket : \mathbb{B} \qquad \mathbf{if} : \mathbb{B} \multimap (\mathbb{B} \,\&\, \mathbb{B}) \multimap \mathbb{B}$$

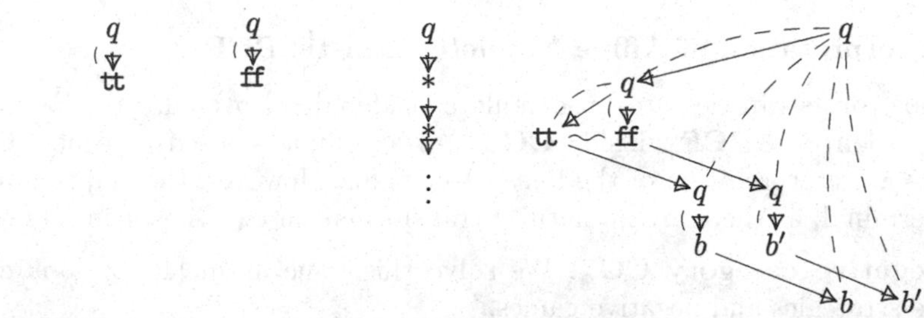

A non-standard point is the interpretation of $\bot$: usually interpreted in game semantics by the minimal strategy simply playing q (as will be done in the next section), our interpretation here reflects the fact that $\bot$ represents an infinite computation that never returns. Conditionals are implemented as usual:

$$[\![\texttt{if}\ M\ N\ N']\!]_{\circledast} = \texttt{if}\ \circledast\ ([\![M]\!]_{\circledast} \parallel \langle [\![N]\!]_{\circledast}, [\![N']\!]_{\circledast}\rangle).$$

Soundness and Adequacy. We now prove adequacy for various notions of convergence. First, we build an uncovered strategy from the operational semantics.

Definition 8 (The operational tree). *Let M be a closed term of type $\mathbb{B}$. We define the pre-strategy $\mathsf{t}(M)$ on $\mathbb{B}$ as follows:*

Events: *An initial event $\bot$ plus one event per derivation $M \to^* M'$.*
Causality: *$\bot$ is below other events, and derivations are ordered by prefix*
Consistency: *A set of events is consistent when its events are comparable.*
Labelling: *$\bot$ has label q, a derivation $M \to^* b$ where $b \in \{\texttt{tt}, \texttt{ff}\}$ is labelled by b. Other derivations are internal.*

As a result, $\mathsf{t}(M)$ is a tree. Our main result of adequacy can now be stated:

Theorem 2. *For a term $\vdash M : \mathbb{B}$, $\mathsf{t}(\texttt{if}\ M\ \texttt{tt}\ \texttt{ff})$ and $[\![M]\!]_{\circledast}$ are weakly bisimilar.*

We need to consider $\mathsf{t}(\texttt{if}\ M\ \texttt{tt}\ \texttt{ff})$ and not simply $\mathsf{t}(M)$ to ensure secrecy. From this theorem, adequacy results for may and fair convergences arise:

Corollary 1. *For any term $\vdash M : \mathbb{B}$, we have:*

May: *$M \to^* \texttt{tt}$ if and only if $[\![M]\!]_{\circledast}$ contains a positive move*
Fair: *For all $M \to^* M'$, M' can converge, if and only if all finite configurations of $[\![M']\!]_{\circledast}$ can be extended to contain a positive move.*

However, we cannot conclude adequacy for must equivalence from Theorem 2. Indeed, must convergence is not generally stable under weak bisimilarity: for instance, (the strategies representing) $\texttt{tt}$ and $\mathrm{Y}\,(\lambda x.\ \texttt{if choice tt}\ x)$ are weakly bisimilar but the latter is not must convergent. To address this in the next section we will refine the interpretation to obtain a closer connection with syntax.

4 Essential Events

The model presented in the previous section is very operational; configurations of $[\![M]\!]_{\circledast}$ can be seen as derivations for an operational semantics. The price, however, is that besides the fact that the interpretation grows dramatically in size, we can only get a category up to weak bisimulation, which can be too coarse (for instance for must convergence). We would like to remove all events that are not relevant to the behaviour of terms up to weak bisimulation. In other words, we want a notion of *essential internal events* that *(1)* suffices to recover all behaviour with respect to weak bisimulation, but which *(2)* is not an obstacle to getting a category up to isomorphism (which amounts to $\text{\ae}_A \circ \sigma \cong \sigma$).

4.1 Definition of Essential Events

As shown before, the loss of behaviours when hiding is due to the disappearance of events participating in a conflict. A neutral event may not have visible consequences but still be relevant if in a minimal conflict; such events are *essential*.

Definition 9. *Let $\sigma : S \rightharpoonup A$ be an uncovered pre-strategy. An* **essential event** *of S is an event s which is either visible, or (internal and) involved in a minimal conflict (that is such that we have $s \rightsquigarrow_x s'$ for some s', x).*

Write E_S for the set of essential events of σ. Any pre-strategy $\sigma : S \rightharpoonup A$ induces another pre-strategy $\mathscr{E}(\sigma) : \mathscr{E}(S) = S \downarrow E_S \rightharpoonup A$ called **the essential part** of σ. The following proves that our definition satisfies *(1)*: no behaviour is lost.

Lemma 7. *An uncovered pre-strategy $\sigma : S \rightharpoonup A$ is weakly bisimilar to $\mathscr{E}(\sigma)$.*

This induces a new notion of (associative) composition only keeping the essential events. For $\sigma : A^\perp \parallel B$ and $\tau : B^\perp \parallel C$, let $\tau \circledcirc \sigma = \mathscr{E}(\tau \circledast \sigma)$. We observe that $\mathscr{E}(\tau \circledast \sigma) \cong \mathscr{E}(\tau) \circledcirc \mathscr{E}(\sigma)$.

Which pre-strategies compose well with copycat with this new composition?

4.2 Essential Strategies

We now can state property *(2)*: the events added by composition with copycat are inessential, hence hidden during composition:

Theorem 3. *Let $\sigma : S \rightharpoonup A$ be an uncovered strategy. Then $\text{æ}_A \circledcirc \sigma \cong \mathscr{E}(\sigma)$.*

This prompts the following definition. An uncovered pre-strategy σ is **essential** when it is a strategy, and if, equivalently: *(1)* all its events are essential, *(2)* $\sigma \cong \mathscr{E}(\sigma)$. We obtain a characterisation of strategies in the spirit of [14]:

Theorem 4. *A pre-strategy $\sigma : S \rightharpoonup A$ is essential if and only if $\text{æ}_A \circledcirc \sigma \cong \sigma$.*

As a result, we get:

Theorem 5. *Race-free games, and essential strategies up to isomorphism form a compact-closed category $\mathbf{CG}_\circledcirc$.*

Relationship Between CG and $\mathbf{CG}_\circledcirc$. Covered strategies can be made into a compact-closed category [5,14]. Remember that the composition of $\sigma : S \to A^\perp \parallel B$ and $\tau : T \to B^\perp \parallel C$ in **CG** is defined as $\tau \odot \sigma = (\tau \circledast \sigma)_\downarrow$.

Lemma 8. *The operation $\sigma \mapsto \sigma_\downarrow$ extends to an identity-on-object functor $\mathbf{CG}_\circledcirc \to \mathbf{CG}$.*

In the other direction, a strategy $\sigma : A$ might not be an essential strategy; in fact it might not even be an uncovered strategy, as it may fail secrecy. Sending σ to $\text{æ}_A \circledcirc \sigma$ delegates the non-deterministic choices to internal events and yields an essential strategy, but this operation is not functorial.

Relationship Between $\mathbf{CG}_{\circledcirc}$ and $\mathbf{CG}_{\circledast}$. The forgetful operation mapping an essential strategy σ to itself, seen as an uncovered strategy, defines a functor $\mathbf{CG}_{\circledcirc} \to \mathbf{CG}_{\circledast}$. Indeed, if two essential strategies are isomorphic, they are also weakly bisimilar. Moreover, we have that $\tau \circledast \sigma \simeq \mathscr{E}(\tau \circledast \sigma) = \tau \circledcirc \sigma$. However the operation $\mathscr{E}(\cdot)$ does not extend to a functor in the other direction even though $\mathscr{E}(\tau) \circledcirc \mathscr{E}(\sigma) \cong \mathscr{E}(\tau \circledast \sigma)$, as it is defined only on concrete representatives, not on equivalence classes for weak bisimilarity.

4.3 Interpretation of $\mathbf{APCF}_{+}$

We now show that this new category also supports a sound and adequate interpretation of $\mathbf{APCF}_{+}$ for various testing equivalences, including must. As before, we need to construct the category of negative games and strategies.

Lemma 9. *There is a cartesian symmetric monoidal category $\mathbf{CG}_{\circledcirc}^{-}$ of negative race-free games and negative essential strategies up to isomorphism. Well-opened negative race-free games form an exponential ideal of $\mathbf{CG}_{\circledcirc}^{-}$.*

We keep the same interpretation of types of affine nondeterministic PCF. Moreover, the strategy **if** is essential. As a result, we let:

$$[\![\bot]\!]_{\circledcirc} = q : \mathbb{B} \qquad [\![\mathtt{if}\ M\,N\,N']\!]_{\circledcirc} = \mathtt{if} \circledcirc ([\![M]\!]_{\circledcirc} \parallel \langle [\![N]\!]_{\circledcirc}, [\![N']\!]_{\circledcirc} \rangle)$$

Using $\mathscr{E}(\sigma \circledast \tau) = \mathscr{E}(\sigma) \circledcirc \mathscr{E}(\tau)$, one can prove by induction that for any term M we have $[\![M]\!]_{\circledcirc} = \mathscr{E}([\![M]\!]_{\circledast})$. Furthermore, this interpretation permits a stronger link between the operational and the denotational semantics:

Theorem 6. *For all terms $\vdash M : \mathbb{B}$, $\mathscr{E}(\mathsf{t}(M)) \cong [\![M]\!]_{\circledcirc}$.*

Theorem 6 implies Theorem 2. It also implies adequacy for must:

Corollary 2. *The interpretation $[\![\cdot]\!]_{\circledcirc}$ is adequate for may, and fair, and must: $\vdash M : \mathbb{B}$ has no infinite derivations if and only if all (possibly infinite) maximal configurations of $[\![M]\!]_{\circledcirc}$ have a positive event.*

This result also implies that $[\![\cdot]\!]_{\circledast}$ is adequate for must.

5 Conclusion

We have described an extension of the games of [14] to uncovered strategies, composed without hiding. It has strong connections with operational semantics, as the interpretations of terms of base type match their tree of reductions. It also forms a compact-closed category up to weak bisimulation, and is adequate for the denotational semantics of programming languages. Identifying the *inessential* events as those responsible for the non-neutrality of copycat, we remove them to yield a compact closed category up to isomorphism. Doing so we obtain our sought-after setting for the denotational semantics of programming languages, one *agnostic w.r.t.* the chosen testing equivalence. The work blends well with the technology of [7] (symmetry, concurrent innocence) dealing with non-affine languages and characterising strategies corresponding to pure programs; these developments appear in the first author's PhD thesis [3].

Acknowledgements. We gratefully acknowledge the support of the ERC Advanced Grant ECSYM, EPSRC grants EP/K034413/1 and EP/K011715/1, and LABEX MILYON (ANR-10-LABX-0070) of Université de Lyon, within the program "Investissements d'Avenir" (ANR-11-IDEX-0007) operated by the ANR.

References

1. Abramsky, S.: Game semantics for programming languages. In: Prívara, I., Ružička, P. (eds.) MFCS 1997. LNCS, vol. 1295, pp. 1–4. Springer, Heidelberg (1997). https://doi.org/10.1007/BFb0029944
2. Brinksma, E., Rensink, A., Vogler, W.: Fair testing. In: Lee, I., Smolka, S.A. (eds.) CONCUR 1995. LNCS, vol. 962, pp. 313–327. Springer, Heidelberg (1995). https://doi.org/10.1007/3-540-60218-6_23
3. Castellan, S.: Concurrent structures in game semantics. Ph.D. thesis, ENS Lyon, France (2017)
4. Castellan, S., Clairambault, P.: Causality vs. interleaving in game semantics. In: CONCUR 2016 - Concurrency Theory (2016)
5. Castellan, S., Clairambault, P., Rideau, S., Winskel, G.: Games and strategies as event structures. Log. Methods Comput. Sci. **13** (2017)
6. Castellan, S., Clairambault, P., Winskel, G.: Symmetry in concurrent games. In: Henzinger, T.A., Miller, D. (eds.) CSL-LICS 2014, Vienna, Austria, July 14–18, 2014, p. 28. ACM (2014)
7. Castellan, S., Clairambault, P., Winskel, G.: The parallel intensionally fully abstract games model of PCF. In: LICS 2015. IEEE Computer Society (2015)
8. Castellan, S., Hayman, J., Lasson, M., Winskel, G.: Strategies as concurrent processes. Electr. Notes Theor. Comput. Sci. **308**, 87–107 (2014)
9. Danos, V., Harmer, R.: Probabilistic game semantics. In: 15th Annual IEEE Symposium on Logic in Computer Science, Santa Barbara, California, USA, 26–29 June 2000, pp. 204–213 (2000)
10. Ghica, D.R., Murawski, A.S.: Angelic semantics of fine-grained concurrency. Ann. Pure Appl. Log. **151**(2–3), 89–114 (2008)
11. Harmer, R., McCusker, G.: A fully abstract game semantics for finite nondeterminism. In: 14th Annual IEEE Symposium on Logic in Computer Science, Trento, Italy, 2–5 July 1999, pp. 422–430 (1999)
12. Hirschowitz, T.: Full abstraction for fair testing in CCS. In: Heckel, R., Milius, S. (eds.) CALCO 2013. LNCS, vol. 8089, pp. 175–190. Springer, Heidelberg (2013). https://doi.org/10.1007/978-3-642-40206-7_14
13. Laird, J.: A game semantics of idealized CSP. Electr. Notes Theor. Comput. Sci. **45**, 232–257 (2001)
14. Rideau, S., Winskel, G.: Concurrent strategies. In: LICS, pp. 409–418. IEEE Computer Society (2011)
15. Tsukada, T., Luke Ong, C.-H.: Innocent strategies are sheaves over plays - deterministic, non-deterministic and probabilistic innocence. CoRR, abs/1409.2764 (2014)
16. Tsukada, T., Luke Ong, C.-H.: Nondeterminism in game semantics via sheaves. In: 30th Annual ACM/IEEE Symposium on Logic in Computer Science, LICS 2015, Kyoto, Japan, 6–10 July 2015, pp. 220–231 (2015)

17. Winskel, G.: Event structures. In: Brauer, W., Reisig, W., Rozenberg, G. (eds.) ACPN 1986. LNCS, vol. 255, pp. 325–392. Springer, Heidelberg (1987). https://doi.org/10.1007/3-540-17906-2_31
18. Yamada, N., Abramsky, S.: Dynamic games and strategies. CoRR, abs/1601.04147 (2016)

A Trace Semantics for System F Parametric Polymorphism

Guilhem Jaber[1] and Nikos Tzevelekos[2](✉)

[1] ENS de Lyon, Université de Lyon, LIP, Lyon, France
[2] Queen Mary University of London, London, England
nikos.tzevelekos@qmul.ac.uk

Abstract. We present a trace model for Strachey parametric polymorphism. The model is built using operational nominal game semantics and captures parametricity by using names. It is used here to prove an operational version of a conjecture of Abadi, Cardelli, Curien and Plotkin which states that Strachey equivalence implies Reynolds equivalence in System F.

1 Introduction

Parametricity was first introduced by Strachey [22] as a way to characterise the behaviour of polymorphic programs as being uniform with respect to the type of the arguments provided. He opposed this notion to ad-hoc polymorphism, where a function can produce arbitrarily different outputs when provided inputs of different types (for example an integer and a boolean). To formalise this notion of parametricity, Reynolds introduced relational parametricity [21]. It is defined using an equivalence on programs, that we call Reynolds equivalence and is a generalisation of logical relations to System F. This equivalence uses arbitrary relations over pairs of types to relate polymorphic programs. So a parametric program that takes related arguments as input will produce related results. Reynolds parametricity has been developed into a fundamental theory for studying polymorphic programs [1,20,23].

Following results of Mitchell on PER-models of polymorphism [18], Abadi, Cardelli, Curien and Plotkin [1,20] introduced another, more intentional notion of equivalence, called Strachey equivalence. Two terms of System F are Strachey equivalent whenever, by removing all their type annotations, we obtain two $\beta\eta$-equivalent untyped terms. The authors conjectured that Strachey equivalence implies Reynolds equivalence (the converse being easily shown to be false).

In this paper we examine a notion of Reynolds equivalence based on operational logical relations, and prove that, for this notion, the conjecture holds. To do so, we introduce a trace model for System F based on operational nominal game semantics [12,14]. Terms in our model are denoted as sets of traces, generated by a labelled transition system, which represent interactions with arbitrary term contexts. In order to abstract away type information from inputs to polymorphic functions, our semantics uses *names* to model such inputs. The idea is

C. Baier and U. Dal Lago (Eds.): FOSSACS 2018, LNCS 10803, pp. 20–38, 2018.
https://doi.org/10.1007/978-3-319-89366-2_2

$$\frac{\Delta;\Gamma,x:\theta\vdash M:\theta'}{\Delta;\Gamma\vdash\lambda x^{\theta}.M:\theta\to\theta'}\quad\frac{\Delta;\Gamma\vdash M:\theta\to\theta'\quad\Delta;\Gamma\vdash N:\theta}{\Delta;\Gamma\vdash MN:\theta'}$$

$$\frac{(x:\theta)\in\Gamma}{\Delta;\Gamma\vdash x:\theta}\quad\frac{\Delta,X;\Gamma\vdash M:\theta}{\Delta;\Gamma\vdash\Lambda X.M:\forall X.\theta}\quad\frac{\Delta;\Gamma\vdash M:\forall X.\theta}{\Delta;\Gamma\vdash M\theta':\theta\{\theta'/X\}}$$

$$(\lambda x.M)N=_{\beta\eta}M\{N/x\}\qquad(\Lambda X.M)\theta=_{\beta\eta}M\{\theta/X\}\qquad\lambda x.Mx=_{\beta\eta}M\qquad\Lambda X.MX=_{\beta\eta}M$$

Fig. 1. Typing rules and $\beta\eta$-equality axioms.

the following: since names have no internal structure, the function has no choice but to act "the same way" on such inputs, i.e. be parametric. Our trace model yields a third notion of equivalence: trace equivalence (i.e. equality of sets of traces). Then, the result is proven by showing that trace equivalence is included in (operational) Reynolds equivalence, while it includes Strachey equivalence.

The traces in our model are formed of *moves*, which represent interactions between the modelled term (the *Player*) and its context (the *Opponent*): either of Player or Opponent can interrogate the terms provided by the other one, or respond to a previous such interrogation. These moves are called *questions* and *answers* respectively. Names enter the scene when calling terms which are of polymorphic type, in which case the calling party would replace the actual argument type θ with a *type name* α, and record locally the correspondence between α and θ. Another use of names in our model is for representing terms that are passed around as arguments to questions. These are called *computation names*, and are typed according to the term they each represent.

2 Definition of System F and Parametricity

We start off by giving the definitions of System F and of the parametric equivalence relations we shall examine on it. The grammar for System F is standard and given by:

$$\begin{aligned}\mathsf{Type}\ni\ &\theta,\theta'::=X\mid\theta\to\theta'\mid\forall X.\theta\\ \mathsf{Term}\ni\ &M,N::=\lambda x^{\theta}.M\mid\Lambda X.M\mid MN\mid M\theta\end{aligned}$$

We write x, etc. for *(term) variables*, sourced from a countable set Var; and X, etc. for *type variables*, taken from TVar. We define substitutions of open variables of either kind in the usual capture-avoiding way. For instance, the term obtained by consecutively applying substitutions $\eta:\mathsf{Var}\rightharpoonup\mathsf{Term}$ and $\delta:\mathsf{TVar}\rightharpoonup\mathsf{Type}$ on M is written $M\{\eta\}\{\delta\}$.

Terms are typed in environments $\Delta;\Gamma$, where Δ is a finite set of type variables, and Γ is a set $\{x_1:\theta_1,\ldots,x_m:\theta_m\}$ of variable-type pairs. The typing rules are given in Fig. 1. The operational semantics we examine is $\beta\eta$-equality, defined as the least syntactic congruence $=_{\beta\eta}$ that includes the axioms given on the RHS part of Fig. 1.

We shall use the following common polymorphic encodings:

- **Bool** $= \forall X.\ X \to X \to X$, **true** $= \Lambda X.\lambda x^X.\lambda y^X.x$ and **false** $= \Lambda X.\lambda x^X.\lambda y^X.y$,
- **Unit** $= \forall X.\ X \to X$ and **id** $= \Lambda X.\lambda x^X.x$.

Reynolds Equivalence. We next introduce logical relations for System F. First, we let Rel be the set of all typed relations between closed terms that are compatible with $=_{\beta\eta}$:

$$\begin{aligned}\mathrm{Rel} = \{(\theta_1,\theta_2,R) \mid R \subseteq \mathsf{Term}\times\mathsf{Term} \wedge \forall (M_1,M_2)\in R.\ \cdot;\cdot \vdash M_i : \theta_i \\ \wedge \forall M_1' =_{\beta\eta} M_1.\forall M_2' =_{\beta\eta} M_2.\ (M_1',M_2')\in R\}\end{aligned}$$

Logical relations $\mathcal{R}[\![\theta]\!]_\delta$ are defined below, indexed by environments $\delta : \mathsf{TVar} \rightharpoonup \mathrm{Rel}$:

$$\begin{aligned}\mathcal{R}[\![X]\!]_\delta &= R \text{ when } \delta(X) = (\_,\_,R)\\ \mathcal{R}[\![\forall X.\theta]\!]_\delta &= \{(M_1,M_2) \mid \forall(\theta_1,\theta_2,R)\in \mathrm{Rel}.\ (M_1\theta_1, M_2\theta_2)\in \mathcal{R}[\![\theta]\!]_{\delta\cdot[X\mapsto(\theta_1,\theta_2,R)]}\}\\ \mathcal{R}[\![\theta_1\to\theta_2]\!]_\delta &= \{(M_1,M_2) \mid \forall (N_1,N_2)\in\mathcal{R}[\![\theta_1]\!]_\delta.\ (M_1N_1, M_2N_2)\in \mathcal{R}[\![\theta_2]\!]_\delta\}\end{aligned}$$

We can now define the first notion of parametric equivalence for System F.

Definition 1. Given terms $\Delta;\Gamma \vdash M_1, M_2 : \theta$, we say that they are ***Reynolds equivalent***, and write $\Delta;\Gamma\vdash M_1 \simeq_{log} M_2 : \theta$, if:

$$\forall \delta\in\mathcal{R}[\![\Delta]\!].\forall(\eta_1,\eta_2)\in\mathcal{R}[\![\Gamma]\!]_\delta.\ (M_1\{\eta_1\}\{\delta_1\}, M_2\{\eta_2\}\{\delta_2\})\in\mathcal{R}[\![\theta]\!]_\delta$$

where $\mathcal{R}[\![\Delta]\!] = \mathrm{dom}(\Delta)\to \mathrm{Rel}$, $\delta_1 = \{(X,\theta_1) \mid \delta(X) = (\theta_1,\_,\_)\}$ (similar for δ_2) and $\mathcal{R}[\![\Gamma]\!]_\delta = \{(\eta_1,\eta_2)\in(\mathrm{dom}(\Gamma)\rightharpoonup \mathsf{Term})^2 \mid \forall(x,\theta')\in\Gamma.\ (\eta_1(x),\eta_2(x))\in\mathcal{R}[\![\theta']\!]_\delta\}$.

The following result is standard [21].

Theorem 2 (Fundamental Property). *If $\Delta;\Gamma\vdash M:\tau$ then $\Delta;\Gamma\vdash M\simeq_{log} M:\theta$.*

Remark 3. Note that our definition of Reynolds equivalence does not coincide with either of the definitions given in [1,20]: therein, parametricity is defined using relational logics (and accompanying proof systems), whereas here we use quantification over concrete relations over closed terms.

Strachey Equivalence. Another notion of parametric equivalence is defined by means of erasing types from terms. We define the *type erasure* **erase**(M) of a term M by:

$$\begin{aligned}&\mathbf{erase}(\Lambda X.M) = \mathbf{erase}(M) &&\mathbf{erase}(MN) = \mathbf{erase}(M)\mathbf{erase}(N)\\ &\mathbf{erase}(\lambda x^\theta.M) = \lambda x.\mathbf{erase}(M) &&\mathbf{erase}(M\theta) = \mathbf{erase}(M)\end{aligned}$$

and **erase**$(x) = x$. Thus, **erase**(M) is an untyped λ-term. Below we overload $=_{\beta\eta}$ to also mean $\beta\eta$-equality in the untyped λ-calculus.

Definition 4. Given terms $\Delta;\Gamma \vdash M_1, M_2 : \theta$, we say that they are ***Strachey equivalent*** if $\mathbf{erase}(M_1) =_{\beta\eta} \mathbf{erase}(M_2)$.

It was conjectured in [1,20] that Reynolds equivalence includes Strachey equivalence. We prove this holds for the version of Reynolds equivalence given in Definition 1.

Theorem 5. *Any two Strachey equivalent terms are also Reynolds equivalent.*

It is interesting to think why a direct approach would not work in order to prove this conjecture. Given Strachey equivalent terms M_1, M_2 of type **Bool**, suppose we want to prove them Reynolds equivalent. We therefore take $(\theta_1, \theta_2, R) \in \text{Rel}$, $(N_{1,1}, N_{2,1}) \in R$, and $(N_{1,2}, N_{2,2}) \in R$, and aim to prove that $(M_1\theta_1 N_{1,1} N_{1,2}, M_2\theta_2 N_{2,1} N_{2,2}) \in R$. Ideally, we would like to prove that there exists $j \in \{1,2\}$ s.t. for all $i \in \{1,2\}$, $M_i\theta_i N_{i,1} N_{i,2} =_{\beta\eta} N_{i,j}$, but that seems overly optimistic. A first trick is to use Theorem 2, to get that M_2 is related with itself. Thus, we get that $(M_2\theta_1 N_{1,1} N_{1,2}, M_2\theta_2 N_{2,1} N_{2,2}) \in R$, and it would suffice to prove $M_1\theta_1 N_{1,1} N_{1,2} =_{\beta\eta} M_2\theta_1 N_{1,1} N_{1,2}$ to conclude. However, our hypothesis is simply that $\mathbf{erase}(M_1) =_{\beta\eta} \mathbf{erase}(M_2)$.

A possible solution to the above could be to *β-reduce* both $M_i\theta_1 N_{1,1} N_{1,2}$, hoping that the distinction between the two terms will vanish. Our trace semantics provides a way to model the interaction between such a term M_i and a context $\bullet\, \theta_j N_{j,1} N_{j,2}$, and to deduce properties about the normal form reached by their application via head reduction.

3 A Nominal Trace Semantics for System F

In this section we introduce a trace semantics for open terms which will be our main vehicle of study for System F. The terms in our semantics will be allowed to contain special constants representing any term that could fill in their open variables (these be term or type variables). The use of names can be seen as a nominal approach to parametricity: parametric types and values are represented in our semantics by names, without internal structure. Thus, e.g. a parametric function is going to behave "the same way" for any input, since the latter will be nothing but a name.

Our approach follows the line of work on nominal techniques [7,19] and nominal operational game semantics [12,14]. We let the set of ***names*** be:

$$\mathsf{N} = \mathsf{TN} \uplus \mathsf{CN}$$

We therefore use two kinds of names: type names $\alpha, \beta \in \mathsf{TN}$; and computation names $c, d \in \mathsf{CN}$. We will range over arbitrary names by a and variants. We extend the syntax of terms and types by including computation and type names as constants, and call the resulting syntax ***namey terms and types***:

$$M, N ::= c \mid x \mid \lambda x^\theta.M \mid \Lambda X.M \mid MN \mid M\theta \qquad \theta, \theta' ::= \alpha \mid X \mid \theta \to \theta' \mid \Lambda X.\theta$$

A namey term or type is *closed* if it contains no free (type/term) variables – but it may contain names. On the other hand, a ***value*** is a closed term in head normal form that contains no names. We range over values with v and variants.

We will use the notation $\hat{M}, \hat{N}$, and variants, to refer jointly to namey terms and namey types. Namey terms are typed with additional typing hypotheses for the added constants. These typings are made explicit in the trace model. By abuse of terminology, we will drop the adjective "namey" and refer to the above simply as "terms" and "types". Formally speaking, namey terms and types form *nominal sets* (cf. Definition 8).

Note 6 (what do c's and α's represent?). A computation name c represents a term that can replace the open variables of a term M. That is, in order to examine the semantics of $\lambda x^\theta.M$, we will look instead at $M\{c/x\}$ where c a computation name of appropriate type. Type names α have a similar purpose, for types.

Our trace semantics is built on top of head reduction, which is reminded next. Moreover, we shall be using types in *extended form*, which determines the number and types of arguments needed in order to fully apply a term of a given type.

Definition 7. The (standard) head reduction rules are given in Fig. 2. Head normal forms are given by the syntax on the LHS below,

$$M_{\mathsf{hnf}} ::= E[x] \mid E[c] \mid \lambda x^\theta.M_{\mathsf{hnf}} \mid \Lambda X.M_{\mathsf{hnf}} \qquad E ::= \bullet \mid EM \mid E\theta$$

where E ranges over ***evaluation contexts*** (defined on the RHS). Evaluation contexts are typed with types of the form $\theta \rightsquigarrow \theta'$. We write $E : \theta \rightsquigarrow \theta'$ if we can derive $\bullet : \theta \vdash E : \theta'$.

An ***extended type form*** is a sequence $(\tau_1, ..., \tau_n, \xi)$ with $\xi \in \mathsf{TVar} \cup \mathsf{TN}$ and, for each i, $\tau_i \in \mathsf{Type} \cup \{\forall X \mid X \in \mathsf{TVar}\}$. Formally, the extended form of a type θ, written $\mathrm{ext}(\theta)$, is defined by:

$$\mathrm{ext}(\forall X.\theta) = (\forall X) :: \mathrm{ext}(\theta) \qquad \mathrm{ext}(\theta \to \theta') = \theta :: \mathrm{ext}(\theta') \qquad \mathrm{ext}(\xi) = (\xi)$$

where we write $h :: t$ for the sequence with head h and tail t (cf. list notation). Elements of the form $\forall X$ in these sequences are binders that bind to their right.

We let $\to^*$ be the reflexive-transitive closure of $\to$. It is a standard result that $\to^*$ preserves typing and (strongly) normalises to head normal forms.

We finally introduce some infrastructure for working with objects with names.

$$\begin{array}{l} (\lambda x.M)N \to M\{N/x\} \\ (\Lambda X.M)\theta \to M\{\theta/X\} \end{array} \qquad \frac{M \to M'}{\lambda x.M \to \lambda x.M'} \qquad \frac{M \to M'}{\Lambda X.M \to \Lambda X.M'} \qquad \frac{M \to M' \quad (*)}{E[M] \to E[M']}$$

Fig. 2. Head reduction rules. Condition $(*)$ stipulates that M be not a Λ/λ-abstraction.

Definition 8. We call a permutation $\pi : \mathsf{N} \to \mathsf{N}$ *finite* if the set $\{a \mid \pi(a) \neq a\}$ is finite, and *component-preserving* if, for all $a \in \mathsf{N}$, $a \in \mathsf{TN}$ iff $\pi(a) \in \mathsf{TN}$.

A ***nominal set*** [7] is a pair $(Z, *)$ of a set Z along with an action $(*)$ from the set of finite component-preserving computations of N on the set Z. For each $z \in Z$, the set of names featuring in z form its ***support***, written $\nu(z)$, which we stipulate to be finite.

In the sequel, when constructing objects with names (such as moves or traces) we shall implicitly assume that these form nominal sets, where the permutation action is defined by taking $\pi * z$ to be the result of applying π to each name in z.

3.1 Trace Semantics Preview

Before formally presenting the trace model, we look at some examples informally, postponing the full details for the next section. Head-reduction brings terms into head normal form. The trace semantics allows us to further 'reduce' terms of the form $E[c\hat{M}_1 \cdots \hat{M}_n]$, where c is some computation name. For such a term, following the game semantics approach [3,11], our model will issue a *move* interrogating the computation c on arguments $\hat{M}_i$, and putting E on top of an *evaluation stack*, denoted $\mathcal{E}$. The move is effectively a call to c, and $\mathcal{E}$ functions as a call stack which registers the calls that have been made and are still pending. This will effectively lead to a labelled transition system in which labels are moves issued by two parties: a *Player (P)*, representing the modelled term, and an *Opponent (O)* representing its enclosing term context.

Traces are sequences of ***moves***, which in turn are tuples of names belonging to one of these four classes, taking $c \in \mathsf{CN}$ and $a_i \in \mathsf{N}$ for each i:

- Player questions $\bar{c}(a_1, ..., a_n)$ (also *P-questions*),
- Opponent questions $c(a_1, ..., a_n)$ (also *O-questions*),
- *PO*-answers $\overline{\mathsf{OK}}\mathsf{OK}$, and *OP*-answers $\mathsf{OK}\overline{\mathsf{OK}}$.

Given a question move as above, we let its *core name* be c. We distinguish a computation name $c_{\text{in}} \in \mathsf{CN}$, and call questions with core name c_{in} ***initial***. We define a ***trace*** T to be a finite sequence of moves. Traces will be restricted to *legal* ones in Definition 12.

In the following examples we give traces produced by simple System F terms. Traces are formally produced by an LTS over configurations whose main component is an evaluation stack. An ***evaluation stack*** is a stack whose elements are typed evaluation contexts, apart from the top element which can also be a typed term:

$$\mathcal{E} ::= \mathcal{E}' \mid (M, \theta) :: \mathcal{E}' \qquad \mathcal{E}' ::= \Diamond \mid (E, \theta \rightsquigarrow \theta') :: \mathcal{E}'$$

We denote the empty stack with $\Diamond$. In the next two examples, for simplicity, configurations shall only contain evaluation stacks.

Example 9. Recall that $\mathbf{id} = \Lambda X.\lambda x^X.x : \mathbf{Unit}$ and $\mathbf{Unit} = \forall X.X \to X$. The extended type of **Unit**, $\mathrm{ext}(\mathbf{Unit}) = (\forall X, X, X)$, indicates that **id** requires two arguments in order to be evaluated: one type and one term of that given type. Thus, the traces produced by **id** will start with an interrogating/calling move $c_{\mathrm{in}}(\alpha, c)$ of O:

- c_{in} is the computation name assigned (by convention) to the term being evaluated (in this case, **id**);
- α, c are names abstracting the actual type and term arguments which **id** is called on. It is assumed that c is of type α.

Starting from the initial move $c_{\mathrm{in}}(\alpha, c)$, a trace of **id** can be produced as follows:

$$\langle\Diamond\rangle \xrightarrow{c_{\mathrm{in}}(\alpha,c)} \langle(\mathbf{id}\,\alpha\, c, \alpha)\rangle \to \langle(c, \alpha)\rangle \xrightarrow{\bar{c}()} \langle(\bullet, \alpha \rightsquigarrow \alpha)\rangle \xrightarrow{\mathsf{OK}\overline{\mathsf{OK}}} \langle\Diamond\rangle$$

Thus, O starts the interaction by interrogating **id** with α, c. This results in $\mathbf{id}\,\alpha\, c$, which gets head reduced to c. At this point, c is a head normal form of type α, and P can answer the initial question $c_{\mathrm{in}}(\alpha, c)$. This is done in two steps. First, P further reduces c by playing a move $\bar{c}()$ (here c takes 0 arguments as $\mathrm{ext}(\alpha) = (\alpha)$), and pushes the current evaluation context $(\bullet, \alpha \rightsquigarrow \alpha)$ on the stack. O then responds by triggering a pair of answers $\mathsf{OK}\overline{\mathsf{OK}}$, which answer both questions played so far. The resulting trace is: $c_{\mathrm{in}}(\alpha, c) \cdot \bar{c}() \cdot \mathsf{OK}\overline{\mathsf{OK}}$.

Note 10 (what are $\mathsf{OK}\overline{\mathsf{OK}}$ *and* $\overline{\mathsf{OK}}\mathsf{OK}$*?).* As System F base types are type variables, there is no real need for answer moves: a type X has no return values. For example, in the game models of Hughes [9] and Laird [15], answer moves were effectively suppressed (either explicitly, or by allowing moves $c(\cdots)$ to function as answers). Here, to give the semantics an operational flavour, we introduce instead explicit 'dummy' answers OK.

Example 11. Consider now $M = \lambda f^{\mathbf{Unit}}.f : \mathbf{Unit} \to \mathbf{Unit}$. We have that $\mathrm{ext}(\mathbf{Unit} \to \mathbf{Unit}) = (\mathbf{Unit}, \forall X, X, X)$, and therefore M requires three arguments for its evaluation: one term of type **Unit**, one type, and one term if that latter type. We can therefore start a trace of M with an initial move $c_{\mathrm{in}}(c_1, \alpha_1, c)$ and continue as follows.

$$\langle\Diamond\rangle \xrightarrow{c_{\mathrm{in}}(c_1,\alpha_1,c_2)} \langle(M c_1 \alpha_1\, c_2, \alpha_1)\rangle \to \langle(c_1\, \alpha_1\, c_2, \alpha_1)\rangle \xrightarrow{\bar{c}_1(\alpha_2,c_3)} \langle(\bullet, \alpha_2 \rightsquigarrow \alpha_1)\rangle$$

Thus, the initial move leads to $M c_1 \alpha_1 c_2$, which in turn reaches the hnf $c_1 \alpha_1 c_2$, with $c_1 : \mathbf{Unit}$, and at that point P needs to invoke c_1 with arguments α_1 and c_2. These are abstracted away by fresh names α_2 and c_3 respectively, which are passed as arguments to c_1. c_3 in particular has type α_2. The result of this invocation will be of type α_2, which is the hole type in $(\bullet : \alpha_2 \rightsquigarrow \alpha_1)$. O can only produce a term of α_2 by simply returning c_3. Similarly to before, this is done in two steps: by O playing $c_3()$, which brings c_2 (the term represented by c_3) at the top of the stack, which in turn triggers a pair of answers $\overline{\mathsf{OK}}\mathsf{OK}$ and brings c_2 inside the context $(\bullet : \alpha_2 \rightsquigarrow \alpha_1)$.

$$\langle(\bullet, \alpha_2 \rightsquigarrow \alpha_1)\rangle \xrightarrow{c_3()} \langle(c_2, \alpha_2) :: (\bullet, \alpha_2 \rightsquigarrow \alpha_1)\rangle \xrightarrow{\overline{\mathsf{OK}}\mathsf{OK}} \langle(c_2, \alpha_1)\rangle \xrightarrow{\bar{c}_2()} \langle(\bullet, \alpha_1 \rightsquigarrow \alpha_1)\rangle \xrightarrow{\mathsf{OK}\overline{\mathsf{OK}}} \langle\Diamond\rangle$$

The latter step leaves us with (c_2, α_1), which reaches $\Diamond$ as in the previous example.

3.2 Definition of the LTS

We now proceed with the formal definition of the trace semantics. We start off with a series of definitions setting the conditions for a trace to be legal.

The names appearing in a trace are owned by whoever introduces them. A move m ***introduces*** a name a in a trace T if m is a question $q(\vec{a})$ with $a_i = a$ for some i. For each $A \in \{O, P\}$, we let the set of names of T that *are owned by* A be:

$$A(T) = \{a \in \mathsf{N} \mid \exists m.\ m \text{ is an } A\text{-question in } T \wedge m \text{ introduces } a\}.$$

We will be referring to the names appearing in $A(T)$ as *A-names*.

Each move in a trace needs to be justified, i.e. depend on an earlier move (unless the move is initial). Justification is defined in different ways for questions and answers. Given a trace T and two moves m, m' in T, we say that m' ***justifies*** m when m' is before m in T and:

- m is a question with core name c and m' introduces c, or
- m is an answer which answers m' (and m' is a question).

Answering of questions is defined as follows. Each answer (occurrence) m answers the pair of question moves (m_1, m_2) containing the last two question moves in T which are before m and have not been answered yet.

We can now define legality conditions for traces. Below, for $A \in \{O, P\}$, we say that a move is *A-starting* if it is an A-question or an $AA^\perp$-answer (where $O^\perp = P$ and $P^\perp = O$). Similarly, a move is *A-ending* if it is either an A-question or an $A^\perp A$-answer.

Definition 12. A trace T is said to be ***legal*** when, for each $A \in \{O, P\}$:

1. A-ending moves can only be followed by $A^\perp$-starting moves;
2. all moves in T are justified, apart from the first move which must be initial;
3. apart from c_{in}, every name of T is introduced exactly once in it;
4. for each A-question with core name $c \neq c_{\text{in}}$, we have $c \in A^\perp(T)$;
5. if an $AA^\perp$-answer answers (m, m') then these are A- and $A^\perp$-questions respectively.

The conditions above can be given names (suggesting their purpose) as follows: 1. *alternation*, 2. *justification*, 3. *well-introduction*, 4. *well-calling*, 5. *well-answering*.

Each trace T has a complement, which we denote $T^\perp$ and is obtained from T by switching O/P in all of its moves (i.e. each $c(\vec{a})$ becomes $\bar{c}(\vec{a})$, $\mathsf{OK}\overline{\mathsf{OK}}$ becomes $\overline{\mathsf{OK}}\mathsf{OK}$, etc). T is legal iff $T^\perp$ is.

Traces are produced by use of a labelled transition system. The LTS comprises moves as labels, and of *configurations* as nodes. Each configuration contains an evaluation stack of terms and environments that need to be evaluated, as well as mappings containing type/term information on names that have appeared so far. We introduced evaluation stacks in the previous section. Here we shall restrict the allowed shapes thereof as follows. We let ***passive*** and ***active*** evaluation stacks be defined by the following two grammars respectively, and take evaluation stacks to be $\mathcal{E} ::= \mathcal{E}_{\mathsf{pass}} \mid \mathcal{E}_{\mathsf{actv}}$,

$$\mathcal{E}_{\mathsf{pass}} ::= \Diamond \mid [(E, \alpha \rightsquigarrow \theta)] \mid (E, \alpha \rightsquigarrow \alpha') :: \mathcal{E}_{\mathsf{pass}}\,, \quad \mathcal{E}_{\mathsf{actv}} ::= [(M, \theta)] \mid (M, \alpha) :: \mathcal{E}_{\mathsf{pass}}\,,$$

where θ ranges over closed types with $\nu(\theta) = \varnothing$, and $\Diamond$ is the empty stack.

The other two components of configurations will be maps γ and ϕ of the shape:

$$\gamma \in (\mathsf{CN} \rightharpoonup (\mathsf{Term} \times \mathsf{Type})) \otimes (\mathsf{TN} \rightharpoonup (\mathsf{Type} \times \{\mathcal{U}\})), \quad \phi \in (\mathsf{CN} \rightharpoonup \mathsf{Type}) \otimes (\mathsf{TN} \rightharpoonup \{\mathcal{U}\}),$$

with $F \otimes G = \{f \cup g \mid f \in F \wedge g \in G\}$. $\mathcal{U}$ is a special "universe" symbol that represents the type of types – it is only used for convenience. Then, in words:

- γ assigns term-type pairs to computation names, and type-$\mathcal{U}$ pairs to type names,
- ϕ assigns types to computation names, and $\mathcal{U}$ to type names.

The role of a map γ is to abstract away terms to computational names, and types to type names. On the other hand, a map ϕ simply types names. In the LTS, when P wants to interrogate an O-computation name c with some arguments, they will abstract away the actual arguments to names, record the abstraction in γ, and call c on these names. On the other hand, when O interrogates a P-computation name c with some move $c(\vec{a})$, we will record in ϕ the types of the (new!) O-names $\vec{a}$.

The abstraction of arguments to names is instrumented by a dedicated operation AVal. This operation assigns to each sequence $((\hat{M}_1, \tau_1), ..., (\hat{M}_n, \tau_n), \xi)$, where $(\tau_1, ..., \tau_n, \xi)$ is an extended type (i.e. the type of the computation name we want to call) and each $\hat{M}_i$ is a closed term or type (the i-th argument), a set of triples of the form $(\vec{a}, \gamma, \beta)$ where:

- $\vec{a}$ is a sequence $(a_1, ..., a_n)$ of names (abstracting each of the arguments $\hat{M}_i$),
- γ is a map as above, with domain $\{a_1, ..., a_n\}$,
- β is the result type one gets after applying each a_i for each τ_i.

The operator is formally defined next. In the same definition we introduce the semantics of types, $[\![\theta]\!]$, as sets of triples of the form $(\vec{a}, \phi, \beta)$, which represent all possible input-output name tuples $(\vec{a}, \beta)$ that are allowed for θ, including their typing ϕ.

(Int) $\langle (M,\theta) :: \mathcal{E}, \gamma, \phi\rangle \to \langle (M',\theta) :: \mathcal{E}, \gamma, \phi\rangle$ when $M \to^* M'$ with M' a head normal form.

(OQ$_0$) $\langle \Diamond, \gamma, \phi\rangle \xrightarrow{c(a_1,\ldots,a_n)} \langle [(M a_1 \cdots a_n, \alpha)], \gamma, \phi\cdot\phi'\rangle$
with $\gamma(c) = (M,\theta)$, $((a_1,\ldots,a_n),\phi',\alpha) \in [\![\theta]\!]$ and $\alpha \in \mathrm{dom}(\phi\cdot\phi')$.

(OQ) $\langle (E, \alpha \rightsquigarrow \theta') :: \mathcal{E}, \gamma, \phi\rangle \xrightarrow{c(a_1,\ldots,a_n)} \langle (M a_1 \cdots a_n, \alpha') :: (E, \alpha \rightsquigarrow \theta') :: \mathcal{E}, \gamma, \phi\cdot\phi'\rangle$
with $\alpha \in \mathrm{dom}(\gamma)$, $\gamma(c) = (M,\theta)$, $((a_1,\ldots,a_n),\phi',\alpha') \in [\![\theta]\!]$ and $\alpha' \in \mathrm{dom}(\phi\cdot\phi') \cup \{\alpha\}$.

(PQ$_0$) $\langle [(E[c\hat{M}_1 \cdots \hat{M}_n],\theta)], \gamma, \phi\rangle \xrightarrow{\bar{c}(a_1,\ldots,a_n)} \langle [(E, \alpha \rightsquigarrow \theta)], \gamma\cdot\gamma', \phi\rangle$
when θ is a closed with empty support, $\mathrm{ext}(\phi(c)) = (\tau_1,\ldots,\tau_n,\xi)$
and $((a_1,\ldots,a_n),\gamma',\alpha) \in \mathsf{AVal}((\hat{M}_1,\tau_1),\ldots,(\hat{M}_n,\tau_n),\xi)$.

(PQ) $\langle (E[c\hat{M}_1 \cdots \hat{M}_n],\alpha') :: \mathcal{E}, \gamma, \phi\rangle \xrightarrow{\bar{c}(a_1,\ldots,a_n)} \langle (E, \alpha \rightsquigarrow \alpha') :: \mathcal{E}, \gamma\cdot\gamma', \phi\rangle$ when $\alpha' \in \mathrm{dom}(\phi)$,
$\mathrm{ext}(\phi(c)) = (\tau_1,\ldots,\tau_n,\xi)$ and $((a_1,\ldots,a_n),\gamma',\alpha) \in \mathsf{AVal}((\hat{M}_1,\tau_1),\ldots,(\hat{M}_n,\tau_n),\xi)$.

(OA) $\langle (\bullet, \alpha \rightsquigarrow \alpha) :: \mathcal{E}, \gamma, \phi\rangle \xrightarrow{\mathsf{OK}\overline{\mathsf{OK}}} \langle \mathcal{E}, \gamma, \phi\rangle$ when $\alpha \in \mathrm{dom}(\phi)$.

(PA) $\langle (M,\alpha) :: (E, \alpha \rightsquigarrow \theta) :: \mathcal{E}, \gamma, \phi\rangle \xrightarrow{\overline{\mathsf{OK}}\mathsf{OK}} \langle (E[M],\theta) :: \mathcal{E}, \gamma, \phi\rangle$ when $\alpha \in \mathrm{dom}(\gamma)$ and M a hnf.

Fig. 3. Reduction rules for the LTS.

Definition 13. Given a closed type θ (which may contain type names), we let its semantics be $[\![\theta]\!] = [\![\mathrm{ext}(\theta)]\!]$, where the latter is defined inductively by:

$$[\![(\alpha)]\!] = \{(\varepsilon, \varepsilon, \alpha)\}$$
$$[\![\theta :: L]\!] = \{((c,\vec{a}), \phi\cdot[c \mapsto \theta], \alpha) \mid c \in \mathsf{CN}, (\vec{a},\phi,\alpha) \in [\![L]\!]\}$$
$$[\![\forall X :: L]\!] = \{((\beta,\vec{a}), \phi\cdot[\beta \mapsto \mathcal{U}], \alpha) \mid \beta \in \mathsf{TN}, (\vec{a},\phi,\alpha) \in [\![L\{\alpha/X\}]\!]\}$$

On the other hand, to each sequence $((\hat{M}_1,\tau_1),\ldots,(\hat{M}_n,\tau_n),\xi)$ we assign a set of *abstract values* $\mathsf{AVal}(((\hat{M}_1,\tau_1),\ldots,(\hat{M}_n,\tau_n),\xi))$ inductively by:

$$\mathsf{AVal}((\alpha)) = \{(\varepsilon, \varepsilon, \alpha)\}$$
$$\mathsf{AVal}((M,\theta) :: L) = \{((c,\vec{a}), \gamma\cdot[c \mapsto (M,\theta)], \alpha) \mid c \in \mathsf{CN}, (\vec{a},\gamma,\alpha) \in \mathsf{AVal}(L)\}$$
$$\mathsf{AVal}((\theta,\forall X) :: L) = \{((\beta,\vec{a}), \gamma\cdot[\beta \mapsto (\theta,\mathcal{U})], \alpha) \mid \beta \in \mathsf{TN}, (\vec{a},\gamma,\alpha) \in \mathsf{AVal}(L\{\beta/X\})\}$$

Both ϕ and γ are finite partial functions whose domains are sets of names. For such maps, the extension notation we used e.g. in $\phi\cdot[c \mapsto z]$ (for appropriate z) means *fresh* extension: $\phi\cdot[c \mapsto z] = \phi \cup \{(c,z)\}$ and given that $c \notin \mathrm{dom}(\phi)$. This notation is extended to whole maps: e.g. $\phi\cdot\phi' = \phi \cup \phi'$ and given that $\mathrm{dom}(\phi) \cap \mathrm{dom}(\phi') = \varnothing$. Moreover, for each map γ we write $\mathsf{fst}(\gamma)$ for its first projection: $\mathsf{fst}(\gamma) = \{(a,\hat{M}) \mid \gamma(a) = (\hat{M},\_)\}$. Similarly, second projection is given by: $\mathsf{snd}(\gamma) = \{(a,Z) \mid \gamma(a) = (\_,Z)\}$.

Definition 14. A ***configuration*** is a triple $\langle \mathcal{E}, \gamma, \phi\rangle$ where $\mathcal{E}$ is an evaluation stack and γ and ϕ are as above. The reduction rules of the LTS are given in Fig. 3. We write $\mathsf{Tr}(C)$ for the set of traces generated by a configuration C.

Given a typed term $\Delta;\Gamma \vdash M : \theta$, with $\Delta = \{X_1,\ldots,X_n\}$, $\Gamma = \{x_1 : \theta_1,\ldots,x_m : \theta_m\}$, we set $\langle \Delta;\Gamma \vdash M : \theta\rangle = \langle \Diamond, [c_{\mathrm{in}} \mapsto (\widetilde{M},\widetilde{\theta})], \varepsilon\rangle$ and

$$[\![\Delta;\Gamma \vdash M : \theta]\!] = \{T \in \mathsf{Tr}(\langle \Delta;\Gamma \vdash M : \theta\rangle) \mid T \text{ has at most one initial move }\}$$

where $\widetilde{\theta} = \forall X_1.\ldots\forall X_n.\theta_1 \to \cdots \to \theta_m \to \theta$ and $\widetilde{M} = \Lambda X_1.\ldots\Lambda X_n.\lambda x_1^{\theta_1}.\ldots\lambda x_m^{\theta_m}.M$.

A configuration is active (resp. passive) if its evaluation stack is so. An active configuration stands for a term being computed and it may only produce P-moves. A passive configuration, on the other hand, stands for a scenario where O is next to play. Moreover, the map ϕ in a configuration contains information on the O-names that have been played, i.e. $\mathrm{dom}(\phi)$ contains O-names, while $\mathrm{dom}(\gamma)$ contains P-names.

To better grasp Fig. 3 let us consider an initial configuration $\langle \Diamond, [c_{\mathrm{in}} \mapsto (M, \theta)], \varepsilon \rangle$ and look at its traces, for some closed term M (so no need for $\widetilde{M}, \widetilde{\theta}$) with empty support.

- At the beginning, the only rule that can be applied is (OQ_0), whereby O interrogates the term M by issuing a move $c_{\mathrm{in}}(\vec{a})$. The names $\vec{a}$ are selected from $[\![\theta]\!]$ and represent arguments that O fully applies the term M on. Since θ has empty support, its extended form is of the shape $(\tau_1, ..., \tau_n, X)$ with X bound by one of the τ_i's. Consequently, when the names $a_1, ..., a_n$ are applied for $\tau_1, ..., \tau_n$, the variable X will be replaced by some type name α. The rule makes this explicit, by requiring that $(\vec{a}, \phi', \alpha) \in [\![\theta]\!]$. Thus, writing ϕ_0 instead of ϕ' and setting $\gamma_0 = [c_{\mathrm{in}} \mapsto (M, \theta)]$, the transition brings us to a configuration $\langle [(M\vec{a}, \alpha)], \gamma_0, \phi_0 \rangle$, where $\mathrm{dom}(\phi_0) = \{a_1, ..., a_n\}$.
- At this point, the term $M\vec{a}$ can be reduced using head reduction and brought to head normal form. Applying the (INT) rule we reach some $\langle [(E[c\hat{M}_1 \cdots \hat{M}_k], \alpha)], \gamma_0, \phi_0 \rangle$.
- We next interrogate the computation name c. The latter must have come from the $a_1, ..., a_n$ that were applied to M, hence is an O-name. To interrogate it, P plays a question $\bar{c}(\vec{a}')$, using the (PQ) rule and assuming $(\vec{a}', \gamma', \alpha') \in \mathsf{AVal}(((\hat{M}_1, \tau_1'), ..., (\hat{M}_k, \tau_k'), \xi))$, $\phi_0(c) = \theta'$, $\mathrm{ext}(\theta') = (\tau_1', ..., \tau_k', \xi)$. This leads to $\langle [(E, \alpha' \rightsquigarrow \alpha)], \gamma_1, \phi_0 \rangle$ $(\gamma_1 = \gamma_0 \cdot \gamma')$.
- We are now at a passive configuration, where E has been stored on the stack and O is required to produce a response of type α'. By definition of AVal, either $\alpha' = \alpha$ or α' is in $a_1', ..., a_k'$ and hence belongs to P. In the latter case, O can only produce such a response by calling back P, using rule (OQ), playing an O-question and adding a new term on the evaluation stack. In the former case, O would directly respond with a hnf of type α, say N. But, since $E : \alpha \rightsquigarrow \alpha$ and therefore $E = \bullet$, P would simply reply back playing N again. To avoid this copycat of hnf's, we simply play an OP-answer and remove the top of the evaluation stack – this is what the (OA) rule achieves.

Example 15. In Fig. 4 we include example traces for terms M_1, M_2 : **Unit** → **Unit** (taken from [1], Instance 3.25) and for the Church numerals M_k : **Nat**. The former pair is an instance of Theorem 21 – Strachey equivalence implies trace equivalence.

In our scenario above we started from a passive configuration with empty stack and a singleton γ. A different way to produce a trace is to start from an active configuration with a stack containing only a term $E[c_{\mathrm{in}}\hat{M}_1 \cdots \hat{M}_n]$, in which case the rule (PQ_0) would commence the trace. More generally, we call a configuration C with stack $\mathcal{E}$:

$M_1 = \lambda f^{\mathbf{Unit}}.\, f\, \mathbf{Unit}\, f : \mathbf{Unit} \to \mathbf{Unit}$, $M_2 = \lambda f^{\mathbf{Unit}}.\Lambda X.\, f(X \to X)(fX) : \mathbf{Unit} \to \mathbf{Unit}$, and $\mathrm{ext}(\mathbf{Unit} \to \mathbf{Unit}) = (\mathbf{Unit}, \forall X, X, X)$. Traces for M_1 (left) and M_2 (right):

$$
\begin{aligned}
&\langle \Diamond, \gamma_0, \varepsilon, \varepsilon \rangle \qquad (\gamma_0 = [c_{\mathrm{in}} \mapsto (M_1, \theta)])\\
&\xrightarrow{c_{\mathrm{in}}(c_1,\alpha_1,c_2)} \langle (M_1\, c_1\, \alpha_1\, c_2, \alpha_1), \gamma_0, \phi_0 \rangle\\
&\to \langle (c_1\, \mathbf{Unit}\, c_1\, \alpha_1\, c_2, \alpha_1), \gamma_0, \phi_0 \rangle\\
&\xrightarrow{\bar{c}_1(\alpha_2,c_3)} \langle (\bullet\, \alpha_1\, c_2, \alpha_2 \rightsquigarrow \alpha_1), \gamma_1, \phi_0 \rangle\\
&\xrightarrow{c_3()} \langle (c_1, \alpha_2) :: (\bullet\, \alpha_1\, c_2, \alpha_2 \rightsquigarrow \alpha_1), \gamma_1, \phi_0 \rangle\\
&\xrightarrow{\overline{\mathsf{OK}}\mathsf{OK}} \langle (c_1\, \alpha_1\, c_2, \alpha_1), \gamma_1, \phi_0 \rangle\\
&\xrightarrow{\bar{c}_1(\alpha_2',c_3')} \langle (\bullet, \alpha_2' \rightsquigarrow \alpha_1), \gamma_2, \phi_0 \rangle\\
&\xrightarrow{c_3'()} \langle (c_2, \alpha_2') :: (\bullet, \alpha_2' \rightsquigarrow \alpha_1), \gamma_2, \phi_0 \rangle\\
&\xrightarrow{\overline{\mathsf{OK}}\mathsf{OK}} \langle (c_2, \alpha_1), \gamma_2, \phi_0 \rangle\\
&\xrightarrow{\bar{c}_2()} \langle (\bullet, \alpha_1 \rightsquigarrow \alpha_1), \gamma_1, \phi_0 \rangle \xrightarrow{\mathsf{OK}\overline{\mathsf{OK}}} \langle \Diamond, \gamma_1, \phi_0 \rangle
\end{aligned}
$$

$$
\begin{aligned}
&\langle \Diamond, \gamma_0', \varepsilon, \varepsilon \rangle \qquad (\gamma_0' = [c_{\mathrm{in}} \mapsto (M_2, \theta)])\\
&\xrightarrow{c_{\mathrm{in}}(c_1,\alpha_1,c_2)} \langle (M_2\, c_1\, \alpha_1\, c_2, \alpha_1), \gamma_0', \phi_0 \rangle\\
&\to \langle (c_1(\alpha_1 \to \alpha_1)(c_1\alpha_1)c_2, \alpha_1), \gamma_0', \phi_0 \rangle\\
&\xrightarrow{\bar{c}_1(\alpha_2,c_3)} \langle (\bullet\, c_2, \alpha_2 \rightsquigarrow \alpha_1), \gamma_1', \phi_0 \rangle\\
&\xrightarrow{c_3()} \langle (c_1\alpha_1, \alpha_2) :: (\bullet\, c_2, \alpha_2 \rightsquigarrow \alpha_1), \gamma_1', \phi_0 \rangle\\
&\xrightarrow{\overline{\mathsf{OK}}\mathsf{OK}} \langle (c_1\, \alpha_1\, c_2, \alpha_1), \gamma_1', \phi_0 \rangle \to \ \dots
\end{aligned}
$$

where:

$$
\begin{aligned}
\phi_0 &= \{c_1 \mapsto \mathbf{Unit}, \alpha_1 \mapsto \mathcal{U}, c_2 \mapsto \alpha_1\}\\
\gamma_1 &= \gamma_0 \cdot [\alpha_2 \mapsto (\mathbf{Unit}, \mathcal{U}), c_3 \mapsto (c_1, \mathbf{Unit})]\\
\gamma_2 &= \gamma_1 \cdot [\alpha_2' \mapsto (\alpha_1, \mathcal{U}), c_3' \mapsto (c_2, \alpha_2')]\\
\gamma_1' &= \gamma_0' \cdot [\alpha_2 \mapsto (\alpha_1 \to \alpha_1, \mathcal{U}), c_3 \mapsto (c_1\alpha_1, \alpha_1 \to \alpha_1)]
\end{aligned}
$$

$$
M_k = \Lambda X.\lambda f^{X \to X}.\, \lambda x^{X}.\, N_{f,x,k} \qquad N_{M_f,M_x,k} = \underbrace{M_f(M_f(\dots(M_f}_{k}\, M_x))\dots)
$$
$$
\mathrm{ext}(\mathbf{Nat}) = (\forall X, X \to X, X, X)
$$

Set $\gamma_0 = [c_{\mathrm{in}} \mapsto (M_k, \mathbf{Nat})]$. Reduction for M_k:

$$
\begin{aligned}
&\langle \Diamond, \gamma_0, \varepsilon, \varepsilon \rangle \xrightarrow{c_{\mathrm{in}}(\alpha_1,c_f,c_x)} \langle (M_k\, \alpha_1\, c_f\, c_x, \alpha_1), \gamma_0, \phi_0 \rangle \to \langle (c_f(N_{c_f,c_x,k-1}), \alpha_1), \gamma_0, \phi_0 \rangle\\
&\xrightarrow{\bar{c}_f(c_1)} \langle (\bullet, \alpha_1 \rightsquigarrow \alpha_1), \gamma_1, \phi_0 \rangle \xrightarrow{\mathsf{OK}\overline{\mathsf{OK}}} \langle \Diamond, \gamma_1, \phi_0 \rangle \xrightarrow{c_1()} \langle (c_f(N_{c_f,c_x,k-2}), \alpha_1), \gamma_1, \phi_0 \rangle\\
&\xrightarrow{\bar{c}_f(c_2)} \langle (\bullet, \alpha_1 \rightsquigarrow \alpha_1), \gamma_2, \phi_0 \rangle \xrightarrow{\mathsf{OK}\overline{\mathsf{OK}}} \langle \Diamond, \gamma_2, \phi_0 \rangle \cdots \xrightarrow{c_{k-1}()} \langle (c_x, \alpha_1), \gamma_{k-1}, \phi_0 \rangle\\
&\xrightarrow{\bar{c}_x()} \langle (\bullet, \alpha_1 \rightsquigarrow \alpha_1), \gamma_{k-1}, \phi_0 \rangle \xrightarrow{\mathsf{OK}\overline{\mathsf{OK}}} \langle \Diamond, \gamma_{k-1}, \phi_0 \rangle
\end{aligned}
$$

where $\phi_0 = \{\alpha_1 \mapsto \mathcal{U}, c_f \mapsto (\alpha_1 \to \alpha_1), c_x \mapsto \alpha_1\}$ and $\gamma_i = \gamma_{i-1} \cdot [c_i \mapsto (N_{c_f,c_x,k-i}, \alpha_1)]$.

Fig. 4. Top: traces for two terms of type **Unit→Unit**. Bottom: traces for Church numeral M_k.

- a ***term configuration***, if $\mathcal{E} = \Diamond$ or the bottom element of $\mathcal{E}$ has type α or $\alpha \rightsquigarrow \alpha'$;
- a ***context configuration***, if the bottom of $\mathcal{E}$ has type θ or $\alpha \rightsquigarrow \theta$, and θ is a closed with empty support.

Each reduction sequence in the LTS can only contain either term or context configurations. In our discussion above and in Example 15 we examine the semantics of terms, and therefore use term configurations. In later sections, when we shall start looking at the semantics of contexts, we will be using context configurations as well.

While we have not defined leaves for our LTS, there is a natural notion of a trace being "completed". In particular, we call a trace T ***complete*** if all its questions have been answered. We write $\mathsf{CTr}(C)$ for the set of complete traces generated from C. Term and context configurations can both produce complete traces. Given a term configuration C and a complete trace T, we write $C \Downarrow_T$ if $C \xrightarrow{T} C'$ and C' has an empty evaluation stack. On the other hand, given a context configuration C, a complete trace T and a value v, we write $C \Downarrow_{T,v}$ if $C \xrightarrow{T} C'$ and C' has an evaluation stack with a single element (v, θ).

Lemma 16. *Given a term configuration C and $T \in \mathsf{Tr}(C)$, then T is complete iff $C \Downarrow_T$.*

We conclude this section by looking at some restrictions characterising actual configurations. We first extend fst to evaluation stacks by: $\mathsf{fst}(\Diamond) = \Diamond$ and $\mathsf{fst}((Z, \_) :: \mathcal{E}) = Z :: \mathsf{fst}(\mathcal{E})$.

Definition 17. A configuration $\langle \mathcal{E}, \gamma, \phi \rangle$ is said to be ***legal*** when:

- $\mathrm{dom}(\gamma) \cap \mathrm{dom}(\phi) = \varnothing$ and $\nu(\mathsf{fst}(\mathcal{E})) \cup \nu(\mathrm{cod}(\mathsf{fst}(\gamma))) \subseteq \mathrm{dom}(\phi)$;
- for all $c \in \mathrm{dom}(\gamma) \cap \mathsf{CN}$, given $\gamma(c) = (M, \theta)$, we have $\Delta_\phi; \Gamma_{\phi,\gamma} \vdash M : \theta\{\gamma_v\}$;
- if the top of $\mathcal{E}$ is (M, θ), then $\Delta_\phi; \Gamma_{\phi,\gamma} \vdash M : \widetilde{\theta}$ with either $\theta = \alpha \in \mathrm{dom}(\gamma)$ and $\gamma(\alpha) = (\widetilde{\theta}, \mathcal{U})$, or $\theta = \alpha \in \mathrm{dom}(\phi)$ and $\widetilde{\theta} = \theta$, or $\theta = \widetilde{\theta}$ is a closed type with empty support and $\mathcal{E} = [(M, \theta)]$;
- If $\mathcal{E} = (M, \alpha_1) :: (E, \alpha_2 \rightsquigarrow \theta) :: \mathcal{E}'$, either $\alpha_1 = \alpha_2$ or $\alpha_1 \in \mathrm{dom}(\phi)$;
- for all $(E, \alpha \rightsquigarrow \theta)$ in $\mathcal{E}$ with $\alpha \in \mathrm{dom}(\gamma)$, $\Delta_\phi; \Gamma_{\phi,\gamma}, \vdash E : \gamma_v(\alpha) \rightsquigarrow \theta$, and either $\theta = \alpha \in \mathrm{dom}(\phi)$ or θ is a closed type with empty support, and $(E, \alpha \rightsquigarrow \theta)$ is at the bottom of $\mathcal{E}$;
- for all $(E, \alpha \rightsquigarrow \theta)$ in $\mathcal{E}$ with $\alpha \in \mathrm{dom}(\phi)$, we have $\theta = \alpha$ and $E = \bullet$;

where $\Delta_\phi = \mathrm{dom}(\phi) \cap \mathsf{TN}$ and $\Gamma_{\phi,\gamma} = \{(x, \theta\{\mathsf{fst}(\gamma)\}) \mid (x, \theta) \in \phi\}$.

Lemma 18. *If C is a legal configuration and $C \xrightarrow{m} C'$ then C' is a legal configuration.*

4 Parametricity in the Trace Model, and Proof of Theorem 5

We next examine the relationship between trace equivalence and the notions of Reynolds and Strachey equivalence. We prove that Strachey equivalence is included in trace equivalence (Theorem 21), which in turn is included in Reynolds equivalence (Theorem 28).

4.1 From Strachey to Trace Equivalence

Definition 19. Let $C_i = \langle \mathcal{E}_i, \gamma_i, \phi_i \rangle$, for $i = 1, 2$, be two configurations. We say that C_1 and C_2 are ***Strachey-equivalent*** when $\mathcal{E}_1$ and $\mathcal{E}_2$ have the same size, $\mathrm{dom}(\gamma_1) = \mathrm{dom}(\gamma_2)$, $\phi_1 = \phi_2$ and:

- for all $c \in \mathrm{dom}(\gamma_1)$, if $\gamma_i(c) = (M_i, \theta_i)$ then $\theta_1 = \theta_2$ and $\mathbf{erase}(M_1) =_{\beta\eta} \mathbf{erase}(M_2)$;
- if (Z_i, α_i) is the j-th element of $\mathcal{E}_i$, then $\alpha_1 = \alpha_2$ and $\mathbf{erase}(Z_1) =_{\beta\eta} \mathbf{erase}(Z_2)$;

where $E_1 =_{\beta\eta} E_2$ just if $E_1[x] =_{\beta\eta} E_2[x]$ for some/all fresh x.

The first inclusion can then be proven as follows.

Lemma 20. *Given two Strachey-equivalent legal configurations C_1, C_2, if $C_1 \xrightarrow{m} C_1'$ for some m, C_1' then there is $C_2 \xrightarrow{m} C_2'$ such that C_1' and C_2' are Strachey-equivalent.*

Theorem 21. *For all Strachey-equivalent $\Delta, \Gamma \vdash M_1, M_2 : \theta$, we have $[\![M_1]\!] = [\![M_2]\!]$.*

Proof. Taking $T \in [\![\Delta; \Gamma \vdash M_1 : \theta]\!]$, we prove that $T \in [\![\Delta; \Gamma \vdash M_2 : \theta]\!]$ by induction on the length of T, using the previous lemma. □

The inclusion above is strict. This is shown, for example, by the following terms $M_{\mathbf{true}}, M_{\mathbf{false}} : \mathbf{Unit} \to \mathbf{Unit}$, which are trace equivalent but not Strachey-equivalent:

$$M_{\mathbf{b}} = \lambda f^{\mathbf{Unit}}.\Lambda X.\lambda x^X.\mathbf{snd}(f(\mathbf{Bool} \times X)\langle \mathbf{b}, x\rangle) \quad (\mathbf{b} = \mathbf{true}, \mathbf{false})$$

Here we use the impredicative encoding of product types [8]: $\theta_1 \times \theta_2 = \forall X.(\theta_1 \to \theta_2 \to X) \to X$, $\langle M, N\rangle = \Lambda X.\lambda f^{\theta_1 \to \theta_2 \to X}.fMN$ and $\mathbf{snd} = \lambda x^{\theta_1 \times \theta_2}.x\theta_2(\lambda y^{\theta_1}.\lambda z^{\theta_2}.z)$. Setting $\gamma_0 = [c_{\mathrm{in}} \mapsto (M_{\mathbf{b}}, \mathbf{Unit} \to \mathbf{Unit})]$ and $C_{\mathbf{b}} = \langle \cdot; \cdot \vdash M_{\mathbf{b}} : \mathbf{Unit} \to \mathbf{Unit}\rangle$, we have:

$$\begin{array}{l} C_{\mathbf{b}} \xrightarrow{c_{\mathrm{in}}(c_f,\alpha,c)} \langle (\mathbf{snd}(c_f(\mathbf{Bool} \times \alpha)\langle \mathbf{b}, c\rangle), \alpha), \gamma_0, \phi_0\rangle \ \ (\phi_0 = [c_f \mapsto \mathbf{Unit}, \alpha \mapsto \mathcal{U}, c \mapsto \alpha]) \\ \xrightarrow{\bar{c}_f(\beta,c')} \langle (\mathbf{snd}\bullet, \beta \rightsquigarrow \alpha), \gamma_1, \phi_0\rangle \qquad (\gamma_1 = \gamma_0 \cdot [\beta \mapsto (\mathbf{Bool} \times \alpha, \mathcal{U}), c' \mapsto (\langle \mathbf{b}, c\rangle, \beta)]) \\ \xrightarrow{c'()} \langle ((\langle \mathbf{b}, c\rangle, \beta) :: (\mathbf{snd}\bullet, \beta \rightsquigarrow \alpha), \gamma_1, \phi_0\rangle \xrightarrow{\overline{\mathsf{OK}}\mathsf{OK}} \langle (\mathbf{snd}\langle \mathbf{b}, c\rangle, \alpha), \gamma_1, \phi_0\rangle \\ \longrightarrow \langle (c, \alpha), \gamma_1, \phi_0\rangle \xrightarrow{\bar{c}()} \langle (\bullet, \alpha \rightsquigarrow \alpha), \gamma_1, \phi_0\rangle \xrightarrow{\mathsf{OK}\overline{\mathsf{OK}}} \langle \diamond, \gamma_1, \phi_0\rangle \end{array}$$

and this is the only complete trace in $[\![M_{\mathbf{b}}]\!]$. Indeed, O cannot interrogate another name, as c_{in} can only be played once, and c' cannot be played with the (OQ_0) rule.

The other inclusion (trace included in Reynolds) is more challenging and requires us to introduce machinery for relating the semantics of terms and semantics of contexts to that of terms and contexts composed.

4.2 Composite LTS

We let a ***composite configuration*** be a tuple $\langle \mathcal{E}_P, \mathcal{E}_O, \gamma_P, \gamma_O\rangle$, where γ_P and γ_O are maps γ as above, $\mathcal{E}_P$ is a term evaluation stack, and $\mathcal{E}_O$ is a context evaluation stack. These configurations represent the interaction between a term

(P-INT) $\langle (M,\alpha) :: \mathcal{E}_P, \mathcal{E}_O, \gamma_P, \gamma_O\rangle \longrightarrow \langle (M',\alpha) :: \mathcal{E}_P, \mathcal{E}_O, \gamma_P, \gamma_O\rangle$ when $M \to^* M'$ (hnf).

(O-INT) $\langle \mathcal{E}_P, (M,\alpha) :: \mathcal{E}_O, \gamma_P, \gamma_O\rangle \longrightarrow \langle (M',\alpha) :: \mathcal{E}_P, \mathcal{E}_O, \gamma_P, \gamma_O\rangle$ when $M \to^* M'$ (hnf).

(PA) $\langle (M,\alpha) :: (E, \alpha \rightsquigarrow \alpha') :: \mathcal{E}_P, (\bullet, \alpha \rightsquigarrow \alpha) :: \mathcal{E}_O, \gamma_P, \gamma_O\rangle \xrightarrow{\overline{\mathsf{OK}}\mathsf{OK}} \langle (E[M],\alpha') :: \mathcal{E}_P, \mathcal{E}_O, \gamma_P, \gamma_O\rangle$ with M a hnf and $\alpha \in \mathrm{dom}(\gamma_P)$.

(OA) $\langle (\bullet, \alpha \rightsquigarrow \alpha) :: \mathcal{E}_P, (M,\alpha) :: (E, \alpha \rightsquigarrow \theta) :: \mathcal{E}_O, \gamma_P, \gamma_O\rangle \xrightarrow{\mathsf{OK}\overline{\mathsf{OK}}} \langle \mathcal{E}_P, (E[M],\theta) :: \mathcal{E}_O, \gamma_P, \gamma_O\rangle$ with M a hnf and $\alpha \in \mathrm{dom}(\gamma_O)$.

(PQ) $\langle (E[c\hat{M}_1 \cdots \hat{M}_n],\alpha') :: \mathcal{E}_P, \mathcal{E}_O, \gamma_P, \gamma_O\rangle$
$\xrightarrow{\bar{c}(\vec{a})} \langle (E, \alpha \rightsquigarrow \alpha') :: \mathcal{E}_P, (M\vec{a},\alpha) :: \mathcal{E}_O, \gamma_P \cdot \gamma', \gamma_O\rangle$ when $\alpha' \in \mathrm{dom}(\gamma_O)$, $\gamma_O(c) = (M,\theta)$, $\mathrm{ext}(\theta) = (\tau_1, \ldots, \tau_n, \xi)$ and $((a_1, \ldots, a_n), \gamma', \alpha) \in \mathsf{AVal}((\hat{M}_1, \tau_1), \ldots, (\hat{M}_2, \tau_n), \xi)$.

(OQ) $\langle \mathcal{E}_P, (E[c\hat{M}_1 \cdots \hat{M}_n],\theta) :: \mathcal{E}_O, \gamma_P, \gamma_O\rangle \xrightarrow{c(\vec{a})} \langle (M\vec{a},\alpha) :: \mathcal{E}_P, (E, \alpha \rightsquigarrow \theta) :: \mathcal{E}_O, \gamma_P, \gamma_O \cdot \gamma'\rangle$ when $\theta = \alpha' \in \mathrm{dom}(\gamma_P)$ or θ a closed type with empty support, with $\gamma_P(c) = (M,\theta)$, $\mathrm{ext}(\theta) = (\tau_1, \cdots, \tau_n, \xi)$ and $((a_1, \ldots, a_n), \gamma', \alpha) \in \mathsf{AVal}((\hat{M}_1, \tau_1), \cdots, (\hat{M}_2, \tau_n), \xi)$.

Fig. 5. Composite LTS.

and a context. The term-part in the interaction is played by $\mathcal{E}_P$ and γ_P, while the context-part by $\mathcal{E}_O$ and γ_O. As with ordinary configurations, we define an LTS for composite ones in Fig. 5. Given a composite configuration C, a trace T and a value v (hnf with empty support) we write $C \Downarrow_{T,v}$ when $C \xrightarrow{T} \langle \Diamond, [(v,\theta)], \gamma_P, \gamma_O\rangle$.

Composite configurations allow us to compose a term and a context semantically: we essentially play the traces of one against the other. Another way to obtain a composite semantics is to work syntactically, i.e. by composing configurations and then executing the resulting term. This is defined next.

Definition 22. Given two evaluation stacks $(\mathcal{E}_P, \mathcal{E}_O)$, we build their ***merge*** (which may not always be defined) $\mathcal{E}_P || \mathcal{E}_O$ inductively by $\Diamond || [(M,\theta)] = M$ and:

$$((M,\alpha) :: \mathcal{E}_P) || ((E, \alpha \rightsquigarrow \theta) :: \mathcal{E}_O) = \mathcal{E}_P || ((E[M],\theta) :: \mathcal{E}_O)$$
$$((E, \alpha \rightsquigarrow \theta) :: \mathcal{E}_P) || ((M,\alpha) :: \mathcal{E}_O) = ((E[M],\theta) :: \mathcal{E}_P) || \mathcal{E}_O$$

When it is defined, we say that $\mathcal{E}_P, \mathcal{E}_O$ are ***compatible***. Then, a composite configuration $C = \langle \mathcal{E}_P, \mathcal{E}_O, \gamma_P, \gamma_O\rangle$ is ***legal*** when $(\mathcal{E}_P, \mathcal{E}_O)$ are compatible and when both $\langle \mathcal{E}_P, \gamma_P, \mathsf{snd}(\gamma_O)\rangle$ and $\langle \mathcal{E}_O, \gamma_O, \mathsf{snd}(\gamma_P)\rangle$ are legal.

We now relate the reduction of a composite configuration with the head reduction of the merge of its two evaluation stacks. First, taking the two environments γ_P, γ_O of a legal composite configuration, we compute their *closure* $(\gamma_P \cdot \gamma_O)^*$ as follows. Setting $\gamma^0 = \mathsf{fst}(\gamma_P \cdot \gamma_O)$, and $\gamma^i = \{(a, \hat{M}\{\gamma\}) \mid (a, \hat{M}) \in \gamma^{i-1}\}$ $(i > 0)$, there is an integer n such that $\nu(\mathrm{cod}(\gamma^n)) = \varnothing$. We write $(\gamma_P \cdot \gamma_O)^*$ for the environment defined as γ^n, for the least n satisfying this latter condition.

Theorem 23. *Given a legal composite configuration $C = \langle \mathcal{E}_P, \mathcal{E}_O, \gamma_P, \gamma_O\rangle$, then $C \Downarrow_{T,v}$ iff $(\mathcal{E}_P || \mathcal{E}_O)\{(\gamma_P \cdot \gamma_O)^*\} \to^* v$.*

Finally, we relate the LTS's for composite configurations and ordinary configurations (Theorem 26). Combined with Theorem 23, this gives us a correlation between the traces of two compatible configurations and the head reduction we obtain once we merge their evaluation stacks.

Definition 24. Given legal configurations $C_P = \langle \mathcal{E}_P, \gamma_P, \phi_P \rangle$ and $C_O = \langle \mathcal{E}_O, \gamma_O, \phi_O \rangle$, we say that they are *compatible* when $\mathcal{E}_P, \mathcal{E}_O$ are compatible, $\mathsf{snd}(\gamma_P) = \phi_O$ and $\mathsf{snd}(\gamma_O) = \phi_P$. For each pair (C_P, C_O) of compatible configurations, we define their merge $C_P \mathbin{\bar\wedge} C_O$ as the composite configuration $\langle \mathcal{E}_P, \mathcal{E}_O, \gamma_P, \gamma_O \rangle$.

Lemma 25. *Taking (C_P, C_O) a pair of compatible configurations, $C_P \mathbin{\bar\wedge} C_O \Downarrow_{T,v}$ iff $C_P \Downarrow_T$ and $C_O \Downarrow_{T^\perp,v}$.*

Theorem 26. *Given $C_{P,1}, C_{P,2}, C_O$ such that $C_{P,1}, C_O$ and $C_{P,2}, C_O$ are pairwise compatible and $\mathsf{Tr}(C_{P,1}) = \mathsf{Tr}(C_{P,2})$, if $C_{P,1} \mathbin{\bar\wedge} C_O \Downarrow_{T,v}$, then $C_{P,2} \mathbin{\bar\wedge} C_O \Downarrow_{T,v}$.*

Proof. From Lemma 25 we get $C_{P,1} \Downarrow_T$ and $C_O \Downarrow_{T^\perp,v}$. Thus, $T \in \mathsf{Tr}(C_{P,1})$ and hence $T \in \mathsf{Tr}(C_{P,2})$. Lemma 16 then yields $C_{P,2} \Downarrow_T$ and, from Lemma 25, $C_{P,2} \mathbin{\bar\wedge} C_O \Downarrow_{T,v}$. □

4.3 Proof of Theorem 5

Theorem 5 follows from Theorems 21 and 28. Theorem 28, which is proved below, shows that any trace equivalent terms are also Reynolds equivalent. This is achieved as follows. In the previous section we saw how to relate reductions of terms-in-context to the semantics of terms and contexts. Given terms M_1, M_2 which are trace equivalent, and fully applying them to related arguments, we obtain head reductions to values. These reductions can be decomposed into LTS reductions producing corresponding traces, for the terms and their argument terms (which form contexts). But, since the terms are trace equivalent, M_2 can simulate the behaviour of M_1 in the context of M_1, and that allows us to show that the two composites reduce to the same value.

We start by extending logical relations to extended types with empty support. We define $\mathcal{R}[\![\mathrm{ext}(\theta)]\!]_\delta$ by:

$$\begin{aligned} \mathcal{R}[\![(X)]\!]_\delta &= \{R \mid \delta(X) = (\_, \_, R)\} \\ \mathcal{R}[\![\theta :: L]\!]_\delta &= \{(M_1, N_1) :: L' \mid (M_1, N_1) \in \mathcal{R}[\![\theta]\!]_\delta \wedge L' \in \mathcal{R}[\![L]\!]_\delta\} \\ \mathcal{R}[\![\forall X :: L]\!]_\delta &= \{(\theta_1, \theta_2) :: L' \mid (\theta_1, \theta_2, R) \in \mathrm{Rel} \wedge L' \in \mathcal{R}[\![L]\!]_{\delta \cdot [X \mapsto (\theta_1, \theta_2, R)]}\} \end{aligned}$$

Lemma 27. $(M_1, M_2) \in \mathcal{R}[\![\theta]\!]_\delta$ *iff for all* $((\hat{N}_1^1, \hat{N}_2^1), \ldots, (\hat{N}_1^n, \hat{N}_2^n), R) \in \mathcal{R}[\![\mathrm{ext}(\theta)]\!]_\delta$, $(M_1 \hat{N}_1^1 \cdots \hat{N}_1^n, M_2 \hat{N}_2^1 \cdots \hat{N}_2^n) \in R$.

Theorem 28. *For all trace equivalent $\Delta; \Gamma \vdash M_1, M_2 : \theta$, we have that $M_1 \simeq_{log} M_2$.*

Proof. Taking $\delta \in \mathcal{R}[\![\Delta]\!]$ and $(\eta_1, \eta_2) \in \mathcal{R}[\![\Gamma]\!]_\delta$, we show $(M_1\{\eta_1\}\{\delta_1\}, M_2\{\eta_2\}\{\delta_2\}) \in \mathcal{R}[\![\theta]\!]_\delta$. Using Lemma 27, we take $((\hat{N}_1^1, \hat{N}_2^1), \ldots, (\hat{N}_1^n, \hat{N}_2^n), R) \in \mathcal{R}[\![\mathrm{ext}(\theta)]\!]_\delta$, and prove that $(M_1\{\eta_1\}\{\delta_1\}\hat{N}_1^1 \cdots \hat{N}_1^n, M_2\{\eta_2\}\{\delta_2\}\hat{N}_2^1 \cdots \hat{N}_2^n) \in R$.

For each $i \in \{1, 2\}$, there exists a value v_i s.t. $M_i\{\eta_i\}\{\delta_i\}\hat{N}_i^1 \cdots \hat{N}_i^n \to^* v_i$. Using the closure of R w.r.t. $=_{\beta\eta}$, it suffices to show that $(v_1, v_2) \in R$. Suppose $\Delta = X_1, \ldots, X_k$ and $\Gamma = x_1 : \theta_1, \ldots, x_m : \theta_m$. We write

C_{P_i} for the configuration $\langle \Delta; \Gamma \vdash M_i : \theta \rangle$, and $C_{O,i}$ for the configuration $\langle c_{\mathrm{in}} \delta_i(X_1) \cdots \delta_i(X_k) \eta_i(x_1) \cdots \eta_i(x_m) \hat{N}_i^1 \cdots \hat{N}_i^n, \varepsilon, [c_{\mathrm{in}} \mapsto \widetilde{\theta}] \rangle$, where $\widetilde{\theta} = \forall X_1 . \ldots . \forall X_n . \theta_1 \to \cdots \to \theta_m \to \theta$.

From Theorem 23, for each $i \in \{1,2\}$ there is a trace T_i such that $C_{P,i} \mathbin{⩕} C_{O,i} \Downarrow_{T_i, v_i}$. M_1, M_2 being trace equivalent, we have that $\mathsf{Tr}(C_{P,1}) = \mathsf{Tr}(C_{P,2})$. So from Theorem 26, we get that $C_{P,2} \mathbin{⩕} C_{O,1} \Downarrow_{T_1, v_1}$, and from Theorem 23 that $M_2\{\eta_1\}\{\delta_1\}\hat{N}_1^1 \cdots \hat{N}_1^n \to^* v_1$. Finally, from Theorem 2, we get that $(M_2\{\eta_1\}\{\delta_1\}\hat{N}_1^1 \cdots \hat{N}_1^n, M_2\{\eta_2\}\{\delta_2\}\hat{N}_2^1 \cdots \hat{N}_2^n) \in R$. Thus, using the closure of R w.r.t. $=_{\beta\eta}$, we have that $(v_1, v_2) \in R$. □

5 Related and Future Work

The literature on parametric polymorphism is vast; here we look at the works closest to ours, which come from the game semantics area. The first game model for System F was introduced by Hughes [9,10]. The model is intentional, in the sense that it is fully complete for $\beta\eta$-equivalence. Starting from that model, de Lataillade [5,6] characterised parametricity categorically via the notion of dinaturality [4]. In [2], Abramsky and Jagadeesan developed a model for System F to characterise genericity, as introduced by Longo et al. [17]. A type θ is said to be *generic* when two terms M_1, M_2 of type $\forall X.\theta'$ are equivalent just if $M_1\theta$ and $M_2\theta$ are equivalent. Their model contains several generic types. More recently, Laird [15] has introduced a game model for System F augmented with mutable variables. His model is closer to ours than the previous ones, and in particular his notion of copycat links can be seen as connected to the use of names for parametricity.

In all of the above models the denotation of terms is built compositionally by induction on the structure of the term. In a different line of work, closer in spirit to our model, Lassen and Levy [16] have introduced normal form bisimulations for a language with parametric polymorphism. These bisimulations are defined on LTSs whose definition has similarities with ours. However, the model is for a CPS-style language which has not only polymorphic but also recursive types. Finally, our own model for a higher-order polymorphic language with general references [13] can be seen as a direct precursor to this work, albeit in a very different setting (call-by-value, with references).

Further on, we would like to study the existence of generic types in our model, as well as its dinaturality properties. We would moreover like to examine coarser notions of trace equivalence that bring us closer to Reynolds polymorphism. Finally, we would like to see if the trace model can be used to prove the original conjecture of [1,20]. While this seems plausible in principle, proving equivalences using definable logical relations requires additional tools, such as restrictions on the LTS, to avoid circular reasoning.

Acknowledgement. Authors supported by the LABEX MILYON (ANR-10-LABX-0070) of Université de Lyon, and the EPSRC (EP/P004172/1) respectively.

References

1. Abadi, M., Cardelli, L., Curien, P.-L.: Formal parametric polymorphism. Theor. Comput. Sci. **121**(1&2), 9–58 (1993)
2. Abramsky, S., Jagadeesan, R.: A game semantics for generic polymorphism. Ann. Pure Appl. Logic **133**(1), 3–37 (2005). Festschrift on the occasion of Helmut Schwichtenberg's 60th birthday
3. Abramsky, S., Jagadeesan, R., Malacaria, P.: Full abstraction for PCF. Inf. Comput. **163**(2), 409–470 (2000)
4. Bainbridge, E.S., Freyd, P.J., Scedrov, A., Scott, P.J.: Functorial polymorphism. Theor. Comput. Sci. **70**(1), 35–64 (1990)
5. de Lataillade, J.: Quantification du second ordre en sémentique des jeux: application aux isomorphismes de types. Ph.D. thesis, Paris 7 (2007)
6. de Lataillade, J.: Second-order type isomorphisms through game semantics. Ann. Pure Appl. Logic **151**(2–3), 115–150 (2008)
7. Gabbay, M., Pitts, A.M.: A new approach to abstract syntax with variable binding. Formal Asp. Comput. **13**(3–5), 341–363 (2002)
8. Girard, J.-Y., Lafont, Y., Taylor, P.: Proofs and Types, vol. 7. Cambridge University Press, Cambridge (1989)
9. Hughes, D.: Hypergame semantics: full completeness for System F. Ph.D. thesis, D. Phil. thesis, Oxford University (2000)
10. Hughes, D.J.D.: Games and definability for System F. In: Proceedings of the 12th Annual IEEE Symposium on Logic in Computer Science, LICS 1997, Washington, DC, USA, 76 pages. IEEE Computer Society (1997)
11. Hyland, J.M.E., Ong, C.L.: On full abstraction for PCF: I, II, and III. Inf. Comput. **163**(2), 285–408 (2000)
12. Jaber, G.: Operational nominal game semantics. In: Pitts, A. (ed.) FoSSaCS 2015. LNCS, vol. 9034, pp. 264–278. Springer, Heidelberg (2015). https://doi.org/10.1007/978-3-662-46678-0_17
13. Jaber, G., Tzevelekos, N.: Trace semantics for polymorphic references. In: Proceedings of the 31st Annual ACM/IEEE Symposium on Logic in Computer Science, LICS 2016, New York, NY, USA, 5–8 July 2016, pp. 585–594 (2016)
14. Laird, J.: A fully abstract trace semantics for general references. In: Arge, L., Cachin, C., Jurdziński, T., Tarlecki, A. (eds.) ICALP 2007. LNCS, vol. 4596, pp. 667–679. Springer, Heidelberg (2007). https://doi.org/10.1007/978-3-540-73420-8_58
15. Laird, J.: Game semantics for a polymorphic programming language. J. ACM **60**(4), 29:1–29:27 (2013)
16. Lassen, S.B., Levy, P.B.: Typed normal form bisimulation for parametric polymorphism. In: Proceedings of the 2008 23rd Annual IEEE Symposium on Logic in Computer Science, LICS 2008, Washington, DC, USA, pp. 341–352. IEEE Computer Society (2008)
17. Longo, G., Milsted, K., Soloviev, S.: The genericity theorem and parametricity in the polymorphic lambda-calculus. Theor. Comput. Sci. **121**(1–2), 323–349 (1993)
18. Mitchell, J.C.: On the equivalence of data representations. In: Lifschitz, V. (ed.) Artificial Intelligence and Mathematical Theory of Computation, pp. 305–329. Academic Press Professional Inc., San Diego (1991)
19. Pitts, A.M.: Nominal Sets: Names and Symmetry in Computer Science. Cambridge University Press, New York (2013)

20. Plotkin, G., Abadi, M.: A logic for parametric polymorphism. In: Bezem, M., Groote, J.F. (eds.) TLCA 1993. LNCS, vol. 664, pp. 361–375. Springer, Heidelberg (1993). https://doi.org/10.1007/BFb0037118
21. Reynolds, J.C.: Types, abstraction and parametric polymorphism. In: IFIP Congress, pp. 513–523 (1983)
22. Strachey, C.: Fundamental concepts in programming languages. Higher-Order Symbolic Comput. **13**(1), 11–49 (2000)
23. Wadler, P.: Theorems for free! In: Proceedings of the Fourth International Conference on Functional Programming Languages and Computer Architecture, FPCA 1989, pp. 347–359. ACM, New York (1989)

Categorical Combinatorics for Non Deterministic Strategies on Simple Games

Clément Jacq(✉) and Paul-André Melliès

Institut de Recherche en Informatique Fondamentale,
Université Paris Diderot, Paris, France
Clement.Jacq@irif.fr

Abstract. The purpose of this paper is to define in a clean and conceptual way a non-deterministic and sheaf-theoretic variant of the category of simple games and deterministic strategies. One thus starts by associating to every simple game a presheaf category of non-deterministic strategies. The bicategory of simple games and non-deterministic strategies is then obtained by a construction inspired by the recent work by Melliès and Zeilberger on type refinement systems. We show that the resulting bicategory is symmetric monoidal closed and cartesian. We also define a 2-comonad which adapts the Curien-Lamarche exponential modality of linear logic to the 2-dimensional and non deterministic framework. We conclude by discussing in what sense the bicategory of simple games defines a model of non deterministic intuitionistic linear logic.

1 Introduction

A new generation of 2-categorical and sheaf-theoretic game semantics is currently emerging in the field of programming language semantics. The games and strategies which determine them are more sophisticated mathematically, and also more difficult to define rigorously, than they were in the deterministic case. For that reason, it is timely to examine more closely the 2-categorical and sheaf-theoretic frameworks available to us in order to formulate these games and strategies in a suitably clean and conceptual way. In this investigation, one benefits from the efforts made in the past twenty-five years to give a clearer mathematical status to the previous generation of game semantics, which was (to a large extent) based on the notion of arena game. We recognize three main lines of work here:

1. the logical approach advocated by Girard, and formulated in ludics [3], polarized linear logic [7] or tensorial logic [12] with its connection to continuations and string diagrams,
2. the combinatorial approach advocated by Hyland, inspired by algebraic topology, and based on the combinatorial description of the structure of pointers in arena games [4],

C. Baier and U. Dal Lago (Eds.): FOSSACS 2018, LNCS 10803, pp. 39–70, 2018.
https://doi.org/10.1007/978-3-319-89366-2_3

3. the concurrent and asynchronous approach advocated by Melliès, based on the description of arena games as asynchronous games, and of strategies as causal concurrent structures playing on them, either in an alternated [9–11] or in a non-alternated way [18].

Interestingly, all the sheaf-theoretic frameworks designed for game semantics today are offsprings of the third approach based on asynchronous games: on the one hand, the notion of concurrent strategy in [19] is a sheaf-theoretic transcription of the notion of receptive ingenuous strategy formulated in [18]; on the other hand, the sheaf-theoretic notion of non-deterministic innocent strategy in [13,17] relies on the diagrammatic and local definition of innocence in alternated asynchronous games [11]. For that reason, our purpose in this paper is to investigate the connection with the second approach, different in spirit and design, and to define a bicategory of simple games and non-deterministic strategies in the sheaf-theoretic style of Harmer et al. [4]. As we will see, our work also integrates a number of elements coming from the first approach, and more specifically, the discovery by Melliès that strategies are presented by generators and relations, and for that reason, are prone to factorisation theorems [14,15]. Since we are interested in sheaf-theoretic models of computations, we should not forget to mention the pioneering work by Hirschowitz and Pous on models of process calculi [5], and its recent connection to game semantics [2].

In the present paper, we start from the category $\mathcal{G}$ of simple games and deterministic strategies between them, and we explain how to turn $\mathcal{G}$ into a bicategory $\mathcal{S}$ of simple games and *non-deterministic* strategies. As we will see, the construction of $\mathcal{S}$ relies on the discovery of a number of elementary but fundamental fibrational properties of the original category $\mathcal{G}$. Since our work is built on [4], let us recall that a simple game A is defined there as a contravariant presheaf $A : \omega^{op} \to \mathbf{Set}$ over the order category $\omega = 0 \to 1 \to 2 \to \cdots$ associated to the infinite countable ordinal ω. A simple game A is thus a family of sets A_n together with a function $\pi_n : A_{n+1} \to A_n$ for all $n \in \mathbb{N}$, depicted as:

$$A_0 \xleftarrow{\pi_0} A_1 \xleftarrow{\pi_1} A_2 \longleftarrow \cdots \quad \longleftarrow A_n \xleftarrow{\pi_n} A_{n+1} \longleftarrow \cdots$$

One requires moreover that A_0 is the singleton set. The intuition is that A is a rooted tree; that A_n contains its plays (or branches) of length n; and that π_n is the prefix function which transports every play of length $n+1$ to its prefix of length n. In particular, every simple game A contains only one play of length 0, which should be thought as the empty play. Every simple game A should be moreover understood as alternating: here, the intuition is that every play of odd length $2n+1$ ends with an Opponent move, and that every play of even length $2n$ ends with a Player move if $n > 0$.

Terminology: An element $a \in A_n$ is called a position of degree n in the game A. The position $a \in A_n$ is called a P-position when its degree n is even, and a O-position when its degree n is odd. Given a position $a \in A_{n+1}$, we write $\pi(a)$ for the position $\pi_n(a)$; similarly, given a position $a \in A_{n+2}$, we write $\pi^2(a)$ for the position $\pi_n \circ \pi_{n+1}(a)$. A simple game A is called O-branching when the function $\pi : A_{2n+2} \rightarrow A_{2n+1}$ is injective, for all $n \in \mathbb{N}$. This means that every Opponent position $a \in A_{2n+1}$ can be extended in at most one way into a Player position $b \in A_{2n+2}$, for all $n \in \mathbb{N}$.

We start the paper by formulating a sheaf-theoretic notion of non-deterministic P-strategy on a simple game A. Recall that a deterministic P-strategy σ of a simple game A is defined in [4] as a family of subsets $\sigma_{2n} \subseteq A_{2n}$ of P-positions, satisfying the following properties, for all $n \in \mathbb{N}$:

(i) Unique empty play — σ_0 is equal to the singleton set A_0,
(ii) Closure under even prefixes — if $a \in \sigma_{2n+2}$ then $\pi^2(a) \in \sigma_{2n}$,
(iii) Determinacy — if $a, b \in \sigma_{2n}$ with $\pi(a) = \pi(b)$, then $a = b$.

In order to generalize this definition to non-deterministic P-strategies, we find convenient to consider the full subcategory ω_P of ω consisting of the strictly positive even numbers, of the form $2n$ for $n > 0$; and the inclusion functor $\iota_P : \omega_P \rightarrow \omega$. Define the presheaf $A_P = A \circ \iota_P$ as the simple game A obtained as the restriction of the presheaf $A : \omega^{op} \rightarrow \mathbf{Set}$ to the subcategory ω_P:

$$A_P \quad = \quad \omega_P^{op} \xrightarrow{\iota_P} \omega^{op} \xrightarrow{A} \mathbf{Set}$$

The collection A_P thus consists of all the Player positions in A, except for the initial one $* \in A(0)$. This leads us to the following definition of (non-deterministic) P-strategy on a simple game A:

Definition 1. *A P-strategy σ on a simple game A is a presheaf $S : \omega_P^{\mathrm{op}} \rightarrow \mathbf{Set}$ over the category ω_P together with a morphism of presheaves $\sigma : S \rightarrow A_P$. We write $\sigma : A$ in that case. The presheaf S is called the* support *of the strategy σ and the elements of S_{2n} are called the runs of degree $2n$ of the strategy, for $n \geq 0$.*

In other words, a P-strategy σ on A is a family of sets S_{2n} indexed by strictly positive numbers $n > 0$, related between them by functions $(\pi_P)_{2n} : S_{2n+2} \rightarrow S_n$ pictured as:

$$S_2 \xleftarrow{\pi_P} S_4 \longleftarrow \quad \cdots \quad \longleftarrow S_{2n} \xleftarrow{\pi_P} S_{2n+2} \longleftarrow \cdots$$

together with a family of functions $\sigma_{2n} : S_{2n} \rightarrow A_{2n}$ making the diagram below commute, for all $n > 0$:

$$\begin{array}{ccccc} S_{2n} & & \xleftarrow{\quad\pi_P\quad} & & S_{2n+2} \\ \downarrow{\scriptstyle \sigma_{2n}} & & & & \downarrow{\scriptstyle \sigma_{2n+2}} \\ A_{2n} & \xleftarrow{\pi} & A_{2n+1} & \xleftarrow{\pi} & A_{2n+2} \end{array}$$

To every simple game A, we associate the category $\mathcal{P}(A)$ of P-strategies over A, defined as the slice category

$$\mathcal{P}(A) = [\omega_P^{op}, \mathbf{Set}] / A_P \tag{1}$$

whose objects are thus the strategies over A, and whose morphisms $\theta : \sigma \to \tau$ between two strategies $\sigma : S \to A$ and $\tau : T \to A$ are the morphisms $\theta : S \to T$ of presheaves satisfying the expected equation: $\sigma = \tau \circ \theta$. We will call those simulations. One main contribution of the paper is the observation that the family of categories $\mathcal{P}(A)$ can be organised into a pseudofunctor

$$\mathcal{P} : \mathcal{G} \longrightarrow \mathbf{Cat}$$

from the category $\mathcal{G}$ of simple games and deterministic strategies. The pseudofunctor $\mathcal{P}$ is moreover monoidal, in the sense that there exists a family of functors

$$m_{A,B} : \mathcal{P}(A) \times \mathcal{P}(B) \longrightarrow \mathcal{P}(A \otimes B)$$

indexed by simple games A, B. As a symmetric monoidal closed category, the category $\mathcal{G}$ is enriched over itself, with the simple game $\mathcal{G}(A, B) = A \multimap B$ constructed from the simple games A and B. Here comes the nice point of the construction: the bicategory $\mathcal{S}$ is simply defined as the bicategory with simple games A, B as objects, and with

$$\mathcal{S}(A, B) = \mathcal{P}(A \multimap B)$$

as category of morphisms between two simple games A and B. In other words, a morphism $\sigma : A \to B$ in $\mathcal{S}$ is a P-strategy $\sigma : A \multimap B$, and a 2-cell $\theta : \sigma \Rightarrow \tau : A \to B$ is a morphism $\theta : \sigma \to \tau$ in the category $\mathcal{P}(A \multimap B)$. At this point, the fact that $\mathcal{S}$ defines a bicategory is easily derived from the lax monoidal structure of the pseudofunctor $\mathcal{P}$. Recall that, as a symmetric monoidal closed category, the category $\mathcal{G}$ is enriched over itself. From a conceptual point of view, the construction of the bicategory $\mathcal{S}$ thus amounts to a change of enrichment category along the lax monoidal pseudofunctor $\mathcal{P} : \mathcal{G} \to \mathbf{Cat}$, transforming the $\mathcal{G}$-enriched category $\mathcal{G}$ into the (weak) $\mathbf{Cat}$-enriched category $\mathcal{S}$.

Besides the construction of $\mathcal{S}$, a great care will be devoted to the analysis of the Curien-Lamarche exponential comonad $!$ on the category $\mathcal{G}$ and to the recipe to turn it into an exponential 2-comonad on the bicategory $\mathcal{S}$. The construction relies on the existence of a family of functors

$$p_A \quad : \quad \mathcal{P}(A) \longrightarrow \mathcal{P}(!A)$$

called "promotion" functors, and natural in the simple game A in the category $\mathcal{G}$. In particular, the functorial part of the exponential 2-comonad $! : \mathcal{S} \to \mathcal{S}$ is defined as the composite:

$$\mathcal{P}(A \multimap B) \xrightarrow{p_{A \multimap B}} \mathcal{P}(!(A \multimap B)) \xrightarrow{\mathcal{P}(n_{A,B})} \mathcal{P}(!A \multimap !B)$$

where $n_{A,B} : !(A \multimap B) \to !A \multimap !B$ is the canonical morphism in $\mathcal{G}$ which provides the structure of a lax monoidal functor to the original comonad $! : \mathcal{G} \to \mathcal{G}$.

2 Non-deterministic *P*-strategies as *P*-cartesian Transductions

As explained in the introduction, a P-strategy $\sigma \in \mathcal{P}(A)$ over a simple game A is defined as an object of the slice category (1) in the category $[\omega_P^{op}, \mathbf{Set}]$ of contravariant presheaves over ω_P. We will use the fact that the slice category is equivalent to the category of contravariant presheaves

$$\mathcal{P}(A) \;=\; [\omega_P^{op}\,, \mathbf{Set}]\,/\,A_P \;\cong\; [\mathbf{tree}(A_P)^{op}, \mathbf{Set}]$$

over the Grothendieck category $\mathbf{tree}(A_P)$ generated by the presheaf $A_P \in [\omega_P^{op}, \mathbf{Set}]$. The category $\mathbf{tree}(A_P)$ has the P-positions of the simple game A as objects, and a morphism $a \to a'$ between $a \in A_{2p}$ and $a' \in A_{2q}$ precisely when $p \leq q$ and $\pi^{2q-2p}(a') = a$. In other words, it is the order category associated to the tree of P-positions of the simple game A.

We find convenient for later purposes to reformulate non-deterministic P-strategies in the following way. This paves the way to a comprehension theorem for the pseudofunctor $\mathcal{P}$, which will be established in the next section. A transduction $\theta : A \to B$ between two simple games $A, B : \omega^{op} \to \mathbf{Set}$ is defined as a natural transformation between the presheaves A and B, given by a family of functions $\theta_n : A_n \to B_n$ making the square $\square_n$ diagram below commute, for all $n \in \mathbb{N}$:

$$\begin{array}{ccc}
A_n & \xleftarrow{\;\pi_n\;} & A_{n+1} \\
{\scriptstyle \theta_n}\downarrow & \square_n & \downarrow{\scriptstyle \theta_{n+1}} \\
B_n & \xleftarrow[\;\pi_n\;]{} & B_{n+1}
\end{array}$$

A transduction $\theta : A \to B$ is called P-cartesian when $\square_{2n}$ is a pullback square for all $n \in \mathbb{N}$; and O-cartesian when $\square_{2n+1}$ is a pullback square for all $n \in \mathbb{N}$. We write $\mathcal{T}$ for the category of simple games and transductions between them, and $\mathcal{T}_P$ (resp. $\mathcal{T}_O$) for the subcategory of P-cartesian (resp. O-cartesian) transductions. Note that the restriction functor

$$(-)_P \quad : \quad [\omega^{op}, \mathbf{Set}] \quad \longrightarrow \quad [\omega_P^{op}\,, \mathbf{Set}]$$

is a fibration, and that a transduction $\theta : A \to B$ between simple games is P-cartesian precisely when it defines a cartesian morphism with respect to the fibration $(-)_P$. For that reason, a P-cartesian transduction $\theta : A \to B$ is entirely characterized by the family of functions $\theta_{2n} : A_{2n} \longrightarrow B_{2n}$ on the P-positions of the simple games A and B, for $n \in \mathbb{N}$. From this follows easily that

Proposition 1. *A P-strategy σ on a simple game A is the same thing as a simple game S together with a P-cartesian transduction $S \to A$. The simple game S is uniquely determined by σ up to isomorphism. It is called the support (or run-tree) of σ, and noted $\{A \,|\, \sigma\}$, while the P-cartesian transduction is noted* $\mathsf{supp}_\sigma : \{A \,|\, \sigma\} \longrightarrow A$.

Note that the definition applies the general principle formulated in [18] that a strategy σ of a game A is a specific kind of map (here a P-cartesian transduction) $S \to A$ from a given game $S = \{A \,|\, \sigma\}$ to the game A of interest. One benefit of this principle is that it unifies the two concepts of game and of strategy, by regarding a strategy σ of a game A as a game S "embedded" in an appropriate way by $S \to A$ inside the simple game A. This insight coming from [18] underlies for instance the construction in [19] of a category of non-deterministic strategies between asynchronous games.

Typically, consider the simple game $A = \mathbb{B}_1 \multimap \mathbb{B}_2$ where $\mathbb{B}$ is the simple boolean game with a unique initial Opponent move q and two Player moves tt for true and ff for false; and where the indices 1, 2 are here to indicate the component of the boolean game $\mathbb{B}$. The simple game A may be represented as the decision tree below:

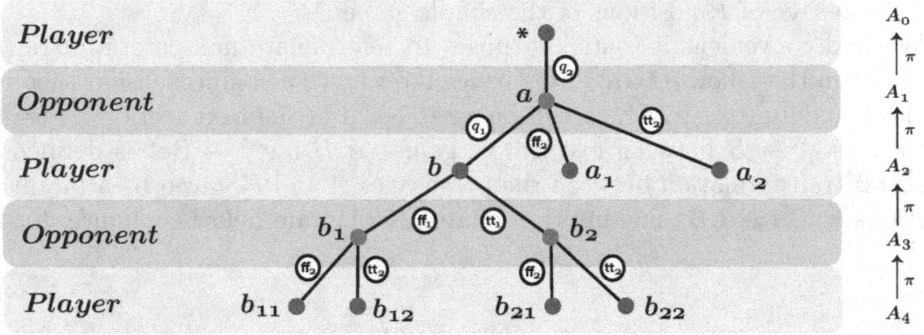

where the sets of positions are defined as:

$$A_1 = \{a\} \qquad A_2 = \{b, a_1, a_2\} \qquad A_3 = \{b_1, b_2\} \qquad A_4 = \{b_{11}, b_{12}, b_{21}, b_{22}\}$$

and where the branches are induced by the prefix functions $\pi_n : A_{n+1} \to A_n$ depicted on the picture above. For the reader's convenience, we label every edge of A by the name of the move which would be used in the more familiar definition of simple games, where plays are defined as sequences of moves [1,6]. Note that every position $a \in A_n$ of degree n is determined by its occurrence, defined as the sequence of n moves from the root $*$ to the position a in the tree A. Typically, the P-position $b \in A_2$ has occurrence $q_2 \cdot q_1$ and the P-position $b_{21} \in A_4$ has occurrence $q_2 \cdot q_1 \cdot \mathsf{tt}_1 \cdot \mathsf{ff}_2$.

By way of illustration, we define the P-strategy $\sigma \in \mathcal{P}(A)$ as the presheaf below

$$* \mapsto \{*\} \quad a_1 \mapsto \emptyset \quad a_2 \mapsto \{x''\}$$
$$b \mapsto \{x'\} \quad b_{11} \mapsto \emptyset \quad b_{12} \mapsto \emptyset \quad b_{21} \mapsto \{z'\} \quad b_{22} \mapsto \{z'', z'''\}$$

on the Grothendieck category $\mathbf{tree}(A_P)$ associated to the presheaf A_P of P-positions in A. As explained in Proposition 1, the P-strategy σ may be equivalently defined as the simple game $S = \{A \,|\, \sigma\}$ below

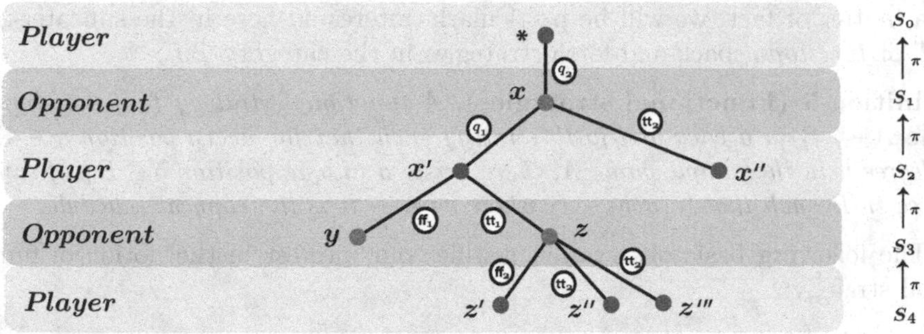

together with the P-cartesian transduction $\mathsf{supp}_\sigma : \{A \,|\, \sigma\} \to A$ described as:

$$x \mapsto a \qquad x' \mapsto b \qquad x'' \mapsto a_2 \qquad y \mapsto b_1 \qquad z' \mapsto b_{21} \qquad z'' \mapsto b_{22} \qquad z''' \mapsto b_{22}$$

It is worth mentioning that the transduction supp_σ may be recovered from the moves labelled on the run-tree $S = \{A \,|\, \sigma\}$. This pictorial description provides a convenient way to describe how the non-deterministic P-strategy σ plays on A. Typically, when questioned by the initial move q_2 of the game, the non-deterministic P-strategy σ answers tt_2 with the run $x'' \in S_2$ or asks the value of the input boolean by playing the move q_1; when the Opponent answers with the move tt_1, the P-strategy reacts by playing the value ff_2 with the run $z' \in S_4$ or by playing the value ff_2 with the runs $z'', z''' \in S_4$. Note in particular that the P-strategy σ is allowed to play two different runs $z'', z''' \in S_4$ of the same play $b_{22} \in A_4$.

3 P-cartesian Transductions as Deterministic Strategies

In the previous section, we have seen how to regard every non-deterministic P-strategy $\sigma \in \mathcal{P}(B)$ as a P-cartesian transduction $\mathsf{supp}_\sigma : \{B \,|\, \sigma\} \to B$ into the simple game B. Our purpose here is to show that every P-cartesian transduction $\theta : A \to B$ can be seen as a particular kind of deterministic strategy of the simple game $A \multimap B$.

Definition 2 (Total strategies). *A deterministic strategy σ of a simple game A is total when for every O-position s such that the P-position $\pi(s)$ is an element of σ, there exists a P-position t in the strategy σ such that $\pi(t) = s$.*

Definition 3 (Back-and-forth strategies). *Given two simple games A and B, a back-and-forth strategy f of the simple game $A \multimap B$ is a deterministic and total strategy whose positions are all of the form (c, a, b) where $c : n \to n$ is a copycat schedule.*

Back-and-forth strategies compose, and thus define a subcategory of $\mathcal{G}$:

Definition 4 (The category $\mathcal{BF}$). *The category $\mathcal{BF}$ of back-and-forth strategies is the subcategory of $\mathcal{G}$ whose objects are the simple games and whose morphisms $f : A \to B$ are the back-and-forth strategies of $A \multimap B$.*

As a matter of fact, we will be particularly interested here in the subcategory $\mathcal{BF}^+$ of *functional* back-and-forth strategies in the category $\mathcal{BF}$.

Definition 5 (Functional strategies). *A functional strategy f of the simple game $A \multimap B$ is a back-and-forth strategy such that for every position $a \in A_n$ of degree n in the simple game A, there exists a unique position $b \in B_n$ of same degree in B such that $(c, a, b) \in f$, where $c : n \to n$ is the copycat schedule.*

The following basic observation justifies our interest in the notion of functional strategy:

Proposition 2. *For all simple games A, B, there is a one-to-one correspondence between the P-cartesian transductions $A \to B$ and the functional strategies in $A \multimap B$.*

Proof. See Appendix E.

For that reason, we will identify P-cartesian transductions and functional strategies from now on. Put together with Proposition 1, this leads us to the following correspondence, which holds for every simple game A:

Proposition 3. *The category $\mathcal{P}(A)$ is equivalent to the slice category $\mathcal{BF}^+/A$.*

The result may be understood as a preliminary form of comprehension: it states that every non-deterministic P-strategy $\sigma \in \mathcal{P}(A)$ may be equivalently seen as a functional P-strategy

$$\mathsf{supp}_\sigma \quad : \quad \{A \,|\, \sigma\} \quad \longrightarrow \quad A \tag{2}$$

in the category $\mathcal{G}$ of simple games and deterministic strategies, obtained by composing the equivalences stated in Propositions 1 and 3. Note that the simple game $\{A \,|\, \sigma\}$ coincides with the run-tree S of the non-deterministic strategy σ formulated in Proposition 1 and that the functional strategy supp_σ coincides with the P-cartesian transduction which "projects" the support S on the simple game A. The property (Proposition 3) is important from a methodological point of view, because it enables us to use the rich toolbox developed for simple games and deterministic strategies, in order to handle non-deterministic strategies *inside* the category $\mathcal{G}$.

4 The Pseudofunctor $\mathcal{P}$

Suppose given a P-strategy $\sigma \in \mathcal{P}(A)$ over the simple game A and a morphism $f : A \to B$ in the category $\mathcal{G}$.

Definition 6. *The P-strategy $\mathcal{P}(f)(\sigma) \in \mathcal{P}(B)$ over the simple game B is defined as the contravariant presheaf over $\mathbf{tree}(B_P)$ which transports every P-position b of the simple game B to the disjoint union defined below:*

$$\mathcal{P}(f)(\sigma) \quad : \quad b \quad \mapsto \quad \coprod_{(e,a,b) \in f} \sigma(a). \tag{3}$$

The fact that (3) defines a presheaf over $\mathcal{P}(B)$ and that $\mathcal{P}$ is a pseudofunctor (see Definition 24) is established in the Appendix F.

This construction equips the family of presheaf categories $\mathcal{P}(A)$ with the structure of a pseudofunctor $\mathcal{P} : \mathcal{G} \longrightarrow \mathbf{Cat}$. Moreover, the pseudo-functor $\mathcal{P}$ has comprehension in the sense of Lawvere [8]. For every simple game B, the *comprehension functor* is defined as the composite

$$\{B \,|\, -\} \quad : \quad \mathcal{P}(B) \quad \longrightarrow \quad \mathcal{BF}^{+}/B \quad \longrightarrow \quad \mathcal{G}/B$$

which transports every non-deterministic P-strategy to the morphism (2) seen as an object of $\mathcal{G}/B$. One establishes that

Theorem 1 (Comprehension). *For every simple game B, the comprehension functor*

$$\{B \,|\, -\} \quad : \quad \mathcal{P}(B) \quad \longrightarrow \quad \mathcal{G}/B$$

has a left adjoint functor

$$\mathsf{image} \quad : \quad \mathcal{G}/B \quad \longrightarrow \quad \mathcal{P}(B).$$

Given a deterministic strategy $f : A \to B$, the contravariant presheaf $\mathsf{image}(f)$ over the category $\mathbf{tree}(B_P)$ transports every P-position b of the game B to the set below:

$$\mathsf{image}(f) \quad : \quad b \quad \mapsto \quad \left\{ \quad (e, a, b) \quad \middle| \quad (e, a, b) \in f \quad \right\}$$

Note that the presheaf $\mathsf{image}(f)$ may be also described by the formula

$$\mathsf{image}(f) \quad = \quad \mathcal{P}(f)(*_A) \quad \in \quad \mathcal{P}(B)$$

where $*_A$ is the terminal object in the category $\mathcal{P}(A)$ of P-strategies over A. Note that the run-tree $\{A \,|\, *_A\}$ of the P-strategy $*_A \in \mathcal{P}(A)$ is the simple game A itself, with supp_{*_A} the identity $i_A : A \to A$. In other words, the P-strategy $*_A$ has exactly one run over each position of the simple game A.

Also note that we will occasionally note positions of $\mathsf{image}(f)$ $b_{(e,a)}$ when there is need to emphasize the fact that $\mathsf{image}(f)$ is a contravariant presheaf over $\mathbf{tree}(B_P)$.

5 The Slender-Functional Factorisation Theorem

In order to establish the comprehension theorem, we prove a factorization theorem in the original category $\mathcal{G}$, which involves slender and functional strategies.

Definition 7. *A deterministic strategy f in a simple game $A \multimap B$ is slender when for every P-position b in the simple game B, there exists exactly one P-position a of the simple game A and exactly one schedule e such that $(e, a, b) \in f$.*

By extension, we say that a morphism $f : A \to B$ in the category $\mathcal{G}$ is slender when the deterministic strategy f is slender in $A \multimap B$. Note that every isomorphism $f : A \to B$ in the category $\mathcal{G}$ is both slender and functional.

Proposition 4. *Suppose that A and B are two simple games and that f is a deterministic strategy of $A \multimap B$. Then, there exists a slender strategy $g : A \to C$ and a functional strategy $h : C \to B$ such that $f = h \circ g$.*

The simple game C is defined as $\{B \,|\, \mathsf{image}(f)\}$ while the slender strategy $g : A \to C$ is defined as

$$g \quad = \quad \left\{ \quad (e, a, (e, a, b)) \quad \middle| \quad (e, a, b) \in f \quad \right\}$$

and $h : C \to B$ is the functional strategy $\mathsf{supp}_{\mathsf{image}(f)}$ associated in Proposition 3 to the P-strategy $\mathsf{image}(f) \in \mathcal{P}(B)$.

Proposition 5. *Suppose that $s : U \to V$ and $f : A \to B$ are two morphisms of the category $\mathcal{G}$. Suppose moreover that s is slender and that f is functional. Then, $s : X \to Y$ is orthogonal to $f : A \to B$ in the sense that for all morphisms $u : X \to A$ and $v : Y \to B$ making the diagram (a) commute, there exists a unique morphism $h : Y \to B$ making the diagram (b) commute in the category $\mathcal{G}$:*

$$(a)\quad \begin{array}{ccc} X & \xrightarrow{u} & A \\ {\scriptstyle s}\downarrow & & \downarrow{\scriptstyle f} \\ Y & \xrightarrow{v} & B \end{array} \qquad\qquad (b)\quad \begin{array}{ccc} X & \xrightarrow{u} & A \\ {\scriptstyle s}\downarrow & \overset{h}{\nearrow} & \downarrow{\scriptstyle f} \\ Y & \xrightarrow{v} & B \end{array}$$

The deterministic strategy $h : Y \to A$ is defined as

$$\begin{aligned} h = & \left\{ (e, y, a) \;\middle|\; \begin{array}{l} \exists x \in X, b \in B, e', e'' \in \Upsilon, \\ (e, y, b) \in v \ \wedge (c, a, b) \in f \ \wedge (e', x, y) \in s \ \wedge (e'', x, a) \in u \end{array} \right\} \\ \uplus & \left\{ (e, y, a) \;\middle|\; \begin{array}{l} \exists x \in X, b \in B, e', e'' \in \Upsilon, \\ (e, y, b) \in v \ \wedge (c, a, b) \in f \ \wedge (e', x, \pi y) \in s \ \wedge (e'', x, \pi a) \in u \end{array} \right\} \end{aligned}$$

Note that the position b is uniquely determined by the position a because f is functional, and that the pair (e', x) is uniquely determined by the position y because s is slender. Moreover, by determinism of $u = h \circ s$, the schedule e'' is entirely determined by the schedules e and e'.

Theorem 2 (Factorization theorem). *The classes $\mathcal{S}$ of slender morphisms and $\mathcal{F}$ of functional morphisms define a factorization system $(\mathcal{S}, \mathcal{F})$ in the category $\mathcal{G}$.*

It is a folklore result that, in that situation, the comprehension theorem (Theorem 1) follows from the factorization theorem. The reason is that the category $\mathcal{P}(B)$ is equivalent (by Proposition 3) to the full subcategory $\mathcal{BF}^+/B$ of functional strategies in the slice category $\mathcal{G}/B$. Seen from that point of view, the comprehension functor $\{B \mid -\}$ coincides with the embedding of $\mathcal{BF}^+/B$ into $\mathcal{G}/B$. It is worth noting that for every P-strategy $\sigma \in \mathcal{P}(A)$, one has an isomorphism

$$\sigma \quad \cong \quad \mathsf{image}(\mathsf{supp}_\sigma)$$

in the category $\mathcal{P}(A)$, and that one has an isomorphism

$$\mathcal{P}(f)(\sigma) \quad \cong \quad \mathsf{image}(f \circ \mathsf{supp}_\sigma) \tag{4}$$

in the category $\mathcal{P}(B)$, for every morphism $f : A \to B$ in the category $\mathcal{G}$. This provides an alternative way to define the pseudofunctor $\mathcal{P}$.

6 The Bicategory $\mathcal{S}$ of Simple Games and Non-deterministic Strategies

In this section, we explain how to construct a bicategory $\mathcal{S}$ of simple games and non-deterministic strategies, starting from the category $\mathcal{G}$. The first step is to equip the pseudofunctor $\mathcal{P}$ with a lax monoidal structure (See Definition 25), based on the definition of tensor product in the category $\mathcal{G}$ formulated in [4], see Appendix B for details. We start by observing that

Proposition 6. *Suppose given two morphisms $f : A \to B$ and $g : C \to D$ in the category $\mathcal{G}$ of simple games and deterministic strategies. The morphism*

$$f \otimes g : A \otimes C \longrightarrow B \otimes D$$

is slender when f and g are slender, and functional when f and g are functional.

Proof. See Appendix G.

Note that the isomorphism $\mathsf{image}(f \otimes g) \cong \mathsf{image}(f) \otimes \mathsf{image}(g)$ follows immediately from this statement and from the factorization theorem (Theorem 2), for every pair of morphisms $f : A \to B$ and $g : C \to D$ in the category $\mathcal{G}$. The tensor product $\sigma \otimes \tau$ of two P-strategies σ and τ is defined in the same spirit, using comprehension:

Definition 8. *Suppose that $\sigma \in \mathcal{P}(A)$ is a P-strategy of a simple game A and that $\tau \in \mathcal{P}(B)$ is a P-strategy of a simple game B. The tensor product $\sigma \otimes \tau$ is the P-strategy of the simple game $A \otimes B$ defined as*

$$\sigma \otimes \tau \quad = \quad \mathsf{image}(\,\mathsf{supp}_\sigma \otimes \mathsf{supp}_\tau\,).$$

Here, the morphism $\mathsf{supp}_\sigma \otimes \mathsf{supp}_\tau : \{A \,|\, \sigma\} \otimes \{B \,|\, \tau\} \to A \otimes B$ denotes the tensor product (computed in the original category $\mathcal{G}$) of the morphisms supp_σ and supp_τ. A direct description of $\sigma \otimes \tau \in \mathcal{P}(A \otimes B)$ is also possible, as the presheaf which transports every position (e, a, b) of the simple game $A \otimes B$ to the set-theoretic product below:

$$\sigma \otimes \tau \quad : \quad (e, a, b) \quad \mapsto \quad \sigma(a) \times \tau(b).$$

As indicated in the introduction, the tensor product of P-strategies defines a family of functors $m_{A,B} : \mathcal{P}(A) \times \mathcal{P}(B) \to \mathcal{P}(A \otimes B)$ which, together with the isomorphism of categories $m_1 : 1 \to \mathcal{P}(1)$, equips the pseudofunctor $\mathcal{P}$ with a lax monoidal structure:

Theorem 3. *The pseudofunctor $\mathcal{P}$ equipped with the family of functors $m_{A,B}$ and m_1 defines a lax monoidal pseudofunctor from $(\mathcal{G}, \otimes, 1)$ to $(\mathbf{Cat}, \times, 1)$.*

Proof. See Appendix H.

The bicategory $\mathcal{S}$ of simple games and non-deterministic strategies is deduced from the lax monoidal pseudofunctor $\mathcal{P}$ in the following generic way, inspired by the idea of monoidal refinement system [16].

Definition 9. *The bicategory $\mathcal{S}$ has simple games A, B, C as objects, with the hom-category $\mathcal{S}(A, B)$ defined as*

$$\mathcal{S}(A, B) \quad = \quad \mathcal{P}(A \multimap B)$$

the composition functor

$$\circ_{A,B,C} : \mathcal{P}(B \multimap C) \times \mathcal{P}(A \multimap B) \longrightarrow \mathcal{P}(A \multimap C)$$

defined as the composite

$$\mathcal{P}(B \multimap C) \times \mathcal{P}(A \multimap B) \xrightarrow{m_{B \multimap C, A \multimap B}} \mathcal{P}((B \multimap C) \otimes (A \multimap B)) \xrightarrow{\mathcal{P}(comp_{A,B,C})} \mathcal{P}(A \multimap C)$$

where $comp_{A,B,C} : (B \multimap C) \otimes (A \multimap B) \longrightarrow (A \multimap C)$ is the morphism which internalizes composition in the symmetric monoidal closed category $\mathcal{G}$. In the same way, the identity in $\mathcal{P}(A \multimap A)$ is defined as the composite

$$1 \xrightarrow{\quad m_1 \quad} \mathcal{P}(1) \xrightarrow{\quad \mathcal{P}(id_A) \quad} \mathcal{P}(A \multimap A)$$

where the morphism $id_A : 1 \to (A \multimap A)$ internalizes the identity morphism in $\mathcal{G}$.

Proposition 7. *The bicategory $\mathcal{S}$ is symmetric monoidal closed in the sense that there exists a family of isomorphisms*

$$\Phi_{A,B,C} \quad : \quad \mathcal{S}(A \otimes B, C) \quad \cong \quad \mathcal{S}(B, A \multimap C).$$

The isomorphism $\Phi_{A,B,C}$ is defined as the image by the pseudofunctor $\mathcal{P}$ of the isomorphism

$$\varphi_{A,B,C} \quad : \quad (A \otimes B) \multimap C \quad \cong \quad B \multimap (A \multimap C)$$

in the category $\mathcal{G}$ between the underlying simple games. One benefit of our conceptual approach is that the monoidal closed structure of $\mathcal{S}$ is neatly deduced from the monoidal closed structure of the original category $\mathcal{G}$.

7 The Exponentional Modality on the Category $\mathcal{G}$

Now that the monoidal bicategory $\mathcal{S}$ has been defined, we analyze how the exponential modality defined in [4] adapts to our sheaf-theoretic framework.

Definition 10. *Let A be a simple game. $!A$ is the simple game whose set $(!A)_n$ of positions of degree n consists of the pairs $(\phi, \overline{a})$ such that:*

- *ϕ is a O-heap over n and $\overline{a} = (a_1, \ldots, a_n)$ is a sequence of positions of A,*
- *for each $k \in \{1, \ldots, n\}$, the sequence of positions in $\overline{a} = (a_1, \ldots, a_n)$ corresponding to the branch of k in ϕ defines a play*

$$\{a_k, a_{\phi(k)}, a_{\phi^2(k)}, \ldots\}$$

of the simple game A.

The predecessor function $\pi_n : (!A)_{n+1} \to (!A)_n$ is defined as $\pi(\phi, \overline{a}) = (\phi \restriction (n), \overline{a} \restriction (n))$.

Definition 11. *Let f be a deterministic strategy of $A \multimap B$. The deterministic strategy $!f$ of $!A \multimap !B$ consists of the positions $(e, (\phi, \overline{a}), (\psi, \overline{b}))$ such that $\phi = e^*\psi$ and, for each branch of (ϕ, e, π), the positions associated to that branch are played by f.*

It is worth observing that the construction of $!f : !A \to !B$ can be decomposed in the following way. Consider the morphism

$$n_{A,B} \quad : \quad !\,(A \multimap B) \quad \longrightarrow \quad !A \multimap !B$$

obtained by currying the composite morphism

$$!\,(A \multimap B) \otimes !A \xrightarrow{\textit{lax monoidal}} !\,((A \multimap B) \otimes A) \xrightarrow{\textit{!evaluation}} !\,B$$

in the symmetric monoidal closed category $\mathcal{G}$, where we use the coercion morphism which provides the exponential modality $! : \mathcal{G} \to \mathcal{G}$ with the structure of a lax monoidal functor.

Definition 12 (#f). *Given a deterministic strategy f of a simple game A, the deterministic strategy $\#f$ of the simple game $!A$ has positions the pairs $(\phi, \overline{a})$ such that for each branch of $(\phi, \overline{a})$, the positions associated to that branch are played by the deterministic strategy f.*

Proposition 8. *Given a morphism $f : A \to B$ of the category $\mathcal{G}$ and its curried form $\lambda a.f : 1 \to A \multimap B$, the composite morphism*

$$1 \xrightarrow{\#\lambda a:A.f} !(A \multimap B) \xrightarrow{n_{A,B}} !A \multimap !B$$

is the curried form $\lambda x : !A.\ !f$ in the category $\mathcal{G}$ of the morphism $!f : !A \longrightarrow !B$.

More details about the original exponential modality in $\mathcal{G}$ will be found in Appendix C. By analogy with Proposition 6, we establish that

Proposition 9. *Suppose that $f : A \to B$ is a morphism in the category $\mathcal{G}$. Then, the morphism*

$$!f \quad : \quad !A \quad \longrightarrow \quad !B$$

is slender when f is slender, and functional when f is functional.

Proof. See Appendix I.

8 The Exponential Modality on the Bicategory $\mathcal{S}$

In this section, we define the linear exponential modality $! : \mathcal{S} \to \mathcal{S}$ on the symmetric monoidal closed bicategory $\mathcal{S}$, in order to define a bicategorical model of intuitionistic linear logic. The construction is inspired by the observation made in the previous section (Proposition 8).

Definition 13. *Given a P-strategy $\sigma \in \mathcal{P}(A)$ of a simple game A, the P-strategy $\#\sigma$ of the simple game $!A$ is defined as the image in $\mathcal{P}(!A)$ of the morphism*

$$!\,\mathsf{supp}_{\sigma} \quad : \quad !\{A \,|\, \sigma\} \quad \longrightarrow \quad !A.$$

Note that the definition of $\#\sigma$ induces a commutative diagram in the category $\mathcal{G}$

$$\begin{array}{ccc} !\{A \,|\, \sigma\} & \xrightarrow{\textit{isomorphism}} & \{!A \,|\, \#\sigma\} \\ & {\scriptstyle !\mathsf{supp}_{\sigma}} \searrow \quad \swarrow {\scriptstyle \mathsf{supp}_{\#\sigma}} & \\ & !A & \end{array}$$

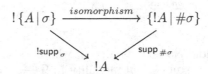

where the top arrow is an isomorphism. Moreover, the definition of $\#\sigma$ coincides with the previous definition (Definition 12) when the P-strategy $\sigma = f$ happens to be deterministic.Consequently, for two games A, B and a deterministic strategy $f : A \multimap B$, we have $\mathsf{image}(!f) \cong \#^{\mathcal{S}}\mathsf{image}(f)$ and $\#^{\mathcal{S}} f = \#f$.

As mentioned in the introduction, this construction $\sigma \mapsto \#\sigma$ defines a functor

$$p_A \quad : \quad \mathcal{P}(A) \quad \longrightarrow \quad \mathcal{P}(!A).$$

Now, remember that a morphism $\sigma : A \to B$ of the bicategory $\mathcal{S}$ is defined as a P-strategy

$$\sigma \in \mathcal{P}(A \multimap B).$$

For that reason, every such morphism $\sigma : A \to B$ induces a P-strategy

$$\#\sigma \in \mathcal{P}(!(A \multimap B)).$$

In order to turn the P-strategy $\#\sigma$ into a P-strategy

$$!\sigma \in \mathcal{P}(\,!A \multimap !B\,)$$

we apply the functor

$$\mathcal{P}(n_{A,B}) \quad : \quad \mathcal{P}(\,!(A \multimap B)\,) \quad \longrightarrow \quad \mathcal{P}(\,!A \multimap !B\,)$$

to the P-strategy $\#\sigma$, where

$$n_{A,B} \quad : \quad !\,(A \multimap B) \quad \longrightarrow \quad !A \multimap !B$$

denotes the structural morphism of $\mathcal{G}$ defined in the previous section. The construction may be summarized as follows:

Definition 14. *The morphism $!\sigma : !A \to !B$ of the bicategory $\mathcal{S}$ associated to the morphism $\sigma : A \to B$ is defined as the P-strategy*

$$\mathcal{P}(n_{A,B})(\#\sigma) \quad \in \quad \mathcal{P}(!A \multimap !B).$$

Theorem 4. *With this definition, $! : \mathcal{S} \to \mathcal{S}$ defines a pseudofunctor from the bicategory $\mathcal{S}$ to itself.*

Proof. See Appendix J.

The family of morphisms

$$\delta_A : !A \to !!\,A \qquad\qquad \varepsilon_A : !A \to A$$

are defined with the same deterministic strategies in $\mathcal{P}(!A \multimap !!\,A)$ and $\mathcal{P}(!A \multimap A)$ as in the original category $\mathcal{G}$. One checks that the families δ and ε define natural transformations between pseudonatural functors on $\mathcal{S}$ (as defined in Definition 26), and that the 2-functor $! : \mathcal{S} \to \mathcal{S}$ defines a 2-comonad in the appropriate bicategorical sense (see Definition 27). The family of morphisms

$$d_A : !A \to !A \otimes !A \qquad\qquad e_A : !A \to 1$$

are defined with the same deterministic strategies in $\mathcal{P}(!A \multimap !A \otimes !A)$ and $\mathcal{P}(!A \multimap 1)$ as in the original category $\mathcal{G}$, and one checks that they define natural transformations between pseudonatural functors on $\mathcal{S}$. One obtains in this way that

Theorem 5. *The bicategory p equipped with the exponential modality* $! : \mathcal{S} \to \mathcal{S}$ *defines a bicategorical model of multiplicative intuitionistic linear logic.*

The formal and rigorous verification of these facts would be extremely tedious if done directly on the bicategory $\mathcal{S}$ of nondeterministic strategies. Our proof relies on the fact that the constructions of the model (Definitions 9, 14) are performed by "push" functors $\mathcal{P}(f)$ above a structural morphism f living in the original category $\mathcal{G}$. The interested reader will find part of the detailed proof in Appendix K.

9 Conclusion

We construct a bicategory $\mathcal{S}$ of simple games and non-deterministic strategies, which is symmetric monoidal closed in the extended 2-dimensional sense. We then equip the bicategory $\mathcal{S}$ with a linear exponential modality $! : \mathcal{S} \to \mathcal{S}$ which defines a bicategorical model of intuitionistic linear logic. This provides, as far as we know, the first sheaf-theoretic and non-deterministic game semantics of intuitionistic linear logic — including, in particular, a detailed description of the exponential modality.

A The Category $\mathcal{G}$ of Simple Games and Deterministic Strategies

We recall the construction of the category Υ of schedules performed in [4] and how we deduce from it the category $\mathcal{G}$ of simple games and deterministic strategies.

Definition 15 (Schedule). *A schedule is defined as a function* $e : \{1, \ldots, n\} \to \{0, 1\}$ *verifying* $e(1) = 1$ *and* $e(2k+1) = e(2k)$ *whenever* $1 \leq 2k \leq n-1$. *The number of* 0*'s and* 1*'s in e are noted* $|e|_0$ *and* $|e|_1$ *respectively. A schedule e is noted* $e : |e|_0 \to |e|_1$.

A schedule $e : p \to q$ may be equivalently seen as a couple $l : (p) \to (p+q)$ and $r : (q) \to (p+q)$ of order-preserving and globally surjective functions, such that $r(1) = 1$ and

$$l(i)\,\text{odd} \Rightarrow l(i+1) = l(i) + 1 \qquad\qquad r(j)\,\text{even} \Rightarrow r(j+1) = r(j) + 1$$

for all $1 \leq i \leq p-1$ and $1 \leq j \leq q-1$, where (n) stands for the finite ordinal $(n) = \{1, \ldots, n\}$.

Definition 16. *The category of schedules Υ has the natural numbers as objects, the schedules $e : p \to q$ as morphisms from p to q.*

The identity morphism $c : p \to p$ is the copycat schedule c characterized by the fact that $c(2k+1) \neq c(2k+2)$ for all $1 \leq 2k \leq 2p$. Details on the composition of two schedules $e : p \to r$ and $e' : r \to q$ as a schedule $e \centerdot e' : p \to q$ can be found in [4]. Now, we explain how we derive the category $\mathcal{G}$ from the category Υ. We start by defining the simple game $A \multimap B$ of linear maps from A to B:

Definition 17. *The simple game $A \multimap B$ is defined as the set $(A \multimap B)_n$ of all the triples (e, a, b) consisting of a schedule $e : p \to q$ with $p + q = n$, a position $a \in A_p$ and $b \in B_q$. The predecessor function π is defined as*

$$\pi(e, a, b) = \begin{cases} (e \restriction (n-1), \pi(a), b) & if\ e(n) = 0 \\ (e \restriction (n-1), a, \pi(b)) & if\ e(n) = 1 \end{cases}$$

Definition 18. *The category $\mathcal{G}$ has simple games A, B as objects, and deterministic P-strategies f, g of $A \multimap B$ as morphisms from A to B. Note that we use latin letters instead of greek letters for deterministic strategies. The identity morphism $i_A : A \to A$ is defined as the P-strategy of $A \multimap A$ whose positions of degree $2n$ are the triples (c, a, a) where $c : n \to n$ is the copycat schedule, and $a \in A_n$. The composite $g \circ f : A \to C$ of two deterministic P-strategies $f : A \to B$ and $g : B \to C$ is the deterministic P-strategy whose set of positions of degree $2n$ is defined as*

$$(g \circ f)_{2n} = \coprod_{\substack{e : p \to r, e' : r \to q \\ p + q = 2n}} \left\{ (e \centerdot e', a, c) \;\middle|\; \exists b \in B_r,\ (e, a, b) \in \sigma_{p+r},\ (e', b, c) \in \tau_{r+q} \right\}$$

B The Tensor Product in the Category $\mathcal{G}$

Definition 19 (Tensorial schedule). *A $\otimes$-schedule is a function $e : \{1, \ldots, n\} \to \{0, 1\}$ verifying $e(2k+1) = e(2k+2)$ whenever $0 \leq 2k \leq n-2$.*

Definition 20 ($A \otimes B$). *The positions of the simple game $A \otimes B$ of degree n are the triples (e, a, b) where $e : p \otimes q$ is a $\otimes$-schedule with $p + q = n$, $a \in A_p$ and $b \in B_q$. The predecessor function π is defined as*

$$\pi(e, a, b) = \begin{cases} (e \restriction (n-1), \pi(a), b) & if\ e(n) = 0 \\ (e \restriction (n-1), a, \pi(b)) & if\ e(n) = 1 \end{cases}$$

The simple game 1 is the simple game with a unique position $*$, of degree 0.

We can also define $\otimes$ on strategies. Intuitively, for $f : A \to B$ and $g : C \to D$ two morphisms of the category $\mathcal{G}$, the plays of the strategy $f \otimes g$ of the simple game $(A \otimes C) \multimap (B \otimes D)$ are obtained by combining through a tensorial schedule plays of f and g.

The intuition is that, once we know the structure of f and g, the structure of plays of $f \otimes g$ is entirely directed by what happens in $B \otimes D$. The only agency that Opponent really has is to decide at some points whether to play on B or D, the rest being handled by the plays of f, g and the structure of $(A \otimes C) \multimap (B \otimes D)$. Formally, this gives the proposition:

Proposition 10. *Let $f : A \multimap B, g : C \multimap D$ be two deterministic strategies. Assuming a valid play of $f \otimes g : A \otimes C \multimap B \otimes D$ and the associated schedules $e : A \otimes C \to B \otimes D, t_1 : A \times C, t_2 : B \times D, e_1 : A \to B, e_2 : C \to D$, the knowledge of t_2, e_1, e_2 is enough to reconstruct e and t_1.*

Proof. The first O move of such a play is in $B \otimes D$ to follow the structure of $A \otimes C \multimap B \otimes D$. This is given to us by t_2. Let us assume it is a move in D (The other case is handled similarly).

The P move after that will necessarily be a move in C or D, as playing a move in A, B would break the structure of $A \multimap B, B \otimes D$ respectively. e_2 gives us the information.

- If it is a move in D, We go back to a situation equivalent to the initial one. We have also started to reconstruct e, which starts by 11.
- If it is a move in C, we start to reconstruct both e which starts by 10 and t_1 which starts by 1.

In this last case, the following O move will be a move in C as a move in A, B, D would break the structure of $A \multimap B, B \otimes D, C \multimap D$ respectively. e is then at 100 and t_1 at 11.

Finally, the following P move will be a move in either C or D as a move in A, B would break the structure of $A \multimap B, B \otimes D$ respectively. e_2 gives us this information.

- If it is a move in D, We go back to a situation equivalent to the initial one. We have also started to reconstruct e, which starts by 1001 and t_1 which starts by 11. We've also played the first two moves of t_2 which is at 11.
- If it is a move in C, we go back to the precedent situation (the one with a fixed O move in C) with e at 1000 and t_1 at 111.

To sum up the described construction, once an opponent move in B or D is played, the play is stuck playing in either $A \multimap B$ or $C \multimap D$ until a player move is played in B, D respectively. t_2 decides whether to play the opponent move in B or D and e_1 guides the play in $A \multimap B$ in the first case, e_2 guides it in $C \multimap D$ in the second. This guides us through the whole play and allows us to reconstruct both e and t_1.

In particular, any compatible plays of $f, g, B \otimes D$ induce a play of $f \otimes g$.

This proposition and its proof are key in several proofs we will make in the rest of the paper.

Proposition 11. *The category $(\mathcal{G}, \otimes, 1, \multimap)$ is symmetric monoidal closed.*

C The Exponential Modality on the Category $\mathcal{G}$

In this section, we recall the combinatorial structures introduced in [4] to construct the linear exponential comonad $! : \mathcal{G} \rightarrow \mathcal{G}$ on the symmetric monoidal closed category $\mathcal{G}$.

Definition 21 (Pointer function). *A pointer function on n is a parity-reversing function*

$$\phi \quad : \quad \{1, \ldots, n\} \quad \longrightarrow \quad \{0, \ldots, n-1\}$$

such that $\phi(i) < i$ for all i. A pointer function ϕ is called an O-heap if $\phi(2k) = 2k - 1$ for all k, and a P-heap if $\phi(2k+1) = 2k$ for all k. The set $\{k, \phi(k), \phi^2(k), ...\}$ will be called the branch of ϕ associated to the integer k. Note that the predecessor function π defined as $\pi(i) = i - 1$ for all i is both an O-heap and a P-heap.

Definition 22. *Suppose that $e : p \rightarrow q$ is a schedule, that ϕ is a O-heap over q and that ψ is a P-heap over p. The O-heap (ϕ, e, ψ) on $p + q$ is defined as follows:*

$$(\phi, e, \psi)(k) \quad = \quad \begin{cases} r(\phi(j)) & if\ k = r(j)\ is\ odd \\ l(\psi(i)) & if\ k = l(i)\ is\ odd \\ k - 1 & otherwise \end{cases}$$

where the schedule e is represented as a pair (l, r) as explained in Appendix A. Intuitively, the O-heap (ϕ, e, ψ) points alongside ϕ when the schedule e is at 1 and alongside ψ otherwise. The fact that (ϕ, e, ψ) defines an O-heap is ensured by the even case.

We recall the partial order over the set of pointer functions introduced in [4].

Definition 23 (Generalization). *Given two pointer functions ϕ, ψ, we say that ϕ is a generalization of ψ, and note $\phi \succeq \psi$, if the branch of ϕ associated to $k \in \{1, .., n\}$ can be injected in the branch of ψ associated to k, or, in other words, if for all k, there exists j such that $\phi(k) = \psi^j(k)$.*

Further in the paper, and in certain proofs, we will also need to look into the structure of $!!A$. Intuitively, positions of $!!A$ are pairs $(\phi, \overline{u})$ where $\overline{u}$ is a sequence of positions of $!A$ and ϕ an O-heap. It is equivalent to another representation using only a sequence of positions of A:

Proposition 12. *A position $(\phi, \overline{u})$ of $!!A$ is equivalent to $(\phi, \psi, \overline{a})$ with $\phi \succeq \psi$, ψ an O-heap, $\overline{a}$ a sequence of positions of A, verifying*

$$\forall i, j \in \{1, \ldots, n\}, (i \neq j) \Rightarrow \exists k, a_{\phi^k(i)} \neq a_{\phi^k(j)}$$

The moves alongside the branches of ψ are then plays of the simple game A.

From this follows a description of the strategy

$$!!f \quad : \quad !!A \quad \longrightarrow \quad !!B$$

for a deterministic strategy $f : A \multimap B$. The positions of $!!f$ are of the form

$$(e, (\phi, \psi, \overline{a}), (\phi', \psi', \overline{b}))$$

where $e^*\phi' = \phi, e^*\psi' = \psi$ and each thread of (ψ, e, π) is a play of the strategy f.

D Some Bicategorical Definitions

In this section, we recall a few definitions required by our bicategorical setting.

Definition 24. *A pseudofunctor is a mapping between bicategories $\mathcal{C}$ and $\mathcal{D}$ where the usual functorial equations $F(f \circ g) = F(f) \circ F(g)$ and $F(Id_A) = Id_{F(A)}$ are only valid up to natural bijectve 2-morphisms in $\mathcal{D}$.*

Definition 25. *Let $(\mathcal{C}, \otimes_\mathcal{C}, 1_\mathcal{C})$ and $(\mathcal{D}, \otimes_\mathcal{D}, 1_\mathcal{D})$ be two monoidal bicategories. A lax monoidal pseudofunctor between them is given by:*

- *a pseudofunctor $F : \mathcal{C} \to \mathcal{D}$*
- *a morphism $\epsilon : 1_\mathcal{D} \to F(1_\mathcal{C})$*
- *for every pair of objects $A, B \in \mathcal{C}$, a natural transformation $\mu_{A,B} : F(A) \otimes_\mathcal{D} F(B) \to F(A \otimes_\mathcal{C} B)$*

satisfying the following conditions:

- *associativity: For every triple of objects $A, B, C \in \mathcal{C}$, the following diagram commutes:*

$$\begin{array}{ccc}
(F(A) \otimes_\mathcal{D} F(B)) \otimes_\mathcal{D} F(C) & \xrightarrow[a^\mathcal{D}_{F(A),F(B),F(C)}]{} & F(A) \otimes_\mathcal{D} (F(B) \otimes_\mathcal{D} F(C)) \\
\Big\downarrow{\scriptstyle \mu_{A,B} \otimes id} & & \Big\downarrow{\scriptstyle id \otimes \mu_{B,C}} \\
F(A \otimes_\mathcal{C} B) \otimes_\mathcal{D} F(C) & & F(A) \otimes_\mathcal{D} F(B \otimes_\mathcal{C} C) \\
\Big\downarrow{\scriptstyle \mu_{A \otimes B,C}} & & \Big\downarrow{\scriptstyle \mu_{A,B \otimes C}} \\
F((A \otimes_\mathcal{C} B) \otimes_\mathcal{C} C) & \xrightarrow[F(a^\mathcal{C}_{A,B,C})]{} & F(A \otimes_\mathcal{C} (B \otimes_\mathcal{C} C))
\end{array}$$

  where the two morphisms $a^\mathcal{C}, a^\mathcal{D}$ denote the associators of the two tensor products.
- *unality: For every object $A \in \mathcal{C}$, the following diagram and its right symmetry both commute:*

$$\begin{array}{ccc}
1_\mathcal{D} \otimes_\mathcal{D} F(A) & \xrightarrow[\epsilon \otimes id]{} & F(1_\mathcal{C}) \otimes_\mathcal{D} F(A) \\
\Big\downarrow{\scriptstyle l^\mathcal{D}_{F(A)}} & & \Big\downarrow{\scriptstyle \mu_{1_\mathcal{C},A}} \\
F(A) & \xleftarrow{F(l^\mathcal{C}{}_A)} & F(1_\mathcal{C} \otimes_\mathcal{C} A)
\end{array}$$

  where $l^\mathcal{C}, l^\mathcal{D}$ denote the left unitors of the two tensor products.

Definition 26. *Let F, G be two pseudofunctors between two bicategories $\mathcal{C}$ and $\mathcal{D}$. A pseudonatural transformation $\phi : F \to G$ is given by:*

- *for every object A of $\mathcal{C}$, a morphism $\phi(A) : F(A) \to G(A)$ of $\mathcal{D}$.*
- *for every morphism $f : A \to B$ of $\mathcal{C}$, a bijective 2−morphism $\phi(f) : \phi(B) \circ F(f) \Rightarrow G(f) \circ \phi(A)$*

such that

- *ϕ respects composition of morphisms, meaning that we have an equivalence between*
$$(\phi(A) \lhd G(f,g)) \cdot (\phi(f) \rhd G(g)) \cdot (F(f) \lhd \phi(g))$$
and
$$\phi(g \circ f) \cdot (F(f,g) \rhd \phi(C)),$$
both being 2-morphisms from
$$\phi(C) \circ F(g) \circ F(f) \Rightarrow G(g \circ f) \circ \phi(A),$$
where $\cdot$ is the vertical composition between 2-morphisms, $\lhd, \rhd$ the two versions of the horizontal composition between a morphism and a 2-morphism, (also called whiskering), anf $F(f,g) : F(g) \circ F(f) \Rightarrow F(g \circ f)$ is the bijective 2-morphism coming from the pseudofunctor F.
- *ϕ respects the identity morphisms, meaning we have an equivalence between*
$$L^{\mathcal{D}}_{\phi(A)} \cdot \epsilon^F_{id_A} \rhd \phi(A)$$
and
$$R^{\mathcal{D}}_{\phi(A)} \cdot \phi(A) \lhd \epsilon^G_{id_A} \cdot \phi(id_A)$$
both being 2-morphisms from
$$\phi(A) \circ F(id_A) \Rightarrow \phi(A)$$
where $L^{\mathcal{D}}_{\phi(A)} : \phi(A) \circ id_{F(A)} \Rightarrow \phi(A)$ is the left unitor coming from the bicategory $\mathcal{D}$ and $\epsilon^F_{id_A} : F(id_A) \Rightarrow id_{F(A)}$ is the bijective 2-morphism coming from the pseudofunctor F.
- *ϕ is natural in the following sense: for every 2-morphism $\psi : f \Rightarrow g$ with $f, g : A \to B$, we have an equivalence between*
$$\phi(g) \cdot F(\psi) \rhd \phi(B)$$
and
$$\phi(A) \lhd G(\psi) \cdot \phi(f).$$

Definition 27. *A fully weak comonad G on a bicategory $\mathcal{C}$ is a pseudofunctor, along with pseudonatural transformations δ and ϵ that satisfy the usual laws of a comonad up to natural bijectiive 2-morphisms in $\mathcal{C}$.*

E Proof of Proposition 2

Proof. Let A, B be two games.

Let σ be a P-cartesian transduction between A and B. The associated deterministic strategy f_σ is simply given by:

$$f_\sigma(2n) = \{(c, a, \sigma(a)) | a \in A(n)\}$$

This definition clearly gives a functional strategy, the determinism being given by the fact that σ is P-cartesian.

Conversely, let f be a functional strategy of $A \multimap B$. The associated P-cartesian transduction σ_f is given by:

$$\sigma_f(2n)(a) = b \text{ s.t. } (c, a, b) \in f(4n)$$

Such a b is unique by functionality of f.

F Proof that $\mathcal{P}$ is a pseudofunctor

Proof. First we need to complete the definition of $\mathcal{P}$ by detailling why, for f a deterministic strategy of $A \to B$ and σ a P-strategy over A, $\mathcal{P}(f)(\sigma)$ is indeed a P-strategy over B, and thus a presheaf over $\mathbf{tree}(B_P)$. For this, we need to define the collection of projector functions $\pi_{2n} : \mathcal{P}(f)(\sigma)(2n) \to \mathcal{P}(f)(\sigma)(2n-2)$ as follows:

For $x \in \mathcal{P}(f)(\sigma)(2n)$ over b (meaning $x \in \mathcal{P}(f)(\sigma)(b)$ and $b \in B_{2n}$), there exists by definition a unique e, a such that $(e, a, b) \in f$ and $x \in \sigma(a)$. From this, we define:

$$\pi_{2n}(x) = \pi_\sigma^k(x), (\pi^{2k+2}(e), \pi_A^{2k}(a), \pi_B^2(b)) \in f.$$

By determinism of f, there is only one such k. Moreover, we also have $\pi_\sigma^k(x) \in \sigma(\pi_A^{2k}(a))$. Consequently, by definition of $\mathcal{P}(f)(\sigma)$, we have $\pi_\sigma^k(x) \in \mathcal{P}(f)(\sigma)(\pi_B^2(b))$ as expected.

Next step is to show that, for a strategy $f : A \to B$, $\mathcal{P}(f)$ is a functor from $\mathcal{P}(A)$ to $\mathcal{P}(B)$. For that, we need to define its effects on simulations. For $\alpha : \sigma \to \tau$, $\mathcal{P}(f)(\alpha) : \mathcal{P}(f)(\sigma) \to \mathcal{P}(f)(\tau)$ is simply defined by applying α to all positions of $\mathcal{P}(f)(\sigma)$, as all those are induced from positions of σ by definition. With this, it is easy to verify that $\mathcal{P}(f)$ preserves identities and composition of simulations.

Finally, let us show that $\mathcal{P}$ is a pseudofunctor.

First, $\mathcal{P}(Id_A)\sigma$ associates to a position a of A the set:

$$\mathcal{P}(Id_A)(\sigma) \quad : \quad a \quad \mapsto \quad \coprod_{(c,a,a) \in Id_A} \quad \sigma(a).$$

which is instantly isomorphic to $\sigma(a)$. Factoring the effect on simulations, it is easy to build a bijective natural natural transformation between $\mathcal{P}(Id_A) \cong Id_{\mathcal{P}(A)}$. Thus $\mathcal{P}(Id_A) \cong Id_{\mathcal{P}(A)}$.

Next, let $f : A \to B$ and $g : B \to C$ two deterministic strategies and σ a P-strategy of A. We have:

$$\mathcal{P}(g)(\mathcal{P}(f)(\sigma) \quad : \quad c \quad \mapsto \quad \coprod_{(e_2,b,c)\in g} \; \coprod_{(e_1,a,b)\in f} \sigma(a).$$

This is easily isomorphic to $\mathcal{P}(g \circ f)\sigma$ which is given by:

$$\mathcal{P}(g \circ f)(\sigma) \quad : \quad c \quad \mapsto \coprod_{(e,a,c)\in g\circ f} \sigma(a).$$

This isomorphism is a consequence of the definition of composition for deterministic strategies, as there is only one triple e_1, e_2, b such that $(e_1, a, b) \in f$, $(e_2, b, c) \in g$ and $e = e_1 \cdot e_2$ for a position $(e, a, c) \in g \circ f$.

This extends into a natural isomorphism between the functors $\mathcal{P}(g \circ f)$ and $\mathcal{P}(g)(\mathcal{P}(f)$, giving us the fact that $\mathcal{P}$ is indeed a pseudofunctor.

G Proof of Proposition 6

Proof. – Let $f : A \multimap B, g : C \multimap D$ be two slender strategies. Let (t_2, b, d)be a player position of $B \otimes D$. Since f and g are slender, there exist unique e_f, a, e_g, c such that $(e_f, a, b) \in f, (e_g, c, d)\in g$. Using t_2, e_f, e_g and Proposition 10, we reconstruct e, t_1 such that $(e, (t_1, a, c), (t_2, b, d))$ is a position of $f \otimes g$. This position is unique as the reconstruction of Proposition 10 is unique, and thus $f \otimes g$ is a slender strategy.

– Let $f : A \multimap B, g : C \multimap D$ be two functional strategies. Let (t_1, a, c) be an opponent position of $A \otimes C$. Since f and g are functional strategies, there exist unique b, d such that $(cp_f, a, b) \in f, (cp_g, c, d) \in g$. The study of $f \otimes g$ done in the proof of Proposition 10 gives us that any valid position of $f \otimes g$ would have a copycat schedule (as the schedule is built from sequences $1.0^k.1$ of cp_f and cp_g. This implies immediately that the only possible position is $cp, (t_1, a, c), (t_1, b, d)$ as no other play would verify the needed structures, and thus $f \otimes g$ is a functional strategy.

H Proof of Theorem 3

Proof. First, we can note that the unit 1 of $\mathcal{G}$ has a unique P-strategy, the empty strategy. Consequently, $\mathcal{P}(1)$ is the singleton category, which is the unit of the cartesian product in **Cat**.

Moreover, to extend $\mathcal{P}$ as a lax monoidal pseudofunctor, we need a transformation $\mu_{A,B} : \mathcal{P}(A) \times \mathcal{P}(B) \to \mathcal{P}(A \otimes B)$ natural in A and B.

Since the morphisms of that transformation live in **Cat**, they are functors. We thus define:

for σ an object of $\mathcal{P}(A)$ and τ an object of $\mathcal{P}(B)$,

$$\mu_{A,B}(\sigma, \tau) = \sigma \otimes \tau$$

for $\alpha : \sigma \to \sigma'$ a morphism of $\mathcal{P}(A)$ and $\beta : \tau \to \tau'$ a morphism of $\mathcal{P}(B)$, $\mu_{A,B}(\alpha, \beta) : \sigma \otimes \tau \to \sigma' \otimes \tau'$ is defined by:

$$\mu_{A,B}(\alpha, \beta)(t, x, y) = (t, \alpha(x), \beta(y))$$

We now need to prove that this transformation is natural in A and B, and that it verifies the two commutative diagrams of a lax monoidal functor (associativity and unitality), up to bijective simulations. Those last two are easy to verify and use similar arguments, so we will focus on the naturality.

We need our transformation to verify the following commutative diagram for A, B, A', B' four games and $f : A \multimap A', g : B \multimap B'$ two deterministic strategies:

$$\begin{array}{ccc}
\mathcal{P}(A) \times \mathcal{P}(B) & \xrightarrow{\mu_{A,B}} & \mathcal{P}(A \otimes B) \\
\downarrow{\scriptstyle \mathcal{P}(f) \times \mathcal{P}(g)} & & \downarrow{\scriptstyle \mathcal{P}(f \otimes g)} \\
\mathcal{P}(A') \times \mathcal{P}(B') & \xrightarrow{\mu_{A',B'}} & \mathcal{P}(A' \otimes B')
\end{array}$$

Let σ be a P-strategy of A and τ a P-strategy of B. Verifying the commutative diagram amounts to finding two reciprocal morphisms between: $\mathcal{P}(f)(\sigma) \otimes \mathcal{P}(g)(\tau)$ and $\mathcal{P}(f \otimes g)(\sigma \otimes \tau)$.

$$\mathcal{P}(f)(\sigma) \otimes \mathcal{P}(g)(\tau) \cong \mathsf{image}(f \circ \mathsf{supp}_\sigma) \otimes \mathsf{image}(g \circ \mathsf{supp}_\tau)$$
$$\mathcal{P}(f)(\sigma) \otimes \mathcal{P}(g)(\tau) \cong \mathsf{image}(f \circ \mathsf{supp}_\sigma \otimes g \circ \mathsf{supp}_\tau) \quad \text{by consequences of prop 6}$$

$$\mathcal{P}(f \otimes g)(\sigma \otimes \tau) \cong \mathsf{image}((f \otimes g) \circ \mathsf{supp}_{\sigma \otimes \tau})$$
$$\mathcal{P}(f \otimes g)(\sigma \otimes \tau) \cong \mathsf{image}((f \otimes g) \circ \mathsf{supp}_\sigma \otimes \mathsf{supp}_\tau) \quad \text{by consequences of prop 6}$$

By bifunctoriality of $\otimes$, we have $f \circ \mathsf{supp}_\sigma \otimes g \circ \mathsf{supp}_\tau \cong (f \otimes g) \circ \mathsf{supp}_\sigma \otimes \mathsf{supp}_\tau$, giving us the equality of the images we need, up to bijective simulations.

I Proof of Proposition 9

Proof. – Let $(\psi, \overline{b} = b_1, ...b_n)$ a P position of $!B$. Since f is slender, for all b_i player positions of $\overline{b}$, there exists a unique pair (e_i, a_i) such that $(e_i, a_i, b_i) \in f$.

We use a method similar to the one used in the proof of Proposition 10. Instead of using the tensorial schedule to guide us in reconstructing the play of $!A \multimap !B$, we use ψ, which indicates us what is the next player move b_i to get to (starting from b_{i-2}, and assuming we have reconstructed e and ϕ so far), and then use the play (e_i, a_i, b_i) to construct the play.

The sequence of moves we add is the suffix of the play (e_i, a_i, b_i) looking like $b_{i-1}a_i^1....a_i^k b_i$ (with $a_i^k = a_i$) as any other move in the play (e_i, a_i, b_i) has already been played (since in particular any b move prior to b_{i-1} has been played.

Player cannot backtrack in the middle of the sequence $b_{i-1}a_i^1....a_i^k b_i$ without breaking the fact that the full play is associated to a O-heap in $!(A \multimap B)$.

This allows us to extend e into $e.1.0^k.1$ and ϕ by linking a_i^1 to its predecessor in A of the play (e_i, a_i, b_i).

This method constructs a valid position of $!f$ as all branches are played following f and ϕ is a O-Heap. It is the only possible position including $\psi, \overline{b}$ as everything we have done was determined by ψ, f and $\overline{b}$. Thus $!f$ is a slender strategy.

- Let $(\phi, \overline{a} = a_1, ...a_n)$ an O position of $!A$. Since f is a functional strategy, for all a_i opponent positions of $\overline{a}$, there exists a unique b_i such that $(c, a_i, b_i) \in f$. By determinism of f, it is also true for all player positions of $\overline{a}$. By using ϕ as a guide, this easily allows us to construct the position of $!f$: $(c, (\phi, \overline{a}), (\phi, \overline{b} = b_1, ...b_n))$.

  It is the unique such position for $(\phi, \overline{a})$ for reasons similar to the ones evoked in the proof for slender strategies. Thus $!f$ is a functional strategy.

J Proof of Theorem 4

Proof. – For a game A, we have by construction:

$$(!_{\mathcal{P}})_{A,B}(Id_A) = \mathcal{P}(n_{A,B}) \circ \#^{\mathcal{S}}(Id_A)$$
$$(!_{\mathcal{P}})_{A,B}(Id_A) = \mathcal{P}(n_{A,B})(\#Id_A) = Id_{!A}$$

- Let A, B, C be three games and σ a P-strategy of $A \multimap B$, τ a P-strategy of $B \multimap C$. We need to prove that there is a natural isomorphic simulation between $!_{\mathcal{P}}(\tau \circ \sigma)$ and $!_{\mathcal{P}}(\tau) \circ !_{\mathcal{P}}(\sigma)$.

  First we will simplify those two strategies through the various properties we have seen so far:
  First $!_{\mathcal{P}}(\tau \circ \sigma)$:

$$!_{\mathcal{P}}(\tau \circ \sigma) = \mathcal{P}(n_{A,C})(\#^{\mathcal{S}}(\tau \circ \sigma))$$
$$!_{\mathcal{P}}(\tau \circ \sigma) \cong \mathsf{image}(n_{A,C} \circ \mathsf{supp}_{\#^{s}(\tau\circ\sigma)}) \text{ by equation 4}$$
$$!_{\mathcal{P}}(\tau \circ \sigma) \cong \mathsf{image}(n_{A,C} \circ \,!\mathsf{supp}_{\tau\circ\sigma}) \text{ by consequence of def 13}$$
$$!_{\mathcal{P}}(\tau \circ \sigma) \cong \mathsf{image}(n_{A,C} \circ \,!\mathsf{supp}_{\mathcal{P}(comp_{A,B,C})(\sigma\otimes\tau)}) \text{ by definition 9}$$
$$!_{\mathcal{P}}(\tau \circ \sigma) \cong \mathsf{image}(n_{A,C} \circ \,!\mathsf{supp}_{\mathsf{image}(comp_{A,B,C}\circ\mathsf{supp}_{\sigma\otimes\tau})}) \text{ by equation 4}$$
$$!_{\mathcal{P}}(\tau \circ \sigma) \cong \mathsf{image}(n_{A,C} \circ \mathsf{supp}_{\mathsf{image}(!(comp_{A,B,C}\circ\mathsf{supp}_{\sigma\otimes\tau}))}) \text{ by consequence of def 13}$$
$$!_{\mathcal{P}}(\tau \circ \sigma) \cong \mathsf{image}(n_{A,C} \circ \,!(comp_{A,B,C} \circ \mathsf{supp}_{\sigma\otimes\tau})) \text{ by theorem 1}$$
$$!_{\mathcal{P}}(\tau \circ \sigma) \cong \mathsf{image}(n_{A,C} \circ \,!comp_{A,B,C} \circ \mathsf{supp}_{\#^{s}(\sigma\otimes\tau)}) \text{ by functoriality of ! and consequence of def 13}$$

Then, $!_{\mathcal{P}}(\tau) \circ\ !_{\mathcal{P}}(\sigma)$:

$$!_{\mathcal{P}}(\tau) \circ\ !_{\mathcal{P}}(\sigma) = \mathcal{P}(n_{B,C})(\#^{\mathcal{S}}\tau) \circ \mathcal{P}(n_{A,B})(\#^{\mathcal{S}}\sigma)$$

$$!_{\mathcal{P}}(\tau) \circ !_{\mathcal{P}}(\sigma) \cong \mathsf{image}(n_{B,C} \circ\ \mathsf{supp}_{\#^{s}(\tau)}) \circ \mathsf{image}(n_{A,B} \circ\ \mathsf{supp}_{\#^{s}(\sigma)}) \text{ by equation 4}$$

$$!_{\mathcal{P}}(\tau) \circ !_{\mathcal{P}}(\sigma) \cong \mathcal{P}(comp_{!A,!B,!C})(\mathsf{image}(n_{A,B} \circ\ \mathsf{supp}_{\#^{s}(\sigma)}) \otimes \mathsf{image}(n_{B,C} \circ\ \mathsf{supp}_{\#^{s}(\tau)})) \text{ by definition 9}$$

$$!_{\mathcal{P}}(\tau) \circ !_{\mathcal{P}}(\sigma) \cong \mathcal{P}(comp_{!A,!B,!C})(\mathsf{image}(n_{A,B} \circ\ \mathsf{supp}_{\#^{s}(\sigma)} \otimes n_{B,C} \circ \mathsf{supp}_{\#^{s}(\tau)})) \text{ by consequence of prop 6}$$

$$!_{\mathcal{P}}(\tau) \circ !_{\mathcal{P}}(\sigma) \cong \mathsf{image}(comp_{!A,!B,!C} \circ (\mathsf{supp}_{\mathsf{image}(n_{A,B} \circ \mathsf{supp}_{\#^{s}(\sigma)} \otimes n_{B,C} \circ \mathsf{supp}_{\#^{s}(\tau)})})) \text{ by equation 4}$$

$$!_{\mathcal{P}}(\tau) \circ !_{\mathcal{P}}(\sigma) \cong \mathsf{image}(comp_{!A,!B,!C} \circ (n_{A,B} \circ\ \mathsf{supp}_{\#^{s}(\sigma)} \otimes\ n_{B,C} \circ\ \mathsf{supp}_{\#^{s}(\tau)})) \text{ by theorem 1}$$

$$!_{\mathcal{P}}(\tau) \circ !_{\mathcal{P}}(\sigma) \cong \mathsf{image}(comp_{!A,!B,!C} \circ n_{A,B} \otimes\ n_{B,C} \circ \mathsf{supp}_{\#^{s}(\sigma)} \otimes \mathsf{supp}_{\#^{s}(\tau)}) \text{ by bifunctoriality of } \otimes$$

$$!_{\mathcal{P}}(\tau) \circ !_{\mathcal{P}}(\sigma) \cong \mathsf{image}(comp_{!A,!B,!C} \circ n_{A,B} \otimes\ n_{B,C} \circ !\mathsf{supp}_{(\sigma)} \otimes !\mathsf{supp}_{(\tau)}) \text{ by consequence of def 13}$$

We intend to prove that those two images are isomorphic. For that, we will make the following remark:
! is lax monoidal in $\mathcal{G}$, meaning that there exists a transformation $\mu_{A,B}$: $!A \otimes !B \to !(A \otimes B)$ natural in A and B. Thus we have the following diagram with the top square commuting by naturality of μ:

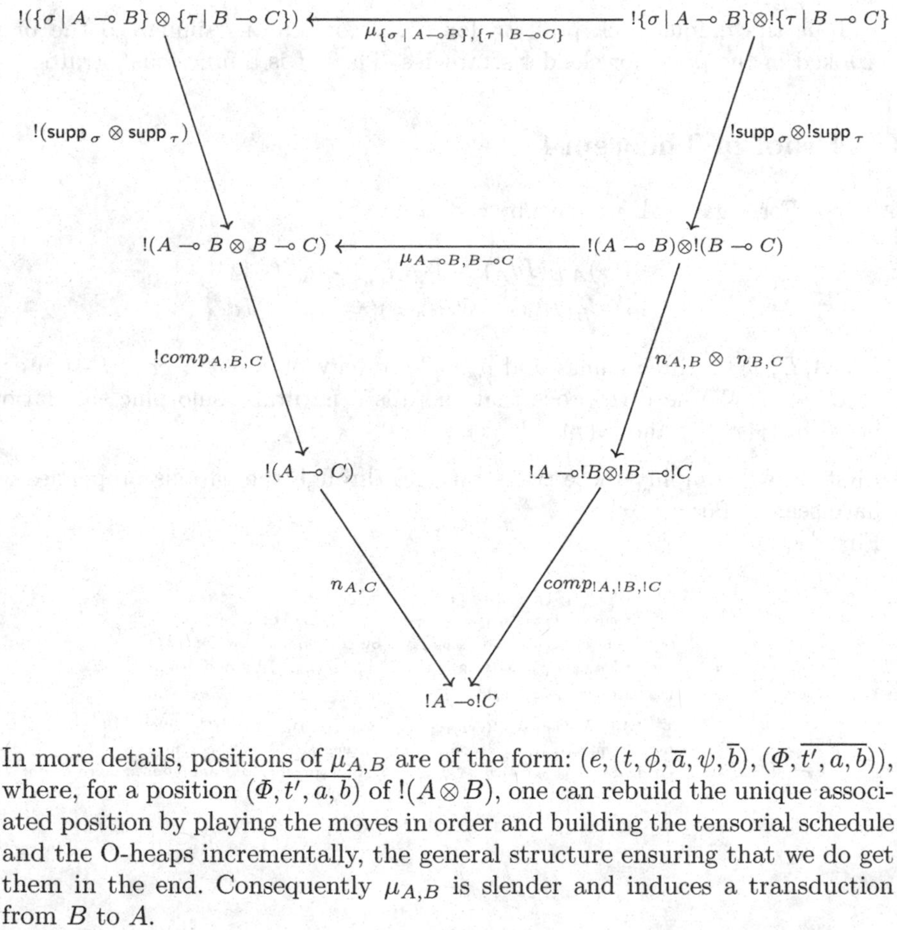

In more details, positions of $\mu_{A,B}$ are of the form: $(e, (t, \phi, \overline{a}, \psi, \overline{b}), (\Phi, \overline{t', a, b}))$, where, for a position $(\Phi, \overline{t', a, b})$ of $!(A \otimes B)$, one can rebuild the unique associated position by playing the moves in order and building the tensorial schedule and the O-heaps incrementally, the general structure ensuring that we do get them in the end. Consequently $\mu_{A,B}$ is slender and induces a transduction from B to A.

Note that it is not bijective as the play of $!(A \otimes B)$ where we play in B, then backtrack to play in A would produce the same play in $!A \otimes !B$ than playing in B then in A without backtracking.
Thus, we have, since $\mu_{\{\sigma \mid A \multimap B\},\{\tau \mid B \multimap C\}}$ is slender:

$$\mathsf{image}(n_{A,C} \circ\ !comp_{A,B,C} \circ\ \mathsf{supp}_{\#^s(\sigma \otimes \tau)}) \cong \mathsf{image}(n_{A,C} \circ\ !comp_{A,B,C} \circ \mathsf{supp}_{\#^s(\sigma \otimes \tau)} \circ \mu_{\{\sigma \mid A \multimap B\},\{\tau \mid B \multimap C\}})$$

Then, by naturality,

$$\mathsf{image}(n_{A,C} \circ\ !comp_{A,B,C} \circ\ \mathsf{supp}_{\#^s(\sigma \otimes \tau)}) \cong \mathsf{image}(n_{A,C} \circ\ !comp_{A,B,C} \circ \mu_{A \multimap B, B \multimap C} \circ\ \mathsf{supp}_{\#^s \sigma \otimes \#^s \tau})$$

Consequently,

$$\mathsf{image}(n_{A,C} \circ\ !comp_{A,B,C} \circ\ \mathsf{supp}_{!(\sigma \otimes \tau)}) \cong \mathsf{image}(comp_{!A,!B,!C} \circ n_{A,B} \otimes\ n_{B,C} \circ !\mathsf{supp}_{(\sigma)} \otimes !\mathsf{supp}_{(\tau)})$$

if and only if

$$\mathsf{image}(n_{A,C} \circ\ !comp_{A,B,C} \circ\ \mu_{A \multimap B, B \multimap C} \circ\ \mathsf{supp}_{!\sigma \otimes !\tau}) \cong \mathsf{image}(comp_{!A,!B,!C} \circ n_{A,B} \otimes\ n_{B,C} \circ !\mathsf{supp}_{(\sigma)} \otimes !\mathsf{supp}_{(\tau)})$$

meaning if and only if

$$\mathsf{image}(n_{A,C} \circ\ !comp_{A,B,C} \circ\ \mu_{A \multimap B, B \multimap C}) \cong \mathsf{image}(comp_{!A,!B,!C} \circ n_{A,B} \otimes\ n_{B,C})$$

An important remark is that $\mu_{A \multimap B, B \multimap C}$ transfers plays p of $(!(A \multimap B) \otimes !(B \multimap C))$ such that there exists $(e, (\phi, \overline{a}), (\psi, \overline{c}))_p \in \mathsf{image}(comp_{!A,!B,!C} \circ n_{A,B} \otimes\ n_{B,C})$ to plays p' of $!(A \multimap B \otimes B \multimap C)$ such that there exists $(e, (\phi, \overline{a}), (\psi, \overline{c}))'_p \in \mathsf{image}(n_{A,C} \circ\ !comp_{A,B,C})$.

In other words μ, when restricted to plays that play a role in the images we outlined, acts as a function from the set of plays of $(!(A \multimap B) \otimes !(B \multimap C))$ to the set of plays of $!(A \multimap B \otimes B \multimap C)$. This can be proved by looking at the respective structures of the plays and induces one half of the isomorphism we need.

We do a similar study by introducing a P-strategy of $!(A \multimap B \otimes B \multimap C) \multimap (!(A \multimap B) \otimes !(B \multimap C))$ that acts as a converse of $\mu_{A \multimap B, B \multimap C}$ for such plays and thus get a converse to our morphism, which will give us the second half of the isomorphism we need. Here is how we proceed:
Let $(t, (\phi, \overline{e, a, b}), (\psi, \overline{f, b, c}))$ be a play of $(!(A \multimap B) \otimes !(B \multimap C))$ such that there exists

$$(e_{!A \multimap !C}, (\phi_{!A}, \overline{a}), (\phi_{!C}, \overline{c}))_{e,(t,\phi,\overline{e,a,b},\psi,\overline{f,b,c})} \in \mathsf{image}(comp_{!A,!B,!C} \circ n_{A,B} \otimes n_{B,C}).$$

In particular, that implies that, since $n_{A,B} \otimes \overline{n_{B,C}}$ doesn't change the order of moves, the sequence of moves of $(t, \overline{(\phi, e, a, b)}, \overline{(\psi, f, b, c)})$ must be able to be the left projection of $comp_{!A,!B,!C}$. This restricts the way the moves can be played.

In particular, B moves from the two components must must answer each other right away, giving sequences without backtrack of the form $c(b_r.b_l.b_l.b_r) * c$, with similar structures for sequences starting and/or finishing with a A move. In addition, there cannot be any backtrack in A or any of the two B component that would not be initiated by a backtrack in a C component.

The idea is that a backtrack in C induces a backtrack in B which is mirrored on the left component and induces a backtrack in A. Those backtracks give us a heap structure and the moves inside a sequence follow a proper tensor schedule, so it can be seen as a play of $!(A \multimap B \otimes B \multimap C)$ and it is easy to verify that this play would produce an element of $\mathsf{image}(n_{A,C} \circ\, !comp_{A,B,C} \circ \mu_{A \multimap B, B \multimap C})$ and that the P-strategy of $!(A \multimap B \otimes B \multimap C) \multimap (!(A \multimap B) \otimes !(B \multimap C))$ built by reorganizing structure without changing order of moves is a converse to $\mu_{A \multimap B, B \multimap C}$.

Consequently, we have the bijection of images we needed and thus an isomorphic simulation between $!_{\mathcal{P}}(\tau \circ \sigma)$ and $!_{\mathcal{P}}(\tau) \circ\, !_{\mathcal{P}}(\sigma)$. It is natural since μ and the isomorphisms involved in the manipulation of images are natural.
The few additional diagrams that must be checked are easy to verify with similar methods, and thus we have that $!_{\mathcal{P}}$ is a pseudofunctor.

K Proof that ! Is a Pseudocomonad

In the following section, we'll detail the construction of the pseudonatural transformations δ and ϵ and prove their naturality. From those definitions, verifying that ! is a pseudocomonad is easy as the morphism part of the two natural transformations coincides with their definition in the deterministic case, making the diagrams commute instantly. After that, we may do a similar study on d, e to give ! the necessary structure to be a linear exponential modality.

We will handle here the case of δ_σ for a P-strategy $\sigma : A \twoheadrightarrow B$. This is, by Definiton 26, a bijective 2-morphism between $!_{\mathcal{P}}!_{\mathcal{P}}\sigma \circ \delta_A$ and $\delta_B \circ\, !_{\mathcal{P}}\sigma$, both being P-strategies of $!A \multimap !!B$.

First note that

$$!_{\mathcal{P}}!_{\mathcal{P}}\sigma \circ \delta_A = \mathsf{image}(comp_{!A,!!A,!!B} \circ\ \mathsf{supp}_{!_{\mathcal{P}}!_{\mathcal{P}}\sigma} \otimes \mathsf{supp}_{\delta_A})$$

and that

$$\delta_B \circ\ !_{\mathcal{P}}\sigma = \mathsf{image}(comp_{!A,!B,!!B} \circ\ \mathsf{supp}_{\delta_B} \otimes \mathsf{supp}_{!_{\mathcal{P}}\sigma}).$$

We want to study the structure of both images to find an isomorphic simulation between them.

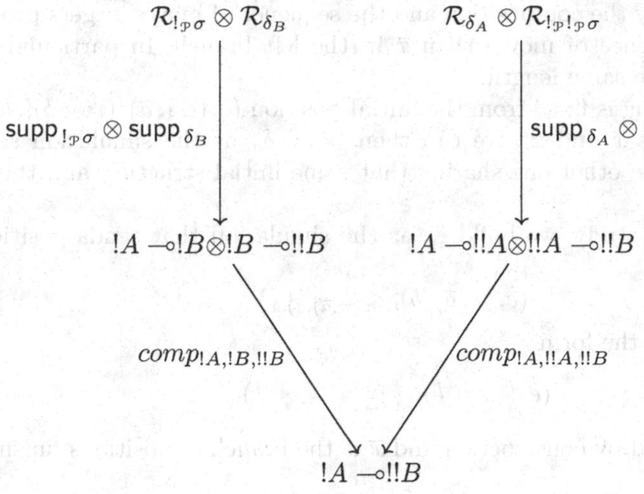

What we will do is start from a position

$$e, (\phi_A, \overline{a}), (\psi_B, \phi_B, \overline{b})$$

of $!A \multimap !!B$ and go back along the arrows to see what structure the positions that produce this position must have.

First, on the left branch, the presence of $comp_{!A,!B,!!B}$ indicates that the position in $!A \multimap !B \otimes !B \multimap !!B$ must be of the form

$$t, (e_1, (\phi_A, \overline{a}), (\varPhi_B, \overline{b'}), (e_2, (\varPhi_B, \overline{b'}), (\psi_B, \phi_B, \overline{b})))$$

for some $t, e_1, e_2, \varPhi_B, \overline{b'}$ such that $e_1 \cdot e_2 = e$.

Since the right component of this position comes from δ_B, we actually have $\overline{b'} = \overline{b}$, $\varPhi_B = \phi_B$, $e_2 = c$ and thus $e_1 = e$ and we actually have the position

$$t, (e, (\phi_A, \overline{a}), (\phi_B, \overline{b}), (c, (\phi_B, \overline{b}), (\psi_B, \phi_B, \overline{b})))$$

for some t which is fixed by the two components for the composition to work.

And thus, this gives us the following position of $\mathcal{R}_{!_{\mathcal{P}}\sigma} \otimes \mathcal{R}_{\delta_B}$:

$$(t, ((\phi_A, e, \pi), \overline{x}), (c, (\phi_B, \overline{b}), (\psi_B, \phi_B, \overline{b})))$$

where $\overline{x}$ is a sequence of moves that gets projected to the sequence of moves of $(e, (\phi_A, \overline{a}), (\phi_B, \overline{b}))$. There is no modification of the order the moves are played in this step, just a reorganization of the structure.

Thus a position of $\mathcal{R}_{\delta_B \circ\, !_{\mathcal{P}}\sigma}$ is of the form

$$(e, (\phi_A, \overline{a}), (\psi_B, \phi_B, \overline{b}))_{(t,((\phi_A,e,\pi),\overline{x}),(c,(\phi_B,\overline{b}),(\psi_B,\phi_B,\overline{b})))}.$$

We apply a similar reasoning to the right branch to obtain the form of a position of $\mathcal{R}_{!_{\mathcal{P}}!_{\mathcal{P}}\sigma \circ \delta_A}$:

$$(e, (\phi_A, \overline{a}), (\psi_B, \phi_B, \overline{b}))_{(t',(c,(\phi_A,\overline{a}),(e*\psi_B,\phi_A,\overline{a})),((e*\psi_B,e,\pi),(\phi_A,e,\pi),\overline{x'}))}$$

where t' is fixed by the composition and the sequence of moves $\overline{x'}$ gets projected to the same sequence of moves than $\overline{x}$ in the left branch. In particular, both sequences have the same length.

Since everything is fixed from the initial position $(e, (\phi_A, \overline{a}), (\psi_B, \phi_B, \overline{b}))$ but the two sequences $\overline{x}$ and $\overline{x'}$, we can then build δ_σ as the simulation sending one position to the other one sharing that same initial structure and the same sequence $\overline{x}$.

With a simlar study, we build ϵ_σ as the simulation that sends positions of the form

$$(e, (\pi, \overline{a}), b)_{t,(c,(\pi,\overline{a}),a),x)}$$

to positions of the form

$$(e, (\pi, \overline{a}), b)_{t',(\pi,\overline{x}),(c,(\pi,\overline{b}),b)}).$$

where t, t' are fixed by construction and $\overline{x}$ is the branch of positions finishing in x in $\mathcal{R}_\sigma$.

Proof. We will now prove the pseudonaturality of ϵ, δ is handled in a similar way. Let us look at the naturality first. Let A, B be two games, σ, τ two P-strategies of $A \multimap B$ and $\alpha : \sigma \to \tau$ a simulation We require that the two following pasting diagrams are equivalent:

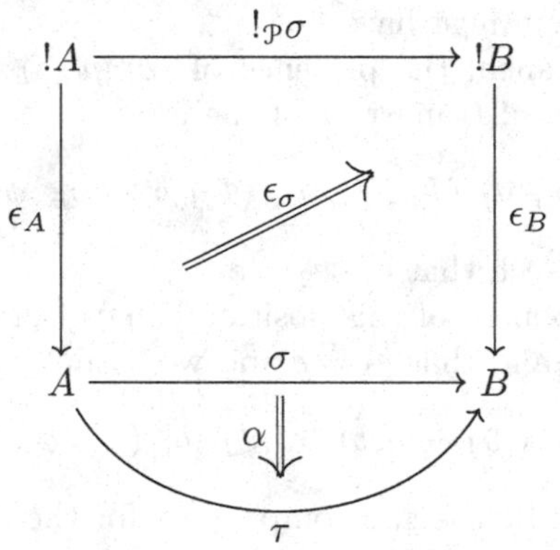

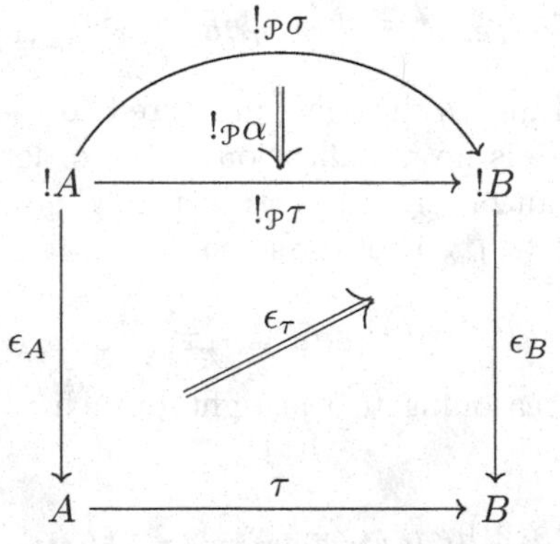

This amounts to the following equality of simulations:

$$(\epsilon_A \lhd \alpha) \cdot \epsilon_\sigma^{-1} = \epsilon_\tau^{-1} \cdot (!_{\mathcal{P}}\alpha \rhd \epsilon_B)$$

where $\lhd, \rhd$ indicate the whiskering that results from the composition of P-strategies and $\cdot$ indicates the vertical composition which is simply the composition of functions. Thus, for a position

$$(e, (\pi, \overline{a}), b)_{t',(\pi,\overline{x}),(c,(\pi,\overline{b}),b)}$$

of $\epsilon_B \circ\ !_{\mathcal{P}}\sigma$, we have:

$$(\epsilon_A \lhd \alpha) \cdot \epsilon_\sigma^{-1} \left((e, (\pi, \overline{a}), b)_{t',(\pi,\overline{x}),(c,(\pi,\overline{b}),b)}\right) = (\epsilon_A \lhd \alpha) \left((e, (\pi, \overline{a}), b)_{t,(c,(\pi,\overline{a}),a),x}\right) \text{ by def of } \epsilon_\sigma$$

$$(\epsilon_A \lhd \alpha) \cdot \epsilon_\sigma^{-1} \left((e, (\pi, \overline{a}), b)_{t',(\pi,\overline{x}),(c,(\pi,\overline{b}),b)}\right) = (e, (\pi, \overline{a}), b)_{t,(c,(\pi,\overline{a}),a),\alpha(x)} \text{ by def of } \mathcal{P}, \epsilon_A$$

On the other hand,

$$\begin{aligned} &\epsilon_\tau^{-1} \cdot (!_{\mathcal{P}}\alpha \rhd \epsilon_B) \left((e, (\pi, \overline{a}), b)_{t',(\pi,\overline{x}),(c,(\pi,\overline{b}),b)}\right) \\ &= \epsilon_\tau^{-1} \left((e, (\pi, \overline{a}), b)_{t',(\pi,\overline{\alpha(x)}),(c,(\pi,\overline{b}),b)}\right) \text{ by def of } \mathcal{P}, \epsilon_B, !_{\mathcal{P}} \end{aligned}$$

$$\begin{aligned} &\epsilon_\tau^{-1} \cdot (!_{\mathcal{P}}\alpha \rhd \epsilon_B) \left((e, (\pi, \overline{a}), b)_{t',(\pi,\overline{x}),(c,(\pi,\overline{b}),b)}\right) \\ &= (e, (\pi, \overline{a}), b)_{t,(c,(\pi,\overline{a}),a),\alpha(x)}) \text{ by def of } \epsilon_\tau \end{aligned}$$

And thus, we have the equivalence we require. The other diagram equalitiies we need to verify are done in a similar way.

The key point to remember from this proof and the similar ones that need to be done, is that, while the form of the positions is a bit heavy, the structures that underly them do most of the work for us, making most of the needed verifications very easy, once the positions have been properly described.

We apply those methods to verify that ! is indeed a pseudocomonad, to define and verify that d_A, e_A are proper pseudonatural transformations and to check that !, along with those transformations, does have the structure of a linear exponential modality.

References

1. Curien, P.-L.: On the symmetry of sequentiality. In: Brookes, S., Main, M., Melton, A., Mislove, M., Schmidt, D. (eds.) MFPS 1993. LNCS, vol. 802, pp. 29–71. Springer, Heidelberg (1994). https://doi.org/10.1007/3-540-58027-1_2
2. Eberhart, C., Hirschowitz, T.: Justified sequences in string diagrams: a comparison between two approached to concurrent game semantics (2016)
3. Girard, J.Y.: Locus Solum: from the rules of logic to the logic of rules. Math. Struct. Comput. Sci. **11**(03), 301–506 (2001)
4. Harmer, R., Hyland, J.M.E., Melliès, P.-A.: Categorical combinatorics for innocent strategies. In: LICS, pp. 379–388 (2007)

5. Hirschowitz, T., Pous, D.: Innocent strategies as presheaves and interactive equivalences for CCS. Sci. Ann. Comput. Sci. (2012)
6. Hyland, M.: Game semantics. In: Semantics of Logics and Computation, Publications of the Newton Institute, pp. 131–184. Cambridge University Press (1997)
7. Laurent, O.: Polarized games. Ann. Pure Appl. Logic **130**(1–3), 79–123 (2004)
8. Lawvere, F.W.: Equality in hyperdoctrines and comprehension schema as an adjoint functor. In: Proceedings of the New York Symposium on Applications of Categorical Algebra, pp. 1–14 (1970)
9. Melliès, P.-A.: Asynchronous games 3: an innocent model of linear logic. In: Category Theory and Computer Science, pp. 171–192 (2004)
10. Melliès, P.-A.: Asynchronous games 4: a fully complete model of propositional linear logic. In: LICS 2005 (2005)
11. Melliès, P.-A.: Asynchronous games 2: the true concurrency of innocence. Theor. Comput. Sci. **358**, 200–228 (2006)
12. Melliès, P.-A.: Game semantics in string diagrams. In: LICS 2012 (2012)
13. Melliès, P.-A.: Tensorial logic with algebraic effects. Talk at the Institut Henri Poincaré, June 2014
14. Melliès, P.-A.: Dialogue categories and chiralities. In: Publications of the Research Institute in Mathematical Sciences (2015)
15. Melliès, P.-A.: Une étude micrologique de la négation. Habilitation thesis (2017)
16. Mellies, P.-A., Zeilberger, N.: Functors are type refinement systems (2015)
17. Ong, C.H.L., Tsukada, T.: Nondeterminism in game semantics via sheaves. In: LICS 2016 (2016)
18. Melliès, P.-A., Mimram, S.: Asynchronous games: innocence without alternation. In: Caires, L., Vasconcelos, V.T. (eds.) CONCUR 2007. LNCS, vol. 4703, pp. 395–411. Springer, Heidelberg (2007). https://doi.org/10.1007/978-3-540-74407-8_27
19. Rideau, S., Winskel, G.: Concurrent strategies. In: LICS 2011 (2011)

A Syntactic View of Computational Adequacy

Marco Devesas Campos(✉) and Paul Blain Levy

School of Computer Science, University of Birmingham, Birmignham, UK
{m.devesascampos,pbl}@cs.bham.ac.uk

Abstract. When presenting a denotational semantics of a language with recursion, it is necessary to show that the semantics is computationally adequate, i.e. that every divergent term denotes the "bottom" element of a domain.

We explain how to view such a theorem as a purely syntactic result. Any theory (congruence) that includes basic laws and is closed under an infinitary rule that we call "rational continuity" has the property that every divergent term is equated with the divergent constant. Therefore, to prove a model adequate, it suffices to show that it validates the basic laws and the rational continuity rule. While this approach was inspired by the categorical, ordered framework of Abramsky et al., neither category theory nor order is needed.

The purpose of the paper is to present this syntactic result for call-by-push-value extended with term-level recursion and polymorphic types. Our account begins with PCF, then includes sum types, then moves to call-by-push-value, and finally includes polymorphic types.

1 Introduction

Models of Recursion. A conventional denotational account of a language with recursion proceeds as follows. First define the syntax and operational semantics. Then give a denotational model. Lastly, prove *soundness*, i.e. if t evaluates to u (written $t \Downarrow u$) then $[\![t]\!] = [\![u]\!]$, and *adequacy*, i.e. if t diverges (written $t \Uparrow$) then $[\![t]\!] = \bot$.

Because it is often convenient to structure a model categorically, Fiore and Plotkin (1994) gave categorical axioms on a model that imply (soundness and) adequacy. Crucially, in their work, as detailed by Fiore (1996), a model is required to be "ω**Cpo**-enriched", meaning that a term denotes an element of a pointed ω-cpo (poset with least element $\bot$ and suprema of all increasing ω-chains), and a term constructor is ω-continuous (preserves suprema of ω-chains). Thus (for a call-by-name language) a term $x : A \vdash t : A$ gives a continuous endofunction f, and the recursion $\mathbf{rec}\, x\,.\, M$ denotes the supremum of $(f^n \bot)_{n \in \mathbb{N}}$, the least (pre)fixpoint of f.

P. B. Levy—Research Supported by UK EPSRC Grant EP/N023757/1.

C. Baier and U. Dal Lago (Eds.): FOSSACS 2018, LNCS 10803, pp. 71–87, 2018.
https://doi.org/10.1007/978-3-319-89366-2_4

However, for the models of Abramsky et al. (2000), Abramsky and McCusker (1997), and McCusker (1998), the requirement of ω**Cpo**-enrichment is too restrictive, because the posets arising do not have suprema of *all* increasing ω-chains (Normann 2006). So these papers use a more relaxed ordered framework where the only suprema that must be preserved are those of chains $(f^n \bot)_{n \in \mathbb{N}}$ of iterated applications. This means that any so called *rational chain* $(g \circ f^n \bot)_{n \in \mathbb{N}}$ has an upper bound given by $g\,(\bigsqcup f^n \bot)$—a property known as *rational continuity* (Wright et al. 1976; cf. also Bloom and Ésik 1993).

Recursion but Rationally. Our goal is to give an even more relaxed version of this "rational" framework for adequacy; one that uses no category theory, order or denotational model. It could be viewed as a purely syntactic result: a property of a *theory* (congruence) $\approx$ rather than of a model. Thus we want $t \Downarrow u$ to imply $t \approx u$, and $t \Uparrow$ to imply $t \approx \Omega$, where Ω is a divergent constant. The benefit of such a result is to modularize the narrative described at the start; we can get adequacy out of the way before we start studying categorical and denotational semantics.

Rational Continuity. Currently we have accomplished this goal for term-level recursion and polymorphic types. (Recursive and existential types are left to future work; see Sect. 6). Our result is that any theory (congruence) $\approx$ will be sound and adequate provided it (a) contains the β-laws, fixpoint law and strictness laws and (b) is closed under an infinitary rule called *rational continuity*. This rule says (for a call-by-name language) that if $C[\mathbf{rec}^n\, x \,.\, t] \approx D[\mathbf{rec}^n\, x \,.\, t]$ for infinitely many $n \in \mathbb{N}$, then $C[\mathbf{rec}\, x \,.\, t] \approx D[\mathbf{rec}\, x \,.\, t]$. Here we write $\mathbf{rec}^n\, x \,.\, t$ for the nth *approximant to recursion*, defined by the clauses $\mathbf{rec}^0\, x \,.\, t := \Omega$ and $\mathbf{rec}^{n+1}\, x \,.\, t := t[\mathbf{rec}^n\, x \,.\, t/x]$.

Plan. To include both call-by-value (CBV) and call-by-name (CBN), we have established our result for call-by-push-value. The latter has both value types and computation types, but the treatment of value types in our proof is more complicated, so we begin in the CBN setting, which has only computation types. Our CBN account itself begins with PCF, which has only base types and function types; we then include sum types, using a proof method adapted from McCusker (1998). Next we move to call-by-push-value, and use *ultimate pattern matching* of values (Lassen and Levy 2008) to treat the value types. Finally we include polymorphic types.

Related Work. Adequacy of topos models has been studied using an internal language (Simpson 2004). Other adequacy results for polymorphic models include realizability semantics (Møgelberg 2009) and game semantics (Laird 2013).

2 PCF

Language. We begin by introducing a version of Plotkin's PCF (1997) that replaces fixpoint combinators with recursion operators and an explicit divergence construct Ω (Table 1). As per usual, terms are taken up to α-equivalence. The set

Table 1. PCF

Types

$$T, U = \mathrm{Bool} \mid \mathrm{Nat} \mid T \to U$$

Typing

$$\frac{(x : T \in \Gamma)}{\Gamma \vdash x : T} \qquad \frac{}{\Gamma \vdash tt : \mathrm{Bool}} \qquad \frac{}{\Gamma \vdash ff : \mathrm{Bool}} \qquad \frac{\Gamma \vdash t : \mathrm{Bool} \quad \Gamma \vdash u : T \quad \Gamma \vdash q : T}{\Gamma \vdash \mathbf{if}\, t\, \mathbf{then}\, u\, \mathbf{else}\, q : T}$$

$$\frac{}{\Gamma \vdash \mathbf{zero} : \mathrm{Nat}} \qquad \frac{\Gamma \vdash t : \mathrm{Nat}}{\Gamma \vdash \mathbf{succ}\, t : \mathrm{Nat}} \qquad \frac{\Gamma \vdash t : \mathrm{Nat}}{\Gamma \vdash \mathbf{pred}\, t : \mathrm{Nat}} \qquad \frac{\Gamma \vdash t : \mathrm{Nat}}{\Gamma \vdash \mathbf{iszero}\, t : \mathrm{Bool}}$$

$$\frac{x : T, \Gamma \vdash t : U}{\Gamma \vdash \lambda x.t : T \to U} \qquad \frac{\Gamma \vdash t : T \to U \quad \Gamma \vdash u : T}{\Gamma \vdash tu : U} \qquad \frac{}{\Gamma \vdash \Omega : T} \qquad \frac{x : T, \Gamma \vdash t : T}{\Gamma \vdash \mathbf{rec}\, x\,.\, t : T}$$

Reduction

$$\frac{}{tt \Downarrow tt} \qquad \frac{}{ff \Downarrow ff} \qquad \frac{t \Downarrow tt \quad u \Downarrow v}{\mathbf{if}\, t\, \mathbf{then}\, u\, \mathbf{else}\, q \Downarrow v} \qquad \frac{t \Downarrow ff \quad q \Downarrow v}{\mathbf{if}\, t\, \mathbf{then}\, u\, \mathbf{else}\, q \Downarrow v}$$

$$\frac{}{\mathbf{zero} \Downarrow \mathbf{zero}} \qquad \frac{t \Downarrow v}{\mathbf{succ}\, t \Downarrow \mathbf{succ}\, v} \qquad \frac{t \Downarrow \mathbf{succ}\, v}{\mathbf{pred}\, t \Downarrow v} \qquad \frac{t \Downarrow \mathbf{zero}}{\mathbf{iszero}\, t \Downarrow tt} \qquad \frac{t \Downarrow \mathbf{succ}\, v}{\mathbf{iszero}\, t \Downarrow ff}$$

$$\frac{}{\lambda x.t \Downarrow \lambda x.t} \qquad \frac{t \Downarrow (\lambda x.t') \quad t'[u/x] \Downarrow v}{tu \Downarrow v} \qquad \frac{t[\mathbf{rec}\, x\,.\, t/x] \Downarrow v}{\mathbf{rec}\, x\,.\, t \Downarrow v}$$

of closed terms of type T will be denoted by CTerms^T and that of normal forms by NF^T. For a closed term t there is at most one v such that $t \Downarrow v$; when there is none we say it diverges and represent this by $t \Uparrow$.

2.1 A Rationally Continuous Theory of PCF

The Theory. A *congruence on terms* is a type-indexed equivalence relation on closed terms of said type satisfying $t \approx t' \implies C[t] \approx C[t']$ for any context $C[-]$ where the hole is closed. (We omit type annotations.) A congruence is a *rationally continuous β-Ω-fix theory* if it also satisfies the rules in Table 2.

The basis for the theory are the obvious β rules that mimic the reduction rules. In a similar vein, the fixpoint rule establishes that each recursive term is the fixpoint of a substitution. These rules alone are enough to establish the soundness of the theory with respect to reduction.

Proposition 1 (Soundness). *Any congruence $\approx$ satisfying the β and fixpoint rules (Table 2) is sound: $t \Downarrow r \implies t \approx r$.*

A Converse. Our sights now turn to proving that divergent terms are identical to Ω. The extra requirement calls for a more refined theory that can more closely mirror the behaviour of reduction. The last two sets of equations in Table 2 fill the gaps in what the reduction rules *don't* say about divergence. The first

Table 2. Rationally continuous β-Ω-fix theory of PCF

Basis An equivalence relation $\approx$ on closed terms satisfying compatibility:

$$\text{for any (closed) } C[-],\ t \approx u \implies C[t] \approx C[u]$$

β Rules

$$\mathbf{if}\, tt\, \mathbf{then}\, u\, \mathbf{else}\, q \approx u \qquad \mathbf{if}\, ff\, \mathbf{then}\, u\, \mathbf{else}\, q \approx q \qquad \mathbf{iszero}\,\mathbf{zero} \approx tt$$

$$\mathbf{iszero}\,\mathbf{succ}^{n+1}\,\mathbf{zero} \approx ff \qquad \mathbf{pred}\,\mathbf{succ}^{n+1}\,\mathbf{zero} \approx \mathbf{succ}^{n}\,\mathbf{zero} \qquad (\lambda x.t)u \approx t[u/x]$$

Fixpoint Rule

$$\mathbf{rec}\, x\,.\,t \approx t[\mathbf{rec}\, x\,.\,t/x]$$

Divergence Rules

$$\Omega u \approx \Omega \qquad \mathbf{if}\, \Omega\, \mathbf{then}\, u\, \mathbf{else}\, q \approx \Omega \qquad \mathbf{iszero}\, \Omega \approx \Omega$$

$$\mathbf{succ}\, \Omega \approx \Omega \qquad \mathbf{pred}\, \Omega \approx \Omega \qquad \mathbf{pred}\,\mathbf{zero} \approx \Omega$$

Rational Continuity for $x : T \vdash t : T$

$$\frac{\exists^\infty n. C[\mathbf{rec}^n\, x\,.\,t] \approx D[\mathbf{rec}^n\, x\,.\,t]}{C[\mathbf{rec}\, x\,.\,t] \approx D[\mathbf{rec}\, x\,.\,t]}$$

where $\mathbf{rec}^0\, x\,.\,t = \Omega$ and $\mathbf{rec}^{n+1}\, x\,.\,t = t[\mathbf{rec}^n\, x\,.\,t/x]$.

relates to the strictness of the operators: divergence of an argument leads to the divergence of the operator, e.g., $\Omega u \approx \Omega$. The second is the rational continuity rule presented in the introduction.

Rational Continuity and Chains. To prove adequacy, one often has to re-write or equate certain terms built with recursion either with some constant or as the unrolling of the recursive term a few times. In cpo models, continuity and compositionality of the interpretations validate the following rule

$$\frac{\forall n \in \mathbb{N}. [\![C[\mathbf{rec}^n\, x\,.\,t]]\!] = [\![D[\mathbf{rec}^n\, x\,.\,t]]\!]}{[\![C[\mathbf{rec}\, x\,.\,t]]\!] = [\![D[\mathbf{rec}\, x\,.\,t]]\!]}$$

But this can be further weakened by requiring only equality *at infinitely many* n, for then one would still be able to define chains with exactly the same least upper bounds. We write $\exists^\infty n.P(n)$ to mean *there exist infinitely many n in $\mathbb{N}$ for which $P(n)$ holds.* This leads us to the syntactic continuity rule in Table 2. Since adequacy refers solely to closed terms, we only require this property for $x : T \vdash t : T$—and therefore $\mathbf{rec}^n\, x\,.\,t$ and $\mathbf{rec}\, x\,.\,t$ are closed. Similarly, by a *rational chain* we mean a chain of the form $C[\mathbf{rec}^n x\,.\,t]$ for infinitely many $n \in \mathbb{N}$, and by its limit we mean the term $C[\mathbf{rec}\, x\,.\,t]$.

2.2 Adequacy

The Claim. We now embark on the syntactic journey towards a proof we have an adequate theory—formally, that $t \Uparrow \implies t \approx \Omega$. By the aforementioned reasons the proof follows the usual approaches by replacing closure under bottom elements and least upper bounds of the relevant chains with closure under divergence and limits of rational chains.

Approximations. First we define abstractly[1] the notion of an approximation candidate between terms and the values they approximate; these are then extended to relations on terms. The concrete relations we use for each type are given by certain actions on approximation candidates (cf., e.g., Pitts 2000). When using the result of an action ϕ on approximation candidates $\triangleleft_1, \dots, \triangleleft_n$ infix, we will sometimes surround the result with brackets, as in $t \; \langle \phi(\triangleleft_1, \dots, \triangleleft_n) \rangle \; u$, to aid readability.

Definition 1 (Approximation Candidates). *An approximation candidate $\triangleleft$ for a type T is a subset of $CTerms^T \times NFs^T$ s.t.:*

1. *$\approx$ Extension: $t \approx t'$ and $t' \triangleleft v \implies t \triangleleft v$*
2. *Rational Admissibility: for $x : T \vdash t : T$*

$$(\exists^\infty n. C[\mathbf{rec}^n \, x \, . \, t] \triangleleft v) \implies C[\mathbf{rec} \, x \, . \, t] \triangleleft v$$

Proposition 2. *If $\triangleleft$ is an approximation candidate for type T, then the binary relation on $CTerms^T$ defined by*

$$t \triangleleft^c u \iff t \approx \Omega \text{ or } (\exists v. u \Downarrow v \text{ and } t \triangleleft v)$$

satisfies the following properties:

1. *Ω Property: $\Omega \triangleleft^c u$, for any $u \in CTerms^T$*
2. *$\approx$ Extension: $t \approx t'$ and $t' \triangleleft^c u \implies t \triangleleft^c u$*
3. *$\Downarrow$ Extension: $t \triangleleft^c u$ and $(\forall v. u \Downarrow v \implies u' \Downarrow v) \implies t \triangleleft^c u'$*
4. *Rational Admissibility: for $x : T \vdash t : T$*

$$(\exists^\infty n. C[\mathbf{rec}^n \, x \, . \, t] \triangleleft^c u) \implies C[\mathbf{rec} \, x \, . \, t] \triangleleft^c u$$

Proof. *To give a taste of how the proofs go using rational admissibility, assume we have $\exists^\infty n. C[\mathbf{rec}^n \, x \, . \, t] \triangleleft^c u$. From the definition, one of two options (possibly both) is true: that an infinite number of terms on the left are identical to Ω; or that for an infinite series of m, $C[\mathbf{rec}^m \, x \, . \, t]$ is related to the value v that u reduces to (determinism of reduction is paramount here). Admissibility then follows by rational continuity in the first case (using the obvious constant context), and by admissibility of $\triangleleft$ (Definition 1) in the second.*

[1] Anticipating our treatment of polymorphism in Sect. 4, we have purposefully set up here a proof structure in the style of Girard (1989).

Proposition 3 (Base Type Actions). *The two binary relations* $\lhd_{Bool} \subseteq CTerms^{Bool} \times NFs^{Bool}$ *and* $\lhd_{Nat} \subseteq CTerms^{Nat} \times NFs^{Nat}$ *defined by*

$$t \lhd_{Bool} v \iff t \approx v \quad \text{and} \quad t \lhd_{Nat} v \iff t \approx v$$

are approximation candidates for Bool and Nat.

Proposition 4 (Arrow Action). *Given approximation candidates* $\lhd_T$ *for* T *and* $\lhd_U$ *for* U, *the binary relation between* $CTerms^{T\to U}$ *and* $NFs^{T\to U}$

$$t\ \langle \lhd_T \to \lhd_U \rangle\ \lambda x.u \iff \forall p \lhd_T^c q\ .\ tp \lhd_U^c u[q/x])$$

is an approximation candidate for $T \to U$.

Definition 2 (Approximation Relation). *The approximation relation* $\lhd_T$ *is the type-indexed family of approximation candidates defined by induction on types, where base types are covered by their respective actions (Proposition 3), and* $\lhd_{T\to U} = \lhd_T \to \lhd_U$ *(Proposition 4).*

Definition 3 (Environments). *Given a typing context* Γ, *an* environment σ *for* Γ *is a substitution that maps each* $x : T \in \Gamma$ *to a closed term of type* $\vdash \sigma(x) : T$. *If* σ_1 *and* σ_2 *are two such, we write* $\sigma_1 \lhd_\Gamma^c \sigma_2$ *to mean* $\sigma_1(x) \lhd_T^c \sigma_2(x)$ *for all* $x : T \in \Gamma$.

Proposition 5. *For any* $\Gamma \vdash t : T$ *and environments* $\sigma_1 \lhd_\Gamma^c \sigma_2$, $t[\sigma_1] \lhd_T^c t[\sigma_2]$.

Corollary 1 (Adequacy). *For every closed* $\vdash t : T$, $t \Uparrow \implies t \approx \Omega$.

Proof. *Applying Proposition* 5 *to* $\vdash t : T$ *(for the empty substitution), we conclude that* $t \lhd_T^c t$; *the definition of* $(-)^c$ *(Proposition* 2*) asserts, then, that either* $t \approx \Omega$ *or* ($t \Downarrow v$ and $t \lhd_T v$); *whereby if* $t \Uparrow$, *it can only be that* $t \approx \Omega$.

3 PCF with Sums

The Extension. Sums provide a slight complication—but one which shows the adaptability of the method. The extension to call-by-name sums is presented in Table 3. With the new reduction rules come new β rules and divergence rules in the theory (Table 4). As before, reduction is deterministic and the theory is sound.

3.1 Adequacy

Action. The action for sums must reflect the structure of its parameters. That is for $\lhd_T$ we expect $t \lhd_{T+U}$ **inl** u exactly when (modulo the theory) t decomposes into some **inl** t' for which $t' \lhd_T u$. The assertion of that existence, though, causes us a small hiccup[2] in proving that $- \lhd_{T+U} v$ is rationally admissible: If we have

[2] A hiccup that will be much amplified in the proof of admissibility for $\lhd_{FA}$ (Sect. 4).

Table 3. Extension of PCF with binary sums

Types $T, U = \ldots \mid T + U$

Typing

$$\frac{\Gamma \vdash t : T}{\Gamma \vdash \mathbf{inl}\, t : T + U} \quad \frac{\Gamma \vdash t : U}{\Gamma \vdash \mathbf{inr}\, t : T + U}$$

$$\frac{\Gamma \vdash t : T + T' \quad x : T, \Gamma \vdash u : U \quad y : T', \Gamma \vdash q : U}{\Gamma \vdash \mathbf{match}\, t\, \mathbf{as}\, \{\mathbf{inl}\, x.u\,,\ \mathbf{inr}\, y.q\} : U}$$

Reduction

$$\frac{}{\mathbf{inl}\, t \Downarrow \mathbf{inl}\, t} \quad \frac{}{\mathbf{inr}\, t \Downarrow \mathbf{inr}\, t} \quad \frac{t \Downarrow \mathbf{inl}\, t' \quad u[t'/x] \Downarrow v}{\mathbf{match}\, t\, \mathbf{as}\, \{\mathbf{inl}\, x.u\,,\ \mathbf{inr}\, y.q\} \Downarrow v}$$

$$\frac{t \Downarrow \mathbf{inr}\, t' \quad q[t'/x] \Downarrow v}{\mathbf{match}\, t\, \mathbf{as}\, \{\mathbf{inl}\, x.u\,,\ \mathbf{inr}\, y.q\} \Downarrow v}$$

Table 4. Extension of the theory in Table 2 with binary sums

β Rules

$$\mathbf{match}\ \mathbf{inl}\, t\, \mathbf{as}\, \{\mathbf{inl}\, x.u\,,\ \mathbf{inr}\, y.q\} \approx u[t/x] \quad \mathbf{match}\ \mathbf{inr}\, t\, \mathbf{as}\, \{\mathbf{inl}\, x.u\,,\ \mathbf{inr}\, y.q\} \approx q[t/x]$$

Divergence Rules $\mathbf{match}\, \Omega\, \mathbf{as}\, \{\mathbf{inl}\, x.u\,,\ \mathbf{inr}\, y.q\} \approx \Omega$

a series of $C[\mathbf{rec}^n\, x\, .\, t] \lhd_{T+U} \mathbf{inl}\, u$, then we know that each of the terms on the left must be identical to some $\mathbf{inl}\, t_n$ with $t_n \lhd_T u$—but do the t_n form a rational chain? It turns out that for every t, simply from the existence of $t \approx \mathbf{inl}\, t'$, and because each type is inhabited by Ω, there is a context that can extract directly the t' (up to equivalence, obviously) from the original term. (An idea we borrowed from McCusker 1998)

Lemma 1. *The contexts*

$$T^l[-] = \mathbf{match}\ -\ \mathbf{as}\, \{\mathbf{inl}\, x.x\,,\ \mathbf{inr}\, y.\Omega\}$$

$$T^r[-] = \mathbf{match}\ -\ \mathbf{as}\, \{\mathbf{inl}\, x.\Omega\,,\ \mathbf{inr}\, y.y\}$$

satisfy $t \approx \mathbf{inl}\, u \implies T^l[t] \approx u$ *and* $t \approx \mathbf{inr}\, u \implies T^r[t] \approx u$.

Proposition 6 (Sum Action). *Given approximation candidates* $\lhd_T$ *for* T *and* $\lhd_U$ *for* U, *the relation between* $CTerms^{T+U}$ *and* NFs^{T+U} *defined by*

$$t\ \langle \lhd_T + \lhd_U \rangle\ \mathbf{inl}\, v \iff (\exists t' \lhd_T^c u.t \approx \mathbf{inl}\, t')$$

$$t \ \langle \lhd_T + \lhd_U \rangle \ \mathbf{inr}\, v \iff (\exists t' \lhd^{\mathrm{c}}_U u.t \approx \mathbf{inr}\, t')$$

is an approximation candidate for $A + B$.

Proof. For rational admissibility, the pre-condition must hold for (at least) one of the two clauses in the definition. Say we have $\exists^\infty n.C[\mathbf{rec}^n\, x\,.\,t] \ \langle \lhd_T + \lhd_U \rangle$ $\mathbf{inl}\, u$ *with each term on the left equivalent to some* $\mathbf{inl}\, t_n$*; rewriting* $t_n \approx \mathcal{T}^l[C[\mathbf{rec}^n\, x\,.\,t]]$ *(Lemma 1) it follows that (Proposition 2)*

$$C[\mathbf{rec}^n\, x\,.\,t] \approx \mathbf{inl}\, \mathcal{T}^l[C[\mathbf{rec}^n\, x\,.\,t]] \text{ and } \mathcal{T}^l[C[\mathbf{rec}^n\, x\,.\,t]] \lhd^{\mathrm{c}}_T u$$

An application of rational continuity of the theory, and one of rational admissibility of $\lhd^{\mathrm{c}}_T$ *(again, Proposition 2) yields* $C[\mathbf{rec}\, x\,.\,t] \approx \mathbf{inl}\, \mathcal{T}^l[C[\mathbf{rec}\, x\,.\,t]]$ *and also* $\mathcal{T}^l[C[\mathbf{rec}\, x\,.\,t]] \lhd^{\mathrm{c}}_T u$ *so that* $C[\mathbf{rec}\, x\,.\,t] \ \langle \lhd_T + \lhd_U \rangle \ \mathbf{inl}\, u$*. (Likewise for the right injection.)*

Adequacy. The rest of the proof of adequacy follows exactly as before. Approximation candidates for sums are derived by induction using the sum action; and with them we can extend Proposition 5.

4 Call-by-Push-Value

Values vs. Computations. We now turn to Call-by-push-value (Levy 2004). This language (Table 5) distinguishes between values and computations, with value types represented by A, A', etc., and computation types by $\underline{B}$, $\underline{B}'$, etc. The set of closed values of type A will be represented by Vals^A; that of closed computations by $\mathrm{Comps}^{\underline{B}}$. Variables always have value type. Here we include value products and sums, products of computation types $\underline{B} \,\Pi\, \underline{B}'$, types FA for computations aiming to return a value, and functions which in CBPV are computations taking values to computations. Central to CBPV, we also include value types $U\underline{B}$ of suspended computations of type $\underline{B}$—which can be of one of two forms.

Recursion. In addition to the usual **thunks** of computations, we also have recursively defined thunks $\mathbf{threc}\, x.t$. An alternative would be to use recursive computations $\Gamma \vdash^{\mathrm{c}} \mathbf{rec}\, x.t : \underline{B}$. Although the two are equivalent via the definitions $\mathbf{rec}\, x.t := \mathbf{force}\, \mathbf{threc}\, x.t$ and $\mathbf{threc}\, x.t := \mathbf{thunk}\, \mathbf{rec}\, x.t$, there are two reasons for preferring **threc**: One is that, in some denotational models (e.g. state or continuation passing), **threc** has a simpler denotation than **rec**. The other is that a treatment based on **threc** would be more easily adapted to call-by-value, where recursion and lambda are combined.

Evaluation. Evaluation (Table 6) pertains only to computations. To those on the co-domain side of the evaluation relation $\Downarrow$, we call the *terminal* computations or, alternatively, the normal forms; and their (typed-indexed) set is represented by $\mathrm{NFs}^{\underline{B}}$. Since we have two forms of thunked computations, the action of forcing one such into execution much act accordingly; this *unthunk*ing (a derived operation on the syntax) returns the computations suspended inside **thunks**, or plucks out the computation from a $\mathbf{threc}\, x\,.\,t$ suitably instantiated by the recursive thunk itself—i. e. $t[\mathbf{threc}\, x\,.\,t/x]$. Note that reduction is deterministic.

Table 5. Call-by-push with recursion-value—syntax

Types

$$A, A', \ldots = 1 \mid A \times A' \mid 0 \mid A + A' \mid U\underline{B} \qquad \underline{B}, \underline{B}', \ldots = FA \mid A \to \underline{B} \mid 1_{\Pi} \mid \underline{B}\,\Pi\,\underline{B}'$$

Typing

$$\frac{(x : A \in \Gamma)}{\Gamma \vdash^{v} x : A} \qquad \frac{\Gamma \vdash^{v} v : A \quad x : A, \Gamma \vdash^{c} t : \underline{B}}{\Gamma \vdash^{c} \mathbf{let}\, v\, \mathbf{be}\, x \,.\, t : \underline{B}} \qquad \frac{\Gamma \vdash^{v} v : A}{\Gamma \vdash^{c} \mathbf{return}\ v : FA}$$

$$\frac{\Gamma \vdash^{c} u : FA \quad x : A, \Gamma \vdash^{c} t : \underline{B}}{\Gamma \vdash^{c} u\, \mathbf{to}\, x \,.\, t : \underline{B}} \qquad \frac{\Gamma \vdash^{c} t : \underline{B}}{\Gamma \vdash^{v} \mathbf{thunk}\, t : U\underline{B}} \qquad \frac{\Gamma \vdash^{v} v : U\underline{B}}{\Gamma \vdash^{c} \mathbf{force}\, v : \underline{B}}$$

$$\frac{x : U\underline{B}, \Gamma \vdash^{c} t : \underline{B}}{\Gamma \vdash^{v} \mathbf{threc}\, x \,.\, t : U\underline{B}} \qquad \frac{}{\Gamma \vdash^{c} \Omega : \underline{B}} \qquad \frac{}{\Gamma \vdash^{v} \langle\rangle : 1}$$

$$\frac{\Gamma \vdash^{v} v : A \quad \Gamma \vdash^{v} v' : A'}{\Gamma \vdash^{v} \langle v, v' \rangle : A \times A'} \qquad \frac{\Gamma \vdash^{v} v : A \times A' \quad x : A, y : A', \Gamma \vdash^{c} t : \underline{B}}{\Gamma \vdash^{c} \mathbf{match}\, v\, \mathbf{as}\, \langle x, y \rangle \,.\, t : \underline{B}}$$

$$\frac{\Gamma \vdash^{v} v : 1 \quad \Gamma \vdash^{c} t : \underline{B}}{\Gamma \vdash^{c} \mathbf{match}\, v\, \mathbf{as}\, \langle\rangle \,.\, t : \underline{B}} \qquad \frac{\Gamma \vdash^{v} v : 0}{\Gamma \vdash^{c} \mathbf{match}\, v\, \mathbf{as}\, \{\} : \underline{B}} \qquad \frac{}{\Gamma \vdash^{c} \lambda\{\} : 1_{\Pi}}$$

$$\frac{\Gamma \vdash^{c} t : \underline{B} \quad \Gamma \vdash^{c} t' : \underline{B}'}{\Gamma \vdash^{c} \lambda\{{}^{l}.t, {}^{r}.t'\} : \underline{B}\,\Pi\,\underline{B}'} \qquad \frac{\Gamma \vdash^{c} t : \underline{B}\,\Pi\,\underline{B}'}{\Gamma \vdash^{c} t^{l} : \underline{B}} \qquad \frac{\Gamma \vdash^{c} t : \underline{B}\,\Pi\,\underline{B}'}{\Gamma \vdash^{c} t^{r} : \underline{B}'}$$

$$\frac{x : A, \Gamma \vdash^{c} t : \underline{B}}{\Gamma \vdash^{c} \lambda x.t : A \to \underline{B}} \qquad \frac{\Gamma \vdash^{v} v : A \quad \Gamma \vdash^{c} t : A \to \underline{B}}{\Gamma \vdash^{c} tv : \underline{B}} \qquad \frac{\Gamma \vdash^{v} v : A}{\Gamma \vdash^{v} \mathbf{inl}\, v : A + A'}$$

$$\frac{\Gamma \vdash^{v} v : A'}{\Gamma \vdash^{v} \mathbf{inr}\, v : A + A'} \qquad \frac{\Gamma \vdash^{v} v : A + A' \quad x : A, \Gamma \vdash^{c} t : \underline{B} \quad y : A', \Gamma \vdash^{c} t' : \underline{B}}{\Gamma \vdash^{c} \mathbf{match}\, v\, \mathbf{as}\, \{\mathbf{inl}\, x.t, \mathbf{inr}\, y.t'\} : \underline{B}}$$

Table 6. Call-by-push-value with recursion—reduction

$$\frac{}{\mathbf{return}\ v \Downarrow \mathbf{return}\ v} \qquad \frac{}{\lambda x.t \Downarrow \lambda x.t} \qquad \frac{}{\lambda\{\} \Downarrow \lambda\{\}} \qquad \frac{}{\lambda\{{}^{l}.t, {}^{r}.t'\} \Downarrow \lambda\{{}^{l}.t, {}^{r}.t'\}}$$

$$\frac{t[v/x] \Downarrow r}{\mathbf{let}\, v\, \mathbf{be}\, x \,.\, t \Downarrow r} \qquad \frac{t \Downarrow r}{\mathbf{match}\ \langle\rangle\ \mathbf{as}\ \langle\rangle \,.\, t \Downarrow r} \qquad \frac{t[v/x, v'/y] \Downarrow r}{\mathbf{match}\ \langle v, v' \rangle\ \mathbf{as}\ \langle x, y \rangle \,.\, t \Downarrow r}$$

$$\frac{t[v/x] \Downarrow r}{\mathbf{match}\ \mathbf{inl}\, v\, \mathbf{as}\ \{\mathbf{inl}\, x.t, \mathbf{inr}\, y.t'\} \Downarrow r} \qquad \frac{t'[v/y] \Downarrow r}{\mathbf{match}\ \mathbf{inr}\, v\, \mathbf{as}\ \{\mathbf{inl}\, x.t, \mathbf{inr}\, y.t'\} \Downarrow r}$$

$$\frac{t \Downarrow \lambda\{{}^{l}.u, {}^{r}.q\} \quad u \Downarrow r}{t^{l} \Downarrow r} \qquad \frac{t \Downarrow \lambda\{{}^{l}.u, {}^{r}.q\} \quad q \Downarrow r}{t^{r} \Downarrow r}$$

$$\frac{u \Downarrow \mathbf{return}\ v \quad t[v/x] \Downarrow r}{u\, \mathbf{to}\, x \,.\, t \Downarrow r} \qquad \frac{t \Downarrow \lambda x.u \quad u[v/x] \Downarrow r}{tv \Downarrow r} \qquad \frac{\mathit{unthunk}\, v \Downarrow r}{\mathbf{force}\, v \Downarrow r}$$

where $\mathit{unthunk}(\mathbf{thunk}\, t) = t$ and $\mathit{unthunk}(\mathbf{threc}\, x \,.\, t) = t[\mathbf{threc}\, x \,.\, t/x]$,

4.1 Theory

Theory. By a (CBPV) congruence on closed terms we mean a type-indexed equivalence relation $\approx$ on closed values and computations such that for all closed terms $t \approx t'$ and (value or computation) context $C[-]$ we have $C[t] \approx C[t']$, respectively. A congruence is a *rationally continuous β-Ω-fix theory* when it satisfies the rules in Table 7. Rational chains are now those built by the application of a context $C[-]$ to the (thunked) approximants $\mathbf{threc}^n$ of recursive thunks and which are defined by the clauses $\mathbf{threc}^0\, x\,.\,t = \mathbf{thunk}\,\Omega$ and $\mathbf{threc}^{n+1}\, x\,.\,t = \mathbf{thunk}\, t[\mathbf{threc}^n\, x\,.\,t/x]$; continuity is defined accordingly. Any congruence including the β and fixpoint rules is easily seen to be sound. We shall show that with the remaining rules it is also adequate.

Table 7. Call-by-push-value with recursion—rationally continuous β-Ω-fix theory

Basis An equivalence relation $\approx$ on closed terms satisfying compatibility:

$$\text{for any } C[-],\ t \approx u \implies C[t] \approx C[u] \qquad \text{for any } C[-],\ v \approx w \implies C[v] \approx C[w]$$

β Rules

$$\mathbf{let}\ v\ \mathbf{be}\ x\,.\,t \approx t[v/x] \qquad \mathbf{match}\ \langle\rangle\ \mathbf{as}\ \langle\rangle\,.\,t \approx t$$

$$\mathbf{match}\ \langle v, v'\rangle\ \mathbf{as}\ \langle x, y\rangle\,.\,t \approx t[v/x, v'/y]$$

$$\mathbf{match}\ \mathbf{inl}\, v\ \mathbf{as}\ \{\mathbf{inl}\, x.t, \mathbf{inr}\, y.t'\} \approx t[v/x]$$

$$\mathbf{match}\ \mathbf{inr}\, v\ \mathbf{as}\ \{\mathbf{inl}\, x.t, \mathbf{inr}\, y.t'\} \approx t'[v/y]$$

$$\left(\lambda\left\{{}^{l}.t, {}^{r}.t'\right\}\right)^{l} \approx t \qquad \left(\lambda\left\{{}^{l}.t, {}^{r}.t'\right\}\right)^{r} \approx t' \qquad \mathbf{return}\ v\ \mathbf{to}\ x.\,t \approx t[v/x]$$

$$\mathbf{force}\,\mathbf{thunk}\, t \approx t \qquad (\lambda x.t)v \approx t[v/x]$$

Fixpoint Rule $\qquad \mathbf{threc}\, x\,.\,t \approx \mathbf{thunk}\, t[\mathbf{threc}\, x\,.\,t/x]$

Divergence Rules $\qquad \Omega\ \mathbf{to}\ x.\,t \approx \Omega \qquad (\Omega)^{l} \approx \Omega \qquad (\Omega)^{r} \approx \Omega \qquad \Omega v \approx \Omega$

Rational Continuity for $x : U\underline{B} \vdash^{c} t : \underline{B}$ and $C[-]$, $D[-]$ computation contexts

$$\frac{\exists^{\infty} n.C[\mathbf{threc}^n\, x\,.\,t] \approx D[\mathbf{threc}^n\, x\,.\,t]}{C[\mathbf{threc}\, x\,.\,t] \approx D[\mathbf{threc}\, x\,.\,t]}$$

where $\mathbf{threc}^0\, x\,.\,t = \mathbf{thunk}\,\Omega$ and $\mathbf{threc}^{n+1}\, x\,.\,t = \mathbf{thunk}\, t[\mathbf{threc}^n\, x\,.\,t/x]$.

4.2 Adequacy

Values: Empty Shells. In the proof of adequacy for PCF with sums we were required to introduce the tests so that we could, metaphorically, peek inside the injections and transform the rational chains there into equivalent ones with the properties we needed (cf. proof of Proposition 6). Here the problem expands to *all value types.* When checking rational admissibility, we need to decompose a value into its *ultimate pattern* and its constituent thunks (Lassen and Levy 2008, following ideas from Abramsky and McCusker 1997; also discernible in the work of Zeilberger 2008) and use those to find equivalent chains that can be used to establish adequacy.

Definition 4 (Ultimate Patterns). *The set of of ultimate patterns* UP^A *for a value type* A *is given by induction on the following rules:* $-_{U\underline{B}} \in UP^{U\underline{B}}$, $\langle\rangle \in UP^1$ *and*

$$\frac{p \in UP^A \quad p' \in UP^{A'}}{\langle p, p'\rangle \in UP^{A\times A'}} \qquad \frac{p \in UP^A}{\mathbf{inl}\, p \in UP^{A+A'}} \qquad \frac{p \in UP^{A'}}{\mathbf{inr}\, p \in UP^{A+A'}}$$

For a given ultimate pattern $p \in UP^A$ *the finite sequence of hole-types in pattern* p *is given by induction by*

$$H(-_{U\underline{B}}) = (U\underline{B}) \qquad H(\langle\rangle) = \epsilon \qquad H(\langle p, p'\rangle) = H(p) +\!\!+ H(p')$$
$$H(\mathbf{inl}\, p) = H(p) \qquad H(\mathbf{inr}\, p) = H(p)$$

Proposition 7 (Value Decomposition). *Given* $\vdash^v v : A$, *there is a unique* $p \in UP^A$ *and a unique sequence* $(\vdash^v v_i : H(p)_i)_{i<|H(p)|}$*—the filling—for which* $v = p @ (v_i)_{i<|H(p)|}$, *using the reassembly function*

$$(-_{U\underline{B}}) @ (v) = v \qquad \langle\rangle @ \epsilon = \langle\rangle$$
$$\mathbf{inl}\, p @ (v_i)_{i<|H(p)|} = \mathbf{inl}(p @ (v_i)_{i<|H(p)|})$$
$$\mathbf{inr}\, p @ (v_i)_{i<|H(p)|} = \mathbf{inr}(p @ (v_i)_{i<|H(p)|})$$
$$\langle p, p'\rangle @ \left((v_i)_{i<|H(p)|} +\!\!+ (v'_i)_{i<|H(p')|}\right) = \left\langle (p @ (v_i)_{i<|H(p)|}), (p' @ (v'_i)_{i<|H(p')|})\right\rangle$$

Tests. Ultimate patterns let us define the tests that extract the computations embedded in a given value. Like in the PCF sum case, we can use them to define values that are equivalent to a given one but make use only of the latter. If the values are derived from some family of contexts for the holes, then we can derive an equivalent context from the respective ultimate pattern.

Definition 5. *For* $p \in UP^A$, *and* $i < |H(p)|$, *we define a context* $\mathcal{T}_i^p[-]$ *by induction on* $p \in UP^A$ *using the rules below. Note that when* $\Gamma \vdash^v - : A$ *the test has type* $\Gamma \vdash^c \mathcal{T}_i^p[-] : \underline{B_i}$ *where* $U\underline{B_i} = H(p)_i$.

$$\mathcal{T}_0^{-_{UB}}[-] = \mathbf{force}\, -$$
$$\mathcal{T}_i^{\mathbf{inl}\, p}[-] = \mathbf{match}\ -\ \mathbf{as}\ \{\mathbf{inl}\, x.\, \mathcal{T}_i^p[x],\ \mathbf{inr}\, y.\Omega\}$$
$$\mathcal{T}_i^{\mathbf{inr}\, p}[-] = \mathbf{match}\ -\ \mathbf{as}\ \{\mathbf{inl}\, x.\Omega,\ \mathbf{inr}\, y.\, \mathcal{T}_i^p[y]\}$$
$$\mathcal{T}_{i<|H(p)|}^{\langle p,p'\rangle}[-] = \mathbf{match} - \mathbf{as} < x, y > .\, \mathcal{T}_i^p[x]$$
$$\mathcal{T}_{i=|H(p)|+i'}^{\langle p,p'\rangle}[-] = \mathbf{match} - \mathbf{as} < x, y > .\, \mathcal{T}_{i'}^{p'}[y]$$

Proposition 8 (Tests Decompose). *Given a pattern* $p \in UP^A$, *a sequence* $(\vdash w_i : H(p)_i)_{i<|H(p)|}$, *and* $i < |H(p)|$, *we have* $\mathcal{T}_i^p[p\,@\,(w_i)_{i<|H(p)|}] \approx \mathbf{force}\, w_i$.

Proposition 9. *For* $\vdash^c t : FA$, *and* $p \in UP^A$, *if* $t \approx \mathbf{return}\, p\,@\,(v_i)_{i<|H(p)|}$ *then, successively:*

1. $\forall i < |H(p)|.\, \mathbf{thunk}(t \text{ } \mathbf{to} \text{ } x.\ \mathcal{T}_i^p[x]) \approx v_i$
2. $p\,@\,(v_i)_{i<|H(p)|} \approx p\,@\,(\mathbf{thunk}(t \text{ } \mathbf{to} \text{ } x.\ \mathcal{T}_i^p[x]))_{i<|H(p)|}$
3. $t \approx \mathbf{return}\, p\,@\,(\mathbf{thunk}(t \text{ } \mathbf{to} \text{ } x.\ \mathcal{T}_i^p[x]))_{i<|H(p)|}$

Approximation Candidates. Unlike PCF where we have computations and normal forms, CBPV has three levels of syntax: values, terminals, and computations. For the purposes of defining the needed approximation candidates, terminals (read: normal forms) and computations, behave like their PCF counterparts and have (now) familiar definitions of approximation candidates. Approximation candidates for value types enforce that: only structurally similar values are related; that they are (left) closed under equivalence of their holes; and that they are closed under the usual chains.

Definition 6 (Approximation Candidates). *Given a value type A, an approximation candidate* $\lhd$ *for A is a subset of* $Vals^A \times Vals^A$ *such that*

1. *Structural Matching:* $p\,@\,(v_i)_i \lhd p'\,@\,(w_i)_i \implies p = p'$
2. *Computational* $\approx$ *Extension: if* $p\,@\,(v'_i)_{i<|H(p)|} \lhd p\,@\,(w_i)_{i<|H(p)|}$ *then*

$$(\forall i < |H(p)|.v_i \approx v'_i) \implies p\,@\,(v_i)_{i<|H(p)|} \lhd p\,@\,(w_i)_{i<|H(p)|}$$

3. *Rational Admissibility: for* $x : U\underline{B} \vdash^c t : \underline{B}$

$$(\exists^\infty n.V[\mathbf{threc}^n x.t] \lhd w) \implies V[\mathbf{threc}\, x\,.\, t] \lhd w$$

Given a computation type $\underline{B}$, *an approximation candidate* $\lhd$ *for* $\underline{B}$ *is a subset of* $Comps^{\underline{B}} \times NF^{\underline{B}}$ *such that*

1. $\approx$ *Extension:* $t \approx t'$ *and* $t' \lhd r \implies t \lhd r$
2. *Rational Admissibility: for* $x : U\underline{B} \vdash^c t : \underline{B}$

$$\exists^\infty n.C[\mathbf{threc}^n\, x\,.\, t] \lhd r \implies C[\mathbf{threc}\, x\,.\, t] \lhd r$$

Proposition 10. *Given a (computation) approximation candidate* $\lhd$ *on* $\underline{B}$, *define its closure as the binary relation* $Comps^{\underline{B}} \times Comps^{\underline{B}}$ *where*

$$t \lhd^c u \iff t \approx \Omega \text{ or } (\exists r.u \Downarrow r \text{ and } t \lhd r)$$

It satisfies the following properties:

1. Ω *Property:* $\Omega \lhd^c u$ *for any* $u \in Comps^{\underline{B}}$
2. $\approx$ *Extension:* $t \approx t'$ *and* $t' \lhd^c u \implies t \lhd^c u$
3. $\Downarrow$ *Extension:* $t \lhd^c u'$ *and* $(\forall r.u' \Downarrow r \implies u \Downarrow r) \implies t \lhd^c u$
4. *Rational Admissibility: for* $x : U\underline{B} \vdash^c t : \underline{B}$

$$(\exists^\infty n.C[\mathbf{threc}^n x\,.\, t] \lhd^c u) \implies C[\mathbf{threc}\, x\,.\, t] \lhd^c u$$

Actions. We can then define the actions on these approximation candidates associated with each type constructor. Mostly this is done by structure (for values) or by use (for computations); the exceptions are U types and F types that we define, respectively, by structure, and by use. Note that it is the existential quantification in the definition of the F action that—very much like PCF sums—requires the use of the tests. Using them, we can easily define, by induction, the approximation relation and thereby establish the adequacy of the theory.

Proposition 11 (Thunk Action). *Let $\lhd$ be an approximation candidate for $\underline{B}$. Then the binary relation*

$$v \langle U(\lhd)\rangle\, w \iff \mathbf{force}\, v \lhd^{c} \mathit{unthunk}\, w$$

is an approximation candidate for $U\underline{B}$.

Proposition 12 (F Action). *Let $\lhd$ be an approximation candidate for A. Then the following is an approximation candidate for FA:*

$$t \langle F(\lhd)\rangle\, \mathbf{return}\, w \iff \exists v \lhd w.t \approx \mathbf{return}\, v$$

Definition 7 (Enviroments). *Given a typing context Γ, an* environment σ *for Γ is a substitution that maps each $x : A \in \Gamma$ to a closed term of type $\vdash^{v} \sigma(x) : A$. If σ_1 and σ_2 are two such, we write $\sigma_1 \lhd_\Gamma \sigma_2$ to mean $\sigma_1(x) \lhd_A \sigma_2(x)$ for all $x : A \in \Gamma$.*

Proposition 13. *For any $\Gamma \vdash^{c} t : B$ (resp. $\Gamma \vdash^{v} v : A$), and environments $\sigma_1 \lhd_\Gamma \sigma_2$ we have $t[\sigma_1] \lhd^{c}_{\underline{B}} t[\sigma_2]$ (resp. $v[\sigma_1] \lhd_A v[\sigma_2]$).*

Corollary 2 (Adequacy). *For any computation $\vdash^{c} t : \underline{B}$, if $t \Uparrow$ then $t \approx \Omega$.*

5 Polymorphic Call-by-Push-Value

Adequacy, Now For All. Our final extension deals with polymorphism. In Call-by-push-value, polymorphic types are computation types. We may quantify over both value and computation types. The extension is presented in Table 8.

We assume two disjoint countable sets of variables, $X, Y, \ldots \in$ VVars and $\underline{X}, \underline{Y}, \ldots \in$ CVars, for value and computation types (resp.). Types are now also considered up to α-equivalence. They will also be considered under context, $\Theta \vdash^{C} \underline{B}$ and $\Theta \vdash^{V} A$, where Θ is some finite subset of VVars $\cup$ CVars that includes the free type variables of the A or $\underline{B}$. (These type judgements have an obvious inductive definition). The proper extension of a type context Θ by a type variable χ will be denoted by χ, Θ. Typing judgements also need to be annotated by a type context, as in $\Theta; \Gamma \vdash^{c} t : \underline{B}$ where Θ includes all the free type variables in the types of Γ and $\underline{B}$. The previous typing rules are extended in the evident way.

Table 8. Polymorphic Call-by-push-value with recursion

Types $A = X \in \text{VVars} \mid \ldots \quad \underline{B} = \underline{X} \in \text{CVars} \mid \ldots \mid \prod X.\underline{B} \mid \prod \underline{X}.\underline{B}$

Typing

$$\frac{X, \Theta; \Gamma \vdash^c t : \underline{B} \quad (X \notin \Theta)}{\Theta; \Gamma \vdash^c \Lambda X.t : \prod X.\underline{B}} \qquad \frac{\Theta; \Gamma \vdash^c t : \prod X.\underline{B} \quad \Theta \vdash^V A}{\Theta; \Gamma \vdash^c tA : \underline{B}[A/X]}$$

$$\frac{\underline{X}, \Theta; \Gamma \vdash^c t : \underline{B} \quad (\underline{X} \notin \Theta)}{\Theta; \Gamma \vdash^c \Lambda \underline{X}.t : \prod \underline{X}.\underline{B}} \qquad \frac{\Theta; \Gamma \vdash^c t : \prod \underline{X}.\underline{B} \quad \Theta \vdash^C \underline{B'}}{\Theta; \Gamma \vdash^c t\underline{B'} : \underline{B}[\underline{B'}/\underline{X}]}$$

Reduction

$$\frac{}{\Lambda X.t \Downarrow \Lambda X.t} \qquad \frac{}{\Lambda \underline{X}.t \Downarrow \Lambda \underline{X}.t}$$

$$\frac{t \Downarrow \Lambda X.u \quad u[A/X] \Downarrow r}{tA \Downarrow r} \qquad \frac{t \Downarrow \Lambda \underline{X}.u \quad u[\underline{B}/\underline{X}] \Downarrow r}{t\underline{B} \Downarrow r}$$

Table 9. Extension of the theory in Table 7 to polymorphism

β Rules	$(\Lambda X.t)A \approx t[A/X]$	$(\Lambda \underline{X}.t)\underline{B} \approx t[\underline{B}/\underline{X}]$
Divergence Rules	$\Omega A \approx \Omega$	$\Omega \underline{B} \approx \Omega$

Reduction and Theory. Reduction—defined only for closed terms of closed type—is still deterministic. On the theory end of things, we equate only closed terms of closed type so that we need only extend the theory of Sect. 4 with the obvious β and divergence rules (Table 9). Unsurprisingly, soundness still stands.

5.1 Adequacy

Approximation Candidates and Actions. Throughout we have worked with approximation candidates—and now we can reap the fruits of that work. The definition of approximation candidates (Definition 6) and of their extension to computations (Proposition 10) can stay exactly the same; as can the actions for non-polymorphic type constructors. The actions of polymorphic types follow.

Proposition 14. *Let $Y \vdash^C \underline{B}$ be a computation type, and ϕ a mapping that assigns to every closed type T and approximation candidate $\lhd \in ACs^T$ an approximation candidate $\phi_{T,\lhd} \in ACs^{\underline{B}[T/Y]}$; then*

$$t \left\langle \prod Y.\phi \right\rangle \Lambda Y.u \iff \textit{for all } \vdash^C T, \lhd \in ACs^T.\, tT \, \langle \phi_{T,\lhd} \rangle^c \, u[T/Y]$$

is an approximation candidate for $\prod Y.\underline{B}$—and likewise for $\prod \underline{Y}.\underline{B}$

Approximations. The approximation relations need to be parametrized by the candidates that will instantiate the type variables so that in the end we arrive at a candidate for a closed type. As usual, we have that it satisfies the weakening and substitution properties that are used in the proof of adequacy for abstractions and type instantiations, respectively.

Definition 8 (Approximation Environment). *An* approximation environment *γ for Θ is a map taking each $\chi \in \Theta$ to a closed type $\gamma^T(\chi)$ of the same kind as χ and an adequacy candidate $\gamma^C(\chi) \in ACs^{\gamma^T(\chi)}$.*

Definition 9 (Parametrized Approximation Relations). *Let $\Theta \vdash^V A$ (resp. $\Theta \vdash^C \underline{B}$) be a (possibly open) type and γ an approximation environment for Θ. The following* parametrized approximation relations, *defined by induction on types, determine an approximation candidate for $A[\gamma^T]$—i.e. A with each type variable χ replaced with $\gamma^T(\chi)$ (resp. $\underline{B}[\gamma^T]$).*

$$
\begin{aligned}
\lhd^{\gamma}_{\Theta \vdash^V X} &= \gamma^C(X) & \lhd^{\gamma}_{\Theta \vdash^C \underline{X}} &= \gamma^C(\underline{X}) \\
\lhd^{\gamma}_{\Theta \vdash^V 1} &= \lhd_1 & \lhd^{\gamma}_{\Theta \vdash^V A \times A'} &= (\lhd^{\gamma}_{\Theta \vdash^V A}) \times (\lhd^{\gamma}_{\Theta \vdash^V A'}) \\
\lhd^{\gamma}_{\Theta \vdash^V 0} &= \lhd_0 & \lhd^{\gamma}_{\Theta \vdash^V A + A'} &= (\lhd^{\gamma}_{\Theta \vdash^V A}) + (\lhd^{\gamma}_{\Theta \vdash^V A'}) \\
\lhd^{\gamma}_{\Theta \vdash^V U\underline{B}} &= U(\lhd^{\gamma}_{\Theta \vdash^C \underline{B}}) & \lhd^{\gamma}_{\Theta \vdash^C FA} &= F(\lhd^{\gamma}_{\Theta \vdash^V A}) \\
\lhd^{\gamma}_{\Theta \vdash^C 1_{\Pi}} &= (\lhd_{1_{\Pi}}) & \lhd^{\gamma}_{\Theta \vdash^C \underline{B} \Pi \underline{B'}} &= (\lhd^{\gamma}_{\Theta \vdash^C \underline{B}}) \, \Pi \, (\lhd^{\gamma}_{\Theta \vdash^C \underline{B'}}) \\
\lhd^{\gamma}_{\Theta \vdash^C A \to \underline{B}} &= (\lhd^{\gamma}_{\Theta \vdash^V A}) \to (\lhd^{\gamma}_{\Theta \vdash^C \underline{B}}) & & \\
\lhd^{\gamma}_{\Theta \vdash^C \prod Y . \underline{B}} &= \prod Y . \left(\lhd^{\gamma[Y \mapsto (-,=)]}_{Y, \Theta \vdash^C \underline{B}} \} \right) & \lhd^{\gamma}_{\Theta \vdash^C \prod \underline{Y} . \underline{B}} &= \prod \underline{Y} . \left(\lhd^{\gamma[\underline{Y} \mapsto (-,=)]}_{\underline{Y}, \Theta \vdash^C \underline{B}} \} \right)
\end{aligned}
$$

Definition 10. *For any Θ and approximation environment γ for Θ, if σ_1 and σ_2 are environments for $\Gamma[\gamma^T]$, we write $\sigma_1 \lhd^{\gamma}_{\Theta;\Gamma} \sigma_2$ to mean $\sigma_1(x) \lhd^{\gamma}_{\Theta \vdash^V A} \sigma_2(x)$ for every $x : A \in \Gamma$.*

Proposition 15. *For any $\Theta; \Gamma \vdash^c t : B$ (resp. $\Theta, \Gamma \vdash^v v : A$), approximation environment γ for Θ, and environments $\sigma_1 \lhd^{\gamma}_{\Theta;\Gamma} \sigma_2$ for Γ*

$$
t[\gamma^T][\sigma_1] \left\langle \lhd^{\gamma}_{\Theta \vdash^C \underline{B}} \right\rangle^c t[\gamma^T][\sigma_2] \qquad (\textit{resp. } v[\gamma^T][\sigma_1] \left\langle \lhd^{\gamma}_{\Theta \vdash^V A} \right\rangle v[\gamma^T][\sigma_2])
$$

6 Concluding Remarks

We have thus seen how, for term-level recursion, the rational continuity rule coupled with β, the fixpoint property of recursion, and strictness of the basic constructors of the language suffices to make a theory adequate. The recipe of the previous sections applies to both call-by-name and call-by-value languages and is compatible with polymorphic types. Along the way we used no category theory; no models were mentioned. We relied only on syntactic constructions and required no external machinery.

Two extensions are conspicuous for their absence: to existential types and to recursive types. In Call-by-push-value, existential types are value types. We conjecture our theorem holds for them but we must find a way to quantify over ultimate patterns. For recursive types, even finding suitable conditions on $\approx$ is challenging. We would like to adapt Pitts' (1996) method of minimal invariant relations but we will need type constructors to be functorial over suitable syntactic categories.

For term-recursion and polymorphism, however, we now know that to prove a model adequate we need only to show that it satisfies the basic laws and rational continuity.

References

Abramsky, S., Jagadeesan, R., Malacaria, P.: Full abstraction for PCF. Inf. Comput. **163**(2), 409–470 (2000)

Abramsky, S., McCusker, G.: Call-by-value games. In: Nielsen, M., Thomas, W. (eds.) CSL 1997. LNCS, vol. 1414, pp. 1–17. Springer, Heidelberg (1998). https://doi.org/10.1007/BFb0028004

Bloom, S.L., Ésik, Z.: Iteration Theories - The Equational Logic of Iterative Processes. EATCS Monographs on Theoretical Computer Science. Springer, Heidelberg (1993). https://doi.org/10.1007/978-3-642-78034-9

Fiore, M.P., Plotkin, G.D.: An axiomatization of computationally adequate domain theoretic models of FPC. In: LICS: IEEE Symposium on Logic in Computer Science (1994)

Fiore, M.P.: Axiomatic Domain Theory in Categories of Partial Maps, Distinguished Dissertations in Computer Science, vol. 14. Cambridge University Press, Cambridge (1996)

Girard, J.Y., Lafont, Y., Taylor, P.: Proofs and Types, Cambridge Tracts in Theoretical Computer Science, vol. 7. Cambridge University Press, Cambridge (1989)

Laird, J.: Game semantics for a polymorphic programming language. J. ACM **60**(4), 29:1–29:27 (2013)

Lassen, S.B., Levy, P.B.: Typed normal form bisimulation for parametric polymorphism. In: Proceedings of the Twenty-Third Annual IEEE Symposium on Logic in Computer Science, LICS 2008, 24–27 June 2008, Pittsburgh, PA, USA, pp. 341–352 (2008)

Levy, P.B.: Call-By-Push-Value: A Functional/Imperative Synthesis, Semantics Structures in Computation, vol. 2. Springer, Dordrecht (2004). https://doi.org/10.1007/978-94-007-0954-6

McCusker, G.: Games and Full Abstraction for a Functional Metalanguage with Recursive Types. CPHC/BCS Distinguished Dissertations. Springer, London (1998). https://doi.org/10.1007/978-1-4471-0615-9

Møgelberg, R.E.: From parametric polymorphism to models of polymorphic FPC. Math. Struct. Comput. Sci. **19**(4), 639–686 (2009)

Normann, D.: On sequential functionals of type 3. Math. Struct. Comput. Sci. **16**(2), 279–289 (2006)

Pitts, A.M.: Relational properties of domains. Inf. Comput. **127**(2), 66–90 (1996)

Pitts, A.M.: Parametric polymorphism and operational equivalence. Math. Struct. Comput. Sci. **10**(3), 321–359 (2000)

Plotkin, G.D.: LCF considered as a programming language. Theor. Comput. Sci. **5**(3), 223–255 (1977)
Simpson, A.K.: Computational adequacy for recursive types in models of intuitionistic set theory. Ann. Pure Appl. Log. **130**(1–3), 207–275 (2004)
Wright, J.B., Thatcher, J.W., Wagner, E.G., Goguen, J.A.: Rational algebraic theories and fixed-point solutions. In: 1976 17th Annual Symposium on Foundations of Computer Science, pp. 147–158. IEEE (1976)
Zeilberger, N.: Focusing and higher-order abstract syntax. In: Proceedings of the 35th Annual ACM SIGPLAN-SIGACT Symposium on Principles of Programming Languages, POPL 2008, pp. 359–369. ACM, New York (2008)

Linearity

A New Linear Logic for Deadlock-Free Session-Typed Processes

Ornela Dardha(✉) and Simon J. Gay

School of Computing Science, University of Glasgow, Glasgow, UK
{Ornela.Dardha,Simon.Gay}@glasgow.ac.uk

Abstract. The π-calculus, viewed as a core concurrent programming language, has been used as the target of much research on type systems for concurrency. In this paper we propose a new type system for deadlock-free session-typed π-calculus processes, by integrating two separate lines of work. The first is the propositions-as-types approach by Caires and Pfenning, which provides a linear logic foundation for session types and guarantees deadlock-freedom by forbidding cyclic process connections. The second is Kobayashi's approach in which types are annotated with priorities so that the type system can check whether or not processes contain genuine cyclic dependencies between communication operations. We combine these two techniques for the first time, and define a new and more expressive variant of classical linear logic with a proof assignment that gives a session type system with Kobayashi-style priorities. This can be seen in three ways: (i) as a new linear logic in which cyclic structures can be derived and a Cycle-elimination theorem generalises Cut-elimination; (ii) as a logically-based session type system, which is more expressive than Caires and Pfenning's; (iii) as a logical foundation for Kobayashi's system, bringing it into the sphere of the propositions-as-types paradigm.

1 Introduction

The Curry-Howard correspondence, or propositions-as-types paradigm, provides a canonical logical foundation for functional programming [42]. It identifies types with logical propositions, programs with proofs, and computation with proof normalisation. It was natural to ask for a similar account of concurrent programming, and this question was brought into focus by the discovery of linear logic [24] and Girard's explicit suggestion that it should have some connection with concurrent computation. Several attempts were made to relate π-calculus processes to the proof nets of classical linear logic [1,8], and to relate CCS-like processes to the $*$-autonomous categories that provide semantics for classical linear logic [2]. However, this work did not result in a convincing propositions-as-types framework for concurrency, and did not continue beyond the 1990s.

Supported by the UK EPSRC grant EP/K034413/1, "From Data Types to Session Types: A Basis for Concurrency and Distribution (ABCD)", and by COST Action IC1201, "Behavioural Types for Reliable Large-Scale Software Systems (BETTY)".

C. Baier and U. Dal Lago (Eds.): FOSSACS 2018, LNCS 10803, pp. 91–109, 2018.
https://doi.org/10.1007/978-3-319-89366-2_5

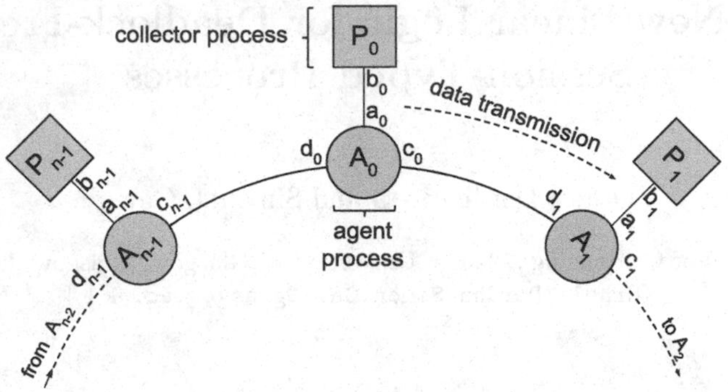

Fig. 1. Cyclic scheduler

Meanwhile, Honda *et al.* [26,27,38] developed *session types* as a formalism for statically checking that messages have the correct types and sequence according to a communication protocol. Research on session types developed and matured over several years, eventually inspiring Caires and Pfenning [12] to discover a Curry-Howard correspondence between dual intuitionistic linear logic [7] and a form of π-calculus with session types [38]. Wadler [41] subsequently gave an alternative formulation based on classical linear logic, and related it to existing work on session types for functional languages [23]. The Caires-Pfenning approach has been widely accepted as a propositions-as-types theory of concurrent programming, as well as providing a logical foundation for session types.

Caires and Pfenning's type system guarantees deadlock-freedom by forbidding cyclic process structures. It provides a logical foundation for deadlock-free session processes, complementing previous approaches to deadlock-freedom in session type systems [9,15,21,22]. The logical approach to session types has been extended in many ways, including features such as dependent types [39], failures and non-determinism [11], sharing and races [6]. All this work relies on the acyclicity condition. However, rejecting cyclic process structures is unnecessarily strict: they are a necessary, but not sufficient, condition for the existence of deadlocked communication operations. As we will show in Example 1 (Fig. 1), there are deadlock-free processes that can naturally be implemented in a cyclic way, but are rejected by Caires and Pfenning's type system.

Our contribution is to define a new logic, *priority-based linear logic* (PLL), and formulate it as a type system for *priority-based* CP (PCP), which is a more expressive class of processes than Wadler's CP [41]. This is the first Curry-Howard correspondence that allows cyclic interconnected processes, while still ensuring deadlock-freedom. The key idea is that PLL includes conditions on inter-channel dependencies based on Kobayashi's type systems [29,30,32]. Our work can be viewed in three ways: (i) as a new linear logic in which cyclic proof structures can be derived; (ii) as an extension of Caires-Pfenning type systems so that they accept more processes, while maintaining the strong logical foundation; (iii) as a logical foundation for Kobayashi-style type systems.

An example of a deadlock-free cyclic process is Milner's well-known scheduler [35], described in the following Example 1.

Example 1 (Cyclic Scheduler, Fig. 1*).* A set of agents $A_0, ..., A_{n-1}$, for $n > 1$, is scheduled to perform a certain task in cyclic order, starting with agent A_0. For all $i \in \{1, ..., n-1\}$, agent A_i sends the result of computation to a collector process P_i, before transmitting further data to agent $A_{(i+1) \bmod n}$. At the end of the round, A_0 sends the final result to P_0. Here we define a finite version of Milner's scheduler, which executes one round of communication.

$$\begin{array}{lll} \mathit{Sched} & \triangleq ...(\boldsymbol{\nu} a_i b_i)...(\boldsymbol{\nu} c_i d_{(i+1) \bmod n})(A_0 \mid A_1 \mid ... \mid A_{n-1} \mid P_0 \mid P_1 \mid ... \mid P_{n-1}) & \\ A_0 & \triangleq c_0[\mathbf{n}_0].d_0(x_0).a_0[\mathbf{m}_0].\mathsf{close}_0 & \\ A_i & \triangleq d_i(x_i).a_i[\mathbf{m}_i].c_i[\mathbf{n}_i].\mathsf{close}_i & i \in \{1, ..., n-1\} \\ P_i & \triangleq b_i(y_i).Q_i & i \in \{0, ..., n-1\} \end{array}$$

Prefix $c_0[\mathbf{n}_0]$ denotes an output on c_0, and $d_0(x_0)$ an input on d_0. For now, let $\mathbf{m}$ and $\mathbf{n}$ denote data. Process close_i closes the channels used by A_i: the details of this closure are irrelevant here (however, they are as in processes Q and R in Example 2). Process Q_i uses the message received from A_i, in internal computation. The construct $(\boldsymbol{\nu} ab)$ creates two channel endpoints a and b and binds them together. The system *Sched* is deadlock-free because $A_1, ..., A_{n-1}$ each wait for a message from the previous A_i before sending, and A_0 sends the initial message.

Sched is not typable in the original type systems by Caires-Pfenning and Wadler. To do that, it would be necessary to break A_0 into two parallel agents $A_0' \triangleq c_0[\mathbf{n}_0].\mathsf{close}_{c_0}$ and $A_0'' \triangleq d_0(x_0).a_0[\mathbf{m}_0].\mathsf{close}_{d_0,a_0}$. This changes the design of the system, yielding a different one. Moreover, if the scheduler continues into a second round of communication, this redesign is not possible because of the potential dependency from the input on d_0 to the next output on c_0. However, *Sched* is typable in PCP; we will show the type assignment at the end of Sect. 2.

There is a natural question at this point: *given that the cyclic scheduler is deadlock-free, is it possible to encode its semantics in* CP, *thus eliminating the need for* PCP? It is possible to define a centralised agent A that communicates with all the collectors P_i, resulting in a system that is semantically equivalent to our *Sched*. However, such an encoding has a global character, and changes the structure of the overall system from distributed to centralised. In programming terms, it corresponds to changing the software design, as we pointed out in Example 1, and ultimately the software architecture, which is not always desirable or even feasible. The aim of PCP is to generalise CP so that deadlock-free processes can be constructed with their natural structure. We would want any encoding of PCP into CP to be structure-preserving, which would mean translating the Cycle rule (given in Fig. 2) homomorphically; this is clearly impossible.

Contributions and Structure of the Paper. In Sect. 2 we define priority-based linear logic (PLL), which extends classical linear logic (CLL) with priorities attached to propositions. These priorities are based on Kobayashi's annotations for deadlock freedom [32]. By following the propositions-as-types paradigm, we define a term assignment for PLL proofs, resulting in priority-based classical processes (PCP), which extends Wadler's CP [41] with MIX and CYCLE rules (Fig. 2). In Sect. 3 we define an operational semantics for PCP. In Sect. 4 we prove CYCLE-elimination (Theorem 1) for PLL, analogous to the standard CUT-elimination theorem for CLL. Consequently, the results for PCP are subject reduction (Theorem 2), top-level deadlock-freedom (Theorem 3), and full deadlock-freedom for closed processes (Theorem 4). In Sect. 5 we discuss related work and conclude the paper.

2 PCP: Classical Processes with MIX and CYCLE

Priority-based CP (PCP) follows the style of Wadler's Classical Processes (CP) [41], with details inspired by Carbone *et al.* [14] and Caires and Pérez [11].

Types. We start with types, which are based on CLL propositions. Let A, B range over types, given in Definition 1. Let $\mathsf{o}, \kappa \in \mathbb{N} \cup \{\omega\}$ range over *priorities*, which are used to annotate types. Let ω be a special element such that $\mathsf{o} < \omega$ for all $\mathsf{o} \in \mathbb{N}$. Often, we will omit ω. We will explain priorities later in this section.

Definition 1 (Types). *Types* (A, B) *are given by:*

$$A, B ::= \bot^{\mathsf{o}} \mid \mathbf{1}^{\mathsf{o}} \mid A \otimes^{\mathsf{o}} B \mid A \mathbin{⅋}^{\mathsf{o}} B \mid \oplus^{\mathsf{o}}\{l_i : A_i\}_{i \in I} \mid \&^{\mathsf{o}}\{l_i : A_i\}_{i \in I} \mid ?^{\mathsf{o}} A \mid !^{\mathsf{o}} A$$

$\bot^{\mathsf{o}}$ and $\mathbf{1}^{\mathsf{o}}$ are associated with channel endpoints that are ready to be closed. $A \otimes^{\mathsf{o}} B$ (respectively, $A \mathbin{⅋}^{\mathsf{o}} B$) is associated with a channel endpoint that first outputs (respectively, inputs) a channel of type A and then proceeds as B. $\oplus^{\mathsf{o}}\{l_i : A_i\}_{i \in I}$ is associated with a channel endpoint over which we can select a label from $\{l_i\}_{i \in I}$, and proceed as A_i. Dually, $\&^{\mathsf{o}}\{l_i : A_i\}_{i \in I}$ is associated with a channel endpoint that can offer a set of labelled types. $?^{\mathsf{o}} A$ types a collection of clients requesting A. Dually, $!^{\mathsf{o}} A$ types a server repeatedly accepting A.

Duality on types is total and is given in Definition 2. It preserves priorities of types.

Definition 2 (Duality). *The* duality *function* $(\cdot)^{\bot}$ *on types is given by:*

$$\begin{array}{rlrl}
(A \mathbin{⅋}^{\mathsf{o}} B)^{\bot} &= A^{\bot} \otimes^{\mathsf{o}} B^{\bot} & (\bot^{\mathsf{o}})^{\bot} &= \mathbf{1}^{\mathsf{o}} \\
(A \otimes^{\mathsf{o}} B)^{\bot} &= A^{\bot} \mathbin{⅋}^{\mathsf{o}} B^{\bot} & (\mathbf{1}^{\mathsf{o}})^{\bot} &= \bot^{\mathsf{o}} \\
(\&^{\mathsf{o}}\{l_i : A_i\}_{i \in I})^{\bot} &= \oplus^{\mathsf{o}}\{l_i : {A_i}^{\bot}\}_{i \in I} & ?^{\mathsf{o}} A^{\bot} &= !^{\mathsf{o}} A^{\bot} \\
(\oplus^{\mathsf{o}}\{l_i : A_i\}_{i \in I})^{\bot} &= \&^{\mathsf{o}}\{l_i : {A_i}^{\bot}\}_{i \in I} & !^{\mathsf{o}} A^{\bot} &= ?^{\mathsf{o}} A^{\bot}
\end{array}$$

Processes. Let P, Q range over processes, given in Definition 3. Let x, y range over channel endpoints, and $\mathbf{m}, \mathbf{n}$ over channel endpoints of type either $\bot^{\mathsf{o}}$ or $\mathbf{1}^{\mathsf{o}}$.

Definition 3 (Processes). *Processes (P, Q) are given by:*

$$\begin{array}{llll} P,Q ::= x[y].P & (output) & \mathbf{0} & (inaction) \\ \quad x(y).P & (input) & P \mid Q & (composition) \\ \quad x \triangleleft l_j.P & (selection) & (\boldsymbol{\nu} x^A y)P & (sessionrestriction) \\ \quad x \triangleright \{l_i : P_i\}_{i \in I} & (branching) & x[\,].\mathbf{0} & (emptyoutput) \\ \quad x \rightarrow y^A & (forwarding) & x().P & (emptyinput) \end{array}$$

Process $x[y].P$ (respectively, $x(y).P$) outputs (respectively, inputs) y on channel endpoint x, and proceeds as P. Process $x \triangleleft l_j.P$ uses x to select l_j from a labelled choice process, typically being $x \triangleright \{l_i : P_i\}_{i \in I}$, and triggers P_j; labels indexed by the finite set I are pairwise distinct. Process $x \rightarrow y^A$ forwards communications from x to y, the latter having type A. Processes also include the inaction process $\mathbf{0}$, the parallel composition of P and Q, denoted $P \mid Q$, and the double restriction constructor $(\boldsymbol{\nu} x^A y)P$: the intention is that x and y denote dual session channel endpoints in P, and A is the type of x. Processes $x[\,].\mathbf{0}$ and $x().P$ are the empty output and empty input, respectively. They denote the closure of a session from the viewpoint of each of the two communicating participants.

Notions of bound/free names in processes are standard; we write $\mathtt{fn}(P)$ to denote the set of free names of P. Also, we write $P\{x/z\}$ to denote the (capture-avoiding) substitution of x for the free occurrences of z in P. Finally, we let $\tilde{x}$, which is different from x, denote a sequence $x_1, \ldots, x_n$ for $n > 0$.

Typing Rules. Typing contexts, ranged over by Γ, Δ, Θ, are sets of typing assumptions $x : A$. We write Γ, Δ for union, requiring the contexts to be disjoint. A typing judgement $P \vdash \Gamma$ means "process P is well typed using context Γ".

Before presenting the typing rules, we need some auxiliary definitions. Our priorities are based on the annotations used by Kobayashi [32], but simplified to single priorities *à la* Padovani [37]. They obey the following laws:

(i) An action of priority o must be prefixed only by actions of priorities *strictly smaller than* o.
(ii) Communication requires *equal priorities* for the complementary actions.

Definition 4 (Priority). *The* priority *function* $\mathsf{pr}(\cdot)$ *on types is given by:*

$$\begin{array}{ll} \mathsf{pr}(A \mathbin{⅋}^{\mathsf{o}} B) = \mathsf{pr}(A \otimes^{\mathsf{o}} B) = \mathsf{o} & \mathsf{pr}(\bot^{\mathsf{o}}) = \mathsf{pr}(\mathbf{1}^{\mathsf{o}}) = \mathsf{o} \\ \mathsf{pr}(\oplus^{\mathsf{o}}\{l_i : A_i\}_{i \in I}) = \mathsf{pr}(\&^{\mathsf{o}}\{l_i : A_i\}_{i \in I}) = \mathsf{o} & \mathsf{pr}(\,?^{\mathsf{o}} A) = \mathsf{pr}(\,!^{\mathsf{o}} A) = \mathsf{o} \end{array}$$

Definition 5 (Lift). *Let $t \in \mathbb{N}$. The* lift *operator* $\uparrow^t(\cdot)$ *on types is given by:*

$$\begin{array}{ll} \uparrow^t(A \mathbin{⅋}^{\mathsf{o}} B) = (\uparrow^t A) \mathbin{⅋}^{(\mathsf{o}+t)} (\uparrow^t B) & \uparrow^t \bot^{\mathsf{o}} = \mathbf{1}^{(\mathsf{o}+t)} \\ \uparrow^t(A \otimes^{\mathsf{o}} B) = (\uparrow^t A) \otimes^{(\mathsf{o}+t)} (\uparrow^t B) & \uparrow^t \mathbf{1}^{\mathsf{o}} = \bot^{(\mathsf{o}+t)} \\ \uparrow^t(\&^{\mathsf{o}}\{l_i : A_i\}_{i \in I}) = \&^{(\mathsf{o}+t)}\{l_i : \uparrow^t A_i\}_{i \in I} & \uparrow^t(\,?^{\mathsf{o}} A) = \,?^{(\mathsf{o}+t)}(\uparrow^t A) \\ \uparrow^t(\oplus^{\mathsf{o}}\{l_i : A_i\}_{i \in I}) = \oplus^{(\mathsf{o}+t)}\{l_i : \uparrow^t A_i\}_{i \in I} & \uparrow^t(\,!^{\mathsf{o}} A) = \,!^{(\mathsf{o}+t)}(\uparrow^t A) \end{array}$$

We assume $\omega + t = \omega$ for all $t \in \mathbb{N}$.
The operator $\uparrow^t$ is extended component-wise to typing contexts: $\uparrow^t \Gamma$.

$$\frac{}{x \to y^A \vdash x{:}A^\perp, y{:}A}\ \textsc{Ax} \qquad \frac{P \vdash \Gamma \quad Q \vdash \Delta}{P \mid Q \vdash \Gamma, \Delta}\ \textsc{Mix} \qquad \frac{P \vdash \Gamma, x{:}A, y{:}A^\perp}{(\nu x^A y)P \vdash \Gamma}\ \textsc{Cycle}$$

$$\frac{}{\mathbf{0} \vdash \emptyset}\ \emptyset \qquad \frac{}{x[\,].\mathbf{0} \vdash x{:}\mathbf{1}^{\mathsf{o}}}\ \mathbf{1} \qquad \frac{P \vdash \Gamma \quad \mathsf{o} < \mathsf{pr}(\Gamma)}{x().P \vdash x{:}\bot^{\mathsf{o}}, \Gamma}\ \bot$$

$$\frac{P \vdash \Gamma, y{:}A, x{:}B \quad \mathsf{o} < \mathsf{pr}(\Gamma)}{x(y).P \vdash \Gamma, x{:}A \mathbin{⅋}^{\mathsf{o}} B}\ ⅋ \qquad \frac{P \vdash \Gamma, y{:}A, x{:}B \quad \mathsf{o} < \mathsf{pr}(\Gamma)}{x[y].P \vdash \Gamma, x{:}A \otimes^{\mathsf{o}} B}\ \otimes$$

$$\frac{\forall i \in I.(P_i \vdash \Gamma, x{:}A_i) \quad \mathsf{o} < \mathsf{pr}(\Gamma)}{x \triangleright \{l_i : P_i\}_{i \in I} \vdash \Gamma, x{:}\&^{\mathsf{o}}\{l_i : A_i\}_{i \in I}}\ \& \qquad \frac{P \vdash \Gamma, x{:}A_j \quad j \in I \quad \mathsf{o} < \mathsf{pr}(\Gamma)}{x \triangleleft l_j.P \vdash \Gamma, x{:}\oplus^{\mathsf{o}}\{l_i : A_i\}_{i \in I}}\ \oplus$$

$$\frac{P \vdash ?\Gamma, y{:}A \quad \mathsf{o} < \mathsf{pr}(?\Gamma)}{!x(y).P \vdash ?\Gamma, x{:}\,!^{\mathsf{o}} A}\ ! \qquad \frac{P \vdash \Gamma, y{:}A \quad \mathsf{o} < \mathsf{pr}(\Gamma)}{?x[y].P \vdash \Gamma, x{:}\,?^{\mathsf{o}} A}\ ?$$

$$\frac{P \vdash \Gamma}{P \vdash \Gamma, x{:}\,?^{\mathsf{o}} A}\ \mathrm{W} \qquad \frac{P \vdash \Gamma, y{:}\,?^{\kappa} A, z{:}\,?^{\kappa'} A \quad \mathsf{o} \leq \kappa \quad \mathsf{o} \leq \kappa' \quad \mathsf{o} < \mathsf{pr}(\Gamma)}{P\{x/y, x/z\} \vdash \Gamma, x{:}\,?^{\mathsf{o}} A}\ \mathrm{C}$$

Fig. 2. Typing rules for PCP.

The typing rules are given in Fig. 2. Ax states that the forwarding process $x \to y^A$ is well typed if x and y have dual types, respectively $A^\perp$ and A. Mix types the parallel composition of two processes P and Q in the union of their disjoint typing contexts. Cycle is our key typing rule; it states that the restriction process is well typed, if the endpoints x and y have dual types, respectively A and $A^\perp$. By Definition 2, A and $A^\perp$ also have the same priorities, enforcing law (ii) above. In classical logic this rule would be unsound, but in PLL it allows deadlock-free cycles. Rule $\emptyset$ states that inaction is well typed in the empty context. Rules $\mathbf{1}$ and $\bot$ type channel closure actions from the viewpoint of each participant. Rule $⅋$ (respectively $\otimes$) types an input process $x(y).P$ (respectively, output process $x[y].P$), with y bound and x of type $A \mathbin{⅋}^{\mathsf{o}} B$ (respectively, $A \otimes^{\mathsf{o}} B$). The priority o is strictly smaller than any priorities in the continuation process P, enforcing law (i) above. This is captured by $\mathsf{o} < \mathsf{pr}(\Gamma)$ in the premises of both rules, abbreviating "for all $z \in \mathtt{dom}(\Gamma), \mathsf{o} < \mathsf{pr}(\Gamma(z))$". Rules & and $\oplus$ type external and internal choice, respectively, and follow the previous two rules. Rule ! types a server and states that if P communicates along y following protocol A, then $!x(y).P$ communicates along x following protocol $!^{\mathsf{o}} A$. The three remaining rules type different numbers of clients. Rule ? is for a single client: if P communicates along y following A, then $?x[y].P$ communicates along x following $?^{\mathsf{o}} A$. Rule W is for no client: if P does not communicate along any channel following A, then it may be regarded as communicating along x following $?^{\mathsf{o}} A$, for some priority o. Rule C is for multiple clients: if P communicates along y following $?^{\kappa} A$, and z following protocol $?^{\kappa'} A$, then $P\{x/y, x/z\}$ communicates along a single channel x following $?^{\mathsf{o}} A$, where $\mathsf{o} \leqslant \kappa$ and $\mathsf{o} \leqslant \kappa'$. The last two conditions are necessary to deal with some cases in the proof of Cycle-elimination (Theorem 1).

Lifting preserves typability, by an easy induction on typing derivations.

Lemma 1. *If* $P \vdash \Gamma$ *then* $P \vdash {\uparrow^t} \Gamma$.

We will use this result in the form of an admissible rule: $\dfrac{P \vdash \Gamma}{P \vdash {\uparrow^t} \Gamma}\,{\uparrow^t}$

The Design of PCP. We have included Mix and Cycle, which allow derivation of both the standard Cut and the Multicut by Abramsky *et al.* [2].

$$\left.\dfrac{\dfrac{\vdash \Gamma, A_1, \dots, A_n \qquad \vdash \Delta, A_1^\perp, \dots, A_n^\perp}{\vdash \Gamma, \Delta, A_1, \dots, A_n, A_1^\perp, \dots, A_n^\perp}\,\textsc{Mix}}{\vdash \Gamma, \Delta}\,\textsc{Cycle}^n\right\}\textsc{Multicut}$$

Conversely, Mix is the nullary case of Multicut, and Cycle can be derived from Ax and Multicut:

$$\left.\dfrac{\vdash \Gamma, A, A^\perp \qquad \dfrac{}{\vdash A^\perp, A}\,\textsc{Ax}}{\vdash \Gamma}\,\textsc{Multicut}\right\}\textsc{Cycle}$$

Having included Mix, we choose Cycle instead of Multicut, as Cycle is more primitive.

In the presence of Mix and Cycle, there is an isomorphism between $A \otimes B$ and $A \parr B$ in CLL. Both $A \otimes B \multimap A \parr B$ and $A \parr B \multimap A \otimes B$, are derivable, where $C \multimap D \triangleq C^\perp \parr D$ in CLL. Equivalently, both $(A^\perp \parr B^\perp) \parr (A \parr B)$ and $(A^\perp \otimes B^\perp) \parr (A \otimes B)$ are derivable. For simplicity, let $\mathsf{pr}(A) = \mathsf{pr}(B) = \omega$; by duality also $\mathsf{pr}(A^\perp) = \mathsf{pr}(B^\perp) = \omega$.

$$\dfrac{\dfrac{\dfrac{\dfrac{\dfrac{}{\vdash A^\perp, A} \qquad \dfrac{}{\vdash B^\perp, B}}{\vdash A^\perp, B^\perp, A, B}\,\textsc{Mix} \quad \mathsf{o}_1 < \omega}{\vdash A^\perp \parr^{\mathsf{o}_1} B^\perp, A, B}\,\parr \quad \mathsf{o}_2 < \mathsf{o}_1}{\vdash A^\perp \parr^{\mathsf{o}_1} B^\perp, A \parr^{\mathsf{o}_2} B}\,\parr}{\vdash (A^\perp \parr^{\mathsf{o}_1} B^\perp) \parr^{\mathsf{o}} (A \parr^{\mathsf{o}_2} B)}\,\parr$$

$$\dfrac{\dfrac{\dfrac{\dfrac{\dfrac{\dfrac{}{\vdash A^\perp, A} \quad \dfrac{}{\vdash B^\perp, B}}{\vdash A^\perp, B^\perp, A, B}\,\textsc{Mix} \quad \mathsf{o}_1 < \omega}{\vdash A^\perp \otimes^{\mathsf{o}_1} B^\perp, A, B}\,\otimes \qquad \dfrac{\dfrac{\dfrac{}{\vdash A^\perp, A} \quad \dfrac{}{\vdash B^\perp, B}}{\vdash A^\perp, B^\perp, A, B}\,\textsc{Mix} \quad \mathsf{o}_2 < \omega}{\vdash A^\perp, B^\perp, A \otimes^{\mathsf{o}_2} B}\,\otimes}{\vdash A^\perp \otimes^{\mathsf{o}_1} B^\perp, A \otimes^{\mathsf{o}_2} B, A^\perp, A, B^\perp, B}\,\textsc{Mix}}{\vdash A^\perp \otimes^{\mathsf{o}_1} B^\perp, A \otimes^{\mathsf{o}_2} B}\,\textsc{Cycle}^2}{\vdash (A^\perp \otimes^{\mathsf{o}_1} B^\perp) \parr^{\mathsf{o}} (A \otimes^{\mathsf{o}_2} B)}\,\parr$$

The above derivations *without* priorities show the isomorphism between $A \otimes B$ and $A \parr B$ in CLL, which does not hold in our PLL, in particular as $\mathsf{o}_1 \neq \mathsf{o}_2$. The distinction between $\otimes$ and $\parr$, preserves the distinction between output and input in the term assignment. However, to simplify derivations, both typing rules (Fig. 2) have the same form. The usual tensor rule, where there are two separate derivations in the premise rather than just one, is derivable by using Mix.

Our type system performs priority-checking. Priorities can be inferred, as in Kobayashi's type system [32] and the tool TyPiCal [28]. We have opted for priority checking over priority inference, as the presentation is more elegant.

The following two examples illustrate the use of priorities. We first establish the structure of the typing derivation, then calculate the priorities. We conclude the section by showing the typing for the cyclic scheduler from Sect. 1.

Example 2 (Cyclic process: deadlock-free). Consider the following process

$$P \triangleq (\boldsymbol{\nu} x_1 y_1)(\boldsymbol{\nu} x_2 y_2)\big[x_1(v).x_2(w).R \mid y_1[\mathbf{n}].y_2[\mathbf{n'}].Q\big]$$

where $R \triangleq x_1().v().x_2().w().\mathbf{0}$ and $Q \triangleq y_1[\,].\mathbf{0} \mid \mathbf{n}[\,].\mathbf{0} \mid y_2[\,].\mathbf{0} \mid \mathbf{n'}[\,].\mathbf{0}$. First, we show the typing derivation for the left-hand side of the parallel, $x_1(v).x_2(w).R$:

$$\dfrac{\dfrac{\dfrac{\dfrac{}{\mathbf{0} \vdash \emptyset}\,\emptyset \qquad \kappa_4 < \kappa_3 < \kappa_2 < \kappa_1}{R \vdash x_1 : \bot^{\kappa_4}\; v : \bot^{\kappa_3}, x_2 : \bot^{\kappa_2}, w : \bot^{\kappa_1} \qquad \mathsf{o}_1 < \kappa_4}\,\bot^4}{x_2(w).R \vdash x_1 : \bot^{\kappa_4}, v : \bot^{\kappa_3}, x_2 : \bot^{\kappa_1} \parr^{\mathsf{o}_1} \bot^{\kappa_2} \qquad \mathsf{o}_2 < \mathsf{o}_1}\,\parr}{x_1(v).x_2(w).R \vdash x_2 : \bot^{\kappa_1} \parr^{\mathsf{o}_1} \bot^{\kappa_2}, x_1 : \bot^{\kappa_3} \parr^{\mathsf{o}_2} \bot^{\kappa_4}}\,\parr \tag{1}$$

Now, the typing derivation for the right-hand side of the parallel, $y_1[\mathbf{n}].y_2[\mathbf{n'}].Q$, and recall that $\kappa_4 < \kappa_3 < \kappa_2 < \kappa_1$:

$$\dfrac{\dfrac{\dfrac{\dfrac{}{y_1[\,].\mathbf{0} \vdash y_1 : \mathbf{1}^{\kappa_4}}\,\mathbf{1} \quad \dfrac{}{\mathbf{n}[\,].\mathbf{0} \vdash \mathbf{n} : \mathbf{1}^{\kappa_3}}\,\mathbf{1} \quad \dfrac{}{y_2[\,].\mathbf{0} \vdash y_1 : \mathbf{1}^{\kappa_2}}\,\mathbf{1} \quad \dfrac{}{\mathbf{n'}[\,].\mathbf{0} \vdash \mathbf{n'} : \mathbf{1}^{\kappa_1}}\,\mathbf{1}}{y_1[\,].\mathbf{0} \mid \mathbf{n}[\,].\mathbf{0} \mid y_2[\,].\mathbf{0} \mid \mathbf{n'}[\,].\mathbf{0} \vdash y_1 : \mathbf{1}^{\kappa_4}, \mathbf{n} : \mathbf{1}^{\kappa_3}, y_2 : \mathbf{1}^{\kappa_2}, \mathbf{n'} : \mathbf{1}^{\kappa_1} \qquad \mathsf{o}_3 < \kappa_4}\,\textsc{Mix}^3}{y_2[\mathbf{n'}].Q \vdash y_1 : \mathbf{1}^{\kappa_4}, \mathbf{n} : \mathbf{1}^{\kappa_3}, y_2 : \mathbf{1}^{\kappa_1} \otimes^{\mathsf{o}_3} \mathbf{1}^{\kappa_2} \qquad \mathsf{o}_4 < \mathsf{o}_3}\,\otimes}{y_1[\mathbf{n}].y_2[\mathbf{n'}].Q \vdash y_2 : \mathbf{1}^{\kappa_1} \otimes^{\mathsf{o}_3} \mathbf{1}^{\kappa_2}, y_1 : \mathbf{1}^{\kappa_3} \otimes^{\mathsf{o}_4} \mathbf{1}^{\kappa_4}}\,\otimes \tag{2}$$

Finally, the typing derivation for process P is as follows:

$$\dfrac{\dfrac{\dfrac{(1) \qquad\qquad (2)}{\begin{array}{l} x_1(v).x_2(w).R \mid y_1[\mathbf{n}].y_2[\mathbf{n'}].Q \vdash \\ \quad x_2 : \bot^{\kappa_1} \parr^{\mathsf{o}_1} \bot^{\kappa_2}, x_1 : \bot^{\kappa_3} \parr^{\mathsf{o}_2} \bot^{\kappa_4}, y_2 : \mathbf{1}^{\kappa_1} \otimes^{\mathsf{o}_3} \mathbf{1}^{\kappa_2}, y_1 : \mathbf{1}^{\kappa_3} \otimes^{\mathsf{o}_4} \mathbf{1}^{\kappa_4} \\ \qquad\qquad \mathsf{o}_1 = \mathsf{o}_3 \end{array}}\,\textsc{Mix}}{\begin{array}{l} (\boldsymbol{\nu} x_2 y_2)\big[x_1(v).x_2(w).R \mid y_1[\mathbf{n}].y_2[\mathbf{n'}].Q\big] \vdash \\ \quad x_1 : \bot^{\kappa_3} \parr^{\mathsf{o}_2} \bot^{\kappa_4}, y_1 : \mathbf{1}^{\kappa_3} \otimes^{\mathsf{o}_4} \mathbf{1}^{\kappa_4} \qquad \mathsf{o}_2 = \mathsf{o}_4 \end{array}}\,\textsc{Cycle}}{(\boldsymbol{\nu} x_1 y_1)(\boldsymbol{\nu} x_2 y_2)\big[x_1(v).x_2(w).R \mid y_1[\mathbf{n}].y_2[\mathbf{n'}].Q\big] \vdash \emptyset}\,\textsc{Cycle}$$

The system of equations

$$\mathsf{o}_2 < \mathsf{o}_1 \qquad \mathsf{o}_4 < \mathsf{o}_3 \qquad \mathsf{o}_1 = \mathsf{o}_3 \qquad \mathsf{o}_2 = \mathsf{o}_4$$

can be solved by the assignment $\mathsf{o}_1 = \mathsf{o}_3 = 1$ and $\mathsf{o}_2 = \mathsf{o}_4 = 0$.

Example 3 (Cyclic process: deadlocked!). Now consider the process

$$P' = (\boldsymbol{\nu} x_1 y_1)(\boldsymbol{\nu} x_2 y_2)\big[x_1(v).x_2(w).R \mid y_2[\mathbf{n'}].y_1[\mathbf{n}].Q\big]$$

where $R = x_1().v().x_2().w().\mathbf{0}$ and $Q = y_1[].\mathbf{0} \mid \mathbf{n}[].\mathbf{0} \mid y_2[].\mathbf{0} \mid \mathbf{n}'[].\mathbf{0}$. Notice that the order of actions on channels y_1 and y_2 is now swapped, thus causing a deadlock! If we tried to construct a typing derivation for process P', we would have for the right-hand side of the parallel the following:

$$\dfrac{\dfrac{\dfrac{\dfrac{}{y_1[].\mathbf{0} \vdash y_1 : \mathbf{1}^{\kappa_4}}\,\mathbf{1} \quad \dfrac{}{\mathbf{n}[].\mathbf{0} \vdash \mathbf{n} : \mathbf{1}^{\kappa_3}}\,\mathbf{1} \quad \dfrac{}{y_2[].\mathbf{0} \vdash y_1 : \mathbf{1}^{\kappa_2}}\,\mathbf{1} \quad \dfrac{}{\mathbf{n}'[].\mathbf{0} \vdash \mathbf{n}' : \mathbf{1}^{\kappa_1}}\,\mathbf{1}}{y_1[].\mathbf{0} \mid \mathbf{n}[].\mathbf{0} \mid y_2[].\mathbf{0} \mid \mathbf{n}'[].\mathbf{0} \vdash y_1 : \mathbf{1}^{\kappa_4}, \mathbf{n} : \mathbf{1}^{\kappa_3}, y_2 : \mathbf{1}^{\kappa_2}, \mathbf{n}' : \mathbf{1}^{\kappa_1}}\,\textsc{Mix}^3 \quad \mathsf{o}_4 < \kappa_4}{y_1[\mathbf{n}].Q \vdash \mathbf{n}' : \mathbf{1}^{\kappa_1}, y_2 : \mathbf{1}^{\kappa_2}, y_1 : \mathbf{1}^{\kappa_3} \otimes^{\mathsf{o}_4} \mathbf{1}^{\kappa_4}}\,\otimes \quad \mathsf{o}_3 < \mathsf{o}_4}{y_2[\mathbf{n}'].y_1[\mathbf{n}].Q \vdash y_1 : \mathbf{1}^{\kappa_3} \otimes^{\mathsf{o}_4} \mathbf{1}^{\kappa_4}, y_2 : \mathbf{1}^{\kappa_1} \otimes^{\mathsf{o}_3} \mathbf{1}^{\kappa_2}}\,\otimes$$

Then, the system of equations

$$\mathsf{o}_2 < \mathsf{o}_1 \qquad \mathsf{o}_3 < \mathsf{o}_4 \qquad \mathsf{o}_1 = \mathsf{o}_3 \qquad \mathsf{o}_2 = \mathsf{o}_4$$

has no solution because it requires $\mathsf{o}_2 < \mathsf{o}_3$ and $\mathsf{o}_3 < \mathsf{o}_2$, which is impossible.

Example 1 continued (Cyclic Scheduler)

$$\begin{array}{ll} \mathit{Sched} \triangleq ...(\boldsymbol{\nu} a_i b_i)...(\boldsymbol{\nu} c_i d_{(i+1) \bmod n})\big(A_0 \mid A_1 \mid ... \mid A_{n-1} \mid P_0 \mid P_1 \mid ... \mid P_{n-1}\big) & \\ \quad A_0 \triangleq c_0[\mathbf{n}_0].d_0(x_0).a_0[\mathbf{m}_0].\mathsf{close}_0 & \\ \quad A_i \triangleq d_i(x_i).a_i[\mathbf{m}_i].c_i[\mathbf{n}_i].\mathsf{close}_i & i \in \{1, ..., n-1\} \\ \quad P_i \triangleq b_i(y_i).Q_i & i \in \{0, ..., n-1\} \end{array}$$

By applying the typing rules in Fig. 2 we can derive $\mathit{Sched} \vdash \emptyset$, since it is a closed process, and assign the following types and priorities:

$$\begin{array}{llll} c_0 : \mathbf{1} \otimes^0 \mathbf{1} & d_0 : \bot \parr^{2(n-1)} \bot & a_0 : \mathbf{1} \otimes^{2(n-1)+1} \mathbf{1} & \text{for } A_0 \\ d_i : \bot \parr^{2i-2} \bot & a_i : \mathbf{1} \otimes^{2i-1} \mathbf{1} & c_i : \mathbf{1} \otimes^{2i} \mathbf{1} & \text{for } A_i,\ 0 < i < n \\ b_0 : \bot \parr^{2(n-1)+1} \bot & b_i : \bot \parr^{2i-1} \bot & & \text{for } P_0 \text{ and } P_i,\ 0 < i < n \end{array}$$

The priorities of types $\bot$ and $\mathbf{1}$ could be easily assigned as Example 2. As the priority of d_{i+1} is $2(i+1) - 2 = 2i$, we can connect it to a_i with a CYCLE.

3 Operational Semantics of PCP

In this section we define structural equivalence, the principal β-reduction rules and commuting conversions. The detailed derivations can be found in [18].

We define structural equivalence to be the smallest congruence relation satisfying the following axioms. SC-AX-SWP allows swapping channels in the forwarding process. SC-AX-CYCLE states that cycle applied to a forwarding process is equivalent to inaction. This allows elimination of unnecessary cycles. Axioms SC-MIX-NIL, SC-MIX-COMM and SC-MIX-ASC state that parallel composition uses the inaction as the neutral element and is commutative and associative. SC-CYCLE-EXT is the standard scope extrusion rule. SC-CYCLE-SWP allows swapping channels and SC-CYCLE-COMM states the commutativity of restriction[1].

[1] Note that associativity of restriction is derived from SC-MIX-COMM and SC-CYCLE-COMM.

SC-Ax-Swp	$x \to y^{A} \vdash x{:}A^{\perp}, y{:}A \;\equiv\; y \to x^{A^{\perp}} \vdash x{:}A^{\perp}, y{:}A$
SC-Ax-Cycle	$(\boldsymbol{\nu} x^{A^{\perp}} y)\, x \to y^{A} \vdash \emptyset \;\equiv\; \mathbf{0} \vdash \emptyset$
SC-Mix-Nil	$\mathbf{0} \mid P \vdash \Gamma \;\equiv\; P \vdash \Gamma$
SC-Mix-Comm	$P \mid Q \vdash \Gamma, \Delta \;\equiv\; Q \mid P \vdash \Gamma, \Delta$
SC-Mix-Asc	$P \mid (Q \mid R) \vdash \Gamma, \Delta, \Theta \;\equiv\; (P \mid Q) \mid R \vdash \Gamma, \Delta, \Theta$
SC-Cycle-Ext	$(\boldsymbol{\nu} x^{A} y)(P \mid Q) \vdash \Gamma, \Delta \;\equiv\; P \mid (\boldsymbol{\nu} x^{A} y) Q \vdash \Gamma, \Delta \quad x, y \notin \mathtt{fn}(P)$
SC-Cycle-Swp	$(\boldsymbol{\nu} x^{A} y) P \vdash \Gamma \;\equiv\; (\boldsymbol{\nu} y^{A^{\perp}} x) P \vdash \Gamma$
SC-Cycle-Comm	$(\boldsymbol{\nu} x^{A} y)(\boldsymbol{\nu} z^{B} w) P \vdash \Gamma \;\equiv\; (\boldsymbol{\nu} z^{B} w)(\boldsymbol{\nu} x^{A} y) P \vdash \Gamma$

The core of the operational semantics consists of β-reductions. In π-calculus terms these are communication steps; in logical terms they are Cycle-elimination steps. $\beta_{\otimes\,\invamp}$ is given in Fig. 3 to illustrate priorities. It simplifies a cycle connecting x of type $A \otimes^{\mathsf{o}} B$ and y of type $A \invamp^{\mathsf{o}} B$, which corresponds to communication between an output on x and an input on y, respectively. Both actions have priority o, which is strictly smaller than any priorities in their typing contexts, respecting the fact that they are top-level prefixes. The remaining β-reductions are summarised below. β_{AxCycle} simplifies a Cycle involving an axiom. $\beta_{\mathbf{1}\perp}$ closes and eliminates channels. $\beta_{\oplus\&}$, similarly to $\beta_{\otimes\,\invamp}$, simplifies a communication between a selection and a branching. $\beta_{!?}$ simplifies a cycle between one server of type $!^{\mathsf{o}} A$ and one client of type $?^{\mathsf{o}} A$. The last two rules differ in the number of clients involved: rule $\beta_{!\mathrm{W}}$ considers no clients, whether $\beta_{!\mathrm{C}}$ considers multiple clients.

$$
\begin{array}{rl}
\beta_{\text{AxCycle}} & (\boldsymbol{\nu} y^{A} z)(x \to y^{A} \mid P) \vdash \Gamma, x{:}A^{\perp} \;\longrightarrow\; P\{x/z\} \vdash \Gamma, x{:}A^{\perp} \\
\beta_{\mathbf{1}\perp} & (\boldsymbol{\nu} x^{A} y)(x[\,].\mathbf{0} \mid y().P) \vdash \Gamma \;\longrightarrow\; P \vdash \Gamma \\
\beta_{\oplus\&} & (\boldsymbol{\nu} x^{\oplus^{\mathsf{o}}\{l_i:B_i\}_{i\in I}} y)\big(x \triangleleft l_j.P \mid y \triangleright \{l_i : Q_i\}_{i\in I}\big) \vdash \Gamma, \Delta \longrightarrow \\
 & \qquad\qquad (\boldsymbol{\nu} x^{B_j} y)\big(P \mid Q_j\big) \vdash \Gamma, \Delta \\
\beta_{!?} & (\boldsymbol{\nu} x^{!^{\mathsf{o}} A} y)\big(!x(v).P \mid ?y[w].Q\big) \vdash ?\Gamma, \Delta \;\longrightarrow\; (\boldsymbol{\nu} v^{A} w)\big(P \mid Q\big) \vdash ?\Gamma, \Delta \\
\beta_{!\mathrm{W}} & (\boldsymbol{\nu} x^{!^{\mathsf{o}} A} y)\big(!x(v).P \mid Q\big) \vdash ?\Gamma, \Delta \;\longrightarrow\; Q \vdash ?\Gamma, \Delta \\
\beta_{!\mathrm{C}} & (\boldsymbol{\nu} x^{!^{\mathsf{o}} A} y)\big(!x(v).P \mid Q\{y/y', y/y''\}\big) \vdash ?\Gamma, \Delta \;\longrightarrow \\
 & (\boldsymbol{\nu} x'^{\,!^{\mathsf{o}} A} y')(!x'(v').P' \mid (\boldsymbol{\nu} x''^{\,!^{\mathsf{o}} A} y'')(!x''(v'').P'' \mid Q)) \vdash ?\Gamma, \Delta
\end{array}
$$

$$
\dfrac{\dfrac{\dfrac{\dfrac{\mathsf{o} < \mathsf{pr}(\Gamma) \qquad P \vdash \Gamma, v{:}A, x{:}B}{x[v].P \vdash \Gamma, x{:}A \otimes^{\mathsf{o}} B}\,\otimes \qquad \dfrac{\mathsf{o} < \mathsf{pr}(\Delta) \qquad Q \vdash \Delta, w{:}A^{\perp}, y{:}B^{\perp}}{y(w).Q \vdash \Delta, y{:}A^{\perp} \invamp^{\mathsf{o}} B^{\perp}}\,\invamp}{x[v].P \mid y(w).Q \vdash \Gamma, \Delta, x{:}A \otimes^{\mathsf{o}} B, y{:}A^{\perp} \invamp^{\mathsf{o}} B^{\perp}}\,\text{Mix}}{(\boldsymbol{\nu} x^{A \otimes^{\mathsf{o}} B} y)\big(x[v].P \mid y(w).Q\big) \vdash \Gamma, \Delta}\,\text{Cycle}}{}
$$

$$
\longrightarrow\quad \dfrac{\dfrac{P \vdash \Gamma, v{:}A, x{:}B \qquad Q \vdash \Delta, w{:}A^{\perp}, y{:}B^{\perp}}{P \mid Q \vdash \Gamma, \Delta, v{:}A, x : B, w{:}A^{\perp}, y : B^{\perp}}\,\text{Mix}}{(\boldsymbol{\nu} v^{A} w)(\boldsymbol{\nu} x^{B} y)(P \mid Q) \vdash \Gamma, \Delta}\,\text{Cycle}^{2}
$$

Fig. 3. β-reduction for $\otimes$ and $\invamp$.

Commuting conversions, following [12,41], allow communication prefixes to be moved to the conclusion of a typing derivation, corresponding to pulling them out of the scope of Cycle rules. In order to account for the sequence of Cycles, here we use $\tilde{\cdot}$. Due to this movement, if a prefix on a channel endpoint x with priority o is pulled out at top level, then to preserve priority conditions in the typing rules in Fig. 2, it is necessary to increase priorities of all actions after the prefix on x. This increase is achieved by using $\uparrow^{o+1}(\cdot)$ in the typing contexts.

$$
\begin{array}{ll}
\kappa_\bot & (\boldsymbol{\nu}\tilde{x}^{\tilde{A}}\tilde{y})(x().P \mid Q) \vdash \Gamma, \Delta, x\colon \bot^{o} \longrightarrow \\
& \qquad x().[(\boldsymbol{\nu}\tilde{x}^{\tilde{A}}\tilde{y})(P \mid Q)] \vdash \uparrow^{o+1}\Gamma, \uparrow^{o+1}\Delta, x\colon \bot^{o} \\
\kappa_\otimes & (\boldsymbol{\nu}\tilde{x}^{\tilde{A}}\tilde{y})(x[v].P \mid Q) \vdash \Gamma, \Delta, x\colon A \otimes^{o} B \longrightarrow \\
& x[v].\big[(\boldsymbol{\nu}\tilde{x}^{\tilde{A}}\tilde{y})(P \mid Q)\big] \vdash (\uparrow^{o+1}\Gamma), (\uparrow^{o+1}\Delta), x\colon(\uparrow^{o+1}A) \otimes^{o} (\uparrow^{o+1}B) \\
\kappa_\parr & (\boldsymbol{\nu}\tilde{x}^{\tilde{A}}\tilde{y})(x(w).P \mid Q) \vdash \Gamma, \Delta, x\colon A \parr^{o} B \longrightarrow \\
& x(w).\big[(\boldsymbol{\nu}\tilde{x}^{\tilde{A}}\tilde{y})(P \mid Q)\big] \vdash (\uparrow^{o+1}\Gamma), (\uparrow^{o+1}\Delta), x\colon(\uparrow^{o+1}A) \parr^{o} (\uparrow^{o+1}B) \\
\kappa_\oplus & (\boldsymbol{\nu}\tilde{x}^{\tilde{A}}\tilde{y})(x \triangleleft l_j.P \mid Q) \vdash \Gamma, \Delta, x\colon \oplus^{o}\{l_i : B_i\}_{i\in I} \longrightarrow \\
& x \triangleleft l_j.[(\boldsymbol{\nu}\tilde{x}^{\tilde{A}}\tilde{y})(P \mid Q)] \vdash (\uparrow^{o+1}\Gamma), (\uparrow^{o+1}\Delta), x\colon \oplus^{o}\{l_i : \uparrow^{o+1}B_i\}_{i\in I} \\
\kappa_\& & (\boldsymbol{\nu}\tilde{x}^{\tilde{A}}\tilde{y})(x \triangleright \{l_i : P_i\}_{i\in I} \mid Q) \vdash \Gamma, \Delta, x\colon \&^{o}\{l_i : B_i\}_{i\in I} \longrightarrow \\
& x \triangleright \{l_i : (\boldsymbol{\nu}\tilde{x}^{\tilde{A}}\tilde{y})(P_i \mid Q)\}_{i\in I} \vdash (\uparrow^{o+1}\Gamma), (\uparrow^{o+1}\Delta), x\colon \&^{o}\{l_i : \uparrow^{o+1}B_i\}_{i\in I} \\
\kappa_? & (\boldsymbol{\nu}\tilde{x}^{\tilde{A}}\tilde{y})(?x[w].P \mid Q) \vdash \Gamma, \Delta, x\colon ?^{o} A \longrightarrow \\
& \qquad ?x[w].\big[(\boldsymbol{\nu}\tilde{x}^{\tilde{A}}\tilde{y})(P \mid Q)\big] \vdash (\uparrow^{o+1}\Gamma), (\uparrow^{o+1}\Delta), x\colon ?^{o}(\uparrow^{o+1}A) \\
\kappa_! & (\boldsymbol{\nu}\tilde{x}^{\widetilde{?^{o}A}}\tilde{y})(!x(v).P \mid Q) \vdash ?\Gamma, \Delta, x\colon !^{o} A \longrightarrow \\
& \qquad !x(v).\big[(\boldsymbol{\nu}\tilde{x}^{\widetilde{?^{o}A}}\tilde{y})(P \mid Q)\big] \vdash (\uparrow^{o+1}\Gamma), (\uparrow^{o+1}\Delta), x\colon !^{o}(\uparrow^{o+1}A)
\end{array}
$$

Finally, we give the following additional reduction rules: closure under structural equivalence, and two congruence rules, for restriction and for parallel.

$$
\begin{array}{ll}
\text{Close-Equiv} & P \equiv Q \quad Q \longrightarrow R \quad R \equiv S \text{ implies } P \longrightarrow S \\
\text{Cong-Cycle} & P \longrightarrow Q \text{ implies } (\boldsymbol{\nu}x^{A}y)P \longrightarrow (\boldsymbol{\nu}x^{A}y)Q \\
\text{Cong-Mix} & P \longrightarrow Q \text{ implies } P \mid R \longrightarrow Q \mid R
\end{array}
$$

4 Results for PLL and PCP

4.1 Cycle-Elimination for PLL

We start with results for Cycle-elimination for PLL; thus here we refer to A, B as propositions, rather than types. The detailed proofs can be found in [18].

Definition 6. *The* degree *function* $\partial(\cdot)$ *on propositions is defined by:*

- $\partial(\mathbf{1}^{o}) = \partial(\bot^{o}) = 1$
- $\partial(A \otimes^{o} B) = \partial(A \parr^{o} B) = \partial(A) + \partial(B) + 1$
- $\partial(\&^{o}\{l_i : A_i\}_{i\in I}) = \partial(\oplus^{o}\{l_i : A_i\}_{i\in I}) = \sum_{i\in I}\{\partial(A_i)\} + 1$
- $\partial(?^{o} A) = \partial(!^{o} A) = \partial(A) + 1$.

Definition 7. *A* Maxicut *is a maximal sequence of* Mix *and* Cycle *rules, ending with a* Cycle *rule.*

Maximality means that the rules applied immediately before a MAXICUT are any rules in Fig. 2, other than MIX or CYCLE. The order in which MIX and CYCLE rules are applied within a MAXICUT is irrelevant. However, Proposition 1, which follows directly from structural equivalence (Sect. 3), allows us to simplify a MAXICUT.

Proposition 1 (Canonical MAXICUT). *Given an arbitrary* MAXICUT, *it is always possible to obtain from it a* canonical MAXICUT *consisting of a sequence of only* MIX *rules followed by a sequence of only* CYCLE *rules.*

Definition 8. *A* single-MIX MAXICUT *contains only one* MIX *rule.*
$A_1, \ldots, A_n, A$ *are* MAXICUT propositions *if they are eliminated by a* MAXICUT.
The degree of a sequence of CYCLEs *is the sum of the degrees of the eliminated propositions.*
The degree of a MAXICUT *is the sum of the degrees of the* CYCLEs *in it.*
The degree of a proof π, $d(\pi)$, *is the* sup *of the degrees of its* MAXICUTs, *implying* $d(\pi) = 0$ *if and only if proof* π *has no* CYCLEs.
The height of a proof π, $h(\pi)$, *is the height of its tree, and it is defined as* $h(\pi) = \sup(h(\pi_i))_{i \in I} + 1$, *where* $\{\pi_i\}_{i \in I}$ *are the subproofs of* π.

MAXICUT has some similarities with the derived MULTICUT: it generalises MULTICUT in the number of MIXes, and a single-MIX MAXICUT is an occurrence of MULTICUT.

The core of CYCLE-elimination for our PLL, as for CUT-elimination for CLL [10,25], is the Principal Lemma (Lemma 3), which eliminates a CYCLE by either (i) replacing it with another CYCLE on simpler propositions, or (ii) pushing it further up the proof tree. Item (i) corresponds to (the logical part of) β-reductions (Sect. 3); and (ii) corresponds to (the logical part of) commuting conversions (Sect. 3).

Exceptionally, $\beta_{!C}$ reduces the original proof in a way that neither (i) nor (ii) are respected. In order to cope with this case, we introduce Lemma 2, which is inspired by Lemma B.1.3 in Bräuner [10], and adapted to our PLL. Lemma 2 allows us to reduce the degree of a proof ending with a single-MIX MAXICUT and having the same degree as the whole proof, and where the last rule applied on the left hand-side immediate subproof is !. Let $[n]$ denote the set $\{1, \ldots, n\}$.

Lemma 2 (Inspired by B.1.3 in Bräuner [10]). *Let* τ *be a proof of the following form, ending with a single-*MIX MAXICUT*:*

$$
\dfrac{\dfrac{\dfrac{\pi}{\vdots}\quad \mathsf{o} < \mathsf{pr}(?\Gamma) \quad \forall i \in [n] : \mathsf{o} < \mathsf{o}_i \quad \vdash ?\Gamma,\ ?^{\mathsf{o}_1} A_1, \ldots,\ ?^{\mathsf{o}_n} A_n, A}{\vdash ?\Gamma,\ ?^{\mathsf{o}_1} A_1, \ldots,\ ?^{\mathsf{o}_n} A_n,\ !^{\mathsf{o}} A}\ ! \qquad \dfrac{\dfrac{\pi'}{\vdots}\quad \mathsf{o} < \mathsf{pr}(\Delta) \quad \forall i \in [n] : \mathsf{o} < \mathsf{o}_i \quad \forall j \in [k] : \mathsf{o} \leqslant \kappa_j \quad \vdash \Delta,\ !^{\mathsf{o}_1} A_1^{\perp}, \ldots,\ !^{\mathsf{o}_n} A_n^{\perp}, (?^{\kappa_j} A^{\perp})_{j \in [k]}}{\vdash \Delta,\ !^{\mathsf{o}_1} A_1^{\perp}, \ldots,\ !^{\mathsf{o}_n} A_n^{\perp},\ ?^{\mathsf{o}} A^{\perp}}\ \mathrm{C}^{k-1}}{\dfrac{\vdash ?\Gamma, \Delta,\ ?^{\mathsf{o}_1} A_1, \ldots,\ ?^{\mathsf{o}_n} A_n,\ !^{\mathsf{o}} A,\ !^{\mathsf{o}_1} A_1^{\perp}, \ldots,\ !^{\mathsf{o}_n} A_n^{\perp},\ ?^{\mathsf{o}} A^{\perp}}{\vdash ?\Gamma, \Delta}\ \textsc{Cycle}}\ \textsc{Mix}
$$

where $d(\pi) < d(\tau)$ and $d(\pi') < d(\tau)$. Then, there is a proof τ' of $\vdash ?\Gamma, \Delta$ such that $d(\tau') < d(\tau)$.

Proof. Induction on $h(\pi')$, with a case-analysis on the last rule applied in π'. □

Lemma 3 (The Principal Lemma). *Let τ be a proof of $\vdash \Gamma$, ending with a canonical* MAXICUT*:*

$$\cfrac{\cfrac{\pi_1 \dots \pi_m}{\vdash \Gamma, A_1, ..., A_n, A, A_1^\perp, ..., A_n^\perp, A^\perp}\ \textsc{Mix}}{\vdash \Gamma}\ \textsc{Cycle}$$

such that for all $i \in [m]$, $d(\pi_i) < d(\tau)$. Then there is a proof τ' of $\vdash \uparrow^t \Gamma$, for some $t \geqslant 0$, such that $d(\tau') < d(\tau)$.

Proof. The proof is by induction on $\sum_{i\in[m]} h(\pi_i)$. Let r_i be the last rule applied in π_i, for $i \in [m]$ and let C_{r_i} be the proposition introduced by r_i. Consider the proposition with the *smallest* priority. If the proposition is not unique, just pick one. Let this proposition be C_{r_k}. Then, π_k is the following proof: $\cfrac{\cdots}{\vdash \Gamma', C_{r_k}}\ r_k$ We proceed by cases on π_k.

– r_k is $\otimes$ on one of the MAXICUT propositions $A_1, \dots, A_n, A$. Without loss of generality, suppose r_k is applied on A, meaning $A = E \otimes^{\mathsf{o}} F$ for some E and F and $\mathsf{o} \geqslant 0$. By $\otimes$ rule in Fig. 2, $\mathsf{o} < \mathsf{pr}(\Gamma')$. Since A is a MAXICUT proposition, by Definition 2, $A^\perp = E^\perp \parr^{\mathsf{o}} F^\perp$. Since $\mathsf{o} < \mathsf{pr}(\Gamma')$ and $\mathsf{pr}(A^\perp) = \mathsf{o}$, it must be that $A^\perp$ is in another proof, say π_h: $\cfrac{\cdots}{\vdash \Gamma'', E^\perp \parr^{\mathsf{o}} F^\perp}\ r_h$

Consider the case where r_h is a multiplicative, additive, exponential or $\perp$ rule in Fig. 2. Suppose r_h is applied on C_{r_h} which is not $A^\perp$. All the mentioned rules require $\mathsf{pr}(C_{r_h}) < \mathsf{pr}(\Gamma'', E^\perp \parr^{\mathsf{o}} F^\perp \setminus C_{r_h})$, implying $\mathsf{pr}(C_{r_h}) < \mathsf{pr}(E^\perp \parr^{\mathsf{o}} F^\perp) = \mathsf{pr}(E \otimes^{\mathsf{o}} F) = \mathsf{o}$. This contradicts the fact that o is the smallest priority. Hence, r_h must be a $\parr$ introducing $A^\perp$.

We construct proof τ_A ending with a single-MIX MAXICUT applied on *at least* A:

$$\cfrac{\cfrac{\cfrac{\cfrac{\pi_\otimes}{\vdots}}{\vdash \Gamma', E, F \quad \mathsf{o} < \mathsf{pr}(\Gamma')}}{\vdash \Gamma', E \otimes^{\mathsf{o}} F}\ \otimes \quad \cfrac{\cfrac{\cfrac{\pi_\parr}{\vdots}}{\vdash \Gamma'', E^\perp, F^\perp \quad \mathsf{o} < \mathsf{pr}(\Gamma'')}}{\vdash \Gamma'', E^\perp \parr^{\mathsf{o}} F^\perp}\ \parr}{\cfrac{\vdash \Gamma', \Gamma'', E \otimes^{\mathsf{o}} F, E^\perp \parr^{\mathsf{o}} F^\perp}{\vdash \Gamma'''}\ \textsc{Cycle}}\ \textsc{Mix}$$

Then, by structural equivalence, we can rewrite τ in terms of τ_A. By applying $\beta_{\otimes\parr}$ on τ_A (only considering the logical part), we obtain a proof τ'_A such that $d(\tau'_A) < d(\tau_A) \leq d(\tau)$, because $\partial(E) + \partial(F) < \partial(E \otimes^{\mathsf{o}} F)$. We can then construct τ' by substituting τ'_A for τ_A in τ, which concludes this case.

– r_k is ! on one of the MAXICUT propositions $A_1, \ldots, A_n, A$. Without loss of generality, suppose r_k introduces A, implying that $A = !^{o} A'$ for some A' and $o \geqslant 0$. Then π_k is the following proof:

$$\dfrac{\begin{array}{c}\pi_! \\ \vdots \\ \vdash ?\Theta, A' \quad o < \mathsf{pr}(?\Theta)\end{array}}{\vdash ?\Theta, !^{o} A'}\ !$$

where $\Gamma' = ?\Theta$. Since A is a MAXICUT proposition, by duality $A^{\perp} = ?^{o} A'^{\perp}$. Since $o < \mathsf{pr}(\Gamma')$ and $\mathsf{pr}(A^{\perp}) = o$, it must be that $A^{\perp}$ is in another proof. Let it be π_h for $h \in [m]$ and $h \neq k$. Then we apply Lemma 2 to π_k and π_h, obtaining a proof which we use to construct τ', as we did in the previous case. □

Lemma 4. *Given a proof τ of $\vdash \Gamma$, such that $d(\tau) > 0$, then for some $t \geqslant 0$ there is a proof τ' of $\vdash \uparrow^t \Gamma$ such that $d(\tau') < d(\tau)$.*

Proof. By induction on $h(\tau)$. We have the following cases.

– If τ ends in a MAXICUT whose degree is *the same as* the degree of τ:

$$\dfrac{\dfrac{\pi_1 \ldots \pi_m}{\vdash \Gamma, A_1, \ldots, A_n, A, A_1^{\perp}, \ldots, A_n^{\perp}, A^{\perp}}\ \text{MIX}^m}{\vdash \Gamma}\ \text{CYCLE}^{n+1}$$

we can apply the induction hypothesis to the subproofs of τ right before the last MIX preceding the sequence of CYCLE. This allows us to reduce their degrees to become smaller than $d(\tau)$. Then we use Lemma 3.

– Otherwise, by using the inductive hypothesis on the immediate subproofs to reduce their degree, we also reduce the degree of the whole proof. □

Theorem 1 (CYCLE-Elimination). *Given any proof of $\vdash \Gamma$, we can construct a* CYCLE*-free proof of $\vdash \uparrow^t \Gamma$, for some $t \geqslant 0$.*

Proof. Iteration on Lemma 4. □

CYCLE-elimination increases the priorities of the propositions in Γ. This is solely due to the (logical part of) our commuting conversions in Sect. 3.

4.2 Deadlock-Freedom for PCP

Theorem 2 (Subject Reduction). *If $P \vdash \Gamma$ and $P \longrightarrow Q$, then $Q \vdash \uparrow^t \Gamma$, for some $t \geqslant 0$.*

Proof. Follows from the β-reductions and commuting conversions in Sect. 3. □

Definition 9. *A process is a* CYCLE *if it is of the form $(\boldsymbol{\nu} x^A y)P$.*

Theorem 3 (Top-Level Deadlock-Freedom). *If $P \vdash \Gamma$ and P is a* Cycle, *then there is some Q such that $P \longrightarrow^* Q$ and Q is not a* Cycle.

Proof. The interpretation of Lemma 3 for PCP is that either (i) a top-level communication occurs, corresponding to a β-reduction, or (ii) commuting conversions are used to push Cycle further inwards in a process. Consequently, iterating Lemma 3 results in eliminating top-level Cycles. □

Eliminating all Cycles, as specified by Theorem 1, would correspond to a semantics in which reduction occurs under prefixes, as discussed by Wadler [41]. In order to achieve this, we would need to introduce additional congruence rules, such as:

$$\frac{P \longrightarrow Q}{x(y).P \longrightarrow x(y).Q}$$

and similarly for other actions. Reductions of this kind are not present in the π-calculus, and we also omit them in our framework.

However, we can eliminate all Cycles in a proof of $\vdash \emptyset$, corresponding to full deadlock-freedom for closed processes. Kobayashi's type system [32] satisfies the same property.

Theorem 4 (Deadlock-Freedom for Closed Processes). *If $P \vdash \emptyset$, then either $P \equiv \mathbf{0}$ or there is Q such that $P \longrightarrow Q$.*

Proof. This follows from Theorems 2 and 3, because if $Q \vdash \emptyset$ and Q is not a Cycle then Q must be a parallel composition of $\mathbf{0}$ processes. □

5 Related Work and Conclusion

Cycle and Multicut rules were explored by Abramsky *et al.* [2–4] in the context of $*$-autonomous categories. That work is not directly comparable with ours, as it only presented a typed semantics for CCS-like processes and did not give a type system for a language or a term assignment for a logical system. Atkey *et al.* [5] added a Multicut rule to CP, producing an isomorphism between $\otimes$ and $⅋$, but they did not consider deadlock-freedom.

In Kobayashi's original type-theoretic approach to deadlock-freedom [29], priorities were abstract tags from a partially ordered set. In later work abstract tags were simplified to natural numbers, and priorities were replaced by pairs of obligations and capabilities [30,32]. The latter change allows more processes to be typed, at the expense of a more complex type system. Padovani [36] adapted Kobayashi's approach to session types, and later on he simplified it to a single priority for linear π-calculus [37]. Then, the single priority technique can be transferred to session types by the encoding of session types into linear types [16,17,19,33]. For simplicity, we have opted for single priorities, as Padovani [37].

The first work on progress for session types, by Dezani-Ciancaglini *et al.* [15,22], guaranteed the property by allowing only one active session at a time. Later work [21] introduced a partial order on channels in Kobayashi-style [29].

Bettini *et al.* [9] applied similar ideas to multiparty session types. The main difference with our work is that we associate priorities with individual communication operations, rather than with entire channels. Carbone *et al.* [13] proved that progress is a compositional form of lock-freedom and introduced a new technique for progress in session types by adopting Kobayashi's type system and the encoding of session types [19]. Vieira and Vasconcelos [40] used single priorities and an abstract partial order in session types to guarantee deadlock-freedom.

The linear logic approach to deadlock-free session types started with Caires and Pfenning [12], based on dual intuitionistic linear logic, and was later formulated for classical linear logic by Wadler [41]. All subsequent work on linear logic and session types enforces deadlock-freedom by forbidding cyclic connections. In their original work, Caires and Pfenning commented that it would be interesting to compare process typability in their system with other approaches including Kobayashi's and Dezani-Ciancaglini's. However, we are aware of only one comparative study of the expressivity of type systems for deadlock-freedom, by Dardha and Pérez [20]. They compared Kobayashi-style typing and CLL typing, and proved that CLL corresponds to Kobayashi's system with the restriction that only single cuts, not multicuts, are allowed.

In this paper, we have presented a new logic, priority-based linear logic (PLL), and a term assignment system, priority-based CP (PCP), that increase the expressivity of deadlock-free session type systems, by combining Caires and Pfenning's linear logic-based approach and Kobayashi's priority-based type system. The novel feature of PLL and PCP is CYCLE, which allows cyclic process structures to be formed if they do not violate ordering conditions on the priorities of prefixes. Following the propositions-as-types paradigm, we prove a CYCLE-elimination theorem analogous to the standard CUT-elimination theorem. As a result of this theorem, we obtain deadlock-freedom for a class of π-calculus processes which is larger than the class typed by Caires and Pfenning. In particular, these are processes that typically share more than one channel in parallel.

There are two main directions for future work. First, develop a type system for a functional language, priority-based GV, and translate it into PCP, along the lines of Lindley and Morris' [34] translation of GV [41] into CP. Second, extend PCP to allow recursion and sharing [6], in order to support more general concurrent programming, while maintaining deadlock-freedom, as well as termination, or typed behavioural equivalence.

Acknowledgements. We are grateful for suggestions and feedback from the anonymous reviewers and colleagues: Wen Kokke, Sam Lindley, Roly Perera, Frank Pfenning, Carsten Schürmann and Philip Wadler.

References

1. Abramsky, S.: Proofs as processes. Theor. Comput. Sci. **135**(1), 5–9 (1994)
2. Abramsky, S., Gay, S.J., Nagarajan, R.: Interaction categories and the foundations of typed concurrent programming. In: Broy, M. (ed.) Proceedings of the NATO Advanced Study Institute on Deductive Program Design, pp. 35–113 (1996)

3. Abramsky, S., Gay, S., Nagarajan, R.: A type-theoretic approach to deadlock-freedom of asynchronous systems. In: Abadi, M., Ito, T. (eds.) TACS 1997. LNCS, vol. 1281, pp. 295–320. Springer, Heidelberg (1997). https://doi.org/10.1007/BFb0014557
4. Abramsky, S., Gay, S.J., Nagarajan, R.: A specification structure for deadlock-freedom of synchronous processes. Theor. Comput. Sci. **222**(1–2), 1–53 (1999)
5. Atkey, R., Lindley, S., Morris, J.G.: Conflation confers concurrency. In: Lindley, S., McBride, C., Trinder, P., Sannella, D. (eds.) A List of Successes That Can Change the World. LNCS, vol. 9600, pp. 32–55. Springer, Cham (2016). https://doi.org/10.1007/978-3-319-30936-1_2
6. Balzer, S., Pfenning, F.: Manifest sharing with session types. In: Proceedings of the ACM on Programming Languages, vol. 1(ICFP), pp. 37:1–37:29 (2017)
7. Barber, A.: Dual intuitionistic linear logic. Technical report ECS-LFCS-96-347, University of Edinburgh (1996). www.lfcs.inf.ed.ac.uk/reports/96/ECS-LFCS-96-347
8. Bellin, G., Scott, P.J.: On the pi-calculus and linear logic. Theor. Comput. Sci. **135**(1), 11–65 (1994)
9. Bettini, L., Coppo, M., D'Antoni, L., De Luca, M., Dezani-Ciancaglini, M., Yoshida, N.: Global progress in dynamically interleaved multiparty sessions. In: van Breugel, F., Chechik, M. (eds.) CONCUR 2008. LNCS, vol. 5201, pp. 418–433. Springer, Heidelberg (2008). https://doi.org/10.1007/978-3-540-85361-9_33
10. Bräuner, T.: Introduction to linear logic. Technical report BRICS LS-96-6, Basic Research Institute in Computer Science, University of Aarhus (1996)
11. Caires, L., Pérez, J.A.: Linearity, control effects, and behavioral types. In: Yang, H. (ed.) ESOP 2017. LNCS, vol. 10201, pp. 229–259. Springer, Heidelberg (2017). https://doi.org/10.1007/978-3-662-54434-1_9
12. Caires, L., Pfenning, F.: Session types as intuitionistic linear propositions. In: Gastin, P., Laroussinie, F. (eds.) CONCUR 2010. LNCS, vol. 6269, pp. 222–236. Springer, Heidelberg (2010). https://doi.org/10.1007/978-3-642-15375-4_16
13. Carbone, M., Dardha, O., Montesi, F.: Progress as compositional lock-freedom. In: Kühn, E., Pugliese, R. (eds.) COORDINATION 2014. LNCS, vol. 8459, pp. 49–64. Springer, Heidelberg (2014). https://doi.org/10.1007/978-3-662-43376-8_4
14. Carbone, M., Lindley, S., Montesi, F., Schürmann, C., Wadler, P.: Coherence generalises duality: a logical explanation of multiparty session types. In: CONCUR. LIPIcs, vol. 59, pp. 33:1–33:15. Schloss Dagstuhl–Leibniz-Zentrum für Informatik (2016)
15. Coppo, M., Dezani-Ciancaglini, M., Yoshida, N.: Asynchronous session types and progress for object oriented languages. In: Bonsangue, M.M., Johnsen, E.B. (eds.) FMOODS 2007. LNCS, vol. 4468, pp. 1–31. Springer, Heidelberg (2007). https://doi.org/10.1007/978-3-540-72952-5_1
16. Dardha, O.: Recursive session types revisited. In: BEAT. EPTCS, vol. 162, pp. 27–34 (2014)
17. Dardha, O.: Type Systems for Distributed Programs: Components and Sessions. Atlantis Studies in Computing, vol. 7. Atlantis Press, Paris (2016). https://doi.org/10.2991/978-94-6239-204-5
18. Dardha, O., Gay, S.J.: A new linear logic for deadlock-free session typed processes. In: 21st International Conference on Foundations of Software Science and Computation Structures, FoSSaCS 2018 (Extended Version). http://www.dcs.gla.ac.uk/~ornela/publications/DG18-Extended.pdf
19. Dardha, O., Giachino, E., Sangiorgi, D.: Session types revisited. In: PPDP, pp. 139–150. ACM (2012)

20. Dardha, O., Pérez, J.A.: Comparing deadlock-free session typed processes. In: EXPRESS/SOS. EPTCS, vol. 190, pp. 1–15 (2015)
21. Dezani-Ciancaglini, M., de'Liguoro, U., Yoshida, N.: On progress for structured communications. In: Barthe, G., Fournet, C. (eds.) TGC 2007. LNCS, vol. 4912, pp. 257–275. Springer, Heidelberg (2008). https://doi.org/10.1007/978-3-540-78663-4_18
22. Dezani-Ciancaglini, M., Mostrous, D., Yoshida, N., Drossopoulou, S.: Session types for object-oriented languages. In: Thomas, D. (ed.) ECOOP 2006. LNCS, vol. 4067, pp. 328–352. Springer, Heidelberg (2006). https://doi.org/10.1007/11785477_20
23. Gay, S.J., Vasconcelos, V.T.: Linear type theory for asynchronous session types. J. Funct. Program. **20**(1), 19–50 (2010)
24. Girard, J.: Linear logic. Theor. Comput. Sci. **50**, 1–102 (1987)
25. Girard, J.-Y., Taylor, P., Lafont, Y.: Proofs and Types. Cambridge University Press, New York (1989)
26. Honda, K.: Types for dyadic interaction. In: Best, E. (ed.) CONCUR 1993. LNCS, vol. 715, pp. 509–523. Springer, Heidelberg (1993). https://doi.org/10.1007/3-540-57208-2_35
27. Honda, K., Vasconcelos, V.T., Kubo, M.: Language primitives and type discipline for structured communication-based programming. In: Hankin, C. (ed.) ESOP 1998. LNCS, vol. 1381, pp. 122–138. Springer, Heidelberg (1998). https://doi.org/10.1007/BFb0053567
28. Kobayashi, N.: TyPiCal: type-based static analyzer for the pi-calculus. www-kb.is.s.u-tokyo.ac.jp/~koba/typical
29. Kobayashi, N.: A partially deadlock-free typed process calculus. ACM Trans. Program. Lang. Syst. **20**(2), 436–482 (1998)
30. Kobayashi, N.: A type system for lock-free processes. Inf. Comput. **177**(2), 122–159 (2002)
31. Kobayashi, N.: Type systems for concurrent programs. In: Aichernig, B.K., Maibaum, T. (eds.) Formal Methods at the Crossroads. From Panacea to Foundational Support. LNCS, vol. 2757, pp. 439–453. Springer, Heidelberg (2003). https://doi.org/10.1007/978-3-540-40007-3_26
32. Kobayashi, N.: A new type system for deadlock-free processes. In: Baier, C., Hermanns, H. (eds.) CONCUR 2006. LNCS, vol. 4137, pp. 233–247. Springer, Heidelberg (2006). https://doi.org/10.1007/11817949_16
33. Kobayashi, N.: Type systems for concurrent programs. Extended version of [31], Tohoku University (2007)
34. Lindley, S., Morris, J.G.: A semantics for propositions as sessions. In: Vitek, J. (ed.) ESOP 2015. LNCS, vol. 9032, pp. 560–584. Springer, Heidelberg (2015). https://doi.org/10.1007/978-3-662-46669-8_23
35. Milner, R.: Communication and Concurrency. Prentice Hall, Upper Saddle River (1989)
36. Padovani, L.: From lock freedom to progress using session types. In: PLACES. EPTCS, vol. 137, pp. 3–19 (2013)
37. Padovani, L.: Deadlock and lock freedom in the linear π-Calculus. In: CSL-LICS, pp. 72:1–72:10. ACM (2014)
38. Takeuchi, K., Honda, K., Kubo, M.: An interaction-based language and its typing system. In: Halatsis, C., Maritsas, D., Philokyprou, G., Theodoridis, S. (eds.) PARLE 1994. LNCS, vol. 817, pp. 398–413. Springer, Heidelberg (1994). https://doi.org/10.1007/3-540-58184-7_118
39. Toninho, B., Caires, L., Pfenning, F.: Dependent session types via intuitionistic linear type theory. In: PPDP, pp. 161–172. ACM (2011)

40. Torres Vieira, H., Thudichum Vasconcelos, V.: Typing progress in communication-centred systems. In: De Nicola, R., Julien, C. (eds.) COORDINATION 2013. LNCS, vol. 7890, pp. 236–250. Springer, Heidelberg (2013). https://doi.org/10.1007/978-3-642-38493-6_17
41. Wadler, P.: Propositions as sessions. In: ICFP, pp. 273–286. ACM (2012)
42. Wadler, P.: Propositions as types. Commun. ACM **58**(12), 75–84 (2015)

A Double Category Theoretic Analysis of Graded Linear Exponential Comonads

Shin-ya Katsumata(✉)

National Institute of Informatics, Tokyo, Japan
s-katsumata@nii.ac.jp

Abstract. Graded linear exponential comonads are an extension of linear exponential comonads wih *grading*, and provide a categorical semantics of resource-sensitive exponential modality in linear logic. In this paper, we propose a concise double-category theoretic formulation of graded linear exponential comonads as a kind of monoid homomorphisms from the multiplicative monoids of semirings to the composition monoids of symmetric monoidal endofunctors. We also exploit this formulation to derive the category of graded comonoid-coalgebras, which decompose graded linear exponential comonads into symmetric monoidal adjunctions plus twists.

1 Introduction

One of the important discoveries in substructural logic is the decomposition of the intuitionistic implication $\phi \Rightarrow \psi$ using the linear implication $\multimap$ and the *exponential modality* !. This discovery was studied by Girard through his *linear logic*, which brought many new ideas and perspectives to logic and programming language semantics.

Inside linear logic proofs, propositions with the exponential modality $!\phi$ can be freely copied or discarded. Later, it was realized that by adding a copy limit to the exponential modality, like $!_r\phi$, linear logic gains fine control of assumption usage. This idea was first implemented in *bounded linear logic* [9], and studied in connection with implicit complexity theory [4,14]. Indexed exponential modalities $!_r$ were then used in wider context: resource management in programming languages [3,7,8,20,23] and control of sensitivity in the metric semantics of programs [5,21].

The categorical structure corresponding to the exponential modality ! was studied by various researchers, and it was identified as a categorical structure called *linear exponential comonad* [1]. One of the celebrated results about linear exponential comonads is that any symmetric lax monoidal adjunction:

$$(\mathbb{D}, 1, \times) \underset{R}{\overset{L}{\underset{\bot}{\rightleftarrows}}} (\mathbb{C}, \mathbf{I}, \otimes) \quad \text{(the monoidal structure } 1, \times \text{ is cartesian)}$$

yields a linear exponential comonad $L \circ R$, and every linear exponential comonad D arises in this way - for $\mathbb{D}$ take the category of Eilenberg-Moore coalgebras of D.

C. Baier and U. Dal Lago (Eds.): FOSSACS 2018, LNCS 10803, pp. 110–127, 2018.
https://doi.org/10.1007/978-3-319-89366-2_6

The categorical structure corresponding to the indexed exponential modality $!_r$ has been proposed as *exponential action* [3] and *graded linear exponential comonad* [7]; they are two different presentations of the same data. Compared to linear exponential comonads, however, categorical understanding of graded linear exponential comonads is not well-established. The aim of this paper is to contribute to this point. Concretely speaking, we show the following categorical results about graded linear exponential comonads:

- We give a new concise formulation of graded linear exponential comonads as vertical monoid homomorphisms from multiplicative monoids of semirings to the composition monoids of symmetric lax monoidal endofunctors. This formulation is given in a rather complex multi-double category of symmetric monoidal categories. The slogan is "to represent a complex structure in a simple category as a simple structure in a complex category".
- In the multi-double category, vertical monoid homomorphisms themselves can be seen as monoids. By considering actions of such monoids, we obtain the concept of *graded comonoid-coalgebras.* They are an extension of Eilenberg-Moore coalgebras to graded linear exponential comonads, and the category of graded comonoid-coalgebras provides a resolution of graded linear exponential comonads by a symmetric lax monoidal adjunction plus a *twist.*

2 Related Work

Graded linear exponential comonads were first introduced as *exponential actions* in [3], and an equivalent definition was given in [7]. This paper adopts the latter definition as the starting point of study. These papers also consider linear type systems with an indexed exponential modality $!_r\phi$, which is directly interpreted by a graded linear exponential comonad. This paper, however, focuses only on the categorical axiomatics of the indexed exponential modality, and omit its syntactic theory. In [2], Breuvart and Pagani gave a construction of graded linear exponential comonads from a set of data called *stratification.* They derived various graded linear exponential comonads on the category of sets and binary relations and the category of coherence spaces. Structures close to, but different from, graded linear exponential comonads were considered in the categorical semantics of the following calculi: *INTML* for interactive computation [23], *coeffect calculus* [20] and *bounded affine types system* [8].

Looking at the dual structure, *graded monads*, first considered in mathematics [6,25], were recently used in the semantic study of logic, systems and programming languages [13,18,19,22]. The resolution of graded monads were studied in [12], mildly extending a classic work by Street [26]. The major difference between graded monads and graded linear exponential comonads is the way how they interact with the monoidal structure. In [13] only *strengths* were considered for graded monads, while graded linear exponential comonads interact with monoidal structures in an intricate manner.

The multicategory of symmetric lax monoidal multifunctors is related to the 2-multicategory of T-algebras for a pseudo-commutative 2-monad T [11].

Hyland and Power studied multifunctors that are symmetric *strong* monoidal in each argument, while in this paper we weaken "strong" to "lax". Yet, we think that by suitably extending their theory, the symmetric lax monoidal multifunctors can also be given in the language of 2-monad theory.

Monoids in the multicategory $\mathbf{MSMC}_l$ in Sect. 5 are similar to the distributivity studied in [15], where Laplaza considered two symmetric *non-strict* monoidal structures together with a *colax* distributivity between them. On the other hand, in this paper, we consider a *strict* monoidal structure on top of the underlying symmetric (non-strict) monoidal structure, and a *lax* distributivity between them.

Preliminaries

For symmetric monoidal categories and symmetric lax monoidal functors, see [16]. In a symmetric monoidal category $\mathbb{C}$, by $\iota : \mathbf{I} \otimes \mathbf{I} \to \mathbf{I}$ we mean the isomorphism $\lambda_{\mathbf{I}} = \rho_{\mathbf{I}}$, and by $\tau : (A \otimes B) \otimes (C \otimes D) \to (A \otimes C) \otimes (B \otimes D)$ we mean the symmetry swapping the second and third component of the tensor product. For functors $F_i : \prod_{j=1}^{m_i} \mathbb{C}_{i,j} \to \mathbb{D}_i$ where $1 \leq i \leq n$, we define $F_1 \times \cdots \times F_n$ to be the composite functor $\prod_{1 \leq i \leq n, 1 \leq j \leq m_i} \mathbb{C}_{i,j} \to \prod_{i=1}^{n} (\prod_{j=1}^{m_i} \mathbb{C}_{i,j}) \to \prod_{i=1}^{n} \mathbb{D}_i$, whose codomain is the product category without the nesting of products.

3 Graded Linear Exponential Comonad

In this paper, comonads are graded by a *partially ordered semiring*. It is a tuple $(R, \leq, 0, +, 1, *)$ such that $(R, 0, +, 1, *)$ is a unital semiring (not necessarily commutative) and $+, *$ are monotone in each argument w.r.t. the partial order $\leq$. The partially ordered monoids of additive and multiplicative parts of R are denoted by $R^+ = (R, \leq, 0, +)$ and $R^* = (R, \leq, 1, *)$, respectively.

Let $\mathbb{C}, \mathbb{D}$ be symmetric monoidal categories. We write $\mathbf{SMC}_l(\mathbb{D}, \mathbb{C})$ for the category of symmetric lax monoidal functors and monoidal natural transformations between them. The following pointwise extension of the tensor unit and tensor product on $\mathbb{C}$ extends to a symmetric monoidal structure on $\mathbf{SMC}_l(\mathbb{D}, \mathbb{C})$:

$$\dot{\mathbf{I}}(D) = \mathbf{I}, \quad (F \mathbin{\dot{\otimes}} G)(D) = FD \otimes GD.$$

(We note that the symmetry in $\mathbb{C}$ is used to make $F \mathbin{\dot{\otimes}} G$ a symmetric lax monoidal functor.) Below by $[\mathbb{D}, \mathbb{C}]_l$ we mean the symmetric monoidal category $(\mathbf{SMC}_l(\mathbb{D}, \mathbb{C}), \dot{\mathbf{I}}, \dot{\otimes})$ of symmetric lax monoidal functors and monoidal natural transformations between them.

3.1 Graded Linear Exponential Comonad

Fix a partially ordered semiring $(R, \leq, 0, +, 1, *)$. We introduce the main subject of this study, *R-graded linear exponential comonad*. This concept first appeared in [3, Definition 13] under the name *exponential action*. We adopt the following definition [7, Sect. 5.2], which is equivalent to the exponential action:

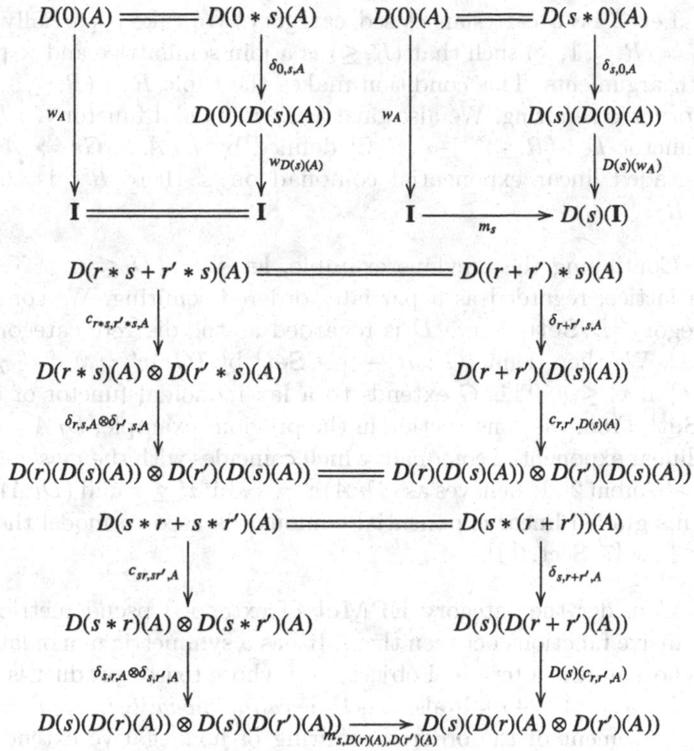

Fig. 1. Four equational axioms related to distributive law

Definition 1. *An* R-graded linear exponential comonad *on a symmetric monoidal category* $\mathbb{C}$ *is a tuple* $(D, w, c, \epsilon, \delta)$ *where*

- $D : (R, \leq) \to \mathbf{SMC}_l(\mathbb{C}, \mathbb{C})$ *is a functor. Below we write* $m_r : \mathbf{I} \to D(r)(\mathbf{I})$ *and* $m_{r,A,B} : D(r)(A) \otimes D(r)(B) \to D(r)(A \otimes B)$ *for the symmetric lax monoidal structure of* $D(r)$.
- $(D, w, c) : R^+ \to [\mathbb{C}, \mathbb{C}]_l$ *is a symmetric colax monoidal functor.*
- $(D, \epsilon, \delta) : R^* \to (\mathbf{SMC}_l(\mathbb{C}, \mathbb{C}), \mathrm{Id}, \circ)$ *is a colax monoidal functor.*

They satisfy four equational axioms in Fig. 1. Moreover, we say that D *is an* R-twist *if* Dr *is strong monoidal for each* $r \in R$*, and* (D, ϵ, δ) *is a strict monoidal functor (hence* $D1 = \mathrm{Id}$ *and* $D(r * r') = Dr \circ Dr'$*).*

When fully expanded, a graded linear exponential comonad specifies one functor $D : (R, \leq) \to [\mathbb{C}, \mathbb{C}]$ and 6 natural transformations:

$$m_r : D(r)(\mathbf{I}) \to \mathbf{I}, \qquad m_{r,A,B} : D(r)(A \otimes B) \to D(r)(A) \otimes D(r)(B)$$
$$w_A : D(0)(A) \to \mathbf{I} \qquad c_{r,r',A} : D(r + r')(A) \to D(r)(A) \otimes D(r')(A)$$
$$\epsilon_A : D(1)(A) \to A \qquad \delta_{r,r',A} : D(r * r')(A) \to D(r)(D(r')(A))$$

satisfying more than 20 equational axioms.

Example 1. Let $\mathbb{C}$ be a cartesian closed category. We take a partially ordered monoid $R^\times = (R, \leq, 1, \times)$ such that $(R, \leq)$ is a join semilattice and $\times$ preserves joins in both arguments. This condition makes the tuple $R = (R, \leq, \bot, \vee, 1, \times)$ a partially ordered semiring. We also take a lax monoidal functor $G : R^\times \to \mathbb{C}$. Then the functor $D : (R, \leq)^{op} \to [\mathbb{C}, \mathbb{C}]$ defined by $DrA = Gr \Rightarrow A$ extends to an R^{op}-graded linear exponential comonad on $\mathbb{C}$ (here R^{op} is the order-opposite of R).

Example 2. Continuing the previous example, let $R = (D, \leq, \bot, \vee, \top, \wedge)$ be a distributive lattice, regarded as a partially ordered semiring. We consider the functor category $[D, \mathbf{Set}]$, where D is regarded as the discrete category of the carrier set D. We then define $G : R \to [D, \mathbf{Set}]$ by $(Gr)r' = \emptyset$ if $r' \not\leq r$, and $(Gr)r' = \{*\}$ if $r' \leq r$. This G extends to a lax monoidal functor of type $G : R^\times \to [D, \mathbf{Set}]$. From the construction in the previous example, $DrA = Gr \Rightarrow A$ is a graded linear exponential comonad, which coincides with the *masking functor* given in [7, Theorem 2]. It behaves as $(DrA)r' = \{*\}$ if $r' \not\leq r$ and $(DrA)r' = Ar'$ if $r' \leq r$. This graded linear exponential comonad is used to model the level of information flow [7, Sect. 6.1].

Example 3. Consider the category **EPMet** of extended pseudometric spaces[1] and nonexpansive functions between them. It has a symmetric monoidal (closed) structure, whose unit is a terminal object, and whose tensor product is given by $(X, d) \otimes (Y, e) = (X \times Y, d + e)$. It also has the *scaling modality* $!_r(X, d) = (X, rd)$, where r is an element of the ordered semiring of nonnegative extended reals, which we denote by $[0, \infty]$. The scaling modality is a $[0, \infty]$-twist with respect to the above symmetric monoidal structure.

The concept of R-graded linear exponential comonad is a generalization of non-graded linear exponential comonad [1, Definition 3]. This was first observed in [3].

Theorem 1. *A* $\mathbf{1}$*-graded linear exponential comonad on a symmetric monoidal category* $\mathbb{C}$ *is exactly a non-graded linear exponential comonad on* $\mathbb{C}$.

On the other hand, $\mathbf{1}$-twists make monoidal structures cartesian:

Theorem 2. *A* $\mathbf{1}$*-twist* D *exists on a symmetric monoidal category* $\mathbb{C}$ *if and only if the symmetric monoidal structure of* $\mathbb{C}$ *is cartesian (i.e.* $\mathbf{I}$ *is terminal and* $\otimes$ *is a binary product).*

Proof. If it exists, the functor part of D must specify the identity functor $\mathrm{Id}_{\mathbb{C}}$ because of the strictness. Next, (Id, w, c) becomes a commutative monoid in $[\mathbb{C}, \mathbb{C}]_l$; especially w, c are monoidal natural transformations. From [17, Corollary 17], the monoidal structure of $\mathbb{C}$ is cartesian. The converse construction is evident.

[1] Here, extended pseudometrics mean the pseudometrics that can return $+\infty$.

4 A Double-Category Theoretic Reformulation of Graded Linear Exponential Comonad

Although it is in a reasonably compact form, the definition of graded linear exponential comonad is yet technical, and it indeed specifies a quite complex structure. The motivation of this study is to have a conceptually clean and compact definition of it.

Particularly, what is less clear in the definition is the extra four axioms related to the distributive law (Fig. 1). In the non-graded setting (i.e. when $R = 1$), these four axioms reduces to simpler axioms, which can be viewed as the following conditions:

- comultiplication δ is a comonoid morphism, (item 4, Sect. 7.4, [17]) and
- weakening w and contraction c are coalgebra morphisms (item 3, Sect. 7.4, [17]).

However, it is not obvious how to upgrade these axioms to the graded setting, because the concept of "graded coalgebra" and "graded comonoid" are not yet defined, at least for graded linear exponential comonads. Especially, the concept of graded coalgebra should be defined after the concept of graded linear exponential comonad, which we are going to define! From this circularity, the above view of the four axioms are not very helpful when upgrading them in the current situation.

It is therefore desirable to have an alternative account on four axioms in Fig. 1, which relies on a notion that already exists *before* graded linear exponential comonads. The key observation of this paper is that these four axioms are an instance of the axioms for 2-cells in the double category **SMC** of symmetric monoidal categories, introduced by Grandis and Paré [10, Sect. 2.3]. In **SMC**, a 2-cell consists of the following data:

$$\begin{array}{ccc} \bullet & \xrightarrow{\;H\;} & \bullet \\ {\scriptstyle V'}\Big\downarrow & \Downarrow a & \Big\downarrow{\scriptstyle V} \\ \bullet & \xrightarrow[\;H'\;]{} & \bullet \end{array}$$

where each $\bullet$ is a (possibly distinct) symmetric monoidal category, horizontal morphisms H, H' are symmetric lax monoidal functors, vertical morphisms V, V' are symmetric colax monoidal functors, and $a : V \circ H \to H' \circ V'$ is a natural transformation (between underlying functors of H, H', V, V') making the following diagrams commute:

$$\begin{array}{ccc} V\mathbf{I} & \longrightarrow & VH\mathbf{I} \\ \Big\downarrow & & \Big\downarrow \\ & & H'V'\mathbf{I} \\ & & \Big\downarrow \\ \mathbf{I} & \longrightarrow & H'\mathbf{I} \end{array} \qquad \begin{array}{ccc} V(HX \otimes HY) & \longrightarrow & VH(X \otimes Y) \\ \Big\downarrow & & \Big\downarrow \\ VHX \otimes VHY & & H'V'(X \otimes Y) \\ \Big\downarrow & & \Big\downarrow \\ H'V'X \otimes H'V'Y & \longrightarrow & H'(V'X \otimes V'Y) \end{array} \tag{1}$$

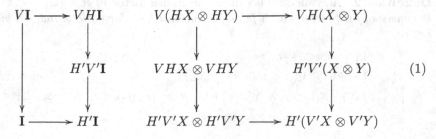

We note that when V, V' (resp. H, H') are identity functors, the above axioms are reduced to the ones for monoidal natural transformations of type $V \to V'$ (resp. $H \to H'$).

Let us see how 2-cell axioms (1) in **SMC** derives the four axioms in Fig. 1.

Proposition 1. *In Definition 1, the four axioms (Fig. 1) can be replaced by the following statement: for each* $r \in R$, *both*

$$\delta_{r,-} : D(r * -) \to Dr \circ D-, \quad \delta_{-,r} : D(- * r) \to D- \circ Dr$$

are 2-cells of the following type in **SMC**:

$$\begin{array}{ccc} R^+ & \xrightarrow{r*-} & R^+ \\ {\scriptstyle D}\downarrow & \Downarrow\delta_{r,-} & \downarrow{\scriptstyle D} \\ [\mathbb{C},\mathbb{C}]_l & \xrightarrow[Dr\circ-]{} & [\mathbb{C},\mathbb{C}]_l \end{array} \qquad \begin{array}{ccc} R^+ & \xrightarrow{-*r} & R^+ \\ {\scriptstyle D}\downarrow & \Downarrow\delta_{-,r} & \downarrow{\scriptstyle D} \\ [\mathbb{C},\mathbb{C}]_l & \xrightarrow[-\circ Dr]{} & [\mathbb{C},\mathbb{C}]_l \end{array}$$

5 Multicategory of Symmetric Lax Monoidal Multifunctors

Proposition 1 says that by fixing one index of the doubly-indexed natural transformation $\delta_{-,=} : D(- * =) \to D- \circ D=$, we obtain a 2-cell in the double category **SMC**. However, δ itself does not live in **SMC**. In order to create a room to accommodate δ as a kind of 2-cell, we extend horizontal morphisms of **SMC** to multi-ary functors that are symmetric lax monoidal in *each argument.* We first study such multi-ary functors in this section.

Let $\mathbb{C}_i$ $(1 \leq i \leq n)$ and $\mathbb{D}$ be symmetric monoidal categories. Intuitively, an n-ary functor $F : \mathbb{C}_1 \times \cdots \times \mathbb{C}_n \to \mathbb{D}$ is symmetric lax monoidal in each argument if it comes with a structure making the functor $F(C_1, .., -_m, .., C_n) : \mathbb{C}_m \to \mathbb{D}$ symmetric lax monoidal for each $m \in \{1, \cdots, n\}$ and $C_i \in \mathbb{C}_i$, $i \in \{1, \cdots, n\} \backslash \{m\}$. Moreover, these symmetric lax monoidal structures commute with each other in a coherent manner.

To formally define such multi-ary symmetric lax monoidal functors, we introduce a notation for sequences. For a sequence $C = C_1, \cdots, C_n$ of mathematical objects, a natural number $1 \leq i \leq n$ and another sequence D, by $C[i : D]$ we mean the sequence obtained by replacing C_i with D. For instance, $(1, 3, 5)[2 : X, Y] = 1, X, Y, 5$. When D is empty, $C[i :]$ stands for the sequence obtained by removing the i-th element of C.

Definition 2. *A* symmetric lax monoidal multifunctor *of type* $(\mathbb{C}_1, \cdots, \mathbb{C}_n) \to \mathbb{D}$ *consists of a functor and a family of natural transformations indexed by* $1 \leq i \leq n$:

$$\begin{aligned} F &: \mathbb{C}_1 \times \cdots \times \mathbb{C}_n \to \mathbb{D} \\ \phi^i_{C[i:]} &: \mathbf{I} \to F(C[i : \mathbf{I}]) \quad (C \in \mathbb{C}_1 \times \cdots \times \mathbb{C}_n) \\ \phi^i_{C[i:X,Y]} &: F(C[i : X]) \otimes F(C[i : Y]) \to F(C[i : X \otimes Y]) \quad (C \in \mathbb{C}_1 \times \cdots \times \mathbb{C}_n, X, Y \in \mathbb{C}_i) \end{aligned}$$

such that:

1. *For each $C \in \mathbb{C}_1 \times \cdots \times \mathbb{C}_n$ and $1 \le i \le n$, The tuple $(F(C[i:-]), \phi^i_{C[i:]}, \phi^i_{C[i:-,=]})$ is a symmetric lax monoidal functor from $\mathbb{C}_i$ to $\mathbb{D}$. We denote it by $F(C/i)$.*
2. *The following equalities hold for each $C \in \mathbb{C}_1 \times \cdots \times \mathbb{C}_n$ and $1 \le i < j \le n$:*

- $\phi^i_{C[j:\mathbf{I}][i:]} = \phi^j_{C[i:\mathbf{I}][j:]}$
- $\phi^j_{C[i:\mathbf{I}][j:P,Q]} \circ (\phi^i_{C[j:P][i:]} \otimes \phi^i_{C[j:Q][i:]}) = \phi^i_{C[j:P\otimes Q][i:]} \circ \iota$
- $\phi^i_{C[j:\mathbf{I}][i:P,Q]} \circ (\phi^j_{C[i:P][j:]} \otimes \phi^j_{C[i:Q][j:]}) = \phi^j_{C[i:P\otimes Q][j:]} \circ \iota$
- $\phi^j_{C[i:X\otimes Y][j:P,Q]} \circ (\phi^i_{C[j:P][i:X,Y]} \otimes \phi^i_{C[j:Q][i:X,Y]}) = \phi^i_{C[j:P\otimes Q][i:X,Y]} \circ (\phi^j_{C[i:X][j:P,Q]} \otimes \phi^j_{C[i:Y][j:P,Q]}) \circ \tau$.

We note that a symmetric lax monoidal multifunctor of type $() \to \mathbb{D}$ is just an object in $\mathbb{D}$, because all natural transformations vanish and only the functor of type $1 \to \mathbb{D}$ remains.

Example 4. Let us see how the definition of a binary symmetric lax monoidal multifunctor $M : (\mathbb{C}, \mathbb{C}) \to \mathbb{C}$ is unfolded. It consists of a functor $M : \mathbb{C} \times \mathbb{C} \to \mathbb{C}$ and the following natural transformations:

$$\phi^1_C : \mathbf{I} \to M(\mathbf{I}, C), \quad \phi^1_{X,Y,C} : M(X,C) \otimes M(Y,C) \to M(X \otimes Y, C)$$
$$\phi^2_C : \mathbf{I} \to M(C, \mathbf{I}), \quad \phi^2_{C,X,Y} : M(C,X) \otimes M(C,Y) \to M(C, X \otimes Y)$$

such that

1. For each $C \in \mathbb{C}$, $(M(-,C), \phi^1_C, \phi^1_{-,=,C})$ and $(M(C,-), \phi^2_C, \phi^2_{C,-,=})$ are symmetric lax monoidal functors of type $\mathbb{C} \to \mathbb{C}$.
2. The following coherence axioms holds:

$$\phi^1_{\mathbf{I}} = \phi^2_{\mathbf{I}}, \quad \phi^1_{C\otimes C'} \circ \iota = \phi^2_{\mathbf{I},C,C'} \circ (\phi^1_C \otimes \phi^1_{C'}), \quad \phi^2_{C\otimes C'} \circ \iota = \phi^1_{C,C',\mathbf{I}} \circ (\phi^2_C \otimes \phi^2_{C'})$$
$$\phi^2_{C\otimes C',D,D'} \circ (\phi^1_{C,C',D} \otimes \phi^1_{C,C',D'}) = \phi^1_{C,C',D\otimes D'} \circ (\phi^2_{C,D,D'} \otimes \phi^2_{C',D,D'}) \circ \tau$$

We will later use the following binary symmetric lax monoidal multifunctors. Let R be a partially ordered semiring and $\mathbb{C}$ be a symmetric monoidal category.

1. The multiplication $(*)$ is a symmetric lax monoidal multifunctor of type $(R^+, R^+) \to R^+$.
2. The evaluation functor $ev : [\mathbb{C}, \mathbb{C}]_l \times \mathbb{C} \to \mathbb{C}$ extends to a symmetric lax monoidal multifunctor of type $([\mathbb{C}, \mathbb{C}]_l, \mathbb{C}) \to \mathbb{C}$.
3. The functor composition $(\circ)$ extends to a symmetric lax monoidal multifunctor of type $([\mathbb{C}, \mathbb{C}]_l, [\mathbb{C}, \mathbb{C}]_l) \to [\mathbb{C}, \mathbb{C}]_l$.

Note that $(*)$ is symmetric strict monoidal in each argument, while $(\circ)$, ev are symmetric strict monoidal in the first argument, and symmetric lax monoidal in the second argument.

Next, for symmetric lax monoidal multifunctors $(F, \phi) : (\mathbb{C}_1, \cdots, \mathbb{C}_n) \to \mathbb{D}$ and $(G_i, \gamma(i)) : (\mathbb{B}_{i,1}, \cdots, \mathbb{B}_{i,m_i}) \to \mathbb{C}_i$ $(1 \le i \le n)$, we define their *multi-composition*. First, we define a bijection $(/) : \{(i,j) \mid 1 \le i \le n, 1 \le j \le m_i\} \to \{1, \cdots, \sum_{1 \le i \le n} m_i\}$, and represent a number in the latter set as the pair of numbers uniquely determined by $(/)$ in the former set. Then the multicomposition is given by the following (H, η):

$$H = F \circ (G_1 \times \cdots \times G_n)$$

$$\eta^{i/j}_{(B_1, \cdots, B_n)[i/j:]} = F((GB_1, \cdots, GB_n)[i : \gamma(i)^j_{B_i[j:]}] \circ \phi^i_{(GB_1, \cdots, GB_n)[i:]}$$

$$\eta^{i/j}_{(B_1, \cdots, B_n)[i/j:X,Y]} = F((GB_1, \cdots, GB_n)[i : \gamma(i)^j_{B_i[j:X,Y]}] \circ \phi^i_{(GB_1, \cdots, GB_n)[i:G(B_i[j:X]), G(B_i[j:Y])]}$$

Theorem 3. *Symmetric monoidal categories, symmetric lax monoidal multifunctors, and the above multi-composition form a multicategory* $\mathbf{MSMC}_l$.

Proof (Proof sketch). To check that symmetric lax monoidal multifunctors are closed under multicomposition, the key case is when $n = 2, m_1 = m_2 = 1$ and $n = 1, m_1 = 2$.

In $\mathbf{MSMC}_l$ we consider *monoids* and *monoid actions*. A *monoid* is a tuple $(\mathbb{C}, U : () \to \mathbb{C}, M : (\mathbb{C}, \mathbb{C}) \to \mathbb{C})$ of a symmetric monoidal category $\mathbb{C}$ and symmetric lax monoidal multifunctors U, M such that

$$\mathrm{Id} = M \circ (\mathrm{Id}, U), \quad \mathrm{Id} = M \circ (U, \mathrm{Id}), \quad M \circ (\mathrm{Id}, M) = M \circ (M, \mathrm{Id}).$$

An *action* of a monoid $(\mathbb{C}, U, M)$ on a symmetric monoidal category $\mathbb{D}$ is a symmetric lax monoidal multifunctor $A : (\mathbb{C}, \mathbb{D}) \to \mathbb{D}$ such that

$$A \circ (U, \mathrm{Id}) = \mathrm{Id}, \quad A \circ (\mathrm{Id}, A) = A \circ (M, \mathrm{Id}).$$

By unfolding the definition, a monoid $(\mathbb{C}, U, M)$ in $\mathbf{MSMC}_l$ equips $\mathbb{C}$ with an additional strict monoidal structure (U, M). The argument-wise symmetric lax monoidal structure on M becomes a lax distributivity (see Example 4). Thus we call a monoid in $\mathbf{MSMC}_l$ a *lax distributive strict rig category*. It has a smaller set of coherence axioms than the one given by Laplaza in [15], thanks to the strictness of (U, M).

Example 5 (Continued from Example 4). $(R^+, 1, *)$ and $([\mathbb{C}, \mathbb{C}]_l, \mathrm{Id}, \circ)$ are both lax distributive strict rig categories. Both monoids acts on themselves. The latter monoid acts on $\mathbb{C}$ with the evaluation functor *ev*.

6 Graded Linear Exponential Comonads as Vertical Monoid Homomorphisms

We now extend the double category **SMC** of Grandis and Paré by replacing horizontal morphisms with symmetric lax monoidal multifunctors. The concept of 2-cells in **SMC** is also replaced by *prisms* — the reason of the name is because they are placed in the middle of the space surrounded by two horizontal multifunctors and vertical morphisms. Such a prism is defined to be a natural transformation that is a 2-cell of **SMC** in *each argument*.

Definition 3. *Let $F : (\mathbb{C}_1, \cdots, \mathbb{C}_n) \to \mathbb{D}$ and $G : (\mathbb{E}_1, \cdots, \mathbb{E}_n) \to \mathbb{F}$ be symmetric lax monoidal multifunctors and $V_i : \mathbb{C}_i \to \mathbb{E}_i$ $(1 \le i \le n)$ and $W : \mathbb{D} \to \mathbb{F}$ be symmetric colax monoidal functors. A* prism *α of type $(V_1, \cdots, V_n) \to W : F \to G$, which is depicted as*

$$\begin{array}{ccc} (\mathbb{C}_1, \cdots, \mathbb{C}_n) & \xrightarrow{\quad F \quad} & \mathbb{D} \\ {\scriptstyle (V_1, \cdots, V_n)} \downarrow & \Downarrow \alpha & \downarrow {\scriptstyle W} \\ (\mathbb{E}_1, \cdots, \mathbb{E}_n) & \xrightarrow[\quad G \quad]{} & \mathbb{F} \end{array}$$

is a natural transformation $\alpha : W \circ F \to G \circ (V_1 \times \cdots \times V_n)$ such that for each $C \in \prod_{i=1}^{n} \mathbb{C}_i$ and $1 \le i \le n$, $\alpha_{C[i:-]}$ is a 2-cell of the following type in the double category **SMC***:*

$$\begin{array}{ccc} \mathbb{C}_i & \xrightarrow{\quad F(C[i:-]) \quad} & \mathbb{D} \\ {\scriptstyle V_i} \downarrow & \Downarrow \alpha_{C[i:-]} & \downarrow {\scriptstyle W} \\ \mathbb{E}_i & \xrightarrow[G((V_1C_1, \cdots, V_nC_n)[i:-])]{} & \mathbb{F} \end{array}$$

We note that when $n = 0$, a prism $\alpha : () \to W : F \to G$ is simply a morphism $\alpha : WF \to G$ in $\mathbb{F}$.

Proposition 2. *Let $D : R^+ \to [\mathbb{C}, \mathbb{C}]_l$ be a symmetric colax monoidal functor and δ be a prism of type $(D, D) \to D : (*) \to (\circ)$, where $(*)$ and $(\circ)$ are symmetric lax monoidal multifunctors appeared in Example 4. Then for each $r \in R$, $\delta_{r,-}$ and $\delta_{-,r}$ are 2-cells of the following type in* **SMC***:*

$$\begin{array}{ccc} R^+ & \xrightarrow{\ r*-\ } & R^+ \\ {\scriptstyle D} \downarrow & \Downarrow \delta_{r,-} & \downarrow {\scriptstyle D} \\ [\mathbb{C}, \mathbb{C}]_l & \xrightarrow[Dr\circ -]{} & [\mathbb{C}, \mathbb{C}]_l \end{array} \qquad \begin{array}{ccc} R^+ & \xrightarrow{\ -*r\ } & R^+ \\ {\scriptstyle D} \downarrow & \Downarrow \delta_{-,r} & \downarrow {\scriptstyle D} \\ [\mathbb{C}, \mathbb{C}]_l & \xrightarrow[-\circ Dr]{} & [\mathbb{C}, \mathbb{C}]_l \end{array}$$

Like double categories, composition of prisms can be done in two directions. Consider the following prisms $(1 \le i \le n)$.

$$\begin{array}{ccc} (\mathbb{B}_{i,1}, \cdots, \mathbb{B}_{i,m_i}) & \xrightarrow{\quad G_i \quad} & \mathbb{C}_i \\ {\scriptstyle (U_{i,1}, \cdots, U_{i,m_i})} \downarrow & \Downarrow \gamma_i & \downarrow {\scriptstyle V_i} \\ (\mathbb{B}'_{i,1}, \cdots, \mathbb{B}'_{i,m_i}) & \xrightarrow{\quad G'_i \quad} & \mathbb{C}'_i \\ {\scriptstyle (U'_{i,1}, \cdots, U'_{i,m_i})} \downarrow & \Downarrow \delta_i & \downarrow {\scriptstyle V_i} \\ (\mathbb{B}''_{i,1}, \cdots, \mathbb{B}''_{i,m_i}) & \xrightarrow[\quad G''_i \quad]{} & \mathbb{C}''_i \end{array} \qquad \begin{array}{ccc} (\mathbb{C}_1, \cdots, \mathbb{C}_n) & \xrightarrow{\quad F \quad} & \mathbb{D} \\ {\scriptstyle (V_1, \cdots, V_n)} \downarrow & \Downarrow \alpha & \downarrow {\scriptstyle W} \\ (\mathbb{C}'_1, \cdots, \mathbb{C}'_n) & \xrightarrow{\quad F' \quad} & \mathbb{D}' \\ {\scriptstyle (V'_1, \cdots, V'_n)} \downarrow & \Downarrow \beta & \downarrow {\scriptstyle W'} \\ (\mathbb{C}''_1, \cdots, \mathbb{C}''_n) & \xrightarrow[\quad F'' \quad]{} & \mathbb{D}'' \end{array}$$

Then define *vertical composition* and *horizontal multicomposition* of prisms by the following (ordinary) natural transformations:

$$\beta \circledcirc \alpha = (\beta \circ (V_1 \times \cdots \times V_n)) \bullet (W' \circ \alpha)$$
$$\alpha \circledast (\gamma_1, \cdots, \gamma_n) = (F' \circ (\gamma_1 \times \cdots \times \gamma_n)) \bullet (\alpha \circ (G_1 \times \cdots \times G_n))$$

where $\bullet$ on the right hand side is the vertical composition of natural transformations.

Proposition 3. *In the above setting,*

1. $\beta \circledcirc \alpha$ *is a prism of type* $(V_1' \circ V_1, \cdots, V_n' \circ V_n) \to W' \circ W : F \to F''$.
2. $\alpha \circledast (\gamma_1, \cdots, \gamma_n)$ *is a prism of type* $(U_{1,1}, \cdots, U_{n,m_n}) \to W : F \circ (G_1, \cdots, G_n) \to F' \circ (G_1', \cdots, G_n')$.
3. *The interchange law holds:*

$$(\beta \circledast (\delta_1, \cdots, \delta_n)) \circledcirc (\alpha \circledast (\gamma_1, \cdots, \gamma_n)) = (\beta \circledcirc \alpha) \circledast (\delta_1 \circledcirc \gamma_1, \cdots, \delta_n \circledcirc \gamma_n).$$

Definition 4. *Let* $(\mathbb{C}, U, M), (\mathbb{D}, U', M')$ *be monoids in* $\mathbf{MSMC}_l$. *A* vertical monoid homomorphism *consists of a symmetric colax monoidal functor* $A : \mathbb{C} \to \mathbb{D}$ *and prisms* $\epsilon : () \to A : U \to U'$ *and* $\delta : (A, A) \to A : M \to M'$*:*

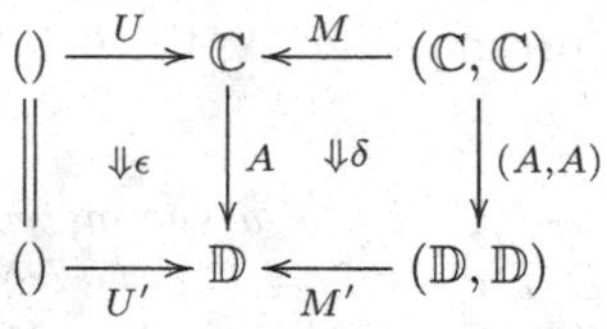

such that the following prism equalities hold:

$$\delta \circledast (\mathrm{id}, \epsilon) = \mathrm{id}, \quad \delta \circledast (\epsilon, \mathrm{id}) = \mathrm{id}, \quad \delta \circledast (\mathrm{id}, \delta) = \delta \circledast (\delta, \mathrm{id}).$$

The above prism equalities amounts to the following equality of natural transformations:

$$M'(AX, \epsilon) \circ \delta_{X,U} = \mathrm{id} \quad M'(\epsilon, AX) \circ \delta_{U,X} = \mathrm{id}$$
$$M'(AX, \delta_{Y,Z}) \circ \delta_{X,M(Y,Z)} = M'(\delta_{X,Y}, AZ) \circ \delta_{M(X,Y),Z}$$

With this concept, we can concisely capture R-graded linear exponential comonads:

Theorem 4. *There is a bijective correspondence between*

1. *A vertical monoid homomorphism* (D, ϵ, δ) *from* $(R^+, 1, *)$ *to* $([\mathbb{C}, \mathbb{C}]_l, \mathrm{Id}, \circ)$.
2. *An* R*-graded linear exponential comonad on* $\mathbb{C}$.

Vertical monoid homomorphisms vertically compose. Therefore we can extend a graded linear exponential comonad (as a vertical monoid homomorphism) by stacking vertical monoid homomorphisms.

Proposition 4. *Let R, S be partially ordered semirings. Then a vertical monoid homomorphism from $(R^+, 1_R, *_R)$ to $(S^+, 1_S, *_S)$ bijectively corresponds to a monotone function $h : (R, \leq_R) \to (S, \leq_S)$ such that $h(\sum_R r_i) \leq \sum_S h(r_i)$ and $h(\prod_R r_i) \leq \prod_S h(r_i)$ (which we call* colax homomorphism*).*

Proposition 5. *Let $F \dashv U : \mathbb{C} \to \mathbb{D}$ be a symmetric lax monoidal adjunction. Then the functor $V^{F\dashv U}$ defined by $V^{F\dashv U}H = F \circ H \circ U$ is a vertical monoid homomorphism from $([\mathbb{C}, \mathbb{C}]_l, \mathrm{Id}, \circ)$ to $([\mathbb{D}, \mathbb{D}]_l, \mathrm{Id}, \circ)$.*

Proof. Let $F \dashv U : \mathbb{C} \to \mathbb{D}$ be a symmetric lax monoidal adjunction. From Kelly's doctrinal adjunction, F is symmetric strong monoidal, hence so is $F \circ -$ in the following diagram:

$$V^{F\dashv U} = [\mathbb{C}, \mathbb{C}]_l \xrightarrow{F\circ -} [\mathbb{C}, \mathbb{D}]_l \xrightarrow{-\circ U} [\mathbb{D}, \mathbb{D}]_l$$

Next, $-\circ U$ above is always symmetric strict monoidal. By composing them, we obtain that $V^{F\dashv U}$ is symmetric strong, hence colax monoidal. We next introduce prisms (ϵ, δ) of the following type:

$$\begin{array}{ccccc}
() & \xrightarrow{\mathrm{Id}} & [\mathbb{C}, \mathbb{C}]_l & \xleftarrow{\circ} & ([\mathbb{C}, \mathbb{C}]_l, [\mathbb{C}, \mathbb{C}]_l) \\
\| & \Downarrow\epsilon & \downarrow V^{F\dashv U} & \Downarrow\delta & \downarrow (V^{F\dashv U}, V^{F\dashv U}) \\
() & \xrightarrow[\mathrm{Id}]{} & [\mathbb{D}, \mathbb{D}]_l & \xleftarrow[\circ]{} & ([\mathbb{D}, \mathbb{D}]_l, [\mathbb{D}, \mathbb{D}]_l)
\end{array}$$

We define ϵ to be the counit of the adjunction $F \dashv U$, which is monoidal natural, and δ be the following natural transformation:

$$\delta_{H_1, H_2} = V^{F\dashv U}(H_1 \circ \eta \circ H_2) : V^{F\dashv U}(H_1 \circ H_2) \to V^{F\dashv U}H_1 \circ V^{F\dashv U}H_2$$

It is routine to check that this satisfies the axioms of prism. □

Theorem 5. *Let R be a partially ordered semiring and D be an R-graded linear exponential comonad on a symmetric monoidal category $\mathbb{C}$. We moreover let S be another partially ordered semiring, $h : S \to R$ be a colax homomorphism and $F \dashv U : \mathbb{C} \to \mathbb{D}$ be a symmetric lax monoidal adjunction. Then the following composite of vertical monoid homomorphisms is an S-graded linear exponential comonad on $\mathbb{D}$.*

$$(S^+, 1_S, *_S) \xrightarrow{h} (R^+, 1_R, *_R) \xrightarrow{D} ([\mathbb{C}, \mathbb{C}]_l, \mathrm{Id}_{\mathbb{C}}, \circ) \xrightarrow{V^{F\dashv U}} ([\mathbb{D}, \mathbb{D}]_l, \mathrm{Id}_{\mathbb{D}}, \circ)$$

We call the above composite the *extension* of D with $F \dashv U$ and h.

7 From Monoid Actions to Graded Comonoid-Coalgebras

Let $(D, \epsilon, \delta) : (R^+, 1, *) \to ([\mathbb{C}, \mathbb{C}]_l, \mathrm{Id}, \circ)$ be an R-graded linear exponential comonad as a vertical monoid homomorphism. The prism equations in

Definition 4 suggests that the vertical monoid homomorphism itself can be seen as a monoid. We can thus consider *monoid actions* of (D, ϵ, δ): it consists of a prism

$$\begin{array}{ccc} (R^+, R^+) & \xrightarrow{\quad * \quad} & R^+ \\ {\scriptstyle (D,A)}\big\downarrow & \Downarrow a & \big\downarrow{\scriptstyle A} \\ ([\mathbb{C}, \mathbb{C}]_l, \mathbb{C}) & \xrightarrow[\quad ev \quad]{} & \mathbb{C} \end{array}$$

such that the following prism equations hold:

$$a \circledast (\delta, \mathrm{id}) = a \circledast (\mathrm{id}, a), \quad a \circledast (\epsilon, \mathrm{id}) = \mathrm{id}.$$

We note that this makes sense because $(*)$ and ev are also monoid actions in $\mathbf{MSMC}_l$; see Example 5. By unfolding this definition, we obtain the following structure, which we name *graded comonoid-coalgebra.*

Definition 5. *Let R be a partially ordered semiring. An R-*graded comonoid-coalgebra *of an R-graded linear exponential comonad $(D, w, c, \epsilon, \delta)$ on a symmetric monoidal category $\mathbb{C}$ is a tuple (A, a, u, o) such that*

- *$(A, u, o) : R^+ \to \mathbb{C}$ is a symmetric colax monoidal functor.*
- *$a_{r,r'} : A(r * r') \to D(r)(A(r'))$ is a natural transformation.*

They satisfy the following six equational axioms:

$$\begin{array}{ccc} A(r * s * t) & \xrightarrow{a_{r,s*t}} & D(r)(A(s * t)) \\ {\scriptstyle a_{r*s,t}}\big\downarrow & & \big\downarrow{\scriptstyle D(r)(a_{s,t})} \\ D(r * s)(A(t)) & \xrightarrow[\delta_{r,s,A(t)}]{} & D(r)(D(s)(A(t))) \end{array} \qquad \begin{array}{ccc} A(1 * t) & \xrightarrow{a_{1,t}} & D(1)(A(t)) \\ \| & & \big\downarrow{\scriptstyle \epsilon_{A(t)}} \\ & & A(t) \end{array}$$

$$\begin{array}{ccc} A(0) & = & A(0 * r) \\ & & \big\downarrow{\scriptstyle a_{0,r}} \\ {\scriptstyle u}\big\downarrow & & D(0)(A(r)) \\ & & \big\downarrow{\scriptstyle w_{A(r)}} \\ \mathbf{I} & = & \mathbf{I} \end{array} \qquad \begin{array}{ccc} A(0) & = & A(r * 0) \\ & & \big\downarrow{\scriptstyle a_{r,0}} \\ {\scriptstyle u}\big\downarrow & & D(r)(A(0)) \\ & & \big\downarrow{\scriptstyle D(r)(u)} \\ \mathbf{I} & \xrightarrow[m_r]{} & D(r)(\mathbf{I}) \end{array}$$

$$\begin{array}{ccc} A(s * r + t * r) & = & A((s + t) * r) \\ {\scriptstyle o_{s*r,t*r}}\big\downarrow & & \big\downarrow{\scriptstyle a_{s+t,r}} \\ A(s * r) \otimes A(t * r) & & D(s + t)(A(r)) \\ {\scriptstyle a_{s,r} \otimes a_{t,r}}\big\downarrow & & \big\downarrow{\scriptstyle c_{s,t,A(r)}} \\ D(s)(A(r)) \otimes D(t)(A(r)) & = & D(s)(A(r)) \otimes D(t)(A(r)) \end{array}$$

$$
\begin{array}{ccc}
A(r*s+r*t) & = & A(r*(s+t)) \\
\Big\downarrow{\scriptstyle o_{r*s,r*t}} & & \Big\downarrow{\scriptstyle a_{r,s+t}} \\
A(r*s)\otimes A(r*t) & & D(r)(A(s+t)) \\
\Big\downarrow{\scriptstyle a_{r,s}\otimes a_{r,t}} & & \Big\downarrow{\scriptstyle D(r)(o_{s,t})} \\
D(r)(A(s))\otimes D(r)(A(t)) & \xrightarrow[m_{r,A(s),A(t)}]{} & D(r)(A(s)\otimes A(t))
\end{array}
$$

A morphism from an R-graded comonoid-coalgebra (A, a, u, o) *to another* (B, b, v, p) *is a monoidal natural transformation* $h : (A, u, o) \to (B, v, p)$ *such that* h *satisfies:*

$$
\begin{array}{ccc}
A(r*s) & \xrightarrow{h_{r*s}} & B(r*s) \\
\Big\downarrow{\scriptstyle a_{r,s}} & & \Big\downarrow{\scriptstyle b_{r,s}} \\
Dr(As) & \xrightarrow[Drh_s]{} & Dr(Bs)
\end{array}
$$

We write $C(\mathbb{C}, D)$ *for the category of R-graded comonoid-coalgebras of* D.

Proposition 6. *Let* R *be a partially ordered semiring and* $(D, w, c, \epsilon, \delta)$ *be an R-graded linear exponential comonad on a symmetric monoidal category* $\mathbb{C}$. *The following gives a symmetric monoidal structure on* $C(\mathbb{C}, D)$:

$\mathbf{I} = (\dot{\mathbf{I}}, (\lambda r, s\ .\ m_r), \mathrm{id}_{\mathbf{I}}, (\lambda r, s\ .\ \iota^{-1}))$

$(A, a, u, o) \otimes (B, b, v, p)$

$= (A \dot{\otimes} B,\ \ \lambda r, r'\ .\ m_{r,Ar',Br'} \circ (a_{r,r'} \otimes b_{r,r'}),\ \ \iota \circ (u \otimes v),\ \ \lambda r, r'\ .\ \tau \circ (o_{r,r'} \otimes p_{r,r'}))$

$(f \otimes g)_r = f_r \otimes g_r$

$(\lambda_A)_r = \lambda_{Ar}, \quad (\rho_A)_r = \rho_{Ar}, \quad (\alpha_{A,B,C})_r = \alpha_{Ar,Br,Cr}, \quad (\sigma_{A,B})_r = \sigma_{Ar,Br}$

When $R = 1$, The category $C(\mathbb{C}, D)$ reduces to the category of Eilenberg-Moore coalgebras of the non-graded linear exponential comonad.

Theorem 6. *Let* $(D, w, c, \epsilon, \delta)$ *be a 1-graded linear exponential comonad on a symmetric monoidal category* $\mathbb{C}$. *Then the category* $C(\mathbb{C}, D)$ *is strong monoidally isomorphic to the category* $\mathbb{C}^D$ *of Eilenberg-Moore coalgebras of the comonad* (D, ϵ, δ).

Like $\mathbb{C}^D$, there is a symmetric lax monoidal adjunction of the following type:

$$
C(\mathbb{C}, D) \underset{U}{\overset{F}{\underset{\longleftarrow}{\overset{\longrightarrow}{\bot}}}} \mathbb{C}
$$

but this itself is not enough to recover D — D takes two arguments, while the composite $F \circ U$ is only equal to the symmetric lax monoidal comonad $D1$ on $\mathbb{C}$.

The category $C(\mathbb{C}, D)$ actually carries an *R-twist* T, which acts on comonoid-coalgebras as follows:

$$Tr(A, \cdots) = (A(- * r), \cdots),$$

and D is recovered as the extension of T with the adjunction $F \dashv U$ (Theorem 5).

Theorem 7. *Let R be a partially ordered semiring and $(D, w, c, \epsilon, \delta)$ be an R-graded linear exponential comonad on a symmetric monoidal category $\mathbb{C}$.*

1. *The functor $F : C(\mathbb{C}, D) \to \mathbb{C}$ given by $F(A, a, u, o) = A1$ and $Fh = h_1$ is symmetric strict monoidal, and has a symmetric lax monoidal right adjoint $U : \mathbb{C} \to C(\mathbb{C}, D)$, whose object part is given by $UA = (\lambda r \,.\, DrA, \lambda r, r' \,.\, \delta_{r,r',A}, w_A, \lambda r, r' \,.\, c_{r,r',A})$.*
2. *The following data give an R-twist T on $C(\mathbb{C}, D)$:*

   $$TrA = (\lambda s \,.\, A(s * r),\ \ \lambda s, s' \,.\, a_{s,s'*r},\ \ u,\ \ \lambda s, s' \,.\, o_{s*r,s'*r}),\quad (Trh)_t = h_{t*r}$$
   $$(m^T_r)_t = \mathrm{id}_{\mathbf{I}},\quad (m^T_{r,A,B})_t = \mathrm{id}_{A(t*r)\otimes B(t*r)},\quad (w^T_A)_t = u,\quad (c^T_{r,s,A})_t = o_{t*r,t*s}.$$

   Here, $A = (A, a, u, o)$ and B are R-graded comonoid coalgebras. From the definition of twists, ϵ^T, δ^T are identities.
3. *The extension of D with $F \dashv U$ (Theorem 5) coincides with the R-graded linear exponential comonad D.*

The following classic result [1, Theorem 6-1] can be reproved by Theorem 7.

Corollary 1. *Let $\mathbb{C}$ be a symmetric monoidal category and Let D be a non-graded linear exponential comonad on $\mathbb{C}$. The canonical symmetric monoidal structure on the category $\mathbb{C}^D$ of Eilenberg-Moore coalgebras of D is cartesian.*

Proof. From Theorem 1, D is a 1-graded linear exponential comonad on $\mathbb{C}$. Therefore $C(\mathbb{C}, D)$ has a 1-twist by Theorem 7-3. Therefore the symmetric monoidal structure of $C(\mathbb{C}, D)$ is cartesian by Theorem 2. Finally, $C(\mathbb{C}, D)$ is strong monoidally isomorphic to $\mathbb{C}^D$ by Theorem 6, hence the symmetric monoidal structure of $\mathbb{C}^D$ is also cartesian. □

We show the finality of the category of graded comonoid-coalgebras. Let R be a partially ordered semiring and D be an R-graded linear exponential comonad on a symmetric monoidal category $\mathbb{C}$. We define a *resolution* of D to be a pair of a symmetric lax monoidal adjunction $J \dashv K : \mathbb{E} \to \mathbb{C}$ and an R-twist (S, w^S, c^S) on $\mathbb{E}$ such that the extension of S with $J \dashv K$ is equal to D. Then the following set of data becomes a strong monoidal functor $(M, m^M, m^M_{E,E'}) : \mathbb{E} \to C(\mathbb{C}, D)$:

$$ME = (\lambda r \,.\, J(Sr)E,\ \ \lambda r, r' \,.\, J(Sr)\eta^{J \dashv K}_{Sr'E},\ \ (m^J)^{-1} \circ w^S_E,\ \ \lambda r, r' \,.\, (m^J_{SrE,Sr'E})^{-1} \circ Jc^S_{r,r',E})$$
$$(Mf)_r = J(Sr)f,\quad (m^M)_r = J(m^S_r) \circ m^J,\quad (m^M_{E,E'})_r = J(m^S_{r,E,E'}) \circ m^J_{SrE,SrE'}$$

(recall that Sr, J are both symmetric strong monoidal).

Theorem 8. *The above M is the unique symmetric strong monoidal functor such that:*

1. *Equality of symmetric lax monoidal functors $M \circ K = U$ and $F \circ M = J$ hold.*
2. *Let $M^* = - \circ M$ and $M_* = M \circ -$ be induced symmetric strict (resp. strong) monoidal functors. Then the following square of symmetric colax monoidal functors commutes.*

$$\begin{array}{ccc} R^+ & \xrightarrow{\quad S \quad} & [\mathbb{E}, \mathbb{E}]_l \\ {\scriptstyle T}\downarrow & & \downarrow{\scriptstyle M_*} \\ [C(\mathbb{C}, D), C(\mathbb{C}, D)]_l & \xrightarrow[\quad M^* \quad]{} & [\mathbb{E}, C(\mathbb{C}, D)]_l \end{array}$$

8 Conclusion

We have given a concise characterization of graded linear exponential comonad as a vertical monoid homomorphism (D, ϵ, δ) from $(R^+, 1, *)$ to $([\mathbb{C}, \mathbb{C}]_l, \mathrm{Id}, \circ)$. This characterization is built upon a combination of the theory of symmetric lax monoidal multifunctors and Grandis and Paré's double category of symmetric monoidal categories. After this characterization, we considered *monoid actions*, and derived the concept of graded comonoid-coalgebras. The category of graded comonoid-coalgebras are shown to give a resolution of the graded linear exponential comonad D. These results are consistent with the theory of non-graded linear exponential comonads developed in [1].

It remains to be seen if the category of graded comonoid-coalgebras can be constructed in a purely double-category theoretic way. In non-graded case, there are other type of categorical models of exponential modality using *Lafont category* and *Seely category* [17]. Graded version of these categories are also an interesting research topic.

Acknowledgment. The author is grateful to Marco Gaboardi, Naohiko Hoshino, Flavien Breuvart, Soichiro Fujii and Paul-Andrè Melliès for many fruitful discussions. This research was supported by JSPS KAKENHI Grant Number JP15K00014 and ERATO Hasuo Metamathematics for Systems Design Project (No. JPMJER1603), JST.

References

1. Benton, N., Bierman, G., de Paiva, V., Hyland, M.: Linear λ-calculus and categorical models revisited. In: Börger, E., Jäger, G., Kleine Büning, H., Martini, S., Richter, M.M. (eds.) CSL 1992. LNCS, vol. 702, pp. 61–84. Springer, Heidelberg (1993). https://doi.org/10.1007/3-540-56992-8_6
2. Breuvart, F., Pagani, M.: Modelling coeffects in the relational semantics of linear logic. In: Kreutzer, S. (ed.) 24th EACSL Annual Conference on Computer Science Logic, CSL 2015, 7–10 September 2015, Berlin, Germany, vol. 41. LIPIcs, pp. 567–581. Schloss Dagstuhl - Leibniz-Zentrum fuer Informatik (2015)

3. Brunel, A., Gaboardi, M., Mazza, D., Zdancewic, S.: A core quantitative coeffect calculus. In: Shao [24], pp. 351–370
4. Dal Lago, U., Schöpp, U.: Functional programming in sublinear space. In: Gordon, A.D. (ed.) ESOP 2010. LNCS, vol. 6012, pp. 205–225. Springer, Heidelberg (2010). https://doi.org/10.1007/978-3-642-11957-6_12
5. de Amorim, A.A., Gaboardi, M., Hsu, J., Katsumata, S., Cherigui, I.: A semantic account of metric preservation. In: Castagna, G., Gordon, A.D. (eds.) Proceedings of the 44th ACM SIGPLAN Symposium on Principles of Programming Languages, POPL 2017, Paris, France, 18–20 January 2017, pp. 545–556. ACM (2017)
6. Durov, N.: New approach to Arakelov geometry. arXiv: 0704.2030 v1 [math AG], April 2008
7. Gaboardi, M., Katsumata, S.-Y., Orchard, D.A., Breuvart, F., Uustalu, T.: Combining effects and coeffects via grading. In: Garrigue, J., Keller, G., Sumii, E. (eds.) Proceedings of the 21st ACM SIGPLAN International Conference on Functional Programming, ICFP 2016, Nara, Japan, 18–22 September 2016, pp. 476–489. ACM (2016)
8. Ghica, D.R., Smith, A.I.: Bounded linear types in a resource semiring. In: Shao [24], pp. 331–350
9. Girard, J.-Y., Scedrov, A., Scott, P.J.: Bounded linear logic: a modular approach to polynomial-time computability. Theoret. Comput. Sci. **97**(1), 1–66 (1992)
10. Grandis, M., Paré, R.: Adjoint for double categories. Cahiers de Topologie et Gomtrie Diffrentielle Catgoriques **45**(3), 193–240 (2004)
11. Hyland, M., Power, J.: Pseudo-commutative monads and pseudo-closed 2-categories. J. Pure Appl. Algebra **175**(1), 141–185 (2002). Special Volume celebrating the 70th birthday of Professor Max Kelly
12. Jacobs, B., Löding, C. (eds.): FoSSaCS 2016. LNCS, vol. 9634. Springer, Heidelberg (2016). https://doi.org/10.1007/978-3-662-49630-5
13. Katsumata, S.: Parametric effect monads and semantics of effect systems. In: Jagannathan, S., Sewell, P. (eds.) The 41st Annual ACM SIGPLAN-SIGACT Symposium on Principles of Programming Languages, POPL 2014, San Diego, CA, USA, 20–21 January 2014, pp. 633–646. ACM (2014)
14. Lago, U.D., Gaboardi, M.: Linear dependent types and relative completeness. In: 2011 IEEE 26th Annual Symposium on Logic in Computer Science, pp. 133–142, June 2011
15. Laplaza, M.L.: Coherence for distributivity. In: Kelly, G.M., Laplaza, M., Lewis, G., Mac Lane, S. (eds.) Coherence in Categories. LNM, vol. 281, pp. 29–65. Springer, Heidelberg (1972). https://doi.org/10.1007/BFb0059555
16. Mac Lane, S.: Categories for the Working Mathematician. GTM, vol. 5. Springer, New York (1978). https://doi.org/10.1007/978-1-4757-4721-8
17. Melliès, P.-A.: Categorical semantics of linear logic. In: Interactive Models of Computation and Program Behaviour, vol. 27. Panoramas et synthses. Société Mathématique de France (2009)
18. Melliès, P.-A.: The parametric continuation monad. Math. Struct. Comput. Sci. **27**(5), 651–680 (2017)
19. Milius, S., Pattinson, D., Schröder, L.: Generic trace semantics and graded monads. In: Moss, L.S., Sobocinski, P. (eds.) 6th Conference on Algebra and Coalgebra in Computer Science, CALCO 2015, 24–26 June 2015, Nijmegen, The Netherlands, vol. 35. LIPIcs, pp. 253–269. Schloss Dagstuhl - Leibniz-Zentrum fuer Informatik (2015)

20. Petricek, T., Orchard, D., Mycroft, A.: Coeffects: unified static analysis of context-dependence. In: Fomin, F.V., Freivalds, R., Kwiatkowska, M., Peleg, D. (eds.) ICALP 2013. LNCS, vol. 7966, pp. 385–397. Springer, Heidelberg (2013). https://doi.org/10.1007/978-3-642-39212-2_35
21. Reed, J., Pierce, B.C.: Distance makes the types grow stronger: a calculus for differential privacy. In: Hudak, P., Weirich, S. (eds.) Proceeding of the 15th ACM SIGPLAN International Conference on Functional Programming, ICFP 2010, Baltimore, Maryland, USA, 27–29 September 2010, pp. 157–168. ACM (2010)
22. Sato, T.: Approximate relational Hoare logic for continuous random samplings. Electr. Notes Theor. Comput. Sci. **325**, 277–298 (2016)
23. Schöpp, U.: Computation-by-interaction with effects. In: Yang, H. (ed.) APLAS 2011. LNCS, vol. 7078, pp. 305–321. Springer, Heidelberg (2011). https://doi.org/10.1007/978-3-642-25318-8_23
24. Shao, Z. (ed.): ESOP 2014. LNCS, vol. 8410. Springer, Heidelberg (2014). https://doi.org/10.1007/978-3-642-54833-8
25. Smirnov, A.L.: Graded monads and rings of polynomials. J. Math. Sci. **151**(3), 3032–3051 (2008)
26. Street, R.: Two constructions on lax functors. Cahiers de Topologie et Géométrie Différentielle Catégoriques (1972)

Depending on Session-Typed Processes

Bernardo Toninho[1,2](✉) and Nobuko Yoshida[2]

[1] NOVA-LINCS, Departamento de Informática, FCT, Universidade Nova de Lisboa, Lisbon, Portugal
[2] Imperial College London, London, UK
b.toninho@imperial.ac.uk

Abstract. This work proposes a dependent type theory that combines functions and session-typed processes (with value dependencies) through a contextual monad, internalising typed processes in a dependently-typed λ-calculus. The proposed framework, by allowing session processes to depend on functions and vice-versa, enables us to specify and statically verify protocols where the choice of the next communication action can depend on specific values of received data. Moreover, the type theoretic nature of the framework endows us with the ability to internally describe and prove predicates on process behaviours. Our main results are type soundness of the framework, and a faithful embedding of the functional layer of the calculus within the session-typed layer, showcasing the expressiveness of dependent session types.

1 Introduction

Session types [14,24] are a typing discipline for communication protocols, whose simplicity provides an extensible framework that allows for integration with a variety of functional type features. One useful instance arising from the proof theoretic exploration of logical quantification is *value dependent session types* [25]. In this work, one can express properties of exchanged data in protocol specifications separately from communication, but *cannot* describe protocols where communication actions depend on the actual exchanged data (e.g. [16, Sect. 2]). Moreover, it does not allow functions or values to depend on protocols (i.e. sessions) or communication, thus preventing reasoning about dependent process behaviours, exploring the proofs-as-programs paradigm of dependent type theory, e.g. [8,17].

Our work addresses the limitations of existing formulations of session types by proposing a type theory that integrates dependent functions *and* session types using a *contextual monad*. This monad internalises a session-typed calculus within a dependently-typed λ-calculus. By allowing session types to depend on λ-terms *and* λ-terms to depend on typed processes (using the monad), we are able to achieve heightened degrees of expressiveness. Exploiting the former direction, we enable writing actual data-dependent communication protocols. Exploiting the latter, we can define and *prove* properties of linearly-typed objects (i.e. processes) within our intuitionistic theory.

C. Baier and U. Dal Lago (Eds.): FOSSACS 2018, LNCS 10803, pp. 128–145, 2018.
https://doi.org/10.1007/978-3-319-89366-2_7

To informally demonstrate how our type theory goes beyond the state of the art in order to represent data-dependent protocols, consider the following session type (we write $\tau \wedge A$ for $\exists x{:}\tau.A$ where x does not occur in A and similarly $\tau \supset A$ for $\forall x{:}\tau.A$ when x is not free in A), $T \triangleq \mathsf{Bool} \supset \oplus\{\mathtt{t} : \mathsf{Nat} \wedge \mathbf{1}, \mathtt{f} : \mathsf{Bool} \wedge \mathbf{1}\}$, representable in existing session typing systems. The type T denotes a protocol which first, inputs a boolean and then either emits the label t, which will be followed by an output of a natural number; or emits the label f and a boolean. The intended protocol described by T is to take the t branch if the received value is t and the f branch otherwise, which we can implement as Q with channel z typed by T as follows:

$$Q \triangleq z(x).\mathsf{case}\ x\ \mathsf{of}\ (\mathsf{true} \Rightarrow z.\mathtt{t}; z\langle 23\rangle.\mathbf{0},\ \mathsf{false} \Rightarrow z.\mathtt{f}; z\langle \mathsf{true}\rangle.\mathbf{0})$$

where $z(x).P$ denotes an input process, $z.\mathtt{t}$ is a process which selects label t and $z\langle 23\rangle.P$ is an output on z. However, since the specification is imprecise, process $z(x).\mathsf{case}\ x\ \mathsf{of}\ (\mathsf{false} \Rightarrow z.\mathtt{t}; z\langle 23\rangle.\mathbf{0},\ \mathsf{true} \Rightarrow z.\mathtt{f}; z\langle \mathsf{true}\rangle.\mathbf{0})$ is also a type-correct implementation of T that does not adhere to the intended protocol. Using our dependent type system, we can narrow the specification to guarantee that the desired protocol is precisely enforced. Consider the following definition of a session-type level conditional where we assume inductive definition and dependent pattern matching mechanisms (stype denotes the *kind* of session types):

$$\mathtt{if} :: \mathsf{Bool} \rightarrow \mathsf{stype} \rightarrow \mathsf{stype} \rightarrow \mathsf{stype}$$
$$\mathtt{if}\ \mathsf{true}\ A\, B\ =\ A \qquad \mathtt{if}\ \mathsf{false}\ A\, B\ =\ B$$

The type-level function above case analyses the boolean and produces its first session type argument if the value is true and the second otherwise. We may now specify a session type that faithfully implements the protocol:

$$T' \triangleq \forall x{:}\mathsf{Bool}.\mathtt{if}\ x\ (\mathsf{Nat} \wedge \mathbf{1})\ (\mathsf{Bool} \wedge \mathbf{1})$$

A process R implementing such a type on channel z is given below:

$$R \triangleq z(x).\mathsf{case}\ x\ \mathsf{of}\ (\mathsf{true} \Rightarrow z\langle 23\rangle.\mathbf{0},\ \mathsf{false} \Rightarrow z\langle \mathsf{true}\rangle.\mathbf{0})$$

Note that if we flip the two branches of the case analysis in R, the session is no longer typable with T', ensuring that the protocol is implemented faithfully.

The example above illustrates a simple yet useful data-dependent protocol. When we further extend our dependent types with a *process* monad [29], where $\{c \leftarrow P \leftarrow \overline{u_j}; \overline{d_i}\}$ is a functional term denoting a process that may be *spawned* by other processes by instantiating the names in $\overline{u_j}$ and $\overline{d_i}$, we can provide more powerful reasoning on processes, enabling refined specifications through the use of type indices (i.e. type families) and an ability to internally specify and verify predicates on process behaviours. We also show that *all* functional types and terms can be faithfully embedded in the process layer using the dependently-typed sessions and process monads.

Contributions. Section 2 introduces our dependent type theory, augmenting the example above by showing how we can reason about process behaviour using

Kinds	K, K'	$::= \mathsf{type} \mid \mathsf{stype} \mid \Pi x{:}\tau.K \mid \Pi t{:}K.K'$
Functional	τ, σ	$::= \Pi x{:}\tau.\sigma \mid \lambda x{:}\tau.\sigma \mid \tau\ M \mid \{\overline{u_j{:}B_j}; \overline{d_i{:}A_i} \vdash c{:}A\} \mid \lambda t :: K.\tau \mid \tau\,\sigma$
Sessions	A, B	$::= {!A} \mid A \multimap B \mid A \otimes B \mid \forall x{:}\tau.A \mid \exists x{:}\tau.A \mid \mathbf{1}$
		$\mid\ \&\{\overline{l_i : A_i}\} \mid \oplus\{\overline{l_i : A_i}\} \mid \lambda x{:}\tau.A \mid A\ M \mid \lambda t{::}K.A \mid A\,B$
Terms	M, N	$::= \lambda x{:}\tau.M \mid \{c \leftarrow P \leftarrow \overline{u_j}; \overline{d_i}\} \mid M\ N \mid x$
Processes	P, Q	$::= \overline{c}\langle d\rangle.P \mid (\nu c)P \mid c(x).P \mid c\langle M\rangle.P \mid {!c(x).P}$
		$\mid\ c.\mathsf{case}\{\overline{l_i \Rightarrow P_i}\} \mid c.l; P \mid [c \leftrightarrow d] \mid \mathbf{0} \mid c \leftarrow M \leftarrow \overline{u_j}; \overline{d_i}; Q$

Fig. 1. Syntax of kinds, types, terms and processes

type families and dependently-typed functions (Sect. 2.3). We then establish the soundness of the theory (Sect. 2.4). Section 3 develops a faithful embedding of the dependent function space in the process layer (Theorem 3.4). Section 4 concludes with related work. Proofs, omitted definitions and additional examples can be found in [32].

2 A Dependent Type Theory of Processes

This section introduces our dependent type theory combining session-typed processes and functions. The theory is a generalisation of the line of work relating linear logic and session types [4,25,29], considering type-level functions and dependent kinds in an intensional type theory with full *mutual* dependencies between functions and processes. This generalisation enables us to express more sophisticated session types (such as those of Sect. 1) and also to define and *prove* properties of processes expressed as type families with proofs as their inhabitants. We focus on the new rules and judgements, pointing the interested reader to [5,25,26] for additional details on the base theory.

2.1 Syntax

The calculus is stratified into two mutually dependent layers of processes and terms, which we often refer to as the *process* and *functional* layers, respectively. The syntax of the theory is given in Fig. 1 (we use x, y for variables ranging over terms and t for variables ranging over types).

Types and Kinds. The process layer is able to refer to terms of the functional layer via appropriate (dependently-typed) communication actions and through a *spawn* construct, allowing for processes encapsulated as functional values to be executed. Dually, the functional layer can refer to the process layer via a *contextual* monad [29] that internalises (open) typed processes as opaque functional values. This mutual dependency is also explicit in the type structure on several axes: process channel usages are typed by a language of session types, which specifies the communication protocols implemented on the used channels, extended with two dependent communication operations $\forall x{:}\tau.A$ and $\exists x{:}\tau.A$, where τ is a functional type and A is a session type in which x may occur. Moreover, we also extend the language of session types with type-level λ-abstraction over terms

$\lambda x{:}\tau.A$ and session types $\lambda t :: K.A$ (with the corresponding elimination forms $A\,M$ and $A\,B$). As we show in Sect. 1, the combination of these features allows for a new degree of expressiveness, enabling us to construct session types whose structure depends on previously communicated values.

The remaining session constructs are standard, following [5]: $!A$ denotes a *shared* session of type A that may be used an arbitrary (finite) number of times; $A \multimap B$ represents a session offering to input a session of type A to then offer the session behaviour B; $A \otimes B$ is the dual operator, denoting a session that outputs A and proceeds as B; $\oplus\{\overline{l_i : A_i}\}$ and $\&\{\overline{l_i : A_i}\}$ represent internal and external labelled choice, respectively; $\mathbf{1}$ denotes the terminated session.

The functional layer is a λ-calculus with dependent functions $\Pi x{:}\tau.\sigma$, type-level λ-abstractions over terms and types (and respective type-level applications) and a *contextual monadic* type $\{\overline{u_j{:}B_j}; \overline{d_i{:}A_i} \vdash c{:}A\}$, denoting a (quoted) process offering session $c{:}A$ by using the *linear* sessions $\overline{d_i{:}A_i}$ and *shared* sessions $\overline{u_j{:}B_j}$ [29]. We often write $\{A\}$ for $\{\cdot;\cdot \vdash c{:}A\}$. The kinding system for our theory contains two base kinds type and stype of functional and session types, respectively. Type-level λ-abstractions require dependent kinds $\Pi x{:}\tau.K$ and $\Pi t :: K.K'$, respectively. We note that the functional connectives form a standard dependent type theory [11,21].

Terms and Processes. Terms include the standard λ-abstractions $\lambda x{:}\tau.M$, applications $M\,N$ and variables x. In order to internalise processes within the functional layer we make use of a monadic process wrapper, written $\{c \leftarrow P \leftarrow \overline{u_j}; \overline{d_i}\}$. In such a construct, the channels c, $\overline{u_j}$ and $\overline{d_i}$ are bound in P, where c is the session channel being offered and $\overline{u_j}$ and $\overline{d_i}$ are the session channels (linear and shared, respectively) being used. We write $\{c \leftarrow P \leftarrow \epsilon\}$ when P does not use any ambient channels, which we abbreviate to $\{P\}$.

The syntax of processes follows that of [5] extended with the monadic elimination form $c \leftarrow M \leftarrow \overline{u_j}; \overline{d_i}; Q$. Such a process construct denotes a term M that is to be evaluated to a monadic value of the form $\{c \leftarrow P \leftarrow \overline{u_j}; \overline{d_i}\}$ which will then be executed in parallel with Q, sharing with it a session channel c and using the provided channels $\overline{u_j}$ and $\overline{d_i}$. We write $c \leftarrow M \leftarrow \epsilon; Q$ when no channels are provided for the execution of M and often abbreviate this to $c \leftarrow M; Q$. The process $\overline{c}\langle d\rangle.P$ denotes the output of the *fresh* channel d along channel c with continuation P, which binds d; $(\nu c)P$ denotes channel hiding, restricting the scope of c to P; $c(x).P$ denotes an input along c, bound to x in P; $c\langle M\rangle.P$ denotes the output of term M along c with continuation P; $!c(x).P$ denotes a replicated input which spawns copies of P; the construct $c.\mathsf{case}\{\overline{l_i \Rightarrow P_i}\}$ codifies a process that waits to receive some label l_j along c, with continuation P_j; dually, $c.l; P$ denotes a process that emits a label l along c and continues as P; $[c \leftrightarrow d]$ denotes a forwarder between c and d, which is operationally implemented as renaming; $P \mid Q$ denotes parallel composition and $\mathbf{0}$ the null process.

2.2 A Dependent Typing System

We now introduce our typing system, defined by a series of mutually inductive judgements, given in Fig. 2. We use Ψ to stand for a typing context for dependent

$\Psi \vdash$	Context Ψ is well-formed.
$\Psi; \Delta \vdash$	Context Δ is well-formed, under assumptions in Ψ.
$\Psi \vdash K$	K is a kind in context Ψ.
$\Psi \vdash \tau :: K$	τ is a (functional) type of kind K in context Ψ.
$\Psi \vdash A :: K$	A is a session type of kind K in context Ψ.
$\Psi \vdash M : \tau$	M has type τ in context Ψ.
$\Psi; \Gamma; \Delta \vdash P :: z{:}A$	P offers session $z{:}A$ when composed with processes offering sessions specified in Γ and Δ in context Ψ.
$\Psi \vdash K_1 = K_2$	Kinds K_1 and K_2 are equal.
$\Psi \vdash \tau = \sigma :: K$	Types τ and σ are equal of kind K.
$\Psi \vdash A = B :: K$	Session types A and B are equal of kind K.
$\Psi \vdash M = N : \tau$	Terms M and N are equal of type τ.
$\Psi \vdash \Delta = \Delta' :: \mathsf{stype}$	Contexts Δ and Δ' are equal, under the assumptions in Ψ.
$\Psi; \Gamma; \Delta \vdash P = Q :: z{:}A$	Processes P and Q are equal with typing $z{:}A$.

Fig. 2. Typing judgements

λ-terms (i.e. assumptions of the form $x{:}\tau$ or $t :: K$, not subject to exchange), Γ for a typing context for *shared* sessions of the form $u{:}A$ (implicitly subject to weakening and contraction) and Δ for a linear context of sessions $x{:}A$. The context well-formedness judgments $\Psi \vdash$ and $\Psi; \Delta \vdash$ require that types and kinds (resp. session types) in Ψ (resp. Δ) are well-formed. The judgments $\Psi \vdash K$, $\Psi \vdash \tau :: K$ and $\Psi \vdash A :: K$ codify well-formedness of kinds, functional and session types (with kind K), respectively. Their rules are standard.

Typing. An excerpt of the typing rules for terms and processes is given in Figs. 3 and 4, respectively, noting that typing enforces types to be of base kind type (respectively stype). The rules for dependent functions are standard, including the type conversion rule which internalises definitional equality of types. We highlight the introduction rule for the monadic construct, which requires the appropriate session types to be well-formed and the process P to offer $c{:}A$ when provided with the appropriate session contexts.

In the typing rules for processes (Fig. 4), presented as a set of right and left rules (the former identifying how to *offer* a session of a given type and the latter how to use such a session), we highlight the rules for dependently-typed communication and monadic elimination (for type-checking purposes we annotate constructs with the respective dependent type – this is akin to functional type theories). To offer a session $c{:}\exists x{:}\tau.A$ we send a term M of type τ and then offer a session $c{:}A\{M/x\}$; dually, to use such a session we perform an input along c, bound to x in Q, warranting a use of c as a session of (open) type A. The rules for the universal are dual. Offering a session $c{:}\forall x{:}\tau.A$ entails receiving on c a term of type τ and offering $c{:}A$. Using a session of such a type requires sending along c a term M of type τ, warranting the use of c as a session of type $A\{M/x\}$.

The rule for the monadic elimination form requires that the term M be of the appropriate monadic type and that the provided channels $\overline{u_j}$ and $\overline{y_i}$ adhere to the typing specified in M's type. Under these conditions, the process Q may then use the session c as session A. The type conversion rules reflect session type definitional equality in typing.

$$\frac{\Psi \vdash \tau :: \mathsf{type} \quad \Psi, x{:}\tau \vdash M : \sigma}{\Psi \vdash \lambda x{:}\tau.M : \Pi x{:}\tau.\sigma}\ (\Pi I) \qquad \frac{\Psi \vdash M : \Pi x{:}\tau.\sigma \quad \Psi \vdash N : \tau}{\Psi \vdash M\,N : \sigma\{N/x\}}\ (\Pi E)$$

$$\frac{\forall i,j.\Psi \vdash A_i, B_j :: \mathsf{stype} \quad \Psi; \overline{u_j{:}B_j}; \overline{d_i{:}A_i} \vdash P :: c{:}A}{\Psi \vdash \{c \leftarrow P \leftarrow \overline{u_j}; \overline{d_i}\} : \{\overline{u_j{:}B_j}; \overline{d_i : A_i} \vdash c{:}A\}}\ (\{\}I) \qquad \frac{\Psi \vdash M : \tau \quad \Psi \vdash \tau = \sigma :: \mathsf{type}}{\Psi \vdash M : \sigma}\ (\mathsf{Conv})$$

Fig. 3. Typing for terms (Excerpt – See [32])

$$\frac{\Psi \vdash M{:}\tau \quad \Psi; \Gamma; \Delta \vdash P :: c{:}A\{M/x\}}{\Psi; \Gamma; \Delta \vdash c\langle M\rangle_{\exists x:\tau.A}.P :: c{:}\exists x{:}\tau.A}\ (\exists\mathsf{R}) \qquad \frac{\Psi \vdash \tau :: \mathsf{type} \quad \Psi, x{:}\tau\ ; \Gamma; \Delta, c{:}A \vdash Q :: d{:}D}{\Psi\ ; \Gamma; \Delta, c{:}\exists x{:}\tau.A \vdash c(x{:}\tau).Q :: d{:}D}\ (\exists\mathsf{L})$$

$$\frac{\Psi \vdash \tau :: \mathsf{type} \quad \Psi, x{:}\tau\ ; \Gamma; \Delta \vdash P :: c{:}A}{\Psi; \Gamma; \Delta \vdash c(x{:}\tau).P :: c{:}\forall x{:}\tau.A}\ (\forall\mathsf{R}) \qquad \frac{\Psi \vdash M{:}\tau \quad \Psi; \Gamma; \Delta, c{:}A\{M/x\} \vdash Q :: d{:}D}{\Psi; \Gamma; \Delta, c{:}\forall x{:}\tau.A \vdash c\langle M\rangle_{\forall x:\tau.A}.Q :: d{:}D}\ (\forall\mathsf{L})$$

$$\frac{\Delta' = \overline{d_i : B_i} \quad \overline{u_j{:}C_j} \subseteq \Gamma \quad \Psi \vdash M : \{\overline{u_j{:}C_j}; \overline{d_i{:}B_i} \vdash c{:}A\} \quad \Psi; \Gamma; \Delta, c{:}A \vdash Q :: z{:}C}{\Psi; \Gamma; \Delta', \Delta \vdash c \leftarrow M \leftarrow \overline{u_j}; \overline{y_i}; Q :: z{:}C}\ (\{\}E)$$

$$\frac{\Psi; \Gamma; \Delta \vdash P :: z{:}A \quad \Psi \vdash A = B :: \mathsf{stype}}{\Psi; \Gamma; \Delta \vdash P :: z{:}B}\ (\mathsf{ConvR}) \qquad \frac{\Psi; \Gamma'; \Delta' \vdash P :: z{:}A \quad \Psi; \Gamma'; \Delta' = \Psi; \Gamma; \Delta}{\Psi; \Gamma; \Delta \vdash P :: z{:}A}\ (\mathsf{ConvL})$$

$$(\mathsf{cut})\ \frac{\Psi; \Gamma; \Delta \vdash P :: c{:}A \quad \Psi; \Gamma; \Delta', c{:}A \vdash Q :: d{:}D}{\Psi; \Gamma; \Delta, \Delta' \vdash (\nu c)(P \mid Q) :: d{:}D}$$

Fig. 4. Typing for processes (Excerpt – See [32])

Definitional Equality. The crux of any dependent type theory lies in its *definitional equality*. Type equality relies on equality of terms which, by including the monadic construct, necessarily relies on a notion of *process* equality.

Our presentation of an intensional definitional equality of terms follows that of [12], where we consider an intrinsically typed relation, including β and η conversion (similarly for type equality which includes β and η principles for the type-level λ-abstractions). An excerpt of the rules for term equality is given in Fig. 5. The remaining rules are congruence rules and closure under symmetry, reflexivity and transitivity. Rule ($\mathsf{TMEq}\beta$) captures the β-reduction, identifying a λ-abstraction applied to an argument with the substitution of the argument in the function body (typed with the appropriately substituted type). We highlight rule ($\mathsf{TMEq}\{\}\eta$), which codifies a general η-like principle for arbitrary terms of monadic type: We form a monadic term that applies the monadic elimination form to M, forwarding the result along the appropriate channel, which becomes a term equivalent to M.

$$\text{(TMEq}\beta\text{)}\ \frac{\Psi \vdash \tau :: \mathsf{type} \quad \Psi, x{:}\tau \vdash M : \sigma \quad \Psi \vdash N : \tau}{\Psi \vdash (\lambda x{:}\tau.M)\, N = M\{N/x\} : \sigma\{N/x\}} \qquad \text{(TMEq}\eta\text{)}\ \frac{\Psi \vdash M : \Pi x{:}\tau.\sigma \quad x \notin fv(M)}{\Psi \vdash \lambda x{:}\tau.M\, x = M : \Pi x{:}\tau.\sigma}$$

$$\text{(TMEq\{\}}\eta\text{)}\ \frac{\Psi \vdash M : \{\overline{u_j{:}B_j}; \overline{d_i{:}A_i} \vdash c{:}A\}}{\Psi \vdash \{c \leftarrow (y \leftarrow M; \overline{u_j}; \overline{d_i}; [y \leftrightarrow c]) \leftarrow \overline{u_j}; \overline{d_i}\} = M : \{\overline{u_j{:}B_j}; \overline{d_i{:}A_i} \vdash c{:}A\}}$$

Fig. 5. Definitional equality of terms (Excerpt – See [32])

$$\text{(PEqRed)}\ \frac{\Psi; \Gamma; \Delta \vdash P :: z{:}A \quad P \rightarrow Q \quad \Psi; \Gamma; \Delta \vdash Q :: z{:}A}{\Psi; \Gamma; \Delta \vdash P = Q :: z{:}A}$$

$$\text{(PEq}\forall\eta\text{)}\ \frac{}{\Psi; \Gamma; d{:}\forall x{:}\tau.A \vdash c(x).d\langle x\rangle.[d \leftrightarrow c] = [d \leftrightarrow c] :: c{:}\forall x{:}\tau.A}$$

$$\text{(PEqCC}\forall\text{)}\ \frac{\Psi; \Gamma; \Delta \vdash P :: d{:}B \quad \Psi, x{:}\tau; \Gamma; \Delta', d{:}B \vdash Q :: c{:}A}{\Psi; \Gamma; \Delta, \Delta' \vdash (\nu d)(P \mid c(x).Q) = c(x).(\nu d)(P \mid Q) :: c{:}\forall x{:}\tau.A}$$

Fig. 6. Definitional equality of processes (Excerpt – See [32])

Definitional equality of processes is summarised in Fig. 6. We rely on process reduction defined below. Definitional equality of processes consists of the usual congruence rules, (typed) reductions and the commutting conversions of linear logic and η-like principles, which allows for forwarding actions to be equated with the primitive syntactic forwarding construct. Commutting conversions amount to sound observational equivalences between processes [22], given that session composition requires name restriction (embodied by the (cut) rule): In rule (PEqCC$\forall$), either process can only be interacted with via channel c and so postponing actions of P to after the input on c (when reading the equality from left to right) cannot impact the process' observable behaviours. While P can in general interact with sessions in Δ (or with Q), these interactions are unobservable due to hiding in the (cut) rule.

Operational Semantics. The operational semantics for the λ-calculus is standard, noting that no reduction can take place inside monadic terms. The operational (reduction) semantics for processes is presented below where we omit closure under structural congruence and the standard congruence rules [4,25,29]. The last rule defines spawning a process in a monadic term.

$$\begin{array}{ll}
c\langle M\rangle.P \mid c(x).Q \rightarrow P \mid Q\{M/x\} & \overline{c}\langle x\rangle.P \mid c(x).Q \rightarrow (\nu x)(P \mid Q) \\
!c(x).P \mid \overline{c}\langle x\rangle.Q \rightarrow !c(x).P \mid (\nu x)(P \mid Q) & c.\mathsf{case}\{\overline{l_i \Rightarrow P_i}\} \mid c.l_j; Q \rightarrow P_j \mid Q \quad (l_j \in \overline{l_i}) \\
(\nu c)(P \mid [c \leftrightarrow d]) \rightarrow P\{d/c\} & c \leftarrow \{c \leftarrow P \leftarrow \overline{u_j}; \overline{d_i}\} \leftarrow \overline{u_j}; \overline{d_i}; Q \rightarrow (\nu c)(P \mid Q)
\end{array}$$

2.3 Example – Reasoning About Processes Using Dependent Types

The use of type indices (i.e. type families) in dependently typed frameworks adds information to types to produce more refined specifications. Our framework enables us to do this at the level of session types.

Consider a session type that "counts down" on a natural number (we assume inductive definitions and dependent pattern matching in the style of [21]):

$$\begin{aligned}
&\mathsf{countDown} && :: \Pi x{:}\mathsf{Nat}.\mathsf{stype}\\
&\mathsf{countDown}\,(\mathsf{succ}(n)) && = \exists y{:}\mathsf{Nat}.\mathsf{countDown}(n)\\
&\mathsf{countDown}\ \mathsf{z} && = \mathbf{1}
\end{aligned}$$

The type family $\mathsf{countDown}(n)$ denotes a session type that emits exactly n numbers and then terminates. We can now write a (dependently-typed) function that produces processes with the appropriate type, given a starting value:

$$\begin{aligned}
&\mathsf{counter} && : \Pi x{:}\mathsf{Nat}.\{\mathsf{countDown}(x)\}\\
&\mathsf{counter}\ (\mathsf{succ}(n)) && = \{c \leftarrow c\langle \mathsf{succ}(n)\rangle.\, d \leftarrow \mathsf{counter}(n); [d \leftrightarrow c]\}\\
&\mathsf{counter}\ \mathsf{z} && = \{c \leftarrow \mathbf{0}\}
\end{aligned}$$

Note how the type of counter, through the type family countDown, allows us to specify exactly the number of times a value is sent. This is in sharp contrast with existing recursive (or inductive/coinductive [18,30]) session types, where one may only specify the general iterative nature of the behaviour (e.g. "send a number and then recurse or terminate").

The example above relies on session type indexing in order to provide additional static guarantees about processes (and the functions that generate them). An alternative way is to consider "simply-typed" programs and then *prove* that they satisfy the desired properties, using the language itself. Consider a simply-typed version of the counter above described as an inductive session type:

$$\begin{aligned}
&\mathsf{simpleCounterT} :: \mathsf{stype}\\
&\mathsf{simpleCounterT} = \oplus\{\mathsf{dec} : \mathsf{Nat} \wedge \mathsf{simpleCounterT}, \mathsf{done} : \mathbf{1}\}
\end{aligned}$$

There are many processes that correctly implement such a type, given that the type merely dictates that the session outputs a natural number and recurses (modulo the dec and done messages to signal which branch of the internal choice is taken). A function that produces processes implementing such a session, mirroring those generated by the counter function above, is:

$$\begin{aligned}
&\mathsf{simpleCounter} && : \mathsf{Nat} \rightarrow \{\mathsf{simpleCounterT}\}\\
&\mathsf{simpleCounter}\ (\mathsf{succ}(n)) && = \{c \leftarrow c.\mathsf{dec}; (\boldsymbol{\nu} d)(d\langle \mathsf{succ}(n)\rangle.\mathbf{0} \mid d(x).c\langle x\rangle.\\
& && \qquad d \leftarrow \mathsf{simpleCounter}(n); [d \leftrightarrow c])\}\\
&\mathsf{simpleCounter}\ \ \mathsf{z} && = \{c \leftarrow c.\mathsf{done}; \mathbf{0}\}
\end{aligned}$$

The process generated by simpleCounter, after emiting the dec label, spawns a process in parallel that sends the appropriate number, which is received by the parallel thread and then sent along the session c. Despite its simplicity, this example embodies a general pattern where a computation is spawned in parallel (itself potentially spawning many other threads) and the main thread then waits for the result before proceeding.

While such a process is typable in most session typing frameworks, our theory enables us to *prove* that the counter implementation above indeed counts down from a given number by defining an appropriate (inductive) type family, indexed by *monadic* values (i.e. processes):

$$\begin{array}{lll} \mathsf{corrCount} & :: & \Pi x{:}\mathsf{Nat}.\Pi y{:}\{\mathsf{simpleCounterT}\}.\mathsf{type} \\ \mathsf{corr}_z & : & \mathsf{corrCount}\,\mathsf{z}\,\{c \leftarrow c.\mathsf{done}; \mathbf{0}\} \\ \mathsf{corr}_n & : & \Pi n{:}\mathsf{Nat}.\Pi P{:}\{\mathsf{simpleCounterT}\}.\mathsf{corrCount}\, n\, P \rightarrow \\ & & \mathsf{corrCount}\,(\mathsf{succ}(n))\,\{c \leftarrow c.\mathsf{dec}; c\langle \mathsf{succ}(n)\rangle.d \leftarrow P; [d \leftrightarrow c]\} \end{array}$$

The type family corrCount, indexed by a natural number and a monadic value implementing the session type simpleCounter, is defined via two constructors: corr_z, which specifies that a correct 0 counter emits the done label and terminates; and corr_n, which given a monadic value P that is a correct n-counter, defines that a correct $(n+1)$-counter emits $n+1$ and then proceeds as P (modulo the label emission bookkeeping).

The proof of correctness of the simpleCounter function above is no more than a function of type $\Pi n{:}\mathsf{Nat}.\mathsf{corrCount}\, n\ (\mathsf{simpleCounter}(n))$, defined below:

$$\begin{array}{lll} \mathsf{prf} & & : \Pi n{:}\mathsf{Nat}.\mathsf{corrCount}\, n\, (\mathsf{simpleCounter}(n)) \\ \mathsf{prf} & \mathsf{z} & = \mathsf{corr}_z \\ \mathsf{prf} & (\mathsf{succ}(n)) & = \mathsf{corr}_n\, n\, (\mathsf{simpleCounter}(n))\, (\mathsf{prf}\; n) \end{array}$$

Note that in this scenario, the processes that index the corrCount type family are not syntactically equal to those generated by simpleCounter, but rather *definitionally* equal.

Typically, the processes that index such correctness specifications tend to be distilled versions of the actual implementations, which often perform some additional internal computation or communication steps. Since our notion of definitional equality of processes includes reduction (and also commuting conversions which account for type-preserving shuffling of internal communication actions [26]), the type conversion mechanism allows us to use the techniques described above to generally reason about specification conformance.

2.4 Type Soundness of the Framework

The main goal of this section is to present type soundness of our framework through a subject reduction result. We also show that our theory guarantees progress for terms and processes. The development requires a series of auxiliary results (detailed in [32]) pertaining to the functional and process layers which are ultimately needed to produce the inversion properties necessary to establish subject reduction. We note that strong normalisation results for linear-logic based session processes are known in the literature [3,26,30], even in the presence of impredicative polymorphism, restricted corecursion and higher-order data.

Such results are directly applicable to our work using appropriate semantics preserving type erasures.

In the remainder we often write $\Psi \vdash \mathcal{J}$ to stand for a well-formedness, typing or definitional equality judgment of the appropriate form. Similarly for $\Psi; \Gamma; \Delta \vdash \mathcal{J}$. We begin with the substitution property, which naturally holds for both layers, noting that the dependently typed nature of the framework requires substitution in both contexts, terms and in types.

Lemma 2.1 (Substitution). *Let $\Psi \vdash M : \tau$:*

1. *If $\Psi, x{:}\tau, \Psi' \vdash \mathcal{J}$ then $\Psi, \Psi'\{M/x\} \vdash \mathcal{J}\{M/x\}$;*
2. *If $\Psi, x{:}\tau, \Psi'; \Gamma; \Delta \vdash \mathcal{J}$ then $\Psi, \Psi'\{M/x\}; \Gamma\{M/x\}; \Delta\{M/x\} \vdash \mathcal{J}\{M/x\}$*

Combining substitution with a form of functionality for typing (i.e. that substitution of equal terms in a well-typed term produces equal terms) and for equality (i.e. that substitution of equal terms in a definitional equality proof produces equal terms), we can establish validity for typing and equality, which is a form of internal soundness of the type theory stating that judgments are consistent across the different levels of the theory.

Lemma 2.2 (Validity for Typing). (1) *If $\Psi \vdash \tau :: K$ or $\Psi \vdash A :: K$ then $\Psi \vdash K$;* (2) *If $\Psi \vdash M : \tau$ then $\Psi \vdash \tau :: \mathsf{type}$; and* (3) *If $\Psi; \Gamma; \Delta \vdash P :: z{:}A$ then $\Psi \vdash A :: \mathsf{stype}$.*

Lemma 2.3 (Validity for Equality)

1. *If $\Psi \vdash M = N : \tau$ then $\Psi \vdash M : \tau$, $\Psi \vdash N : \tau$ and $\Psi \vdash \tau :: \mathsf{type}$*
2. *If $\Psi \vdash \tau = \sigma :: K$ then $\Psi \vdash \tau :: K$, $\Psi \vdash \sigma :: K$ and $\Psi \vdash K$*
3. *If $\Psi \vdash A = B :: K$ then $\Psi \vdash A :: K$, $\Psi \vdash B :: K$ and $\Psi \vdash K$*
4. *If $\Psi \vdash K = K'$ then $\Psi \vdash K$ and $\Psi \vdash K'$*
5. *If $\Psi; \Gamma; \Delta \vdash P = Q :: z{:}A$ then $\Psi; \Gamma; \Delta \vdash P :: z{:}A$, $\Psi; \Gamma; \Delta \vdash Q :: z{:}A$ and $\Psi \vdash A :: \mathsf{stype}$.*

With these results we establish the appropriate inversion and injectivity properties which then enable us to show unicity of types (and kinds).

Theorem 2.4 (Unicity of Types and Kinds)

1. *If $\Psi \vdash M : \tau$ and $\Psi \vdash M : \tau'$ then $\Psi \vdash \tau = \tau' :: \mathsf{type}$*
2. *If $\Psi \vdash \tau :: K$ and $\Psi \vdash \tau :: K'$ then $\Psi \vdash K = K'$*
3. *If $\Psi; \Gamma; \Delta \vdash P :: z{:}A$ and $\Psi; \Gamma; \Delta \vdash P :: z{:}A'$ then $\Psi \vdash A = A' :: \mathsf{stype}$*
4. *If $\Psi \vdash A :: K$ and $\Psi \vdash A :: K'$ then $\Psi \vdash K = K'$.*

All the results above, combined with the process-level properties established in [5,26,27] enable us to show the following:

Theorem 2.5 (Subject Reduction – Terms). *If $\Psi \vdash M : \tau$ and $M \rightarrow M'$ then $\Psi \vdash M' : \tau$.*

Theorem 2.6 (Subject Reduction – Processes). *If $\Psi; \Gamma; \Delta \vdash P :: z{:}A$ and $P \rightarrow P'$ then $\exists Q$ such that $P' \equiv Q$ and $\Psi; \Gamma; \Delta \vdash Q :: z{:}A$.*

Theorem 2.7 (Progress – Terms). *If $\Psi \vdash M : \tau$ then either M is a value or $M \rightarrow M'$.*

As common in logical-based session type theories, typing enforces a strong notion of *global* progress which states that closed processes that are waiting to perform communication actions cannot get stuck (this relies on a notion of *live* process, defined as $\mathsf{live}(P)$ iff $P \equiv (\boldsymbol{\nu}\tilde{n})(\pi.Q \mid R)$ for some process R, sequence of names $\tilde{n}$ and a non-replicated guarded process $\pi.Q$). We note that the restricted typing for P is without loss of generality, due to the (cut) rule.

Theorem 2.8 (Progress – Processes). *If $\Psi; \cdot; \cdot \vdash P :: c{:}\mathbf{1}$ and $\mathsf{live}(P)$ then $\exists Q$ such that $P \rightarrow Q$.*

3 Embedding the Functional Layer in the Process Layer

Having introduced our type theory and showcased some of its informal expressiveness in terms of the ability to specify and *statically* verify true data dependent protocols, as well as the ability to prove properties of processes, we now develop a formal expressiveness result for our theory, showing that the process level type constructs are able to encode the dependently-typed functional layer, faithfully preserving type dependencies.

Specifically, we show that (1) the type-level constructs in the functional layer can be represented by those in the process layer combined with the contextual monad type, and (2) all term level constructs can be represented by session-typed processes that exchange monadic values. Thus, we show that both λ-abstraction and application can be eliminated while still preserving non-trivial type dependencies. Crucially, we note that the monadic construct *cannot* be fully eliminated due to the cross-layer nature of session type dependencies: In the process layer, simply-kinded dependent types (i.e. types with kind stype) are of the form $\forall x{:}\tau.A$ where τ is of kind type and A of kind stype (where x may occur). Operationally, such a session denotes an input of some term M of type τ with a continuation of type $A\{M/x\}$. Thus, to faithfully encode type dependencies we cannot represent such a type with a non-dependently typed input (e.g. a type of the form $A \multimap B$).

3.1 The Embedding

A first attempt. Given the observation above, a seemingly reasonable option would be to attempt an encoding that maintains monadic objects solely at the level of type indices and then exploits Girard's encoding [9] of function types $\tau \rightarrow \sigma$ as $![\![\tau]\!] \rightarrow [\![\sigma]\!]$, which is adequate for session-typed processes [28]. Thus a candidate encoding for the type $\Pi x{:}\tau.\sigma$ would be $\forall x{:}\{[\![\tau]\!]\}.![\![\tau]\!] \multimap [\![\sigma]\!]$, where $[\![-]\!]$ denotes our encoding on types. If we then consider the encoding at the level of terms, typing dictates the following (we write $[\![M]\!]_z$ for the process encoding of $M : \tau$, where z is the session channel along which one may observe the "result" of the encoding, typed with $[\![\tau]\!]$):

$$[\![\lambda x{:}\tau.M]\!]_z \triangleq z(x).z(x').[\![M]\!]_z$$
$$[\![M\,N]\!]_z \triangleq (\boldsymbol{\nu} x)([\![M]\!]_x \mid x\langle\{[\![N]\!]_y\}\rangle.\overline{x}\langle x'\rangle.(!x'(y).[\![N]\!]_y \mid [x \leftrightarrow z])$$

However, this candidate encoding breaks down once we consider definitional equality. Specifically, compositionality (i.e. the relationship between $[\![M\{N/x\}]\!]_z$ and the encoding of N substituted in that of M) requires us to relate $[\![M\{N/x\}]\!]_z$ with $(\boldsymbol{\nu} x)([\![M]\!]_z\{\{[\![N]\!]_y\}/x\} \mid !x'(y).[\![N]\!]_y)$, which relies on reasoning up-to *observational equivalence* of processes, a much stronger relation than our notion of definitional equality. Therefore it is *fundamentally* impossible for such an encoding to preserve our definitional equality, and thus it cannot preserve typing in the general case.

A faithful embedding. We now develop our embedding of the functional layer into the process layer which is compatible with definitional equality. Our target calculus is reminiscent of a higher-order (in the sense of higher-order processes [23]) session calculus [19]. Our encoding $[\![-]\!]$ is inductively defined on kinds, types, session types, terms and processes. As usual in process encodings of the λ-calculus, the encoding of a term M is indexed by a result channel z, written $[\![M]\!]_z$, where the behaviour of M may be observed.

$$[\![\lambda x{:}\tau.M]\!]_z \triangleq z(x).z(x').[\![M]\!]_z$$
$$[\![M\,N]\!]_z \triangleq (\boldsymbol{\nu} x)([\![M]\!]_x \mid x\langle\{[\![N]\!]_y\}\rangle.\overline{x}\langle x'\rangle.(!x'(y).[\![N]\!]_y \mid [x \leftrightarrow z])$$

Kind:

$[\![\mathsf{type}]\!] \triangleq \mathsf{stype}$ $\quad$ $[\![\mathsf{stype}]\!] \triangleq \mathsf{stype}$

$[\![\Pi x{:}\tau.K]\!] \triangleq \Pi x{:}\{[\![\tau]\!]\}.[\![K]\!]$ $\quad$ $[\![\Pi t :: K_1.K_2]\!] \triangleq \Pi t{::}[\![K_1]\!].[\![K_2]\!]$

Functional:

$[\![\Pi x{:}\tau.\sigma]\!] \triangleq \forall x{:}\{[\![\tau]\!]\}.[\![\sigma]\!]$ $\quad$ $[\![\{\overline{u_j{:}B_j}; \overline{d_i{:}B_i} \vdash c{:}A\}]\!] \triangleq \overline{![\![B_j]\!]} \multimap \overline{[\![B_i]\!]} \multimap [\![A]\!]$

$[\![\lambda x{:}\tau.\sigma]\!] \triangleq \lambda x{:}\{[\![\tau]\!]\}.[\![\sigma]\!]$ $\quad$ $[\![\tau\, M]\!] \triangleq [\![\tau]\!]\,\{[\![M]\!]_c\}$

$[\![\lambda t{::}K.\tau]\!] \triangleq \lambda t{::}[\![K]\!].[\![\tau]\!]$ $\quad$ $[\![\tau\,\sigma]\!] \triangleq [\![\tau]\!]\,[\![\sigma]\!]$

Session:

$[\![\forall x{:}\tau.A]\!] \triangleq \forall x{:}\{[\![\tau]\!]\}.[\![A]\!]$ $\quad$ $[\![\exists x{:}\tau.A]\!] \triangleq \exists x{:}\{[\![\tau]\!]\}.[\![A]\!]$

$[\![\lambda x{:}\tau.A]\!] \triangleq \lambda x{:}\{[\![\tau]\!]\}.[\![A]\!]$ $\quad$ $[\![A\,M]\!] \triangleq [\![A]\!]\,\{[\![M]\!]_c\}$

Terms:

$[\![\lambda x{:}\tau.M]\!]_z \triangleq z(x{:}\{[\![\tau]\!]\}).[\![M]\!]_z$ $\quad$ $[\![M\,N]\!]_z \triangleq (\boldsymbol{\nu} x)([\![M]\!]_x \mid x\langle\{[\![N]\!]_y\}\rangle.[x \leftrightarrow z])$

$[\![x]\!]_z \triangleq y \leftarrow x; [y \leftrightarrow z]$ $\quad$ $[\![\{z \leftarrow P \leftarrow \overline{u_j}; \overline{d_i}\}]\!]_z \triangleq z(u_0).\ldots.z(u_j).z(d_0).\ldots.z(d_n).[\![P]\!]$

Processes:

$[\![(\boldsymbol{\nu} x)(P \mid Q)]\!] \triangleq (\boldsymbol{\nu} x)([\![P]\!] \mid [\![Q]\!])$ $\quad$ $[\![\mathbf{0}]\!] \triangleq \mathbf{0}$ $\quad$ $[\![\overline{x}\langle y\rangle.(P \mid Q)]\!] \triangleq \overline{x}\langle y\rangle.([\![P]\!] \mid [\![Q]\!])$

$[\![x\langle M\rangle.P]\!] \triangleq x\langle\{[\![M]\!]_y\}\rangle.[\![P]\!]$ $\quad$ $[\![x(y).P]\!] \triangleq x(y).[\![P]\!]$

$$[\![c \leftarrow M \leftarrow \overline{u_j}; \overline{y_i}; Q]\!] \triangleq (\boldsymbol{\nu} c)([\![M]\!]_c \mid \overline{c}\langle v_1\rangle.(\overline{u_1}\langle a_1\rangle.[a_1 \leftrightarrow v_1] \mid \cdots \mid$$
$$\overline{c}\langle d_1\rangle.([y_1 \leftrightarrow d_1] \mid \cdots \mid \overline{c}\langle d_n\rangle.([y_n \leftrightarrow d_n] \mid [\![Q]\!])\ldots)$$

Fig. 7. An embedding of dependent functions into processes

The embedding is presented in Fig. 7, noting that the encoding extends straightforwardly to typing contexts, where functional contexts $\Psi, x{:}\tau$ are mapped to $\{\llbracket\Psi\rrbracket\}, x{:}\{\llbracket\tau\rrbracket\}$. The mapping of base kinds is straightforward. Dependent kinds $\Pi x{:}\tau.K$ rely on the monad for well-formedness and are encoded as (session) kinds of the form $\Pi x{:}\{\llbracket\tau\rrbracket\}.\llbracket K\rrbracket$. The higher-kinded types in the functional layer are translated to the corresponding type-level constructs of the process layer where all objects that must be type-kinded rely on the monad to satisfy this constraint. For instance, $\lambda x{:}\tau.\sigma$ is mapped to the session-type abstraction $\lambda x{:}\{\llbracket\tau\rrbracket\}.\llbracket\sigma\rrbracket$ and the type-level application $\tau\, M$ is translated to $\llbracket\tau\rrbracket\,\{\llbracket M\rrbracket_c\}$. Given the observation above on embedding the dependent function type $\Pi x{:}\tau.\sigma$, we translate it directly to $\forall x{:}\{\llbracket\tau\rrbracket\}.\llbracket\sigma\rrbracket$, that is, functions from τ to σ are mapped to sessions that input *processes* implementing $\llbracket\tau\rrbracket$ and then behave as $\llbracket\sigma\rrbracket$ accordingly. The encoding for monadic types simply realises the contextual nature of the monad by performing a sequence of inputs of the appropriate types (with the shared sessions being of ! type).

The mutually dependent nature of the framework requires us to extend the mapping to the process layer. Session types are mapped homomorphically (e.g. $\llbracket A \multimap B\rrbracket \triangleq \llbracket A\rrbracket \multimap \llbracket B\rrbracket$) with the exception of dependent inputs and outputs which rely on the monad, similarly for type-level functions and application.

The encoding of λ-terms is guided by the embedding for types: the abstraction $\lambda x{:}\tau.M$ is mapped to an input of a term of type $\{\llbracket\tau\rrbracket\}$ with continuation $\llbracket M\rrbracket_z$; application $M\,N$ is mapped to the composition of the encoding of M on a fresh name x with the corresponding output of $\{\llbracket N\rrbracket_y\}$, which is then forwarded to the result channel z; monadic expressions are translated to the appropriate sequence of inputs, as dictated by the translation of the monadic type; and, the translation of variables makes use of the monadic elimination form (since the encoding enforces variables to always be of monadic type) combined with forwarding to the appropriate result channel.

The mapping for processes is mostly homomorphic, using the monad constructor as needed. The only significant exception is the encoding for monadic elimination which must provide the encoded monadic term $\llbracket M\rrbracket_c$ with the necessary channels. Since the session calculus does not support communication of free names this is achieved by a sequence of outputs of fresh names combined with forwarding of the appropriate channel. To account for replicated sessions we must first trigger the replication via an output which is then forwarded accordingly.

We can illustrate our encoding via a simple example of an encoded function (we omit type annotations for conciseness):

$$\begin{array}{l}\llbracket(\lambda x.x)\,(\lambda x.\lambda y.y)\rrbracket_z = (\boldsymbol{\nu} c)(\llbracket\lambda x.x\rrbracket_c \mid c\langle\{\llbracket\lambda x.\lambda y.y\rrbracket_w\}\rangle.[c \leftrightarrow z]) \\ = \;(\boldsymbol{\nu} c)(c(x).y \leftarrow x; [y \leftrightarrow c] \mid c\langle\{w(x).w(y).d \leftarrow y; [d \leftrightarrow w]\}\rangle.[c \leftrightarrow z]) \\ \to^{+} z(x).z(y).d \leftarrow y; [d \leftrightarrow z] \;=\; \llbracket\lambda x.\lambda y.y\rrbracket_z\end{array}$$

3.2 Properties of the Embedding

We now state the key properties satisfied by our embedding, ultimately resulting in type preservation and operational correspondence. For conciseness, in the statements below we list only the cases for terms and processes, omitting those for types and kinds (see [32]). The key property that is needed is a notion of compositionality, which unlike in the sketch above no longer falls outside of definitional equality.

Lemma 3.1 (Compositionality)

1. $\Psi; \Gamma; \Delta \vdash [\![M\{N/x\}]\!]_z = [\![M]\!]_z\{\{[\![N]\!]_y\}/x\} :: z{:}[\![A\{N/x\}]\!]$
2. $\Psi; \Gamma; \Delta \vdash [\![P\{M/x\}]\!] :: z{:}[\![A\{M/x\}]\!]$ *iff* $\Psi; \Gamma; \Delta \vdash [\![P]\!]\{\{[\![M]\!]_c\}/x\} :: z{:}[\![A]\!]\{\{[\![M]\!]_c\}/x\}$.

Given the dependently typed nature of the framework, establishing the key properties of the encoding must be done simultaneously (relying on some auxiliary results – see [32]).

Theorem 3.2 (Preservation of Equality)

1. If $\Psi \vdash M = N : \tau$ *then* $\{[\![\Psi]\!]\}; \cdot; \cdot \vdash [\![M]\!]_z = [\![N]\!]_z :: z{:}[\![\tau]\!]$
2. If $\Psi; \Gamma; \Delta \vdash P = Q :: z{:}A$ *then* $\{[\![\Psi]\!]\}; [\![\Gamma]\!]; [\![\Delta]\!] \vdash [\![P]\!] = [\![Q]\!] :: z{:}[\![A]\!]$.

Theorem 3.3 (Preservation of Typing)

1. If $\Psi \vdash M : \tau$ *then* $\{[\![\Psi]\!]\}; \cdot; \cdot \vdash [\![M]\!]_z :: z{:}[\![\tau]\!]$
2. If $\Psi; \Gamma; \Delta \vdash P :: z{:}A$ *then* $\{[\![\Psi]\!]\}; [\![\Gamma]\!]; [\![\Delta]\!] \vdash [\![P]\!] :: z{:}[\![A]\!]$.

Theorem 3.4 (Operational Correspondence). *If* $\Psi; \Gamma; \Delta \vdash P :: z{:}A$ *and* $\Psi \vdash M : \tau$ *then:*

1. (a) If $P \to P'$ *then* $[\![P]\!] \to^+ Q$ *with* $\{[\![\Psi]\!]\}; [\![\Gamma]\!]; [\![\Delta]\!] \vdash Q = [\![P']\!] :: z{:}[\![A]\!]$ *and*
(b) if $[\![P]\!] \to P'$ *then* $P \to^+ Q$ *with* $\{[\![\Psi]\!]\}; [\![\Gamma]\!]; [\![\Delta]\!] \vdash P' = [\![Q]\!] :: z{:}[\![A]\!]$
2. (a) If $M \to M'$ *then* $[\![M]\!]_z \to^+ N$ *with* $\{[\![\Psi]\!]\}; \cdot; \cdot \vdash N = [\![M']\!]_z :: z{:}[\![\tau]\!]$ *and*
(b) if $[\![M]\!]_z \to P$ *then* $M \to N$ *with* $\{[\![\Psi]\!]\}; \cdot; \cdot \vdash [\![N]\!]_z = P :: z{:}[\![\tau]\!]$.

In Theorem 3.4, (a) is commonly referred to as operational completeness, with (b) establishing soundness. As exemplified above, our encoding satisfies a very precise operational correspondence with the original λ-terms.

4 Related and Future Work

Enriching Session Types via Type Structure. Exploiting the linear logical foundations of session types, [25] considers a form of value dependencies where session types can state properties of exchanged data values, while the work [29] introduces the contextual monad in a simply-typed setting. Our development not only subsumes these two works, but goes beyond simple value dependencies by extending to a richer type structure and integrating dependencies with the contextual monad. Recently, [1] considers a non-conservative extension of linear logic-based session types with sharing, allowing true non-determinism. Their work includes dependent quantifications with shared channels, but their type syntax does *not* include free type variables, so the actual type dependencies do not arise (see [1, 37:8]). Thus none of the examples in this paper can be represented in [1]. The work [16] studies gradual session types. To the best of our knowledge, the main example in [1, Sect. 2] is *statically* representable in our framework as in the example of Sect. 1, where protocol actions depend on values that are communicated (or passed as function arguments).

In the context of multiparty session types, the theory of multiparty indexed session types is studied in [7], and implemented in a protocol description language [20]. The main aim of these works is to use indexed types to represent an arbitrary number of session *participants*. The work [31] extends [25] to multiparty sessions in order to treat value dependency across multiple participants. Extending our framework to multiparty [15] or non-logic based session types [14] is an interesting future topic.

Combining Linear and Dependent Types. Many works have studied the various challenges of integrating linearity in dependent functional type theories. We focus on the most closely related works. The work [6] introduced the Linear Logical Framework (LLF), integrating linearity with the LF [11] type theory, which was later extended to the Concurrent Logical Framework (CLF) [33], accounting for further linear connectives. Their theory is representable in our framework through the contextual monad (encompassing full intuitionistic linear logic), depending on linearly-typed processes that can express dependently typed functions (Sect. 3).

The work of [17] integrates linearity with type dependencies by extending LNL [2]. Their work is aimed at reasoning about imperative programs using a form of Hoare triples, requiring features that we do not study in this work such has proof irrelevance and computationally irrelevant quantification. Formally, their type theory is extensional which introduces significant technical differences from our intensional type theory, such as a realisability model in the style of NuPRL [10] to establish consistency.

Recently, [8] proposed an extension of LLF with first-class contexts (which may contain both linear and unrestricted hypotheses). While the contextual aspects of their theory are reminiscent of our contextual monad, their framework differs significantly from ours, since it is designed to enable higher-order abstract syntax (commonplace in the LF family of type theories), focusing on a type system for canonical LF objects with a meta-language that includes contexts and context manipulation. They do not consider additives since their integration with first-class contexts can break canonicity.

While none of the above works considers processes as primitive, their techniques should be useful for, e.g. developing algorithmic type-checking and integrating inductive and coinductive session types based on [18,26,30].

Dependent Types and Higher-Order π-calculus. The work [35] studies a form of dependent types where the type of processes takes the form of a mapping Δ from channels x to channel types T representing an interface of process P. The dependency is specified as $\Pi(x{:}T)\Delta$, representing a channel abstraction of the environment. This notion is extended to an existential channel dependency type $\Sigma(x{:}T)\Delta$ to address fresh name creation [13,34]. Combining our process monad with dependent types can be regarded as an "interface" which describes explicit channel usages for processes. The main differences are (1) our dependent types are more general, treating full dependent families including terms and processes in types, while [13,34,35] study only channel dependency to environments (i.e. neither terms nor processes appear in types, only channels); and (2) our calculus emits only fresh names, not needing to handle the complex scoping mechanism treated in [13,34]. In this sense, the process monad provides an elegant framework to handle higher-order computations and assign non-trivial types to processes.

Acknowledgements. The authors would like to thank the anonymous reviews for their comments and suggestions. This work is partially supported by EPSRC EP/K034413/1, EP/K011715/1, EP/L00058X/1, EP/N027833/1, EP/N028201/1 and NOVA LINCS (UID/CEC/04516/2013).

References

1. Balzer, S., Pfenning, F.: Manifest sharing with session types. PACMPL **1**(ICFP), 37:1–37:29 (2017)
2. Benton, P.N.: A mixed linear and non-linear logic: proofs, terms and models. In: Pacholski, L., Tiuryn, J. (eds.) CSL 1994. LNCS, vol. 933, pp. 121–135. Springer, Heidelberg (1995). https://doi.org/10.1007/BFb0022251
3. Caires, L., Pérez, J.A., Pfenning, F., Toninho, B.: Behavioral polymorphism and parametricity in session-based communication. In: Felleisen, M., Gardner, P. (eds.) ESOP 2013. LNCS, vol. 7792, pp. 330–349. Springer, Heidelberg (2013). https://doi.org/10.1007/978-3-642-37036-6_19
4. Caires, L., Pfenning, F.: Session types as intuitionistic linear propositions. In: Gastin, P., Laroussinie, F. (eds.) CONCUR 2010. LNCS, vol. 6269, pp. 222–236. Springer, Heidelberg (2010). https://doi.org/10.1007/978-3-642-15375-4_16
5. Caires, L., Pfenning, F., Toninho, B.: Linear logic propositions as session types. Math. Struct. Comput. Sci. **26**(3), 367–423 (2016)
6. Cervesato, I., Pfenning, F.: A linear logical framework. Inf. Comput. **179**(1), 19–75 (2002)
7. Deniélou, P., Yoshida, N., Bejleri, A., Hu, R.: Parameterised multiparty session types. Log. Methods Comput. Sci. **8**(4), 1–46 (2012). https://doi.org/10.2168/LMCS-8(4:6)2012
8. Georges, A.L., Murawska, A., Otis, S., Pientka, B.: **LINCX**: a linear logical framework with first-class contexts. In: Yang, H. (ed.) ESOP 2017. LNCS, vol. 10201, pp. 530–555. Springer, Heidelberg (2017). https://doi.org/10.1007/978-3-662-54434-1_20
9. Girard, J.: Linear logic. Theor. Comput. Sci. **50**, 1–102 (1987)
10. Harper, R.: Constructing type systems over an operational semantics. J. Symbolic Comput. **14**(1), 71–84 (1992)
11. Harper, R., Honsell, F., Plotkin, G.D.: A framework for defining logics. J. ACM **40**(1), 143–184 (1993)
12. Harper, R., Pfenning, F.: On equivalence and canonical forms in the LF type theory. ACM Trans. Comput. Log. **6**(1), 61–101 (2005)
13. Hennessy, M., Rathke, J., Yoshida, N.: safeDpi: a language for controlling mobile code. Acta Inf. **42**(4–5), 227–290 (2005)

14. Honda, K., Vasconcelos, V.T., Kubo, M.: Language primitives and type discipline for structured communication-based programming. In: Hankin, C. (ed.) ESOP 1998. LNCS, vol. 1381, pp. 122–138. Springer, Heidelberg (1998). https://doi.org/10.1007/BFb0053567
15. Honda, K., Yoshida, N., Carbone, M.: Multiparty asynchronous session types. J. ACM **63**(1), 9:1–9:67 (2016). https://doi.org/10.1145/2827695
16. Igarashi, A., Thiemann, P., Vasconcelos, V.T., Wadler, P.: Gradual session types. PACMPL **1**(ICFP), 38:1–38:28 (2017)
17. Krishnaswami, N.R., Pradic, P., Benton, N.: Integrating linear and dependent types. In: POPL 2015, pp. 17–30 (2015)
18. Lindley, S., Morris, J.G.: Talking bananas: structural recursion for session types. In: ICFP 2016, pp. 434–447 (2016)
19. Mostrous, D., Yoshida, N.: Two session typing systems for higher-order mobile processes. In: Della Rocca, S.R. (ed.) TLCA 2007. LNCS, vol. 4583, pp. 321–335. Springer, Heidelberg (2007). https://doi.org/10.1007/978-3-540-73228-0_23
20. Ng, N., Yoshida, N.: Pabble: parameterised scribble. Serv. Oriented Comput. Appl. **9**(3–4), 269–284 (2015)
21. Norell, U.: Towards a practical programming language based on dependent type theory. Ph.D. thesis, Department of Computer Science and Engineering, Chalmers University of Technology (2007)
22. Pérez, J.A., Caires, L., Pfenning, F., Toninho, B.: Linear logical relations for session-based concurrency. In: Seidl, H. (ed.) ESOP 2012. LNCS, vol. 7211, pp. 539–558. Springer, Heidelberg (2012). https://doi.org/10.1007/978-3-642-28869-2_27
23. Sangiorgi, D., Walker, D.: The Pi-calculus: A Theory of Mobile Processes. C.U.P, Cambridge (2001)
24. Takeuchi, K., Honda, K., Kubo, M.: An interaction-based language and its typing system. In: Halatsis, C., Maritsas, D., Philokyprou, G., Theodoridis, S. (eds.) PARLE 1994. LNCS, vol. 817, pp. 398–413. Springer, Heidelberg (1994). https://doi.org/10.1007/3-540-58184-7_118
25. Toninho, B., Caires, L., Pfenning, F.: Dependent session types via intuitionistic linear type theory. In: PPDP 2011, pp. 161–172 (2011)
26. Toninho, B.: A logical foundation for session-based concurrent computation. Ph.D. thesis, Carnegie Mellon University and New University of Lisbon (2015)
27. Toninho, B., Caires, L., Pfenning, F.: Dependent session types via intuitionistic linear type theory. Technical report CMU-CS-11-139, School of Computer Science, Carnegie Mellon University (2011)
28. Toninho, B., Caires, L., Pfenning, F.: Functions as session-typed processes. In: Birkedal, L. (ed.) FoSSaCS 2012. LNCS, vol. 7213, pp. 346–360. Springer, Heidelberg (2012). https://doi.org/10.1007/978-3-642-28729-9_23
29. Toninho, B., Caires, L., Pfenning, F.: Higher-order processes, functions, and sessions: a monadic integration. In: Felleisen, M., Gardner, P. (eds.) ESOP 2013. LNCS, vol. 7792, pp. 350–369. Springer, Heidelberg (2013). https://doi.org/10.1007/978-3-642-37036-6_20
30. Toninho, B., Caires, L., Pfenning, F.: Corecursion and non-divergence in session-typed processes. In: Maffei, M., Tuosto, E. (eds.) TGC 2014. LNCS, vol. 8902, pp. 159–175. Springer, Heidelberg (2014). https://doi.org/10.1007/978-3-662-45917-1_11
31. Toninho, B., Yoshida, N.: Certifying data in multiparty session types. J. Log. Algebraic Methods Program. **90**(C), 61–83 (2017)

32. Toninho, B., Yoshida, N.: Depending on session-typed processes. CoRR abs/1801.08114 (2017). https://arxiv.org/abs/1801.08114
33. Watkins, K., Cervesato, I., Pfenning, F., Walker, D.: A concurrent logical framework: the propositional fragment. In: Berardi, S., Coppo, M., Damiani, F. (eds.) TYPES 2003. LNCS, vol. 3085, pp. 355–377. Springer, Heidelberg (2004). https://doi.org/10.1007/978-3-540-24849-1_23
34. Yoshida, N.: Channel dependent types for higher-order mobile processes. In: Proceedings of the 31st ACM SIGPLAN-SIGACT Symposium on Principles of Programming Languages, POPL 2004, Venice, Italy, 14–16 January 2004, pp. 147–160 (2004)
35. Yoshida, N., Hennessy, M.: Assigning types to processes. Inf. Comput. **174**(2), 143–179 (2002)

FabULous Interoperability for ML and a Linear Language

Gabriel Scherer[1,2(✉)], Max New[1], Nick Rioux[1], and Amal Ahmed[1,3]

[1] Northeastern University, Boston, USA
maxnew@ccs.neu.edu, rioux.n@husky.neu.edu, A.Ahmed@northeastern.edu
[2] Inria Saclay, Palaiseau, France
gabriel.scherer@inria.fr
[3] Inria Paris, Paris, France

Abstract. Instead of a monolithic programming language trying to cover all features of interest, some programming systems are designed by combining together simpler languages that cooperate to cover the same feature space. This can improve usability by making each part simpler than the whole, but there is a risk of *abstraction leaks* from one language to another that would break expectations of the users familiar with only one or some of the involved languages.

We propose a formal specification for what it means for a given language in a multi-language system to be usable without leaks: it should embed into the multi-language in a *fully abstract* way, that is, its contextual equivalence should be unchanged in the larger system.

To demonstrate our proposed design principle and formal specification criterion, we design a multi-language programming system that combines an ML-like statically typed functional language and another language with linear types and linear state. Our goal is to cover a good part of the expressiveness of languages that mix functional programming and linear state (ownership), at only a fraction of the complexity. We prove that the embedding of ML into the multi-language system is fully abstract: functional programmers should not fear abstraction leaks. We show examples of combined programs demonstrating in-place memory updates and safe resource handling, and an implementation extending OCaml with our linear language.

1 Introduction

Feature accretion is a common trend among mature but actively evolving programming languages, including C++, Haskell, Java, OCaml, Python, and Scala. Each new feature strives for generality and expressiveness, and may provide a large usability improvement to users of the particular problem domain or programming

Note: Due to severe space restrictions, many details have been omitted from this presentation of our work. We strongly encourage the reader to consult the complete version at https://arxiv.org/pdf/1707.04984.

C. Baier and U. Dal Lago (Eds.): FOSSACS 2018, LNCS 10803, pp. 146–162, 2018.
https://doi.org/10.1007/978-3-319-89366-2_8

style it was designed to empower (e.g., XML documents, asynchronous communication, staged evaluation). But feature creep in general-purpose languages may also make it harder for programmers to master the language as a whole, degrade the user experience (e.g., leading to more cryptic error messages), require additional work on the part of tooling providers, and lead to fragility in language implementations.

A natural response to increased language complexity is to define subsets of the language designed for a better programming experience. For instance, a subset can be easier to teach (e.g., "Core" ML[1], Haskell 98 as opposed to GHC Haskell, Scala mastery levels[2]); it can facilitate static analysis or decrease the risk of programming errors, while remaining sufficiently expressive for the target users' needs (e.g., MISRA C, Spark/Ada); it can enforce a common style within a company; or it can be designed to encourage a transition to deprecate some ill-behaved language features (e.g., strict Javascript).

Once a subset has been selected, it may be the case that users write whole programs purely in the subset (possibly using tooling to enforce that property), but programs will commonly rely on other libraries that are not themselves implemented in the same subset of the language. If users stay in the subset while using these libraries, they will only interact with the part of the library whose interface is expressible in the subset. But does the behavior of the library respect the expectations of users who only know the subset? When calling a function from within the subset breaks subset expectations, it is a sign of *leaky abstraction*.

How should we design languages with useful subsets that manage complexity and avoid abstraction leaks?

We propose to look at this question from a different, but equivalent, angle: instead of designing a single big monolithic language with some nicer subsets, we propose to consider *multi-language* programming systems where several smaller programming languages interact together to cover the same feature space. Each language or sub-combination of languages is a subset, in the above sense, of the multi-language, and there is a clear definition of *abstraction leaks* in terms of user experience: a user who only knows some of the languages of the system should be able to use the multi-language system, interacting with code written in the other languages, without have their expectations violated. If we write a program in Java and call a function that, internally, is implemented in Scala, there should be no surprises—our experience should be the same as when calling a pure Java function. Similarly, consider the subset of Haskell that does not contain `IO` (input-output as a type-tracked effect): the expectations of a user of this language, for instance in terms of valid equational reasoning, should not be violated by adding `IO` back to the language—in the absence of the abstraction-leaking `unsafePerformIO`.

We propose a *formal specification* for a "no abstraction leaks" guarantee that can be used as a design criterion to design new multi-language systems, with graceful interoperation properties. It is based on the formal notion of *full abstraction* which has previously been used to study the denotational semantics

[1] https://caml.inria.fr/pub/docs/u3-ocaml/ocaml-ml.html.
[2] http://www.scala-lang.org/old/node/8610.

of programming languages (Meyer and Sieber 1988; Milner 1977; Cartwright and Felleisen 1992; Jeffrey and Rathke 2005; Abramsky, Jagadeesan, and Malacaria 2000), and the formal property of compilers (Ahmed and Blume 2008, 2011; Devriese et al. 2016; New et al. 2016; Patrignani et al. 2015), but not for user-facing languages. A compiler C from a source language S to a target language T is *fully abstract* if, whenever two source terms s_1 and s_2 are indistinguishable in S, their translations $C(s_1)$ and $C(s_2)$ are indistinguishable in T. In a multi-language $G + E$ formed of a general-purpose, user-friendly language G and a more advanced language E—one that provides an escape hatch for experts to write code that can't be implemented in G—we say that E does not *leak* into G if the embedding of G into the multi-language $G + E$ is fully abstract.

To demonstrate that our formal specification is reasonable, we design a novel multi-language programming system that satisfies it. Our multi-language λ^{UL} combines a general-purpose functional programming language λ^{U} (unrestricted) of the ML family with an advanced language λ^{L} (linear) with *linear types* and linear state. It is less convient to program in λ^{L}'s restrictive type system, but users can write programs in λ^{L} that could not be written in λ^{U}: they can use linear types, locally, to enforce resource usage protocols (typestate), and they can use linear state and the linear ownership discipline to write programs that do in-place update to allocate less memory, yet remain observationally pure.

Consider for example the following mixed-language program. The blue fragments are written in the general-purpose, user-friendly functional language, while the red fragments are written in the linear language. The boundaries UL and LU allow switching between languages. The program reads all lines from a file, accumulating them in a list, and concatenating it into a single string when the end-of-file (EOF) is reached.

```
let concat_lines path : String = UL(
 loop (open LU(path)) LU(Nil)
 where rec loop handle LU(acc : List String) =
  match line handle with
  | Next line LU(handle) -> loop handle LU(Cons line acc)
  | EOF handle -> close handle; LU(rev_concat "\n" acc))
```

The linear type system ensures that the file handle is properly closed: removing the close handle call would give a type error. On the other hand, only the parts concerned with the resource-handling logic need to be written in the red linear language; the user can keep all general-purpose logic (here, how to accumulate lines and what to do with them at the end) in the more convenient general-purpose blue language—and call this function from a blue-language program. Fine-grained boundaries allow users to rely on each language's strength and to use the advanced features only when necessary.

In this example, the file-handle API specifies that the call to line, which reads a line, returns the data at type ![String]. The latter represents how U values of type String can be put into a *lump* type to be passed to the linear world where they are treated as opaque blackboxes that must be passed back to the ML world for consumption. For other examples, such as in-place list manipulation or transient operations on an persistent data structure, we will need a deeper form of interoperability where the linear world creates, dissects or manipulates U values. To enable this, our multi-language supports translation of types from one language to the other, using a *type compatibility* relation $\sigma \simeq \sigma$ between λ^{U} types σ and λ^{L} types σ.

We claim the following contributions:

1. We propose a formal specification of what it means for advanced language features to be introduced in a (multi-)language system without introducing a class of abstraction leaks that break equational reasoning. This specification captures a useful *usability* property, and we hope it will help us and others design more usable programming languages, much like the formal notion of *principal types* served to better understand and design type inference systems.
2. We design a simple linear language, λ^{L}, that supports linear state (Sect. 2). This simple design for linear state is a contribution of its own. A nice property of the language (shared by some other linear languages) is that the code has both an imperative interpretation—with in-place memory update, which provides resource guarantees—and a functional interpretation—which aids program reasoning. The imperative and functional interpretations have different resource usage, but the same input/output behavior.
3. We present a multi-language programming system λ^{UL} combining a core ML language, λ^{U} (U for Unrestricted, as opposed to Linear) with λ^{L} and prove that the embedding of the ML language λ^{U} in λ^{UL} is fully abstract (Sect. 3). Moreover, the multi-language is designed to ensure that our full abstraction result is stable under extension of the embedded ML language λ^{U}.

2 The λ^{U} and λ^{L} Languages

The unrestricted language λ^{U} is a run-of-the-mill idealized ML language with functions, pairs, sums, iso-recursive types and polymorphism. It is presented in its explicitly typed form—we will not discuss type inference in this work. The full syntax is described in Fig. 1, and the typing rules in Fig. 2. The dynamic semantics is completely standard. Having binary sums, binary products and iso-recursive types lets us express algebraic datatypes in the usual way.

The novelty lies in the linear language λ^{L}, which we present in several steps. As is common in λ-calculi with references, the small-step operational semantics is given for a language that is not exactly the surface language in which programs

Types $\sigma ::= \alpha \mid \sigma_1 \times \sigma_2 \mid 1 \mid \sigma_1 \to \sigma_2 \mid \sigma_1 + \sigma_2 \mid \mu\alpha.\sigma \mid \forall\alpha.\sigma$

Expr. $\mathsf{e} ::= \mathsf{x} \mid \langle \mathsf{e}_1, \mathsf{e}_2 \rangle \mid \pi_1\,\mathsf{e} \mid \pi_2\,\mathsf{e} \mid \langle\rangle \mid \mathsf{e}_1; \mathsf{e}_2 \mid \lambda(\mathsf{x}:\sigma).\,\mathsf{e} \mid \mathsf{e}_1\,\mathsf{e}_2 \mid$
$\quad \mathsf{inj}_i\,\mathsf{e} \mid \mathsf{case}\,\mathsf{e}'\,\mathsf{of}\,\mathsf{x}_1.\,\mathsf{e}_1 \mid \mathsf{x}_2.\,\mathsf{e}_2 \mid \mathsf{fold}_{\mu\alpha.\sigma}\,\mathsf{e} \mid \mathsf{unfold}\,\mathsf{e} \mid \Lambda\alpha.\,\mathsf{e} \mid \mathsf{e}\,[\sigma]$

Values $\mathsf{v} ::= \mathsf{x} \mid \langle \mathsf{v}_1, \mathsf{v}_2 \rangle \mid \langle\rangle \mid \lambda(\mathsf{x}:\sigma).\,\mathsf{e} \mid \mathsf{inj}_1\,\mathsf{v} \mid \mathsf{inj}_2\,\mathsf{v} \mid \mathsf{fold}_{\mu\alpha.\sigma}\,\mathsf{v} \mid \Lambda\alpha.\,\mathsf{v}$

Contexts $\Gamma ::= \cdot \mid \Gamma, \mathsf{x}:\sigma \mid \Gamma, \alpha$

Fig. 1. Unrestricted language: syntax

$$\boxed{\Gamma \vdash_{\mathsf{U}} \mathsf{e} : \sigma}$$

$$\frac{\mathsf{x}{:}\sigma \in \Gamma}{\Gamma \vdash_{\mathsf{U}} \mathsf{x} : \sigma} \qquad \frac{}{\Gamma \vdash_{\mathsf{U}} \langle\rangle : 1} \qquad \frac{\Gamma \vdash_{\mathsf{U}} \mathsf{e} : 1 \qquad \Gamma \vdash_{\mathsf{U}} \mathsf{e}' : \sigma}{\Gamma \vdash_{\mathsf{U}} \mathsf{e}; \mathsf{e}' : \sigma}$$

$$\frac{\Gamma \vdash_{\mathsf{U}} \mathsf{e}_1 : \sigma_1 \qquad \Gamma \vdash_{\mathsf{U}} \mathsf{e}_2 : \sigma_2}{\Gamma \vdash_{\mathsf{U}} \langle \mathsf{e}_1, \mathsf{e}_2 \rangle : \sigma_1 \times \sigma_2} \qquad \frac{\Gamma \vdash_{\mathsf{U}} \mathsf{e} : \sigma_1 \times \sigma_2}{\Gamma \vdash_{\mathsf{U}} \pi_i\, \mathsf{e} : \sigma_i}$$

$$\frac{\Gamma, \mathsf{x}{:}\sigma \vdash_{\mathsf{U}} \mathsf{e} : \sigma'}{\Gamma \vdash_{\mathsf{U}} \lambda(\mathsf{x}{:}\sigma).\, \mathsf{e} : \sigma \to \sigma'} \qquad \frac{\Gamma \vdash_{\mathsf{U}} \mathsf{e} : \sigma' \to \sigma \qquad \Gamma \vdash_{\mathsf{U}} \mathsf{e}' : \sigma'}{\Gamma \vdash_{\mathsf{U}} \mathsf{e}\ \mathsf{e}' : \sigma}$$

$$\frac{\Gamma \vdash_{\mathsf{U}} \mathsf{e} : \sigma_i}{\Gamma \vdash_{\mathsf{U}} \mathsf{inj}_i\, \mathsf{e} : \sigma_1 + \sigma_2} \qquad \frac{\Gamma \vdash_{\mathsf{U}} \mathsf{e} : \sigma_1 + \sigma_2 \qquad \begin{array}{c}\Gamma, \mathsf{x}_1 : \sigma_1 \vdash_{\mathsf{U}} \mathsf{e}_1 : \sigma \\ \Gamma, \mathsf{x}_2 : \sigma_2 \vdash_{\mathsf{U}} \mathsf{e}_2 : \sigma\end{array}}{\Gamma \vdash_{\mathsf{U}} \mathsf{case}\, \mathsf{e}\, \mathsf{of}\, \mathsf{x}_1.\, \mathsf{e}_1 \mid \mathsf{x}_2.\, \mathsf{e}_2 : \sigma}$$

$$\frac{\Gamma \vdash_{\mathsf{U}} \mathsf{e} : \sigma[\mu\alpha.\, \sigma/\alpha]}{\Gamma \vdash_{\mathsf{U}} \mathsf{fold}_{\mu\alpha.\sigma}\, \mathsf{e} : \mu\alpha.\, \sigma} \qquad \frac{\Gamma \vdash_{\mathsf{U}} \mathsf{e} : \mu\alpha.\, \sigma}{\Gamma \vdash_{\mathsf{U}} \mathsf{unfold}\, \mathsf{e} : \sigma[\mu\alpha.\, \sigma/\alpha]}$$

$$\frac{\Gamma, \alpha \vdash_{\mathsf{U}} \mathsf{v} : \sigma}{\Gamma \vdash_{\mathsf{U}} \Lambda\alpha.\, \mathsf{v} : \forall\alpha.\, \sigma} \qquad \frac{\Gamma \vdash_{\mathsf{U}} \mathsf{e} : \forall\alpha.\, \sigma \qquad \Gamma \vdash \sigma'}{\Gamma \vdash_{\mathsf{U}} \mathsf{e}\, [\sigma'] : \sigma[\sigma'/\alpha]}$$

Fig. 2. Unrestricted language: static semantics

are written, because memory allocation returns *locations* ℓ that are not in the grammar of surface terms. Reductions are defined on *configurations*, a local store paired with a term in a slightly larger *internal* language. We have two type systems, a type system on surface terms, that does not mention locations and stores—which is the one a programmer needs to know—and a type system on configurations, which contains enough static information to reason about the dynamics of our language and prove subject reduction. Again, this follows the standard structure of syntactic soundness proofs for languages with a mutable store.

2.1 The Core of λ^{L}

Figure 3 presents the surface syntax of our linear language λ^{L}. For the syntactic categories of types σ, and expressions e, the last line contains the constructions related to the linear store that we only discuss in Sect. 2.2.

In technical terms, our linear type system is exactly propositional intuitionistic linear logic, extended with iso-recursive types. For simplicity and because we did not need them, our current system also does not have polymorphism or additive/lazy pairs $\sigma_1 \,\&\, \sigma_2$. Additive pairs would be a trivial addition, but polymorphism would require more work when we define the multi-language semantics in Sect. 3.

In less technical terms, our type system can enforce that values be used *linearly*, meaning that they cannot be duplicated or erased, they have to be deconstructed

Types	$\sigma ::= \sigma_1 \otimes \sigma_2 \mid 1 \mid \sigma_1 \multimap \sigma_2 \mid \sigma_1 \oplus \sigma_2 \mid \mu\alpha.\sigma \mid \alpha \mid !\sigma \mid \mathsf{Box}\ 1\ \sigma \mid \mathsf{Box}\ 0$
Expr.	$\mathsf{e} ::= \mathsf{x} \mid \langle \mathsf{e}_1, \mathsf{e}_2 \rangle \mid \mathsf{let}\ \langle \mathsf{v}_1, \mathsf{v}_2 \rangle = \mathsf{e}_1\ \mathsf{in}\ \mathsf{e}_2 \mid \langle\rangle \mid \mathsf{e}_1; \mathsf{e}_2 \mid \lambda(\mathsf{x}:\sigma).\,\mathsf{e} \mid \mathsf{e}_1\ \mathsf{e}_2 \mid$ $\mathsf{inj}_1\,\mathsf{e} \mid \mathsf{inj}_2\,\mathsf{e} \mid \mathsf{case}\,\mathsf{e}'\ \mathsf{of}\ \mathsf{x}_1.\,\mathsf{e}_1 \mid \mathsf{x}_2.\,\mathsf{e}_2 \mid \mathsf{fold}_{\mu\alpha.\sigma}\,\mathsf{e} \mid \mathsf{unfold}\,\mathsf{e} \mid$ $\mathsf{share}\,\mathsf{e} \mid \mathsf{copy}\,\mathsf{e} \mid \mathsf{new}\,\mathsf{e} \mid \mathsf{free}\,\mathsf{e} \mid \mathsf{box}\,\mathsf{e} \mid \mathsf{unbox}\,\mathsf{e}$
Values	$\mathsf{v} ::= \mathsf{x} \mid \langle \mathsf{v}_1, \mathsf{v}_2 \rangle \mid \langle\rangle \mid \lambda(\mathsf{x}:\sigma).\,\mathsf{e} \mid \mathsf{inj}_1\,\mathsf{v} \mid \mathsf{inj}_2\,\mathsf{v} \mid \mathsf{fold}_{\mu\alpha.\sigma}\,\mathsf{v} \mid \mathsf{share}\,\mathsf{v}$
Contexts	$\Gamma ::= \cdot \mid \Gamma, \mathsf{x}:\sigma$

Fig. 3. Linear language: surface syntax

exactly once. Only some types have this linearity restriction; others allow duplication and sharing of values at will. We can think of linear values as *resources* to be spent wisely; for any linear value somewhere in a term, there can be only one way to access this value, so we can interpret the language as enforcing an *ownership* discipline where whoever points to a linear value owns it.

In particular, linear functions of type $\sigma_1 \multimap \sigma_2$ must be called exactly once, and their results must in turn be consumed – they can safely capture linear resources. On the other hand, the non-linear, duplicable values are those at types of the form $!\sigma$ — the *exponential* modality of linear logic. If the term e has duplicable type $!\sigma$, then the term copy e has type σ: this creates a local copy of the value that is uniquely-owned by its receiver and must be consumed linearily.

This resource-usage discipline is enforced by the surface typing rules of λ^{L}, presented in Fig. 4. They are exactly the standard (two-sided) logical rules of intuitionistic linear logic, annotated with program terms. The non-duplicability of linear values is enforced by the way contexts are merged by the inference rules: if e_1 is type-checked in the context Γ_1 and e_2 in Γ_2, then the linear pair $\langle \mathsf{e}_1, \mathsf{e}_2 \rangle$ is only valid in the combined context $\Gamma_1 \curlyvee \Gamma_2$. The $(\curlyvee)$ operation is partial; this combined context is defined only if the variables shared by Γ_1 and Γ_2 are duplicable—their type is of the form $!\sigma$. In other words, a variable at a non-duplicable type in $\Gamma_1 \curlyvee \Gamma_2$ cannot possibly appear in both Γ_1 and Γ_2: it must appear exactly once[3].

The expression share e takes a term at some type σ and creates a "shared" term, whose value will be duplicable. Its typing rule uses a context of the form $!\Gamma$, which is defined as the pointwise application of the (!) connectives to all the types in Γ. In other words, the context of this rule must only have duplicable types: a term can only be made duplicable if it does not depend on linear resources from the context. Otherwise, duplicating the shared value could break the unique-ownership discipline on these linear resources.

Finally, the linear isomorphism notation for fold and unfold in Fig. 4 defines them as primitive functions, at the given linear function type, in the empty context – using them does not consume resources. This notation also means that, operationally, these two operations shall be inverses of each other. The rules for the linear store type Box 1 σ and Box 0 are described in Sect. 2.2.

[3] Standard presentations of linear logic force contexts to be completely distinct, but have a separate rule to duplicate linear variables, which is less natural for programming.

$$\boxed{\Gamma_1 \curlyvee \Gamma_2}$$

$$
\begin{array}{lll}
(\Gamma_1, \mathsf{x}{:}!\sigma) \curlyvee (\Gamma_2, \mathsf{x}{:}!\sigma) & \overset{\text{def}}{=} (\Gamma_1 \curlyvee \Gamma_2), \mathsf{x}{:}!\sigma & \\
(\Gamma_1, \mathsf{x}{:}\sigma) \curlyvee \Gamma_2 & \overset{\text{def}}{=} (\Gamma_1 \curlyvee \Gamma_2), \mathsf{x}{:}\sigma & (\mathsf{x} \notin \Gamma_2) \\
\Gamma_1 \curlyvee (\Gamma_2, \mathsf{x}{:}\sigma) & \overset{\text{def}}{=} (\Gamma_1 \curlyvee \Gamma_2), \mathsf{x}{:}\sigma & (\mathsf{x} \notin \Gamma_1)
\end{array}
$$

$$\boxed{\Gamma \vdash_{\mathsf{L}} \mathsf{e} : \sigma}$$

$$
\frac{}{!\Gamma, \mathsf{x}{:}\sigma \vdash_{\mathsf{L}} \mathsf{x} : \sigma}
\qquad
\frac{\Gamma_1 \vdash_{\mathsf{L}} \mathsf{e}_1 : \sigma_1 \qquad \Gamma_2 \vdash_{\mathsf{L}} \mathsf{e}_2 : \sigma_2}{\Gamma_1 \curlyvee \Gamma_2 \vdash_{\mathsf{L}} \langle \mathsf{e}_1, \mathsf{e}_2 \rangle : \sigma_1 \otimes \sigma_2}
$$

$$
\frac{\Gamma \vdash_{\mathsf{L}} \mathsf{e} : \sigma_1 \otimes \sigma_2 \qquad \Gamma', \mathsf{x}_1{:}\sigma_1, \mathsf{x}_2{:}\sigma_2 \vdash_{\mathsf{L}} \mathsf{e}' : \sigma}{\Gamma \curlyvee \Gamma' \vdash_{\mathsf{L}} \mathsf{let}\ \langle \mathsf{x}_1, \mathsf{x}_2 \rangle = \mathsf{e}\ \mathsf{in}\ \mathsf{e}' : \sigma}
$$

$$
\frac{}{!\Gamma \vdash_{\mathsf{L}} \langle\rangle : 1}
\qquad
\frac{\Gamma \vdash_{\mathsf{L}} \mathsf{e} : 1 \qquad \Gamma' \vdash_{\mathsf{L}} \mathsf{e}' : \sigma}{\Gamma \curlyvee \Gamma' \vdash_{\mathsf{L}} \mathsf{e}; \mathsf{e}' : \sigma}
\qquad
\frac{\Gamma, \mathsf{x}{:}\sigma \vdash_{\mathsf{L}} \mathsf{e} : \sigma'}{\Gamma \vdash_{\mathsf{L}} \lambda(\mathsf{x}{:}\sigma).\, \mathsf{e} : \sigma \multimap \sigma'}
$$

$$
\frac{\Gamma \vdash_{\mathsf{L}} \mathsf{e} : \sigma' \multimap \sigma \qquad \Gamma' \vdash_{\mathsf{L}} \mathsf{e}' : \sigma'}{\Gamma \curlyvee \Gamma' \vdash_{\mathsf{L}} \mathsf{e}\ \mathsf{e}' : \sigma}
\qquad
\frac{\Gamma \vdash_{\mathsf{L}} \mathsf{e} : \sigma_i}{\Gamma \vdash_{\mathsf{L}} \mathsf{inj}_i\, \mathsf{e} : \sigma_1 \oplus \sigma_2}
$$

$$
\frac{\Gamma \vdash_{\mathsf{L}} \mathsf{e} : \sigma_1 \oplus \sigma_2 \qquad \Gamma', \mathsf{x}_1 : \sigma_1 \vdash_{\mathsf{L}} \mathsf{e}_1 : \sigma \qquad \Gamma', \mathsf{x}_2 : \sigma_2 \vdash_{\mathsf{L}} \mathsf{e}_2 : \sigma}{\Gamma \curlyvee \Gamma' \vdash_{\mathsf{L}} \mathsf{case}\ \mathsf{e}\ \mathsf{of}\ \mathsf{x}_1.\, \mathsf{e}_1 \mid \mathsf{x}_2.\, \mathsf{e}_2 : \sigma}
\qquad
\frac{!\Gamma \vdash_{\mathsf{L}} \mathsf{e} : \sigma}{!\Gamma \vdash_{\mathsf{L}} \mathsf{share}\ \mathsf{e} : !\sigma}
\qquad
\frac{\Gamma \vdash_{\mathsf{L}} \mathsf{e} : !\sigma}{\Gamma \vdash_{\mathsf{L}} \mathsf{copy}\ \mathsf{e} : \sigma}
$$

$$
\mu\alpha.\,\sigma \ \underset{\mathsf{fold}_{\mu\alpha.\sigma}}{\overset{\mathsf{unfold}}{\rightleftarrows}}\ \sigma[\mu\alpha.\,\sigma/\alpha]
\qquad
1 \ \underset{\mathsf{free}}{\overset{\mathsf{new}}{\rightleftarrows}}\ \mathsf{Box}\ 0
\qquad
\mathsf{Box}\ 1\ \sigma \ \underset{\mathsf{box}}{\overset{\mathsf{unbox}}{\rightleftarrows}}\ (\mathsf{Box}\ 0) \otimes \sigma
$$

Fig. 4. Linear language: surface static semantics

head reduction $\boxed{\mathsf{e} \overset{\mathsf{L}}{\rightsquigarrow} \mathsf{e}'}$ $\boxed{(s \mid \mathsf{e}) \overset{\mathsf{L}}{\rightsquigarrow} (s' \mid \mathsf{e}')}$

$$
(\emptyset \mid \langle\rangle) \ \underset{\mathsf{free}}{\overset{\mathsf{new}}{\rightleftarrows}}\ ([\ell \mapsto \cdot] \mid \ell)
\qquad
(s[\ell \mapsto \cdot] \mid \langle \ell, \mathsf{v} \rangle) \ \underset{\mathsf{unbox}}{\overset{\mathsf{box}}{\rightleftarrows}}\ ([\ell \mapsto (s \mid \mathsf{v})] \mid \ell)
$$

$$
(\lambda(\mathsf{x}{:}\sigma).\, \mathsf{e})\ \mathsf{v} \overset{\mathsf{L}}{\rightsquigarrow} \mathsf{e}[\mathsf{v}/\mathsf{x}]
\qquad
\mathsf{copy}\ (\mathsf{share}(s : \Psi).\, \mathsf{inj}_i\, \mathsf{v}) \overset{\mathsf{L}}{\rightsquigarrow} \mathsf{inj}_i\, \mathsf{copy}\ (\mathsf{share}(s : \Psi).\, \mathsf{v})
$$

$$
(\emptyset \mid \mathsf{copy}\ (\mathsf{share}(s : \Psi).\, \lambda(\mathsf{x}{:}\sigma).\, \mathsf{e})) \overset{\mathsf{L}}{\rightsquigarrow} (s \mid \lambda(\mathsf{x}{:}\sigma).\, \mathsf{e})
$$

$$
\mathsf{copy}\ (\mathsf{share}([\ell \mapsto \cdot] : (\cdot; \cdot \vdash \ell : \mathsf{Box}\ 0)).\, \ell) \overset{\mathsf{L}}{\rightsquigarrow} \mathsf{new}\ \langle\rangle
$$

$$
\begin{array}{l}
\mathsf{copy}\ (\mathsf{share}([\ell \mapsto (s \mid \mathsf{v})] : (\Psi; !\Gamma \vdash \ell : \mathsf{Box}\ 1\ \sigma)).\, \ell) \\
\qquad \overset{\mathsf{L}}{\rightsquigarrow} \mathsf{box}\ \langle \mathsf{new}\ \langle\rangle, \mathsf{copy}\ (\mathsf{share}(s : \Psi).\, \mathsf{v}) \rangle
\end{array}
$$

Fig. 5. Internal linear language: typing and reduction (excerpt)

2.2 Linear Memory in λ^{L}

The surface typing rules for the linear store are given at the end of Fig. 4. The linear type $\mathsf{Box}\ 1\ \sigma$ represents a memory location that holds a value of type σ. The type $\mathsf{Box}\ 0$ represents a location that has been allocated, but does not currently hold a value. The primitive operations to act on this type are given as linear isomorphisms: new allocates, turning a unit value into an empty location; conversely, free reclaims an empty location. Putting a value into the location and taking it out are expressed by box and unbox, which convert between a pair of an empty location and a value, of type $(\mathsf{Box}\ 0) \otimes \sigma$, and a full location, of type $\mathsf{Box}\ 1\ \sigma$.

For example, the following program takes a full reference and a value, and swaps the value with the content of the reference:

$$\lambda(\mathsf{p} : (\mathsf{Box}\ 1\ \sigma) \otimes \sigma).\ \mathsf{let}\ \langle \mathsf{r}, \mathsf{x} \rangle = \mathsf{p}\ \mathsf{in}\ \mathsf{let}\ \langle \mathsf{l}, \mathsf{x}_l \rangle = \mathsf{unbox}\ \mathsf{r}\ \mathsf{in}\ \langle \mathsf{box}\ \langle \mathsf{l}, \mathsf{x} \rangle, \mathsf{x}_l \rangle$$

The programming style following from this presentation of linear memory is functional, or applicative, rather than imperative. Rather than insisting on the mutability of references—which is allowed by the linear discipline—we may think of the type $\mathsf{Box}\ 1 \sigma$ as representing the indirection through the heap that is implicit in functional programs. In a sense, we are not writing imperative programs with a mutable store, but rather making explicit the allocations and dereferences happening in higher-level purely functional language. In this view, empty cells allow memory reuse.

This view that $\mathsf{Box}\ 1\ \sigma$ represents indirection through the memory suggests we can encode lists of values of type σ by the type $\mathsf{LinList}\, \sigma \stackrel{\text{def}}{=} \mu\alpha.1 \oplus \mathsf{Box}\ 1\ (\sigma \otimes \alpha)$. The placement of the box inside the sum mirrors the fact that empty list is represented as an immediate value in functional languages. From this type definition, one can write an in-place reverse function on lists of σ as follows:

```
fix λ(rev_into : LinList σ ⊸ LinList σ ⊸ LinList σ).
        λ(xs : LinList σ). λ(acc : LinList σ).
  case unfold xs of
  | y. (y; acc)
  | y. let ⟨l, p⟩ = unbox y in
        let ⟨xs, x⟩ = p in
        rev_into xs (fold (inj₂ (box ⟨l, ⟨x, acc⟩⟩)))
```

Our linear language λ^{L} is a formal language that is not terribly convenient to program directly. We will not present a full surface language in this work, but one could easily define syntactic sugar to write the exact same function as follows:

```
rev_into Nil                acc = acc
rev_into (Cons ⟨x, xs⟩@ l) acc = rev_into xs (Cons ⟨x, acc⟩@ l)
```

One can read this function as the usual functional `rev_append` function on lists, annotated with memory reuse information: if we assume we are the unique owner of the input list and won't need it anymore, we can reuse the memory of its cons cells (given in this example the name l) to store the reversed list.

On the other hand, if you read the box and unbox as imperative operations, this code expresses the usual imperative pointer-reversal algorithm.

This double view of linear state occurs in other programming systems with linear state. It was recently emphasized in O'Connor et al. (2016), where the functional point of view is seen as easing formal verification, while the imperative view is used as a compilation technique to produce efficient C code from linear programs.

2.3 Internal λ^{L} Syntax and Typing

To give a dynamic semantics for λ^{L} and prove it sound, we need to extend the language with explicit stores and store locations. Indeed, the allocating term $\mathsf{new}\,\langle\rangle$ should reduce to a "fresh location" ℓ allocated in some store s, and neither are part of the surface-language syntax. The corresponding internal typing judgment is more complex, but note that users do not need to know about it to reason about correctness of surface programs. The internal typing is essential for the soundness proof, but also useful for defining the multi-language semantics in Sect. 3.

We work with *configurations* $(s \mid \mathsf{e})$, which are pairs of a store s and a term e. Our internal typing judgment $\Psi; \Gamma \vdash_{\mathsf{L}} s \mid \mathsf{e} : \sigma$ checks configurations, not just terms, and relies not only on a typing context for variables Γ but also on a *store typing* Ψ, which maps the locations of the configuration to typing assumptions.

Unfortunately, due to space limits, we will not present this part of the type system – which is not directly exposed to users of the language. See some examples of reduction rules in Fig. 5, and the long version of this work.

2.4 Reduction of Internal Terms

In the long version of this work we give a reduction relation between linear configurations $(s \mid \mathsf{e}) \overset{\mathsf{L}}{\hookrightarrow} (s' \mid \mathsf{e}')$ and prove a subject reduction result.

Theorem 1 (Subject reduction for λ^{L}). *If $\Psi; \Gamma \vdash_{\mathsf{L}} s \mid \mathsf{e} : \sigma$ and $(s \mid \mathsf{e}) \overset{\mathsf{L}}{\hookrightarrow} (s' \mid \mathsf{e}')$, then there exists a (unique) Ψ' such that $\Psi'; \Gamma \vdash_{\mathsf{L}} s' \mid \mathsf{e}' : \sigma$.*

3 Multi-language Semantics

To formally define our multi-language semantics we create a combined language λ^{UL} which lets us compose term fragments from both λ^{U} and λ^{L} together, and we give an operational semantics to this combined language. Interoperability is enabled by specifying how to transport values across the language boundaries.

Multi-language systems in the wild are not defined in this way: both languages are given a semantics, by interpretation or compilation, in terms of a shared lower-level language (C, assembly, the JVM or CLR bytecode, or Racket's core forms), and the two languages are combined at that level. Our formal multi-language description can be seen as a model of such combinations, that gives a specification of the expected observable behavior of this language combination.

Another difference from multi-languages in the wild is our use of very fine-grained language boundaries: a term written in one language can have its sub-terms written in the other, provided the type-checking rules allow it. Most multi-language systems, typically using Foreign Function Interfaces, offer coarser-grained composition at the level of compilation units. Fine-grained composition of existing languages, as done in the Eco project (Barrett et al. 2016), is difficult because of semantic mismatches. In the full version of this work we demonstrate that fine-grained composition is a rewarding language design, enabling new programming patterns.

3.1 Lump Type and Language Boundaries

The core components the multi-language semantics are shown Fig. 6—the communication of values from one language to the other will be described in the next section. The multi-language λ^{UL} has two distinct syntactic categories of types, values, and expressions: those that come from λ^{U} and those that come from λ^{L}. Contexts, on the other hand, are mixed, and can have variables of both sorts. For a mixed context Γ, the notation $!\Gamma$ only applies (!) to its linear variables.

The typing rules of λ^{U} and λ^{L} are imported into our multi-language system, working on those two separate categories of program. They need to be extended to handle mixed contexts Γ instead of their original contexts Γ and Γ. In the linear case, the rules look exactly the same. In the ML case, the typing rules implicitly duplicate all the variables in the context. It would be unsound to extend them to arbitrary linear variables, so they use a duplicable context $!\Gamma$.

To build interesting multi-language programs, we need a way to insert a fragment coming from a language into a term written in another. This is done using *language boundaries*, two new term formers $\mathcal{LU}(\mathsf{e})$ and $\mathcal{UL}(s{:}\Psi \mid \mathsf{e})$ that inject an ML term into the syntactic category of linear terms, and a linear configuration into the syntactic category of ML terms.

Of course, we need new typing rules for these term-level constructions, clarifying when it is valid to send a value from λ^{U} into λ^{L} and vice versa. It would be incorrect to allow sending any type from one language into the other—for instance, by adding the counterpart of our language boundaries in the syntax of types—since values of linear types must be uniquely owned so they cannot possibly be sent to the ML side as the ML type system cannot enforce unique ownership.

On the other hand, any ML value could safely be sent to the linear world. For closed types, we could provide a corresponding linear type (1 maps to $!1$, etc.), but an ML value may also be typed by an abstract type variable α, in which case we can't know what the linear counterpart should be. Instead of trying to provide translations, we will send any ML type σ to the *lump type* $[\sigma]$, which embeds ML types into linear types. A lump is a blackbox, not a type translation: the linear language does not assume anything about the behavior of its values—the values of $[\sigma]$ are of the form $[\mathsf{v}]$, where $\mathsf{v} : \sigma$ is an ML value that the linear world cannot use. More precisely, we only propagate the information that ML values are all duplicable by sending σ to $![\sigma]$.

The typing rules for language boundaries insert lumps when going from λ^{U} to λ^{L}, and remove them when going back from λ^{L} to λ^{U}. In particular, arbitrary linear types cannot occur at the boundary, they must be of the form $![\sigma]$.

Types $\sigma \mid \sigma$

σ (unchanged from Figure 1)

$\sigma \quad +::= \cdots \mid [\sigma]$

Values $\mathsf{v} \mid \mathsf{v}$

v (unchanged from Figure 1)

$\mathsf{v} \quad +::= \cdots \mid [\mathsf{v}]$

Expressions $\mathsf{e} \mid \mathsf{e}$

$\mathsf{e} \quad +::= \cdots \mid \mathcal{UL}(s{:}\Psi \mid \mathsf{e})$

with $\mathcal{UL}(\mathsf{e}) \stackrel{\text{def}}{=} \mathcal{UL}(\emptyset{:}\cdot \mid \mathsf{e})$

$\mathsf{e} \quad +::= \cdots \mid \mathcal{LU}(\mathsf{e})$

Contexts $\Gamma ::= \cdot \mid \Gamma, \mathsf{x}{:}\sigma \mid \Gamma, \alpha \mid \Gamma, \mathsf{x}{:}\sigma$

Typing rules $\boxed{\Gamma \vdash_{\mathsf{LU}} \mathsf{e} : \sigma}$ $\boxed{\Psi \mid \Gamma \vdash_{\mathsf{UL}} s \mid \mathsf{e} : \sigma}$

with $\Gamma \vdash_{\mathsf{UL}} \mathsf{e} : \sigma \stackrel{\text{def}}{=} \cdot \mid \Gamma \vdash_{\mathsf{UL}} \emptyset \mid \mathsf{e} : \sigma$

(Typing rules of $\Gamma \vdash_{\mathsf{U}} \mathsf{e} : \sigma$ reused, with mixed context $!\Gamma$)
(Typing rules of $\Psi; \Gamma \vdash_{\mathsf{L}} s \mid \mathsf{e} : \sigma$ reused, with mixed context Γ)

$$\frac{!\Gamma \vdash_{\mathsf{LU}} \mathsf{e} : \sigma}{\cdot \mid !\Gamma \vdash_{\mathsf{UL}} \emptyset \mid \mathcal{LU}(\mathsf{e}) : ![\sigma]} \qquad \frac{\Psi \mid !\Gamma \vdash_{\mathsf{UL}} s \mid \mathsf{e} : ![\sigma]}{!\Gamma \vdash_{\mathsf{LU}} \mathcal{UL}(s{:}\Psi \mid \mathsf{e}) : \sigma}$$

Reduction rules

(Reduction rules of λ^{U} and λ^{L} reused unchanged)
$$\frac{\mathsf{e} \stackrel{\mathsf{U}}{\hookrightarrow} \mathsf{e}'}{\mathcal{LU}(\mathsf{e}) \stackrel{\mathsf{L}}{\hookrightarrow} \mathcal{LU}(\mathsf{e}')}$$

$$\frac{}{\mathcal{LU}(\mathsf{v}) \stackrel{\mathsf{L}}{\rightsquigarrow} [\mathsf{v}]} \qquad \frac{}{\mathcal{UL}(\emptyset{:}\cdot \mid \mathsf{share}\,[\mathsf{v}]) \stackrel{\mathsf{U}}{\hookrightarrow} \mathsf{v}}$$

$$\frac{\Psi \mid \Gamma \vdash_{\mathsf{UL}} s \mid \mathsf{e} : \sigma \qquad (s \mid \mathsf{e}) \stackrel{\mathsf{L}}{\hookrightarrow} (s' \mid \mathsf{e}') \qquad \Psi' \mid \Gamma \vdash_{\mathsf{UL}} s' \mid \mathsf{e}' : \sigma}{\mathcal{UL}(s{:}\Psi \mid \mathsf{e}) \stackrel{\mathsf{U}}{\hookrightarrow} \mathcal{UL}(s'{:}\Psi' \mid \mathsf{e}')}$$

Fig. 6. Multi-language: lump and boundaries

static compatibility $\boxed{\Sigma \vdash_{\text{UL}} \sigma \simeq \sigma}$

$$\frac{}{\Sigma \vdash_{\text{UL}} \sigma \simeq ![\sigma]} \qquad \frac{\Sigma \vdash_{\text{UL}} \sigma_1 \simeq !\sigma_1 \qquad \Sigma \vdash_{\text{UL}} \sigma_2 \simeq !\sigma_2}{\Sigma \vdash_{\text{UL}} \sigma_1 + \sigma_2 \simeq !(\sigma_1 \oplus \sigma_2)}$$

$$\frac{\Sigma \vdash_{\text{UL}} \sigma \simeq !\sigma \qquad \Sigma \vdash_{\text{UL}} \sigma' \simeq !\sigma'}{\Sigma \vdash_{\text{UL}} \sigma \to \sigma' \simeq !(!\sigma \multimap !\sigma')} \qquad \frac{\Sigma \vdash_{\text{UL}} \sigma \simeq !\sigma}{\Sigma \vdash_{\text{UL}} \sigma \simeq !!\sigma} \qquad \frac{\Sigma \vdash_{\text{UL}} \sigma \simeq !\sigma}{\Sigma \vdash_{\text{UL}} \sigma \simeq !(\mathsf{Box}\ 1\ \sigma)}$$

value conversion $\boxed{\mathsf{v} \leftrightarrow^{\sigma} \mathsf{v}}$

$$\frac{}{\mathsf{v} \leftrightarrow^{![\sigma]} \mathsf{share}\,[\mathsf{v}]} \qquad \frac{\mathsf{v} \leftrightarrow^{!\sigma_i} \mathsf{share}(s:\Psi).\,\mathsf{v}}{\mathsf{inj_i}\,\mathsf{v} \leftrightarrow^{!(\sigma_1 \oplus \sigma_2)} \mathsf{share}(s:\Psi).\,\mathsf{inj_i}\,\mathsf{v}}$$

$$\frac{\sigma \simeq \sigma \qquad \sigma' \simeq \sigma'}{\mathsf{e} \to^{!(!\sigma \multimap !\sigma')} \mathsf{share}\,\lambda(\mathsf{x}:!\sigma).\,{}^{\sigma'}\mathcal{LU}(\mathsf{e}\ \mathcal{UL}^{\sigma}(\mathsf{x}))} \qquad \frac{\sigma \simeq \sigma \qquad \sigma' \simeq \sigma'}{\lambda(\mathsf{x}:\sigma).\,\mathcal{UL}^{\sigma'}(\mathsf{copy}\ \mathsf{e}\ {}^{\sigma}\mathcal{LU}(\mathsf{x})) \leftarrow^{!(\sigma \multimap \sigma')} \mathsf{e}}$$

$$\frac{\mathsf{v} \leftrightarrow^{!\sigma} \mathsf{share}\,\mathsf{v}}{\mathsf{v} \leftrightarrow^{!!\sigma} \mathsf{share}\,(\mathsf{share}\,\mathsf{v})} \qquad \frac{\mathsf{v} \leftrightarrow^{!\sigma} \mathsf{share}(s:\Psi).\,\mathsf{v}}{\mathsf{v} \leftrightarrow^{!\mathsf{Box}\ 1\ \sigma} \mathsf{share}([\ell \mapsto (s \mid \mathsf{v})] : (\cdot; \ell \vdash \Psi : \mathsf{Box}\ 1\ \sigma)).\,\ell}$$

$$\frac{\mathsf{v} \leftrightarrow^{!\sigma[\mu\alpha.\sigma/\alpha]} \mathsf{share}(s:\Psi).\,\mathsf{v}}{\mathsf{fold}_{\mu\alpha.\sigma}\,\mathsf{v} \leftrightarrow^{!\mu\alpha.\sigma} \mathsf{share}(s:\Psi).\,(\mathsf{fold}_{\mu\alpha.\sigma}\,\mathsf{v})}$$

Fig. 7. Interoperability: static and dynamic semantics (excerpt)

Finally, boundaries have reduction rules: a term or configuration inside a boundary in reduction position is reduced until it becomes a value, and then a lump is added or removed depending on the boundary direction. Note that because the v in $\mathcal{UL}(s:\Psi \mid \mathsf{v})$ is at a duplicable type $![\sigma]$, we know by inversion that the store is empty.

3.2 Interoperability: Static Semantics

If the linear language could not interact with lumped values at all, our multi-language programs would be rather boring, as the only way for the linear extension to provide a value back to ML would be to have received it from λ^{U} and pass it back unchanged (as in the lump embedding of Matthews and Findler (2009)). To provide a real interaction, we provide a way to extract values out of a lump $![\sigma]$, use it at some linear type σ, and put it back in before sending the result to λ^{U}.

The correspondence between intuitionistic types σ and linear types σ is specified by a heterogeneous *compatibility relation* $\sigma \simeq \sigma$ – defined in full in Fig. 7. The specification of this relation is that if $\sigma \simeq \sigma$ holds, then the space of values of $![\sigma]$ and σ are isomorphic: we can convert back and forth between them. When this relation holds, the term-formers lump^{σ} and ${}^{\sigma}\mathsf{unlump}$ perform the conversion.

The term $\mathcal{LU}(\mathsf{e})$ turns a $\mathsf{e} : \sigma$ into a lumped type $![\sigma]$, and we need to unlump it with some ${}^{\sigma}\mathsf{unlump}$ for a compatible $\sigma \simeq \sigma$ to interact with it on the linear side. It is common to combine both operations and we provide syntactic sugar for it: ${}^{\sigma}\mathcal{LU}(\mathsf{e})$. Similarly $\mathcal{UL}^{\sigma}(\mathsf{e})$ first lumps a linear term then sends the result to the ML world.

3.3 Interoperability: Dynamic Semantics

When the relation $\sigma \simeq \sigma$ holds, we can define a relation $\mathsf{v} \leftrightarrow^{\sigma} \mathsf{v}$ between the values of σ and the values of σ – see the long version of this work. It is functional in both direction: with our definition v is uniquely determined from v and conversely. We then define the reduction rule for (un)lumping: if $\mathsf{v} \leftrightarrow^{\sigma} \mathsf{v}$, then

$$(\emptyset \mid {}^{\sigma}\mathsf{unlump}\ (\mathsf{share}\,[\mathsf{v}])) \overset{\mathsf{L}}{\hookrightarrow} (\emptyset \mid \mathsf{v}) \qquad (\emptyset \mid \mathsf{lump}^{\sigma}\ \mathsf{v}) \overset{\mathsf{L}}{\hookrightarrow} (\emptyset \mid \mathsf{share}\,[\mathsf{v}])$$

3.4 Full Abstraction from λ^{U} into λ^{UL}

We can now state the major meta-theoretical result of this work, which is the proposed multi-language design extends the simple language λ^{U} in a way that provably has, in a certain sense, "no abstraction leaks".

Definition 1 (Contextual equivalence in λ^{U}). *We say that* e, e' *such that* $\Gamma \vdash_{\mathrm{U}} \mathsf{e}, \mathsf{e}' : \sigma$ *are* contextually equivalent, *written* $\mathsf{e} \approx^{ctx}_{\mathrm{U}} \mathsf{e}'$, *if, for any expression context* $C[\square]$ *such that* $\cdot \vdash_{\mathrm{U}} C[\mathsf{e}] : 1$, *the closed terms* $C[\mathsf{e}]$ *and* $C[\mathsf{e}']$ *are equi-terminating.*

Definition 2 (Contextual equivalence in λ^{UL}). *We say that* e, e' *such that* $\Gamma \vdash_{\mathrm{LU}} \mathsf{e}, \mathsf{e}' : \sigma$ *are* contextually equivalent, *written* $\mathsf{e} \approx^{ctx}_{\mathrm{LU}} \mathsf{e}'$, *if, for any expression context* $C[\square]$ *such that* $\cdot \vdash_{\mathrm{LU}} C[\mathsf{e}] : 1$, *the closed terms* $C[\mathsf{e}]$ *and* $C[\mathsf{e}']$ *are equi-terminating.*

Theorem 2 (Full Abstraction). *The embedding of* λ^{U} *into* λ^{UL} *is fully-abstract:*

$$\Gamma \vdash_{\mathsf{U}} \mathsf{e} \approx^{ctx}_{\mathsf{U}} \mathsf{e}' : \sigma \qquad \Longrightarrow \qquad \Gamma \vdash_{\mathsf{LU}} \mathsf{e} \approx^{ctx}_{\mathsf{LU}} \mathsf{e}' : \sigma$$

4 Conclusion and Related Work

Having a stack of usable, interoperable languages, extensions or dialects is at the forefront of the Racket approach to programming environments, in particular for teaching (Felleisen et al. 2004).

Our multi-language semantics builds on the seminal work by Matthews and Findler (2009), who gave a formal semantics of interoperability between a dynamically and a statically typed language. Others have followed the Matthews-Findler approach of designing multi-language systems with fine-grained boundaries—for instance, formalizing interoperability between a simply and dependently typed language (Osera et al. 2012); between a functional and typed assembly language (Patterson et al. 2017); between an ML-like and an affinely typed language, where linearity is enforced at runtime on the ML side using stateful contracts (Tov and Pucella 2010); and between the source and target languages of compilation to specify compiler correctness (Perconti and Ahmed 2014). However, all these papers address only the question of soundness of the multi-language; we propose a formal treatment of *usability* and absence of abstraction leaks.

The only work to establish that a language embeds into a multi-language in a fully abstract way is the work on fully abstract compilation by Ahmed and Blume (2011) and New et al. (2016) who show that their compiler's source language embeds into their source-target multi-language in a fully abstract way. But the focus of this work was on fully abstract compilation, not on usability of user-facing languages.

The Eco project (Barrett et al. 2016) is studying multi-language systems where user-exposed languages are combined in a very fine-grained way; it is closely related in that it studies the user experience in a multi-language system. The choice of an existing dynamic language creates delicate interoperability issues (conflicting variable scoping rules, etc.) as well as performance challenges. We propose a different approach, to design new multi-languages from scratch with interoperability in mind to avoid legacy obstacles.

We are not aware of existing systems exploiting the simple idea of using promotion to capture uniquely-owned state and dereliction to copy it—common formulations would rather perform copies on the contraction rule.

The general idea that linear types can permit reuse of unused allocated cells is not new. In Wadler (1990), a system is proposed with both linear and non-linear types to attack precisely this problem. It is however more distant from standard linear logic and somewhat ad-hoc; for example, there is no way to permanently turn a uniquely-owned value into a shared value, it provides instead a local *borrowing* construction that comes with ad-hoc restrictions necessary for safety. (The inability to *give up* unique ownership, which is essential in our list-programming examples, seems to also be missing from Rust, where one would need to perform a costly operation of traversing the graph of the value to turn all pointers into `Arc` nodes.)

The RAML project (Hoffmann et al. 2012) also combines linear logic and memory reuse: its *destructive match* operator will implicitly reuse consumed cells in new allocations occurring within the match body. Multi-languages give us the option to explore more explicit, flexible representations of those low-level concern, without imposing the complexity to all programmers.

A recent related work is the Cogent language (O'Connor et al. 2016), in which linear state is also viewed as both functional and imperative – the latter view enabling memory reuse. The language design is interestingly reversed: in Cogent, the linear layer is the simple language that everyone uses, and the non-linear language is a complex but powerful language that is used when one really has to, named C.

Our linear language λ^{L} is sensibly simpler, and in several ways less expressive, than advanced programming languages based on linear logic (Tov and Pucella 2011), separation logic (Balabonski et al. 2016), fine-grained permissions (Garcia et al. 2014): it is not designed to stand on its own, but to serve as a useful sidekick to a functional language, allowing safer resource handling.

One major simplification of our design compared to more advanced linear or separation-logic-based languages is that we do not separate physical locations from the logical capability/permission to access them (e.g., as in Ahmed et al. (2007)). This restricts expressiveness in well-understood ways (Fahndrich and DeLine 2002): shared values cannot point to linear values.

Alms (Tov and Pucella 2011), Quill (Morris 2016) and Linear Haskell (Bernardy et al. 2018) add linear types to a functional language, trying hard not to lose desirable usability property, such as type inference or the genericity of polymorphic higher-order functions. This is very challenging; for example, Linear Haskell gives up on principality of inference[4]. Our multi-language design side-steps this issue as the general-purpose language remains unchanged. Language boundaries are more rigid than an ideal no-compromise language, as they force users to preserve the distinction between the general-purpose and the advanced features; it is precisely this compromise that gives a design of reduced complexity.

Finally, on the side of the semantics, our system is related to LNL (Benton 1994), a calculus for linear logic that, in a sense, is itself built as a multi-language system where (non-duplicable) linear types and (duplicable) intuitionistic types interact through a boundary. It is not surprising that our design contains an instance of this adjunction: for any σ there is a unique σ such that $\sigma \simeq !\sigma$, and converting a σ value to this σ and back gives a $!\sigma$ and is provably equivalent, by boundary cancellation, to just using share.

Acknowledgments. We thank our anonymous reviewers for their feedback, as well as Neelakantan Krishnaswami, François Pottier, Jennifer Paykin, Sylvie Boldot and Simon Peyton-Jones for our discussions on this work.

This work was supported in part by the National Science Foundation under grants CCF-1422133 and CCF-1453796, and the European Research Council under ERC Starting Grant SECOMP (715753). Any opinions, findings, and conclusions expressed in this material are those of the authors and do not necessarily reflect the views of our funding agencies.

References

Abramsky, S., Jagadeesan, R., Malacaria, P.: Full abstraction for PCF. Inf. Comput. **163**(2), 409–470 (2000)

Ahmed, A., Blume, M.: Typed closure conversion preserves observational equivalence. In: International Conference on Functional Programming (ICFP), Victoria, British Columbia, Canada, pp. 157–168, September 2008

[4] Thanks to Stephen Dolan for pointing out that $\lambda f.\lambda x.\, f\ x$ has several incompatible Linear Haskell types.

Ahmed, A., Blume, M.: An equivalence-preserving CPS translation via multi-language semantics. In: International Conference on Functional Programming (ICFP), Tokyo, Japan, pp. 431–444, September 2011

Ahmed, A., Fluet, M., Morrisett, G.: L3: a linear language with locations. Fundamenta Informaticae **77**(4), 397–449 (2007)

Balabonski, T., Pottier, F., Protzenko, J.: The design and formalization of Mezzo, a permission-based programming language. ACM Trans. Program. Lang. Syst. **38**(4), 14:1–14:94 (2016)

Barrett, E., Bolz, C.F., Diekmann, L., Tratt, L.: Fine-grained language composition: a case study. In: ECOOP (2016)

Benton, P.N.: A mixed linear and non-linear logic: proofs, terms and models. In: Pacholski, L., Tiuryn, J. (eds.) CSL 1994. LNCS, vol. 933, pp. 121–135. Springer, Heidelberg (1995). https://doi.org/10.1007/BFb0022251

Bernardy, J.-P., Boespflug, M., Newton, R.R., Jones, S.P., Spiwack, A.: Linear haskell: practical linearity in a higher-order polymorphic language. PACMPL **2**(POPL), 5:1–5:29 (2018). https://doi.org/10.1145/3158093

Cartwright, R., Felleisen, M.: Observable sequentiality and full abstraction. In: ACM Symposium on Principles of Programming Languages (POPL), Albuquerque, New Mexico, pp. 328–342 (1992)

Devriese, D., Patrignani, M., Piessens, F.: Fully-abstract compilation by approximate back-translation. In: ACM Symposium on Principles of Programming Languages (POPL), St. Petersburg, Florida (2016)

Fahndrich, M., DeLine, R.: Adoption and focus: practical linear types for imperative programming. In: PLDI 2002 (2002)

Felleisen, M., Findler, R.B., Flatt, M., Krishnamurthi, S.: The teachscheme! project: computing and programming for every student. Comput. Sci. Educ. **14**(1), 55–77 (2004)

Garcia, R., Tanter, É., Wolff, R., Aldrich, J.: Foundations of typestate-oriented programming. TOPLAS **36**, 12 (2014)

Hoffmann, J., Aehlig, K., Hofmann, M.: Resource aware ML. In: Madhusudan, P., Seshia, S.A. (eds.) CAV 2012. LNCS, vol. 7358, pp. 781–786. Springer, Heidelberg (2012). https://doi.org/10.1007/978-3-642-31424-7_64

Jeffrey, A., Rathke, J.: Java JR: fully abstract trace semantics for a Core Java Language. In: Sagiv, M. (ed.) ESOP 2005. LNCS, vol. 3444, pp. 423–438. Springer, Heidelberg (2005). https://doi.org/10.1007/978-3-540-31987-0_29

Matthews, J., Findler, R.B.: Operational semantics for multi-language programs. ACM Trans. Program. Lang. Syst. (TOPLAS) **31**(3), 12 (2009)

Meyer, A.R., Sieber, K.: Towards fully abstract semantics for local variables. In: ACM Symposium on Principles of Programming Languages (POPL), San Diego, California, pp. 191–203 (1988)

Milner, R.: Fully abstract models of typed lambda calculi. Theor. Comput. Sci. **4**(1), 1–22 (1977)

Morris, J.G.: The best of both worlds: linear functional programming without compromise. In: ICFP (2016)

New, M.S., Bowman, W.J., Ahmed, A.: Fully abstract compilation via universal embedding. In: International Conference on Functional Programming (ICFP), Nara, Japan, September 2016

O'Connor, L., Chen, Z., Rizkallah, C., Amani, S., Lim, J., Murray, T., Nagashima, Y., Sewell, T., Klein, G.: Refinement through restraint: bringing down the cost of verification. In: ICFP (2016)

Osera, P.M., Sjöberg, V., Zdancewic, S.: Dependent interoperability. In: Programming Languages Meets Program Verification (PLPV), January 2012

Patrignani, M., Agten, P., Strackx, R., Jacobs, B., Clarke, D., Piessens, F.: Secure compilation to protected module architectures. ACM Trans. Program. Lang. Syst. **37**(2), 6:1–6:50 (2015)

Patterson, D., Perconti, J., Dimoulas, C., Ahmed, A.: FunTAL: reasonably mixing a functional language with assembly. In: ACM SIGPLAN Conference on Programming Language Design and Implementation (PLDI), Barcelona, Spain, June 2017. http://www.ccs.neu.edu/home/amal/papers/funtal.pdf

Perconti, J.T., Ahmed, A.: Verifying an open compiler using multi-language semantics. In: Shao, Z. (ed.) ESOP 2014. LNCS, vol. 8410, pp. 128–148. Springer, Heidelberg (2014). https://doi.org/10.1007/978-3-642-54833-8_8

Tov, J.A., Pucella, R.: Stateful contracts for affine types. In: Gordon, A.D. (ed.) ESOP 2010. LNCS, vol. 6012, pp. 550–569. Springer, Heidelberg (2010). https://doi.org/10.1007/978-3-642-11957-6_29

Tov, J.A., Pucella, R.: Practical affine types. In: POPL (2011)

Wadler, P.: Linear types can change the world! In: Programming Concepts and Methods (1990)

Concurrency

Automata for True Concurrency Properties

Paolo Baldan(✉) and Tommaso Padoan

Dipartimento di Matematica, Università di Padova, Padua, Italy
{baldan,padoan}@math.unipd.it

Abstract. We present an automata-theoretic framework for the model checking of true concurrency properties. These are specified in a fixpoint logic, corresponding to history-preserving bisimilarity, capable of describing events in computations and their dependencies. The models of the logic are event structures or any formalism which can be given a causal semantics, like Petri nets. Given a formula and an event structure satisfying suitable regularity conditions we show how to construct a parity tree automaton whose language is non-empty if and only if the event structure satisfies the formula. The automaton, due to the nature of event structure models, is usually infinite. We discuss how it can be quotiented to an equivalent finite automaton, where emptiness can be checked effectively. In order to show the applicability of the approach, we discuss how it instantiates to finite safe Petri nets. As a proof of concept we provide a model checking tool implementing the technique.

1 Introduction

Behavioural logics with the corresponding verification techniques are a cornerstone of automated verification. For concurrent and distributed systems, so called true concurrent models can be an appropriate choice, since they describe not only the possible steps in the evolution of the system but also their causal dependencies. A widely used foundational model in this class is given by Winskel's event structures [1]. They describe the behaviour of a system in terms of events in computations and two dependency relations: a partial order modelling causality and an additional relation modelling conflict. A survey on the use of such causal models can be found in [2]. Recently they have been used in the study of concurrency in weak memory models [3,4], for process mining and differencing [5], in the study of atomicity [6] and of information flow [7] properties.

Operational models can be abstracted by considering true concurrent equivalences that range from hereditary history preserving bisimilarity to the coarser pomset and step equivalences (see, e.g., [8]) and behavioural logics expressing causal properties (see, e.g., [9–14] for a necessarily partial list and [15–19] for some related verification techniques).

Event-based logics have been recently introduced [20,21], capable of uniformly characterising the equivalences in the true concurrent spectrum. Their formulae include variables which are bound to events in computations and describe

C. Baier and U. Dal Lago (Eds.): FOSSACS 2018, LNCS 10803, pp. 165–182, 2018.
https://doi.org/10.1007/978-3-319-89366-2_9

their dependencies. While the relation between operational models, behavioural equivalences and event-based true concurrent logics is well understood, the corresponding model checking problem has received limited attention.

We focus on the logic referred to as $\mathcal{L}_{hp}$ in [20], corresponding to a classical equivalence in the spectrum, i.e., history preserving (hp-)bisimilarity [22–24].

Decidability of model checking is not obvious since event structure models are infinite even for finite state systems and the possibility of expressing properties that depends on the past often leads to undecidability [25]. In a recent paper [26] we proved the decidability of the problem for the alternation free fragment of the logic $\mathcal{L}_{hp}$ over a class of event structures satisfying a suitable regularity condition [27] referred to as strong regularity. The proof relies on a tableau-based model checking procedure. Despite the infiniteness of the model, a suitable stop condition can be identified, ensuring that a successful finite tableau can be generated if and only if the formula is satisfied by the model.

Besides the limitation to the alternation free fragment of $\mathcal{L}_{hp}$, a shortcoming of the approach is that a direct implementation of the procedure can be extremely inefficient. Roughly speaking, the problem is that in the search of a successful tableau, branches which are, in some sense, equivalent are explored several times.

In this paper we devise an automata-theoretic technique, in the style of [28], for model checking $\mathcal{L}_{hp}$ that works for the full logic, without constraints on the alternation depth. Besides providing an alternative approach for model-checking $\mathcal{L}_{hp}$, amenable of a more efficient implementation, this generalises the decidability result of [26] to the full logic $\mathcal{L}_{hp}$. Given a formula in $\mathcal{L}_{hp}$ and a strongly regular event structure, the procedure generates a parity tree automaton. Satisfiability is reduced to emptiness in the sense that the event structure satisfies the formula if and only if the automaton accepts a non-empty language.

The result is not directly usable for practical purposes since the automaton is infinite for any non-trivial event structure. However an equivalence on states can be defined such that the quotiented automaton accepts the same language as the original one. Whenever such equivalence is of finite index the quotiented automaton is finite, so that satisfaction of the formula can be checked effectively on the quotient. We show that for all strongly regular event structures a canonical equivalence always exists that is of finite index.

The procedure is developed abstractly on event structures. A concrete algorithm on some formalism requires the effectiveness of the chosen equivalence on states. We develop a concrete instantiation of the algorithm on finite safe Petri nets. It is implemented in a tool, wishfully called *True concurrency workbench* (TCWB), written in Haskell. Roughly, the search of an accepting run in the automaton can be seen as an optimisation of the procedure for building a successful tableau in [26] where the graph structure underlying the automaton helps in the reuse of the information discovered. Some tests reveal that the TCWB is way more efficient than the direct implementation of the tableau-based procedure (which could not manage most of the examples in the TCWB repository).

The rest of the paper is structured as follows. In Sect. 2 we review event structures, strong regularity and the logic $\mathcal{L}_{hp}$ of interest in the paper. In Sect. 3

we introduce (infinite state) parity tree automata and we show how the model checking problem for $\mathcal{L}_{hp}$ on strongly regular PES can be reduced to the non-emptiness of the language of such automata. In Sect. 4 we discuss the instantiation of the approach to Petri nets. Finally, in Sect. 5 we discuss some related work and outline directions of future research. Due to space limitations, proofs are only sketched.

2 Event Structures and True Concurrent Logic

We introduce prime event structures [1] and the subclass of strongly regular event structures on which our model checking approach will be developed. Then we present the logic for true concurrency of interest in the paper.

2.1 Prime Event Structures and Regularity

Throughout the paper $\mathbb{E}$ is a fixed countable set of events, Λ a finite set of labels ranged over by $\mathsf{a}, \mathsf{b}, \mathsf{c} \ldots$ and $\lambda : \mathbb{E} \to \Lambda$ a labelling function.

Definition 1 (prime event structure). *A (Λ-labelled)* prime event structure *(PES) is a tuple $\mathcal{E} = \langle E, \leq, \# \rangle$, where $E \subseteq \mathbb{E}$ is the set of* events *and $\leq$, $\#$ are binary relations on E, called* causality *and* conflict *respectively, such that:*
1. $\leq$ is a partial order and $\lceil e \rceil = \{e' \in E \mid e' \leq e\}$ is finite for all $e \in E$;
2. $\#$ is irreflexive, symmetric and inherited along $\leq$, i.e., for all $e, e', e'' \in E$, if $e \# e' \leq e''$ then $e \# e''$.

The PES $\mathcal{E}_1 = \langle E_1, \leq_1, \#_1 \rangle$, $\mathcal{E}_2 = \langle E_2, \leq_2, \#_2 \rangle$ are isomorphic, *written $\mathcal{E}_1 \sim \mathcal{E}_2$, when there is a bijection $\iota : E_1 \to E_2$ such that for all $e_1, e_1' \in E_1$, it holds $e_1 \leq_1 e_1'$ iff $\iota(e_1) \leq_2 \iota(e_1')$ and $e_1 \#_1 e_1'$ iff $\iota(e_1) \#_2 \iota(e_1')$ and $\lambda(e_1) = \lambda(\iota(e_1))$.*

In the following, we will assume that the components of a PES $\mathcal{E}$ are named as in the definition above, possibly with subscripts. The concept of concurrent computation for PESs is captured by the notion of configuration.

Definition 2 (configuration). *A* configuration *of a PES $\mathcal{E}$ is a finite set of events $C \subseteq E$* consistent *(i.e., $\neg(e \# e')$ for all $e, e' \in C$) and* causally closed *(i.e., $\lceil e \rceil \subseteq C$ for all $e \in C$). We denote by $\mathcal{C}(\mathcal{E})$ the set of configurations of $\mathcal{E}$.*

The evolution of a PES can be represented by a transition system over configurations, with the empty configuration as initial state.

Definition 3 (transition system). *Let $\mathcal{E}$ be a PES and let $C \in \mathcal{C}(\mathcal{E})$. Given $e \in E \setminus C$ such that $C \cup \{e\} \in \mathcal{C}(\mathcal{E})$, and $X, Y \subseteq C$ with $X \subseteq \lceil e \rceil$, $Y \cap \lceil e \rceil = \emptyset$ we write $C \xrightarrow{X, \overline{Y} < e}_{\lambda(e)} C \cup \{e\}$. The set of* enabled *events at a configuration C is defined as $en(C) = \{e \in E \mid C \xrightarrow{e} C'\}$. The PES is called k-*bounded *for some $k \in \mathbb{N}$ (or simply* bounded*) if $|en(C)| \leq k$ for all $C \in \mathcal{C}(\mathcal{E})$.*

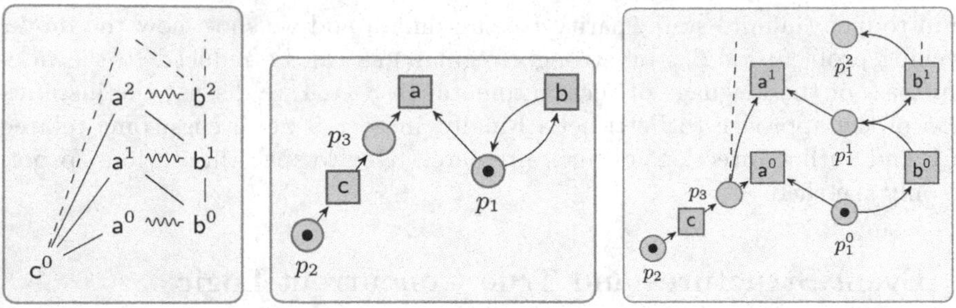

(a) $\mathcal{E}_{\mathcal{N}}$ (b) $\mathcal{N}$ (c) $\mathcal{U}(\mathcal{N})$

Fig. 1. (a) A PES $\mathcal{E}_{\mathcal{N}}$ associated with the net $\mathcal{N}$ in (b) via its unfolding (c).

Transitions are labelled by the executed event e. In addition, they report its label $\lambda(e)$, a subset of causes X and a set of events $Y \subseteq C$ concurrent with e. When X or Y are empty they are normally often, i.e., e.g., we write $C \xrightarrow{X<e}_{\lambda(e)} C'$ for $C \xrightarrow{\emptyset<e}_{\lambda(e)} C'$ and $C \xrightarrow{e}_{\lambda(e)} C'$ for $C \xrightarrow{\emptyset,\overline{\emptyset}<e}_{\lambda(e)} C'$.

The PES modelling a non-trivial system is normally infinite. We will work on a subclass identified by finitarity requirements on the possible substructures.

Definition 4 (residual). *Let $\mathcal{E}$ be a* PES. *For a configuration $C \in \mathcal{C}(\mathcal{E})$, the* residual *of $\mathcal{E}$ after C, is defined as $\mathcal{E}[C] = \{e \mid e \in E \setminus C \wedge C \cup \{e\}$ consistent$\}$.*

The residual of $\mathcal{E}$ can be seen as a PES, endowed with the restriction of causality and conflict of $\mathcal{E}$. Intuitively, it represents the PES that remains to be executed after the computation expressed by C. Given $C \in \mathcal{C}(\mathcal{E})$ and $X \subseteq C$, we denote by $\mathcal{E}[C] \cup X$ the PES obtained from $\mathcal{E}[C]$ by adding the events in X with the causal dependencies they had in the original PES $\mathcal{E}$.

Definition 5 (strong regularity). *A* PES *$\mathcal{E}$ is called* strongly regular *when it is bounded and for each $k \in \mathbb{N}$ the set $\{\mathcal{E}[C] \cup \{e_1, \ldots, e_k\} \mid C \in \mathcal{C}(\mathcal{E}) \wedge e_1, \ldots, e_k \in C\}$ is finite up to isomorphism of* PESs.

Strong regularity [26] is obtained from the notion of regularity in [27], by replacing residuals with residuals extended with a bounded number of events from the past. Intuitively, this is important since we are interested in history dependent properties. We will later show in Sect. 4 that the PESs associated with finite safe Petri nets, i.e., the regular trace PESs [27], are strongly regular.

A simple PES is depicted in Fig. 1a. Graphically, curly lines represent immediate conflicts and the causal partial order proceeds upwards along the straight lines. Events are denoted by their labels, possibly with superscripts. For instance, in $\mathcal{E}_{\mathcal{N}}$, the events a^0 and b^0, labelled by a and b, respectively, are in conflict. Event c^0 causes the events a^i and it is concurrent with b^i for all $i \in \mathbb{N}$. It is an infinite PES associated with the Petri net $\mathcal{N}$ in Fig. 1b in a way that will be discussed in Sect. 4.1, hence it is strongly regular by Corollary 1. It has five (equivalence classes of) residuals extended with an event from the past $\mathcal{E}_{\mathcal{N}}[\{\mathsf{b}^0\}] \cup \{\mathsf{b}^0\}$, $\mathcal{E}_{\mathcal{N}}[\{\mathsf{c}^0, \mathsf{b}^0\}] \cup \{\mathsf{b}^0\}$, $\mathcal{E}_{\mathcal{N}}[\{\mathsf{c}^0, \mathsf{a}^0\}] \cup \{\mathsf{c}^0\}$, $\mathcal{E}_{\mathcal{N}}[\{\mathsf{c}^0, \mathsf{a}^0\}] \cup \{\mathsf{a}^0\}$, and $\mathcal{E}_{\mathcal{N}}[\{\mathsf{c}^0, \mathsf{b}^0, \mathsf{a}^1\}] \cup \{\mathsf{b}^0\}$.

2.2 True Concurrent Logic

The logic of interest for this paper, originally defined in [20], is a Hennessy-Milner style logic that allows one to specify the dependencies (causality and concurrency) between events in computation.

Logic formulae include event variables, from a fixed denumerable set *Var*, denoted by $x, y, \ldots$. Tuples of variables like $x_1, \ldots, x_n$ will be denoted by a corresponding boldface letter $\mathbf{x}$ and, abusing the notation, tuples will be often used as sets. The logic includes diamond and box modalities. The formula $\langle\!| \mathbf{x}, \overline{\mathbf{y}} < \mathsf{a}\, z |\!\rangle\, \varphi$ holds in a configuration when an a-labelled event e is enabled which causally depends on the events bound to $\mathbf{x}$ and is concurrent with those in $\mathbf{y}$. Event e is executed and then the formula φ must hold, with e bound to variable z. Dually, $[\![\mathbf{x}, \overline{\mathbf{y}} < \mathsf{a}\, z]\!]\, \varphi$ is satisfied when all a-labelled events causally dependent on $\mathbf{x}$ and concurrent with $\mathbf{y}$ bring to a configuration where φ holds.

For dealing with fixpoint operators we fix a denumerable set $\mathcal{X}^a$ of *abstract propositions*, ranged over by $X, Y, \ldots$. Each abstract proposition X has an arity $ar(X)$ and it represents a formula with $ar(X)$ (unnamed) free event variables. Then, for $\mathbf{x}$ such that $|\mathbf{x}| = ar(X)$, we write $X(\mathbf{x})$ to indicate the abstract proposition X whose free event variables are named $\mathbf{x}$.

Definition 6 (syntax). *The syntax of $\mathcal{L}_{hp}$ over the sets of event variables Var, abstract propositions $\mathcal{X}^a$ and labels Λ is defined as follows:*

$$\begin{aligned} \varphi \ ::= \ & X(\mathbf{x}) \mid \mathsf{T} \mid \varphi \wedge \varphi \mid \langle\!| \mathbf{x}, \overline{\mathbf{y}} < \mathsf{a}\, z |\!\rangle\, \varphi \mid \nu X(\mathbf{x}).\varphi \\ & \mid \mathsf{F} \mid \varphi \vee \varphi \mid [\![\mathbf{x}, \overline{\mathbf{y}} < \mathsf{a}\, z]\!]\, \varphi \mid \mu X(\mathbf{x}).\varphi \end{aligned}$$

For a formula φ we denote by $fv(\varphi)$ its free event variables, defined in the obvious way. Just note that the modalities act as binders for the variable representing the event executed, hence $fv(\langle\!| \mathbf{x}, \overline{\mathbf{y}} < \mathsf{a}\, z |\!\rangle\, \varphi) = fv([\![\mathbf{x}, \overline{\mathbf{y}} < \mathsf{a}\, z]\!]\, \varphi) = (fv(\varphi) \setminus \{z\}) \cup \mathbf{x} \cup \mathbf{y}$. For formulae $\nu X(\mathbf{x}).\varphi$ and $\mu X(\mathbf{x}).\varphi$ we require that $fv(\varphi) = \mathbf{x}$. The free propositions in φ not bound by μ or ν, are denoted by $fp(\varphi)$. When both $fv(\varphi)$ and $fp(\varphi)$ are empty we say that φ is *closed*. When $\mathbf{x}$ or $\mathbf{y}$ are empty are omitted, e.g., we write $\langle\!| \mathsf{a}\, z |\!\rangle\, \varphi$ for $\langle\!| \emptyset, \overline{\emptyset} < \mathsf{a}\, z |\!\rangle\, \varphi$.

For example, the formula $\varphi_1 = \langle\!| \mathsf{c}\, x |\!\rangle (\langle\!| x < \mathsf{a}\, y |\!\rangle \mathsf{T} \wedge \langle\!| \overline{x} < \mathsf{b}\, z |\!\rangle \mathsf{T})$ requires that, after the execution of a c-labelled event, one can choose between a causally dependent a-labelled event and a concurrent b-labelled event. It is satisfied by $\mathcal{E}_{\mathcal{N}}$ in Fig. 1a. Instead $\varphi_2 = \langle\!| \mathsf{c}\, x |\!\rangle (\langle\!| \overline{x} < \mathsf{a}\, y |\!\rangle \mathsf{T} \wedge \langle\!| \overline{x} < \mathsf{b}\, z |\!\rangle \mathsf{T})$ requiring both events to be concurrent would be false. Moving to infinite computations, consider $\varphi_3 = [\![\mathsf{b}\, x]\!] \nu Z(x). \langle\!| \mathsf{c}\, z |\!\rangle \langle\!| \overline{z} < \mathsf{b}\, y |\!\rangle \mathsf{T} \wedge [\![x < \mathsf{b}\, y]\!] Z(y)$, expressing that all non-empty causal chains of b-labelled events reach a state where it is possible to execute two concurrent events labelled c and b, respectively. Then φ_3 holds in $\mathcal{E}_{\mathcal{N}}$. Another formula satisfied by $\mathcal{E}_{\mathcal{N}}$ is $\varphi_4 = \langle\!| \mathsf{c}\, x |\!\rangle \langle\!| \overline{x} < \mathsf{b}\, y |\!\rangle \nu X(x, y). \langle\!| y, \overline{x} < \mathsf{b}\, z |\!\rangle X(x, z)$

requiring the existence of an infinite causal chain of b-labelled events, concurrent with a c-labelled event.

The logic $\mathcal{L}_{hp}$ is interpreted over PESs. The satisfaction of a formula is defined with respect to a configuration C and a (total) function $\eta : Var \to E$, called an *environment*, that binds free variables in φ to events in C. Namely, if $Env_{\mathcal{E}}$ denotes the set of environments, the semantics of a formula will be a set of pairs in $\mathcal{C}(\mathcal{E}) \times Env_{\mathcal{E}}$. The semantics of $\mathcal{L}_{hp}$ also depends on a *proposition environment* $\pi : \mathcal{X} \to 2^{\mathcal{C}(\mathcal{E}) \times Env_{\mathcal{E}}}$ which provides an interpretation for propositions. In order to ensure that the semantics of a formula only depends on the events associated with its free variables and is independent on the naming of the variables, it is required that if $(C, \eta) \in \pi(X(\mathbf{x}))$ and $\eta'(\mathbf{y}) = \eta(\mathbf{x})$ pointwise, then $(C, \eta') \in \pi(X(\mathbf{y}))$. We denote by $PEnv_{\mathcal{E}}$ the set of proposition environments, ranged over by π.

We can now give the semantics of logic $\mathcal{L}_{hp}$. Given an event environment η and an event e we write $\eta[x \mapsto e]$ for the updated environment which maps x to e. Similarly, for a proposition environment π and $S \subseteq \mathcal{C}(\mathcal{E}) \times Env_{\mathcal{E}}$, we write $\pi[Z(\mathbf{x}) \mapsto S]$ for the corresponding update.

Definition 7 (semantics). *Let $\mathcal{E}$ be a* PES. *The denotation of a formula φ in $\mathcal{L}_{hp}$ is given by the function $\{\!|\cdot|\!\}^{\mathcal{E}} : \mathcal{L}_{hp} \to PEnv_{\mathcal{E}} \to 2^{\mathcal{C}(\mathcal{E}) \times Env_{\mathcal{E}}}$ defined inductively as follows, where we write $\{\!|\varphi|\!\}^{\mathcal{E}}_{\pi}$ instead of $\{\!|\varphi|\!\}^{\mathcal{E}}(\pi)$:*

$$\{\!|\mathsf{T}|\!\}^{\mathcal{E}}_{\pi} = \mathcal{C}(\mathcal{E}) \times Env_{\mathcal{E}} \qquad \{\!|\mathsf{F}|\!\}^{\mathcal{E}}_{\pi} = \emptyset \qquad \{\!|Z(\mathbf{y})|\!\}^{\mathcal{E}}_{\pi} = \pi(Z(\mathbf{y}))$$

$$\{\!|\varphi_1 \wedge \varphi_2|\!\}^{\mathcal{E}}_{\pi} = \{\!|\varphi_1|\!\}^{\mathcal{E}}_{\pi} \cap \{\!|\varphi_2|\!\}^{\mathcal{E}}_{\pi} \qquad \{\!|\varphi_1 \vee \varphi_2|\!\}^{\mathcal{E}}_{\pi} = \{\!|\varphi_1|\!\}^{\mathcal{E}}_{\pi} \cup \{\!|\varphi_2|\!\}^{\mathcal{E}}_{\pi}$$

$$\{\!|\langle\!|\mathbf{x}, \overline{\mathbf{y}} < \mathsf{a}\, z|\!\rangle\, \varphi|\!\}^{\mathcal{E}}_{\pi} = \{(C, \eta) \mid \exists e.\, C \xrightarrow{\eta(\mathbf{x}), \overline{\eta(\mathbf{y})} < e}_{\mathsf{a}} C' \;\wedge\; (C', \eta[z \mapsto e]) \in \{\!|\varphi|\!\}^{\mathcal{E}}_{\pi}\}$$

$$\{\!|[\![\mathbf{x}, \overline{\mathbf{y}} < \mathsf{a}\, z]\!]\, \varphi|\!\}^{\mathcal{E}}_{\pi} = \{(C, \eta) \mid \forall e.\, C \xrightarrow{\eta(\mathbf{x}), \overline{\eta(\mathbf{y})} < e}_{\mathsf{a}} C' \;\Rightarrow\; (C', \eta[z \mapsto e]) \in \{\!|\varphi|\!\}^{\mathcal{E}}_{\pi}\}$$

$$\{\!|\nu Z(\mathbf{x}).\varphi|\!\}^{\mathcal{E}}_{\pi} = gfp(f_{\varphi, Z(\mathbf{x}), \pi}) \qquad \{\!|\mu Z(\mathbf{x}).\varphi|\!\}^{\mathcal{E}}_{\pi} = lfp(f_{\varphi, Z(\mathbf{x}), \pi})$$

where $f_{\varphi, Z(\mathbf{x}), \pi} : 2^{\mathcal{C}(\mathcal{E}) \times Env_{\mathcal{E}}} \to 2^{\mathcal{C}(\mathcal{E}) \times Env_{\mathcal{E}}}$ is defined by $f_{\varphi, Z(\mathbf{x}), \pi}(S) = \{\!|\varphi|\!\}^{\mathcal{E}}_{\pi[Z(\mathbf{x}) \mapsto S]}$ and $gfp(f_{\varphi, Z(\mathbf{x}), \pi})$ (resp. $lfp(f_{\varphi, Z(\mathbf{x}), \pi})$) denotes the corresponding greatest (resp. least) fixpoint. We say that a PES *$\mathcal{E}$ satisfies a formula φ and write $\mathcal{E} \models \varphi$ if $(\emptyset, \eta) \in \{\!|\varphi|\!\}^{\mathcal{E}}_{\pi}$ for all environments η and π.*

The semantics of boolean operators is standard. The formula $\langle\!|\mathbf{x}, \overline{\mathbf{y}} < \mathsf{a}\, z|\!\rangle\, \varphi$ holds in (C, η) when configuration C enables an a-labelled event e that causally depends on (at least) the events bound to the variables in $\mathbf{x}$ and concurrent with (at least) those bound to the variables in $\mathbf{y}$ and, once executed, it produces a new configuration $C' = C \cup \{e\}$ which, paired with the environment $\eta' = \eta[z \mapsto e]$, satisfies the formula φ. Dually, $[\![\mathbf{x}, \overline{\mathbf{y}} < \mathsf{a}\, z]\!]\, \varphi$ holds when all a-labelled events executable from C, caused by $\mathbf{x}$ and concurrent with $\mathbf{y}$ bring to a configuration where φ is satisfied.

The fixpoints corresponding to the formulae $\nu Z(\mathbf{x}).\varphi$ and $\mu Z(\mathbf{x}).\varphi$ are guaranteed to exist by Knaster-Tarski theorem, since the set $2^{\mathcal{C}(\mathcal{E}) \times Env_{\mathcal{E}}}$ ordered by subset inclusion is a complete lattice and the functions $f_{\varphi, Z(\mathbf{x}), \pi}$ are monotonic.

3 Automata-Based Model Checker

We introduce nondeterministic parity tree automata and we show how the model checking problem for $\mathcal{L}_{hp}$ on strongly regular PESs can be reduced to the non-emptiness of the language of such automata. The automaton naturally generated from a PES and a formula has an infinite number of states. We discuss how the automaton can be quotiented to a finite one accepting the same language and thus potentially useful for model checking purposes.

3.1 Infinite Parity Tree Automata

Automata on infinite trees revealed to be a powerful tool to various problems in the setting of branching temporal logics. Here we focus on nondeterministic parity tree automata [29], with some (slightly) non-standard features. We work on k-trees (rather than on binary trees), a choice that will simplify the presentation, and we allow for possibly infinite state automata.

When automata are used for model checking purposes it is standard to restrict to unlabelled trees. A *k-bounded branching tree* or *k-tree*, for short, is a subset $\mathcal{T} \subseteq [1,k]^\star$, such that

1. $\mathcal{T}$ is prefix closed, i.e., if $wv \in \mathcal{T}$ then $w \in \mathcal{T}$
2. $w1 \in \mathcal{T}$ for all $w \in \mathcal{T}$
3. for all $i \in [2,k]$ if $wi \in \mathcal{T}$ then $w(i-1) \in \mathcal{T}$.

Elements of $\mathcal{T}$ are the nodes of the tree. The empty string ϵ corresponds to the root. A string of the form wi corresponds to the i-th child of w. Hence by (2) each branch is infinite and by (3) the presence of the i-th child implies the presence of the j-th children for $j \leq i$.

Definition 8 (nondeterministic parity automaton). *A k-bounded nondeterministic parity tree automaton (NPA) is a tuple $\mathcal{A} = \langle Q, \rightarrow, q_0, \mathcal{F} \rangle$ where Q is a set of states, $\rightarrow \subseteq Q \times \bigcup_{i=1}^{k} Q^k$ is the* transition relation, *$q_0 \in Q$ is the* initial state, *and $\mathcal{F} = (F_0, \ldots, F_h)$ is the* acceptance condition, *where $F_0, \ldots, F_h \subseteq Q$ are mutually disjoint subsets of states.*

Transitions are written as $q \rightarrow (q_1, \ldots, q_m)$ instead of $(q, (q_1, \ldots, q_m)) \in \rightarrow$.

Given a k-tree $\mathcal{T}$, a *run* of $\mathcal{A}$ on $\mathcal{T}$ is a labelling of $\mathcal{T}$ over the states $r : \mathcal{T} \rightarrow Q$ consistent with the transition relation, i.e., such that $r(\epsilon) = q_0$ and for all $u \in \mathcal{T}$, with m children, there is a transition $r(u) \rightarrow (r(u1), \ldots, r(um))$ in $\mathcal{A}$. A *path* in the run r is an infinite sequence of states $p = (q_0, q_1, \ldots)$ labelling a complete path from the root in the tree. It is called *accepting* if there exists an even number $l \in [0,h]$ such that the set $\{j \mid q_j \in F_l\}$ is infinite and the set $\{j \mid q_j \in \bigcup_{l<i\leq h} F_i\}$ is finite. The run r is *accepting* if all paths are accepting.

Definition 9 (language of an NPA). *Let $\mathcal{A}$ be an NPA. The language of $\mathcal{A}$, denoted by $L(\mathcal{A})$, consists of the trees $\mathcal{T}$ which admit an accepting run.*

Observe that for a k-bounded NPA, the language $L(\mathcal{A})$ is a set of k-trees.

The possibility of having an infinite number of states and the associated acceptance condition are somehow non-standard. However, it is easy to see that whenever an NPA is finite, the acceptance condition coincides with the standard one requiring a single state with maximal even priority to occur infinitely often.

Since NPAs are nondeterministic, different runs (possibly infinitely many) can exist for the same input tree. Still, the non-emptiness problem, also for our k-ary variant, is decidable when the number of states is finite (and solvable by a corresponding parity game [30]).

3.2 Infinite NPAs for Model Checking

We show how, given a PES and a closed formula in $\mathcal{L}_{hp}$, we can build an NPA in a way that, for strongly regular PESs, the satisfaction of φ in $\mathcal{E}$ reduces to the non-emptiness of the automaton language. The construction is inspired by that in [28] for the mu-calculus.

The acceptance condition for the automaton will refer to the fixpoint alternation in the formulae of $\mathcal{L}_{hp}$. We adapt a definition from [28]. A fixpoint formula $\alpha X(\mathbf{y}).\varphi'$, for $\alpha \in \{\nu, \mu\}$, is called an α-formula. Hereafter α ranges over $\{\nu, \mu\}$. Given an α-formula $\varphi = \alpha X(\mathbf{y}).\varphi'$, we say that a subformula ψ of φ is a *direct active subformula*, written $\psi \sqsubseteq_d \varphi$, if the abstract proposition X appears free in ψ. The transitive closure of $\sqsubseteq_d$ is a partial order and when $\psi \sqsubseteq_d^* \varphi$ we say that ψ is an *active subformula* of φ. We denote by $\mathsf{sf}(\varphi)$ the set of subformulae of a formula φ and by $\mathsf{sf}_\alpha(\varphi)$ the set of active α-subformulae.

The *alternation depth* of a formula φ in $\mathcal{L}_{hp}$, written $\mathsf{ad}(\varphi)$, is defined, for a ν-formula φ, as $\mathsf{ad}(\varphi) = \max\{1 + \mathsf{ad}(\psi) \mid \psi \in \mathsf{sf}_\mu(\varphi)\}$ and dually, for a μ-formula φ, as $\mathsf{ad}(\varphi) = \max\{1 + \mathsf{ad}(\psi) \mid \psi \in \mathsf{sf}_\nu(\varphi)\}$. For any other formula φ, $\mathsf{ad}(\varphi) = max\{\mathsf{ad}(\psi) \mid \psi \in \mathsf{sf}(\varphi) \setminus \{\varphi\}\}$. It is intended that $\max \emptyset = 0$. E.g., by the first clause above, the alternation depth of $\nu X(\mathbf{x}).\,\varphi$ is 0 in absence of active μ-subformulae.

Hereafter we assume that in every formula different bound propositions have different names, so that we can refer to the fixpoint subformula quantifying an abstract proposition. This requirement can always be fulfilled by alpha-renaming.

Hereafter, if X and X' are abstract propositions quantified in α-subformulae $\alpha X(\mathbf{x}).\,\varphi$ and $\alpha' X'(\mathbf{x}').\,\varphi'$, we will write $\mathsf{ad}(X)$ for $\mathsf{ad}(\alpha X(\mathbf{x}).\,\varphi)$ and $X \sqsubseteq_d X'$ for $\alpha X(\mathbf{x}).\,\varphi \sqsubseteq_d \alpha' X'(\mathbf{x}').\,\varphi'$. Moreover, given a PES $\mathcal{E}$, for a pair $(C, \eta) \in \mathcal{C}(\mathcal{E}) \times Env_{\mathcal{E}}$ and variables $\mathbf{x}$, $\mathbf{y}$, z, we define $(\mathbf{x}, \overline{\mathbf{y}} < \mathsf{a}z)$-successors of (C, η), as

$$\mathsf{Succ}^{\mathbf{x},\overline{\mathbf{y}}<\mathsf{a}z}(C, \eta) = \{(C', \eta[z \mapsto e]) \mid C \xrightarrow{\eta(\mathbf{x}), \overline{\eta(\mathbf{y})} < e}_{\mathsf{a}} C'\}.$$

We can now illustrate the construction of the NPA for a formula and a PES.

Definition 10 (NPA for a formula). *Let $\mathcal{E}$ be a bounded* PES *and let $\varphi \in \mathcal{L}_{hp}$ be a closed formula. The NPA for $\mathcal{E}$ and φ is $\mathcal{A}_{\mathcal{E},\varphi} = \langle Q, \to, q_0, \mathcal{F} \rangle$ defined* *as follows. The set of states* $Q \subseteq \mathcal{C}(\mathcal{E}) \times Env_{\mathcal{E}} \times \mathsf{sf}(\varphi)$ *is* $Q = \{(C, \eta, \psi) \mid \eta(fv(\psi)) \subseteq C\}$. *The initial state* $q_0 = (\emptyset, \eta, \varphi)$, *for some chosen* $\eta \in Env_{\mathcal{E}}$. *The transition relation is defined, for any state* $q = (C, \eta, \psi) \in Q$, *by:*

- *if* $\psi = \mathsf{T}$ *or* $\psi = \mathsf{F}$, *then* $q \to (q)$;
- *if* $\psi = \psi_1 \wedge \psi_2$, *then* $q \to (q_1, q_2)$ *where* $q_i = (C, \eta, \psi_i)$, $i \in \{1, 2\}$;
- *if* $\psi = \psi_1 \vee \psi_2$, *then* $q \to (q_1)$ *and* $q \to (q_2)$ *where* $q_i = (C, \eta, \psi_i)$, $i \in \{1, 2\}$;
- *if* $\psi = [\![\mathbf{x}, \overline{\mathbf{y}} < \mathsf{a}\, z]\!]\psi'$ *and* $\mathsf{Succ}^{\mathbf{x}, \overline{\mathbf{y}} < \mathsf{a}z}(C, \eta) = \{(C_1, \eta_1), \ldots, (C_n, \eta_n)\} \neq \emptyset$ *then* $q \to (q_1, \ldots, q_n)$ *where* $q_i = (C_i, \eta_i, \psi')$ *for* $i \in [1, n]$, *otherwise* $q \to (q)$;
- *if* $\psi = \langle\!|\mathbf{x}, \overline{\mathbf{y}} < \mathsf{a}\, z|\!\rangle\psi'$ *and* $\mathsf{Succ}^{\mathbf{x}, \overline{\mathbf{y}} < \mathsf{a}z}(C, \eta) = \{(C_1, \eta_1), \ldots, (C_n, \eta_n)\} \neq \emptyset$ *then* $q \to (q_i)$ *where* $q_i = (C_i, \eta_i, \psi')$ *for* $i \in [1, n]$, *otherwise* $q \to (q)$;
- *if* $\psi = \alpha X(\mathbf{x}).\psi'$ *then* $q \to (q')$ *where* $q' = (C, \eta, X(\mathbf{x}))$;
- *if* $\psi = X(\mathbf{y})$ *and* $\psi' \in \mathsf{sf}(\varphi)$ *is the unique subformula such that* $\psi' = \alpha X(\mathbf{x}).\psi''$ *then* $q \to (q')$ *where* $q' = (C, \eta[\mathbf{x} \mapsto \eta(\mathbf{y})], \psi'')$.

The acceptance condition is $\mathcal{F} = (F_0, \ldots, F_h)$ *where* $h = \mathsf{ad}(\varphi) + 1$ *and the* F_i *are as follows. Consider* $A_0, \ldots, A_h \subseteq \mathsf{sf}(\varphi)$ *such that for* $i \in [0, h]$, *if* i *is even (odd) then* A_i *contains exactly all propositions quantified in* ν*-subformulae (*μ*-subformulae) with alternation depth* i *or* $i - 1$. *Then* $F_0 = (\mathcal{C}(\mathcal{E}) \times Env_{\mathcal{E}} \times (A_0 \cup \{\mathsf{T}\})) \cup B$ *where* $B = \{(C, \eta, [\![\mathbf{x}, \overline{\mathbf{y}} < \mathsf{a}\, z]\!]\psi) \mid \mathsf{Succ}^{\mathbf{x}, \overline{\mathbf{y}} < \mathsf{a}z}(C, \eta) = \emptyset\}$ *is the set of all subformulae of* φ *in a context where they are trivially true, and* $F_i = \mathcal{C}(\mathcal{E}) \times Env_{\mathcal{E}} \times A_i$, *for* $i \in [1, h]$.

States of $\mathcal{A}_{\mathcal{E},\varphi}$ are triples (C, η, φ) consisting of a configuration C, an environment η and a subformula ψ of the original formula φ. The intuition is that a transition reduces the satisfaction of a formula in a state to that of subformulae in possibly updated states. It can just decompose the formula, as it happens for $\wedge$ or $\vee$, check the satisfaction of a modal operator, thus changing the state consequently, or unfold a fixpoint.

The automaton $\mathcal{A}_{\mathcal{E},\varphi}$ is bounded but normally infinite (whenever the PES $\mathcal{E}$ is infinite and the formula φ includes some non-trivial fixpoint).

We next show that for a strongly regular PES the satisfaction of the formula φ on the PES $\mathcal{E}$ reduces to the non-emptiness of the language of $\mathcal{A}_{\mathcal{E},\varphi}$.

Theorem 1 (model checking via non-emptiness). *Let* $\mathcal{E}$ *be a strongly regular* PES *and let* $\check{\varphi}$ *be a closed formula in* $\mathcal{L}_{hp}$. *Then* $L(\mathcal{A}_{\mathcal{E},\check{\varphi}}) \neq \emptyset$ *iff* $\mathcal{E} \models \check{\varphi}$.

We next provide an outline of the proof. A basic ingredient is an equivalence that can be defined on the NPA. As a first step we introduce a generalised notion of residual in which the relation with some selected events in the past is kept.

Definition 11 (pointed residual). *Given a* PES $\mathcal{E}$ *and a set* X, *a* X-pointed configuration *is a pair* $\langle C, \zeta \rangle$ *where* $C \in \mathcal{C}(\mathcal{E})$ *and* $\zeta : X \to C$ *is a function. We say that the* X*-pointed configurations* $\langle C, \zeta \rangle$, $\langle C', \zeta' \rangle$ *have isomorphic pointed residuals, written* $\mathcal{E}[\langle C, \zeta \rangle] \approx \mathcal{E}[\langle C', \zeta' \rangle]$ *if there is an isomorphism of* PESs $\iota : \mathcal{E}[C] \to \mathcal{E}[C']$ *such that for all* $x \in X$, $e \in \mathcal{E}[C]$ *we have* $\zeta(x) \leq e$ *iff* $\zeta'(x) \leq \iota(e)$.

Then two states are deemed equivalent if they involve the same subformula (up to renaming of the event variables) and the configurations, pointed by the free variables in the formulae, have isomorphic residuals. This resembles the notion of contextualised equivalence used on tableau judgments in [26].

Definition 12 (future equivalence). *Let $\mathcal{E}$ be a* PES*, φ be a formula and let $q_i = (C_i, \eta_i, \psi_i)$, $i \in \{1, 2\}$ be two states of the NPA $\mathcal{A}_{\mathcal{E},\varphi}$. We say that q_1 and q_2 are* future equivalent*, written $q_1 \approx_f q_2$, if there exists a formula ψ and substitutions $\sigma_i : fv(\psi) \to fv(\psi_i)$ such that $\psi\sigma_i = \psi_i$, for $i \in \{1, 2\}$, and the $fv(\psi)$-pointed configurations $\langle C_i, \eta_i \circ \sigma_i \rangle$ have isomorphic pointed residuals.*

It can be shown that, given $q_i = (C_i, \eta_i, \psi_i)$, $i \in \{1, 2\}$ as above, for all proposition environments π (satisfying a technical property of saturation) we have that $(C_1, \eta_1) \in \{\!|\psi_1|\!\}^{\mathcal{E}}_{\pi}$ if and only if $(C_2, \eta_2) \in \{\!|\psi_2|\!\}^{\mathcal{E}}_{\pi}$. Additionally, using strong regularity, one can prove that the semantics of fixpoint formulae is properly captured by finite approximants and that equivalence $\approx_f$ is of finite index. These are fundamental building bricks in the proof of Theorem 1 which, roughly, proceeds as follows.

Assume that the language $L(\mathcal{A}_{\mathcal{E},\varphi}) \neq \emptyset$. Then there is an accepting run r over some k-tree $\mathcal{T}$. Since φ is finite, in each infinite path there are infinitely many states $q_{i_h} = (C_{i_h}, \eta_{i_h}, \psi_{i_h})$ where ψ_{i_h} is the same subformula, up to renaming. Since $\approx_f$ is of finite index, infinitely many such states are equivalent. Then one deduces that, for some h, the subformula ψ_{i_h} is satisfied in (C_{i_h}, η_{i_h}). For fixpoint subformulae, this requires to show that, since the run is accepting, the subformula of maximal alternation depth that repeats infinitely often is a ν-formula and use the fact that, as mentioned before, its semantics can be finitely approximated. Then, by a form of backward soundness of the transitions, we get that all the nodes, including the root, contain formulae which are satisfied.

For the converse implication, assume that $\mathcal{E} \models \varphi$. Starting from the initial state $q_0 = (\emptyset, \eta, \varphi)$ where the formula is satisfied, and using the automaton transitions, we can build a k-tree $\mathcal{T}$ and a run where for each state (C', η', ψ) the subformula ψ is satisfied in (C', η') and such run can be proved to be accepting.

3.3 Quotienting the Automaton

In order to have an effective procedure for checking the satisfaction of a formula we need to build a suitable quotient of the NPA, with respect to an equivalence which preserves emptiness. A simple but important observation is that it is sufficient to require that the equivalence is a bisimulation in the following sense. An analogous notion is studied in [31] in the setting of nondeterministic tree automata over finite trees.

Definition 13 (bisimulation). *Given an NPA $\mathcal{A}$, a symmetric relation $R \subseteq Q \times Q$ over the set of states is a bisimulation if for all $(q, q') \in R$*

1. *for all $i \in [0, h]$, $q \in F_i \iff q' \in F_i$;*
2. *if $q \to (q_1, \ldots, q_m)$ then $q' \to (q'_1, \ldots, q'_m)$ with $(q_i, q'_i) \in R$ for $i \in [1, m]$.*

Given an NPA $\mathcal{A}$ and an equivalence $\equiv$ on the set of states which is a bisimulation, we define the quotient as $\mathcal{A}_{/\equiv} = \langle Q_{/\equiv}, \rightarrow_{/\equiv}, [q_0]_{\equiv}, \mathcal{F}_{/\equiv}\rangle$ where $[q]_{\equiv} \rightarrow_{/\equiv} ([q_1]_{\equiv}, \ldots, [q_m]_{\equiv})$ if $q \rightarrow (q_1, \ldots, q_m)$ and $\mathcal{F}_{/\equiv} = (F_{0/\equiv}, \ldots, F_{h/\equiv})$. An NPA and its quotient accept exactly the same language.

Theorem 2 (language preservation). *Let $\mathcal{A}$ be an NPA and let $\equiv$ be an equivalence on the set of states which is a bisimulation. Then $L(\mathcal{A}_{/\equiv}) = L(\mathcal{A})$.*

When $\equiv$ is of finite index, the quotient $\mathcal{A}_{\mathcal{E},\varphi/\equiv}$ is finite and, exploiting Theorems 1 and 2, we can verify whether $\mathcal{E} \models \varphi$ by checking the emptiness of the language accepted by $\mathcal{A}_{\mathcal{E},\varphi/\equiv}$. Clearly a concrete algorithm will not first generate the infinite state NPA and then take the quotient, but it rather performs the quotient on the fly: whenever a new state would be equivalent to one already generated, the transition loops back to the existing state.

Whenever $\mathcal{E}$ is strongly regular, the future equivalence on states (see Definition 12) provides a bisimulation equivalence of finite index over $\mathcal{A}_{\mathcal{E},\varphi}$.

Lemma 1 ($\approx_f$ is a bisimulation). *Let $\mathcal{E}$ be a strongly regular* PES *and let φ be a closed formula in $\mathcal{L}_{hp}$. Then the future equivalence $\approx_f$ on $\mathcal{A}_{\mathcal{E},\varphi}$ is a bisimulation and it is of finite index.*

An obstacle towards the use of the quotiented NPA for model checking purposes is the fact that the future equivalence could be hard to compute (or even undecidable). In order to make the construction effective we need a decidable bisimulation equivalence on the NPA and the effectiveness of the set of successors of a state. This is further discussed in the next section.

4 Model Checking Petri Nets

We show how the model checking approach outlined before can be instantiated on finite safe Petri nets, a classical model of concurrency and distribution [32], by identifying a suitable effective bisimulation equivalence on the NPA.

4.1 Petri Nets and Their Event Structure Semantics

A *Petri net* is a tuple $\mathcal{N} = (P, T, F, M_0)$ where P, T are disjoint sets of *places* and *transitions*, respectively, $F : (P \times T) \cup (T \times P) \rightarrow \{0, 1\}$ is the *flow function*, and M_0 is the initial marking, i.e., the initial state of the net. We assume that the set of transitions is a subset of a fixed set $\mathbb{T}$ with a labelling $\lambda_N : \mathbb{T} \rightarrow \Lambda$.

A *marking* of $\mathcal{N}$ is a function $M : P \rightarrow \mathbb{N}$, indicating for each place the number of tokens in the place. A transition $t \in T$ is *enabled* at a marking M if $M(p) \geq F(p, t)$ for all $p \in P$. In this case it can be *fired* leading to a new marking M' defined by $M'(p) = M(p) + F(t, p) - F(p, t)$ for all places $p \in P$. This is written $M[t\rangle M'$. We denote by $\mathcal{R}(\mathcal{N})$ the set of markings reachable in $\mathcal{N}$ via a sequence of firings starting from the initial marking. We say that a marking M is *coverable* if there exists $M' \in \mathcal{R}(\mathcal{N})$ such that $M \leq M'$, pointwise. A net

$\mathcal{N}$ is *safe* if for every reachable marking $M \in \mathcal{R}(\mathcal{N})$ and all $p \in P$ we have $M(p) \leq 1$. Hereafter we will consider only safe nets. Hence markings will be often confused with the corresponding subset of places $\{p \mid M(p) = 1\} \subseteq P$. For $x \in P \cup T$ the *pre-set* and *post-set* are defined $^{\bullet}x = \{y \in P \cup T \mid F(y, x) = 1\}$ and $x^{\bullet} = \{y \in P \cup T \mid F(x, y) = 1\}$ respectively.

An example of Petri net can be found in Fig. 1b. Graphically places and transitions are drawn as circles and rectangles, respectively, while the flow function is rendered by means of directed arcs connecting places and transitions. Markings are represented by inserting tokens (black dots) in the corresponding places.

The concurrent behaviour of a Petri net can be represented by its unfolding $\mathcal{U}(\mathcal{N})$, an acyclic net constructed inductively starting from the initial marking of $\mathcal{N}$ and then adding, at each step, an occurrence of each enabled transition.

Definition 14 (unfolding). *Let $\mathcal{N} = (P, T, F, m_0)$ be a safe net. Define the net $U^{(0)} = (P^{(0)}, T^{(0)}, F^{(0)})$ as $T^{(0)} = \emptyset$, $P^{(0)} = \{(p, \bot) \mid p \in m_0\}$ and $F^{(0)} = \emptyset$, where $\bot$ is an element not belonging to P, T or F. The unfolding is the least net $\mathcal{U}(\mathcal{N}) = (P^{(\omega)}, T^{(\omega)}, F^{(\omega)})$ containing $U^{(0)}$ and such that*

- *if $t \in T$, the set of places $X \subseteq P^{(\omega)}$ is coverable and $\pi_1(X) = {}^{\bullet}t$, then $e = (t, X) \in T^{(\omega)}$;*
- *for any $e = (t, X) \in T^{(\omega)}$, the set $Z = \{(p, e) \mid p \in \pi_1(e)^{\bullet}\} \subseteq P^{(\omega)}$ where $\pi_1(u, v) = u$; moreover ${}^{\bullet}e = X$ and $e^{\bullet} = Z$.*

Places and transitions in the unfolding represent tokens and firing of transitions, respectively, of the original net. The projection π_1 over the first component maps places and transitions of the unfolding to the corresponding items of the original net $\mathcal{N}$. The initial marking is implicitly identified as the set of minimal places. For historical reasons transitions and places in the unfolding are also called *events* and *conditions*, respectively.

One can define *causality* $\leq_N$ over the unfolding as the transitive closure of the flow relation. *Conflict* is the relation $e\#e'$ if ${}^{\bullet}e \cap {}^{\bullet}e' \neq \emptyset$, inherited along causality. The events $T^{(\omega)}$ of the unfolding of a finite safe net, endowed with causality and conflict, form a PES, denoted $\mathcal{E}(\mathcal{N})$. The transitions of a configuration $C \in \mathcal{C}(\mathcal{E}(\mathcal{N}))$ can be fired in any order compatible with causality, producing a marking $C^{\circ} = (P^{(0)} \cup \bigcup_{t \in C} t^{\bullet}) \setminus (\bigcup_{t \in C} {}^{\bullet}t)$ in $\mathcal{U}(\mathcal{N})$; in turn, this corresponds to a reachable marking of $\mathcal{N}$ given by $\mathsf{M}(C) = \pi_1(C^{\circ})$. As an example, the unfolding $\mathcal{U}(\mathcal{N})$ of the running example net $\mathcal{N}$ and the corresponding PES can be found in Figs. 1c and a.

4.2 Automata Model Checking for Petri Nets

The PES associated with a safe Petri net is known to be regular [27]. We next prove that it is also strongly regular and thus we can apply the theory developed so far for model checking $\mathcal{L}_{hp}$ over safe Petri nets.

Let $\mathcal{N} = \langle S, T, F, M_0 \rangle$ be a safe Petri net. A basic observation is that the residual of the PES $\mathcal{E}(\mathcal{N})$ with respect to a configuration $C \in \mathcal{C}(\mathcal{E}(\mathcal{N}))$ is uniquely determined by the marking produced by C. This correspondence can be extended to pointed configurations by considering markings which additionally record, for the events of interest in the past, the places in the marking which are caused by such events. This motivates the definition below.

Definition 15 (pointed marking). *Let $\mathcal{N} = \langle S, T, F, M_0 \rangle$ be a safe Petri net. Given a set X, a X-*pointed marking *is a pair $\langle M, r \rangle$ with $r : X \to 2^M$.*

A X-pointed configuration $\langle C, \zeta \rangle$ induces an X-pointed marking $\mathsf{M}(\langle C, \zeta \rangle) = \langle \mathsf{M}(C), r \rangle$ where $r(x) = \{\pi_1(b) \mid b \in C^\circ \ \wedge \ \zeta(x) < b\}$. Pointed configurations producing the same pointed marking have isomorphic pointed residuals.

Proposition 1 (pointed markings vs residuals). *Let $\mathcal{N} = \langle S, T, F, M_0 \rangle$ be a safe Petri net. Given a set X and two X-pointed configurations $\langle C_1, \zeta_1 \rangle$, $\langle C_2, \zeta_2 \rangle$ in $\mathcal{U}(\mathcal{N})$, if $\mathsf{M}(\langle C_1, \zeta_1 \rangle) = \mathsf{M}(\langle C_2, \zeta_2 \rangle)$ then $\mathcal{E}(\mathcal{N})[\langle C_1, \zeta_1 \rangle] \approx \mathcal{E}(\mathcal{N})[\langle C_2, \zeta_2 \rangle]$.*

By the previous result the PES associated with a finite safe Petri net is strongly regular. Indeed, the number of residuals of X-pointed configurations, up to isomorphism, by Proposition 1, is smaller than the number of X-pointed markings, which is clearly finite since the net is safe.

Corollary 1 (strong regularity). *Let $\mathcal{N}$ be finite safe Petri net. Then the corresponding* PES *$\mathcal{E}(\mathcal{N})$ is strongly regular.*

In order to instantiate the model checking framework to finite safe Petri nets, the idea is to take an equivalence over the infinite NPA by abstracting the (pointed) configurations associated with its states to pointed markings.

Definition 16 (pointed-marking equivalence on NPA). *Let $\mathcal{N}$ be a finite safe Petri net and let φ be a closed formula in $\mathcal{L}_{hp}$. Two states q_1, q_2 in the NPA $\mathcal{A}_{\mathcal{E}(\mathcal{N}),\varphi}$ are* pointed-marking equivalent, *written $q_1 \approx_m q_2$, if $q_i = \langle C_i, \eta_i, \psi \rangle$, $i \in \{1, 2\}$, for some $\psi \in \mathsf{sf}(\varphi)$ and $\mathsf{M}(\langle C_1, \eta_{1|fv(\psi)} \rangle) = \mathsf{M}(\langle C_2, \eta_{2|fv(\psi)} \rangle)$.*

Using Proposition 1 we can immediately prove that $\approx_m$ refines $\approx_f$. Moreover we can show that $\approx_m$ is a bisimulation in the sense of Definition 13.

Proposition 2 (marking equivalence is a bisimulation). *Let $\mathcal{N}$ be a finite safe Petri net and let φ be a closed formula in $\mathcal{L}_{hp}$. The equivalence $\approx_m$ on the automaton $\mathcal{A}_{\mathcal{E}(\mathcal{N}),\varphi}$ is a bisimulation and it is of finite index.*

Relying on Propositions 1 and 2 we provide an explicit construction of the quotient automaton $\mathcal{A}_{\mathcal{E}(\mathcal{N}),\varphi_{/\approx_m}}$. We introduce a convenient notation for transitions between pointed markings. Given the variables $\mathbf{x}$, $\mathbf{y}$, a set X such that $\mathbf{x} \cup \mathbf{y} \subseteq X$ and an X-pointed marking $\langle M, r \rangle$, we write $\langle M, r \rangle \xrightarrow{\mathbf{x}, \overline{\mathbf{y}} < t}_{\mathsf{a}, z} \langle M', r' \rangle$ if $M[t\rangle M'$, $\lambda_N(t) = \mathsf{a}$, for all $x \in \mathbf{x}$ we have $r(x) \cap {}^\bullet t \neq \emptyset$ and for all $y \in \mathbf{y}$ it holds $r(y) \cap {}^\bullet t = \emptyset$ and r' is defined by $r'(z) = t^\bullet$ and $r'(w) = (r(w) \cap M') \cup \{s \mid r(w) \cap {}^\bullet t \neq \emptyset \ \wedge \ s \in t^\bullet\}$, for $w \neq z$. In words, from the pointed marking $\langle M, r \rangle$ transition t is fired and "pointed" by variable z. Transition t is required to consume tokens caused by $\mathbf{x}$ and not to consume tokens caused by $\mathbf{y}$, in order to be itself caused by $\mathbf{x}$ and independent from $\mathbf{y}$. After the firing, variables which were causes of some $p \in {}^\bullet t$ become causes of the places in $t^\bullet$ and, clearly, z causes $t^\bullet$.

Construction 1 (quotient NPA). *Let $\mathcal{N}$ be a finite safe Petri net and let $\varphi \in \mathcal{L}_{hp}$ be a closed formula. The quotient NPA $\mathcal{A}_{\mathcal{E}(\mathcal{N}),\varphi_{/\approx_m}}$ is defined as follows. The set of states $Q = \{(M, r, \psi) \mid M \in \mathcal{R}(\mathcal{N}) \wedge r : fv(\psi) \to 2^M \wedge \psi \in \mathsf{sf}(\varphi)\}$. The initial state $q_0 = (M_0, \emptyset, \varphi)$. The transition relation is defined, for any state $q = (M, r, \psi) \in Q$, by:*

- *if $\psi = \mathsf{T}$ or $\psi = \mathsf{F}$, then $q \to (q)$*
- *if $\psi = \psi_1 \wedge \psi_2$, then $q \to (q_1, q_2)$ where $q_i = (M, r, \psi_i)$, $i \in \{1, 2\}$*
- *if $\psi = \psi_1 \vee \psi_2$, then $q \to (q_1)$ and $q \to (q_2)$ where $q_i = (M, r, \psi_i)$, $i \in \{1, 2\}$*
- *if $\psi = [\![\mathbf{x}, \overline{\mathbf{y}} < \mathsf{a}\, z]\!]\psi'$, let $S = \{(M', r'_{|fv(\psi')}) \mid \langle M, r\rangle \xrightarrow{\mathbf{x}, \overline{\mathbf{y}} < t}_{a,z} \langle M', r'\rangle\}$; if $S = \{(M_1, r_1), \ldots, (M_n, r_n)\} \neq \emptyset$ then $q \to (q_1, \ldots, q_n)$ where $q_i = (M_i, r_i, \psi')$ for $i \in [1, n]$, otherwise $q \to (q)$;*
- *if $\psi = \langle\!| \mathbf{x}, \overline{\mathbf{y}} < \mathsf{a}\, z |\!\rangle\psi'$, let $S = \{(M', r'_{|fv(\psi')}) \mid \langle M, r\rangle \xrightarrow{\mathbf{x}, \overline{\mathbf{y}} < t}_{a,z} \langle M', r'\rangle\}$; if $S = \{(M_1, r_1), \ldots, (M_n, r_n)\} \neq \emptyset$ then $q \to (q_i)$ where $q_i = (M_i, r_i, \psi')$ for $i \in [1, n]$, otherwise $q \to (q)$;*
- *if $\psi = \alpha X(\mathbf{x}).\psi'$ then $q \to (q')$ where $q' = (M, r, X(\mathbf{x}))$;*
- *if $\psi = X(\mathbf{y})$ and $\psi' \in \mathsf{sf}(\varphi)$ is the subformula such that $\psi' = \alpha X(\mathbf{x}).\psi''$ then $q \to (q')$ where $q' = (M, r[\mathbf{x} \mapsto r(\mathbf{y})], \psi'')$.*

The acceptance condition is as in Definition 10.

4.3 A Prototype Tool

The algorithm for model checking Petri nets outlined before is implemented in the prototype tool TCWB (*True Concurrency Workbench*) [33], written in Haskell. The tool inputs a safe Petri net $\mathcal{N}$ and a closed formula φ of $\mathcal{L}_{hp}$ and outputs the truth value of the formula on the initial marking of $\mathcal{N}$. The algorithm builds the quotient NPA $\mathcal{A}_{\mathcal{E}(\mathcal{N}),\varphi_{/\approx_m}}$ "on demand", i.e., the states of the automaton are generated when they are explored in the search of an accepting run. A path is recognised as successful when it includes a loop where a $\sqsubseteq_d^*$-maximal subformula is T, a $[\![\,]\!]$-subformula or a ν-subformula. In this way only the fragment of $\mathcal{A}_{\mathcal{E}(\mathcal{N}),\varphi_{/\approx_m}}$ relevant to decide the satisfaction of φ is built.

Given a net $\mathcal{N} = (P, T, F, M_0)$ and a formula φ, the number of states in the quotient automaton $\mathcal{A}_{\mathcal{E}(\mathcal{N}),\varphi_{/\approx_m}}$ can be bounded as follows. Recall that a state consists of a triple (M, r, ψ) where $\psi \in \mathsf{sf}(\varphi)$, M is a reachable marking and $r : fv(\psi) \to 2^M$ is a function. This leads to an upper bound $O(|\mathsf{sf}(\varphi)| \cdot |\mathcal{R}(\mathcal{N})| \cdot 2^{|P| \cdot v})$, where $v = max\{|fv(\psi)| : \psi \in \mathsf{sf}(\varphi)\}$ is the largest number of event variables appearing free in a subformula of φ. In turn, since $|\mathcal{R}(\mathcal{N})| \leq 2^{|P|}$, this is bounded by $O(|\mathsf{sf}(\varphi)| \cdot 2^{|P| \cdot (v+1)})$. The size of the automaton is thus exponential in the size of the net and linear in the size of the formula. Moving from the interleaving fragment of the logic (where $v = 0$) to formulae capable of expressing true concurrent properties thus causes an exponential blow up. However, note that the worst case scenario requires all transitions to be related by causality and concurrency to all places in any possible way, something that should be quite unlikely in practice. Indeed, despite the fact that the tool is very preliminary and more tweaks and optimisations could improve its efficiency, for the practical tests we performed the execution time seems to be typically well below than the theoretical worst case upper bound.

5 Conclusions

We introduced an automata-theoretic framework for the model checking of the logic for true concurrency $\mathcal{L}_{hp}$, representing the logical counterpart of a classical true concurrent equivalence, i.e., history preserving bisimilarity. The approach is developed abstractly for strongly regular PESs, that include regular trace PESs. A concrete model-checking procedure requires the identification of an effective bisimulation equivalence for the construction of the quotient automaton. We showed how this can be done for finite safe Petri nets. The technique is implemented in a proof-of-concept tool.

We proved that the class of regular trace PESs is included in that of strongly regular PESs which in turn is included in the class of regular PESs. The precise relation of strongly regular PESs with the other two classes is still unclear and interesting in view of [34] that recently showed that regular trace PESs are strictly included in regular PESs, disproving Thiagarajan's conjecture.

Several other papers deal with model checking for logics on event structures. In [35] a technique is proposed for model checking a CTL-style logic with modalities for immediate causality and conflict on a subclass of PESs. The logic is quite different from ours as formulae are satisfied by single events, the idea being that an event, with its causes, represents the local state of a component. The procedure involves the construction of a finite representation of the PES associated with a program which has some conceptual relation with our quotienting phase. In [19] the author shows that first order logic and Monadic Trace Logic (MTL), a restricted form of monadic second order (MSO) logic are decidable on regular trace event structures. The possibility of directly observing conflicts in MTL and thus of distinguishing behaviourally equivalent PESs (e.g., the PESs consisting of a single or two conflicting copies of an event), and the presence in $\mathcal{L}_{hp}$ of propositions which are non-monadic with respect to event variables, make these logics not immediate to compare. Still, a deeper investigation is definitively worth to pursue, especially in view of the fact that, in the propositional case, the mu-calculus corresponds to the bisimulation invariant fragment of MSO logic [36].

The work summarised in [18] develops a game theoretic approach for model-checking a concurrent logic over partial order models. It has been observed in [20] that such logic is incomparable to $\mathcal{L}_{hp}$. Preliminary investigations shows that our model-checking framework could be adapted to such a logic and, more generally, to a logic joining the expressive power of the two. Moreover, further exploring the potentialities of a game theoretic approach in our setting represents an interesting venue of further research.

Compared to our previous work [26], we extended the range of the technique to the full logic $\mathcal{L}_{hp}$, without limitations concerning the alternation depth of formulae. Relaxing the restriction to strongly regular PESs, instead, appears to be quite problematic unless one is willing to deal with transfinite runs which, however, would be of very limited practical interest.

The tool is still very preliminary. As suggested by its (wishful) name (inspired by the classical Edinburgh Concurrency Workbench [37]) we would like to bring the TCWB to a more mature stage, working on optimisations and adding an interface that gives access to a richer set of commands.

Acknowledgements. We are grateful to Perdita Stevens for insightful hints and pointers to the literature and to the anonymous reviewers for their comments.

References

1. Winskel, G.: Event structures. In: Brauer, W., Reisig, W., Rozenberg, G. (eds.) ACPN 1986. LNCS, vol. 255, pp. 325–392. Springer, Heidelberg (1987). https://doi.org/10.1007/3-540-17906-2_31
2. Winskel, G.: Events, causality and symmetry. Comput. J. **54**(1), 42–57 (2011)
3. Pichon-Pharabod, J., Sewell, P.: A concurrency semantics for relaxed atomics that permits optimisation and avoids thin-air executions. In: Bodík, R., Majumdar, R. (eds.) Proceedings of POPL 2016, pp. 622–633. ACM (2016)
4. Jeffrey, A., Riely, J.: On thin air reads towards an event structures model of relaxed memory. In: Grohe, M., Koskinen, E., Shankar, N. (eds.) Proceedings of LICS 2016, pp. 759–767. ACM (2016)
5. Dumas, M., García-Bañuelos, L.: Process mining reloaded: event structures as a unified representation of process models and event logs. In: Devillers, R., Valmari, A. (eds.) PETRI NETS 2015. LNCS, vol. 9115, pp. 33–48. Springer, Cham (2015). https://doi.org/10.1007/978-3-319-19488-2_2
6. Farzan, A., Madhusudan, P.: Causal atomicity. In: Ball, T., Jones, R.B. (eds.) CAV 2006. LNCS, vol. 4144, pp. 315–328. Springer, Heidelberg (2006). https://doi.org/10.1007/11817963_30
7. Baldan, P., Carraro, A.: A causal view on non-intereference. Fundamenta Informaticae **140**(1), 1–38 (2015)
8. van Glabbeek, R., Goltz, U.: Refinement of actions and equivalence notions for concurrent systems. Acta Informatica **37**(4/5), 229–327 (2001)
9. De Nicola, R., Ferrari, G.L.: Observational logics and concurrency models. In: Nori, K.V., Veni Madhavan, C.E. (eds.) FSTTCS 1990. LNCS, vol. 472, pp. 301–315. Springer, Heidelberg (1990). https://doi.org/10.1007/3-540-53487-3_53
10. Bednarczyk, M.A.: Hereditary history preserving bisimulations or what is the power of the future perfect in program logics. Technical report, Polish Academy of Sciences (1991)
11. Pinchinat, S., Laroussinie, F., Schnoebelen, P.: Logical characterization of truly concurrent bisimulation. Technical report 114, LIFIA-IMAG, Grenoble (1994)
12. Penczek, W.: Branching time and partial order in temporal logics. In: Time and Logic: A Computational Approach, pp. 179–228. UCL Press (1995)
13. Nielsen, M., Clausen, C.: Games and logics for a noninterleaving bisimulation. Nord. J. Comput. **2**(2), 221–249 (1995)
14. Bradfield, J., Fröschle, S.: Independence-friendly modal logic and true concurrency. Nord. J. Comput. **9**(1), 102–117 (2002)
15. Alur, R., Peled, D., Penczek, W.: Model-checking of causality properties. In: Proceedings of LICS 1995, pp. 90–100. IEEE Computer Society (1995)

16. Gutierrez, J., Bradfield, J.: Model-checking games for fixpoint logics with partial order models. In: Bravetti, M., Zavattaro, G. (eds.) CONCUR 2009. LNCS, vol. 5710, pp. 354–368. Springer, Heidelberg (2009). https://doi.org/10.1007/978-3-642-04081-8_24
17. Gutierrez, J.: Logics and bisimulation games for concurrency, causality and conflict. In: de Alfaro, L. (ed.) FoSSaCS 2009. LNCS, vol. 5504, pp. 48–62. Springer, Heidelberg (2009). https://doi.org/10.1007/978-3-642-00596-1_5
18. Gutierrez, J.: On bisimulation and model-checking for concurrent systems with partial order semantics. Ph.D. thesis, University of Edinburgh (2011)
19. Madhusudan, P.: Model-checking trace event structures. In: Proceedings of LICS 2013, pp. 371–380. IEEE Computer Society (2003)
20. Baldan, P., Crafa, S.: A logic for true concurrency. J. ACM **61**(4), 24:1–24:36 (2014)
21. Phillips, I., Ulidowski, I.: Event identifier logic. Math. Struct. Comput. Sci. **24**(2), 1–51 (2014)
22. Best, E., Devillers, R., Kiehn, A., Pomello, L.: Fully concurrent bisimulation. Acta Informatica **28**, 231–261 (1991)
23. Rabinovich, A.M., Trakhtenbrot, B.A.: Behaviour structures and nets. Fundamenta Informaticae **11**, 357–404 (1988)
24. Degano, P., De Nicola, R., Montanari, U.: Partial orderings descriptions and observations of nondeterministic concurrent processes. In: de Bakker, J.W., de Roever, W.-P., Rozenberg, G. (eds.) REX 1988. LNCS, vol. 354, pp. 438–466. Springer, Heidelberg (1989). https://doi.org/10.1007/BFb0013030
25. Jurdzinski, M., Nielsen, M., Srba, J.: Undecidability of domino games and hhp-bisimilarity. Inf. Comput. **184**(2), 343–368 (2003)
26. Baldan, P., Padoan, T.: Local model checking in a logic for true concurrency. In: Esparza, J., Murawski, A.S. (eds.) FoSSaCS 2017. LNCS, vol. 10203, pp. 407–423. Springer, Heidelberg (2017). https://doi.org/10.1007/978-3-662-54458-7_24
27. Thiagarajan, P.S.: Regular event structures and finite Petri nets: a conjecture. In: Brauer, W., Ehrig, H., Karhumäki, J., Salomaa, A. (eds.) Formal and Natural Computing. LNCS, vol. 2300, pp. 244–253. Springer, Heidelberg (2002). https://doi.org/10.1007/3-540-45711-9_14
28. Emerson, E.A., Jutla, C.S., Sistla, A.P.: On model checking for the μ-calculus and its fragments. Theor. Comput. Sci. **258**(1–2), 491–522 (2001)
29. Mostowski, A.W.: Regular expressions for infinite trees and a standard form of automata. In: Skowron, A. (ed.) SCT 1984. LNCS, vol. 208, pp. 157–168. Springer, Heidelberg (1985). https://doi.org/10.1007/3-540-16066-3_15
30. Klauck, H.: Algorithms for parity games. In: Grädel, E., Thomas, W., Wilke, T. (eds.) Automata Logics, and Infinite Games. LNCS, vol. 2500, pp. 107–129. Springer, Heidelberg (2002). https://doi.org/10.1007/3-540-36387-4_7
31. Abdulla, P.A., Kaati, L., Högberg, J.: Bisimulation minimization of tree automata. In: Ibarra, O.H., Yen, H.-C. (eds.) CIAA 2006. LNCS, vol. 4094, pp. 173–185. Springer, Heidelberg (2006). https://doi.org/10.1007/11812128_17
32. Petri, C.: Kommunikation mit Automaten. Schriften des Institutes für Instrumentelle Matematik, Bonn (1962)
33. Padoan, T.: True concurrency workbench. http://github.com/tpadoan/TCWB
34. Chalopin, J., Chepoi, V.: A counterexample to Thiagarajan's conjecture on regular event structures. In: Chatzigiannakis, I., Indyk, P., Kuhn, F., Muscholl, A. (eds.) Proceedings of ICALP 2017, LIPIcs, vol. 80, pp. 101:1–101:14, Schloss Dagstuhl - Leibniz-Zentrum fuer Informatik (2017)

35. Penczek, W.: Model-checking for a subclass of event structures. In: Brinksma, E. (ed.) TACAS 1997. LNCS, vol. 1217, pp. 145–164. Springer, Heidelberg (1997). https://doi.org/10.1007/BFb0035386
36. Janin, D., Walukiewicz, I.: On the expressive completeness of the propositional mu-calculus with respect to monadic second order logic. In: Montanari, U., Sassone, V. (eds.) CONCUR 1996. LNCS, vol. 1119, pp. 263–277. Springer, Heidelberg (1996). https://doi.org/10.1007/3-540-61604-7_60
37. Stevens, P., Stirling, C.: Practical model-checking using games. In: Steffen, B. (ed.) TACAS 1998. LNCS, vol. 1384, pp. 85–101. Springer, Heidelberg (1998). https://doi.org/10.1007/BFb0054166

A Theory of Encodings and Expressiveness (Extended Abstract)

Rob van Glabbeek[1,2(✉)]

[1] Data61, CSIRO, Sydney, Australia
rvg@cs.stanford.edu

[2] Computer Science and Engineering, University of New South Wales, Sydney, Australia

Abstract. This paper proposes a definition of what it means for one system description language to encode another one, thereby enabling an ordering of system description languages with respect to expressive power. I compare the proposed definition with other definitions of encoding and expressiveness found in the literature, and illustrate it on a well-known case study: the encoding of the synchronous in the asynchronous π-calculus.

1 Introduction

This paper, like [16,21], aims at answering the question what it means for one language to encode another one, and making the resulting definition applicable to order system description languages like CCS, CSP and the π-calculus with respect to their expressive power.

To this end it proposes a unifying concept of valid translation between two languages *up to* a semantic equivalence or preorder. It applies to languages whose semantics interprets the operators and recursion constructs as operations on a set of values, called a *domain*. Languages can be partially ordered by their expressiveness up to the chosen equivalence or preorder according to the existence of valid translations between them.

The concept of a [valid] translation between system description languages (or *process calculi*) was first formally defined by Boudol [3]. There, and in most other related work in this area, the domain in which a system description language is interpreted consists of the closed expressions from the language itself. In [14] I have reformulated Boudol's definition, while dropping the requirement that the domain of interpretation is the set of closed terms. This allows (but does not enforce) a clear separation of syntax and semantics, in the tradition of universal algebra. Nevertheless, the definition employed in [14] only deals with the case that all (relevant) elements in the domain are denotable as the interpretations of closed terms. In [16] situations are described where such a restriction is undesirable. In addition, both [3,14] require the semantic equivalence $\sim$ under which two languages are compared to be a congruence for both of them. This is too severe a restriction to capture many recent encodings [1,2,7,30,31,33,38,43].

C. Baier and U. Dal Lago (Eds.): FOSSACS 2018, LNCS 10803, pp. 183–202, 2018.
https://doi.org/10.1007/978-3-319-89366-2_10

In [16] I alleviated these two restrictions by proposing two notions of encoding: *correct* and *valid* translations up to $\sim$. Each of them generalises the proposals of [3,14]. The former drops the restriction on denotability as well as $\sim$ being a congruence for the whole target language, but it requires $\sim$ to be a congruence for the source language, as well as for the source's image within the target. The latter drops both congruence requirements (and allows $\sim$ to be a preorder rather than an equivalence), but at the expense of requiring denotability by closed terms. In situations where $\sim$ is a congruence for the source language's image within the target language *and* all semantic values are denotable, the two notions agree.

The current paper further generalises the work of [16] by proposing a new notion of a valid translation that incorporates the correct and valid translations of [16] as special cases. It drops the congruence requirements as well as the restriction on denotability.

As in [16], my aim is to generalise the concept of a valid translation as much as possible, so that it is uniformly applicable in many situations, and not just in the world of process calculi. Also, it needs to be equally applicable to encodability and separation results, the latter saying that an encoding of one language in another does not exists. At the same time, I try to derive this concept from a unifying principle, rather than collecting a set of criteria that justify a number of known encodability and separation results that are intuitively justified.

Overview of the Paper. Section 2 defines my new concept of a valid translation up to a semantic equivalence or preorder $\mathrel{\bullet\!\!\sim}$. Roughly, a valid translation of one language into another is a mapping from the expressions in the first language to those in the second that preserves their meaning, i.e. such that the meaning of a translated expression is semantically equivalent to the meaning of the original.

Section 3 shows that this concept generalises the notion of a correct translation from [16]: a translation is correct up to a semantic equivalence $\sim$ iff it is valid up to $\sim$ and $\sim$ is a congruence for the source language as well as for the image of the source language within the target language.

Likewise, [18]—the full version of this paper—establishes the coincidence of my validity-based notion of expressiveness with the one from [16] when applying both to languages for which all semantic values are denotable by closed terms.

One language is said to be at least as expressive as another up to $\mathrel{\bullet\!\!\sim}$ iff there exists a valid translation up to $\mathrel{\bullet\!\!\sim}$ of the latter language into the former. Section 4 shows that "being at least as expressive as" is a preorder on languages. This expressiveness preorder depends on the choice of $\mathrel{\bullet\!\!\sim}$, and a coarser choice (making less distinctions) yields a richer preorder of expressiveness inclusions.

Section 6 illustrates the framework on a well-known case study: the encoding of the synchronous in the asynchronous π-calculus.

Section 7 discusses the *congruence closure* of a semantic equivalence for a given language, and remarks that in the presence of operators with infinite arity it is not always a congruence. Section 8 states a useful congruence closure property for valid translations: if a translation between two languages exists that is valid up a semantic equivalence $\sim$, then it is even valid up to an equivalence that

- on the source language coincides with the congruence closure of $\sim$
- on the image of the source within the target language also coincides with the congruence closure of $\sim$
- melts each equivalence class of the source with exactly one of the target.

Section 9 concludes that the framework established thus far is great for comparing the expressiveness of languages, but falls short for the purpose of combining language features. This requires a congruence reflection theorem, provided in Sect. 12, for languages satisfying postulates formulated in Sects. 5, 10 and 11.

Section 12 defines when a translation is *compositional*, and shows that any valid translation up to $\mathrel{\bullet\!\!\sim}$ can be modified into a compositional translation valid up to $\mathrel{\bullet\!\!\sim}$. This requires restricting attention to languages and preorders $\mathrel{\bullet\!\!\sim}$ that satisfy some mild sanity requirements—the postulates of Sects. 10 and 11. Hence, for the purpose of comparing the expressive power of languages, valid translations between them may be presumed compositional.

Section 13 compares my approach with the one of Gorla [21], and concludes. Omitted proofs and counterexamples (marked by ¶) can be found in [18].

2 Languages, Valid Translations, and Expressiveness

A language consists of *syntax* and *semantics*. The syntax determines the valid expressions in the language. The semantics is given by a mapping $[\![\ \ ]\!]$ that associates with each valid expression its meaning, which can for instance be an object, concept or statement.

Following [16], I represent a language $\mathcal{L}$ as a pair $(\mathbb{T}_\mathcal{L}, [\![\ \ ]\!]_\mathcal{L})$ of a set $\mathbb{T}_\mathcal{L}$ of valid expressions in $\mathcal{L}$ and a mapping $[\![\ \ ]\!]_\mathcal{L} : \mathbb{T}_\mathcal{L} \rightarrow \mathcal{D}_\mathcal{L}$ from $\mathbb{T}_\mathcal{L}$ in some set of meanings $\mathcal{D}_\mathcal{L}$.

Definition 1 **([16]).** A *translation* from a language $\mathcal{L}$ into a language $\mathcal{L}'$ is a mapping $\mathcal{T} : \mathbb{T}_\mathcal{L} \rightarrow \mathbb{T}_{\mathcal{L}'}$.

In this paper, I consider single-sorted languages $\mathcal{L}$ in which *expressions* or *terms* are built from variables (taken from a set $\mathcal{X}$) by means of operators (including constants) and possibly recursion constructs. For such languages the meaning $[\![E]\!]_\mathcal{L}$ of an $\mathcal{L}$-expression E is a function of type $(\mathcal{X} \rightarrow \mathbf{V}) \rightarrow \mathbf{V}$ for a given sets of *values* $\mathbf{V}$. It associates a value $[\![E]\!]_\mathcal{L}(\rho) \in \mathbf{V}$ to E that depends on the choice of a *valuation* ρ: $\mathcal{X} \rightarrow \mathbf{V}$. The valuation associates a value from $\mathbf{V}$ with each variable.

Since normally the names of variables are irrelevant and the cardinality of the set of variables satisfies only the requirement that it is "sufficiently large", no generality is lost by insisting that two (system description) languages whose expressiveness is being compared employ the same set of (process) variables. On the other hand, two languages $\mathcal{L}$ and $\mathcal{L}'$ may be interpreted in different domains of values $\mathbf{V}$ and $\mathbf{V}'$.

Let $\mathcal{L}$ and $\mathcal{L}'$ be languages as considered above, with semantic mappings

$$[\![\ ]\!]_{\mathcal{L}} : \mathbb{T}_{\mathcal{L}} \rightarrow ((\mathcal{X} \rightarrow \mathbf{V}) \rightarrow \mathbf{V}) \quad \text{and} \quad [\![\ ]\!]_{\mathcal{L}'} : \mathbb{T}_{\mathcal{L}'} \rightarrow ((\mathcal{X} \rightarrow \mathbf{V}') \rightarrow \mathbf{V}').$$

In order to compare these languages w.r.t. their expressive power I need a semantic equivalence or preorder $\mathrel{\bullet\!\!\sim}$ that is defined on a unifying domain of interpretation $\mathbf{Z}$, with $\mathbf{V}, \mathbf{V}' \subseteq \mathbf{Z}$.[1] Intuitively, $v' \mathrel{\bullet\!\!\sim} v$ with $v \in \mathbf{V}$ and $v' \in \mathbf{V}'$ means that values v and v' are sufficiently alike for our purposes, so that one can accept a translation of an expression with meaning v into an expression with meaning v'. Below, target values of a translation (in $\mathbf{V}'$) are written on the left.

Correct and a *valid* translations up to a semantic equivalence or preorder $\mathrel{\bullet\!\!\sim}$ were introduced in [16]. Here I redefine these concepts in terms of a new concept of *correctness w.r.t. a semantic translation.*

Definition 2. Let $\mathbf{V}$ and $\mathbf{V}'$ be domains of values in which two languages $\mathcal{L}$ and $\mathcal{L}'$ are interpreted. A *semantic translation* from $\mathbf{V}$ into $\mathbf{V}'$ is a relation $\mathbf{R} \subseteq \mathbf{V}' \times \mathbf{V}$ such that $\forall v \in \mathbf{V}. \exists v' \in \mathbf{V}'. v' \mathbf{R} v$.

Thus every semantic value in $\mathbf{V}$ needs to have a counterpart in $\mathbf{V}'$—possibly multiple ones. For valuations $\eta : \mathcal{X} \rightarrow \mathbf{V}'$, $\rho : \mathcal{X} \rightarrow \mathbf{V}$ I write $\eta \,\mathbf{R}\, \rho$ iff $\eta(X) \,\mathbf{R}\, \rho(X)$ for each $X \in \mathcal{X}$.

Definition 3. A translation $\mathcal{T} : \mathbb{T}_{\mathcal{L}} \rightarrow \mathbb{T}_{\mathcal{L}'}$ is *correct* w.r.t. a semantic translation $\mathbf{R}$ if $[\![\mathcal{T}(E)]\!]_{\mathcal{L}'}(\eta) \,\mathbf{R}\, [\![E]\!]_{\mathcal{L}}(\rho)$ for all expressions $E \in \mathbb{T}_{\mathcal{L}}$ and all valuations $\eta : \mathcal{X} \rightarrow \mathbf{V}'$ and $\rho : \mathcal{X} \rightarrow \mathbf{V}$ with $\eta \,\mathbf{R}\, \rho$.

Thus $\mathcal{T}$ is correct iff the meaning of the translation of an expression E is a counterpart of the meaning of E, no matter what values are filled in for the variables, provided that the value filled in for a given variable X occurring in the translation $\mathcal{T}(E)$ is a counterpart of the value filled in for X in E.

Definition 4. A translation $\mathcal{T} : \mathbb{T}_{\mathcal{L}} \rightarrow \mathbb{T}_{\mathcal{L}'}$ is *correct* up to $\mathrel{\bullet\!\!\sim}$ iff $\mathrel{\bullet\!\!\sim}$ is an equivalence, the restriction $\mathbf{R}$ of $\mathrel{\bullet\!\!\sim}$ to $\mathbf{V}' \times \mathbf{V}$ is a semantic translation, and $\mathcal{T}$ is correct w.r.t. $\mathbf{R}$.

Definition 5. A translation $\mathcal{T}$ is *valid* up to $\mathrel{\bullet\!\!\sim}$ iff it is correct w.r.t. some semantic translation $\mathbf{R} \subseteq \mathrel{\bullet\!\!\sim}$. Language $\mathcal{L}'$ is at least as *expressive* as $\mathcal{L}$ up to $\mathrel{\bullet\!\!\sim}$ if a translation valid up to $\mathrel{\bullet\!\!\sim}$ from $\mathcal{L}$ into $\mathcal{L}'$ exists.

Example 4 in [18] illustrates both notions and shows their difference.

[1] I will be chiefly interested in the case that $\mathrel{\bullet\!\!\sim}$ is an equivalence—hence the choice of a symbol that looks like $\sim$. However, to establish Observation 2 and Theorem 2 below, it suffices to know that $\mathrel{\bullet\!\!\sim}$ is reflexive and transitive. My convention is that the dotted end of $\mathrel{\bullet\!\!\sim}$ points to a translation and the other end to an original—without offering an intuition for the possible asymmetry.

3 Correct = Valid + Congruence

In [16] the concept of a correct translation up to $\sim$ was defined, for $\sim$ a semantic equivalence on $\mathbf{Z}$. Here two valuations $\eta, \rho : \mathcal{X} \to \mathbf{Z}$ are called *$\sim$-equivalent*, $\eta \sim \rho$, if $\eta(X) \sim \rho(X)$ for each $X \in \mathcal{X}$. In case there exists a $v \in \mathbf{V}$ for which there is no $\sim$-equivalent $v' \in \mathbf{V}'$, there is no correct translation from $\mathcal{L}$ into $\mathcal{L}'$ up to $\sim$. Namely, the semantics of $\mathcal{L}$ describes, among others, how any $\mathcal{L}$-operator evaluates the argument value v, and this aspect of the language has no counterpart in $\mathcal{L}'$. Therefore, [16] requires

$$\forall v \in \mathbf{V}.\ \exists v' \in \mathbf{V}'.\ v' \sim v. \tag{1}$$

This implies that for any valuation $\rho : \mathcal{X} \to \mathbf{V}$ there is an $\eta : \mathcal{X} \to \mathbf{V}'$ with $\eta \sim \rho$.

Definition 6 ([16]). A translation $\mathcal{T}$ from $\mathcal{L}$ into $\mathcal{L}'$ is *correct up to* $\sim$ iff (1) holds and $[\![\mathcal{T}(E)]\!]_{\mathcal{L}'}(\eta) \sim [\![E]\!]_{\mathcal{L}}(\rho)$ for all $E \in \mathbb{T}_{\mathcal{L}}$ and all valuations $\eta : \mathcal{X} \to \mathbf{V}'$ and $\rho : \mathcal{X} \to \mathbf{V}$ with $\eta \sim \rho$.

Note that this definition agrees completely with Definition 4. Requirement (1) above corresponds to $\mathbf{R}$ being a semantic translation in Definition 4.

If a correct translation up to $\sim$ from $\mathcal{L}$ into $\mathcal{L}'$ exists, then $\sim$ must be a congruence for $\mathcal{L}$.

Definition 7. An equivalence relation $\sim$ is a *congruence* for a language $\mathcal{L}$ interpreted in a semantic domain $\mathbf{V}$ if $[\![E]\!]_{\mathcal{L}}(\nu) \sim [\![E]\!]_{\mathcal{L}}(\rho)$ for any $\mathcal{L}$-expression E and any valuations $\nu, \rho : \mathcal{X} \to \mathbf{V}$ with $\nu \sim \rho$.[2]

Proposition 1 ([16]). If $\mathcal{T}$ is a correct translation up to $\sim$ from $\mathcal{L}$ into $\mathcal{L}'$, then $\sim$ is a congruence for $\mathcal{L}$

The existence of a correct translation up to $\sim$ from $\mathcal{L}$ into $\mathcal{L}'$ does not imply that $\sim$ is a congruence for $\mathcal{L}'$. However, $\sim$ has the properties of a congruence for those expressions of $\mathcal{L}'$ that arise as translations of expressions of $\mathcal{L}$, when restricting attention to valuations into $\mathbf{U} := \{v \in \mathbf{V}' \mid \exists v \in \mathbf{V}.\ v' \sim v\}$. In [16] this called a *congruence for* $\mathcal{T}(\mathcal{L})$.

Definition 8. Let $\mathcal{T} : \mathbb{T}_{\mathcal{L}} \to \mathbb{T}_{\mathcal{L}'}$ be a translation from $\mathcal{L}$ into $\mathcal{L}'$. An equivalence $\sim$ on $\mathbf{V}'$ is a *congruence for* $\mathcal{T}(\mathcal{L})$ if $[\![\mathcal{T}(E)]\!]_{\mathcal{L}'}(\theta) \sim [\![\mathcal{T}(E)]\!]_{\mathcal{L}'}(\eta)$ for any $E \in \mathbb{T}_{\mathcal{L}}$ and $\theta, \eta : \mathcal{X} \to \mathbf{U}$ with $\theta \sim \eta$.

Proposition 2 ([16]). If $\mathcal{T}$ is a correct translation up to $\sim$ from $\mathcal{L}$ into $\mathcal{L}'$, then $\sim$ is a congruence for $\mathcal{T}(\mathcal{L})$.

The following theorem tells that the notion of validity proposed in Sect. 2 can be seen as a generalisation of the notion of correctness from [16] that applies to equivalences (and preorders) $\sim$ that need not be congruences for $\mathcal{L}$ or $\mathcal{T}(\mathcal{L})$.

Theorem 1. A translation $\mathcal{T}$ from $\mathcal{L}$ into $\mathcal{L}'$ is correct up to a semantic equivalence $\sim$ iff it is valid up to $\sim$ and $\sim$ is a congruence for $\mathcal{T}(\mathcal{L})$. ¶

[2] This is called a *lean* congruence in [17]; in the presence of recursion, stricter congruence requirements are common. Those are not needed in this paper.

4 A Hierarchy of Expressiveness Preorders

An equivalence or preorder $\sim$ on a class $\mathbf{Z}$ is said to be *finer*, *stronger*, or *more discriminating* than another equivalence or preorder $\approx$ on $\mathbf{Z}$ if $v \sim w \Rightarrow v \approx w$ for all $v, w \in \mathbf{Z}$.

Observation 1. Let $\mathcal{T} : \mathbb{T}_{\mathcal{L}} \rightarrow \mathbb{T}_{\mathcal{L}'}$ be a translation from $\mathcal{L}$ into $\mathcal{L}'$, and let $\sim$ be finer than $\approx$. If $\mathcal{T}$ is valid up to $\sim$, then it is also valid up to $\approx$.

The quality of a translation depends on the choice of the equivalence or preorder up to which it is valid. Any two languages are equally expressive up to the universal equivalence, relating any two processes. Hence, the equivalence or preorder needs to be chosen carefully to match the intended applications of the languages under comparison. In general, as shown by Observation 1, using a finer equivalence or preorder yields a stronger claim that one language can be encoded in another. On the other hand, when separating two languages $\mathcal{L}$ and $\mathcal{L}'$ by showing that $\mathcal{L}$ *cannot* be encoded in $\mathcal{L}'$, a coarser equivalence yields a stronger claim.

Observation 2. The identity is a valid translation up to any preorder from any language into itself.

Theorem 2. If valid translations up to $\sim$ exists from $\mathcal{L}_1$ into $\mathcal{L}_2$ and from $\mathcal{L}_2$ into $\mathcal{L}_3$, then there is a valid translation up to $\sim$ from $\mathcal{L}_1$ into $\mathcal{L}_3$. ¶

Theorem 2 and Observation 2 show that the relation "being at least as expressive as up to $\sim$" is a preorder on languages.

5 Closed-Term Languages

The languages considered in this paper feature *variables*, *operators* of *arity* $n \in \mathbb{N}$, and/or other constructs. The set $\mathbb{T}_{\mathcal{L}}$ of $\mathcal{L}$-expressions is inductively defined by:

- $X \in \mathbb{T}_{\mathcal{L}}$ for each variable $X \in \mathcal{X}$,
- $f(E_1, \ldots, E_n) \in \mathbb{T}_{\mathcal{L}}$ for each n-ary operator f and expressions $E_i \in \mathbb{T}_{\mathcal{L}}$,
- and clauses for the other constructs, if any.

Examples of other constructs are the infinite summation operator $\sum_{i \in I} E_i$ of CCS, which takes arbitrary many arguments, or the recursion construct $\mu X.E$, that has one argument, but *binds* all occurrences of X in that argument.

In general a construct has a number (possibly infinite) of argument expressions and it may bind certain variables within some of its arguments—the *scope* of the binding. An occurrence of a variable X in an expression is *bound* if it occurs within the scope of a construct that binds X, and *free* otherwise.

The semantics of such a language is given, in part, by a domain of values $\mathbf{V}$, and an interpretation of each n-ary operator f of $\mathcal{L}$ as an n-ary operation $f^{\mathbf{V}} : \mathbf{V}^n \rightarrow \mathbf{V}$ on $\mathbf{V}$. Using the equations

$$[\![X]\!]_{\mathcal{L}}(\rho) = \rho(X) \quad \text{and} \quad [\![f(E_1, \ldots, E_n)]\!]_{\mathcal{L}}(\rho) = f^{\mathbf{V}}([\![E_1]\!]_{\mathcal{L}}(\rho), \ldots, [\![E_n]\!]_{\mathcal{L}}(\rho))$$

this allows an inductive definition of the meaning $[\![E]\!]_{\mathcal{L}}$ of an $\mathcal{L}$-expression E. Moreover, $[\![E]\!]_{\mathcal{L}}(\rho)$ only depends on the restriction of ρ to the set $fv(E)$ of variables occurring free in E.

The set $\mathrm{T}_{\mathcal{L}} \subseteq \mathbb{T}_{\mathcal{L}}$ of *closed terms* of $\mathcal{L}$ consists of those $\mathcal{L}$-expressions $E \in \mathbb{T}_{\mathcal{L}}$ with $fv(E) = \emptyset$. If $P \in \mathrm{T}_{\mathcal{L}}$ and $\mathbf{V} \neq \emptyset$ then $[\![P]\!]_{\mathcal{L}}(\rho)$ is independent of the choice of $\rho : \mathcal{X} \to \mathbf{V}$, and therefore denoted $[\![P]\!]_{\mathcal{L}}$.

Definition 9. A *substitution* in $\mathcal{L}$ is a partial function $\sigma : \mathcal{X} \rightharpoonup \mathbb{T}_{\mathcal{L}}$ from the variables to the $\mathcal{L}$-expressions. For a given $\mathcal{L}$-expression $E \in \mathbb{T}_{\mathcal{L}}$, $E[\sigma] \in \mathbb{T}_{\mathcal{L}}$ denotes the $\mathcal{L}$-expression E in which each free occurrence of a variable $X \in dom(\sigma)$ is replaced by $\sigma(X)$, while renaming bound variables in E so as to avoid a free variable Y occurring in an expression $\sigma(X)$ ending up being bound in $E[\sigma]$. A substitution is *closed* if it has the form $\sigma : \mathcal{X} \to \mathrm{T}_{\mathcal{L}}$.

An important class of languages used in concurrency theory are the ones where the distinction between syntax and semantic is effectively dropped by taking $\mathbf{V} = \mathrm{T}_{\mathcal{L}}$, i.e. where the domain of values where the language is interpreted in consists of the closed terms of the language. Here a valuation is the same as a closed substitution, and $[\![E]\!]_{\mathcal{L}}(\rho)$ for $E \in \mathbb{T}_{\mathcal{L}}$ and $\rho : \mathcal{X} \to \mathrm{T}_{\mathcal{L}}$ is defined to be $E[\rho] \in \mathrm{T}_{\mathcal{L}}$. I will call such languages *closed-term* languages.

6 Translating a Synchronous into an Asynchronous π

As an illustration of the concepts introduced above, consider the π-calculus as presented in [28], i.e., the one of [44] without matching, τ-prefixing, and choice.

Given a set of *names* $\mathcal{N}$, the set $\mathbb{T}_\pi$ of *process expressions* or *terms* E of the calculus is given by

$$E ::= X \ \mid\ \mathbf{0} \ \mid\ \bar{x}y.E \ \mid\ x(z).E \ \mid\ E|E' \ \mid\ (\nu z)E \ \mid\ !E$$

with x, y, z ranging over $\mathcal{N}$, and X over $\mathcal{X}$, the set of *process variables*. Process variables are not considered in [44], although they are common in languages like CCS [27] that feature a recursion construct. Since process variables form a central part of my notion of a valid or correct translation, here they have simply been added. This works generally. In Sect. 12 I show that for the purpose of accessing whether one language is as expressive as another, translations between them can be assumed to be compositional. This important result would be lost if process variables were dropped from the language. In that case compositionality would need to be stated as a separate requirement for valid translations.

Closed process expressions are called *processes*. The π-calculus is usually presented as a closed-term language, in that the semantic value associated with a closed term is simply itself. Yet, the real semantics is given by a reduction relation between processes, defined below.

Definition 10. An occurrence of a name z in π-calculus process $P \in \mathrm{T}_\pi$ is *bound* if it occurs within a subexpression $x(z).P'$ or $(\nu z)P'$ of P; otherwise it is *free*. Let $\mathsf{n}(P)$ (resp. $\mathsf{bn}(P)$) be the set of names occurring (bound) in $P \in \mathrm{T}_\pi$. *Structural congruence*, $\equiv$, is the smallest congruence relation on processes satisfying

$$\begin{array}{rclrclrcl} P_1|(P_2|P_3) & \equiv & (P_1|P_2)|P_3 & !P & \equiv & P|!P & (\nu w)(P|Q) & \equiv & P|(\nu w)Q \\ P_1|P_2 & \equiv & P_2|P_1 & (\nu z)\mathbf{0} & \equiv & \mathbf{0} & x(z).P & \equiv & x(w).P\{^w\!/\!_z\} \\ P|\mathbf{0} & \equiv & P & (\nu z)(\nu w)P & \equiv & (\nu w)(\nu z)P & (\nu z)P & \equiv & (\nu w)P\{^w\!/\!_z\}\ . \end{array}$$

Here the rightmost column only holds when $w \notin \mathsf{n}(P)$, and $P\{^w\!/\!_z\}$ denotes the process obtained by replacing each free occurrence of z in P by w.

Definition 11. The *reduction relation*, $\rightarrow\ \subseteq \mathrm{T}_\pi \times \mathrm{T}_\pi$, is generated by the following rules.

$$\frac{}{\bar{x}z.P|x(y).Q \rightarrow P|Q\{^z\!/\!_y\}}\ (z \notin \mathsf{bn}(Q))$$

$$\frac{P \rightarrow P'}{P|Q \rightarrow P'|Q} \qquad \frac{P \rightarrow P'}{(\nu z)P \rightarrow (\nu z)P'} \qquad \frac{Q \equiv P \quad P \rightarrow P' \quad P' \equiv Q'}{Q \rightarrow Q'}$$

Let $\Longrightarrow$ be the reflexive and transitive closure of $\rightarrow$. The observable behaviour of π-calculus processes is often stated in terms of the outputs they can produce (abstracting from the value communicated on an output channel).

Definition 12. Let $x \in \mathcal{N}$. A process P has a *strong output barb* on x, notation $P{\downarrow_{\bar{x}}}$, if P can perform an output action $\bar{x}z$. This is defined inductively:

$$(\bar{x}z.(P)){\downarrow_{\bar{x}}} \qquad \frac{P{\downarrow_{\bar{x}}}}{(P|Q){\downarrow_{\bar{x}}}} \qquad \frac{Q{\downarrow_{\bar{x}}}}{(P|Q){\downarrow_{\bar{x}}}} \qquad \frac{P{\downarrow_{\bar{x}}} \quad x \neq z}{((\nu z)P){\downarrow_{\bar{x}}}} \qquad \frac{P{\downarrow_{\bar{x}}}}{(!P){\downarrow_{\bar{x}}}}$$

A process P has a *weak output barb* on x, $P{\Downarrow_{\bar{x}}}$, if there is a P' with $P \Longrightarrow P'{\downarrow_{\bar{x}}}$.

A common semantic equivalence applied in the π-calculus is *weak barbed congruence* [29,44].

Definition 13. *Weak (output) barbed bisimilarity* is the largest symmetric relation $\stackrel{\bullet}{\approx}\ \subseteq \mathrm{T}_\pi \times \mathrm{T}_\pi$ such that

- $P \stackrel{\bullet}{\approx} Q$ and $P{\downarrow_{\bar{x}}}$ implies $Q{\Downarrow_{\bar{x}}}$, and
- $P \stackrel{\bullet}{\approx} Q$ and $P \Longrightarrow P'$ implies $Q \Longrightarrow Q'$ for some Q' with $P' \stackrel{\bullet}{\approx} Q'$.

Weak barbed congruence, $\cong^c$, is the largest congruence included in $\stackrel{\bullet}{\approx}$.

Often *input barbs*, defined similarly, are included in the definition of weak barbed bisimilarity [44]. This is known to induce the same notion of weak barbed congruence [44]. Another technique for defining weak barbed congruence is to use a barb, or set of barbs, external to the language under investigation, that are added to the language as constants [21], similar to the theory of testing of [9]. This method is useful for languages with a reduction semantics that do not feature a clear notion of barb, or where there is ambiguity in which barbs should be counted and which not, or for comparing languages with different kinds of barb.

Example 1. $\bar{x}z.\mathbf{0} \not\cong^c (\nu u)(\bar{x}u.\mathbf{0}|u(v).\bar{v}z.\mathbf{0})$.
For let $E := X|x(u).\bar{u}v.\mathbf{0}$ with $\rho(X) = \bar{x}z.\mathbf{0}$ and $\zeta(X) = (\nu u)(\bar{x}u.\mathbf{0}|u(v).\bar{v}z.\mathbf{0})$. Then $E[\zeta] \rightarrow (\nu u)(u(v).\bar{v}z.\mathbf{0}|\bar{u}v.\mathbf{0}) \rightarrow (\bar{v}z.\mathbf{0})\downarrow_{\bar{v}}$ but $(E[\rho])\not\Downarrow_{\bar{v}}$.

The asynchronous π-calculus, as introduced by Honda and Tokoro in [24] and by Boudol in [4], is the sublanguage $\mathrm{a}\pi$ of the fragment π of the π-calculus presented above where all subexpressions $\bar{x}y.E$ have the form $\bar{x}y.\mathbf{0}$. *Asynchronous barbed congruence*, $\cong^c_\mathrm{a}$, is the largest congruence *for the asynchronous π-calculus* included in $\dot{\approx}$. Since $\mathrm{a}\pi$ is a sublanguage of π, $\cong^c_\mathrm{a}$ is at least as coarse an equivalence as $\cong^c$, i.e. $\cong^c \subseteq \cong^c_\mathrm{a}$. The inclusion is strict, since $!x(z).\bar{x}z.\mathbf{0} \cong^c_\mathrm{a} \mathbf{0}$, yet $!x(z).\bar{x}z.\mathbf{0} \not\cong^c \mathbf{0}$ [44]. Since all expressions used in Example 1 belong to $\mathrm{a}\pi$, one even has $\bar{x}z.\mathbf{0} \not\cong^c_\mathrm{a} (\nu u)(\bar{x}u.\mathbf{0}|u(v).\bar{v}z.\mathbf{0})$.

Boudol [4] defined a translation $\mathcal{T}$ from π to $\mathrm{a}\pi$ inductively as follows:

$$\begin{aligned}
\mathcal{T}(X) &= X && \text{for } X \in \mathcal{X}\\
\mathcal{T}(\mathbf{0}) &= \mathbf{0}\\
\mathcal{T}(\bar{x}z.P) &= (u)(\bar{x}u|u(v).(\bar{v}z|\mathcal{T}(P))) && \text{choosing } u,v \notin \mathsf{n}(P),\ u \neq v\\
\mathcal{T}(x(y).P) &= x(u).(v)(\bar{u}v|v(y).\mathcal{T}(P)) && \text{choosing } u,v \notin \mathsf{n}(P),\ u \neq v\\
\mathcal{T}(P|Q) &= (\mathcal{T}(P)|\mathcal{T}(Q))\\
\mathcal{T}(!P) &= !\mathcal{T}(P)\\
\mathcal{T}((\nu x)P) &= (\nu x)\mathcal{T}(P)
\end{aligned}$$

Example 1 shows that $\mathcal{T}$ is not valid up to $\cong^c$. In fact, it is not even valid up to $\cong^c_\mathrm{a}$. However, as shown in [25], it is valid up to $\dot{\approx}$. Since $\dot{\approx}$ is not a congruence (for π or $\mathrm{a}\pi$) it is not correct up to $\dot{\approx}$.

7 Congruence Closure

Definition 14. An equivalence relation $\sim$ is a *1-hole congruence* for a language $\mathcal{L}$ interpreted in a semantic domain $\mathbf{V}$ if $[\![E]\!]_\mathcal{L}(\nu) \sim [\![E]\!]_\mathcal{L}(\rho)$ for any $\mathcal{L}$-expression E and any valuations $\nu,\rho : \mathcal{X} \rightarrow \mathbf{V}$ with $\nu \sim^1 \rho$. Here ν,ρ are *$\sim^1$-equivalent*, $\nu \sim^1 \rho$, if $\nu(X) \sim \rho(X)$ for some $X \in \mathcal{X}$ and $\nu(Y) = \rho(Y)$ for all variables $Y \neq X$.

An *n-hole congruence* for any finite $n \in \mathbb{N}$ can be defined in the same vain, and it is well known and easy to check that a 1-hole congruence $\sim$ is also an n-hole congruence, for any $n \in \mathbb{N}$. However, in the presence of operators with infinitely many arguments, a 1-hole congruence need not be a congruence.

Example 2. Let $\mathbf{V}$ be $(\mathbb{N} \times \mathbb{N}) \cup \{\infty\}$, with the well-order $\leq$ on $\mathbf{V}$ inherited lexicographically from the default order on $\mathbb{N}$ and ∞ the largest element. So $(n,m) \leq (n',m')$ iff $n \leq n' \vee (n = m \wedge m \leq m')$. Consider the language $\mathcal{L}$ with constants 0, 1 and (1), interpreted in $\mathbf{V}$ as $(0,0)$, $(1,0)$ and $(0,1)$, respectively, the binary operator $+$, interpreted by $(n_1,m_1) +^\mathbf{V} (n_2,m_2) = (n_1{+}n_2, m_1{+}m_2)$ and $\infty{+}E = E{+}\infty = \infty$, and the construct $\sup(E_i)_{i\in I}$ that takes any number of arguments (dependent on the set of the index sets I). The interpretation of sup in $\mathbf{V}$ is to take the supremum of its arguments w.r.t. the well-order $\leq$. In case sup is given finitely many arguments, it simply returns the largest. However $\sup((n,i))_{i\in\mathbb{N}} = (n{+}1, 0)$.

Now let the equivalence relation $\sim$ on $\mathbf{V}$ be defined by $(n,m) \sim (n',m')$ iff $n = n'$, leaving ∞ in an equivalence class of its own. This relation is a 1-hole congruence on $\mathcal{L}$. Hence, it is also a 2-hole congruence, so one has

$$((n_1,m_1) \sim (n_1',m_1') \wedge (n_2,m_2) \sim (n_2',m_2')) \Rightarrow (n_1,m_1) + (n_2,m_2) \sim (n_1',m_1') + (n_2',m_2').$$

Yet it fails to be a congruence: $(n,i) \sim (n,0)$ for all $i \in \mathbb{N}$, but

$$(n{+}1,0) = \sup((n,i))_{i\in\mathbb{N}} \not\sim \sup((n,0))_{i\in\mathbb{N}} = (n,0).$$

It is well known and easy to check that the collection of equivalence relations on any domain $\mathbf{V}$, ordered by inclusion, forms a complete lattice—namely the intersection of arbitrary many equivalence relations is again an equivalence relation. Likewise, the collection of 1-hole congruences for $\mathcal{L}$ is also a complete lattice, and moreover a complete sublattice of the complete lattice of equivalence relations on $\mathbf{V}$. The latter implies that for any collection C of 1-hole congruence relations, the least equivalence relation that contains all elements of C (exists and) happens to be a 1-hole congruence relation. Again, this is a property that is well known [22] and easy to prove. It follows that for any equivalence relation $\sim$ there exists a largest 1-hole congruence for $\mathcal{L}$ contained in $\sim$. I will denote this 1-hole congruence by $\sim_{\mathcal{L}}^{1c}$, and call it the *congruence closure* of $\sim$ w.r.t. $\mathcal{L}$. One has $v_1 \sim_{\mathcal{L}}^{1c} v_2$ for $v_1, v_2 \in \mathbf{V}$ iff $[\![E]\!]_{\mathcal{L}}(\nu) \sim [\![E]\!]_{\mathcal{L}}(\rho)$ for any $\mathcal{L}$-expression E and any valuations $\nu, \rho : \mathcal{X} \rightarrow \mathbf{V}$ with $\nu(X) = v_1$ and $\rho(X) = v_2$ for some $X \in \mathcal{X}$ and $\nu(Y) = \rho(Y)$ for all $Y \neq X$. Such results do not generally hold for congruences.

Example 3. Continue Example 2, but skipping the operator $+$. Let $\sim_k$ be the equivalence on $\mathbf{V}$ defined by $(n,m) \sim_k (n',m')$ iff $n = n' \wedge (m = m' \vee m, m' \leq k)$. It is easy to check that all $\sim_k$ for $k \in \mathbb{N}$ are congruences on the reduced $\mathcal{L}$, and contained in $\sim$. Yet their least upper bound (in the lattice of equivalence relations on $\mathbf{V}$) is $\sim$, which is not a congruence itself. In particular, there is no largest congruence contained in $\sim$.

When dealing with languages $\mathcal{L}$ in which all operators and other constructs have a finite arity, so that each $E \in \mathbb{T}_{\mathcal{L}}$ contains only finitely many variables, there is no difference between a congruence and a 1-hole congruence, and thus $\sim_{\mathcal{L}}^{1c}$ is a congruence relation for any equivalence $\sim$. I will apply the theory of expressiveness presented in this paper also to languages like CCS that have operators (such as $\sum_{i\in I} E_i$) of infinite arity. However, in all such cases I'm currently aware of, the relevant choices of $\mathcal{L}$ and $\sim$ have the property that $\sim_{\mathcal{L}}^{1c}$ is in fact a congruence relation. As an example, consider weak bisimilarity [27]. This equivalence relation fails to be a congruence for $\sum$. However, the coarsest 1-hole congruence contained in this relation, often called *rooted* weak bisimilarity, happens to be a congruence. In fact, when congruence-closing weak bisimilarity w.r.t. the binary sum, the result [15] is also a congruence for the infinitary sum, as well as for all other operators of CCS [27].

Definition 15. Let $\mathcal{T}$ be a translation from $\mathcal{L}$ into $\mathcal{L}'$. A subset $\mathbf{W}$ of $\mathbf{V}'$ is *closed* under $\mathcal{T}(\mathcal{L})$ if $[\![\mathcal{T}(E)]\!](\eta) \in \mathbf{W}$ for any expression $E \in \mathbb{T}_{\mathcal{L}}$ and valuation $\eta : \mathcal{X} \to \mathbf{W}$. An equivalence $\sim$ on $\mathbf{W}$ is a *congruence* (respectively *1-hole congruence*) for $\mathcal{T}(\mathcal{L})$ on $\mathbf{W}$ if for any $E \in \mathbb{T}_{\mathcal{L}}$ and $\theta, \eta : \mathcal{X} \to \mathbf{W}$ with $\theta \sim \eta$ (respectively $\theta \sim^1 \eta$) one has $[\![\mathcal{T}(E)]\!]_{\mathcal{L}'}(\theta) \sim [\![\mathcal{T}(E)]\!]_{\mathcal{L}'}(\eta)$.

Proposition 3. Let $\mathcal{T}$ be a translation from $\mathcal{L}$ into $\mathcal{L}'$ that is correct w.r.t. a semantic translation $\mathbf{R} \subseteq \mathbf{V}' \times \mathbf{V}$. Let $\mathbf{R}(\mathbf{V}) := \{v' \in \mathbf{V}' \mid \exists v \in \mathbf{V}.\ v'\mathbf{R}v\}$. Then $\mathbf{R}(\mathbf{V})$ is closed under $\mathcal{T}(\mathcal{L})$.

Proof: Let $E \in \mathbb{T}_{\mathcal{L}}$ and $\eta : \mathcal{X} \to \mathbf{R}(\mathbf{V})$. Take $\rho : \mathcal{X} \to \mathbf{V}$ with $\rho\mathbf{R}\eta$. Then $[\![\mathcal{T}(E)]\!]_{\mathcal{L}'}(\eta)\mathbf{R}[\![E]\!]_{\mathcal{L}}(\rho)$. Since $[\![E]\!]_{\mathcal{L}}(\rho) \in \mathbf{V}$ one has $[\![\mathcal{T}(E)]\!]_{\mathcal{L}'}(\eta) \in \mathbf{R}(\mathbf{V})$. □

Proposition 4. Let the translation $\mathcal{T}$ from $\mathcal{L}$ into $\mathcal{L}'$ be correct w.r.t. the semantic translation $\mathbf{R} \subseteq\, \sim$. Then $\sim$ is a (1-hole) congruence for $\mathcal{L}$ iff it is a (1-hole) congruence for $\mathcal{T}(\mathcal{L})$ on $\mathbf{R}(\mathbf{V})$.

Proof: First suppose $\sim$ is a congruence for $\mathcal{L}$. Let $E \in \mathbb{T}_{\mathcal{L}}$ and $\theta, \eta : \mathcal{X} \to \mathbf{R}(\mathbf{V})$ with $\theta \sim \eta$. By the definition of $\mathbf{R}(\mathbf{V})$ there are valuations $\nu, \rho : \mathcal{X} \to \mathbf{V}$ with $\theta\ \mathbf{R}\ \nu$ and $\eta\ \mathbf{R}\ \rho$. Now $\nu \sim \theta \sim \eta \sim \rho$, so

$$[\![\mathcal{T}(E)]\!]_{\mathcal{L}'}(\theta)\mathbf{R}[\![E]\!]_{\mathcal{L}}(\nu) \sim [\![E]\!]_{\mathcal{L}}(\rho)\mathbf{R}^{-1}[\![\mathcal{T}(E)]\!]_{\mathcal{L}'}(\eta)$$

and hence $[\![\mathcal{T}(E)]\!]_{\mathcal{L}'}(\theta) \sim [\![\mathcal{T}(E)]\!]_{\mathcal{L}'}(\eta)$. The other direction proceeds in the same way.

Now suppose $\sim$ is a 1-hole congruence for $\mathcal{L}$. Let $E \in \mathbb{T}_{\mathcal{L}}$ and $\theta, \eta : \mathcal{X} \to \mathbf{R}(\mathbf{V})$ with $\theta \sim^1 \eta$. Then $\theta(X) \sim \eta(X)$ for some $X \in \mathcal{X}$ and $\theta(Y) = \eta(Y)$ for all $Y \neq X$. So there must be $\nu, \rho : \mathcal{X} \to \mathbf{V}$ with $\theta\ \mathbf{R}\ \nu$, $\eta\ \mathbf{R}\ \rho$ and $\nu(Y) = \rho(Y)$ for all $Y \neq X$. Since $\nu(X) \sim \theta(X) \sim \eta(X) \sim \rho(X)$ it follows that $\nu \sim^1 \rho$. The conclusion proceeds as above, and the other direction goes likewise. □

The requirement of being a congruence for $\mathcal{T}(\mathcal{L})$ on $\mathbf{R}(\mathbf{V})$ is slightly weaker than that of being a congruence for $\mathcal{T}(\mathcal{L})$—cf. Definition 8—for it proceeds by restricting attention to valuations into $\mathbf{R}(\mathbf{V}) \subseteq \mathbf{U}$. ¶

8 A Congruence Closure Property for Valid Translations

In many applications, semantic values in the domain of interpretation of a language $\mathcal{L}$ are only meaningful up to a semantic equivalence $\sim^c$, and the intended semantic domain could just as well be seen as the set of $\sim^c$-equivalence classes of values. For this purpose it is essential that $\sim^c$ is a congruence for $\mathcal{L}$. Often $\sim^c$ is the congruence closure of a coarser semantic equivalence $\sim$, so that two values end up being identified iff they are $\sim$-equivalent in every context. An example of this occurred in Sect. 6, with $\stackrel{\bullet}{\approx}$ in the rôle of $\sim$ and $\cong^c$ in the rôle of $\sim^c$. Now Theorem 4, contributed in this section, says that if a translation from $\mathcal{L}$ into $\mathcal{L}'$ is valid up to $\sim$, then it is even valid up to an equivalence $\sim^{1c}_{\mathcal{L},\mathbf{R}}$ that extends $\sim^c$ from $\mathbf{V}$ to a subdomain $\mathbf{W}$ of $\mathbf{V}'$ that suffices for the interpretation of translated expressions from $\mathcal{L}$. This equivalence $\sim^{1c}_{\mathcal{L},\mathbf{R}}$ coincides with the congruence closure of $\sim$ on $\mathcal{L}$, as well as on $\mathcal{T}(\mathcal{L})$, and melts each equivalence class of $\mathbf{V}$ with exactly one of $\mathbf{W}$, and vice versa.

Let $\mathcal{L}$ and $\mathcal{L}'$ be languages with $[\![\ ]\!]_{\mathcal{L}} : \mathbb{T}_{\mathcal{L}} \rightarrow ((\mathcal{X} \rightarrow \mathbf{V}) \rightarrow \mathbf{V})$ and $[\![\ ]\!]_{\mathcal{L}'} : \mathbb{T}_{\mathcal{L}'} \rightarrow ((\mathcal{X} \rightarrow \mathbf{V}') \rightarrow \mathbf{V}')$. In this section I assume that $\mathbf{V} \cap \mathbf{V}' = \emptyset$. To apply the results to the general case, just adapt $\mathcal{L}'$ by using a copy of $\mathbf{V}'$—any preorder $\sim$ on $\mathbf{V} \cup \mathbf{V}'$ extends to this copy by considering each copied element $\sim$-equivalent to the original.

Definition 16. Given any semantic translation $\mathbf{R}$, let $\equiv_{\mathbf{R}} \subseteq (\mathbf{V} \cup \mathbf{V}')^2$ be the smallest equivalence relation on $\mathbf{V} \cup \mathbf{V}'$ containing $\mathbf{R}$.

Theorem 3. If a translation $\mathcal{T}$ is correct w.r.t. the semantic translation $\mathbf{R}$, then $\equiv_{\mathbf{R}}$ is a 1-hole congruence for $\mathcal{L}$. ¶

By Proposition 4 $\equiv_{\mathbf{R}}$ also is a 1-hole congruence for $\mathcal{T}(\mathcal{L})$ on $\mathbf{R}(\mathbf{V})$. Only the subset $\mathbf{R}(\mathbf{V})$ of $\mathbf{V}'$ matters for the purpose of translating $\mathcal{L}$ into $\mathcal{L}'$. On $\mathbf{V}' \backslash \mathbf{R}(\mathbf{V})$ the equivalence $\equiv_{\mathbf{R}}$ is the identity.

Theorem 4. Let $\mathcal{T}$ be a translation from a language $\mathcal{L}$, with semantic domain $\mathbf{V}$, into a language $\mathcal{L}'$, with domain $\mathbf{V}'$, that is valid up to a semantic equivalence $\sim$. Then $\mathcal{T}$ is even valid up to a semantic equivalence $\sim^{1c}_{\mathcal{L},\mathbf{R}}$, contained in $\sim$, such that (1) the restriction of $\sim^{1c}_{\mathcal{L},\mathbf{R}}$ to $\mathbf{V}$ is the largest 1-hole congruence for $\mathcal{L}$ contained in $\sim$, (2) the set $\mathbf{W} := \{v \in \mathbf{V}' \mid \exists v \in \mathbf{V}.\ v' \sim^{1c}_{\mathcal{L},\mathbf{R}} v\}$ is closed under $\mathcal{T}(\mathcal{L})$, and (3) the restriction of $\sim^{1c}_{\mathcal{L},\mathbf{R}}$ to $\mathbf{W}$ is the largest 1-hole congruence for $\mathcal{T}(\mathcal{L})$ on $\mathbf{W}$ that is contained in $\sim$. ¶

Note that each equivalence class of $\sim^{1c}_{\mathcal{L},\mathbf{R}}$ on $\mathbf{V} \cup \mathbf{W}$ melts an equivalence class of $\sim^{1c}_{\mathcal{L},\mathbf{R}}$ on $\mathbf{V}$ with one of $\sim^{1c}_{\mathcal{L},\mathbf{R}}$ on $\mathbf{W}$. Moreover, on $\mathbf{V}$ the relation is completely determined by $\mathcal{L}$ and $\sim$. However, in general the whole relation $\sim^{1c}_{\mathcal{L},\mathbf{R}}$ is not completely determined by $\mathcal{L}$ and $\sim$. ¶

Corollary 1. Let $\mathcal{T}$ be a translation from a language $\mathcal{L}$, with semantic domain $\mathbf{V}$, into a language $\mathcal{L}'$, with domain $\mathbf{V}'$, valid up to a semantic equivalence $\sim$, and suppose the congruence closure $\sim^{1}_{\mathcal{L}}$ of $\sim$ w.r.t. $\mathcal{L}$ is in fact a congruence. Then $\mathcal{T}$ is correct up to the equivalence $\sim^{1c}_{\mathcal{L},\mathbf{R}}$ described in Theorem 4. ¶

The languages π and $\mathrm{a}\pi$ of Sect. 6 do not feature operators (or other constructs) of infinite arity. Hence the congruence closure $\sim^{1c}_{\pi}$ or $\sim^{1c}_{\mathrm{a}\pi}$ of an equivalence $\sim$ on π or $\mathrm{a}\pi$ is always a congruence. So by Corollary 1 Boudol's translation $\mathcal{T}$ is correct up to an equivalence $\dot{\approx}^{c}_{\pi,\mathbf{R}}$, defined on the disjoint union of the domains T_{π} and $\mathrm{T}_{\mathrm{a}\pi}$ on which the two languages are interpreted. This equivalence is contained in $\dot{\approx}$, and on the source domain T_{π} coincides with $\cong^{c}$. By Theorem 4, the restriction of $\dot{\approx}^{c}_{\pi,\mathbf{R}}$ to a subdomain $\mathbf{W} \subseteq \mathrm{T}_{\mathrm{a}\pi}$ is the largest congruence for $\mathcal{T}(\pi)$ on $\mathbf{W}$ that is contained in $\sim$. As $\cong^{c}_{\mathrm{a}}$ is a congruence for all of $\mathrm{a}\pi$ on all of $\mathrm{T}_{\mathrm{a}\pi}$, and contained in $\dot{\approx}$, it is certainly a congruence for $\mathcal{T}(\pi)$ on $\mathbf{W}$, and thus contained in $\dot{\approx}^{c}_{\pi,\mathbf{R}}$. This inclusion turns out to be strict. As an illustration of that, note that $\bar{x}z.\mathbf{0}|\bar{x}z.\mathbf{0} \cong^{c} \bar{x}z.\bar{x}z\mathbf{0}$. (This follows since these processes are strong (early) bisimilar [44] and thus strong full bisimilar by [44, Definition 2.2.2].) Consequently, their translations must be related by $\stackrel{\bullet}{\approx}{}^c_{\pi,\mathbf{R}}$. So, for distinct $u, v, y, w, x, z \in \mathcal{N}$,

$$(u)(\bar{x}u|u(v).(\bar{v}z|\mathbf{0}))\big|(u)(\bar{x}u|u(v).(\bar{v}z|\mathbf{0})) \stackrel{\bullet}{\approx}{}^c_{\pi,\mathbf{R}} (y)(\bar{x}y|u(w).(\bar{w}z|(u)(\bar{x}u|u(v).(\bar{v}z|\mathbf{0})))).$$

Yet, these processes are not $\cong^c_{\mathrm{a}}$-equivalent, as can be seen by putting them in a context $x(y).x(y).\bar{r}(s)|X$. There, only the left-hand side has a weak barb $\Downarrow_{\bar{r}}$.

9 Integrating Language Features Through Translations

The results of the previous section show how valid translations are satisfactory for comparing the expressiveness of languages. If there is a valid translation $\mathcal{T}$ from $\mathcal{L}$ to $\mathcal{L}'$ up to $\sim$, and (as usual) $\sim^{1c}_{\mathcal{L}}$ is a congruence, then all truths that can be expressed in terms of $\mathcal{L}$ can be mimicked in $\mathcal{L}'$. For the congruence classes of $\sim^{1c}_{\mathcal{L}}$ translate bijectively to congruence classes of an induced equivalence relation on the domain of $\mathcal{T}(\mathcal{L})$ (within the domain of $\mathcal{L}'$), and all operations on those congruence classes that can be performed by contexts of $\mathcal{L}$ have a perfect counterpart in terms of contexts of $\mathcal{T}(\mathcal{L})$. This state of affairs was illustrated on Boudol's translation from a synchronous to an asynchronous π-calculus.

There is however one desirable property of translations between languages that has not yet been achieved, namely to combine the powers of two languages into one unified language. If both languages $\mathcal{L}_1$ and $\mathcal{L}_2$ have valid translations into a language $\mathcal{L}'$, then all that can be done with $\mathcal{L}_1$ can be mimicked in a fragment of $\mathcal{L}'$, and all that can be done with $\mathcal{L}_2$ can be mimicked in another fragment of $\mathcal{L}'$. In order for these two fragments to combine, one would like to employ a single congruence relation on $\mathcal{L}'$ that specialises to congruence relations for $\mathcal{T}_1(\mathcal{L}_1)$ and $\mathcal{T}_2(\mathcal{L}_2)$, which form the counterparts of relevant congruence relations for the source languages $\mathcal{L}_1$ and $\mathcal{L}_2$.

In terms of the translation $\mathcal{T}$ from π to aπ, the equivalence $\cong^c_{\mathrm{a}}$ on $\mathrm{T}_{\mathrm{a}\pi}$ would be the right congruence relation to consider for aπ. Ideally, this congruence would extend to an equivalence $\cong^c_{\pi,\mathrm{a}\pi}$ on the disjoint union $\mathrm{T}_\pi \uplus \mathrm{T}_{\mathrm{a}\pi}$, such that the restriction of $\cong^c_{\pi,\mathrm{a}\pi}$ to T_π is a congruence for π. Necessarily, this congruence on T_π would have to distinguish the terms $\bar{x}z.\mathbf{0}|\bar{x}z.\mathbf{0}$ and $\bar{x}z.\bar{x}z\mathbf{0}$, since their translations are distinguished by $\cong^c_{\mathrm{a}}$. One therefore expects $\cong^c_{\pi,\mathrm{a}\pi}$ on T_π to be strictly finer than $\cong^c$. Here it is important that the union of T_π and $\mathrm{T}_{a\pi}$ on which this congruence is defined is required to be disjoint. For if one considers $\mathrm{T}_{a\pi}$ as a subset of T_π, then we obtain that the restriction of $\cong^c_{\pi,\mathrm{a}\pi}$ to that subset (1) coincides with $\cong^c_{\mathrm{a}}$ and (2) is strictly finer than $\cong^c$. This contradicts the fact that $\cong^c$ is strictly finer than $\cong^c_{\mathrm{a}}$.

In Sect. 12 I will show that such a congruence $\cong^c_{\pi,\mathrm{a}\pi}$ indeed exists. In fact, under a few very mild conditions this result holds generally, provided that the source language $\mathcal{L}$ is a closed-term language. ¶

10 A Unique Decomposition of Terms

The results of Sect. 12 apply only to languages satisfying two postulates, formulated below, and to preorders $\sim$ that "respect $\stackrel{\alpha}{=}$", defined in Sect. 11.

Definition 17. *α-conversion* is the act of renaming all occurrences of a bound variable X within the scope of its binding into another variable, say Y, while avoiding capture of free variables. Here one speaks of *capture* when a free occurrence of Y turns into a bound one.

Write $E \stackrel{\alpha}{=} F$ if expression E can be converted into F by acts of α-conversion.

In languages where there are multiple types of bound variables, $\stackrel{\alpha}{=}$ allows conversion of all of them. In a π-calculus with recursion, for instance, there could be bound process variables $X \in \mathcal{X}$ as well as bound names $x \in \mathcal{N}$. The last two conversions in the right column of Definition 10 define α-conversion for names.

Postulate 1 ([16], paraphrased). There exists a class of expressions called *standard heads*, and a class of substitutions called *standard substitutions*, such that for each expression E, if not a variable, there are unique standard heads H and substitutions σ such that $E \stackrel{\alpha}{=} H[\sigma]$.

A term $f(c, g(c))$, for instance, can be written as $H[\sigma]$ where $H = f(X_1, X_2)$ is a head, and $\sigma : \{X_1, X_2\} \to \mathbb{T}_{\mathcal{L}}$ is given by $\sigma(X_1) = c$ and $\sigma(X_2) = g(c)$. The head H is standardised by means of a particular (arbitrary) choice for its argument variables X_1 and X_2. σ is standardised through a particular choice of the bound variables that may occur in the expressions $\sigma(X)$. A head for a recursive expression $\mu X.f(g(c), g(g(X)))$ is $\mu X.f(Y, g(g(X)))$. See [16] for further detail.

This postulate is easy to show for each common type of system description language, and I am not aware of any counterexamples. However, while striving for maximal generality, I consider languages with (recursion-like) constructs that are yet to be invented, and in view of those, this principle has to be postulated rather than derived.

11 Invariance of Meaning Under α-conversion

Write $v \stackrel{\alpha}{=}_{\mathcal{L}} w$, with $v, w \in \mathbf{V}$, iff there are terms $E, F \in \mathbb{T}_{\mathcal{L}}$ with $E \stackrel{\alpha}{=} F$, and a valuation $\zeta : \mathcal{X} \to \mathbf{V}$ such that $[\![E]\!]_{\mathcal{L}}(\zeta) = v$ and $[\![F]\!]_{\mathcal{L}}(\zeta) = w$. This relation is reflexive and symmetric.

In [16] I limited attention to languages satisfying

$$\text{if } E \stackrel{\alpha}{=} F \text{ then } [\![E]\!]_{\mathcal{L}} = [\![F]\!]_{\mathcal{L}}. \tag{2}$$

This postulate says that the meaning of an expression is invariant under α-conversion. It can be reformulated as the requirement that $\stackrel{\alpha}{=}_{\mathcal{L}}$ is the identity relation. This postulate is satisfied by all my intended applications, except for the important class of closed-term languages. Languages like CCS and the π-calculus can be regarded as falling in this class (although it is also possible to declare the meaning of a term under a valuation to be an $\stackrel{\alpha}{=}$-equivalence class of closed terms). To bring this type of application within the scope of my theory, here I weaken this postulate by requiring merely that $\stackrel{\alpha}{=}_{\mathcal{L}}$ is an equivalence.

Postulate 2. $\stackrel{\alpha}{=}_{\mathcal{L}}$ is an equivalence relation.

This postulate is needed in Sect. 12. I also need to restrict attention to preorders $\sim$ with $\stackrel{\alpha}{=}_{\mathcal{L}} \subseteq \sim$. When that holds I say that the preorder $\sim$ *respects* $\stackrel{\alpha}{=}_{\mathcal{L}}$. If (2) holds—which strengthens of Postulate 2—then *any* preorder respects $\stackrel{\alpha}{=}_{\mathcal{L}}$.

12 Compositionality

An important property of translations, defined below, is *compositionality*. In this section show I that any valid translation up to a preorder $\sim$ can be modified into such a translation that moreover is compositional, provided one restricts attention to languages that satisfy Postulates 1 and 2, and preorders $\sim$ that respect $\stackrel{\alpha}{=}$.

Definition 18. A translation $\mathcal{T}$ from $\mathcal{L}$ into $\mathcal{L}'$ is *compositional* if

(1) $\mathcal{T}(E[\sigma]) \stackrel{\alpha}{=} \mathcal{T}(E)[\mathcal{T} \circ \sigma]$ for each $E \in \mathbb{T}_{\mathcal{L}}$ and $\sigma : fv(E) \rightarrow \mathbb{T}_{\mathcal{L}}$,
(2) $E \stackrel{\alpha}{=} F$ implies $\mathcal{T}(E) \stackrel{\alpha}{=} \mathcal{T}(F)$ for all $E, F \in \mathbb{T}_{\mathcal{L}}$,
(3) and moreover $\mathcal{T}(X) = X$ for each $X \in \mathcal{X}$.

In case $E = f(t_1, \ldots, t_n)$ for certain $t_i \in \mathbb{T}_{\mathcal{L}}$ this amounts to $\mathcal{T}(f(t_1, \ldots, t_n)) \stackrel{\alpha}{=} E_f(\mathcal{T}(t_1), \ldots, \mathcal{T}(t_n))$, where $E_f := \mathcal{T}(f(X_1, \ldots, X_n))$ and $E_f(u_1, \ldots, u_n)$ denotes the result of the simultaneous substitution in this expression of the terms $u_i \in \mathbb{T}_{\mathcal{L}'}$ for the free variables X_i, for $i = 1, \ldots, n$. The first requirement of Definition 18 is more general and covers language constructs other than functions, such as recursion. Requiring equality rather than $\stackrel{\alpha}{=}$ is too demanding. ¶

Lemma 1. If $\mathcal{T}_1 : \mathbb{T}_{\mathcal{L}_1} \rightarrow \mathbb{T}_{\mathcal{L}_2}$ and $\mathcal{T}_2 : \mathbb{T}_{\mathcal{L}_2} \rightarrow \mathbb{T}_{\mathcal{L}_3}$ are compositional translations, then so is their composition $\mathcal{T}_2 \circ \mathcal{T}_1 : \mathbb{T}_{\mathcal{L}_1} \rightarrow \mathbb{T}_{\mathcal{L}_3}$, defined by $\mathcal{T}_2 \circ \mathcal{T}_1(E) := \mathcal{T}_2(\mathcal{T}_1(E))$ for all $E \in \mathcal{L}_1$.

Proof: (1) $\mathcal{T}_2(\mathcal{T}_1(E[\sigma])) \stackrel{\alpha}{=} \mathcal{T}_2(\mathcal{T}_1(E)[\mathcal{T}_1 \circ \sigma]) \stackrel{\alpha}{=} \mathcal{T}_2(\mathcal{T}_1(E))[\mathcal{T}_2 \circ \mathcal{T}_1 \circ \sigma])$ for each $\sigma : \mathcal{X} \rightharpoonup \mathbb{T}_{\mathcal{L}_1}$ and $E \in \mathbb{T}_{\mathcal{L}_1}$. Here the derivation of the first $\stackrel{\alpha}{=}$ uses Property (2) of Definition 18—and this is the reason for requiring that property.

(2) $E \stackrel{\alpha}{=} F$ implies $\mathcal{T}_1(E) \stackrel{\alpha}{=} \mathcal{T}_1(F)$ and $\mathcal{T}_2(\mathcal{T}_1(E)) \stackrel{\alpha}{=} \mathcal{T}_2(\mathcal{T}_1(F))$ for all $E, F \in \mathbb{T}_{\mathcal{L}}$.

(3) $\mathcal{T}_2(\mathcal{T}_1(X)) = \mathcal{T}_2(X) = X$ for each $X \in \mathcal{X}$. □

Theorem 5. Let $\mathcal{L}$ and $\mathcal{L}'$ be languages that satisfy Postulates 1 and 2, and $\sim$ a preorder that respects $\stackrel{\alpha}{=}_{\mathcal{L}}$ and $\stackrel{\alpha}{=}_{\mathcal{L}'}$. If any valid (or correct) translation from $\mathcal{L}$ into $\mathcal{L}'$ up to $\sim$ exists, then there exists a compositional translation that is valid (or correct) up to $\sim$. ¶

Hence, for the purpose of comparing the expressive power of languages, valid translations between them can be assumed to be compositional. For correct translations this was already established in [16], but assuming (2), a stronger version of Postulate 2.

I can now establish the theorem promised in Sect. 9. In view of Theorem 5, no great sacrifices are made by assuming that the translation $\mathcal{T}$ is compositional. Other "mild conditions" needed are Postulate 2 for $\mathcal{L}'$ and $\approx$ respecting $\stackrel{\alpha}{=}_{\mathcal{L}'}$.

Theorem 6. Let $\mathcal{L}$ be a closed-term language and $\mathcal{L}'$ a language that satisfies Postulate 2. Let $\mathcal{T}$ be a compositional translation from $\mathcal{L}$ into $\mathcal{L}'$ that is valid up to $\sim$. Let $\approx$ be any congruence for $\mathcal{L}'$ containing $\stackrel{\alpha}{=}_{\mathcal{L}'}$ and contained in $\sim$. Then $\mathcal{T}$ is correct up to an equivalence $\approx_{\mathcal{T}}$ on $\mathbf{V} \cup \mathbf{V}'$, contained in $\sim$, that on $\mathbf{V}'$ coincides with $\approx$. ¶

13 Related Work

The concept of *full abstraction* stems from Milner [26]. It indicates a satisfactory connection between a denotational and an operational semantics of a language. Riecke [42] and Shapiro [45] adapt this notion to translations between languages.

Definition 19. A translation $\mathcal{T}: \mathbb{T}_{\mathcal{L}_S} \rightarrow \mathbb{T}_{\mathcal{L}_T}$ is *fully abstract* w.r.t. the equivalences $\sim_S \subseteq \mathrm{T}^2_{\mathcal{L}_S}$ and $\sim_T \subseteq \mathrm{T}^2_{\mathcal{L}_T}$ if, for all $P, Q \in \mathrm{T}_{\mathcal{L}_S}$, $P \sim_S Q \Leftrightarrow \mathcal{T}(P) \sim_T \mathcal{T}(Q)$.

In [42,45], $\sim_S$ and $\sim_T$ are required to be congruence closures—see [18] for more detail. The simplified definition above was used in [1,30,31]. Fu [10] bases a theory of expressiveness on full abstraction, with a divergence-preserving form of barbed branching bisimilarity [19] in the rôle of $\sim_S$ and $\sim_T$. A comparison of full abstraction with the approach of the present paper appears in [18].

In the last twenty years, a great number of encodability and separation results have appeared, comparing CCS, Mobile Ambients, and several versions of the π-calculus (with and without recursion; with mixed choice, separated choice or asynchronous) [1,2,5–8,11–13,23,30–34,38–41,43,46]; see [20,21] for an overview. Many of these results employ different and somewhat ad-hoc criteria on what constitutes a valid encoding, and thus are hard to compare with each other. Several of these criteria are discussed and compared in [35,36]. Gorla [21] collected some essential features of these approaches and integrated them in a proposal for a valid encoding that justifies most encodings and some separation results from the literature.

Like Boudol [3] and the present paper, Gorla requires a compositionality condition for encodings. However, his criterion is weaker than mine (cf. Definition 18) in that the expression E_f encoding an operator f may be dependent on the set of names occurring freely in the expressions given as arguments of f. This issue is further discussed in [16]. It is an interesting topic for future research to see if there are any valid encodability results à la [21] that suffer from my proposed strengthening of compositionality.

The second criterion of [21] is a form of invariance under name-substitution. It serves to partially undo the effect of making the compositionality requirement name-dependent. In my setting I have not yet found the need for such a condition. In [16] I argue that this criterion as formalised in [21] is too restrictive.

The remaining three requirements of Gorla (the 'semantic' requirements) are very close to an instantiation of mine with a particular preorder $\mathrel{\bullet\!\!\sim}$. If one takes $\mathrel{\bullet\!\!\sim}$ to be weak barbed bisimilarity with explicit divergence (i.e. relating divergent states with divergent states only), using barbs external to the language, as discussed in Sect. 6, then an valid translation in my sense satisfies Gorla's semantic criteria, provided that the equivalence $\equiv$ on the target language that acts as a parameter in Gorla's third criterion is also taken to be weak barbed bisimilarity with explicit divergence. The precise relationships between the proposals of [16,21] are further discussed in [37].

Further work is needed to sort out to what extent the two approaches have relevant differences when evaluating encoding and separation results from the literature. Another topic for future work is to sort out how dependent known encoding and separation results are on the chosen equivalence or preorder.

References

1. Baldamus, M., Parrow, J., Victor, B.: A fully abstract encoding of the *pi*-calculus with data terms. In: Caires, L., Italiano, G.F., Monteiro, L., Palamidessi, C., Yung, M. (eds.) ICALP 2005. LNCS, vol. 3580, pp. 1202–1213. Springer, Heidelberg (2005). https://doi.org/10.1007/11523468_97
2. Boreale, M.: On the expressiveness of internal mobility in name-passing calculi. TCS **195**(2), 205–226 (1998). https://doi.org/10.1016/S0304-3975(97)00220-X
3. Boudol, G.: Notes on algebraic calculi of processes. In: Apt, K. (ed.) Logics and Models of Concurrent Systems. NATO ASI Series, vol. 13, pp. 261–303. Springer, Heidelberg (1985). https://doi.org/10.1007/978-3-642-82453-1_9
4. Boudol, G.: Asynchrony and the π-calculus (Note). Technical report 1702, INRIA (1992)
5. Busi, N., Gabbrielli, M., Zavattaro, G.: On the expressive power of recursion, replication and iteration in process calculi. Math. Struct. Comput. Sci. **19**(6), 1191–1222 (2009). https://doi.org/10.1017/S096012950999017X
6. Cacciagrano, D., Corradini, F., Aranda, J., Valencia, F.D.: Linearity, persistence and testing semantics in the asynchronous pi-calculus. ENTCS **194**(2), 59–84 (2008). https://doi.org/10.1016/j.entcs.2007.11.006
7. Carbone, M., Maffeis, S.: On the expressive power of polyadic synchronisation in pi-calculus. Nord. J. Comput. **10**(2), 70–98 (2003)
8. Cardelli, L., Gordon, A.D.: Mobile ambients. TCS **240**(1), 177–213 (2000). https://doi.org/10.1016/S0304-3975(99)00231-5
9. De Nicola, R., Hennessy, M.: Testing equivalences for processes. TCS **34**, 83–133 (1984). https://doi.org/10.1016/0304-3975(84)90113-0
10. Fu, Y.: Theory of interaction. Theor. Comput. Sci. **611**, 1–49 (2016). https://doi.org/10.1016/j.tcs.2015.07.043
11. Given-Wilson, T.: Expressiveness via intensionality and concurrency. In: Ciobanu, G., Méry, D. (eds.) ICTAC 2014. LNCS, vol. 8687, pp. 206–223. Springer, Cham (2014). https://doi.org/10.1007/978-3-319-10882-7_13

12. Given-Wilson, T.: On the expressiveness of intensional communication. In: Proceedings of EXPRESS/SOS 2014. EPTCS, vol. 160, pp. 30–46 (2014). https://doi.org/10.4204/EPTCS.160.4
13. Given-Wilson, T., Legay, A.: On the expressiveness of joining. In: Proceedings of ICE 2015. EPTCS, vol. 189, pp. 99–113 (2015). https://doi.org/10.4204/EPTCS.189.9
14. van Glabbeek, R.J.: On the expressiveness of ACP (extended abstract). In: Ponse, A., Verhoef, C., van Vlijmen, S.F.M. (eds.) ACP 1994. Workshops in Computing, pp. 188–217. Springer, London (1994). https://doi.org/10.1007/978-1-4471-2120-6_8. http://theory.stanford.edu/ rvg/abstracts.html#31
15. Glabbeek, R.J.: A characterisation of weak bisimulation congruence. In: Middeldorp, A., van Oostrom, V., van Raamsdonk, F., de Vrijer, R. (eds.) Processes, Terms and Cycles: Steps on the Road to Infinity. LNCS, vol. 3838, pp. 26–39. Springer, Heidelberg (2005). https://doi.org/10.1007/11601548_4
16. van Glabbeek, R.J.: Musings on encodings and expressiveness. In: Proceedings of EXPRESS/SOS 2012. EPTCS, vol. 89, pp. 81–98 (2012). https://doi.org/10.4204/EPTCS.89.7
17. van Glabbeek, R.J.: Lean and full congruence formats for recursion. In: Proceedings of LICS 2017 (2017). https://doi.org/10.1109/LICS.2017.8005142
18. van Glabbeek, R.J.: A theory of encodings and expressiveness (2018). http://theory.stanford.edu/~rvg/abstracts.html#tra. Full version of current paper
19. van Glabbeek, R.J., Luttik, B., Trčka, N.: Branching bisimilarity with explicit divergence. Fund. Inform. **93**(4), 371–392 (2009). https://doi.org/10.3233/FI-2009-109
20. Gorla, D.: A taxonomy of process calculi for distribution and mobility. Distrib. Comput. **23**(4), 273–299 (2010). https://doi.org/10.1007/s00446-010-0120-6
21. Gorla, D.: Towards a unified approach to encodability and separation results for process calculi. I&C **208**(9), 1031–1053 (2010). https://doi.org/10.1016/j.ic.2010.05.002
22. Grätzer, G.: Lattice theory: foundation (2010). https://doi.org/10.1007/978-3-0348-0018-1
23. Hatzel, M., Wagner, C., Peters, K., Nestmann, U.: Encoding CSP into CCS. In: Proceedings of EXPRESS/SOS 2015. EPTCS, vol. 190, pp. 61–75 (2015). https://doi.org/10.4204/EPTCS.190.5
24. Honda, K., Tokoro, M.: An object calculus for asynchronous communication. In: America, P. (ed.) ECOOP 1991. LNCS, vol. 512, pp. 133–147. Springer, Heidelberg (1991). https://doi.org/10.1007/BFb0057019
25. Lippert, C., Mennicke, S., van Glabbeek, R.J., Goltz, U.: A case study on evaluating encodings between process calculi. Technical report (2018). http://theory.stanford.edu/~rvg/abstracts.html#129
26. Milner, R.: Processes: a mathematical model for computing agents. Stud. Log. Found. Math. **80**, 157–173 (1975). https://doi.org/10.1016/S0049-237X(08)71948-7. Logic Colloquium 1973
27. Milner, R.: Operational and algebraic semantics of concurrent processes. In: Handbook of Theoretical Computer Science, Chap. 19, pp. 1201–1242 (1990)
28. Milner, R.: Functions as processes. Math. Struct. Comput. Sci. **2**(2), 119–141 (1992). https://doi.org/10.1017/S0960129500001407
29. Milner, R., Sangiorgi, D.: Barbed bisimulation. In: Kuich, W. (ed.) ICALP 1992. LNCS, vol. 623, pp. 685–695. Springer, Heidelberg (1992). https://doi.org/10.1007/3-540-55719-9_114

30. Nestmann, U.: What is a "Good" encoding of guarded choice? I&C **156**(1–2), 287–319 (2000). https://doi.org/10.1006/inco.1999.2822
31. Nestmann, U., Pierce, B.C.: Decoding choice encodings. I&C **163**(1), 1–59 (2000). https://doi.org/10.1006/inco.2000.2868. An earlier version appeared in Proc. CONCUR '96
32. Palamidessi, C.: Comparing the expressive power of the synchronous and asynchronous pi-calculi. Math. Struct. Comput. Sci. **13**(5), 685–719 (2003). https://doi.org/10.1017/S0960129503004043
33. Palamidessi, C., Saraswat, V.A., Valencia, F.D., Victor, B.: On the expressiveness of linearity vs persistence in the asychronous pi-calculus. In: Proceedings of LICS 2006, pp. 59–68. IEEE Computer Society Press (2006). https://doi.org/10.1109/LICS.2006.39
34. Parrow, J.: Trios in concert. In: Proof, Language, and Interaction, Essays in Honour of Robin Milner, pp. 623–638. The MIT Press (2000)
35. Parrow, J.: Expressiveness of process algebras. In: ENTCS, vol. 209, pp. 173–186 (2008). https://doi.org/10.1016/j.entcs.2008.04.011
36. Peters, K.: Translational expressiveness. Comparing process calculi using encodings. Ph.D. thesis, TU Berlin (2012). https://doi.org/10.14279/depositonce-3416
37. Peters, K., van Glabbeek, R.J.: Analysing and comparing encodability criteria. In: EXPRESS/SOS 2015. EPTCS, vol. 190, pp. 46–60 (2015). https://doi.org/10.4204/EPTCS.190.4
38. Peters, K., Nestmann, U.: Is it a "Good" encoding of mixed choice? In: Birkedal, L. (ed.) FoSSaCS 2012, vol. 7213, pp. 210–224. Springer, Heidelberg (2012). https://doi.org/10.1007/978-3-642-28729-9_14
39. Peters, K., Nestmann, U.: Breaking symmetries. Math. Struct. Comput. Sci. **26**(6), 1054–1106 (2016). https://doi.org/10.1017/S0960129514000346
40. Peters, K., Nestmann, U., Goltz, U.: On distributability in process calculi. In: Felleisen, M., Gardner, P. (eds.) ESOP 2013. LNCS, vol. 7792, pp. 310–329. Springer, Heidelberg (2013). https://doi.org/10.1007/978-3-642-37036-6_18
41. Phillips, I., Vigliotti, M.G.: Leader election in rings of ambient processes. TCS **356**(3), 468–494 (2006). https://doi.org/10.1016/j.tcs.2006.02.004
42. Riecke, J.G.: Fully abstract translations between functional languages. In: Proceedings of POPL 1991, pp. 245–254. ACM Press (1991). https://doi.org/10.1145/99583.99617
43. Sangiorgi, D.: From π-calculus to higher-order π-calculus — and back. In: Gaudel, M.-C., Jouannaud, J.-P. (eds.) CAAP 1993. LNCS, vol. 668, pp. 151–166. Springer, Heidelberg (1993). https://doi.org/10.1007/3-540-56610-4_62
44. Sangiorgi, D., Walker, D.: The π-calculus: A Theory of Mobile Processes. Cambridge University Press, Cambridge (2001)
45. Shapiro, E.Y.: Separating concurrent languages with categories of language embeddings. In: STOC 1991, pp. 198–208. ACM (1991). https://doi.org/10.1145/103418.103423
46. Vigliotti, M.G., Phillips, I., Palamidessi, C.: Tutorial on separation results in process calculi via leader election problems. TCS **388**(1–3), 267–289 (2007). https://doi.org/10.1016/j.tcs.2007.09.001

A Framework for Parameterized Monitorability

Luca Aceto[1,2], Antonis Achilleos[2(✉)], Adrian Francalanza[3], and Anna Ingólfsdóttir[2]

[1] Gran Sasso Science Institute, L'Aquila, Italy
[2] School of Computer Science, Reykjavik University, Reykjavik, Iceland
{luca,antonios,annai}@ru.is
[3] Department of Computer Science, ICT, University of Malta, Msida, Malta
adrian.francalanza@um.edu.mt

Abstract. We introduce a general framework for Runtime Verification, parameterized with respect to a set of conditions. These conditions are encoded in the trace generated by a monitored process, which a monitor can observe. We present this parameterized framework in its general form and prove that it corresponds to a fragment of HML with recursion, extended with these conditions. We then show how this framework can be applied to a number of instantiations of the set of conditions.

1 Introduction

Runtime Verification (RV) is a lightweight verification technique that checks whether a system satisfies a correctness property by analysing the *current execution* of the system [20,29], expressed as a trace of execution events. Using the additional information obtained at runtime, the technique can often mitigate state explosion problems typically associated with more traditional verification techniques. At the same time, limiting the verification analysis to the current execution trace hinders the expressiveness of RV when compared to more exhaustive approaches. In fact, there are correctness properties that cannot be satisfactorily verified at runtime (*e.g.* the finiteness of the trace considered up to the current execution point prohibits the verification of liveness properties). Because of this reason, RV is often used as part of a multi-pronged approach towards ensuring system correctness [5,6,8,14,15,25], *complementing* other verification techniques such as model checking, testing and type checking.

In order to attain an effective verification strategy consisting of multiple verification techniques that include RV, it is crucial to understand the expressive power of each technique: one can then determine how to best decompose the verification burden into subtasks that can then be assigned to the most appropriate verification technique. *Monitorability* concerns itself with identifying the

This research was supported by the project "TheoFoMon: Theoretical Foundations for Monitorability" (grant number: 163406-051) of the Icelandic Research Fund.

C. Baier and U. Dal Lago (Eds.): FOSSACS 2018, LNCS 10803, pp. 203–220, 2018.
https://doi.org/10.1007/978-3-319-89366-2_11

properties that are analysable by RV. In [21,22] (and subsequently in [2]), the problem of monitorability was studied for properties expressed in a variant of the modal μ-calculus [26] called μHML [28]. The choice of the logic was motivated by the fact that it can embed widely used logics such as CTL and LTL, and by the fact that it is agnostic of the underlying verification method used—this leads to better separation of concerns and guarantees a good level of generality for the results obtained. The main result in [2,21,22] is the identification of a monitorable syntactic subset of the logic μHML (*i.e.*, a set of logical formulas for which monitors carrying out the necessary runtime analysis exist) that is shown to be maximally expressive (*i.e.*, any property that is monitorable in the logic may be expressed in terms of this syntactic subset). We are unaware of other maximality results of this kind in the context of RV.

In this work we strive towards extending the monitorability limits identified in [2,21,22] for μHML. Particularly, for any logic or specification language, monitorability is a function of the underlying monitoring setup. In [2,21,22], the framework assumes a *classical* monitoring setup, whereby a (single) monitor incrementally analyses an ordered trace of events describing the computation steps that were executed by the system. A key observation made by this paper is that, in general, execution traces need *not* be limited to the reporting of events *that happened.* For instance, they may describe events that *could not have happened* at specific points in the execution of a system. Alternatively, they may also include descriptions for depth-bounded trees of computations that *were possible* at specific points in an execution. We conjecture that there are instances where this additional information can be feasibly encoded in a trace, either dynamically or by way of a pre-processing phase (based, *e.g.*, on the examination of logs of previous system executions, or on the full static checking of sub-components making up the system). More importantly, this additional information could, in principle, permit the verification of more properties at runtime.

The contribution of this paper is a study of how the aforementioned augmented monitoring setups may affect the monitorability of μHML, potentially extending the maximality limits identified in [2,21,22]. More concretely:

1. We show how these aspects can be expressed and studied in a general monitoring framework with (abstract) conditions, Theorems 3 and 4 *resp.* in Sects. 3 and 5.
2. We instantiate the general framework with trace conditions that describe the inability to perform actions, amounting to refusals [31], Propositions 1 and 5.
3. We also instantiate the framework with conditions describing finite execution graphs, amounting to the recursion-free fragment of the logic [24], Propositions 2 and 3.
4. Finally, we instantiate the framework with trace conditions that record information from previous monitored runs of the system, Proposition 4. This, in turn, leads us to a notion of alternating monitoring that allows monitors to aggregate information over monitored runs. We show that this extends the monitorable fragment of our logic in a natural and significant way.

The remainder of the paper is structured as follows. After outlining the necessary preliminaries in Sect. 2, we develop our parameterized monitoring framework with conditions in Sect. 3 for a monitoring setup that allows monitors to observe both silent and external actions of systems. The two condition instantiations for this strong setting are presented in Sect. 4. In Sect. 5 we extend the parameterized monitoring framework with conditions to a weak monitoring setup that abstracts from internal moves, followed by two instantiations similar to those presented in Sect. 4. Section 6 concludes by discussing related and future work.

2 Background

Labelled Transition Systems. We assume a set of *external* actions $\textsc{Act}$ and a distinguished *silent* action τ. We let α range over $\textsc{Act}$ and μ over $\textsc{Act} \cup \{\tau\}$. A *Labelled Transition System* (LTS) on $\textsc{Act}$ is a triple

$$L = \langle P, \textsc{Act}, \rightarrow_L \rangle,$$

where P is a nonempty set of system states referred to as *processes* $p, q, \ldots$, and $\rightarrow_L \subseteq P \times (\textsc{Act} \cup \{\tau\}) \times P$ is a transition relation. We write $p \xrightarrow{\mu}_L q$ instead of $(p, \mu, q) \in \rightarrow_L$. By $p \xrightarrow{\mu}_L$ we mean that there is some q such that $p \xrightarrow{\mu}_L q$. We use $p \stackrel{\mu}{\Rightarrow}_L q$ to mean that, in L, p can derive q using a single μ action and any number of silent actions, *i.e.*, $p(\xrightarrow{\tau}_L)^* \xrightarrow{\mu}_L (\xrightarrow{\tau}_L)^* q$. We distinguish between (general) traces $s = \mu_1\mu_2 \ldots \mu_r \in (\textsc{Act} \cup \{\tau\})^*$ and *external traces* $t = \alpha_1\alpha_2 \ldots \alpha_r \in \textsc{Act}^*$. For a general trace $s = \mu_1\mu_2 \ldots \mu_r \in (\textsc{Act} \cup \{\tau\})^*$, $p \xrightarrow{s}_L q$ means $p \xrightarrow{\mu_1}_L \xrightarrow{\mu_2}_L \ldots \xrightarrow{\mu_r}_L q$; and for an external trace $t = \alpha_1\alpha_2 \ldots \alpha_r \in \textsc{Act}^*$, $p \stackrel{t}{\Rightarrow}_L q$ means $p \stackrel{\alpha_1}{\Rightarrow}_L \stackrel{\alpha_2}{\Rightarrow}_L \ldots \stackrel{\alpha_r}{\Rightarrow}_L q$ when $r \geq 1$ and $p(\xrightarrow{\tau})^* q$ when $t = \varepsilon$ is the empty trace. We occasionally omit the subscript L when it is clear from the context.

Example 1. The (standard) regular fragment of CCS [30] with grammar:

$$p, q \in \textsc{Proc} ::= \mathsf{nil} \quad \mid \mu.p \quad \mid p + q \quad \mid \mathsf{rec}\ x.p \quad \mid x,$$

where $x, y, z, \ldots$ are from some countably infinite set of variables $\textsc{Var}$, and the transition relation defined as:

$$\textsc{Act}\frac{}{\mu.p \xrightarrow{\mu} p} \qquad \textsc{Rec}\frac{p[\mathsf{rec}\ x.p/x] \xrightarrow{\mu} q}{\mathtt{rec}x.p \xrightarrow{\mu} q} \qquad \textsc{SelL}\frac{p \xrightarrow{\mu} p'}{p + q \xrightarrow{\mu} p'} \qquad \textsc{SelR}\frac{q \xrightarrow{\mu} q'}{p + q \xrightarrow{\mu} q'}$$

constitutes the LTS $\langle \textsc{Proc}, \textsc{Act}, \rightarrow \rangle$. We often use the CCS notation above to describe processes. ■

Specification Logic. Properties about the behaviour of processes may be specified via the logic μHML [4,28], a reformulation of the modal μ-calculus [26].

Definition 1. *μHML formulae on* Act *are defined by the grammar:*

$$\begin{aligned}\varphi,\psi \in \mu\text{HML} ::= \; & \mathsf{tt} & \mid \; & \mathsf{ff} & \mid \; & \varphi \wedge \psi & \mid \; & \varphi \vee \psi \\ \mid \; & \langle\mu\rangle\varphi & \mid \; & [\mu]\varphi & \mid \; & \mathsf{min}\, X.\varphi & \mid \; & \mathsf{max}\, X.\varphi & \mid \; X\end{aligned}$$

where $X, Y, Z, \ldots$ come from a countably infinite set of logical variables LVar*. For a given LTS $L = \langle P, \text{Act}, \rightarrow\rangle$, an environment ρ is a function $\rho : \text{LVar} \rightarrow 2^P$. Given an environment ρ, $X \in \text{LVar}$, and $S \subseteq P$, $\rho[x \mapsto S]$ denotes the environment where $\rho[X \mapsto S](X) = S$ and $\rho[X \mapsto S](Y) = \rho(Y)$, for all $Y \neq X$. The semantics of a μHML formula φ over an LTS L relative to an environment ρ, denoted as $[\![\varphi, \rho]\!]_L$, is defined as follows:*

$$[\![\mathsf{tt}, \rho]\!]_L = P \qquad [\![\mathsf{ff}, \rho]\!]_L = \emptyset \qquad [\![X, \rho]\!]_L = \rho(X)$$

$$[\![\varphi_1 \wedge \varphi_2, \rho]\!]_L = [\![\varphi_1, \rho]\!]_L \cap [\![\varphi_2, \rho]\!]_L \qquad [\![\varphi_1 \vee \varphi_2, \rho]\!]_L = [\![\varphi_1, \rho]\!]_L \cup [\![\varphi_2, \rho]\!]_L$$

$$[\![[\mu]\varphi, \rho]\!]_L = \left\{p \mid \forall q.\; p \xrightarrow{\mu} q \text{ implies } q \in [\![\varphi, \rho]\!]_L\right\}$$

$$[\![\langle\mu\rangle\varphi, \rho]\!]_L = \left\{p \mid \exists q.\; p \xrightarrow{\mu} q \text{ and } q \in [\![\varphi, \rho]\!]_L\right\}$$

$$[\![\mathsf{min}\, X.\varphi, \rho]\!]_L = \bigcap \{S \mid S \supseteq [\![\varphi, \rho[X \mapsto S]]\!]_L\}$$

$$[\![\mathsf{max}\, X.\varphi, \rho]\!]_L = \bigcup \{S \mid S \subseteq [\![\varphi, \rho[X \mapsto S]]\!]_L\}$$

Formulas φ and ψ are equivalent, denoted as $\varphi \equiv \psi$, when $[\![\varphi, \rho]\!]_L = [\![\psi, \rho]\!]_L$ for every environment ρ and LTS L. We often consider closed formulae and simply write $[\![\varphi]\!]_L$ for $[\![\varphi, \rho]\!]_L$ when the semantics of φ is independent of ρ. ■

The logic μHML is very expressive. It is also agnostic of the technique to be employed for verification. The property of monitorability, however, fundamentally relies on the monitoring setup considered.

Monitoring Systems. A *monitoring setup* on Act is a triple $\langle M, I, L\rangle$, where L is a system LTS on Act, M is a monitor LTS on Act, and I is the instrumentation describing how to compose L and M into an LTS, denoted by $I(M, L)$, on Act. We call the pair (M, I) a *monitoring system* on Act. For $M = \langle \text{Mon}, \text{Act}, \rightarrow_M\rangle$, Mon is set of monitor states (ranged over by m) and $\rightarrow_M$ is the *monitor semantics* described in terms of the behavioural state transitions a monitor takes when it analyses trace events $\mu \in \text{Act} \cup \{\tau\}$. The states of the composite LTS $I(M, L)$ are written as $m \triangleleft p$, where m is a monitor state and p is a system state; the monitored-system transition relation is denoted here by $\rightarrow_{I(M,L)}$. We present our results with a focus on *rejection* monitors, *i.e.*, monitors with a designated rejection state no, and hence safety fragments of the logic μHML. However, our results and arguments apply dually to acceptance monitors (with a designated acceptance state yes) and co-safety properties; see [21, 22] for details.

Definition 2. *Fix a monitoring setup* $\langle M, I, L\rangle$ *on* Act *and let* m *be a monitor state of* M *and* φ *a closed formula of* μ*HML on* Act. *We say that* m (M, I)-rejects *(or simply* rejects, *if* M, I *are evident) a process* p *in* L, *written as* $\mathbf{rej}_{\langle M,I,L\rangle}(m, p)$, *when there are a process* q *in* L *and a trace* $s \in (\text{Act} \cup \{\tau\})^*$ *such that* $m \triangleleft p \xrightarrow{s}_{I(M,L)} \mathsf{no} \triangleleft q$. *We say that* m (M, I)-monitors for φ *on* L *whenever*

$$\text{for each process } p \text{ of } L, \mathbf{rej}_{\langle M,I,L\rangle}(m, p) \text{ if and only if } p \notin [\![\varphi]\!]_L.$$

(Subscripts are omitted when they are clear from the context.) Finally, m (M, I)-monitors for φ *when* m (M, I)*-monitors for* φ *on* L *for every LTS* L *on* Act. *The monitoring system* (M, I) *is often omitted when evident.* ■

We define monitorability for μHML in terms of monitoring systems (M, I).

Definition 3. *Fix a monitoring system* (M, I) *and a fragment* Λ *of* μ*HML. We say that* (M, I) *rejection-monitors for* Λ *whenever:*

- *For all closed* $\varphi \in \Lambda$, *there exists an* m *from* M *that* (M, I)*-monitors for* φ.
- *For all* m *of* M, *there exists a closed* $\varphi \in \Lambda$ *that is* (M, I)*-monitored by* m. ■

We note that if a monitoring system and a fragment Λ of μHML satisfy the conditions of Definition 3, then Λ is the largest fragment of μHML that is monitored by the monitoring system. Stated otherwise, any other logic fragment Λ' that satisfies the conditions of Definition 3 must be equally expressive to Λ, *i.e.*, $\forall \varphi' \in \Lambda' \cdot \exists \varphi \in \Lambda \cdot \varphi \equiv \varphi'$ and vice versa. Definition 3 can be dually given for acceptance-monitorability, when considering acceptance monitors. We next review two monitoring systems that respectively rejection-monitor for two different fragments of μHML. We omit the corresponding monitoring systems for acceptance-monitors, that monitor for the dual fragments of μHML.

The Basic Monitoring Setup. The following monitoring system, presented in [2], does *not* distinguish between silent actions and external actions.

Definition 4. *A* basic *monitor on* Act *is defined by the grammar:*

$$m, n \in \text{Mon}_b ::= \mathsf{end} \quad | \ \mathsf{no} \quad | \ \mu.m \quad | \ m + n \quad | \ \mathsf{rec}\, x.m \quad | \ x,$$

where x *comes from a countably infinite set of monitor variables. Constant* no *denotes the* rejection verdict *state whereas* end *denotes the* inconclusive verdict *state. The basic monitor LTS* M_b *is the one whose states are the closed monitors of* Mon_b *and whose transition relation is defined by the (standard) rules in Table 1 (we elide the symmetric rule for* $m + n$*).* ■

Note that by rule mVrd in Table 1, verdicts are irrevocable and monitors can only describe suffix-closed behaviour.

Table 1. Behaviour and instrumentation rules for monitored systems ($v \in \{\mathsf{end}, \mathsf{no}\}$).

Monitor semantics			
$\textsc{mRec}\frac{m[\mathsf{rec}\ x.m/x]\xrightarrow{\mu}m'}{\mathsf{rec}\ x.m\xrightarrow{\mu}m'}$	$\textsc{mSel}\frac{m\xrightarrow{\mu}m'}{m+n\xrightarrow{\mu}m'}$	$\textsc{mAct}\frac{}{\mu.m\xrightarrow{\mu}m}$	$\textsc{mVrd}\frac{}{v\xrightarrow{\mu}v}$
Instrumentation semantics			
$\textsc{iMon}\frac{p\xrightarrow{\mu}_L q \quad m\xrightarrow{\mu}_M n}{m\triangleleft p\xrightarrow{\mu}_{I(M,L)} n\triangleleft q}$	$\textsc{iTer}\frac{p\xrightarrow{\mu}_L q \quad m\not\xrightarrow{\mu}_M}{m\triangleleft p\xrightarrow{\mu}_{I(M,L)} \mathsf{end}\triangleleft q}$	$\textsc{iAbs}\frac{p\xrightarrow{\tau}_L q}{m\triangleleft p\xrightarrow{\tau}_{I(M,L)} m\triangleleft q}$	

Definition 5. *Given a system LTS L and a monitor LTS M that agree on* Act, *the* basic *instrumentation LTS, denoted by $I_b(M, L)$, is defined by the rules* iMon *and* iTer *in Table 1. (We do* not *consider rule* iAbs *for now.)* ■

Instrumentation often relegates monitors to a passive role, whereby a monitored system transitions only when the system itself can. In rule iMon, when the system produces a trace event μ that the monitor is able to analyse (and transition from m to n), the constituent components of a monitored system $m \triangleleft p$ move in lockstep. Conversely, when the system produces an event μ that the monitor is *unable* to analyse, the monitored system still executes, according to iTer, but the monitor transitions to the inconclusive state, where it remains for the rest of the computation.

We refer to the pair (M_b, I_b) from Definitions 4 and 5 as the *basic monitoring system.* For each system LTS L that agrees with the full monitoring system on Act, we can show a correspondence between the respective monitoring setup $\langle M_b, I_b, L\rangle$ and the following syntactic subset of μHML.

Definition 6. *The* safety μHML *is defined by the grammar:*

$$\theta, \chi \in \text{sHML} ::= \mathsf{tt} \quad | \ \mathsf{ff} \quad | \ [\mu]\theta \quad | \ \theta \wedge \chi \quad | \ \mathsf{max}\ X.\theta \quad | \ X \quad ■$$

Theorem 1 ([2]). *The* basic *monitoring system (M_b, I_b) monitors for the logical fragment* sHML. □

The proof of Theorem 1 relies on a monitor synthesis and a formula synthesis function. The monitor synthesis function, $(\!|-|\!) : \text{sHML} \to \textsc{Mon}_b$, is defined on the structure of the input formula and assumes a bijective mapping between formula variables and monitor recursion variables:

$$(\!|\mathsf{tt}|\!) = \mathsf{end} \qquad (\!|\mathsf{ff}|\!) = \mathsf{no} \qquad (\!|X|\!) = x$$

$$(\!|[\mu]\psi|\!) = \begin{cases} \mathsf{end} & \text{if } (\!|\psi|\!) = \mathsf{end} \\ \mu.(\!|\psi|\!) & \text{otherwise} \end{cases} \qquad (\!|\mathsf{max}\ X.\psi|\!) = \begin{cases} \mathsf{end} & \text{if } (\!|\psi|\!) = \mathsf{end} \\ \mathsf{rec}\ x.(\!|\psi|\!) & \text{otherwise} \end{cases}$$

$$(\!|\psi_1 \wedge \psi_2|\!) = \begin{cases} (\!|\psi_1|\!) & \text{if } (\!|\psi_2|\!) = \mathsf{end} \\ (\!|\psi_2|\!) & \text{if } (\!|\psi_1|\!) = \mathsf{end} \\ (\!|\psi_1|\!) + (\!|\psi_2|\!) & \text{otherwise} \end{cases}$$

The case analyses in the above synthesis procedure handle some of the redundancies that may be present in formula specifications. For instance, it turns out that $\mathsf{max}\ X.[\mu]\mathsf{tt} \equiv \mathsf{tt}$ and, accordingly, $(\!|\mathsf{max}\ X.[\mu]\mathsf{tt}|\!) = (\!|\mathsf{tt}|\!) = \mathsf{end}$. The formula synthesis function is defined analogously (see [2,22] for more details).

Monitoring for External Actions. The results obtained in [21,22] can be expressed and recovered within our more general framework. We can express a weak version of the modalities employed in [3,21,22] as follows:

$$[[\mu]]\varphi \equiv \mathsf{max}\ X.([\tau]X \wedge [\mu]\mathsf{max}\ Y.(\varphi \wedge [\tau]Y)) \text{ and}$$
$$\langle\langle\mu\rangle\rangle\varphi \equiv \mathsf{min}\ X.(\langle\tau\rangle X \vee \langle\mu\rangle\mathsf{min}\ Y.(\varphi \vee \langle\tau\rangle Y)).$$

Definition 7. Weak safety μHML, *presented in [21,22], is defined by the grammar:*

$$\pi, \kappa \in \mathrm{WsHML} ::= \ tt \quad |\ ff \quad |\ [[\alpha]]\pi \quad |\ \pi \wedge \kappa \quad |\ max\ X.\pi \quad |\ X. \qquad \blacksquare$$

Definition 8. *The set* MON_e *of external monitors on* ACT *contains all the basic monitors that do not use the silent action* τ. *The corresponding external monitor LTS* M_e, *is defined similarly to* M_b, *but with the closed monitors in* MON_e *as its states.* External instrumentation, *denoted by* I_e, *is defined by the* three *rules* IMON, ITER *and* IABS *in Table 1, where in the case of* IMON *and* ITER, *action* μ *is substituted by the external action* α. *We refer to the pair* (M_e, I_e) *as the* external monitoring system, *amounting to the setup in [21,22].* ■

Theorem 2 ([22]). *The* external *monitoring system* (M_e, I_e) *rejection-monitors for the logical fragment* WsHML. □

3 Monitors that Detect Conditions

Given a set of processes P, a pair (C, r) is a condition framework when C is a non-empty set of *conditions* and $r : C \to 2^P$ is a valuation function. We assume a fixed condition framework (C, r) and we extend the syntax and semantics of μHML so that for every condition $c \in C$, both c and $\neg c$ are formulas and for every LTS L on set of processes P, $[\![c]\!] = r(c)$ and $[\![\neg c]\!] = P \setminus r(c)$. We call the extended logic $\mu\mathrm{HML}^{(C,r)}$. Since, in all the instances we consider, r is easily inferred from C, it is often omitted and we simply write C instead of (C, r) and $\mu\mathrm{HML}^{(C,r)}$ as $\mu\mathrm{HML}^{C}$. We say that process p satisfies c when $p \in [\![c]\!]$. We assume that C is closed under negation, meaning that for every $c \in C$, there is some $c' \in C$, such that $[\![c']\!] = [\![\neg c]\!]$. Conditions represent certain properties of processes that the instrumentation is able to report.

We extend the syntax of monitors, so that if m is a monitor and c a condition, then $c.m$ is a monitor. The idea is that if $c.m$ detects that the process satisfies c, then it can transition to m.

Definition 9. *A* basic C-monitor *on* ACT *is defined by the grammar:*

$$m, n \in \textsc{Mon}_b^C ::= \mathsf{end} \quad | \ \mathsf{no} \quad | \ \mu.m \quad | \ c.m \quad | \ m+n \quad | \ \mathsf{rec}\ x.m \quad | \ x,$$

where x comes from a countably infinite set of monitor variables and $c \in C$. Basic C-monitor behaviour is defined as in Table 1, but allowing μ to range over $\textsc{Act} \cup C \cup \{\tau\}$. *We call the resulting monitor LTS M_b^C.* ■

A monitor detects the satisfaction of condition c when the monitored system has transitioned to a process that satisfies c. To express this intuition, we add rule ICON to the instrumentation rules of Table 1:

$$\textsc{iCon}\ \frac{p \in [\![c]\!] \ \text{ and } \ m \xrightarrow{c}_M n}{m \triangleleft p \xrightarrow{\tau}_{I(M,L)} n \triangleleft p}.$$

We call the resulting instrumentation I_b^C. We observe that the resulting monitor setup is transparent with respect to external actions: an external trace of the monitored system results in exactly the same external trace of the instrumentation LTS. However, the general traces are not preserved, as the rule ICON may introduce additional silent transitions for the instrumentation trace. However, we argue that this is an expected consequence of the instrumentation verifying the conditions of C. C-monitors monitor for sHML^C:

Definition 10. *The strong safety fragment of μHML^C is defined as:*

$$\varphi, \psi \in \text{sHML}^C ::= \mathsf{tt} \quad | \ \mathsf{ff} \quad | \ [\mu]\varphi \quad | \ \neg c \vee \varphi \quad | \ \varphi \wedge \psi \quad | \ \mathsf{max}\ X.\varphi \quad | \ X,$$

where $c \in C$. We note that $\neg c \vee \varphi$ can be viewed as an implication $c \rightarrow \varphi$ asserting that if c holds, then φ must also hold. ■

It is immediate to see that sHML^C is a fragment of μHML^C and when $C \subseteq \mu\text{HML}$, it is also a fragment of μHML. Finally, if C is closed under negation, then $\neg c \vee \varphi$ can be rewritten as $c' \vee \varphi$, where $[\![c']\!] = [\![\neg c]\!]$, and in the following we often take advantage of this equivalence to simplify the syntax of sHML^C.

Theorem 3. *The monitoring system (M_b^C, I_b^C) monitors for* sHML^C. □

We note that Theorem 3 implies that sHML^C is the largest monitorable fragment of μHML^C, relative to C.

4 Instantiations

We consider two possible instantiations for parameter C in the framework presented in Sect. 3. Since each of these instantiations consists of a fragment from the logic μHML itself, they both show how monitorability for μHML can be extended when using certain augmented traces.

4.1 The Inability to Perform an Action

The monitoring framework of [2,22] (used also in other works such as [18,19]), is based on the idea that, while a system is executing, it performs discrete computational steps called events (actions) that are recorded and relayed to the monitor for analysis. Based on the analysed events, the monitor then transitions from state to state. One may however also consider instrumentations that record a system's *inability* to perform a certain action. Examples of this arise naturally in situations where actions are requested unsuccessfully by an external entity on a system, or whenever the instrumentation is able to report system stability (*i.e.*, the inability of performing internal actions). For instance, such observations were considered in [1,31], in the context of testing preorders.

In our setting, a process is unable to perform action μ exactly when it satisfies $[\mu]\mathsf{ff}$. For monitors that are able to detect the inability or failure of a process to perform actions, we set $F_{\text{ACT}} = \{[\mu]\mathsf{ff} \mid \mu \in \text{ACT} \cup \{\tau\}\}$ as the set of conditions. By Theorem 3, the resulting maximal monitorable fragment of μHML is given by the grammar:

$$\varphi, \psi \in \text{sHML}^{F_{\text{ACT}}} ::= \mathsf{tt} \quad \mid \mathsf{ff} \quad \mid [\mu]\varphi \quad \mid \langle\mu\rangle\mathsf{tt} \vee \varphi \quad \mid \varphi \wedge \psi \mid \mathsf{max}\, X.\varphi \quad \mid X.$$

We note the fact that μHML is closed under negation, where $\neg[\mu]\mathsf{ff} = \langle\mu\rangle\mathsf{tt}$.

Proposition 1. *The monitoring system* $(M_b^{F_{\text{ACT}}}, I_b^{F_{\text{ACT}}})$ *monitors for the logical fragment* $\text{sHML}^{F_{\text{ACT}}}$. □

A special case of interest are monitors that can detect process stability, *i.e.*, processes satisfying $[\tau]\mathsf{ff}$. Such monitors monitor for $\text{sHML}^{\{[\tau]\mathsf{ff}\}}$, namely sHML from Definition 6 extended with formulas of the form $\langle\tau\rangle\mathsf{tt} \vee \varphi$.

4.2 Depth-Bounded Static Analysis

In multi-pronged approaches using a combination of verification techniques, one could statically verify parts of a program (from specific execution points) with respect to certain behavioural properties using techniques such as Bounded Model Checking [11] and Partial Model Checking [7]. Typical examples arise in component-based software using modules, objects or agents that can be verified in isolation. This pre-computed verification can then be recorded as annotations to a component and subsequently reported by the instrumentation as part of the execution trace. This strategy would certainly be feasible for depth-bounded static analysis for which the original logic HML [24]—the recursion-free fragment of μHML given below—is an ideal fit.

$$\eta, \chi \in \text{HML} ::= \mathsf{tt} \quad \mid \mathsf{ff} \mid \eta \wedge \chi \quad \mid \eta \vee \chi \mid [\mu]\eta \quad \mid \langle\mu\rangle\eta.$$

Again, HML is closed under negation [4]. If we allow monitors to detect the satisfaction of these kinds of conditions, then, according to Theorem 3, the maximal fragment of μHML that we can monitor for, with HML as a condition framework, is sHML^{HML}, defined by the following grammar:

$$\varphi, \psi ::= \mathsf{tt} \mid \mathsf{ff} \mid [\mu]\varphi \mid \eta \vee \varphi \mid \varphi \wedge \psi \mid \max X.\varphi \mid X,$$

where $\eta \in \text{HML}$. Another way to describe sHML^{HML} is as the μHML fragment that includes all formulas whereby for every subformula of the form $\varphi \vee \psi$, at most *one* of the constituent subformulas φ, ψ uses recursion.

Proposition 2. *The monitoring system* $(M_b^{\text{HML}}, I_b^{\text{HML}})$ *monitors for the logical fragment* sHML^{HML}. □

Instead of HML, we can alternatively use a fragment HML^d of HML that only allows formulas with nesting depth for the modalities of at most d. Since the complexity of checking HML formulas is directly dependent on this modal depth, there are cases where the overheads of checking such formulas are deemed to be low enough to be adequately checked for at runtime instead of checking for them statically.

5 Extending External Monitorability

We explore the impact of considering traces that encode conditions from Sect. 3 on the monitorability of the weak version of the logic used in [21,22]:

$$\begin{aligned}\varphi, \psi \in \text{W}\mu\text{HML} ::= \ & \mathtt{tt} && \mid \mathtt{ff} && \mid \varphi \wedge \psi && \mid \varphi \vee \psi \\ & \mid \langle\langle \alpha \rangle\rangle \varphi && \mid [[\alpha]]\varphi && \mid \min X.\varphi && \mid \max X.\varphi && \mid X.\end{aligned}$$

This version of the logic abstracts away from internal moves performed by the system—note that the weak modality formulas are restricted to external actions α as opposed to the general ones, μ. The semantics follows that presented in Sect. 2, but can alternatively be given a more direct inductive definition, *e.g.*

$$[\![[[\alpha]]\varphi, \rho]\!] = \{p \mid \forall q.\ p \overset{\alpha}{\Rightarrow} q \text{ implies } q \in [\![\varphi, \rho]\!]\}.$$

The main aim of this section is to extend the maximally-expressive monitorable subset of μHML that was identified in [21,22] using the framework developed in Sect. 3.

5.1 External Monitoring with Conditions

We define the external monitoring system with conditions similarly to Sect. 3. The syntax of Definition 8 is extended so that, for any instance of C, if m is a monitor and c a condition from C, then $c.m$ is a monitor.

Definition 11. *An* external C-monitor *on* Act *is defined by the grammar:*

$$m, n \in \textsc{Mon}_e^C ::= \mathsf{end} \mid \mathsf{no} \mid \alpha.m \mid c.m \mid m+n \mid \mathsf{rec}\, x.m \mid x,$$

where $c \in C$. C-monitor behaviour is defined as in Table 1, but extending rule mAct *to condition prefixes that generate condition actions (i.e., μ ranges over* Act $\cup$ C*). We call the resulting monitor LTS M_e^C.*

For the instrumentation relation called I_e^C, we consider the rules iMon, iTer *from Table 1 for external actions α instead of the general action μ, rule* iAbs *from the same table, and rule* iCon *from Sect. 3.* ■

Note that the monitoring system (M_e^C, I_e^C) may be used to detect τ-transitions *implicitly*—we conjecture that this cannot be avoided in general. Consider two conflicting conditions c_1 and c_2, *i.e.*, $[\![c_1]\!] \cap [\![c_2]\!] = \emptyset$. Definition 11 permits monitors of the form $c_1.c_2.m$ that encode the fact that state m can only be reached when the system under scrutiny performs a non-empty sequence of τ-moves to transition from a state satisfying c_1 to another state satisfying c_2. This, in some sense, is also related to obscure silent action monitoring studied in [2].

We identify the grammar for the maximally-expressive monitorable syntactic subset of the logic WμHML. It uses the formula $[[\varepsilon]]\varphi$ defined as:

$$[[\varepsilon]]\varphi \equiv \mathsf{max}\, X.(\varphi \wedge [\tau]X)$$

The modality $[[\varepsilon]]\varphi$ quantifies universally over the set of processes that can be reached from a given one via any number of silent steps. Together with its dual $\langle\langle\varepsilon\rangle\rangle\varphi$ modality, $[[\varepsilon]]\varphi$ is used in the modal characterisation of weak bisimilarity [30,34], in which τ transitions from one process may be matched by a (possibly empty) sequence of τ transitions from another.

Definition 12. *The weak safety fragment of* WμHML *with C is defined as:*

$$\varphi, \psi \in \mathrm{WsHML}^C ::= \mathit{tt} \mid \mathit{ff} \mid [[\alpha]]\varphi \mid [[\varepsilon]](\neg c \vee \varphi) \mid \varphi \wedge \psi \mid \mathsf{max}\, X.\varphi \mid X,$$

where $c \in C$. ■

Theorem 4. *The monitoring system (M_e^C, I_e^C) monitors for* WsHML^C. □

We highlight the need to insulate the appearance of the implication $\neg c \vee \varphi$ from internal system behaviour by using the modality $[[\varepsilon]]$ in Definition 12. For conditions that are invariant under τ-transitions, this modality is not required but it cannot be eliminated otherwise; we revisit this point in Example 2.

5.2 Instantiating External Monitors with Conditions

We consider three different instantiations to our parametric external monitoring system of Sect. 5.1.

Recursion-Free Formulas. The weak version of HML, denoted by wHML, is the recursion-free fragment of WμHML. Similarly to what was argued earlier in Sect. 4.2, it is an appropriate set of conditions to instantiate set C in WsHML^C, and the maximal monitorable fragment of WμHML with conditions from wHML is WsHML$^{w\text{HML}}$, defined by the following grammar, where $\eta \in w$HML:

$$\varphi, \psi ::= \mathsf{tt} \quad | \ \mathsf{ff} \quad | \ [[\alpha]]\varphi \quad | \ [[\varepsilon]](\eta \vee \varphi) \quad | \ \varphi \wedge \psi \quad | \ \mathsf{max}\ X.\varphi \quad | \ X.$$

Proposition 3. *The monitoring system* $(M_e^{w\text{HML}}, I_e^{w\text{HML}})$ *monitors for the logical fragment* WsHML$^{w\text{HML}}$. □

An important observation (that is perhaps surprising) is that WsHML$^{w\text{HML}}$ is *not* a fragment of WμHML, as the following example demonstrates.

Example 2. Although for any (closed) WsHML formula φ we have the logical equivalence $[[\varepsilon]]\varphi \equiv \varphi$ (notice that the monitor for φ that is guaranteed by Theorem 2 also monitors for $[[\varepsilon]]\varphi$), this logical equivalence does not hold for a formula φ from WμHML. Consider the formula φ_ϵ below that may be expressed using a formula from WsHML$^{w\text{HML}}$:

$$\varphi_\epsilon = [[\varepsilon]]\langle\langle\alpha\rangle\rangle\mathsf{tt} \equiv [[\varepsilon]](\langle\langle\alpha\rangle\rangle\mathsf{tt} \vee \mathsf{ff}) \in \text{WsHML}^{w\text{HML}}.$$

Formula φ_ϵ is not equivalent to $\langle\langle\alpha\rangle\rangle\mathsf{tt}$ (*e.g.* the process $\alpha.\mathsf{nil} + \tau.\mathsf{nil}$ satisfies $\langle\langle\alpha\rangle\rangle\mathsf{tt}$, but not φ_ϵ) meaning that $[[\varepsilon]]$ plays a discerning role in the context of WμHML. Furthermore, φ_ϵ holds for process $\tau.\alpha.\mathsf{nil}$, but not for $\alpha.\mathsf{nil}+\tau.\mathsf{nil}$, even though these two processes cannot be distinguished by *any* WμHML formula. In fact, it turns out that they are bisimilar with respect to *weak external transitions* and this bisimulation characterises the satisfaction of WμHML formulas [24]. Thus, there is no formula in WμHML that is equivalent to φ_ϵ. ■

Previous Runs and Alternating Monitoring. A monitoring system could reuse information from previous system runs, perhaps recorded as execution logs, and whenever (sub)traces can be associated with specific states of the system, these can also be used as an instantiation for our parametric framework. More concretely, in [21,22] it is shown that traces can be used to characterise the violation of WsHML formulas, or the satisfaction of formulas from the dual fragment, WcHML, defined below.

Definition 13. *The* co-safety WμHML *is defined by the grammar:*

$$\pi, \kappa \in \text{WcHML} ::= \mathit{tt} \quad | \ \mathit{ff} \quad | \ \langle\langle\alpha\rangle\rangle\theta \quad | \ \theta \vee \chi \quad | \ \mathit{min}\ X.\theta \quad | \ X \ ■$$

The witnessed rejection and acceptance traces can in turn be used as part of an augmented trace for an instantiation for C to obtain the monitorable dual logics WsHML^{WcHML} and WcHML^{WsHML} that alternate between rejection monitoring and acceptance monitoring. The logic $\text{WsHML}^{\text{WcHML}}$ is defined by the following grammar, where $\theta \in \text{WsHML}$:

$$\varphi, \psi ::= \mathsf{tt} \quad | \; \mathsf{ff} \; | \; [[\alpha]]\varphi \quad | \; [[\varepsilon]](\theta \vee \varphi) \quad | \; \varphi \wedge \psi \quad | \; \max X.\varphi \quad | \; X;$$

and $\text{WcHML}^{\text{WsHML}}$ is defined by the following grammar, where $\chi \in \text{WcHML}$:

$$\pi, \kappa ::= \mathsf{tt} \quad | \; \mathsf{ff} \quad | \; \langle\langle\alpha\rangle\rangle\pi \quad | \; \langle\langle\varepsilon\rangle\rangle(\chi \wedge \pi) \quad | \; \pi \vee \kappa \quad | \; \min X.\varphi \quad | \; X.$$

Proposition 4. *The monitoring system* $(M_e^{\text{WcHML}}, I_e^{\text{WcHML}})$ *rejection-monitors for the logical fragment* $\text{WsHML}^{\text{WcHML}}$. □

One should observe that in this case, $\text{WsHML}^{\text{WcHML}}$ *is* a fragment of WμHML, in contrast to the previous instantiation $\text{WsHML}^{w\text{HML}}$ from Sect. 5.2.

Lemma 1. *For every* $[[\varepsilon]](\eta \vee \varphi) \in \text{WsHML}^{\text{WcHML}}$ *(where* $\eta \in \text{WsHML}$*), we have* $[[\varepsilon]](\eta \vee \varphi) \equiv \eta \vee \varphi$. □

Corollary 1. *For every formula in* $\text{WsHML}^{\text{WcHML}}$*, there is a logically equivalent formula in* WμHML. □

This entails that $\text{WsHML}^{\text{WcHML}}$ can be reformulated using the following, simpler, grammar (here $\eta \in \text{WsHML}$) which is clearly a fragment of WμHML:

$$\varphi, \psi ::= \mathsf{tt} \quad | \; \mathsf{ff} \quad | \; [[\alpha]]\varphi \quad | \; \eta \vee \varphi \quad | \; \varphi \wedge \psi \quad | \; \max X.\varphi \quad | \; X.$$

If the monitoring system can use such information from previous runs, there is no reason to limit this information to just one previous run. If the instrumentation mechanism can record up to i prior runs, the monitorable logic may be described as WsHML^{i+1}, defined inductively in the following way:

- $\text{WsHML}^1 = \text{WsHML}$ and $\text{WcHML}^1 = \text{WcHML}$; and
- $\text{WsHML}^{i+1} = \text{WsHML}^{\text{WcHML}^i}$ and $\text{WcHML}^{i+1} = \text{WcHML}^{\text{WsHML}^i}$.

Whenever this setup can be extended to unlimited prior runs, the resulting rejection-monitorable fragment would be $\text{WsHML}^\omega = \bigcup_i \text{WsHML}^i$, which is also described by the following grammar:

$$\varphi, \psi ::= \mathsf{tt} \quad | \; \mathsf{ff} \quad | \; [[\alpha]]\varphi \quad | \; \varphi \vee \psi \quad | \; \varphi \wedge \psi \quad | \; \max X.\varphi \quad | \; X.$$

WsHML^ω is a non-trivial extension of WsHML *which is still within* WμHML.

Failure to Execute an Action and Refusals. In Subsect. 4.1, we instantiated the condition set C as the set of formulas from μHML that assert the inability of a process to perform an action. These formulas are of the form $[\alpha]\mathsf{ff}$. We recast this approach in the setting of weak monitorability. In this setting where the monitoring system and the specification formulas ignore any silent transitions, the inability of a process to perform an α-transition acquires a different meaning from the one used for the basic system. In particular, we consider a stronger version of these conditions that incorporates stability; this makes them invariant over τ-transitions. We say that p *refuses* α when $p \not\xrightarrow{\tau}$ and $p \not\xrightarrow{\alpha}$. In [31], a very similar notion is used for refusal testing (see also [1]). Thus, much in line with [31], we use the following definition.

Definition 14. *A process p of an LTS L* refuses *action $\alpha \in$ ACT and write p **ref** α when $p \not\xrightarrow{\tau}_L$ and $p \not\xrightarrow{\alpha}_L$. The set of conditions that corresponds to refusals is thus $R_{\text{ACT}} = \{[\tau]ff \wedge [\alpha]ff \mid \alpha \in \text{ACT}\}$.* ■

According to Theorem 4, the largest fragment of μHML that we can monitor for, using monitors that can detect refusals, is $\text{WsHML}^{R_{\text{ACT}}}$, given by the following grammar:

$$\begin{aligned}\varphi, \psi ::= \;& \mathsf{tt} && \mid \mathsf{ff} && \mid [[\alpha]]\varphi && \mid [[\varepsilon]](\langle\tau\rangle\mathsf{tt} \vee \langle\alpha\rangle\mathsf{tt} \vee \varphi) \\ & \mid \varphi \wedge \psi && \mid \mathsf{max}\, X.\varphi && \mid X. &&\end{aligned}$$

Again, $\langle\tau\rangle\mathsf{tt} \vee \langle\alpha\rangle\mathsf{tt} \vee \varphi$ is best read as the implication $([\tau]\mathsf{ff} \wedge [\alpha]\mathsf{ff}) \rightarrow \varphi$: if the process is stable and cannot perform an α-transition, then φ must hold.

Proposition 5. *The monitoring system $(M_e^{R_{\text{ACT}}}, I_e^{R_{\text{ACT}}})$ monitors for the logical fragment $\text{WsHML}^{R_{\text{ACT}}}$.* □

Example 3. Consider the formula

$$\varphi_s = [[\varepsilon]](\langle\tau\rangle\mathsf{tt} \vee \langle\alpha\rangle\mathsf{tt} \vee [[\beta]]\mathsf{ff}) \in \text{WsHML}^{R_{\text{ACT}}}.$$

Formula φ_s claims that at every stable state that the system can reach, if action α is impossible, then action β should also be impossible. We can see that φ_s is true for $\tau.\mathsf{nil} + \beta.\mathsf{nil}$, but not for $\beta.\mathsf{nil}$. However, the two processes cannot be distinguished by WμHML, as they have the same weak external transitions. Therefore, $\text{WsHML}^{R_{\text{ACT}}}$ is not a fragment of WμHML—but, as we have seen, it is a fragment of μHML. Here we have a part of the formula that clearly is not part of WμHML. That is $\langle\tau\rangle\mathsf{tt}$, which asserts that the process can perform a silent transition. ■

Example 4. Let us consider an LTS L_0 of stable processes—that is, L_0 is an LTS without any silent transitions. L_0 offers a simplified setting to cast our observations. In this case, the $[[\varepsilon]]$, $[\tau]$, and $\langle\tau\rangle$ modalities can be eliminated from our formulas, and weak modalities are equivalent to strong modalities. This allows us to simplify the grammar for $\text{WsHML}^{F_{\text{ACT}}}$ as follows:

$$\begin{aligned}\varphi, \psi ::= \;& \mathsf{tt} && \mid \mathsf{ff} && \mid [\alpha]\varphi && \mid \langle\alpha\rangle\mathsf{tt} \vee \varphi \\ & \mid \varphi \wedge \psi && \mid \mathsf{max}\, X.\varphi && \mid X. &&\end{aligned}$$

Perhaps unsurprisingly, this grammar yields the same formulas as the restriction of grammar of Subsect. 4.1 on external actions. An instance of a specification that can be formalized in this fragment is the following. Consider a simple server-client system, where the client can request a resource, which is represented by action rq, and the server may give a positive response, represented by action rs, after which it needs to allocate said resource to the client, represented by action al. A reasonable specification for the server is that if it is impossible at the moment to provide a resource, then it should not give a positive response to the client. In the above simplification of $\text{WsHML}^{F_{\text{ACT}}}$, this specification can be formalized as $[\mathtt{rq}](\langle\mathtt{al}\rangle\mathtt{tt} \vee [\mathtt{rs}]\mathtt{ff})$. If the LTS includes silent transitions, the corresponding specification would be written as

$$\varphi_r = [\mathtt{rq}][[\varepsilon]](\langle\tau\rangle\mathtt{tt} \vee \langle\mathtt{al}\rangle\mathtt{tt} \vee [[\mathtt{rs}]]\mathtt{ff}).$$

In other words, after a request, if the server cannot provide a resource and it is stable—so, there is no possibility that after some time the resource will be available—then the server should not give a positive response to the client. ■

6 Conclusions

In order to devise effective verification strategies that straddle between the pre- and post-deployment phases of software production, one needs to understand better the monitorability aspects of the correctness properties that are to be verified. We have presented a general framework that allows us to determine maximal monitorable fragments of an expressive logic that is agnostic of the verification technique employed, namely μHML. By way of a number of instantiations, we also show how the framework can be used to reason about the monitorability induced by various forms of augmented traces. Our next immediate concern is to validate the proposed instantiations empirically by constructing monitoring systems and tools that are based on these results, as we did already for the original monitorability results of [21,22] in [9,10,12].

Related Work. Monitorability for μHML was first examined in [21,22]. This work introduced the external monitoring system and identified WsHML as the largest monitorable fragment of μHML, with respect to that system. The ensuring work in [2] focused on monitoring setups that can distinguish silent actions to a varying degree, and introduced the basic monitoring system, showing analogous monitorability results for μHML.

Monitorability has also been examined for languages defined over traces, such as LTL. Pnueli and Zaks in [32] define a notion of monitorability over traces, although they do not attempt maximal monitorability results. Diekert and Leuckert revisited monitorability from a topological perspective in [16]. Falcone *et al.* in [17] extended the work in [32] to incorporate enforcement and introduced a notion of monitorability on traces that is parameterized with respect to a truth domain that corresponds to our separation to acceptance- and rejection-monitorable properties. In [13], the authors use a monitoring system that can generate derivations of satisfied formulas from a fragment of LTL. However, they do not argue that this fragment is somehow maximal. There is a significant body of work on synthesizing monitors from LTL formulas, *e.g.* [13,23,33,35], and it would be worth investigating whether our general techniques for monitor synthesis can be applied effectively in these cases.

Phillips introduced *refusal testing* in [31] as a way to extend the capabilities of testing (see [18] for a discussion on how our monitoring setup relates to testing preorders). The meaning of refusals in [31] is very close to the one in Definition 14 and it is interesting to note how Phillips' use of tests for refusal formulas is similar to our monitoring mechanisms for refusals. Abramsky [1] uses refusals in the context of a much more powerful testing machinery, in order to identify the kind of testing power that is required for distinguishing non-bisimilar processes.

The decomposition of the verification burden across verification techniques, or across iterations of alternating monitoring runs as presented in Sect. 5, can be seen as a method for *quotienting*. In [7] Andersen studies quotienting of the specification logics discussed in this paper to reduce the state-space during model checking and thus increase its efficiency (see also [27] for a more recent treatment). The techniques used rely heavily on the model's concurrency constructs and may produce formulas that are larger in size than the original, but which can be checked against a smaller component of the model. In multi-pronged approaches to verification one would expect to encounter similar difficulties occasionally.

References

1. Abramsky, S.: Observation equivalence as a testing equivalence. Theor. Comput. Sci. **53**(2–3), 225–241 (1987)
2. Aceto, L., Achilleos, A., Francalanza, A., Ingólfsdóttir, A.: Monitoring for silent actions. In: 37th IARCS Annual Conference on Foundations of Software Technology and Theoretical Computer Science, FSTTCS 2017 (2017, to appear)
3. Aceto, L., Achilleos, A., Francalanza, A., Ingólfsdóttir, A., Kjartansson, S.Ö.: Determinizing monitors for HML with recursion. CoRR abs/1611.10212 (2016)
4. Aceto, L., Ingólfsdóttir, A., Larsen, K.G., Srba, J.: Reactive Systems: Modelling, Specification and Verification. Cambridge University Press, New York (2007)
5. Ahrendt, W., Chimento, J.M., Pace, G.J., Schneider, G.: A specification language for static and runtime verification of data and control properties. In: Bjørner, N., de Boer, F. (eds.) FM 2015. LNCS, vol. 9109, pp. 108–125. Springer, Cham (2015). https://doi.org/10.1007/978-3-319-19249-9_8
6. Aktug, I., Naliuka, K.: ConSpec - a formal language for policy specification. Sci. Comput. Programm. **74**(1–2), 2–12 (2008)
7. Andersen, H.R.: Partial model checking (extended). In: Proceedings of Tenth Annual IEEE Symposium on Logic in Computer Science, pp. 398–407. IEEE (1995)
8. Artho, C., Barringer, H., Goldberg, A., Havelund, K., Khurshid, S., Lowry, M.R., Pasareanu, C.S., Rosu, G., Sen, K., Visser, W., Washington, R.: Combining test case generation and runtime verification. Theor. Comput. Sci. **336**(2–3), 209–234 (2005)
9. Attard, D.P., Francalanza, A.: A monitoring tool for a branching-time logic. In: Falcone, Y., Sánchez, C. (eds.) RV 2016. LNCS, vol. 10012, pp. 473–481. Springer, Cham (2016). https://doi.org/10.1007/978-3-319-46982-9_31

10. Attard, D.P., Francalanza, A.: Trace partitioning and local monitoring for asynchronous components. In: Cimatti, A., Sirjani, M. (eds.) SEFM 2017. LNCS, vol. 10469, pp. 219–235. Springer, Cham (2017). https://doi.org/10.1007/978-3-319-66197-1_14
11. Biere, A., Cimatti, A., Clarke, E., Zhu, Y.: Symbolic model checking without BDDs. In: Cleaveland, W.R. (ed.) TACAS 1999. LNCS, vol. 1579, pp. 193–207. Springer, Heidelberg (1999). https://doi.org/10.1007/3-540-49059-0_14
12. Cassar, I., Francalanza, A.: On implementing a monitor-oriented programming framework for actor systems. In: Ábrahám, E., Huisman, M. (eds.) IFM 2016. LNCS, vol. 9681, pp. 176–192. Springer, Cham (2016). https://doi.org/10.1007/978-3-319-33693-0_12
13. Cini, C., Francalanza, A.: An LTL proof system for runtime verification. In: Baier, C., Tinelli, C. (eds.) TACAS 2015. LNCS, vol. 9035, pp. 581–595. Springer, Heidelberg (2015). https://doi.org/10.1007/978-3-662-46681-0_54
14. Decker, N., Leucker, M., Thoma, D.: jUnit^{RV}–adding runtime verification to jUnit. In: Brat, G., Rungta, N., Venet, A. (eds.) NFM 2013. LNCS, vol. 7871, pp. 459–464. Springer, Heidelberg (2013). https://doi.org/10.1007/978-3-642-38088-4_34
15. Desai, A., Dreossi, T., Seshia, S.A.: Combining model checking and runtime verification for safe robotics. In: Lahiri, S., Reger, G. (eds.) RV 2017. LNCS, vol. 10548, pp. 172–189. Springer, Cham (2017). https://doi.org/10.1007/978-3-319-67531-2_11
16. Diekert, V., Leucker, M.: Topology, monitorable properties and runtime verification. Theor. Comput. Sci. **537**, 29–41 (2014)
17. Falcone, Y., Fernandez, J.C., Mounier, L.: What can you verify and enforce at runtime? Int. J. Softw. Tools Technol. Trans. **14**(3), 349–382 (2012)
18. Francalanza, A.: A theory of monitors. In: Jacobs, B., Löding, C. (eds.) FoSSaCS 2016. LNCS, vol. 9634, pp. 145–161. Springer, Heidelberg (2016). https://doi.org/10.1007/978-3-662-49630-5_9
19. Francalanza, A.: Consistently-detecting monitors. In: Meyer, R., Nestmann, U. (eds.) 28th International Conference on Concurrency Theory (CONCUR 2017). LIPIcs, vol. 85, pp. 8:1–8:19. Schloss Dagstuhl, Dagstuhl (2017)
20. Francalanza, A., Aceto, L., Achilleos, A., Attard, D.P., Cassar, I., Della Monica, D., Ingólfsdóttir, A.: A foundation for runtime monitoring. In: Lahiri, S., Reger, G. (eds.) RV 2017. LNCS, vol. 10548, pp. 8–29. Springer, Cham (2017). https://doi.org/10.1007/978-3-319-67531-2_2
21. Francalanza, A., Aceto, L., Ingolfsdottir, A.: On verifying Hennessy-Milner logic with recursion at runtime. In: Bartocci, E., Majumdar, R. (eds.) RV 2015. LNCS, vol. 9333, pp. 71–86. Springer, Cham (2015). https://doi.org/10.1007/978-3-319-23820-3_5
22. Francalanza, A., Aceto, L., Ingolfsdottir, A.: Monitorability for the Hennessy-Milner logic with recursion. Formal Meth. Syst. Des. (FMSD) **51**(1), 87–116 (2017)
23. Geilen, M.: On the construction of monitors for temporal logic properties. Electron. Notes Theor. Comput. Sci. **55**(2), 181–199 (2001)
24. Hennessy, M., Milner, R.: Algebraic laws for nondeterminism and concurrency. J. ACM **32**(1), 137–161 (1985)
25. Kejstová, K., Ročkai, P., Barnat, J.: From model checking to runtime verification and back. In: Lahiri, S., Reger, G. (eds.) RV 2017. LNCS, vol. 10548, pp. 225–240. Springer, Cham (2017). https://doi.org/10.1007/978-3-319-67531-2_14
26. Kozen, D.: Results on the propositional μ-calculus. Theor. Comput. Sci. **27**(3), 333–354 (1983)

27. Lang, F., Mateescu, R.: Partial model checking using networks of labelled transition systems and boolean equation systems. Log. Meth. Comput. Sci. **9**(4), 1–32 (2013)
28. Larsen, K.G.: Proof systems for satisfiability in Hennessy-Milner logic with recursion. Theor. Comput. Sci. **72**(2), 265–288 (1990)
29. Leucker, M., Schallhart, C.: A brief account of runtime verification. J. Log. Algebraic Program. **78**(5), 293–303 (2009)
30. Milner, R.: Communication and Concurrency. Prentice-Hall Inc, Upper Saddle River (1989)
31. Phillips, I.: Refusal testing. Theor. Comput. Sci. **50**(3), 241–284 (1987)
32. Pnueli, A., Zaks, A.: PSL model checking and run-time verification via testers. In: Misra, J., Nipkow, T., Sekerinski, E. (eds.) FM 2006. LNCS, vol. 4085, pp. 573–586. Springer, Heidelberg (2006). https://doi.org/10.1007/11813040_38
33. Sen, K., Roşu, G., Agha, G.: Generating optimal linear temporal logic monitors by coinduction. In: Saraswat, V.A. (ed.) ASIAN 2003. LNCS, vol. 2896, pp. 260–275. Springer, Heidelberg (2003). https://doi.org/10.1007/978-3-540-40965-6_17
34. Stirling, C.: Modal and Temporal Properties of Processes. Springer, New York (2001)
35. Vardi, M.Y.: An automata-theoretic approach to linear temporal logic. In: Moller, F., Birtwistle, G. (eds.) Logics for Concurrency. LNCS, vol. 1043, pp. 238–266. Springer, Heidelberg (1996). https://doi.org/10.1007/3-540-60915-6_6

Logics for Bisimulation and Divergence

Xinxin Liu, Tingting Yu(✉), and Wenhui Zhang

State Key Laboratory of Computer Science, Institute of Software, CAS, University of Chinese Academy of Sciences, Beijing, China
{xinxin,yutt,zwh}@ios.ac.cn

Abstract. The study of modal logics and various bisimulation equivalences so far shows the following progression: 1. weak bisimilarity is characterized by Hennessy-Milner logic (HML), a simple propositional modal logic with a weak possibility modality, and 2. extending HML by refining the weak possibility modality one obtains a logic which characterizes branching bisimilarity, a refinement of weak bisimilarity, and 3. further extending the logic with a divergence modality one obtains a logic which characterizes branching bisimilarity with explicit divergence, a refinement of branching bisimilarity. In this paper, we explore the development by exchanging the above 2 and 3, i.e. by first extending HML with a divergence modality and then refining the weak possibility modality in the extended logic. We have the following findings: A. extending HML with a new divergence modality one obtains a new logic which characterizes complete weak bisimilarity, an equivalence relation with distinguishing power in between weak bisimilarity and branching bisimilarity with explicit divergence; B. further extending the obtained logic by refining the weak possibility modality in it one obtains another logic which characterizes branching bisimilarity with explicit divergence. As main results of the paper, the logic in A. provides a modal characterization for complete weak bisimilarity, and moreover the two new logics in A. and B. are both sub-logics of the known logic obtained in above 3.

1 Introduction

Weak bisimilarity is a popular equivalence relation introduced by Milner [9]. It is defined through the notion of weak bisimulation which was proposed by Milner [9] based on an idea independently discovered by van Benthem [4] and Park [8]. The importance of weak bisimulation is that it not only defines an equivalence relation but also provides a verification technique for the equality. A well-known theoretical result for weak bisimilarity is that the equivalence is characterized by a modal logic which is known as Hennessy-Milner logic (HML) [2] in the following sense: two processes are equivalent with respect to weak bisimilarity if and only if they satisfy exactly the same set of HML formulas.

Supported by the CAS-INRIA major project No. GJHZ1844.

C. Baier and U. Dal Lago (Eds.): FOSSACS 2018, LNCS 10803, pp. 221–237, 2018.
https://doi.org/10.1007/978-3-319-89366-2_12

Because weak bisimilarity does not preserve divergence, i.e. it is possible for two equivalent processes that one of them is capable of endless internal computations while the other is not, various divergence preserving versions of weak bisimulation equivalences and pre-orders are studied later [1,3,5,13]. Complete weak bisimilarity is a newly proposed divergence preserving weak bisimulation equivalence [10]. Like weak bisimilarity, complete weak bisimilarity is supported by a bisimulation verification technique called inductive weak bisimulation, which can be very helpful in practical verification that concerns divergence. One of the main aims of this paper is to find a modal logic which characterizes complete weak bisimilarity just as HML characterizes weak bisimilarity.

We will put our study into a more general context. The study of modal logics and various bisimulation equivalences so far shows the following progression which reveals the co-related increase for the expressive power of the logics and the distinguishing power of the equivalences:

1. Weak bisimilarity is characterized by HML which is a simple propositional modal logic with a weak possibility modality [2];
2. Extending HML by refining the weak possibility modality one obtains a logic which characterizes branching bisimilarity [5,6], a refinement of weak bisimilarity proposed in [12],
3. Further extending the logic with a divergence modality one obtains a logic which characterizes branching bisimilarity with explicit divergence [13], a refinement of branching bisimilarity proposed in [12].

In this paper, we explore the development by exchanging the order of 2 and 3, i.e. by first extending HML with a divergence modality and then refining the weak possibility modality in the extended logic. We have the following findings:

A. Extending HML with a new divergence modality one obtains a logic which characterizes complete weak bisimilarity, an equivalence relation with distinguishing power in between weak bisimilarity and branching bisimilarity with explicit divergence;
B. Further extending the obtained logic by refining the weak possibility modality in it one obtains another logic which characterizes branching bisimilarity with explicit divergence.

To summarize the results of the paper:

- The above A. is the wanted result of modal characterization of complete weak bisimilarity.
- The two new logics in A. and B. are both sub-logics of the known logic mentioned in above 3, hence showing a clear picture of the sub-logic relationships of the corresponding characterization results.
- For finite-state systems, we also use the modal characterization to show a reduction from the problem of checking equality of complete weak bisimilarity to the problem of checking equality of ordinary weak bisimilarity, thus provide a decision procedure for the problem of checking equality of finite-state systems with respect to complete weak bisimilarity.

The rest of the paper is organized as follows. Section 2 presents the definitions of the equalities, i.e. weak bisimilarity, complete weak bisimilarity, branching bisimilarity, and branching bisimilarity with explicit divergence. Section 3 studies the relationships of the modal logic characterizations of the equalities. Section 4 studies reductions for decision problems concerning finite-state processes. Section 5 concludes.

2 Bisimulations and Divergence

In this section, after settling some necessary preliminaries, we introduce the main equivalence relation, i.e. complete weak bisimilarity, together with some related equivalences like branching bisimilarity and branching bisimilarity with explicit divergence.

Definition 1 (Labeled transition systems). *A* labeled transition system *(or LTS) is a triple* $\mathcal{A} = \langle S, A, \longrightarrow \rangle$ *where:*

- *S is a set of* states, *A is a set of* actions, *$\longrightarrow \subseteq S \times (A \cup \{\tau\}) \times S$ is the* transition relation. *τ is the* silent action *which is assumed not in A. An element (s, α, t) of $\longrightarrow$, usually written as $s \xrightarrow{\alpha} t$, is called a* transition*;*
- *A finite run of $\mathcal{A}$ is a finite, nonempty alternating sequence of states and actions: $\rho = s_0\alpha_0 s_1\alpha_1 \ldots s_{n-1}\alpha_{n-1}s_n$ which begins with a state and ends with a state, such that for $0 \leq i < n$, $s_i \xrightarrow{\alpha_i} s_{i+1}$. We also say that ρ is a finite run from s_0 to s_n;*
- *For $\rho = s_0\alpha_0 s_1\alpha_1 \ldots s_{n-1}\alpha_{n-1}s_n$, define $\mathrm{Act}(\rho) = \alpha_0\alpha_1 \ldots \alpha_{n-1}$, and* $\mathrm{length}(\rho) = n$;
- *An infinite run of $\mathcal{A}$ is an infinite, alternating sequence of states and actions: $\rho = s_0\alpha_0 s_1\alpha_1 \ldots$ which begins with a state, such that for all for $i = 0, 1, \ldots,$ $s_i \xrightarrow{\alpha_i} s_{i+1}$. We also say that ρ is an infinite run from s_0;*
- *A (finite or infinite) τ-run of $\mathcal{A}$ is a (finite or infinite) run of $\mathcal{A}$ in which all actions are τ's.*

For a finite sequence of actions $l \in (A \cup \{\tau\})^*$, let $\widehat{l} \in A^*$ be the sequence obtained by deleting all τ's from l.

We use standard notations for multi-step τ transitions, and the so-called double-arrow transitions: write $s \Longrightarrow s'$ if there is a finite τ-run from s to s'; write $s \stackrel{\alpha}{\Longrightarrow} s'$ if there exist t, t' such that $s \Longrightarrow t, t \xrightarrow{\alpha} t', t' \Longrightarrow s'$. Note the important difference between $s \Longrightarrow s'$ and $s \stackrel{\tau}{\Longrightarrow} s'$: the former means that from s to s' there is a finite τ-run (could be a τ-run with zero length), while the latter means that from s to s' there is a finite τ-run with non-zero length. Thus $s \Longrightarrow s$ holds for all $s \in S$, while $s \stackrel{\tau}{\Longrightarrow} s$ holds only when s is on a τ-loop consisting of one or more τ-transitions. Also for $l \in (A \cup \{\tau\})^*$ we will write $s \stackrel{\widehat{l}}{\Longrightarrow} s'$ if there is a finite run ρ from s to s' with $Act(\rho) = l$. Note that $s \stackrel{\epsilon}{\Longrightarrow} s'$ means exactly $s \Longrightarrow s'$, where ϵ is the empty string.

Next, we review the well-known notions of weak bisimulation, weak bisimilarity [9], and branching bisimulation, branching bisimilarity [12].

Definition 2 (Weak and branching bisimulations). *Let $\mathcal{A} = \langle S, A, \longrightarrow \rangle$ be an LTS. A binary relation $R \subseteq S \times S$ is a* weak bisimulation *if it is symmetric and moreover for all $(s,t) \in R$ the following holds:*

whenever $s \xrightarrow{\alpha} s'$, then there exists t' such that $t \stackrel{\hat{\alpha}}{\Longrightarrow} t'$ and $(s',t') \in R$.

A binary relation $R \subseteq S \times S$ is a branching bisimulation *if it is symmetric and moreover for all $(s,t) \in R$ the following holds:*

whenever $s \xrightarrow{\alpha} s'$, then either $\alpha = \tau$, and there exists t' such that $t \Longrightarrow t'$ and $(s,t'), (s',t') \in R$, or there exist t', t'' such that $t \Longrightarrow t', t' \xrightarrow{\alpha} t''$ and $(s,t'), (s',t'') \in R$.

Now define two relations $\approx, \approx_b$ as follows:

$$\approx \; = \bigcup \{R \mid R \text{ is a weak bisimulation}\},$$
$$\approx_b = \bigcup \{R \mid R \text{ is a branching bisimulation}\}.$$

The notions of weak and branching bisimulations enjoy some nice properties as stated in the following Lemmas 1 and 2, which then lead to the important Theorem 1 that justifies Definition 2.

Lemma 1. *If $\{R_i \mid i \in I\}$ is a set of weak bisimulations, then $\bigcup\{R_i \mid i \in I\}$ is a weak bisimulation. If $\{R_i \mid i \in I\}$ is a set of branching bisimulations, then $\bigcup\{R_i \mid i \in I\}$ is a branching bisimulation.*

For two binary relations R_1, R_2, we write $R_1 \cdot R_2$ for the composition of R_1 and R_2, i.e. $R_1 \cdot R_2 = \{(s,t) \mid \exists u.(s,u) \in R_1, (u,t) \in R_2\}$.

Lemma 2. *If R_1, R_2 are weak bisimulations, then $R_1 \cdot R_2 \cup R_2 \cdot R_1$ is also a weak bisimulation. If R_1, R_2 are branching bisimulations, then $R_1 \cdot R_2 \cup R_2 \cdot R_1$ is also a branching bisimulation.*

The proofs of the above two lemmas directly follow from Definition 2 (Note that we modified the conditions for branching bisimulation as in [11]). With the above two lemmas, it is routine to prove the following theorem, which justifies the definitions of $\approx$ and $\approx_b$.

Theorem 1. *$\approx$ is an equivalence relation, and it is the largest weak bisimulation. $\approx_b$ is an equivalence relation, and it is the largest branching bisimulation.*

With Theorem 1, $\approx$ and $\approx_b$ are usually called *weak bisimilarity* and *branching bisimilarity* respectively.

It is well-known that neither $\approx$ nor $\approx_b$ preserves divergence, i.e. it is possible for two states s and t such that $s \approx t$ while there is an infinite τ-run from s but no infinite τ-run from t.

In order to obtain divergence preserving relations, we can adopt the approach used in [12] by introducing the following definition.

Definition 3 *(*Weak and branching bisimulation with explicit divergence*). Let $\mathcal{A} = \langle S, A, \longrightarrow \rangle$ be an LTS. A state $s \in S$ is said* divergent with respect to *an equivalence relation $\equiv$, written $s \Uparrow_{\equiv}$, if from s there is an infinite τ-run ρ such that all the states on ρ are $\equiv$-equivalent to s.*

An equivalence relation $\equiv$ on S is called a weak bisimulation with explicit divergence *if $\equiv$ is a weak bisimulation and moreover whenever $s \equiv t$ it holds that $s \Uparrow_{\equiv}$ if and only if $t \Uparrow_{\equiv}$.*

An equivalence relation $\equiv$ on S is called a branching bisimulation with explicit divergence *if $\equiv$ is a branching bisimulation and moreover whenever $s \equiv t$ it holds that $s \Uparrow_{\equiv}$ if and only if $t \Uparrow_{\equiv}$.*

Now define two relations $\approx^{\triangle}, \approx_b^{\triangle}$ as follows:

$$\approx^{\triangle} = \bigcup\{\equiv \mid \equiv \text{ is a weak bisimulation with explicit divergence}\},$$
$$\approx_b^{\triangle} = \bigcup\{\equiv \mid \equiv \text{ is a branching bisimulation with explicit divergence}\}.$$

$\approx^{\triangle}$ and $\approx_b^{\triangle}$ are called weak bisimilarity with explicit divergence *and* branching bisimilarity with explicit divergence *respectively.*

At this point, let us see a non-trivial example of branching bisimulation with explicit divergence. Define $\equiv_{sc}$, the *strongly connected* relation, such that $s \equiv_{sc} t$ if and only if $s \Longrightarrow t$ and $t \Longrightarrow s$. That is $s \equiv_{sc} t$ just in case s and t can reach each other by performing τ actions. It only takes a second to check that $\equiv_{sc}$ is an equivalence relation. Moreover we have:

Proposition 1. *$\equiv_{sc}$ is a branching bisimulation with explicit divergence.*

The following lemma is easy to prove.

Lemma 3. *If $\equiv$ is a weak bisimulation with explicit divergence, then $\equiv$ preserves divergence, i.e. whenever $s \equiv t$ then there is an infinite τ-run from s if and only if there is one from t.*

With this lemma, we can show that $\approx^{\triangle}$ preserves divergence as follows. If ρ is an infinite τ-run from s and $s \approx^{\triangle} t$, then there is a weak bisimulation with explicit divergence $\equiv$ such that $s \equiv t$, then by Lemma 3 there is an infinite τ-run from t, thus $\approx^{\triangle}$ preserves divergence. One is tempting to say that with Lemma 3, $\approx^{\triangle}$ obviously preserves divergence, since $\approx^{\triangle}$ is a weak bisimulation with explicit divergence. However, to apply Lemma 3 in this way, we first have to prove that $\approx^{\triangle}$ is a weak bisimulation with explicit divergence, and at least for the moment we do not know if this is indeed the case.

Thus, as the definitions of $\approx$ and $\approx_b$ are justified by Theorem 1, the definitions of $\approx^{\triangle}$ and $\approx_b^{\triangle}$ also need justification. That is to say we need to confirm that $\approx^{\triangle}$ as defined is indeed the largest weak bisimulation with explicit divergence and, $\approx_b^{\triangle}$ the largest branching bisimulation with explicit divergence (as it is stated in the definition we even do not know whether $\approx^{\triangle}$ and $\approx_b^{\triangle}$ are equivalence relations!). But this time the task is not as easy, since we no longer have the corresponding lemmas available as Lemmas 1 and 2 for Theorem 1. As a matter of fact this implies that we do not know whether the notion of weak bisimulation with explicit divergence is a fixed-point of some monotonic functions on the complete lattice of equivalence relations, and hence the Knaster-Tarski fixed-point theorem is not applicable in this case. Thus we need to find a different way to justify Definition 3. For the time being we have the following obvious lemma, which clarifies the justification task.

Lemma 4. $\approx^{\triangle}$ *(*$\approx_b^{\triangle}$*) is the largest weak (branching) bisimulation with explicit divergence if and only if the largest weak (branching) bisimulation with explicit divergence exists.*

Justification of the definition of $\approx_b^{\triangle}$ can be found in [13,14], while not in [12] where it was introduced the first time. While a justification for $\approx_b^{\triangle}$ might be taken as granted, a justification for $\approx^{\triangle}$ may seem to be more necessary. This is because in a weak bisimulation equivalence relation, unlike branching bisimulation, an infinite τ-run from a process may be matched by an infinite τ-run from a related process in a way that the sequences of equivalence classes passed through by the two runs may not be the same. So one needs to be more careful in dealing with $\approx^{\triangle}$. According to Lemma 4, in order to prove that $\approx^{\triangle}$ is a weak bisimulation with explicit divergence we only need to show that the largest weak bisimulation with explicit divergence exists. This approach was taken in [10], where two relations called *complete weak bisimilarity* and *complete branching bisimilarity* were constructed and proved to be the largest weak bisimulation with explicit divergence and largest branching bisimulation with explicit divergence respectively. In this paper, for self containment we will present a justification of the definition of $\approx^{\triangle}$ in the next section, by using the logical characterization result. For the convenience of names, in the paper we will freely use the name of complete weak (branching) bisimilarity as synonym for weak (branching) bisimilarity with explicit divergence.

3 Modal Characterization

The main aims of this section is to look for a modal logic characterization of complete weak bisimilarity $\approx^{\triangle}$, and study its relationship with logic characterizations of other bisimulation equivalences. For that, we first review some of the existing logic characterization results.

In [2] a modal logic, later known as Hennessy-Milner logic (HML), was introduced and proved that two given processes are equivalent under weak bisimularity $\approx$ if and only if they satisfy the same set of HML formulas. This is the so-called Hennessy-Milner theorem. The key constructor in HML is the weak possibility modality $\langle\langle u\rangle\rangle F$, which asserts that after the observation of u some state with property F is reached. In [6], the weak possibility modality was refined to an *until* modality in the form of $F_1\langle\alpha\rangle F_2$, meaning that there is a finite τ-run such that all the states on it satisfy F_1, and the last state can perform an α action and arrives at a state satisfying F_2, and it was proved that the refined logic characterizes branching bisimilarity $\approx_b$, just as HML characterizes weak bisimilarity.

In [5] the weak possibility modality was refined to a *just-before* modality in the form of $F_1\{\alpha\}F_2$, meaning that there is a finite τ-run such that the last state satisfies F_1 and can perform an α action and arrives at a state satisfying F_2, and it was proved that the refined logic, named Φ_{jb}, also characterizes branching bisimilarity $\approx_b$. In [13], Φ_{jb} was further extended to the logic $\Phi_{jb}^{\triangle}$ with a divergence modality in the form of ΔF, meaning that there is an infinite τ-run on which eventually all the states satisfy F, and it was proved that $\Phi_{jb}^{\triangle}$ characterizes branching bisimilarity with explicit divergence $\approx_b^{\triangle}$.

As the starting point of the work of this paper, we describe a modal logic HMLbΔ which is basically $\Phi_{jb}^{\triangle}$ with a derived operator $\langle\langle u\rangle\rangle$. The set of formulas of HMLbΔ is defined by the following syntax of BNF rules:

$$F ::= \bigwedge_{i\in I} F_i | \neg F | F_1\{u\}F_2 | \langle\langle u\rangle\rangle F | \Delta F$$

where I is an index set which could be infinite, $\{u\}$ (with $u \in A \cup \{\epsilon\}$) is the *just-before* modality introduced in [5], $\langle\langle u\rangle\rangle$ is the usual *weak possibility* modality as in [9], and Δ is the *divergence* modality introduced in [13].

Definition 4. *Let $\mathcal{A} = \langle S, A, \longrightarrow\rangle$ be an LTS. The satisfaction relation $\models$ between states and formulas of HMLbΔ is defined by induction on the structure of formulas as follows:*

1. $s \models \bigwedge_{i\in I} F_i$ *if, for all* $i \in I$, $s \models F_i$;
2. $s \models \neg F$ *if* $s \models F$ *does not hold;*
3. $s \models F_1\{u\}F_2$ *if there exist* $t, t' \in S$ *such that* $t \models F_1, t' \models F_2$, $s \Longrightarrow t$ *and* ($t \xrightarrow{u} t'$ *(when* $u \in A$*) or* $t \xrightarrow{\tau} t'$ *(when* $u = \epsilon$*)) or there is* $t \in S$ *such that* $t \models F_1, t \models F_2$, $s \Longrightarrow t$ *and* $u = \epsilon$;
4. $s \models \langle\langle u\rangle\rangle F$ *if there is* $t \in S$ *such that* $s \overset{u}{\Longrightarrow} t$ *and* $t \models F$;
5. $s \models \Delta F$ *if there is an infinite* τ*-run* σ *from* s *such that* $\sigma = s\tau s_1 \tau s_2 \ldots s_i \tau \ldots$ *and there is* $n > 0$ *such that* $s_i \models F$ *for all* $i \geq n$ *(in other words, there are only finitely many positions on* σ *where* F *does not hold).*

First note that this logic can express some interesting properties of infinite behaviours of processes. For example, Δ**true** asserts the existence of an infinite τ-run, where **true** is a short hand for $\bigwedge_{i\in\emptyset} F_i$ (which is the first formula of HMLbΔ according to the BNF rules). The logic is basic, however it might be more expressive than one expect due to the use of infinite conjunction with the construction $\bigwedge_{i\in I} F_i$ when I is an infinite set.

As usual we will write binary conjunction $F_1 \wedge F_2$ for $\bigwedge_{i\in\{1,2\}} F_i$, and binary disjunction $F_1 \vee F_2$ for $\neg \bigwedge_{i\in\{1,2\}} \neg F_i$. For two HMLb$\Delta$ formulas F_1, F_2, we say that F_1 and F_2 are equivalent logic formulas, written $F_1 \Leftrightarrow F_2$, if for any process s of any LTS it holds that $s \models F_1$ if and only if $s \models F_2$.

The following proposition shows that $\langle\langle u\rangle\rangle$ is a derived operator in the sense that it can be defined in terms of the just-before operator $\{u\}$.

Proposition 2. *For any HMLbΔ formula F and $a \neq \tau$, the following equivalences hold:*

1. $\langle\langle \epsilon \rangle\rangle F \Leftrightarrow \mathbf{true}\{\epsilon\}F$;
2. $\langle\langle a \rangle\rangle F \Leftrightarrow \mathbf{true}\{a\}(\mathbf{true}\{\epsilon\}F)$.

Proof. Immediately follows from Definition 4. □

We write HMLb for the sub-logic of HMLbΔ which consists of formulas constructed without the divergence modality Δ. Then HML, the normal Hennessy-Milner logic, is a sub-logic of HMLb consisting of formulas constructed without the just-before modality $\{u\}$. With the result in Proposition 2 that $\langle\langle u \rangle\rangle$ is a derived operator of $\{u\}$, then the following is a theorem which immediately follows from the characterization result for $\Phi_{jb}^{\triangle}$ in [13].

Theorem 2 *(HMLbΔ characterization of $\approx_b^{\triangle}$). Let s, t be two states. Then $s \approx_b^{\triangle} t$ if and only if s and t satisfy the same set of HMLbΔ formulas.*

Likewise, the following is a theorem immediately follows from the characterization result for Φ_{jb} in [5].

Theorem 3 *(HMLb characterization of $\approx_b$). Let s, t be two states. Then $s \approx_b t$ if and only if s and t satisfy the same set of HMLb formulas.*

The following is the famous Hennessy-Milner theorem, which can be found in Chap. 10 of [9].

Theorem 4 *(HML characterization of $\approx$). Let s, t be two states. Then $s \approx t$ if and only if s and t satisfy the same set of HML formulas.*

The last three theorems give modal logic characterizations for $\approx_b^{\triangle}, \approx_b$ and $\approx$ respectively, still missing is a modal logic characterization for $\approx^{\triangle}$. Considering that HMLb is the extension of HML by the just-before modality and that HMLbΔ is the extension of HML by the just-before *and* the divergence modality, an obvious attempt is to extend HML with the divergence modality and hopefully that will give us a logic which characterizes $\approx^{\triangle}$. However it turns out that the divergence construction ΔF is not preserved by $\approx^{\triangle}$, as the following example shows.

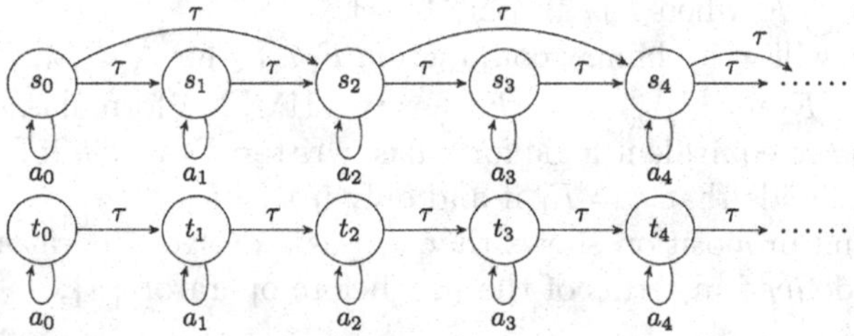

Example 1. The drawing shows an LTS $\mathcal{P} = \langle S, A, \longrightarrow\rangle$ where $A = \{a_i \mid i \geq 0\}$, $S = \{s_i \mid i \geq 0\} \cup \{t_i \mid i \geq 0\}$, and the transition relation is as follows:

- for each $i \geq 0$, if i is even then there are exactly three transitions out of s_i: $s_i \xrightarrow{a_i} s_i, s_i \xrightarrow{\tau} s_{i+1}, s_i \xrightarrow{\tau} s_{i+2}$;
  if i is odd then there are exactly two transitions out of s_i: $s_i \xrightarrow{a_i} s_i, s_i \xrightarrow{\tau} s_{i+1}$.
- for each $i \geq 0$, there are exactly two transitions out of t_i: $t_i \xrightarrow{a_i} t_i, t_i \xrightarrow{\tau} t_{i+1}$.

Now define $\equiv$ to be the following relation:

$$\{(s_i, s_i) \mid i \geq 0\} \cup \{(t_i, t_i) \mid i \geq 0\} \cup \{(s_i, t_i) \mid i \geq 0\} \cup \{(t_i, s_i) \mid i \geq 0\}.$$

The following facts about $\equiv$ are easy to verify:

1. $\equiv$ is an equivalence relation;
2. $\equiv$ is a weak bisimulation;
3. for every $s \in S$, whenever $s \xrightarrow{\tau} t$ then $s \not\equiv t$. Hence whenever $s \equiv t$ then $s \Uparrow_{\equiv}$ if and only if $t \Uparrow_{\equiv}$.

Thus $\equiv$ is a weak bisimulation with explicit divergence, and $s_0 \approx^{\triangle} t_0$. In the following we show that there is an HML formula F such that ΔF is satisfied by s_0 and not by t_0.

Let F_k be the following formula:

$$(\langle\langle a_{2k}\rangle\rangle\mathbf{true} \wedge \langle\langle a_{2k+1}\rangle\rangle\mathbf{true}) \vee (\neg\langle\langle a_{2k}\rangle\rangle\mathbf{true} \wedge \neg\langle\langle a_{2k+1}\rangle\rangle\mathbf{true}).$$

That is, F_k asserts that the pair of actions a_{2k} and a_{2k+1} are either both enabled or both disabled. It is clear that F_k holds for every state of S except s_{2k+1} and t_{2k+1}. Thus $\bigwedge\{F_k \mid k \geq 0\}$ holds on every even numbered position (i.e. $s_0, t_0, s_2, t_2, \ldots$) while does not hold on every odd numbered position (i.e. $s_1, t_1, s_3, t_3, \ldots$).

Now $\Delta \bigwedge\{F_k \mid k \geq 0\}$ is satisfied by s_0 but not by t_0. To see that, note that from s_0 there is an infinite τ-run $\sigma = s_0\tau s_2\tau \ldots s_{2k}\tau \ldots$ and every state on σ satisfies $\bigwedge\{F_k \mid k \geq 0\}$, while the only infinite τ-run from t_0 is $t_0\tau t_1\tau \ldots$, on which there are infinitely many states that do not satisfy $\bigwedge\{F_k \mid k \geq 0\}$. □

Thus, we need to find a different divergence modality. For that we introduce the *weak divergence modality* Δ_ϵ into HMLbΔ, by extending the BNF rules as follows:

$$F ::= \ldots \mid \Delta_\epsilon F.$$

And then add the following interpretation into Definition 4.

6. $s \models \Delta_\epsilon F$ if there is an infinite τ-run σ from s such that for every state s' on σ it holds that $s' \Longrightarrow t$ for some $t \models F$.

The following is a depiction of the condition for $s \models \Delta_\epsilon F$.

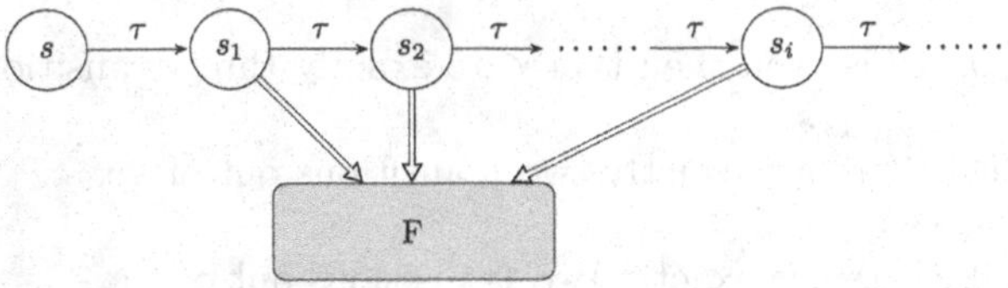

Proposition 3. *For any HMLbΔ formula F, the following equivalence holds:*

$$\Delta_\epsilon F \Leftrightarrow \Delta \langle\langle \epsilon \rangle\rangle F.$$

Proof. Immediately follows from Definition 4 together with the above interpretation for $\Delta_\epsilon F$. □

This proposition shows that Δ_ϵ is a derived operator of Δ and $\langle\langle \epsilon \rangle\rangle$, and that with Δ_ϵ added into HMLbΔ the expressiveness of the extended logic does not increase. So we still call the logic HMLbΔ after extending with Δ_ϵ, and we write HMLΔ_ϵ for the sub-logic where the only modalities allowed are the weak possibility modality $\langle\langle u \rangle\rangle$ and the weak divergence modality Δ_ϵ. With the new divergence modality we can obtain another sub-logic HMLbΔ_ϵ in which Δ_ϵ is allowed but not Δ.

Given a sub-logic L of HMLbΔ, it induces an equivalence relation $\equiv_L$ on states such that $s \equiv_L t$ if and only if s and t satisfy the same set of formulas in the sub-logic. We call $\equiv_L$ the equivalence induced by L. The following is a summary of the sub-logics of HMLbΔ that we concerned about and the corresponding induced equivalences:

1. Let $\equiv_b^\triangle$ be the equivalence induced by HMLbΔ;
2. Let $\equiv_b$ be the equivalence induced by HMLb;
3. Let $\equiv_w$ be the equivalence induced by HML;
4. Let $\equiv_w^{\triangle_\epsilon}$ be the equivalence induced by HMLΔ_ϵ;
5. Let $\equiv_b^{\triangle_\epsilon}$ be the equivalence induced by HMLbΔ_ϵ.

In the rest of this section we will show that HMLΔ_ϵ characterizes $\approx^\triangle$, i.e. $\approx^\triangle$ coincides with $\equiv_w^{\triangle_\epsilon}$. To prove $\approx^\triangle \subseteq \equiv_w^{\triangle_\epsilon}$, we show that for every weak bisimulation with explicit divergence $\equiv$ it holds that $\equiv \subseteq \equiv_w^{\triangle_\epsilon}$ (Lemma 5). To prove $\equiv_w^{\triangle_\epsilon} \subseteq \approx^\triangle$, we show that $\equiv_w^{\triangle_\epsilon}$ is a weak bisimulation with explicit divergence (Lemma 8).

Example 1 shows what ΔF is not preserved by $\approx^\triangle$, while the following lemma guarantees that $\Delta_\epsilon F$ is preserved by $\approx^\triangle$. Here we omit the proof.

Lemma 5. *Let $\equiv$ be a weak bisimulation with explicit divergence, F be an HMLΔ_ϵ formula. If $s \equiv t$ and $s \models F$, then $t \models F$. Thus if $\equiv$ is a weak bisimulation with explicit divergence then $\equiv \subseteq \equiv_w^{\triangle_\epsilon}$.*

Lemma 6. *Let $s \Longrightarrow t$. Then*

1. *whenever $t \models F_1\{u\}F_2$ then $s \models F_1\{u\}F_2$;*
2. *whenever $t \models \langle\langle u\rangle\rangle F$ then $s \models \langle\langle u\rangle\rangle F$;*
3. *whenever $t \models \Delta_\epsilon F$ then $s \models \Delta_\epsilon F$.*

Proof. We only prove 3. With the similar idea we can prove 1 and 2.

Suppose $t \models \Delta_\epsilon F$. Thus from t there is an infinite τ-run ρ such that for each state t' on ρ there exists t'' with $t' \Longrightarrow t''$ and $t'' \models F$. Now since $s \Longrightarrow t$, by adding a prefix to ρ we can easily obtain an infinite run ρ' with starting state s such that for each state t' on ρ' there exists t'' with $t' \Longrightarrow t''$ and $t'' \models F$, hence $s \models \Delta_\epsilon F$. □

The following is the so-called *stuttering lemma* for $\equiv_w^{\triangle_\epsilon}$.

Lemma 7. *If $s \Longrightarrow s', s' \Longrightarrow t$, and $s \equiv_w^{\triangle_\epsilon} t$ then $s \equiv_w^{\triangle_\epsilon} s'$.*

Proof. In this case we only need to prove the following: for any HMLΔ_ϵ formula F, it holds that $s \models F$ if and only if $s' \models F$. We carry out the proof by induction on the structure of F.

For $\bigwedge_{i\in I} F_i$, we have the following sequence of equivalences: $s \models \bigwedge_{i\in I} F_i$ iff $s \models F_i$ for every $i \in I$ (by definition of $\models$) iff $s' \models F_i$ for every $i \in I$ (by induction hypothesis) iff $s' \models \bigwedge_{i\in I} F_i$ (by definition of $\models$). In the same way we can prove it for the case $\neg F$.

For $\langle\langle u\rangle\rangle F$, suppose $s \models \langle\langle u\rangle\rangle F$. Then $t \models \langle\langle u\rangle\rangle F$ by $s \equiv_w^{\triangle_\epsilon} t$, then it immediately follows that $s' \models \langle\langle u\rangle\rangle F$ by $s' \Longrightarrow t$ and Lemma 6. On the other hand, suppose $s' \models \langle\langle u\rangle\rangle F$, then $s \models \langle\langle u\rangle\rangle F$ immediately follows by $s \Longrightarrow s'$ and Lemma 6. In the same way we can prove it for the case $\Delta_\epsilon F$. □

Lemma 8. *$\equiv_w^{\triangle_\epsilon}$ is a weak bisimulation with explicit divergence.*

Proof. To prove that $\equiv_w^{\triangle_\epsilon}$ is a weak bisimulation with explicit divergence, we need to establish the following:

1. $\equiv_w^{\triangle_\epsilon}$ is an equivalence relation;
2. $\equiv_w^{\triangle_\epsilon}$ is a weak bisimulation;
3. if $s \equiv_w^{\triangle_\epsilon} t$, then $s \Uparrow_{\equiv_w^{\triangle_\epsilon}}$ iff $t \Uparrow_{\equiv_w^{\triangle_\epsilon}}$.

It is obvious that $\equiv_w^{\triangle_\epsilon}$ is an equivalence relation. The way to prove that $\equiv_w^{\triangle_\epsilon}$ is a weak bisimulation is exactly the same as the way to prove that $\equiv_w$ is a weak bisimulation [9]. We prove 3. in the following.

First, let us note that for a pair of states s, t with $s \not\equiv_w^{\triangle_\epsilon} t$, by the definition of $\equiv_w^{\triangle_\epsilon}$ there exists an HMLΔ_ϵ formula F_t^s, which is often called a *distinguishing formula* of s and t, such that $s \models F_t^s$ and $t \not\models F_t^s$.

Suppose $s \equiv_w^{\triangle_\epsilon} t$, and $s \Uparrow_{\equiv_w^{\triangle_\epsilon}}$, then there is an infinite τ-run ρ from s with all the states on it $\equiv_w^{\triangle_\epsilon}$-equivalent to s. We construct the following formula F^s

$$\bigwedge\{F_u^s \mid t \overset{\tau}{\Longrightarrow} u, u \not\equiv_w^{\triangle_\epsilon} s\}.$$

Clearly $s \models F^s$. Moreover $s \models \Delta_\epsilon F^s$, since for any state s' on ρ, there is s'' such that $s' \Longrightarrow s''$ and $s'' \models F^s$ (just take s'' to be s', thus $s' \Longrightarrow s'$, and $s' \models F^s$ by $s' \equiv_w^{\triangle_\epsilon} s$). Now because $t \equiv_w^{\triangle_\epsilon} s$, thus $t \models \Delta_\epsilon F^s$. In the following we will show that $t \models \Delta_\epsilon F^s$ implies $t \Uparrow_{\equiv_w^{\triangle_\epsilon}}$.

Since $t \models \Delta_\epsilon F^s$, there is an infinite τ-run σ from t such that for any state t' on σ there exists t'' with $t' \Longrightarrow t''$ and $t'' \models F^s$. Now we will show that if t' is a state on ρ then $t' \equiv_w^{\triangle_\epsilon} t$.

Note that the construction of F^s guarantees the following property:

$$\text{if } t \Longrightarrow t' \text{ and } t' \models F^s \text{ then } t' \equiv_w^{\triangle_\epsilon} t.$$

To see that, let $t \stackrel{\tau}{\Longrightarrow} t'$. Suppose $t' \not\equiv_w^{\triangle_\epsilon} t$, then $t' \not\equiv_w^{\triangle_\epsilon} s$, which implies $t' \not\models F^s$ because in this case $F^s_{t'}$, which is a distinguishing formula of s and t', is one of the conjuncts of F^s, and $t' \not\models F^s_{t'}$.

Now for any state t' on σ, since $t \Longrightarrow t'$ and $t' \Longrightarrow t''$ for some t'' with $t'' \models F^s$, and by the above property of F^s we know that $t'' \equiv_w^{\triangle_\epsilon} t$, then by Lemma 7 $t' \equiv_w^{\triangle_\epsilon} t$, thus σ is the infinite τ-run that we are looking for. □

At last, we can state the modal characterization theorem for $\approx^\triangle$.

Theorem 5. *(HMLΔ_ϵ characterization of $\approx^\triangle$) $\equiv_w^{\triangle_\epsilon}$ coincides with $\approx^\triangle$, that is for any pair of states s and t, $s \approx^\triangle t$ if and only if s and t satisfy the same set of HMLΔ_ϵ formulas.*

Proof. By Lemma 5, $\approx^\triangle \subseteq \equiv_w^{\triangle_\epsilon}$, and by Lemma 8 $\equiv_w^{\triangle_\epsilon}$ is a weak bisimulation with explicit divergence, hence $\equiv_w^{\triangle_\epsilon} \subseteq \approx^\triangle$. □

And at the same time we obtain the following theorem which justifies the definition of $\approx^\triangle$.

Theorem 6. $\approx^\triangle$ *is a weak bisimulation with explicit divergence, and it is the largest weak bisimulation with explicit divergence.*

Proof. By Lemmas 5 and 8, $\equiv_w^{\triangle_\epsilon}$ is the largest weak bisimulation with explicit divergence. By Theorem 5 $\approx^\triangle$ is the same as $\equiv_w^{\triangle_\epsilon}$, hence $\approx^\triangle$ is the largest weak bisimulation with explicit divergence. □

Perhaps a little surprise is the following new modal characterization result for branching bisimilarity with explicit divergence $\approx_b^\triangle$.

Theorem 7 *(HMLbΔ_ϵ characterization of $\approx_b^\triangle$). Let s, t be two states. Then $s \approx_b^\triangle t$ if and only if s and t satisfy the same set of HMLbΔ_ϵ formulas.*

Proof. Here we give the following sketch.

Suppose $s \approx_b^\triangle t$ and $s \models F$ for some HMLbΔ_ϵ formula F, just note that by Proposition 3 there is an HMLbΔ formula F' with $F' \Leftrightarrow F$, then $s \models F'$ and by Theorem 2 $t \models F'$ thus $t \models F$.

For the other direction, we prove that $\equiv_b^{\triangle_\epsilon}$ is a branching bisimulation with explicit divergence. We can prove that $\equiv_b^{\triangle_\epsilon}$ is a branching bisimulation in the same way to prove that $\equiv_b$ is a branching bisimulation as the proof of Theorem 3 in [5]. Suppose $s \equiv_b^{\triangle_\epsilon} t$ and there is an infinite τ-run from s with all the states on the run in the same $\equiv_b^{\triangle_\epsilon}$-equivalence class of s, we can prove that there is an infinite τ-run from t with all the states on the run in the same $\equiv_b^{\triangle_\epsilon}$-equivalence class of t as we prove it for $\equiv_w^{\triangle_\epsilon}$ in Lemma 8, with the help of a lemma similar to Lemma 7 with $\equiv_b^{\triangle_\epsilon}$ in place of $\equiv_w^{\triangle_\epsilon}$. □

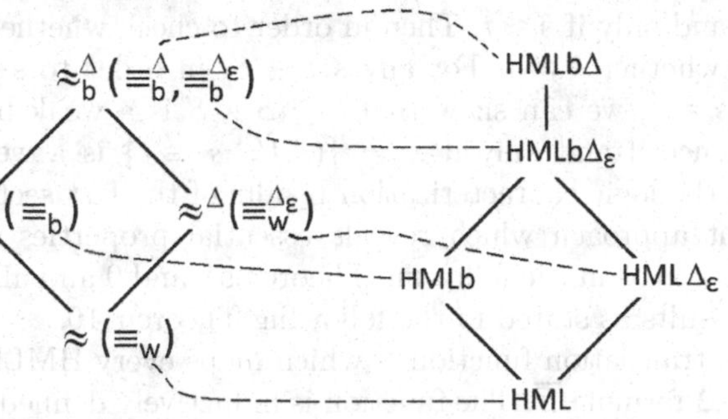

By Theorems 2 and 7, HMLbΔ and HMLbΔ_ϵ both characterize $\approx_b^\triangle$. Now the results about the relationships of various bisimulation equivalence relations and the logics can be summarized as the above lattice shaped diagrams, where on the left the equality on the higher end of an edge is included in the equality on the lower end of the edge, and on the right the logic on the lower end of an edge is a sub-logic of the one on the higher end of the edge, and the dotted lines represent the logic characterization results.

4 Divergence in Finite State Systems

The motivating problem of this section is the problem of checking complete weak bisimilarity for finite-state processes:

> given an LTS $\langle S, A, \longrightarrow \rangle$ and two states $s, t \in S$, where S and A are finite sets, decide whether $s \approx^\triangle t$.

We will show that this problem can be solved by reducing it to the problem of checking weak bisimilarity for finite-state processes which can be solved by a well-known partition algorithm [7]:

> given an LTS $\langle S, A, \longrightarrow \rangle$ and two states $s, t \in S$, where S and A are finite sets, decide whether $s \approx t$.

The reduction is as follows. Let $\mathcal{P} = \langle S, A, \longrightarrow \rangle$ be a finite-state labeled transition system, i.e. both S and A are finite sets, δ be an action not in A. Then we can construct a new finite-state LTS $\mathcal{P}_\delta = \langle \widehat{S}, \widehat{A}, \longrightarrow' \rangle$ where $\widehat{S} = \{\hat{s} \mid s \in S\}$, $\widehat{A} = A \cup \{\delta\}$, $\longrightarrow' = \{(\hat{s}, \alpha, \hat{s}') \mid s \xrightarrow{\alpha} s'\} \cup \{(\hat{s}, \delta, \hat{s}) \mid s \stackrel{\tau}{\Longrightarrow} s\}$.

The idea of the reduction is pretty straightforward: in a finite-state system, the existence of an infinite τ-run from a state s is equivalent to the existence of a so-called *looping state* s' such that $s \Longrightarrow s'$ and $s' \stackrel{\tau}{\Longrightarrow} s'$, and then the looping states can be marked by a particular new action δ. Thus the transitions of the constructed system $\mathcal{P}_\delta$ is like the original system $\mathcal{P}$ except that every looping state s is indicated by a new transition $\hat{s} \stackrel{\delta}{\longrightarrow}{}' \hat{s}$. In the following when there will cause no confusion we will simply write $\hat{s} \stackrel{\alpha}{\longrightarrow} \hat{t}$ instead of $\hat{s} \stackrel{\alpha}{\longrightarrow}{}' \hat{t}$ for $s, t \in S$.

Now to complete the reduction, we will show that for any $s, t \in S$, it holds that $s \approx^{\triangle} t$ if and only if $\hat{s} \approx \hat{t}$. Then in order to check whether $s \approx^{\triangle} t$ we only need to check whether $\hat{s} \approx \hat{t}$. For any $s, t \in S$, in order to show that $s \approx^{\triangle} t$ if and only if $\hat{s} \approx \hat{t}$, we can show that $\equiv \subseteq S \times S$ is a weak bisimulation with explicit divergence if and only if $\widehat{\equiv} = \{(\hat{s}, \hat{t}) \mid s \equiv t\}$ is a weak bisimulation. However, with the logic characterization results of the last section, here we will take a different approach which reveals essential properties of the reduction construction as stated in the following Theorems 8 and 9 and allows us to obtain more general results as stated in the following Theorem 10.

We define a translation function $\overline{}$ which maps every HMLbΔ formula F to another HMLbΔ formula $\overline{F}$. The function is inductively defined on the structure of the formula as follows:

$$
\begin{array}{rl|rl}
\overline{\bigwedge_{i\in I} F_i} = \bigwedge_{i\in I} \overline{F_i} & & \overline{\neg F} = \neg \overline{F} \\
\overline{F_1\{u\}F_2} = \overline{F_1}\{u\}\overline{F_2} & (u \neq \delta) & \overline{F_1\{\delta\}F_2} = \neg\mathbf{true} \\
\overline{\langle\langle u\rangle\rangle F} = \langle\langle u\rangle\rangle \overline{F} & (u \neq \delta) & \overline{\langle\langle \delta\rangle\rangle F} = \neg\mathbf{true} \\
\overline{\Delta F} = \mathbf{true}\{\delta\}\overline{F} & & \overline{\Delta_\epsilon F} = \langle\langle \delta\rangle\rangle \overline{F}
\end{array}
$$

Theorem 8. *If F is an HMLbΔ formula, then $\overline{F}$ is an HMLb formula. Moreover if F is an HMLΔ_ϵ formula, then $\overline{F}$ is an HML formula.*

For a finite-state LTS $\mathcal{P} = \langle S, A, \longrightarrow\rangle$, let $\mathcal{P}_\delta = \langle \widehat{S}, A, \longrightarrow'\rangle$ be the finite-state LTS constructed above, $s \in S$. Then for any HMLbΔ formula F, it holds that $s \models F$ if and only if $\hat{s} \models \overline{F}$.

The proof, which is omitted here, is a routine induction on the structure of the formulas. Here we just explain the idea behind the translation function $\overline{}$ from which one can see the rationale behind Theorem 8. The key is to understand why $F_1\{\delta\}F_2$ is translated to $\neg\mathbf{true}$. As we have pointed out above, δ is an action which is not in A and which is used in the reduction to mark divergence. That implies that any process s from $\mathcal{P}$ is not capable of an δ action, hence the property $F_1\{\delta\}F_2$ will never be satisfied by any process from $\mathcal{P}$. That is why $F_1\{\delta\}F_2$ is translated to $\neg\mathbf{true}$. For the same reason $\langle\langle \delta\rangle\rangle F$ is also translated to $\neg\mathbf{true}$.

Also, we can define a translation function $\underline{}$ which maps every HMLb formula F to an HMLbΔ formula $\underline{F}$. The function is inductively defined on the structure of the formula as follows:

$$
\begin{array}{rl|rl}
\underline{\bigwedge_{i\in I} F_i} = \bigwedge_{i\in I} \underline{F_i} & & \underline{\neg F} = \neg \underline{F} \\
\underline{F_1\{u\}F_2} = \underline{F_1}\{u\}\underline{F_2} & (u \neq \delta) & \underline{F_1\{\delta\}F_2} = \Delta(F_1 \wedge F_2) \\
\underline{\langle\langle u\rangle\rangle F} = \langle\langle u\rangle\rangle \underline{F} & (u \neq \delta) & \underline{\langle\langle \delta\rangle\rangle F} = \Delta_\epsilon \underline{F}
\end{array}
$$

Theorem 9. *If F is an HMLb formula, then $\underline{F}$ is an HMLbΔ formula. Moreover if F is an HML formula, then $\underline{F}$ is an HMLΔ_ϵ formula.*

For a finite-state LTS $\mathcal{P} = \langle S, A, \longrightarrow \rangle$, let $\mathcal{P}_\delta = \langle \widehat{S}, A, \longrightarrow' \rangle$ be the finite-state LTS constructed above, $s \in S$. Then for any HMLb formula F, it holds that $s \models \underline{F}$ if and only if $\hat{s} \models F$.

Now we obtain the following theorem which guarantees the correctness of our reduction.

Theorem 10. *For a finite-state LTS $\mathcal{P} = \langle S, A \longrightarrow \rangle$, let $\mathcal{P}_\delta = \langle \widehat{S}, A, \longrightarrow' \rangle$ be the finite-state LTS constructed above. Then for $s, t \in S$:*

1. $s \approx^{\triangle} t$ if and only if $\hat{s} \approx \hat{t}$;
2. $s \approx_b^{\triangle} t$ if and only if $\hat{s} \approx_b \hat{t}$.

Proof. Here we only prove 1. The way to prove 2. is the same. Since $\approx^{\triangle}$ coincides with $\equiv_w^{\triangle_\epsilon}$ and $\approx$ coincides with $\equiv_w$, to prove 1. we only need to prove that $s \equiv_w^{\triangle_\epsilon} t$ if and only if $\hat{s} \equiv_w \hat{t}$.

Suppose $s \equiv_w^{\triangle_\epsilon} t$. If $\hat{s} \models F$ for some HML formula F, then by Theorem 9, $\underline{F}$ is an HMLΔ_ϵ formula and $s \models \underline{F}$. Then by the condition that $s \equiv_w^{\triangle_\epsilon} t$, we have $t \models \underline{F}$, and again by Theorem 9, $\hat{t} \models F$. Thus $\hat{s} \equiv_w \hat{t}$.

Suppose $\hat{s} \equiv_w \hat{t}$. If $s \models F$ for some HMLΔ_ϵ formula F, then by Theorem 8, $\overline{F}$ is an HML formula and $\hat{s} \models F$. Then by the condition that $\hat{s} \equiv \hat{t}$, we have $\hat{t} \models \overline{F}$, and again by Theorem 8, $t \models F$. Thus $s \equiv_w^{\triangle_\epsilon} t$. □

Theorem 8 also suggests a simple solution to the model checking problem for HMLbΔ (which can have many solutions). The model checking problem here is to ask, for any given state s of a fnite-state LTS $\mathcal{P}$ and any given finite HMLbΔ formula F (finite in the sense that only finite conjunctions are allowed in F), how to decide whether $s \models F$ holds or not. By Theorem 8, this problem can be reduced to the problem of deciding if $\hat{s} \models \overline{F}$ holds or not, which comes with simple decision procedures because here $\hat{s}$ is a state in the finite-state LTS $\mathcal{P}_\delta$ and $\overline{F}$ is a finite HMLb formula.

5 Conclusion

To summarize, by introducing a new divergence modality, the weak divergence modality Δ_ϵ, we obtain logic characterization results for two divergence sensitive bisimulation equivalence relations. One is the first modal logic characterization for complete weak bisimilarity $\approx^{\triangle}$, and the other is a new modal logic characterization for branching bisimilarity with explicit divergence $\approx_b^{\triangle}$. With these new characterization results we showed a clear picture of the sub-logic relationships of various logic characterization results. By using these new characterization results, we provide reductions from the divergence sensitive equality checking problems and model checking problems to the divergence blind equality checking problems and model checking problems respectively for finite-state systems.

Complete weak bisimilarity $\approx^{\triangle}$ was first defined in [10], which is a refinement of weak bisimilarity $\approx$ [9] by taking divergence behavior into account. Since this is a relatively new equivalence relation, the logic characterization problem and equality checking problem for finite-state systems have not been treated before this paper. The relation $\approx_b^{\triangle}$ was defined in [12] which is a refinement of branching bisimilarity $\approx_b$ [12]. In [15], the equality checking problem of *stutter equivalence* on Kripke structures is solved by a reduction to the equality checking problem of *divergence blind* stutter equivalence problem. Stutter equivalence and divergence blind stutter equivalence are the Kripke structure versions of branching bisimilarity with explicit divergence and branching bisimilarity. The reduction presented in Sect. 4 is inspired by the reduction in [15].

The study of modal logic characterization of bisimulation equivalence relations was initiated by Hennessy and Milner in [2]. For branching bisimilarity, modal characterization results were studied in [5,6], where different modalities for branching structures were used. In [6], besides the extension of Hennessy-Milner logic with the until operator mentioned earlier in the paper, two other logics were proposed to characterize branching bisimilarity. One is another extension of Hennessy-Milner logic which exploits the power of backward modalities. The other is CTL* without the next-time operator interpreted over all paths, not just over maximal ones. In [13] a modal logic was proposed to characterize branching bisimilarity with explicit divergence by combining modalities for branching bisimilarity in [5] and a divergence modality Δ. In [14], an extension of CTL* without the next operator is proposed which also characterizes branching bisimilarity with explicit divergence.

References

1. Hennessy, M.C.B., Plotkin, G.D.: A term model for CCS. In: Dembiński, P. (ed.) MFCS 1980. LNCS, vol. 88, pp. 261–274. Springer, Heidelberg (1980). https://doi.org/10.1007/BFb0022510
2. Hennessy, M., Milner, R.: Algebraic laws for non-determinism and concurrency. J. ACM **32**(1), 131–161 (1985)
3. Walker, D.J.: Bisimulation and divergence. Inf. Comput. **85**, 212–241 (1990)
4. van Benthem, J.: Modal Logic and Classical Logic. Bibliopolis, Naples (1983)
5. Glabbeek, R.J.: The linear time—branching time spectrum II. In: Best, E. (ed.) CONCUR 1993. LNCS, vol. 715, pp. 66–81. Springer, Heidelberg (1993). https://doi.org/10.1007/3-540-57208-2_6
6. de Nicola, R., Vaandrager, F.: Three logics for branching bisimulation. J. ACM **42**(2), 458–487 (1995)
7. Kannelakis, P., Smolka, S.: CCS expressions, finite state processes and three problems of equivalence. Inf. Comput. **86**, 43–68 (1990)
8. Park, D.: Concurrency and automata on infinite sequences. In: Deussen, P. (ed.) GI-TCS 1981. LNCS, vol. 104, pp. 167–183. Springer, Heidelberg (1981). https://doi.org/10.1007/BFb0017309
9. Milner, R.: Communication and Concurrency. Prentice-Hall, Upper Saddle River (1989)

10. Liu, X., Yu, T., Zhang, W.: Analyzing divergence in bisimulation semantics. In: Proceedings of POPL (2017)
11. Basten, T.: Branching bisimularity is an equivalence indeed!. Inf. Process. Lett. **58**(3), 141–147 (1996)
12. van Glabbeek, R.J., Weijland, P.: Branching time and abstraction in bisimulation semantics. J. ACM **43**(3), 555–600 (1996)
13. van Glabbeek, R.J., Luttik, B., Trcka, N.: Branching bisimilarity with explicit divergence. Fundam. Inf. **93**(4), 371–392 (2009)
14. van Glabbeek, R.J., Luttik, B., Trcka, N.: Computation tree logic with deadlock detection. Log. Methods Comput. Sci. **5**(4) (2009)
15. Groote, J.F., Vaandrager, F.: An efficient algorithm for branching bisimulation and stuttering equivalence. In: Paterson, M.S. (ed.) ICALP 1990. LNCS, vol. 443, pp. 626–638. Springer, Heidelberg (1990). https://doi.org/10.1007/BFb0032063

Lambda-Calculi and Types

Call-by-Need, Neededness and All That

Delia Kesner[1], Alejandro Ríos[2], and Andrés Viso[2,3](✉)

[1] IRIF, CNRS and Univ. Paris-Diderot, Paris, France
[2] Universidad de Buenos Aires, Buenos Aires, Argentina
aeviso@dc.uba.ar
[3] CONICET, Buenos Aires, Argentina

Abstract. We show that call-by-need is observationally equivalent to weak-head needed reduction. The proof of this result uses a semantical argument based on a (non-idempotent) intersection type system called $\mathcal{V}$. Interestingly, system $\mathcal{V}$ also allows to syntactically identify all the weak-head needed redexes of a term.

1 Introduction

One of the fundamental notions underlying this paper is the one of *needed reduction* in λ-calculus, which is to be used here to understand (lazy) evaluation of functional programs. Key notions are those of reducible and non-reducible programs: the former are programs (represented by λ-terms) containing non-evaluated subprograms, called reducible expressions (redexes), whereas the latter can be seen as definitive results of computations, called normal forms. It turns out that every reducible program contains a special kind of redex known as needed or, in other words, every λ-term not in normal form contains a needed redex. A redex r is said to be *needed* in a λ-term t if r has to be contracted (*i.e.* evaluated) sooner or later when reducing t to *normal form*, or, informally said, if there is no way of avoiding r to reach a normal form.

The needed strategy, which always contracts a needed redex, is normalising [8], *i.e.* if a term can be reduced (in any way) to a normal form, then contraction of needed redexes necessarily terminates. This is an excellent starting point to design an evaluation strategy, but unfortunately, neededness of a redex is not decidable [8]. As a consequence, real implementations of functional languages cannot be directly based on this notion.

Our goal is, however, to establish a clear connection between the semantical notion of neededness and different implementations of lazy functional languages (*e.g.* Miranda or Haskell). Such implementations are based on *call-by-need calculi*, pioneered by Wadsworth [20], and extensively studied *e.g.* in [3]. Indeed, call-by-need calculi fill the gap between the well-known operational semantics of the call-by-name λ-calculus and the actual implementations of lazy functional languages. While call-by-name re-evaluates an argument each time it is used –an

This work was partially founded by LIA INFINIS.

C. Baier and U. Dal Lago (Eds.): FOSSACS 2018, LNCS 10803, pp. 241–257, 2018.
https://doi.org/10.1007/978-3-319-89366-2_13

operation which is quite expensive– call-by-need can be seen as a *memoized* version of call-by-name, where the value of an argument is stored the first time it is evaluated for subsequent uses. For example, if $t = \Delta\,(I\,I)$, where $\Delta = \lambda x.x\,x$ and $I = \lambda z.z$, then call-by-name duplicates the argument $I\,I$, while lazy languages first reduce $I\,I$ to the value I so that further uses of this argument do not need to evaluate it again.

While the notion of needed reduction is defined with respect to (full strong) *normal forms*, call-by-need calculi evaluate programs to special values called *weak-head normal forms*, which are either abstractions or arbitrary applications headed by a variable (*i.e.* terms of the form $x\,t_1 \ldots t_n$ where $t_1 \ldots t_n$ are arbitrary terms). To overcome this shortfall, we first adapt the notion of needed redex to terms that are not going to be fully reduced to *normal forms* but only to *weak-head normal forms*. Thus, informally, a redex r is *weak-head needed* in a term t if r has to be contracted sooner or later when reducing t to a weak-head normal form. The derived notion of strategy is called a *weak-head needed strategy*, which always contracts a weak-head needed redex.

This paper introduces two independent results about weak-head neededness, both obtained by means of (non-idempotent) intersection types [12,13] (a survey can be found in [9]). We consider, in particular, typing system $\mathcal{V}$ [14] and show that it allows to identify all the weak-head needed redexes of a weak-head normalising term. This is done by adapting the classical notion of *principal type* [17] and proving that a redex in a weak-head normalising term t is weak-head needed iff it is typed in a principally typed derivation for t in $\mathcal{V}$.

Our second goal is to show observational equivalence between call-by-need and weak-head needed reduction. Two terms are observationally equivalent when all the empirically testable computations on them are identical. This means that a term t can be evaluated to a weak-head normal form using the call-by-need machinery if and only if the weak-head needed reduction normalises t.

By means of system $\mathcal{V}$ mentioned so far we use a technique to reason about observational equivalence that is flexible, general and easy to verify or even certify. Indeed, system $\mathcal{V}$ provides a semantic argument: first showing that a term t is typable in system $\mathcal{V}$ iff it is normalising for the weak-head needed strategy ($t \in \mathcal{WN}_{\mathtt{whnd}}$), then by resorting to some results in [14], showing that system $\mathcal{V}$ is complete for call-by-name, *i.e.* a term t is typable in system $\mathcal{V}$ iff t is normalising for call-by-name ($t \in \mathcal{WN}_{\mathtt{name}}$); and that t is normalising for call-by-name iff t is normalising for call-by-need ($t \in \mathcal{WN}_{\mathtt{need}}$). Thus completing the following chain of equivalences:

$$t \in \mathcal{WN}_{\mathtt{whnd}} \Longleftrightarrow t \text{ typable in } \mathcal{V} \Longleftrightarrow t \in \mathcal{WN}_{\mathtt{name}} \Longleftrightarrow t \in \mathcal{WN}_{\mathtt{need}}$$

This leads to the observational equivalence between call-by-need, call-by-name and weak-head needed reduction.

Structure of the paper: Sect. 2 introduces preliminary concepts while Sect. 3 defines different notions of needed reduction. The type system $\mathcal{V}$ is studied in Sect. 4. Section 5 extends β-reduction to derivation trees. We show in Sect. 6 how system $\mathcal{V}$ identifies weak-head needed redexes, while Sect. 7 gives a characterisation of normalisation for the weak-head needed reduction. Sect. 8 is devoted to define call-by-need. Finally, Sect. 9 presents the observational equivalence result.

2 Preliminaries

This section introduces some standard definitions and notions concerning the reduction strategies studied in this paper, that is, call-by-name, head and weak-head reduction, and neededness, this later notion being based on the *theory of residuals* [7].

2.1 The Call-by-Name Lambda-Calculus

Given a countable infinite set $\mathcal{X}$ of variables $x, y, z, \ldots$ we consider the following grammar:

$$\begin{array}{rl} \textbf{(Terms)} & t, u ::= x \in \mathcal{X} \mid t\,u \mid \lambda x.t \\ \textbf{(Values)} & v ::= \lambda x.t \\ \textbf{(Contexts)} & \mathtt{C} ::= \Box \mid \mathtt{C}\,t \mid t\,\mathtt{C} \mid \lambda x.\mathtt{C} \\ \textbf{(Name contexts)} & \mathtt{E} ::= \Box \mid \mathtt{E}\,t \end{array}$$

The set of λ-terms is denoted by $\mathcal{T}_{\mathtt{a}}$. We use I, K and Ω to denote the terms $\lambda x.x$, $\lambda x.\lambda y.x$ and $(\lambda x.x\,x)\,(\lambda x.x\,x)$ respectively. We use $\mathtt{C}\langle t\rangle$ (resp. $\mathtt{E}\langle t\rangle$) for the term obtained by replacing the hole $\Box$ of $\mathtt{C}$ (resp. $\mathtt{E}$) by t. The sets of ***free*** and ***bound variables*** of a term t, written respectively $\mathtt{fv}(t)$ and $\mathtt{bv}(t)$, are defined as usual [7]. We work with the standard notion of ***α-conversion***, *i.e.* renaming of bound variables for abstractions; thus for example $\lambda x.x\,y =_\alpha \lambda z.z\,y$.

A term of the form $(\lambda x.t)\,u$ is called a ***β-redex*** (or just ***redex*** when β is clear from the context) and λx is called the ***anchor*** of the redex. The ***one-step reduction relation*** $\rightarrow_\beta$ (resp. $\rightarrow_{\mathtt{name}}$) is given by the closure by contexts $\mathtt{C}$ (resp. $\mathtt{E}$) of the rewriting rule $(\lambda x.t)\,u \mapsto_\beta t\,\{x/u\}$, where $\_\{\_/\_\}$ denotes the capture-free standard higher-order substitution. Thus, call-by-name forbids reduction inside arguments and λ-abstractions, *e.g.* $(\lambda x.II)\,(II) \rightarrow_\beta (\lambda x.II)\,I$ and $(\lambda x.II)\,(II) \rightarrow_\beta (\lambda x.I)\,(II)$ but neither $(\lambda x.II)\,(II) \rightarrow_{\mathtt{name}} (\lambda x.II)\,I$ nor $(\lambda x.II)\,(II) \rightarrow_{\mathtt{name}} (\lambda x.I)\,(II)$ holds. We write $\twoheadrightarrow_\beta$ (resp. $\twoheadrightarrow_{\mathtt{name}}$) for the reflexive-transitive closure of $\rightarrow_\beta$ (resp. $\rightarrow_{\mathtt{name}}$).

2.2 Head, Weak-Head and Leftmost Reductions

In order to introduce different notions of reduction, we start by formalising the general mechanism of reduction which consists in contracting a redex at some specific occurrence. ***Occurrences*** are finite words over the alphabet $\{0, 1\}$. We use ϵ to denote the empty word and notation a^{n} for $\mathsf{n} \in \mathbb{N}$ concatenations of some letter a of the alphabet. The set of ***occurrences*** of a given term is defined by induction as follows: $\mathsf{oc}(x) \overset{def}{=} \{\epsilon\}$; $\mathsf{oc}(t\,u) \overset{def}{=} \{\epsilon\} \cup \{0\mathsf{p} \mid \mathsf{p} \in \mathsf{oc}(t)\} \cup \{1\mathsf{p} \mid \mathsf{p} \in \mathsf{oc}(u)\}$; $\mathsf{oc}(\lambda x.t) \overset{def}{=} \{\epsilon\} \cup \{0\mathsf{p} \mid \mathsf{p} \in \mathsf{oc}(t)\}$.

Given two occurrences p and q, we use the notation $\mathsf{p} \leq \mathsf{q}$ to mean that p is a ***prefix*** of q, *i.e.* there is p′ such that $\mathsf{pp'} = \mathsf{q}$. We denote by $t|_\mathsf{p}$ the ***subterm of*** t ***at occurrence*** p, defined as expected [4], thus for example $((\lambda x.y)\, z)|_{00} = y$. The set of ***redex occurrences*** of t is defined by $\mathtt{roc}(t) \stackrel{def}{=} \{\mathsf{p} \in \mathsf{oc}(t) \mid t|_\mathsf{p} = (\lambda x.s)\, u\}$. We use the notation $\mathsf{r} : t \rightarrow_\beta t'$ to mean that $\mathsf{r} \in \mathtt{roc}(t)$ and t reduces to t' by ***contracting*** the redex at occurrence r, *e.g.* $000 : (\lambda x.(\lambda y.y)\, x\, x)\, z \rightarrow_\beta (\lambda x.x\, x)\, z$. This notion is extended to reduction sequences as expected, and noted $\rho : t \twoheadrightarrow_\beta t'$, where ρ is the list of all the redex occurrences contracted along the reduction sequence. We use *nil* to denote the empty reduction sequence, so that $nil : t \twoheadrightarrow_\beta t$ holds for every term t.

Any term t has exactly one of the following forms: $\lambda x_1 \ldots . \lambda x_n . y\, t_1 \ldots t_m$ or $\lambda x_1 \ldots . \lambda x_n . (\lambda y.s)\, u\, t_1 \ldots t_m$ with $n, m \geq 0$. In the latter case we say that $(\lambda y.s)\, u$ is the ***head redex*** of t, while in the former case there is no head redex. Moreover, if $n = 0$, we say that $(\lambda y.s)\, u$ is the ***weak-head redex*** of t. In terms of occurrences, the *head redex* of t is the *minimal* redex occurrence of the form 0^n with $n \geq 0$. In particular, if it satisfies that $t|_{0^k}$ is not an abstraction for every $k \leq n$, it is the *weak-head redex* of t. A reduction sequence contracting at each step the head redex (resp. weak-head redex) of the corresponding term is called the ***head reduction*** (resp. ***weak-head reduction***).

Given two redex occurrences $\mathsf{r}, \mathsf{r}' \in \mathtt{roc}(t)$, we say that r is ***to-the-left of*** r′ if the anchor of r is to the left of the anchor of r′. Thus for example, the redex occurrence 0 is to-the-left of 1 in the term $(I\, x)\, (I\, y)$, and ϵ is to-the-left of 00 in $(\lambda x.(I\, I))\, z$. Alternatively, the relation *to-the-left* can be understood as a dictionary order between redex occurrences, *i.e.* r is *to-the-left of* r′ if either $\mathsf{r}' = \mathsf{rq}$ with $\mathsf{q} \neq \epsilon$ (*i.e.* r is a proper prefix of r′); or $\mathsf{r} = \mathsf{p0q}$ and $\mathsf{r}' = \mathsf{p1q}'$ (*i.e.* they share a common prefix and r is on the left-hand side of an application while r′ is on the right-hand side). Notice that in any case this implies $\mathsf{r}' \not\leq \mathsf{r}$. Since this notion defines a total order on redexes, every term not in normal form has a unique ***leftmost redex***. The term t ***leftmost reduces*** to t' if t reduces to t' and the reduction step contracts the leftmost redex of t. For example, $(I\, x)\, (I\, y)$ leftmost reduces to $x\, (I\, y)$ and $(\lambda x.(I\, I))\, z$ leftmost reduces to $I\, I$. This notion extends to reduction sequences as expected.

3 Towards Neededness

Needed reduction is based on two fundamental notions: that of residual, which describes how a given redex is traced all along a reduction sequence, and that of normal form, which gives the form of the expected result of the reduction sequence. This section extends the standard notion of needed reduction [8] to those of head and weak-head needed reductions.

3.1 Residuals

Given a term t, $\mathsf{p} \in \mathsf{oc}(t)$ and $\mathsf{r} \in \mathtt{roc}(t)$, the ***descendants of* p *after* r *in*** t, written p/r, is the set of *occurrences* defined as follows:

$$\begin{array}{c} \varnothing \text{ if } \mathsf{p} = \mathsf{r} \text{ or } \mathsf{p} = \mathsf{r0} \\ \{\mathsf{p}\} \text{ if } \mathsf{r} \not\leq \mathsf{p} \\ \{\mathsf{rq}\} \text{ if } \mathsf{p} = \mathsf{r00q} \\ \{\mathsf{rkq} \mid s|_{\mathsf{k}} = x\} \text{ if } \mathsf{p} = \mathsf{r1q} \text{ with } t|_{\mathsf{r}} = (\lambda x.s)\, u \end{array}$$

For instance, given $t = (\lambda x.(\lambda y.x)\, x)\, z$, then $\mathsf{oc}(t) = \{\epsilon, 0, 1, 00, 000, 001, 0000\}$, $\mathtt{roc}(t) = \{\epsilon, 00\}$, $00/00 = \varnothing$, $\epsilon/00 = \{\epsilon\}$, $00/\epsilon = \{\epsilon\}$ and $1/\epsilon = \{1, 00\}$.

Notice that $\mathsf{p}/\mathsf{r} \subseteq \mathsf{oc}(t')$ where $\mathsf{r} : t \to_\beta t'$. Furthermore, if p is the occurrence of a redex in t (*i.e.* $\mathsf{p} \in \mathtt{roc}(t)$), then $\mathsf{p}/\mathsf{r} \subseteq \mathtt{roc}(t')$, and each position in p/r is called a ***residual*** of p after reducing r. This notion is extended to sets of redex occurrences, indeed, the ***residuals of*** $\mathcal{P}$ ***after*** r ***in*** t are $\mathcal{P}/\mathsf{r} \stackrel{\text{def}}{=} \bigcup_{\mathsf{p}\in\mathcal{P}} \mathsf{p}/\mathsf{r}$. In particular $\varnothing/\mathsf{r} = \varnothing$. Given $\rho : t \twoheadrightarrow_\beta t'$ and $\mathcal{P} \subseteq \mathtt{roc}(t)$, the ***residuals of*** $\mathcal{P}$ ***after the sequence*** ρ are: $\mathcal{P}/nil \stackrel{\text{def}}{=} \mathcal{P}$ and $\mathcal{P}/\mathsf{r}\rho' \stackrel{\text{def}}{=} (\mathcal{P}/\mathsf{r})/\rho'$.

Stability of the to-the-left relation makes use of the notion of residual:

Lemma 1. *Given a term* t*, let* $\mathsf{l}, \mathsf{r}, \mathsf{s} \in \mathtt{roc}(t)$ *such that* l *is to-the-left of* r, $\mathsf{s} \not\leq \mathsf{l}$ *and* $\mathsf{s} : t \to_\beta t'$*. Then,* $\mathsf{l} \in \mathtt{roc}(t')$ *and* l *is to-the-left of* r' *for every* $\mathsf{r}' \in \mathsf{r}/\mathsf{s}$.

Proof. By case analysis using the definition of *to-the-left* [15]. □

Notice that this result does not only implies that the leftmost redex is preserved by reduction of other redexes, but also that the residual of the leftmost redex occurs in exactly the same occurrence as the original one.

Corollary 1. *Given a term* t*, and* $\mathsf{l} \in \mathtt{roc}(t)$ *the leftmost redex of* t*, if the reduction* $\rho : t \twoheadrightarrow_\beta t'$ *contracts neither* l *nor any of its residuals, then* $\mathsf{l} \in \mathtt{roc}(t')$ *is the leftmost redex of* t'.

Proof. By induction on the length of ρ using Lemma 1. □

3.2 Notions of Normal Form

The expected result of evaluating a program is specified by means of some appropriate notion of normal form. Given any relation $\to_\mathcal{R}$, a term t is said to be in $\mathcal{R}$***-normal form*** ($\mathcal{NF}_\mathcal{R}$) iff there is no t' such that $t \to_\mathcal{R} t'$. A term t is $\mathcal{R}$***-normalising*** ($\mathcal{WN}_\mathcal{R}$) iff there exists $u \in \mathcal{NF}_\mathcal{R}$ such that $t \twoheadrightarrow_\mathcal{R} u$. Thus, given an $\mathcal{R}$-normalising term t, we can define the set of $\mathcal{R}$-normal forms of t as $\mathtt{nf}_\mathcal{R}(t) \stackrel{\text{def}}{=} \{t' \mid t \twoheadrightarrow_\mathcal{R} t' \wedge t' \in \mathcal{NF}_\mathcal{R}\}$.

In particular, it turns out that a term in ***weak-head*** β***-normal form*** ($\mathcal{WHNF}_\beta$) is of the form $x\, t_1 \ldots t_n$ ($n \geq 0$) or $\lambda x.t$, where $t, t_1, \ldots, t_n$ are arbitrary terms, *i.e.* it has no weak-head redex. The set of weak-head β-normal forms of t is $\mathtt{whnf}_\beta(t) \stackrel{\text{def}}{=} \{t' \mid t \twoheadrightarrow_\beta t' \wedge t' \in \mathcal{WHNF}_\beta\}$.

Similarly, a term in ***head β-normal form*** ($\mathcal{HNF}_\beta$) turns out to be of the form $\lambda x_1.\ldots\lambda x_n.x\,t_1\ldots t_m$ $(n, m \geq 0)$, *i.e.* it has no head redex. The set of head β-normal forms of t is given by $\mathtt{hnf}_\beta(t) \stackrel{\text{def}}{=} \{t' \mid t \twoheadrightarrow_\beta t' \wedge t' \in \mathcal{HNF}_\beta\}$.

Last, any term in ***β-normal form*** ($\mathcal{NF}_\beta$) has the form $\lambda x_1.\ldots\lambda x_n.x\,t_1\ldots t_m$ $(n, m \geq 0)$ where $t_1, \ldots, t_m$ are themselves in β-normal form. It is well-known that the set $\mathtt{nf}_\beta(t)$ is a singleton, so we may use it either as a set or as its unique element.

It is worth noticing that $\mathcal{NF}_\beta \subset \mathcal{HNF}_\beta \subset \mathcal{WHNF}_\beta$. Indeed, the inclusions are strict, for instance $\lambda x.(\lambda y.y)\,z$ is in weak-head but not in head β-normal form, while $x\,((\lambda y.y)\,x)\,z$ is in head but not in β-normal form.

3.3 Notions of Needed Reduction

The different notions of normal form considered in Sect. 3.2 suggest different notions of needed reduction, besides the standard one in the literature [8]. Indeed, consider $\mathsf{r} \in \mathtt{roc}(t)$. We say that r is ***used*** in a reduction sequence ρ iff ρ reduces r or some residual of r. Then:

1. r is ***needed*** in t if every reduction sequence from t to β-normal form uses r;
2. r is ***head needed*** in t if every reduction sequence from t to head β-normal form uses r;
3. r is ***weak-head needed*** in t if every reduction sequence of t to weak-head β-normal form uses r.

Notice in particular that $\mathtt{nf}_\beta(t) = \varnothing$ (resp. $\mathtt{hnf}_\beta(t) = \varnothing$ or $\mathtt{whnf}_\beta(t) = \varnothing$) implies every redex in t is needed (resp. head needed or weak-head needed).

A ***one-step reduction*** $\rightarrow_\beta$ is ***needed*** (resp. ***head*** or ***weak-head needed***), noted $\rightarrow_{\mathtt{nd}}$ (resp. $\rightarrow_{\mathtt{hnd}}$ or $\rightarrow_{\mathtt{whnd}}$), if the contracted redex is needed (resp. head or weak-head needed). A ***reduction sequence*** $\twoheadrightarrow_\beta$ is ***needed*** (resp. ***head*** or ***weak-head needed***), noted $\twoheadrightarrow_{\mathtt{nd}}$ (resp. $\twoheadrightarrow_{\mathtt{hnd}}$ or $\twoheadrightarrow_{\mathtt{whnd}}$), if every reduction step in the sequence is needed (resp. head or weak-head needed).

For instance, consider the reduction sequence:

$$(\lambda y.\lambda x.I\,x\,(\underline{I\,I}_{\mathsf{r}_1}))\,(I\,I) \rightarrow_{\mathtt{nd}} (\lambda y.\lambda x.\underline{I\,x}_{\mathsf{r}_2}\,I)\,(I\,I) \rightarrow_{\mathtt{nd}} \underline{(\lambda y.\lambda x.x\,I)\,(I\,I)}_{\mathsf{r}_3} \rightarrow_{\mathtt{nd}} \lambda x.x\,I$$

which is needed but not head needed, since redex r_1 might not be contracted to reach a head normal form:

$$(\lambda y.\lambda x.\underline{I\,x}_{\mathsf{r}_2}\,(I\,I))\,(I\,I) \rightarrow_{\mathtt{hnd}} \underline{(\lambda y.\lambda x.x\,(I\,I))\,(I\,I)}_{\mathsf{r}_3} \rightarrow_{\mathtt{hnd}} \lambda x.x\,(I\,I)$$

Moreover, this second reduction sequence is head needed but not weak-head needed since only redex r_3 is needed to get a weak-head normal form:

$$\underline{(\lambda y.\lambda x.I\,x\,(I\,I))\,(I\,I)}_{\mathsf{r}_3} \rightarrow_{\mathtt{whnd}} \lambda x.I\,x\,(I\,I)$$

Notice that the following equalities hold: $\mathcal{NF}_{\mathtt{nd}} = \mathcal{NF}_\beta$, $\mathcal{NF}_{\mathtt{hnd}} = \mathcal{HNF}_\beta$ and $\mathcal{NF}_{\mathtt{whnd}} = \mathcal{WHNF}_\beta$.

Leftmost redexes and reduction sequences are indeed needed:

Lemma 2. *The leftmost redex in any term not in normal form (resp. head or weak-head normal form) is needed (resp. head or weak-head needed).*

Proof. By contradiction using the definition of *needed* [15]. □

Theorem 1. *Let* $\mathsf{r} \in \mathtt{roc}(t)$ *and* $\rho : t \twoheadrightarrow_\beta t'$ *be the leftmost reduction (resp. head reduction or weak-head reduction) starting with* t *such that* $t' = \mathtt{nf}_\beta(t)$ *(resp.* $t' \in \mathtt{hnf}_\beta(t)$ *or* $t' \in \mathtt{whnf}_\beta(t)$*). Then,* r *is needed (resp. head or weak-head needed) in* t *iff* r *is used in* ρ*.*

Proof. By definition of *needed* using Lemma 2 [15]. □

Notice that the weak-head reduction is a prefix of the head reduction, which is in turn a prefix of the leftmost reduction to normal form. As a consequence, it is immediate to see that every weak-head needed redex is in particular head needed, and every head needed redex is needed as well. For example, consider:

$$\underline{(\lambda y.\lambda x.\overline{I\,x}^{\mathsf{r}_2}\,(\overline{I\,I}^{\mathsf{r}_3}))\,(\overline{I\,I}^{\mathsf{r}_4})}_{\mathsf{r}_1}$$

where r_3 is a needed redex but not head needed nor weak-head needed. However, r_2 is both needed and head needed, while r_1 is the only weak-head needed redex in the term, and r_4 is not needed at all.

4 The Type System $\mathcal{V}$

In this section we recall the (non-idempotent) intersection type system $\mathcal{V}$ [14] –an extension of those in [12,13]– used here to characterise normalising terms w.r.t. the weak-head strategy. More precisely, we show that t is typable in system $\mathcal{V}$ if and only if t is normalising when only weak-head needed redexes are contracted. This characterisation is used in Sect. 9 to conclude that the weak-head needed strategy is observationally equivalent to the call-by-need calculus (to be introduced in Sect. 8).

Given a constant type a that denotes *answers* and a countable infinite set $\mathcal{B}$ of base type variables $\alpha, \beta, \gamma, \ldots$, we define the following sets of types:

$$\begin{array}{rl} \textbf{(Types)} & \tau, \sigma ::= \mathsf{a} \mid \alpha \in \mathcal{B} \mid \mathcal{M} \to \tau \\ \textbf{(Multiset types)} & \mathcal{M}, \mathcal{N} ::= \{\!\{\tau_i\}\!\}_{i \in I} \qquad \text{where } I \text{ is a finite set} \end{array}$$

The empty multiset is denoted by $\{\!\{\}\!\}$. We remark that types are *strict* [18], *i.e.* the right-hand sides of functional types are never multisets. Thus, the general form of a type is $\mathcal{M}_1 \to \ldots \to \mathcal{M}_n \to \tau$ with τ being the constant type or a base type variable.

Typing contexts (or just ***contexts***), written Γ, Δ, are functions from variables to multiset types, assigning the empty multiset to all but a finite set of variables. The domain of Γ is given by $\mathtt{dom}(\Gamma) \stackrel{def}{=} \{x \mid \Gamma(x) \neq \{\!\{\}\!\}\}$. The ***union of contexts***, written $\Gamma + \Delta$, is defined by $(\Gamma + \Delta)(x) \stackrel{def}{=} \Gamma(x) \sqcup \Delta(x)$, where $\sqcup$ denotes multiset union. An example is $(x : \{\!\{\sigma\}\!\}, y : \{\!\{\tau\}\!\}) + (x : \{\!\{\sigma\}\!\}, z : \{\!\{\tau\}\!\}) = (x : \{\!\{\sigma, \sigma\}\!\}, y : \{\!\{\tau\}\!\}, z : \{\!\{\tau\}\!\})$. This notion is extended to several contexts as expected, so that $+_{i \in I} \Gamma_i$ denotes a finite union of contexts (when $I = \emptyset$ the notation is to be understood as the empty context). We write $\Gamma \setminus\!\!\setminus x$ for the context $(\Gamma \setminus\!\!\setminus x)(x) = \{\!\{\}\!\}$ and $(\Gamma \setminus\!\!\setminus x)(y) = \Gamma(y)$ if $y \neq x$.

Type judgements have the form $\Gamma \vdash t : \tau$, where Γ is a typing context, t is a term and τ is a type. The intersection type system $\mathcal{V}$ for the λ-calculus is given in Fig. 1.

$$\frac{}{x : \{\!\{\tau\}\!\} \vdash x : \tau}\,(\mathtt{ax}) \qquad \frac{\Gamma \vdash t : \tau}{\Gamma \setminus\!\!\setminus x \vdash \lambda x.t : \Gamma(x) \to \tau}\,(\to\mathtt{i})$$

$$\frac{}{\vdash \lambda x.t : \mathtt{a}}\,(\mathtt{val}) \qquad \frac{\Gamma \vdash t : \{\!\{\sigma_i\}\!\}_{i \in I} \to \tau \quad (\Delta_i \vdash u : \sigma_i)_{i \in I}}{\Gamma +_{i \in I} \Delta_i \vdash t\,u : \tau}\,(\to\mathtt{e})$$

Fig. 1. The non-idempotent intersection type system $\mathcal{V}$.

The constant type $\mathtt{a}$ in rule $(\mathtt{val})$ is used to type values. The axiom $(\mathtt{ax})$ is relevant (there is no weakening) and the rule $(\to\mathtt{e})$ is multiplicative. Note that the argument of an application is typed $\#(I)$ times by the premises of rule $(\to\mathtt{e})$. A particular case is when $I = \emptyset$: the subterm u occurring in the typed term $t\,u$ turns out to be untyped.

A ***(type) derivation*** is a tree obtained by applying the (inductive) typing rules of system $\mathcal{V}$. The notation $\rhd_{\mathcal{V}} \Gamma \vdash t : \tau$ means there is a derivation of the judgement $\Gamma \vdash t : \tau$ in system $\mathcal{V}$. The term t is typable in system $\mathcal{V}$, or $\mathcal{V}$-typable, iff t is the ***subject*** of some derivation, *i.e.* iff there are Γ and τ such that $\rhd_{\mathcal{V}} \Gamma \vdash t : \tau$. We use the capital Greek letters $\Phi, \Psi, \ldots$ to name type derivations, by writing for example $\Phi \rhd_{\mathcal{V}} \Gamma \vdash t : \tau$. For short, we usually denote with Φ_t a derivation with subject t for some type and context. The ***size of the derivation*** Φ, denoted by $\mathtt{sz}(\Phi)$, is defined as the number of nodes of the corresponding derivation tree. We write $\mathtt{RULE}(\Phi) \in \{(\mathtt{ax}), (\to\mathtt{i}), (\to\mathtt{e})\}$ to access the last rule applied in the derivation Φ. Likewise, $\mathtt{PREM}(\Phi)$ is the *multiset* of proper maximal subderivations of Φ. For instance, given

$$\Phi = \frac{\Phi_t \quad (\Phi_u^i)_{i \in I}}{\Gamma \vdash t\,u : \tau}\,(\to\mathtt{e})$$

we have $\mathtt{RULE}(\Phi) = (\to\mathtt{e})$ and $\mathtt{PREM}(\Phi) = \{\!\{\Phi_t\}\!\} \sqcup \{\!\{\Phi_u^i \mid i \in I\}\!\}$. We also use functions $\mathtt{CTXT}(\Phi)$, $\mathtt{SUBJ}(\Phi)$ and $\mathtt{TYPE}(\Phi)$ to access the context, subject and type of the judgement in the root of the derivation tree respectively. For short, we also use notation $\Phi(x)$ to denote the type associated to the variable x in the typing environment of the conclusion of Φ (*i.e.* $\Phi(x) \stackrel{def}{=} \mathtt{CTXT}(\Phi)(x)$).

Intersection type systems can usually be seen as models [11], *i.e.* typing is stable by convertibility: if t is typable and $t =_\beta t'$, then t' is typable too. This property splits in two different statements known as *subject reduction* and *subject expansion* respectively, the first one giving stability of typing by reduction, the second one by expansion. In the particular case of *non-idempotent types*, subject reduction refines to *weighted subject-reduction*, stating that not only typability is stable by reduction, but also that the size of type derivations is decreasing. Moreover, this decrease is strict when reduction is performed on special occurrences of redexes, called *typed occurrences*. We now introduce all these concepts.

Given a type derivation Φ, the set $\mathtt{TOC}(\Phi)$ of ***typed occurrences*** of Φ, which is a subset of $\mathtt{oc}(\mathtt{SUBJ}(\Phi))$, is defined by induction on the last rule of Φ.

- If $\mathtt{RULE}(\Phi) \in \{(\mathtt{ax}), (\mathtt{val})\}$, then $\mathtt{TOC}(\Phi) \stackrel{def}{=} \{\epsilon\}$.
- If $\mathtt{RULE}(\Phi) = (\rightarrow\mathtt{i})$ with $\mathtt{SUBJ}(\Phi) = \lambda x.t$ and $\mathtt{PREM}(\Phi) = \{\!\{\Phi_t\}\!\}$, then $\mathtt{TOC}(\Phi) \stackrel{def}{=} \{\epsilon\} \cup \{0\mathsf{p} \mid \mathsf{p} \in \mathtt{TOC}(\Phi_t)\}$.
- If $\mathtt{RULE}(\Phi) = (\rightarrow\mathtt{e})$ with $\mathtt{SUBJ}(\Phi) = t\,u$ and $\mathtt{PREM}(\Phi) = \{\!\{\Phi_t\}\!\} \sqcup \{\!\{\Phi^i_u \mid i \in I\}\!\}$, then $\mathtt{TOC}(\Phi) \stackrel{def}{=} \{\epsilon\} \cup \{0\mathsf{p} \mid \mathsf{p} \in \mathtt{TOC}(\Phi_t)\} \cup (\bigcup_{i\in I} \{1\mathsf{p} \mid \mathsf{p} \in \mathtt{TOC}(\Phi^i_u)\})$.

Remark that there are two kind of untyped occurrences, those inside untyped arguments of applications, and those inside untyped bodies of abstractions. For instance consider the following type derivations:

$$\Phi_K = \cfrac{\cfrac{\cfrac{}{x : \{\!\{\mathtt{a}\}\!\} \vdash x : \mathtt{a}}\,(\mathtt{ax})}{x : \{\!\{\mathtt{a}\}\!\} \vdash \lambda y.x : \{\!\{\}\!\} \rightarrow \mathtt{a}}\,(\rightarrow\mathtt{i})}{\vdash K : \{\!\{\mathtt{a}\}\!\} \rightarrow \{\!\{\}\!\} \rightarrow \mathtt{a}}\,(\rightarrow\mathtt{i}) \qquad \begin{array}{l} \Phi_{KI} = \cfrac{\Phi_K \quad \cfrac{}{\vdash I : \mathtt{a}}\,(\mathtt{val})}{\vdash K I : \{\!\{\}\!\} \rightarrow \mathtt{a}}\,(\rightarrow\mathtt{e}) \\ \Phi_{KI\Omega} = \cfrac{}{\vdash K I \Omega : \mathtt{a}}\,(\rightarrow\mathtt{e}) \end{array}$$

Then, $\mathtt{TOC}(\Phi_{KI\Omega}) = \{\epsilon, 0, 00, 01, 000, 0000\} \subseteq \mathtt{oc}(KI\Omega)$.

Remark 1. The weak-head redex of a typed term is always a typed occurrence.

Given Φ and $\mathsf{p} \in \mathtt{TOC}(\Phi)$, we define $\Phi|_\mathsf{p}$ as the *multiset* of ***all the subderivations of Φ at occurrence*** p (a formal definition can be found in [15]). Note that $\Phi|_\mathsf{p}$ is a multiset since the subterm of $\mathtt{SUBJ}(\Phi)$ at position p may be typed several times in Φ, due to rule $(\rightarrow\mathtt{e})$.

We can now state the two main properties of system $\mathcal{V}$, whose proofs can be found in Sect. 7 of [9].

Theorem 2 (Weighted Subject Reduction). *Let $\Phi \rhd_\mathcal{V} \Gamma \vdash t : \tau$. If $\mathsf{r} : t \rightarrow_\beta t'$, then there exists Φ' s.t. $\Phi' \rhd_\mathcal{V} \Gamma \vdash t' : \tau$. Moreover,*

1. *If $\mathsf{r} \in \mathtt{TOC}(\Phi)$, then $\mathtt{sz}(\Phi) > \mathtt{sz}(\Phi')$.*
2. *If $\mathsf{r} \notin \mathtt{TOC}(\Phi)$, then $\mathtt{sz}(\Phi) = \mathtt{sz}(\Phi')$.*

Theorem 3 (Subject Expansion). *Let $\Phi' \rhd_\mathcal{V} \Gamma \vdash t' : \tau$. If $t \rightarrow_\beta t'$, then there exists Φ s.t. $\Phi \rhd_\mathcal{V} \Gamma \vdash t : \tau$.*

Note that weighted subject reduction implies that reduction of typed redex occurrences turns out to be normalising.

5 Substitution and Reduction on Derivations

In order to relate typed redex occurrences of convertible terms, we now extend the notion of β-reduction to derivation trees, by making use of a natural and basic concept of typed substitution. In contrast to substitution and β-reduction on *terms*, these operations are now both non-deterministic on derivation trees (see [19] for discussions and examples). Given a variable x and type derivations Φ_t and $(\Phi_u^i)_{i\in I}$, the ***typed substitution*** of x by $(\Phi_u^i)_{i\in I}$ in Φ_t, written $\Phi_t\left\{x/(\Phi_u^i)_{i\in I}\right\}$ by making an abuse of notation, is a type derivation inductively defined on Φ_t, only if $\Phi_t(x) = \{\!\{\mathtt{TYPE}(\Phi_u^i)\}\!\}_{i\in I}$. This non-deterministic construction may be non-trivial but it can be naturally formalised in a quite straightforward way (full details can be found in [15]). Intuitively, the typed substitution replaces typed occurrences of x in Φ_t by a corresponding derivation Φ_u^i matching the same type, where such a matching is chosen in a non-deterministic way. Moreover, it also substitutes all untyped occurrences of x by u, where this untyped operation is completely deterministic. Thus, for example, consider the following substitution, where Φ_{KI} is defined in Sect. 4:

$$\left(\dfrac{\dfrac{}{x:\{\!\{\{\!\{\}\!\}\to\mathtt{a}\}\!\}\vdash x:\{\!\{\}\!\}\to\mathtt{a}}\,(\mathtt{ax})}{x:\{\!\{\{\!\{\}\!\}\to\mathtt{a}\}\!\}\vdash x\,x:\mathtt{a}}\,(\to\mathtt{e})\right)\{x/\Phi_{KI}\} = \dfrac{\Phi_{KI}}{\vdash (KI)\,(KI):\mathtt{a}}\,(\to\mathtt{e})$$

The following lemma relates the typed occurrences of the trees composing a substitution and those of the substituted tree itself:

Lemma 3. *Let Φ_t and $(\Phi_u^i)_{i\in I}$ be derivations such that $\Phi_t\left\{x/(\Phi_u^i)_{i\in I}\right\}$ is defined, and $\mathsf{p}\in\mathsf{oc}(t)$. Then,*

1. $\mathsf{p}\in\mathtt{TOC}(\Phi_t)$ *iff* $\mathsf{p}\in\mathtt{TOC}(\Phi_t\left\{x/(\Phi_u^i)_{i\in I}\right\})$.
2. $\mathsf{q}\in\mathtt{TOC}(\Phi_u^k)$ *for some* $k\in I$ *iff there exists* $\mathsf{p}\in\mathtt{TOC}(\Phi_t)$ *such that* $t|_{\mathsf{p}}=x$ *and* $\mathsf{pq}\in\mathtt{TOC}(\Phi_t\left\{x/(\Phi_u^i)_{i\in I}\right\})$.

Proof. By induction on Φ_t. □

Based on the previous notion of substitutions on derivations, we are now able to introduce (non-deterministic) reduction on derivation trees. The ***reduction relation*** $\to_\beta$ on derivation trees is then defined by first considering the following basic rewriting rules.

1. For typed β-redexes:

$$\dfrac{\dfrac{\Phi_t \rhd_{\mathcal{V}} \Gamma; x:\{\!\{\sigma_i\}\!\}_{i\in I}\vdash t:\tau}{\Gamma\vdash\lambda x.t:\{\!\{\sigma_i\}\!\}_{i\in I}\to\tau}\qquad (\Phi_u^i\rhd_{\mathcal{V}}\Delta_i\vdash u:\sigma_i)_{i\in I}}{\Gamma +_{i\in I}\Delta_i\vdash(\lambda x.t)\,u:\tau}\mapsto_\beta \Phi_t\left\{x/(\Phi_u^i)_{i\in I}\right\}$$

2. For β-redexes in untyped occurrences, with $u \to_\beta u'$:

$$\frac{\Gamma \vdash t : \{\!\!\{\}\!\!\} \to \tau}{\Gamma \vdash t\,u : \tau} \mapsto_\nu \frac{\Gamma \vdash t : \{\!\!\{\}\!\!\} \to \tau}{\Gamma \vdash t\,u' : \tau} \qquad \frac{}{\vdash \lambda x.u : \mathsf{a}} \mapsto_\xi \frac{}{\vdash \lambda x.u' : \mathsf{a}}$$

As in the case of the λ-calculus, where reduction is closed under usual *term* contexts, we need to close the previous relation under *derivation tree* contexts. However, a one-step reduction on a given subterm causes many one-step reductions in the corresponding derivation tree (recall $\Phi|_{\mathsf{p}}$ is defined to be a multiset). Then, informally, given a redex occurrence r of t, a type derivation Φ of t, and the multiset of minimal subderivations of Φ containing r, written $\mathscr{M}$, we apply the reduction rules $\mapsto_{\beta,\nu,\xi}$ to all the elements of $\mathscr{M}$, thus obtaining a multiset $\mathscr{M}'$, and we recompose the type derivation of the reduct of t (see [15] for a formal definition). This gives the reduction relation $\to_\beta$ on trees. A reduction sequence on derivation trees contracting only redexes in typed positions is dubbed a ***typed reduction sequence***.

Note that typed reductions are normalising by Theorem 2, yielding a special kind of derivation. Indeed, given a type derivation $\Phi \rhd_{\mathcal{V}} \Gamma \vdash t : \tau$, we say that Φ is ***normal*** iff $\mathtt{TOC}(\Phi) \cap \mathtt{roc}(t) = \varnothing$. Reduction on trees induces reduction on terms: when $\rho : \Phi \twoheadrightarrow_\beta \Phi'$, then $\mathtt{SUBJ}(\Phi) \twoheadrightarrow_\beta \mathtt{SUBJ}(\Phi')$. By abuse of notation we may denote both sequences with the same letter ρ.

6 Weak-Head Neededness and Typed Occurrences

This section presents one of our main results. It establishes a connection between weak-head needed redexes and typed redex occurrences. More precisely, we first show in Sect. 6.1 that every weak-head needed redex occurrence turns out to be a typed occurrence, whatever its type derivation is. The converse does not however hold. But, we show in Sect. 6.2 that any typed occurrence in a special kind of typed derivation (that we call principal) corresponds to a weak-head needed redex occurrence. We start with a technical lemma.

Lemma 4. *Let* $\mathsf{r} : \Phi_t \to_\beta \Phi_{t'}$ *and* $\mathsf{p} \in \mathsf{oc}(t)$ *such that* $\mathsf{p} \neq \mathsf{r}$ *and* $\mathsf{p} \neq \mathsf{r}0$. *Then,* $\mathsf{p} \in \mathtt{TOC}(\Phi_t)$ *iff there exists* $\mathsf{p}' \in \mathsf{p}/\mathsf{r}$ *such that* $\mathsf{p}' \in \mathtt{TOC}(\Phi_{t'})$.

Proof. By induction on r using Lemma 3. □

6.1 Weak-Head Needed Redexes Are Typed

In order to show that every weak-head needed redex occurrence corresponds to a typed occurrence in some type derivation we start by proving that typed occurrences do not come from untyped ones.

Lemma 5. *Let* $\rho : \Phi_t \twoheadrightarrow_\beta \Phi_{t'}$ *and* $\mathsf{p} \in \mathsf{oc}(t)$. *If there exists* $\mathsf{p}' \in \mathsf{p}/\rho$ *such that* $\mathsf{p}' \in \mathtt{TOC}(\Phi_{t'})$, *then* $\mathsf{p} \in \mathtt{TOC}(\Phi_t)$.

Proof. Straightforward induction on ρ using Lemma 4. □

Theorem 4. *Let* r *be a weak-head needed redex in* t. *Let* Φ *be a type derivation of* t. *Then,* $\mathsf{r} \in \mathtt{TOC}(\Phi)$.

Proof. By Theorem 1, r is used in the weak-head reduction from t to $t' \in \mathcal{WHNF}_\beta$. By Remark 1, the weak-head reduction contracts only typed redexes. Thus, r or some of its residuals is a typed occurrence in its corresponding derivation tree. Finally, we conclude by Lemma 5, $\mathsf{r} \in \mathtt{TOC}(\Phi)$. □

6.2 Principally Typed Redexes Are Weak-Head Needed

As mentioned before, the converse of Theorem 4 does not hold: there are some typed occurrences that do not correspond to any weak-head needed redex occurrence. This can be illustrated in the following examples (recall $\Phi_{KI\Omega}$ defined in Sect. 4):

$$\dfrac{\Phi_{KI\Omega}}{\vdash \lambda y.KI\Omega : \{\!\{\}\!\} \to \mathtt{a}}\,(\to\mathtt{i}) \qquad \dfrac{\dfrac{}{y : \{\!\{\{\!\{\mathtt{a}\}\!\} \to \mathtt{a}\}\!\} \vdash y : \{\!\{\mathtt{a}\}\!\} \to \mathtt{a}}\,(\mathtt{ax}) \qquad \Phi_{KI\Omega}}{y : \{\!\{\{\!\{\mathtt{a}\}\!\} \to \mathtt{a}\}\!\} \vdash y\,(KI\Omega) : \mathtt{a}}\,(\to\mathtt{e})$$

Indeed, the occurrence 0 (resp 1) in the term $\lambda y.KI\Omega$ (resp. $y\,(KI\Omega)$) is typed but not weak-head needed, since both terms are already in weak-head normal form. Fortunately, typing relates to redex occurrences if we restrict type derivations to *principal* ones: given a term t in weak-head β-normal form, the derivation $\Phi \rhd_{\mathcal{V}} \Gamma \vdash t : \tau$ is ***normal principally typed*** if:

- $t = x\,t_1 \ldots t_n$ $(n \geq 0)$, and $\Gamma = \{x : \{\!\{\overbrace{\{\!\{\}\!\} \to \ldots \to \{\!\{\}\!\}}^{n \text{ times}} \to \tau\}\!\}\}$ and τ is a type variable α (*i.e.* none of the t_i are typed), or
- $t = \lambda x.t'$, and $\Gamma = \varnothing$ and $\tau = \mathtt{a}$.

Given a weak-head normalising term t such that $\Phi_t \rhd_{\mathcal{V}} \Gamma \vdash t : \tau$, we say that Φ_t is ***principally typed*** if $\Phi_t \twoheadrightarrow_\beta \Phi_{t'}$ for some $t' \in \mathtt{whnf}_\beta(t)$ implies $\Phi_{t'}$ is normal principally typed.

Note in particular that the previous definition does not depend on the chosen weak-head normal form t': suppose $t'' \in \mathtt{whnf}_\beta(t)$ is another weak-head normal form of t, then t' and t'' are convertible terms by the Church-Rosser property [7] so that t' can be normal principally typed iff t'' can, by Theorems 2 and 3.

Lemma 6. *Let* Φ_t *be a type derivation with subject* t *and* $\mathsf{r} \in \mathtt{roc}(t) \cap \mathtt{TOC}(\Phi_t)$. *Let* $\rho : \Phi_t \twoheadrightarrow_\beta \Phi_{t'}$ *such that* $\Phi_{t'}$ *is normal. Then,* r *is used in* ρ.

Proof. Straightforward induction on ρ using Lemma 4. □

The notions of leftmost and weak-head needed reductions on (untyped) terms naturally extends to *typed* reductions on tree derivations. We thus have:

Lemma 7. *Let* t *be a weak-head normalising term and* Φ_t *be principally typed. Then, a leftmost typed reduction sequence starting at* Φ_t *is weak-head needed.*

Proof. By induction on the leftmost typed sequence (called ρ). If ρ is empty the result is immediate. If not, we show that t has a typed weak-head needed redex (which is leftmost by definition) and conclude by inductive hypothesis. Indeed, assume $t \in \mathcal{WHNF}_\beta$. By definition Φ_t is normal principally typed and thus it has no typed redexes. This contradicts ρ being non-empty. Hence, t has a weak-head redex r (*i.e.* $t \notin \mathcal{WHNF}_\beta$). Moreover, r is both typed (by Remark 1) and weak-head needed (by Lemma 2). Thus, we conclude. □

Theorem 5. *Let t be a weak-head normalising term, Φ_t be principally typed and $\mathsf{r} \in \mathtt{roc}(t) \cap \mathtt{TOC}(\Phi_t)$. Then, r is a weak-head needed redex in t.*

Proof. Let $\rho : \Phi_t \twoheadrightarrow_\beta \Phi_{t'}$ be the leftmost typed reduction sequence where $\Phi_{t'}$ is normal. Note that $\Phi_{t'}$ exists by definition of *principally typed*. By Lemma 7, ρ is a weak-head needed reduction sequence. Moreover, by Lemma 6, r is used in ρ. Hence, r is a weak-head needed redex in t. □

As a direct consequence of Theorems 4 and 5, given a weak-head normalising term t, the typed redex occurrences in its principally typed derivation (which always exists) correspond to its weak-head needed redexes. Hence, system $\mathcal{V}$ allows to identify all the weak-head needed redexes of a weak-head normalising term.

7 Characterising Weak-Head Needed Normalisation

This section presents one of the main pieces contributing to our observational equivalence result. Indeed, we relate typing with weak-head neededness by showing that any typable term in system $\mathcal{V}$ is normalising for weak-head needed reduction. This characterisation highlights the power of intersection types. We start by a technical lemma.

Lemma 8. *Let $\Phi \rhd_\mathcal{V} \Gamma \vdash t : \tau$. Then, Φ normal implies $t \in \mathcal{WHNF}_\beta$.*

Proof. By induction on Φ analysing the last rule applied. □

Let $\rho : t_1 \twoheadrightarrow_\beta t_n$. We say that ρ is a ***left-to-right*** reduction sequence iff for every $i < n$ if $\mathsf{r}_\mathsf{i} : t_i \rightarrow_\beta t_{i+1}$ and l_i is to the left of r_i then, for every $j > i$ such that $\mathsf{r}_\mathsf{j} : t_j \rightarrow_\beta t_{j+1}$ we have that $\mathsf{r}_\mathsf{j} \notin \{\mathsf{l}_\mathsf{i}\}/\rho_{ij}$ where $\rho_{ij} : t_i \twoheadrightarrow_\beta t_j$ is the corresponding subsequence of ρ. In other words, for every j and every $i < j$, r_j is not a residual of a redex to the left of r_i (relative to the given reduction subsequence from t_i to t_j) [7].

Left-to-right reductions define in particular standard strategies, which give canonical ways to construct reduction sequences from one term to another:

Theorem 6 ([7]). *If $t \twoheadrightarrow_\beta t'$, there exists a left-to-right reduction from t to t'.*

Theorem 7. *Let $t \in \mathcal{T}_\mathsf{a}$. Then, $\Phi \rhd_\mathcal{V} \Gamma \vdash t : \tau$ iff $t \in \mathcal{WN}_{\mathtt{whnd}}$.*

Proof. ⇒) By Theorem 2 we know that the strategy reducing only typed redex occurrences is normalising, *i.e.* there exist t' and Φ' such that $t \twoheadrightarrow_\beta t'$, $\Phi' \rhd_{\mathcal{V}} \Gamma \vdash t' : \tau$ and Φ' normal. Then, by Lemma 8, $t' \in \mathcal{WHNF}_\beta$. By Theorem 6, there exists a left-to-right reduction $\rho : t \twoheadrightarrow_\beta t'$. Let us write

$$\rho : t = t_1 \twoheadrightarrow_\beta t_n \twoheadrightarrow_\beta t'$$

such that $t_1, \ldots, t_{n-1} \notin \mathcal{WHNF}_\beta$ and $t_n \in \mathcal{WHNF}_\beta$.

We claim that all reduction steps in $t_1 \twoheadrightarrow_\beta t_n$ are leftmost. Assume towards a contradiction that there exists $k < n$ such that $\mathsf{r} : t_k \rightarrow_\beta t_{k+1}$ and r is not the leftmost redex of t_k (written l_k). Since ρ is a left-to-right reduction, no residual of l_k is contracted after the k-th step. Thus, there is a reduction sequence from $t_k \notin \mathcal{WHNF}_\beta$ to $t_n \in \mathcal{WHNF}_\beta$ such that l_k is not used in it. This leads to a contradiction with l_k being weak-head needed in t_k by Lemma 2.

As a consequence, there is a leftmost reduction sequence $t \twoheadrightarrow_\beta t_n$. Moreover, by Lem. 2, $t \twoheadrightarrow_{\mathtt{whnd}} t_n \in \mathcal{WHNF}_\beta = \mathcal{NF}_{\mathtt{whnd}}$. Thus, $t \in \mathcal{WN}_{\mathtt{whnd}}$.

⇐) Consider the reduction $\rho : t \twoheadrightarrow_{\mathtt{whnd}} t'$ with $t' \in \mathtt{whnf}_\beta(t)$. Let $\Phi' \rhd_{\mathcal{V}} \Gamma \vdash t' : \tau$ be the normal principally typed derivation for t' as defined in Sect. 6.2. Finally, we conclude by induction in ρ using Theorem 3, $\Phi \rhd_{\mathcal{V}} \Gamma \vdash t : \tau$. □

8 The Call-by-Need Lambda-Calculus

This section describes the syntax and the operational semantics of the call-by-need lambda-calculus introduced in [1]. It is more concise than previous specifications of call-by-need [2,3,10,16], but it is operationally equivalent to them [6], so that our results could also be presented by using alternative specifications.

Given a countable infinite set $\mathcal{X}$ of variables $x, y, z, \ldots$ we define different syntactic categories for terms, values, list contexts, answers and need contexts:

$$\begin{array}{rl}
\textbf{(Terms)} & t, u ::= x \in \mathcal{X} \mid t\,u \mid \lambda x.t \mid t[x\backslash u] \\
\textbf{(Values)} & v ::= \lambda x.t \\
\textbf{(List contexts)} & \mathtt{L} ::= \Box \mid \mathtt{L}[x\backslash t] \\
\textbf{(Answers)} & a ::= \mathtt{L}\langle \lambda y.t \rangle \\
\textbf{(Need contexts)} & \mathtt{M}, \mathtt{N} ::= \Box \mid \mathtt{N}\,t \mid \mathtt{N}[x\backslash t] \mid \mathtt{N}\langle\!\langle x \rangle\!\rangle[x\backslash \mathtt{M}]
\end{array}$$

We denote the set of terms by $\mathcal{T}_{\mathtt{e}}$. Terms of the form $t[x\backslash u]$ are ***closures***, and $[x\backslash u]$ is called an ***explicit substitution*** (ES). The set of $\mathcal{T}_{\mathtt{e}}$-terms without ES is the set of *terms of the* λ*-calculus*, *i.e.* $\mathcal{T}_{\mathtt{a}}$. The notions of ***free*** and ***bound*** variables are defined as expected, in particular, $\mathtt{fv}(t[x\backslash u]) \stackrel{def}{=} \mathtt{fv}(t) \setminus \{x\} \cup \mathtt{fv}(u)$, $\mathtt{fv}(\lambda x.t) \stackrel{def}{=} \mathtt{fv}(t) \setminus \{x\}$, $\mathtt{bv}(t[x\backslash u]) \stackrel{def}{=} \mathtt{bv}(t) \cup \{x\} \cup \mathtt{bv}(u)$ and $\mathtt{bv}(\lambda x.t) \stackrel{def}{=} \mathtt{bv}(t) \cup \{x\}$. We extend the standard notion of ***α-conversion*** to ES, as expected.

We use the special notation $\mathtt{N}\langle\!\langle u \rangle\!\rangle$ or $\mathtt{L}\langle\!\langle u \rangle\!\rangle$ when the free variables of u are not captured by the context, *i.e.* there are no abstractions or explicit substitutions in the context that binds the free variables of u. Thus for example, given $\mathtt{N} = (\Box\, x)[x\backslash z]$, we have $(y\, x)[x\backslash z] = \mathtt{N}\langle y \rangle = \mathtt{N}\langle\!\langle y \rangle\!\rangle$, but $(x\, x)[x\backslash z] = \mathtt{N}\langle x \rangle$ cannot be written as $\mathtt{N}\langle\!\langle x \rangle\!\rangle$. Notice the use of this special notation in the last case of needed contexts, an example of such case being $(x\,y)[y\backslash t][x\backslash\Box]$.

The ***call-by-need calculus***, introduced in [1], is given by the set of terms $\mathcal{T}_{\mathtt{e}}$ and the ***reduction relation*** $\rightarrow_{\mathtt{need}}$, the ***union*** of $\rightarrow_{\mathtt{dB}}$ and $\rightarrow_{\mathtt{lsv}}$, which are, respectively, the closure by *need contexts* of the following rewriting rules:

$$\mathtt{L}\langle \lambda x.t \rangle\, u \mapsto_{\mathtt{dB}} \mathtt{L}\langle t[x\backslash u] \rangle$$
$$\mathtt{N}\langle\!\langle x \rangle\!\rangle[x\backslash \mathtt{L}\langle v \rangle] \mapsto_{\mathtt{lsv}} \mathtt{L}\langle \mathtt{N}\langle\!\langle v \rangle\!\rangle[x\backslash v] \rangle$$

These rules avoid capture of free variables. An example of need-reduction sequence is the following, where the redex of each step is underlined for clearness:

$$\begin{array}{llll}
\underline{(\lambda x_1.I\,(x_1\,I))\,(\lambda y.I\,y)} & \rightarrow_{\mathtt{dB}} & (\underline{I\,(x_1\,I)})[x_1\backslash\lambda y.I\,y] & \rightarrow_{\mathtt{dB}} \\
\underline{x_2[x_2\backslash x_1\,I][x_1\backslash\lambda y.I\,y]} & \rightarrow_{\mathtt{lsv}} & x_2[x_2\backslash\underline{(\lambda x_3.I\,x_3)\,I}][x_1\backslash\lambda y.I\,y] & \rightarrow_{\mathtt{dB}} \\
x_2[x_2\backslash(\underline{I\,x_3})[x_3\backslash I]][x_1\backslash\lambda y.I\,y] & \rightarrow_{\mathtt{dB}} & x_2[x_2\backslash\underline{x_4[x_4\backslash x_3][x_3\backslash I]}][x_1\backslash\lambda y.I\,y] & \rightarrow_{\mathtt{lsv}} \\
x_2[x_2\backslash\underline{x_4[x_4\backslash I]}[x_3\backslash I]][x_1\backslash\lambda y.I\,y] & \rightarrow_{\mathtt{lsv}} & \underline{x_2[x_2\backslash I[x_4\backslash I][x_3\backslash I]]}[x_1\backslash\lambda y.I\,y] & \rightarrow_{\mathtt{lsv}} \\
I[x_2\backslash I][x_4\backslash I][x_3\backslash I][x_1\backslash\lambda y.I\,y] & & &
\end{array}$$

As for call-by-name, reduction preserves free variables, *i.e.* $t \rightarrow_{\mathtt{need}} t'$ implies $\mathtt{fv}(t) \supseteq \mathtt{fv}(t')$. Notice that call-by-need reduction is also weak, so that answers are not need-reducible.

9 Observational Equivalence

The results in Sect. 7 are used here to prove soundness and completeness of call-by-need w.r.t weak-head neededness, our second main result. More precisely, a call-by-need interpreter stops in a value if and only if the weak-head needed reduction stops in a value. This means that call-by-need and call-by-name are observationally equivalent.

Formally, given a reduction relation $\mathcal{R}$ on a term language $\mathcal{T}$, and an associated notion of context for $\mathcal{T}$, we define t to be ***observationally equivalent*** to u, written $t \cong_{\mathcal{R}} u$, iff $\mathtt{C}\langle t \rangle \in \mathcal{WN}_{\mathcal{R}} \Leftrightarrow \mathtt{C}\langle u \rangle \in \mathcal{WN}_{\mathcal{R}}$ for every context $\mathtt{C}$. In order to show our final result we resort to the following theorem:

Theorem 8 ([14]).
1. *Let* $t \in \mathcal{T}_{\mathtt{a}}$. *Then,* $\Phi \rhd_{\mathcal{V}} \Gamma \vdash t : \tau$ *iff* $t \in \mathcal{WN}_{\mathtt{name}}$.
2. *For all terms* t *and* u *in* $\mathcal{T}_{\mathtt{a}}$, $t \cong_{\mathtt{name}} u$ *iff* $t \cong_{\mathtt{need}} u$.

These observations allows us to conclude:

Theorem 9. *For all terms* t *and* u *in* $\mathcal{T}_{\mathtt{a}}$, $t \cong_{\mathtt{whnd}} u$ *iff* $t \cong_{\mathtt{need}} u$.

Proof. By Theorem 8:2 it is sufficient to show $t \cong_{\mathtt{whnd}} u$ iff $t \cong_{\mathtt{name}} u$. The proof proceeds as follows:

$$\begin{array}{rclll}
 & t \cong_{\mathtt{name}} u & & \text{iff} & \text{(definition)} \\
\mathtt{C}\langle t \rangle \in \mathcal{WN}_{\mathtt{name}} & \Leftrightarrow & \mathtt{C}\langle u \rangle \in \mathcal{WN}_{\mathtt{name}} & \text{iff} & \text{(Theorem 8:1)} \\
\mathtt{C}\langle t \rangle \text{ typable in } \mathcal{V} & \Leftrightarrow & \mathtt{C}\langle u \rangle \text{ typable in } \mathcal{V} & \text{iff} & \text{(Theorem 7)} \\
\mathtt{C}\langle t \rangle \in \mathcal{WN}_{\mathtt{whnd}} & \Leftrightarrow & \mathtt{C}\langle u \rangle \in \mathcal{WN}_{\mathtt{whnd}} & \text{iff} & \text{(definition)} \\
 & t \cong_{\mathtt{whnd}} u & & &
\end{array}$$

□

10 Conclusion

We establish a clear connection between the semantical standard notion of neededness and the syntactical concept of call-by-need. The use of non-idempotent types –a powerful technique being able to characterise different operational properties– provides a simple and natural tool to show observational equivalence between these two notions. We refer the reader to [5] for other proof techniques (not based on intersection types) used to connect semantical notions of neededness with syntactical notions of lazy evaluation.

An interesting (and not difficult) extension of our result in Sect. 6 is that call-by-need reduction (defined on λ-terms with explicit substitutions) contracts only dB weak-head needed redexes, for an appropriate (and very natural) notion of weak-head needed redex for λ-terms with explicit substitutions. A technical tool to obtain such a result would be the type system $\mathcal{A}$ [14], a straightforward adaptation of system $\mathcal{V}$ to call-by-need syntax.

Given the recent formulation of *strong call-by-need* [6] describing a deterministic call-by-need strategy to normal form (instead of weak-head normal form), it would be natural to extend our technique to obtain an observational equivalence result between the standard notion of needed reduction (to full normal forms) and the strong call-by-need strategy. This remains as future work.

References

1. Accattoli, B., Barenbaum, P., Mazza, D.: Distilling abstract machines. In: Jeuring, J., Chakravarty, M.M.T. (eds.) Proceedings of the 19th ACM SIGPLAN International Conference on Functional Programming, Gothenburg, Sweden, 1–3 September 2014, pp. 363–376. ACM (2014)
2. Ariola, Z.M., Felleisen, M.: The call-by-need lambda calculus. J. Funct. Program. **7**(3), 265–301 (1997)
3. Ariola, Z.M., Felleisen, M., Maraist, J., Odersky, M., Wadler, P.: The call-by-need lambda calculus. In: Cytron, R.K., Lee, P. (eds.) Conference Record of POPL 1995: 22nd ACM SIGPLAN-SIGACT Symposium on Principles of Programming Languages, San Francisco, California, USA, 23–25 January 1995, pp. 233–246. ACM Press (1995)
4. Baader, F., Nipkow, T.: Term Rewriting and all That. Cambridge University Press, New York (1998)
5. Balabonski, T.: La pleine paresse, une certaine optimalité. Ph.D. Thesis, Université Paris-Diderot (2012)
6. Balabonski, T., Barenbaum, P., Bonelli, E., Kesner, D.: Foundations of strong call by need. PACMPL **1**(ICFP), 20:1–20:29 (2017)
7. Barendregt, H.P.: The Lambda Calculus Its Syntax and Semantics, vol. 103, revised edition, North Holland (1984)
8. Barendregt, H.P., Kennaway, J.R., Klop, J.W., Sleep, M.R.: Needed reduction and spine strategies for the lambda calculus. Inf. Comput. **75**(3), 191–231 (1987)
9. Bucciarelli, A., Kesner, D., Ventura, D.: Non-idempotent intersection types for the lambda-calculus. Logic J. IGPL **25**(4), 431–464 (2017)

10. Chang, S., Felleisen, M.: The call-by-need lambda calculus, revisited. In: Seidl, H. (ed.) ESOP 2012. LNCS, vol. 7211, pp. 128–147. Springer, Heidelberg (2012). https://doi.org/10.1007/978-3-642-28869-2_7
11. Coppo, M., Dezani-Ciancaglini, M.: An extension of the basic functionality theory for the λ-calculus. Notre Dame J. Formal Log. **21**(4), 685–693 (1980)
12. de Carvalho, D.: Sémantiques de la logique linéaire et temps de calcul. Ph.D. thesis, Université Aix-Marseille II (2007)
13. Gardner, P.: Discovering needed reductions using type theory. In: Hagiya, M., Mitchell, J.C. (eds.) TACS 1994. LNCS, vol. 789, pp. 555–574. Springer, Heidelberg (1994). https://doi.org/10.1007/3-540-57887-0_115
14. Kesner, D.: Reasoning about call-by-need by means of types. In: Jacobs, B., Löding, C. (eds.) FoSSaCS 2016. LNCS, vol. 9634, pp. 424–441. Springer, Heidelberg (2016). https://doi.org/10.1007/978-3-662-49630-5_25
15. Kesner, D., Ríos, A., Viso, A.: Call-by-need, neededness and all that. Extended report (2017). https://arxiv.org/abs/1801.10519
16. Maraist, J., Odersky, M., Wadler, P.: The call-by-need lambda calculus. J. Funct. Program. **8**(3), 275–317 (1998)
17. Rocca, S.R.D.: Principal type scheme and unification for intersection type discipline. Theor. Comput. Sci. **59**, 181–209 (1988)
18. van Bakel, S.: Complete restrictions of the intersection type discipline. Theor. Comput. Sci. **102**(1), 135–163 (1992)
19. Vial, P.: Non-idempotent intersection types, beyond lambda-calculus. Ph.D. thesis, Université Paris-Diderot (2017)
20. Wadsworth, C.P.: Semantics and pragmatics of the lambda calculus. Ph.D. thesis, Oxford University (1971)

Fitch-Style Modal Lambda Calculi

Ranald Clouston(✉)

Department of Computer Science, Aarhus University, Aarhus, Denmark
ranald.clouston@cs.au.dk

Abstract. Fitch-style modal deduction, in which modalities are eliminated by opening a subordinate proof, and introduced by shutting one, were investigated in the 1990s as a basis for lambda calculi. We show that such calculi have good computational properties for a variety of intuitionistic modal logics. Semantics are given in cartesian closed categories equipped with an adjunction of endofunctors, with the necessity modality interpreted by the right adjoint. Where this functor is an idempotent comonad, a coherence result on the semantics allows us to present a calculus for intuitionistic S4 that is simpler than others in the literature. We show the calculi can be extended à la tense logic with the left adjoint of necessity, and are then complete for the categorical semantics.

Keywords: Intuitionistic modal logic · Typed lambda calculi
Categorical semantics

1 Introduction

The Curry-Howard propositions-as-types isomorphism [21,39,41] provides a correspondence between natural deduction and typed lambda calculus of interest to both logicians and computer scientists. For the logician, term assignment offers a convenient notation to express and reason about syntactic properties such as proof normalisation, and, especially in the presence of dependent types, allows proofs of non-trivial mathematical theorems to be checked by computer programs. For the computer scientist, logics have been repurposed as typing disciplines to address problems in computing in sometimes surprising ways. Following Lambek [25], categories form a third leg of the isomorphism. Categorical semantics can be used to prove the consistency of a calculus, and they are crucial if we wish to prove or program in some particular mathematical setting. For example, see the use of the topos of trees as a setting for both programming with guarded recursion, and proof by Löb induction, by Clouston et al. [11].

This work involved two functors, 'later' and 'constant'. Where functors interact appropriately with finite products they correspond to necessity modalities in

We gratefully acknowledge discussions with Patrick Bahr, Lars Birkedal, Aleš Bizjak, Christian Uldal Graulund, G.A. Kavvos, Bassel Mannaa, Rasmus Ejlers Møgelberg, Andrew M. Pitts, and Bas Spitters, and the comments of the anonymous referees. This research was supported by a research grant (12386) from Villum Fonden.

C. Baier and U. Dal Lago (Eds.): FOSSACS 2018, LNCS 10803, pp. 258–275, 2018.
https://doi.org/10.1007/978-3-319-89366-2_14

intuitionistic normal modal logic, usually written $\Box$. Such modalities have been extensively studied by logicians, and the corresponding type-formers are widely applicable in computing, for example to monads [32], staged programming [13], propositional truncation [2], and recent work in homotopy type theory [37]. There is hence a need to develop all sides of the Curry-Howard-Lambek isomorphism for necessity modalities. Approaches to modal lambda calculi are diverse; see the survey by Kavvos [23], and remarks in the final section of this paper. This paper focuses on *Fitch-style* modal lambda calculi as first proposed by Borghuis [9] and (as the "two-dimensional" approach) by Martini and Masini [29].

Fitch-style modal lambda calculi[1] adapt the proof methods of Fitch [19] in which given a formula $\Box A$ we may open a '(strict) subordinate proof' in which we eliminate the $\Box$ to get premise A. Such a subordinate proof with conclusion B can then be shut by introducing a $\Box$ to conclude $\Box B$. Different modal logics can be encoded by tweaking the open and shut rules; for example we could shut the proof to conclude merely B, if we had the T axiom $\Box B \to B$. Normal modal logics are usually understood with respect to Kripke's possible worlds semantics (for the intuitionistic version, see e.g. Simpson [38, Sect. 3.3]). In this setting Fitch's approach is highly intuitive, as opening a subordinate proof corresponds to travelling to a generic related world, while shutting corresponds to returning to the original world. See Fitting [20, Chap. 4] for a lengthier discussion of this approach to natural deduction.

Borghuis [9] kept track of subordinate proofs in a sequent presentation by introducing a new structural connective to the context when a $\Box$ is eliminated, and removing it from the context when one is introduced, in a style reminiscent of the treatment of modal logic in display calculus [42], or for that matter of the standard duality between implication and comma. To the category theorist, this suggests an operation on contexts *left adjoint* to $\Box$. This paper exploits this insight by presenting categorical semantics for Fitch-style modal calculi for the first time, answering the challenge of de Paiva and Ritter [33, Sect. 4], by modelling necessity modalities as right adjoints. This is logically sound and complete, yet less general than modelling modalities as monoidal functors as done for example by Bellin et al. [4]. For example, truncation in sets is monoidal but has no right adjoint. Nonetheless adjunctions are ubiquitous, and in their presence we argue that the case for Fitch-style calculi is compelling. Examples of right adoints of interest to type theorists include the aforementioned modalities of guarded recursion, the closure modalities of (differential) cohesive ∞-toposes [36, Sect. 3], and atom-abstraction in nominal sets [31].

In Sect. 2 we present Borghuis's calculus for the logic Intuitionistic K, the most basic intuitionistic modal logic of necessity. To the results of confluence, subject reduction, and strong normalisation already shown by Borghuis we add canonicity and the subformula property, with the latter proof raising a subtle issue with sums not previously observed. We give categorical semantics for this style of calculus for the first time and prove soundness. In Sect. 3 we introduce the

[1] 'Fitch-style' deduction can also be used to mean the linear presentation of natural deduction with subordinate proofs for implication.

left adjoint as a first-class type former à la intuitionistic tense logic [17], in which the "everywhere in the future" modality is paired with "somewhere in the past". To our knowledge this is the first natural deduction calculus, let alone lambda calculus, for any notion of tense logic. It is not entirely satisfactory as it lacks the subformula property, but it does allow us to prove categorical completeness. In Sect. 4 we show how the basic techniques developed for Intuitionistic K extend to Intuitionistic S4, one of the most-studied intuitionistic modal logics. Instead of working with known Fitch-style calculi for this logic [13,34] we explore a new, particularly simple, calculus where the modality is *idempotent*, i.e. $\Box A$ and $\Box\Box A$ are not merely logically equivalent, but isomorphic. Our semantics for this calculus rely on an unusual 'coherence' proof. In Sect. 5 we present a calculus corresponding to the logic Intuitionistic R. In Sect. 6 we conclude with a discussion of related and further work.

2 Intuitionistic K

This section presents results for the calculus of Borghuis [9] for the most basic modal logic for necessity, first identified to our knowledge by Božić et al. [10] as $HK_\Box$; following Yokota [43] we use the name Intuitionistic K (IK). This logic extends intuitionistic logic with a new unary connective $\Box$, one new axiom

K: $\Box(A \to B) \to \Box A \to \Box B$

and one new inference rule

Necessitation: if A is a theorem, then so is $\Box A$.

2.1 Type System

Contexts are defined by the grammar

$$\Gamma \triangleq \cdot \mid \Gamma, x : A \mid \Gamma, 🔓$$

where x is a variable not in Γ, A is a formula of intuitionistic modal logic, and 🔓 is called a *lock*. The open lock symbol is used to suggest that a box has been opened, allowing access to its contents.

Ignoring variables and terms, sequents $\Gamma \vdash A$ may be interpreted as intuitionistic modal formulae by the translation

- $[\![\cdot \vdash A]\!] = A$;
- $[\![B, \Gamma \vdash A]\!] = B \to [\![\Gamma \vdash A]\!]$;
- $[\![🔓, \Gamma \vdash A]\!] = \Box[\![\Gamma \vdash A]\!]$.

This interpretation will suffice to confirm the soundness and completeness of our calculus, considered as a natural deduction calculus, with respect to IK. It is however not a satisfactory basis for a categorical semantics, because it does not interpret the context as an object. In Sect. 2.3 we shall see that 🔓 may instead by interpreted as a *left adjoint* of $\Box$, applied to the context to its left.

Figure 1 presents the typing rules. Rules for the product constructions 1, $A \times B$, $\langle\rangle$, $\langle t, u\rangle$, $\pi_1\, t$, $\pi_2\, t$ are as usual and so are omitted, while sums are discussed at the end of Sect. 2.2. Note that variables can only be introduced or abstracted if they do not appear to the left of a lock. In the variable rule the context Γ' builds in variable exchange, while in the open rule Γ' builds in variable weakening. Exchange of variables with locks, and weakening for locks, are not admissible.

$$\frac{}{\Gamma, x : A, \Gamma' \vdash x : A}\ \text{🔓} \notin \Gamma' \qquad \frac{\Gamma, x : A \vdash t : B}{\Gamma \vdash \lambda x.t : A \to B} \qquad \frac{\Gamma \vdash t : A \to B \qquad \Gamma \vdash u : A}{\Gamma \vdash t\, u : B}$$

$$\frac{\Gamma, \text{🔓} \vdash t : A}{\Gamma \vdash \mathsf{shut}\, t : \Box A} \qquad \frac{\Gamma \vdash t : \Box A}{\Gamma, \text{🔓}, \Gamma' \vdash \mathsf{open}\, t : A}\ \text{🔓} \notin \Gamma'$$

Fig. 1. Typing rules for Intuitionistic K

Theorem 2.1 (Logical Soundness and Completeness). *A formula is a theorem of IK if and only if it is an inhabited type in the empty context.*

We can for example show that the K axiom is inhabited:

$$\dfrac{\dfrac{f : \Box(A \to B), x : \Box A, \text{🔓} \vdash \mathsf{open}\, f : A \to B \qquad f, x, \text{🔓} \vdash \mathsf{open}\, x : A}{f : \Box(A \to B), x : \Box A, \text{🔓} \vdash (\mathsf{open}\, f)(\mathsf{open}\, x) : B}}{f : \Box(A \to B), x : \Box A \vdash \mathsf{shut}((\mathsf{open}\, f)(\mathsf{open}\, x)) : \Box B}$$

2.2 Computation

We extend the usual notion of β-reduction on untyped terms with the rule

$$\mathsf{open}\, \mathsf{shut}\, t \ \mapsto\ t$$

We write $\rightsquigarrow$ for the reflexive transitive closure of $\mapsto$. This relation is plainly confluent. Two lemmas, proved by easy inductions on the derivation of the terms t, then allow us to prove subject reduction:

Lemma 2.2 (Variable Weakening). *If $\Gamma, \Gamma' \vdash t : B$ then $\Gamma, x : A, \Gamma' \vdash t : B$.*

Lemma 2.3 (Substitution). *If $\Gamma, x : A, \Gamma' \vdash t : B$ and $\Gamma \vdash u : A$ then $\Gamma, \Gamma' \vdash t[u/x] : B$.*

Theorem 2.4 (Subject Reduction). *If $\Gamma \vdash t : A$ and $t \mapsto u$ then $\Gamma \vdash u : A$.*

Proof. β-reduction for $\to$ requires Lemma 2.3, and for $\Box$ requires Lemma 2.2.

A term t is *normalisable* if there exists an integer $\nu(t)$ bounding the length of any reduction sequence starting with t, and *normal* if $\nu(t)$ is 0. By standard techniques we prove the following theorems:

Theorem 2.5 (Strong Normalisation). *Given $\Gamma \vdash t : A$, the term t is normalisable.*

Theorem 2.6 (Canonicity). *If Γ is a context containing no variable assignments, $\Gamma \vdash t : A$, and t is normal, then the main term-former of t is the introduction for the main type-former of A.*

Concretely, if A is some base type then t is a value of that type.

Theorem 2.7 (Subformula Property). *Given $\Gamma \vdash t : A$ with t normal, all subterms of t have as their type in the derivation tree a subtype of A, or a subtype of a type assigned in Γ.*

To attain this final theorem we need to take some care with sums. It is well known that lambda calculi with sums do not enjoy the subformula property unless they have additional reductions called commuting conversions [21, Chap. 10]. However the commuting conversions for the $\Box$ type

$$\mathsf{open}\,\mathsf{case}\, s \,\mathsf{of}\, x.t; y.u \mapsto \mathsf{case}\, s \,\mathsf{of}\, x.\mathsf{open}\, t; y.\mathsf{open}\, u$$
$$\mathsf{open}\,\mathsf{abort}\, t \mapsto \mathsf{abort}\, t$$

do not obviously enjoy subject reduction because open might change the context. However if we tweak the definitions of the elimination term-formers for sums according to Fig. 2 then all results of this section indeed hold.

$$\frac{\Gamma \vdash s : A+B \qquad \Gamma, x : A, \Gamma' \vdash t : C \qquad \Gamma, y : B, \Gamma' \vdash u : C}{\Gamma, \Gamma' \vdash \mathsf{case}\, s \,\mathsf{of}\, x.t; y.u : C} \qquad \frac{\Gamma \vdash t : 0}{\Gamma, \Gamma' \vdash \mathsf{abort}\, t : A}$$

Fig. 2. Elimination term-formers for sums

Finally, while we will not explore computational aspects of η-equivalence in this paper, we do note that

$$\mathsf{shut}\,\mathsf{open}\, t \;=\; t$$

obeys subject reduction in both directions (provided, in the expansion case, that the type of t has $\Box$ as its main type-former).

2.3 Categorical Semantics

This section goes beyond Theorem 2.1 to establish the soundness of the type system with respect to a *categorical semantics*, in cartesian closed categories $\mathcal{C}$ equipped with an endofunctor $\Box$ that has a *left adjoint*, which we write $\blacklozenge$.

We interpret types as $\mathcal{C}$-objects via the structure of $\mathcal{C}$ in the obvious way. We then interpret contexts as $\mathcal{C}$-objects by

- $[\![\cdot]\!] \triangleq 1$;
- $[\![\Gamma, x : A]\!] \triangleq [\![\Gamma]\!] \times A$;
- $[\![\Gamma, \text{🔓}]\!] \triangleq \blacklozenge[\![\Gamma]\!]$.

We omit the brackets $[\![\cdots]\!]$ where no confusion is possible, and usually abuse notation by omitting the left-most '$1\times$' where the left of the context is a variable.

We will also sometimes interpret contexts Γ as endofunctors, abusing notation to also write them as $[\![\Gamma]\!]$, or merely Γ, by taking $[\![\cdot]\!]$ as the identity, $[\![\Gamma, x : A]\!] = [\![\Gamma]\!] \times A$, and $[\![\Gamma, \text{🔓}]\!] = \blacklozenge[\![\Gamma]\!]$.

We interpret $\Gamma \vdash t : A$ as a $\mathcal{C}$-arrow $[\![\Gamma \vdash t : A]\!] : [\![\Gamma]\!] \to A$, often abbreviated to $[\![t]\!]$, or merely t, by induction on the derivation of t as follows.

Standard constructions such as variables, abstraction and application are interpreted as usual. To interpret the rules for sums of Fig. 2 we use the fact that $\blacklozenge$, as a left adjoint, preserves colimits.

shut: we simply apply the isomorphism $\mathcal{C}(\blacklozenge[\![\Gamma]\!], A) \to \mathcal{C}([\![\Gamma]\!], \Box A)$ given by the $\blacklozenge \dashv \Box$ adjunction.

open: We apply the isomorphism $\mathcal{C}([\![\Gamma]\!], \Box A) \to \mathcal{C}(\blacklozenge[\![\Gamma]\!], A)$ to the arrow interpreting the premise, then compose with the projection $[\![\Gamma, \text{🔓}, \Gamma']\!] \to [\![\Gamma, \text{🔓}]\!]$.

Theorem 2.8 (Categorical Soundness). *If $\Gamma \vdash t : A$ and $t \mapsto t'$ then* $[\![t]\!] = [\![t']\!]$.

We also have that η-equivalent terms have the same denotation.

3 Left Adjoints and Categorical Completeness

In this section we extend the calculus to include the left adjoint $\blacklozenge$ as a first-class type-former, and hence prove categorical completeness. The underlying logic is the fragment of intuitionistic tense logic [17] with just one pair of modalities, studied by Dzik et al. [15] as 'intuitionistic logic with a Galois connection'; we use the name $IK_{\blacklozenge}$. We have two new axioms

η^m: $A \to \Box\blacklozenge A$

ε^m: $\blacklozenge\Box A \to A$

We use the superscript m to identify these as the unit as the unit and counit of the *modal* adjunction $\blacklozenge \dashv \Box$, to differentiate them from other (co)units used elsewhere in the paper. We have one new inference rule:

Monotonicity: if $A \to B$ is a theorem, then so is $\blacklozenge A \to \blacklozenge B$.

3.1 Type System and Computation

We extend the type system of Fig. 1 with the new rules for $\blacklozenge$ presented in Fig. 3. $\blacklozenge$, unlike $\Box$, need not commute with products, so does not interact well with contexts. Hence the subterms of a $\mathsf{let\,dia}$ term may not share variables.

$$\frac{\Gamma \vdash t : A}{\Gamma, \text{🔓}, \Gamma' \vdash \mathsf{dia}\, t : \blacklozenge A}\ \text{🔓} \notin \Gamma' \qquad\qquad \frac{\Gamma \vdash t : \blacklozenge A \qquad x : A, \text{🔓} \vdash u : B}{\Gamma \vdash \mathsf{let\,dia}\, x\, \mathsf{be}\, t\, \mathsf{in}\, u : B}$$

Fig. 3. Additional typing rules for logic $IK_{\blacklozenge}$

We can construct the axioms of $IK_{\blacklozenge}$:

$$\frac{x : A, \text{🔓} \vdash \mathsf{dia}\, x : \blacklozenge A}{x : A \vdash \mathsf{shut\,dia}\, x : \Box\blacklozenge A} \qquad \frac{x : \blacklozenge\Box A \vdash x : \blacklozenge\Box A \qquad y : \Box A, \text{🔓} \vdash \mathsf{open}\, y : A}{x : \blacklozenge\Box A \vdash \mathsf{let\,dia}\, y\, \mathsf{be}\, x\, \mathsf{in\,open}\, y : A}$$

and given a closed term $f : A \to B$ we have the monotonicity construction

$$\frac{x : \blacklozenge A \vdash x : \blacklozenge A \qquad y : A, \text{🔓} \vdash \mathsf{dia}(f\, y) : \blacklozenge B}{x : \blacklozenge A \vdash \mathsf{let\,dia}\, y\, \mathsf{be}\, x\, \mathsf{in\,dia}(f\, y) : \blacklozenge B}$$

To this we add the new β rule

$$\mathsf{let\,dia}\, x\, \mathsf{be\,dia}\, t\, \mathsf{in}\, u \ \mapsto\ u[t/x]$$

We can hence extend the syntactic results of the previous section to the logic $IK_{\blacklozenge}$, with the exception of the subformula property. Consider the term

$$\frac{x : \blacklozenge A \vdash \mathsf{let\,dia}\, y\, \mathsf{be}\, x\, \mathsf{in}\, \lambda z.\mathsf{dia}\, y : \blacklozenge A \to \blacklozenge A \qquad x : \blacklozenge A \vdash x : \blacklozenge A}{x : \blacklozenge A \vdash (\mathsf{let\,dia}\, y\, \mathsf{be}\, x\, \mathsf{in}\, \lambda z.\mathsf{dia}\, y)x : \blacklozenge A}$$

This term is normal but evidently fails the subformula property. One might expect, as with sums, that a commuting conversion would save the day by reducing the term to $\mathsf{let\,dia}\, y\, \mathsf{be}\, x\, \mathsf{in}\, ((\lambda z.\mathsf{dia}\, y)x)$, but this term sees the free variable x appear in the second subterm of a $\mathsf{let\,dia}$ expression, which is not permitted.

We now turn to η-equivalence, and an equivalence which we call *associativity*:

$$\begin{aligned} \mathsf{let\,dia}\, x\, \mathsf{be}\, t\, \mathsf{in\,dia}\, x &= t \\ \mathsf{let\,dia}\, x\, \mathsf{be}\, s\, \mathsf{in}\, (t[u/y]) &= t[\mathsf{let\,dia}\, x\, \mathsf{be}\, s\, \mathsf{in}\, u/y] \text{if } t\text{'s context contains } y \text{ only} \end{aligned}$$

For example, under associativity the counter-example to the subformula property equals $(\lambda z.\mathsf{let\,dia}\, y\, \mathsf{be}\, x\, \mathsf{in\,dia}\, y)x$, which reduces to $\mathsf{let\,dia}\, y\, \mathsf{be}\, x\, \mathsf{in\,dia}\, y$, which is η-equal to x. The equivalences enjoy subject reduction in both directions (requiring, as usual, that t has the right type for η-expansion).

3.2 Categorical Semantics

We interpret the new term-formers in the same categories as used in Sect. 2.3. For dia, given $t : \Gamma \to A$ we compose $\blacklozenge t$ with the projection $\Gamma, 🔓, \Gamma' \to \Gamma, 🔓$. The denotation of let dia x be t in u is simply $u \circ t$. We may then confirm the soundness of β-reduction, η-equivalence, and associativity; we call these equivalences collectively *definitional equivalence.*

We extend standard techniques for proving completeness [25], constructing a *term model*, a category with types as objects and, as arrows $A \to B$, terms of form $x : A \vdash t : B$ modulo definitional equivalence. This is a category by taking identity as the term x and composition $u \circ t$ as $u[t/x]$. It is a cartesian closed category using the type- and term-formers for products and function spaces.

The modalities $\blacklozenge$ and $\Box$ act on types; they also act on terms by, for $\blacklozenge$, the monotonicity construction, and for $\Box$, mapping $x : A \vdash t : B$ to $x : \Box A \vdash$ shut t[open x/x] $: \Box B$. One can check these constructions are functorial, and that the terms for η^m and ε^m are natural and obey the triangle equalities for the adjunction $\blacklozenge \dashv \Box$.

Given a context Γ we define the *context term* $\Gamma \vdash c_\Gamma : [\![\Gamma]\!]$ by

- $c_\cdot \triangleq \langle\rangle$;
- $c_{\Gamma, x:A} \triangleq \langle c_\Gamma, x \rangle$;
- $c_{\Gamma, 🔓} \triangleq$ dia c_Γ.

Lemma 3.1. *Given $\Gamma \vdash t : A$, t is definitionally equal to $[\![\Gamma \vdash t : A]\!][c_\Gamma/x]$.*

Theorem 3.2 (Categorical Completeness). *If $\Gamma \vdash t : A$ and $\Gamma \vdash u : A$ are equal in all models then they are definitionally equal.*

Proof. t and u have equal denotations in the term model, so their denotations are definitionally equal. Definitional equality is preserved by substitution, so $[\![\Gamma \vdash t : A]\!][c_\Gamma/x] = [\![\Gamma \vdash u : A]\!][c_\Gamma/x]$, so by Lemma 3.1, $t = u$.

4 Intuitionistic S4 for Idempotent Comonads

Intuitionistic S4 (IS4) is the extension of IK with the axioms

T: $\Box A \to A$

4: $\Box A \to \Box\Box A$

To the category theorist IS4 naturally suggests the notion of a *comonad.* IS4 is one of the most studied and widely applied intuitionistic modal logics; in particular there exist two Fitch-style calculi [13, 34]. We conjecture that similar results to the previous sections could be developed for these calculi. Instead of pursuing such a result, we here show that a simpler calculus is possible if we restrict to *idempotent* comonads, where $\Box A$ and $\Box\Box A$ are isomorphic. This restriction picks out an important class of examples – see for example the discussion of Rijke et al. [35] – and relies on a novel 'coherence' proof.

4.1 Type System and Computation

A calculus for IS4 is obtained by replacing the open rule of Fig. 1 by

$$\frac{\Gamma \vdash t : \Box A}{\Gamma, \Gamma' \vdash \mathsf{open}\, t : A}$$

The T and 4 axioms are obtained by

$$\frac{x : \Box A \vdash x : \Box A}{x : \Box A \vdash \mathsf{open}\, x : A} \qquad \frac{\dfrac{x : \Box A, \text{🔓}, \text{🔓} \vdash \mathsf{open}\, x : A}{x : \Box A, \text{🔓} \vdash \mathsf{shut}\,\mathsf{open}\, x : \Box A}}{x : \Box A \vdash \mathsf{shut}\,\mathsf{shut}\,\mathsf{open}\, x : \Box\Box A}$$

This confirms logical completeness; once can also easily check soundness.

Subject reduction for the β-reduction $\mathsf{open}\,\mathsf{shut}\, t \mapsto t$ requires a new lemma, proved by an easy induction on t:

Lemma 4.1 (Lock Replacement). *If* $\Gamma, \text{🔓}, \Gamma'' \vdash t : A$ *then* $\Gamma, \Gamma', \Gamma'' \vdash t : A$.

The key syntactic Theorems 2.5, 2.6, and 2.7 then follow easily.

η-expansion obeys subject reduction as before, but it is not the case, for example, that the term presented above for the 4 axiom reduces to $\mathsf{shut}\, x$. We may however accept a notion of η-reduction on typed terms-in-context:

$$\Gamma \vdash \mathsf{shut}\,\mathsf{open}\, t \mapsto t : \Box A \text{ provided that } \Gamma \vdash t : \Box A$$

This equivalence is more powerful than it might appear; it allows us to derive the idempotence of $\Box$, as the 4 axiom is mutually inverse with the instance $\Box\Box A \to \Box A$ of the T axiom. That is, $\lambda x.\mathsf{open}\,\mathsf{shut}\,\mathsf{shut}\,\mathsf{open}\, x$ reduces to the identity on $\Box A$, and $\lambda x.\mathsf{shut}\,\mathsf{shut}\,\mathsf{open}\,\mathsf{open}\, x$ reduces to the identity on $\Box\Box A$.

4.2 Categorical Semantics

We give semantics to our type theory in a cartesian closed category with an adjunction of endofunctors $\blacklozenge \dashv \Box$ in which $\Box$ is a *comonad*. Equivalently [16, Sect. 3], $\blacklozenge$ is a monad, equipped with a unit η and multiplication μ. To confirm the *coherence* of these semantics, discussed in the next subsection, and the soundness of η-equivalence, we further require that $\Box$ is idempotent, or equivalently that all $\mu_A : \blacklozenge\blacklozenge A \to \blacklozenge A$ are isomorphisms with inverses $\eta_{\blacklozenge A} = \blacklozenge \eta_A$.

To define the semantics we define *lock replacement* natural transformations $l_\Gamma : [\![\Gamma]\!] \to \blacklozenge$, corresponding to Lemma 4.1, by induction on Γ:

- $l_{.}$ is the unit η of the monad;
- $l_{\Gamma, x:A}$ is the projection composed with l_Γ;
- $l_{\Gamma, \text{🔓}}$ is $\blacklozenge l_\Gamma$ composed with μ.

Note that $l_{\text{🔓}}$ is the identity by the monad laws.

We may now define the interpretation of open: given $t : \Gamma \to \Box A$ we apply the adjunction to get an arrow $\blacklozenge\Gamma \to A$, then compose with $l_{\Gamma'} : \Gamma, \Gamma' \to \Gamma, \text{🔓}$.

Lemma 4.2. *If we replace part of a context with a lock, then replace part of the new context that includes the new lock, we could have done this in one step:*

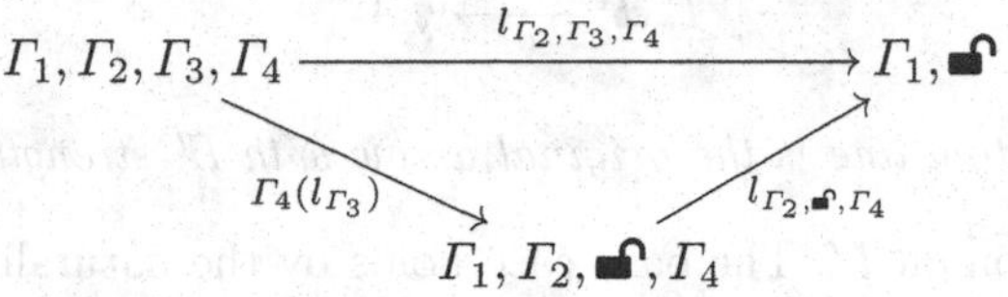

Proof. By induction on Γ_4, with the base case following by induction on Γ_3.

Lemma 4.3. $[\![\Gamma, \text{🔓}, \Gamma'' \vdash t : A]\!] \circ [\![\Gamma'']\!](l_{\Gamma'}) = [\![\Gamma, \Gamma', \Gamma'' \vdash t : A]\!]$.

Proof. By induction on the derivation of t.

Now open shut t, where the open has weakening Γ', has denotation $\varepsilon^m \circ \blacklozenge\Box t \circ \blacklozenge\eta^m \circ l_{\Gamma'}$, which is $t \circ l_{\Gamma'}$ by the naturality of ε^m, and the adjunction. This is what is required by Lemma 4.3, so β-reduction for $\Box$ is soundly modelled.

4.3 Coherence

Because the open rule involves a weakening, and does not explicitly record in the term what that weakening is, the same typed term-in-context can be the root of multiple derivation trees, for example:

$$\frac{\dfrac{x : \Box\Box A \vdash x : \Box\Box A}{x : \Box\Box A, \text{🔓} \vdash \mathsf{open}\, x : \Box A}}{x : \Box\Box A, \text{🔓}, \text{🔓} \vdash \mathsf{open}\,\mathsf{open}\, x : A} \qquad \frac{\dfrac{x : \Box\Box A \vdash x : \Box\Box A}{x : \Box\Box A \vdash \mathsf{open}\, x : \Box A}}{x : \Box\Box A, \text{🔓}, \text{🔓} \vdash \mathsf{open}\,\mathsf{open}\, x : A}$$

The categorical semantics of the previous section is defined by induction on derivations, and so does not truly give semantics to *terms* unless any two trees with the same root must have the same denotation. In this section we show that this property, here called *coherence*, indeed holds. We make crucial use of the idempotence of the comonad $\Box$.

We first observe that if $\Gamma, \Gamma', \Gamma'' \vdash t : A$ and all variables of Γ' are not free in t, then $\Gamma, \Gamma'' \vdash t : A$. The following lemma, proved by easy inductions, describes how the denotations of these derivations are related:

Lemma 4.4. *1. If x is not free in t then $\Gamma, x : A, \Gamma' \vdash t : B$ has the same denotation as $\Gamma, \Gamma' \vdash t : B \circ \Gamma'(pr)$.*
2. $\Gamma, \Gamma' \vdash t : B$ has denotation $\Gamma, \text{🔓}, \Gamma' \vdash t : B \circ \Gamma'(\eta)$.

The technical lemma below is the only place where idempotence is used.

Lemma 4.5. *Given $\Gamma, \Gamma' \vdash t : A$ with Γ' not free in t, we have*

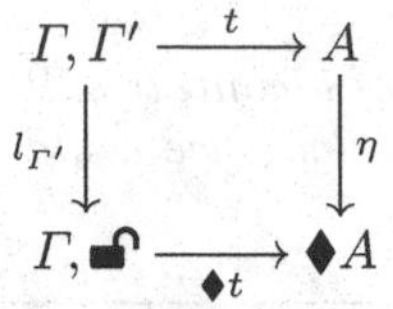

where t on the bottom line is the original arrow with Γ' strengthened away.

Proof. By induction on Γ'. The base case holds by the naturality of η.

We present only the lock case: $\eta \circ t = \blacklozenge t \circ \eta$ by the naturality of η. But by **idempotence**, $\eta : \Gamma, \Gamma',$ 🔓 $\to \Gamma, \Gamma',$ 🔓, 🔓 equals $\blacklozenge\eta$. Then by Lemma 4.4 $\blacklozenge t \circ \blacklozenge\eta$ is $\blacklozenge[\![\Gamma, \Gamma' \vdash t : A]\!]$, i.e. we have strengthened the lock away and can hence use our induction hypothesis, making the top trapezium commute in:

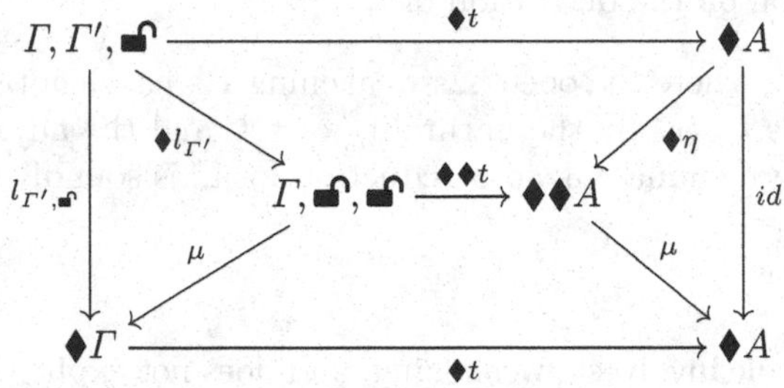

The left triangle commutes by definition, the bottom trapezium commutes by the naturality of μ, and the right triangle commutes by the monad laws.

Lemma 4.6. *Given $\Gamma, \Gamma' \vdash t : A$ with Γ' not free in t, we have*

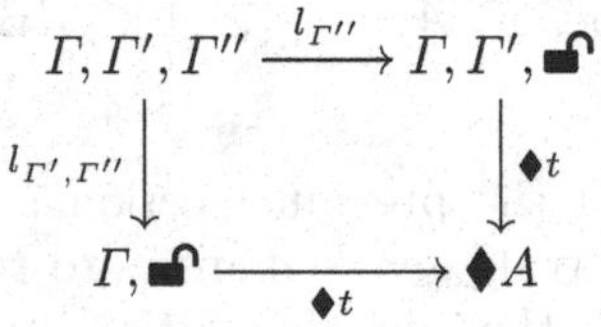

where the bottom t is obtained via strengthening.

Proof. By induction on Γ''. The base case follows by Lemma 4.5.

Lemma 4.7. *Given $\Gamma, \Gamma' \vdash t : \Box A$ with the variables of Γ' not free in t, the following arrows are equal:*

- *$\Gamma, \Gamma', \Gamma'' \vdash$ $\mathsf{open}\, t : A$ where the weakening is Γ'';*
- *obtaining an arrow $\Gamma \to \Box A$ via Lemma 4.4, then applying* open *with weakening Γ, Γ''.*

Proof. Immediate from Lemma 4.6, i.e.

$$\begin{array}{ccccc} \Gamma, \Gamma', \Gamma'' & \xrightarrow{l_{\Gamma''}} & \Gamma, \Gamma', \text{🔓} & & \\ {\scriptstyle l_{\Gamma',\Gamma''}}\downarrow & & \downarrow{\scriptstyle \blacklozenge t} & & \\ \Gamma, \text{🔓} & \xrightarrow[\blacklozenge t]{} & \blacklozenge \Box A & \xrightarrow[\varepsilon^m]{} & A \end{array}$$

Theorem 4.8 (Coherence). *Given two different derivation trees of a term, their denotation is equal.*

Proof. By induction on the number of nodes in the trees. The base case with one node is trivial. Suppose we have $n+1$ nodes. Then the induction hypothesis immediately completes the proof unless the nodes above the roots are non-equal. Then the final construction must be an instance of open, i.e. we have

$$\frac{\Gamma \vdash t : \Box A}{\Gamma, \Gamma', \Gamma'' \vdash \mathsf{open}\, t : A} \qquad \frac{\Gamma, \Gamma' \vdash t : \Box A}{\Gamma, \Gamma', \Gamma'' \vdash \mathsf{open}\, t : A}$$

Clearly any variables in Γ' are not free in t, so we can use Lemma 4.4 on the top line of the right hand tree to derive $\Gamma \vdash t : \Box A$. By induction hypothesis this has the same denotation as the top line of the left hand tree. But Lemma 4.7 tells us that applying this strengthening and then opening with Γ', Γ'' is the same as opening with Γ'' only.

We can now demonstrate the soundness of η-equivalence: given $\Gamma \vdash t : \Box A$ and $\Gamma \vdash \mathsf{shut\,open}\, t : \Box A$ by any derivations, we can by coherence safely assume that open used one lock only as its weakening, and so the arrows are equal by the $\blacklozenge \dashv \Box$ adjunction.

4.4 Left Adjoints and Categorical Completeness

Following Sect. 3 we can add $\blacklozenge$ to the type theory; we need only modify the dia rule to

$$\frac{\Gamma \vdash t : A}{\Gamma, \Gamma' \vdash \mathsf{dia}\, t : \blacklozenge A}$$

to retain Lemma 4.1. The results of the previous sections, apart once more for the subformula property, still hold, where we define the denotation of $\Gamma, \Gamma' \vdash \mathsf{dia}\, t$ as $\blacklozenge t$ composed with $l_{\Gamma'}$. In particular, we must confirm that Lemma 3.1 extends to the new definitions of open and dia, for which we need the lemma below:

Lemma 4.9. *Given the term* $x : [\![\Gamma, \Gamma']\!] \vdash l_{\Gamma'} : \blacklozenge [\![\Gamma]\!]$ *defined in the term model,* $l_{\Gamma'}[c_{\Gamma,\Gamma'}/x]$ *is definitionally equal to* $\mathsf{dia}\, c_{\Gamma}$.

Now $[\![\mathsf{open}\, t]\!][c_{\Gamma,\Gamma'}/x]$ is $\mathsf{let\,dia}\, x\, \mathsf{be}\, (\mathsf{let\,dia}\, x\, \mathsf{be}\, l_{\Gamma'}[c_{\Gamma,\Gamma'}/x]\, \mathsf{in\,dia}[\![t]\!])\, \mathsf{in\,open}\, x$, which by the lemma above is $\mathsf{let\,dia}\, x\, \mathsf{be}\, (\mathsf{let\,dia}\, x\, \mathsf{be\,dia}\, c_{\Gamma}\, \mathsf{in\,dia}[\![t]\!])\, \mathsf{in\,open}\, x \mapsto \mathsf{open}[\![t]\!][c_{\Gamma}/x]$, which equals $\mathsf{open}\, t$ by induction. The proof for dia is similar.

5 Intuitionistic R

One can readily imagine how the calculus for IS4 could be modified for logics with only one of the T and 4 axioms. In this section we instead illustrate the flexibility of Fitch-style calculi by defining a calculus for the rather different logic Intuitionistic R (IR), which extends IK with the axiom

R: $A \to \Box A$

This axiom was first studied for intuitionistic necessity modalities by Curry [12], along with the axiom M, $\Box\Box A \to \Box A$, to develop a logic for monads. The importance of the logic with R but without M was established by McBride and Paterson [30] who showed that it captured the useful programming abstraction of *applicative functors*. We take the name R for the axiom from Fairtlough and Mendler [18], and for the logic from Litak [28].

We modify Figs. 1 and 3 simply by removing the side-conditions $\text{🔓} \notin \Gamma$ from the variable, open, and dia rules. We can then derive R:

$$\frac{x : A, \text{🔓} \vdash x : A}{x : A \vdash \mathsf{shut}\, x : \Box A}$$

For substitution and subject reduction we require the following lemma, easily proved by induction on the derivation of t:

Lemma 5.1 (Lock Weakening). *If* $\Gamma, \Gamma' \vdash t : A$ *then* $\Gamma, \text{🔓}, \Gamma' \vdash t : A$.

We can also observe that η-equivalence preserves types in both directions.

We give semantics for this calculus in a cartesian closed category equipped with an adjunction of endofunctors $\blacklozenge \dashv \Box$ and a 'point' natural transformation $r : Id \to \Box$ preserved by $\Box$, i.e. $\Box r = r : \Box A \to \Box\Box A$. This last property makes this model slightly less general than the notion of tensorial strength used for categorical semantics by McBride and Paterson [30], but is needed for coherence and the soundness of η-equivalence. We will use the arrow $\blacklozenge A \to A$ defined by applying the adjunction to r; we call this q and note the property:

Lemma 5.2. $q = \blacklozenge q : \blacklozenge\blacklozenge A \to \blacklozenge A$.

The *weakening* natural transformation $w_\Gamma : \Gamma \to Id$ is defined by induction on Γ via projection and q. Variables are then denoted by projection composed with weakening, and weakening is used similarly for open and dia. We can hence show the soundness of β-reduction for $\Box$ and $\blacklozenge$. For the soundness of η-equivalence for $\Box$ we need the following lemma:

Lemma 5.3. $w_{\Gamma', \text{🔓}} = \blacklozenge w_{\text{🔓}, \Gamma'} : \Gamma, \text{🔓}, \Gamma', \text{🔓} \to \Gamma, \text{🔓}$.

The denotation of $\Gamma, \text{🔓}, \Gamma' \vdash \mathsf{shut}\,\mathsf{open}\, t$ is $\Box\varepsilon^m \circ \Box\blacklozenge t \circ \Box w_{\Gamma', \text{🔓}} \circ \eta^m$. By the above lemma we replace $\Box w_{\Gamma', \text{🔓}}$ with $\Box\blacklozenge w_{\text{🔓}, \Gamma'}$, so by the naturality of η^m we have $\Box\varepsilon^m \circ \eta^m \circ t \circ w_{\text{🔓}, \Gamma'}$, which is $t \circ w_{\text{🔓}, \Gamma'}$ by the monad laws.

Moving to coherence, we conduct a similar induction to Theorem 4.8, considering the case

$$\frac{\Gamma \vdash t : \Box A}{\Gamma, \text{🔓}, \Gamma', \text{🔓}, \Gamma'' \vdash \mathsf{open}\, t : A} \qquad \frac{\Gamma, \text{🔓}, \Gamma' \vdash t : \Box A}{\Gamma, \text{🔓}, \Gamma', \text{🔓}, \Gamma'' \vdash \mathsf{open}\, t : A}$$

The top line on the left weakens to the top line on the right, with denotation $t \circ w_{\text{🔓},\Gamma'}$. By induction this equals the denotation of the top line of the right. Then the right hand term has denotation $\varepsilon^m \circ \blacklozenge t \circ \blacklozenge w_{\text{🔓},\Gamma'} \circ w_{\Gamma''}$. But by Lemma 5.3 $\blacklozenge w_{\text{🔓},\Gamma'} = w_{\Gamma',\text{🔓}}$. It is clear that $w_{\Gamma',\text{🔓}} \circ w_{\Gamma''} = w_{\Gamma',\text{🔓},\Gamma''}$, which is exactly the weakening used on the left. Coherence for dia follows similarly.

Moving finally to categorical completeness, in the term model $\Box t \circ r$ is $\mathsf{shut}\, t[\mathsf{open}\,\mathsf{shut}\, x/x]$, which reduces to $\mathsf{shut}\, t$, so r is natural. $\Box r : \Box A \to \Box\Box A$ is $\mathsf{shut}\,\mathsf{shut}\,\mathsf{open}\, x$, which is indeed η-equal to $\mathsf{shut}\, x$.

We finally need to update Lemma 3.1 for our new definitions. We do this via a lemma similar to Lemma 4.9:

Lemma 5.4. *Given the term $x : [\![\Gamma, \Gamma']\!] \vdash w_{\Gamma'} : [\![\Gamma]\!]$ defined in the term model, $w_{\Gamma'}[c_{\Gamma,\Gamma'}/x]$ is definitionally equal to c_Γ.*

Now the denotation of $\Gamma, x : A, \Gamma' \vdash x : A$ is $\pi_2 w_{\Gamma'}$. Therefore we have $\pi_2 w_{\Gamma'}[c_{\Gamma,A,\Gamma'}/x]$, which is $\pi_2 c_{\Gamma,A}$ by the lemma above. This is $\pi_2\langle c_\Gamma, x\rangle$, which reduces to x.

The denotation of $\Gamma, \text{🔓}, \Gamma' \vdash \mathsf{open}\, t : A$ is $\mathsf{let\, dia}\, x\, \mathsf{be}\, w_{\Gamma'}\, \mathsf{in\, open}[\![t]\!]$. Applying the substitution $[c_{\Gamma,\text{🔓},\Gamma'}/x]$ along with the lemma above yields the term $\mathsf{let\, dia}\, x\, \mathsf{be\, dia}\, c_\Gamma\, \mathsf{in\, open}[\![t]\!] \mapsto \mathsf{open}[\![t]\!][c_\Gamma/x]$, and induction completes. The calculations for dia follow similarly.

6 Related and Further Work

Conventional contexts. Lambda calculi with conventional contexts containing typed variables only have been proposed for the logic of monads [32], for IS4 [5], for IK [4], and for a logic with 'Löb induction' [6], from which one can extract a calculus for IR. In previous work [11] we developed the *guarded lambda calculus* featuring two modalities, where one ('constant') was an (idempotent) comonad, and the other ('later') supported a notion of guarded recursion corresponding to Löb induction. We therefore used the existing work [5,6] 'off the shelf'.

Problems arose when we attempted to extend our calculus with dependent types [7]. Neither of the calculi with conventional contexts we had used scaled well to this extension. The calculus for IS4 [5], whose terms involved explicit substitutions, turned out to require these substitutions on types also, which added a level of complexity that made it difficult to write even quite basic dependently typed programs. The constant modality was therefore jettisoned in favour of an approach based on clock quantification [1], of which more below. The calculus for later employed a connective $\circledast$ (from McBride and Patterson [30]) which acted on function spaces under the modality. However with dependent types we need to act not merely on function spaces, but on Π-types, and $\circledast$ was unable to be used. Instead a novel notion of 'delayed substitution' was introduced. These were given an equational theory, but some of these equations could not be directed, so they did not give rise to a useful notion of computation.

Modalities as quantifiers. The suggestive but formally rather underdeveloped paper of De Queiroz and Gabbay [14] proposed that necessity modalities should be treated as universal quantifiers, inspired by the standard semantics of necessity as 'for all possible worlds'. This is one way to understand the relationship between the constant modality and clock quantification [1]. However clock quantification is more general than a single constant modality because we can identify multiple free clock variables with multiple 'dimensions' in which a type may or may not be constant. This gap in generality can probably be bridged by using multiple independent constant modalities. More problematically, while it is clear what the denotational semantics of the constant modality are, the best model for clock quantifiers yet found [8] is rather complicated and still leaves open some problems with coherence in the presence of a universe.

Previous Fitch-style calculi. The Fitch-style approach was pioneered, apparently independently, by Martini and Masini [29] and Borghuis [9]. Martini and Masini's work is rather notationally heavy, and weakening appears not to be admissible. Borghuis's calculus for IK is excellent, but his calculi for stronger logics are not so compelling, as each different axiom is expressed with another version of the open or shut rules, not all of which compute when combined. The calculus for IS4 of Pfenning and Wong [34], refined by Davies and Pfenning [13, Sect. 4], provide the basis of the IS4 calculus of this paper, but involve some complications which appear to correlate to not assuming idempotence. We have extended this previous work by investigating the subformula property, introducing categorical semantics, and showing how left adjoints to necessity modalities à la tense logic can be used as types. Finally, the recent clocked type theory of Bahr et al. [3] independently gave a treatment of the later modality that on inspection is precisely Fitch-style (albeit with named 'locks'), and which has better computational properties than the delayed substitution approach.

Dual contexts. Davies and Pfenning [13] use a pair of contexts $\Delta;\Gamma$ with intended meaning $\Box\Delta \wedge \Gamma$. This is quite different from the semantics of Fitch-style sequents, where structure in the context denotes the *left adjoint* of $\Box$. In recent work Kavvos [24] has shown that dual contexts may capture a number of different modal logics, and the approach has been used as a foundation for both pen-and-paper mathematics [37] and, via an Agda fork [40], formalisation [26]. We support this work but there is reason to explore other options. First, writing programs with dual context calculi was described by Davies and Pfenning themselves as 'somewhat awkward', and in the same paper they suggest the Fitch-style approach as a less awkward alternative. Indeed, Fitch's approach was exactly designed to capture 'natural' modal deduction. Second, any application with multiple interacting modalities is unlikely to be accommodated in a mere two zones; the *mode theories* of Licata et al. [27] extend the dual zone approach to a richer setting in which interacting modalities, substructural contexts, and even Fitch-style natural deduction can be expressed[2], but the increase in complexity is considerable and much work remains to be done.

Further logics and algorithmic properties. We wish to bring more logics into the Fitch-style framework, in particular the logic of the later modality, extending IR with the strong Löb axiom $(\Box A \to A) \to A$. The obvious treatment of this axiom does not terminate. but Bahr et al. [3] suggest that this can be managed by giving names to locks. We would further like to develop calculi with multiple modalities. This is easy to do by assigning each modality its own lock; two IK modalities give exactly the intuitionistic tense logic of Goré et al. [22]. The situation is rather more interesting where the modalities interact, as with the later and constant modalities. Finally, we would like to further investigate algorithmic properties of Fitch-style calculi such as type checking, type inference, and η-expansion and other notions of computation. In particular, we wonder if a notion of commuting conversion can be defined so that the calculi with ⧫ enjoy the subformula property.

References

1. Atkey, R., McBride, C.: Productive coprogramming with guarded recursion. In: ICFP (2013)
2. Awodey, S., Bauer, A.: Propositions as [types]. J. Log. Comput. **14**(4), 447–471 (2004)
3. Bahr, P., Grathwohl, H.B., Møgelberg, R.E.: The clocks are ticking: no more delays! In: LICS, pp. 1–12 (2017)
4. Bellin, G., De Paiva, V., Ritter, E.: Extended Curry-Howard correspondence for a basic constructive modal logic. In: M4M (2001)
5. Bierman, G.M., de Paiva, V.C.: On an intuitionistic modal logic. Stud. Logica. **65**(3), 383–416 (2000)
6. Birkedal, L., Møgelberg, R.E.: Intensional type theory with guarded recursive types qua fixed points on universes. In: LICS (2013)
7. Bizjak, A., Grathwohl, H.B., Clouston, R., Møgelberg, R.E., Birkedal, L.: Guarded dependent type theory with coinductive types. In: Jacobs, B., Löding, C. (eds.) FoSSaCS 2016. LNCS, vol. 9634, pp. 20–35. Springer, Heidelberg (2016). https://doi.org/10.1007/978-3-662-49630-5_2
8. Bizjak, A., Møgelberg, R.E.: Denotational semantics for guarded dependent type theory. Math. Struct. Comput. Sci. (2018, to appear)
9. Borghuis, V.A.J.: Coming to terms with modal logic: on the interpretation of modalities in typed lambda-calculus. Ph.D. thesis, Technische Universiteit Eindhoven (1994)
10. Božić, M., Došen, K.: Models for normal intuitionistic modal logics. Stud. Logica. **43**(3), 217–245 (1984)
11. Clouston, R., Bizjak, A., Grathwohl, H.B., Birkedal, L.: The guarded lambda-calculus: programming and reasoning with guarded recursion for coinductive types. LMCS **12**(3) (2016). https://lmcs.episciences.org/2019

[2] We are grateful to an anoymous reviewer for this last observation.

12. Curry, H.B.: A Theory of Formal Deducibility. University of Notre Dame Press, Indiana (1957)
13. Davies, R., Pfenning, F.: A modal analysis of staged computation. JACM **48**(3), 555–604 (2001)
14. De Queiroz, R.J., Gabbay, D.M.: The functional interpretation of modal necessity. In: de Rijike, M. (eds.) Advances in Intensional Logic, pp. 61–91. Springer, Dordrecht (1997). https://doi.org/10.1007/978-94-015-8879-9_3
15. Dzik, W., Järvinen, J., Kondo, M.: Intuitionistic propositional logic with Galois connections. Log. J. IGPL **18**(6), 837–858 (2009)
16. Eilenberg, S., Moore, J.C.: Adjoint functors and triples. Illinois J. Math. **9**(3), 381–398 (1965)
17. Ewald, W.: Intuitionistic tense and modal logic. J. Symb. Log. **51**(1), 166–179 (1986)
18. Fairtlough, M., Mendler, M.: Propositional lax logic. Inform. Comput. **137**(1), 1–33 (1997)
19. Fitch, F.B.: Symbolic Logic, An Introduction. Ronald Press Co., New York (1952)
20. Fitting, M.: Proof Methods for Modal and Intuitionistic Logics. D. Reidel Publishing Co., Dordrecht (1983)
21. Girard, J.Y., Taylor, P., Lafont, Y.: Proofs and Types. Cambridge University Press, Cambridge (1989)
22. Goré, R., Postniece, L., Tiu, A.: Cut-elimination and proof search for bi-intuitionistic tense logic. In: AiML (2010)
23. Kavvos, G.A.: The many worlds of modal λ-calculi: I. Curry-Howard for necessity, possibility and time. arXiv:1605.08106 (2016)
24. Kavvos, G.: Dual-context calculi for modal logic. In: LICS (2017)
25. Lambek, J., Scott, P.J.: Introduction to Higher-Order Categorical Logic. Cambridge University Press, Cambridge (1986)
26. Licata, D.R., Orton, I., Pitts, A.M., Spitters, B.: Internal universes in models of homotopy type theory (2018, unpublished)
27. Licata, D.R., Shulman, M., Riley, M.: A fibrational framework for substructural and modal logics. In: FSCD (2017)
28. Litak, T.: Constructive modalities with provability smack. In: Bezhanishvili, G. (ed.) Leo Esakia on Duality in Modal and Intuitionistic Logics. OCL, vol. 4, pp. 187–216. Springer, Dordrecht (2014). https://doi.org/10.1007/978-94-017-8860-1_8
29. Martini, S., Masini, A.: A computational interpretation of modal proofs. In: Wansing, H. (ed.) Proof Theory of Modal Logic, pp. 213–241. Springer, Dordrecht (1996). https://doi.org/10.1007/978-94-017-2798-3_12
30. McBride, C., Paterson, R.: Applicative programming with effects. J. Funct. Program. **18**(1), 1–13 (2008)
31. Menni, M.: About И-quantifiers. Appl. Categ. Struct. **11**(5), 421–445 (2003)
32. Moggi, E.: Computational lambda-calculus and monads. In: LICS, pp. 14–23 (1989)
33. de Paiva, V., Ritter, E.: Basic constructive modality. In: Logic without Frontiers: Festschrift for Walter Alexandre Carnielli on the occasion of his 60th Birthday, pp. 411–428. College Publication (2011)
34. Pfenning, F., Wong, H.C.: On a modal λ-calculus for S4. In: MFPS (1995)
35. Rijke, E., Shulman, M., Spitters, B.: Modalities in homotopy type theory. arXiv:1706.07526 (2017)
36. Schreiber, U.: Differential cohomology in a cohesive infinity-topos. arXiv:1310.7930 (2013)

37. Shulman, M.: Brouwer's fixed-point theorem in real-cohesive homotopy type theory. Math. Struct. Comput. Sci. (2017). https://www.cambridge.org/core/journals/mathematical-structures-in-computer-science/article/brouwers-fixedpoint-theorem-in-realcohesive-homotopy-type-theory/8270C2EAC4EE5D5CDBA17EEB3FF6B19E
38. Simpson, A.K.: The proof theory and semantics of intuitionistic modal logic. Ph.D. thesis, University of Edinburgh (1994)
39. Sørensen, M.H., Urzyczyn, P.: Lectures on the Curry-Howard Isomorphism. Elsevier, New York (2006)
40. Vezzosi, A.: Agda-flat (2017). https://github.com/agda/agda/tree/flat. GitHub repository
41. Wadler, P.: Propositions as types. Commun. ACM **58**(12), 75–84 (2015)
42. Wansing, H.: Sequent calculi for normal modal propositional logics. J. Log. Comput. **4**(2), 125–142 (1994)
43. Yokota, S.: General characterization results on intuitionistic modal propositional logics. Commentarii Mathematici Universitatis Sancti Pauli **34**(2), 177–199 (1985)

Realizability Interpretation and Normalization of Typed Call-by-Need λ-calculus with Control

Étienne Miquey[1,2(✉)] and Hugo Herbelin[2]

[1] Équipe Gallinette, Inria, LS2N (CNRS), Université de Nantes, Nantes, France
etienne.miquey@inria.fr
[2] Équipe πr^2, Inria, IRIF (CNRS), Université Paris-Diderot, Paris, France
herbelin@inria.fr

Abstract. We define a variant of Krivine realizability where realizers are pairs of a term and a substitution. This variant allows us to prove the normalization of a simply-typed call-by-need λ-calculus with control due to Ariola *et al.* Indeed, in such call-by-need calculus, substitutions have to be delayed until knowing if an argument is really needed. We then extend the proof to a call-by-need λ-calculus equipped with a type system equivalent to classical second-order predicate logic, representing one step towards proving the normalization of the call-by-need classical second-order arithmetic introduced by the second author to provide a proof-as-program interpretation of the axiom of dependent choice.

1 Introduction

1.1 Realizability-Based Normalization

Normalization by realizability is a standard technique to prove the normalization of typed λ-calculi. Originally introduced by Tait [36] to prove the normalization of System T, it was extended by Girard to prove the normalization of System F [11]. This kind of techniques, also called normalization by reducibility or normalization by logical relations, works by interpreting each type by a set of typed or untyped terms seen as realizers of the type, then showing that the way these sets of realizers are built preserve properties such as normalization. Over the years, multiple uses and generalization of this method have been done, for a more detailed account of which we refer the reader to the work of Gallier [9].

Realizability techniques were adapted to the normalization of various calculi for classical logic (see e.g. [3,32]). A specific framework tailored to the study of realizability for classical logic has been designed by Krivine [19] on top of a λ-calculus with control whose reduction is defined in terms of an abstract machine. In such a machinery, terms are evaluated in front of stacks; and control (thus classical logic) is made available through the possibility of saving and restoring stacks. During the last twenty years, Krivine's classical realizability turned out to be fruitful both from the point of view of logic, leading to the construction of

C. Baier and U. Dal Lago (Eds.): FOSSACS 2018, LNCS 10803, pp. 276–292, 2018.
https://doi.org/10.1007/978-3-319-89366-2_15

new models of set theory, and generalizing in particular the technique of Cohen's forcing [20–22]; and on its computational facet, providing alternative tools to the analysis of the computational content of classical programs[1].

Noteworthily, Krivine realizability is one of the approaches contributing to advocating the motto that through the Curry-Howard correspondence, with new programming instructions come new reasoning principles[2]. Our original motivation for the present work is actually in line with this idea, in the sense that our long-term purpose is to give a realizability interpretation to dPA^ω, a call-by-need calculus defined by the second author [15]. In this calculus, the lazy evaluation is indeed a fundamental ingredient in order to obtain an executable proof term for the axiom of dependent choice.

1.2 Contributions of the Paper

In order to address the normalization of typed call-by-need λ-calculus, we design a variant of Krivine's classical realizability, where the realizers are closures (a term with a substitution for its free variables). The call-by-need λ-calculus with control that we consider is the $\overline{\lambda}_{[lv\tau\star]}$-calculus. This calculus, that was defined by Ariola *et al.* [2], is syntactically described in an extension with explicit substitutions of the $\lambda\mu\tilde{\mu}$-calculus [6,14,29]. The syntax of the $\lambda\mu\tilde{\mu}$-calculus itself refines the syntax of the λ-calculus by syntactically distinguishing between *terms* and *evaluation contexts*. It also contains *commands* which combine terms and evaluation contexts so that they can interact together. Thinking of evaluation contexts as stacks and commands as states, the $\lambda\mu\tilde{\mu}$-calculus can also be seen as a syntax for abstract machines. As for a proof-as-program point of view, the $\lambda\mu\tilde{\mu}$-calculus and its variants can be seen as a term syntax for proofs of Gentzen's sequent calculus. In particular, the $\lambda\mu\tilde{\mu}$-calculus contains control operators which give a computational interpretation to classical logic.

We give a proof of normalization first for the simply-typed $\overline{\lambda}_{[lv\tau\star]}$-calculus[3], then for a type system with first-order and second-order quantification. While we only apply our technique to the normalization of the $\overline{\lambda}_{[lv\tau\star]}$-calculus, our interpretation incidentally suggests a way to adapt Krivine realizability to other call-by-need settings. This paves the way to the computational interpretation of classical proofs using lazy evaluation or shared memory cells, including the case of the call-by-need second order arithmetic dPA^ω [15].

[1] See for instance [27] about witness extraction or [12,13] about specification problems.

[2] For instance, one way to realize the axiom of dependent choice in classical realizability is by means of an extra instruction `quote` [18].

[3] Even though it has not been done formally, the normalization of the $\overline{\lambda}_{lv}$-calculus presented in [2] should also be derivable from Polonowski's proof of strong normalization of the non-deterministic $\lambda\mu\tilde{\mu}$-calculus [35]. The $\overline{\lambda}_{lv}$-calculus (a big-step variant of the $\overline{\lambda}_{[lv\tau\star]}$-calculus introduced in Ariola *et al.*) is indeed a particular evaluation strategy for the $\lambda\mu\tilde{\mu}$-calculus, so that the strong normalization of the non-deterministic variant of the latter should imply the normalization of the former as a particular case.

2 The $\overline{\lambda}_{[lv\tau\star]}$-calculus

2.1 The Call-by-Need Evaluation Strategy

The call-by-need evaluation strategy of the λ-calculus evaluates arguments of functions only when needed, and, when needed, shares their evaluations across all places where the argument is required. The call-by-need evaluation is at the heart of a functional programming language such as Haskell. It has in common with the call-by-value evaluation strategy that all places where a same argument is used share the same value. Nevertheless, it observationally behaves like the call-by-name evaluation strategy (for the pure λ-calculus), in the sense that a given computation eventually evaluates to a value if and only if it evaluates to the same value (up to inner reduction) along the call-by-name evaluation. In particular, in a setting with non-terminating computations, it is not observationally equivalent to the call-by-value evaluation. Indeed, if the evaluation of a useless argument loops in the call-by-value evaluation, the whole computation loops, which is not the case of call-by-name and call-by-need evaluations.

These three evaluation strategies can be turned into equational theories. For call-by-name and call-by-value, this was done by Plotkin through continuation-passing-style (CPS) semantics characterizing these theories [34]. For the call-by-need evaluation strategy, a specific equational theory reflecting the intensional behavior of the strategy into a semantics was proposed independently by Ariola and Felleisen [1], and by Maraist et al. [26]. A continuation-passing-style semantics was proposed in the 90s by Okasaki et al. [30]. However, this semantics does not ensure normalization of simply-typed call-by-need evaluation, as shown in [2], thus failing to ensure a property which holds in the simply-typed call-by-name and call-by-value cases.

Continuation-passing-style semantics *de facto* gives a semantics to the extension of λ-calculus with control operators[4]. In particular, even though call-by-name and call-by-need are observationally equivalent on pure λ-calculus, their different intentional behaviors induce different CPS semantics, leading to different observational behaviors when control operators are considered. On the other hand, the semantics of calculi with control can also be reconstructed from an analysis of the duality between programs and their evaluation contexts, and the duality between the `let` construct (which binds programs) and a control operator such as Parigot's μ (which binds evaluation contexts). Such an analysis can be done in the context of the $\lambda\mu\tilde{\mu}$-calculus [6,14].

In the call-by-name and call-by-value cases, the approach based on $\lambda\mu\tilde{\mu}$-calculus leads to continuation-passing style semantics similar to the ones given by Plotkin or, in the call-by-name case, also to the one by Lafont et al. [23]. As for call-by-need, in [2] is defined the $\overline{\lambda}_{lv}$-calculus, a call-by-need version of the $\lambda\mu\tilde{\mu}$-calculus. A continuation-passing style semantics is then defined via a calculus called $\overline{\lambda}_{[lv\tau\star]}$ [2]. This semantics, which is different from Okasaki, Lee and Tarditi's one [30], is the object of study in this paper.

[4] That is to say with operators such as Scheme's `callcc`, Felleisen's $\mathcal{C}$, $\mathcal{K}$, or $\mathcal{A}$ operators [8], Parigot's μ and [ ] operators [31], Crolard's `catch` and `throw` operators [5].

2.2 Explicit Environments

While the results presented in this paper could be directly expressed using the $\overline{\lambda}_{lv}$-calculus, the realizability interpretation naturally arises from the decomposition of this calculus into a different calculus with an explicit *environment*, the $\overline{\lambda}_{[lv\tau\star]}$-calculus [2]. Indeed, as we shall see in the sequel, the decomposition highlights different syntactic categories that are deeply involved in the type system and in the definition of the realizability interpretation.

The $\overline{\lambda}_{[lv\tau\star]}$-calculus is a reformulation of the $\overline{\lambda}_{lv}$-calculus with explicit environments, called *stores* and denoted by τ. Stores consists of a list of bindings of the form $[x := t]$, where x is a term variable and t a term, and of bindings of the form $[\alpha := e]$ where α is a context variable and e a context. For instance, in the closure $c\tau[x := t]\tau'$, the variable x is bound to t in c and τ'. Besides, the term t might be an unevaluated term (*i.e.* lazily stored), so that if x is eagerly demanded at some point during the execution of this closure, t will be reduced in order to obtain a value. In the case where t indeed produces a value V, the store will be updated with the binding $[x := V]$. However, a binding of this form (with a value) is fixed for the rest of the execution. As such, our so-called stores somewhat behave like lazy explicit substitutions or mutable environments.

To draw the comparison between our structures and the usual notions of stores and environments, two things should be observed. First, the usual notion of store refers to a structure of list that is fully mutable, in the sense that the cells can be updated at any time and thus values might be replaced. Second, the usual notion of environment designates a structure in which variables are bounded to closures made of a term and an environment. In particular, terms and environments are duplicated, *i.e.* sharing is not allowed. Such a structure resemble to a tree whose nodes are decorated by terms, as opposed to a machinery allowing sharing (like ours) whose underlying structure is broadly a directed acyclic graph. See for instance [24] for a Krivine abstract machine with sharing.

2.3 Syntax and Reduction Rules

The lazy evaluation of terms allows for the following reduction rule: us to reduce a command $\langle \mu\alpha.c \| \tilde{\mu}x.c' \rangle$ to the command c' together with the binding $[x := \mu\alpha.c]$.

$$\langle \mu\alpha.c \| \tilde{\mu}x.c' \rangle \to c'[x := \mu\alpha.c]$$

In this case, the term $\mu\alpha.c$ is left unevaluated ("frozen") in the store, until possibly reaching a command in which the variable x is needed. When evaluation reaches a command of the form $\langle x \| F \rangle \tau[x := \mu\alpha.c]\tau'$, the binding is opened and the term is evaluated in front of the context $\tilde{\mu}[x].\langle x \| F \rangle \tau'$:

$$\langle x \| F \rangle \tau[x := \mu\alpha.c]\tau' \to \langle \mu\alpha.c \| \tilde{\mu}[x].\langle x \| F \rangle \tau' \rangle \tau$$

The reader can think of the previous rule as the "defrosting" operation of the frozen term $\mu\alpha.c$: this term is evaluated in the prefix of the store τ which predates it, in front of the context $\tilde{\mu}[x].\langle x \| F \rangle \tau'$ where the $\tilde{\mu}[x]$ binder is waiting for a value.

Strong values	$v ::= \lambda x.t \mid \mathbf{k}$		
Weak values	$V ::= v \mid x$		
Terms	$t, u ::= V \mid \mu\alpha.c$		
Forcing contexts	$F ::= t \cdot E \mid \boldsymbol{\kappa}$		
Catchable contexts	$E ::= F \mid \alpha \mid \tilde{\mu}[x].\langle x \| F\rangle\tau$		
Evaluation contexts	$e ::= E \mid \tilde{\mu}x.c$		
Stores	$\tau ::= \varepsilon \mid \tau[x := t] \mid \tau[\alpha := E]$		
Commands	$c ::= \langle t \| e\rangle$		
Closures	$l ::= c\tau$		
(BETA)	$\langle \lambda x.t \| u \cdot E\rangle\tau$	$\rightarrow$	$\langle u \| \tilde{\mu}x.\langle t \| E\rangle\rangle\tau$
(LET)	$\langle t \| \tilde{\mu}x.c\rangle\tau$	$\rightarrow$	$c\tau[x := t]$
(CATCH)	$\langle \mu\alpha.c \| E\rangle\tau$	$\rightarrow$	$c\tau[\alpha := E]$
(LOOKUP$_\alpha$)	$\langle V \| \alpha\rangle\tau[\alpha := E]\tau'$	$\rightarrow$	$\langle V \| E\rangle\tau[\alpha := E]\tau'$
(LOOKUP$_x$)	$\langle x \| F\rangle\tau[x := t]\tau'$	$\rightarrow$	$\langle t \| \tilde{\mu}[x].\langle x \| F\rangle\tau'\rangle\tau$
(RESTORE)	$\langle V \| \tilde{\mu}[x].\langle x \| F\rangle\tau'\rangle\tau$	$\rightarrow$	$\langle V \| F\rangle\tau[x := V]\tau'$

Fig. 1. Syntax and reduction rules of the $\overline{\lambda}_{[lv\tau\star]}$-calculus

This context keeps trace of the part of the store τ' that was originally located after the binding $[x := ...]$. This way, if a value V is indeed furnished for the binder $\tilde{\mu}[x]$, the original command $\langle x \| F\rangle$ is evaluated in the updated full store:

$$\langle V \| \tilde{\mu}[x].\langle x \| F\rangle\tau'\rangle\tau \rightarrow \langle V \| F\rangle\tau[x := V]\tau'$$

The brackets in $\tilde{\mu}[x].c$ are used to express the fact that the variable x is forced at top-level (unlike contexts of the shape $\tilde{\mu}x.C[\langle x \| F\rangle]$ in the $\overline{\lambda}_{lv}$-calculus). The reduction system resembles the one of an abstract machine. Especially, it allows us to keep the standard redex at the top of a command and avoids searching through the meta-context for work to be done.

Note that our approach slightly differ from [2] since we split values into two categories: strong values (v) and weak values (V). The strong values correspond to values strictly speaking. The weak values include the variables which force the evaluation of terms to which they refer into shared strong value. Their evaluation may require capturing a continuation. The syntax of the language, which includes constants $\mathbf{k}$ and co-constants $\boldsymbol{\kappa}$, is given in Fig. 1. As for the reduction $\rightarrow$, we define it as the compatible reflexive transitive closure of the rules given in Fig. 1.

The different syntactic categories can be understood as the different levels of alternation in a context-free abstract machine (see [2]): the priority is first given to contexts at level e (lazy storage of terms), then to terms at level t (evaluation of $\mu\alpha$ into values), then back to contexts at level E and so on until level v. These different categories are directly reflected in the definition of the abstract machine defined in [2], and will thus be involved in the definition of our realizability interpretation. We chose to highlight this by distinguishing different types of sequents already in the typing rules that we shall now present.

$$\frac{(\mathbf{k}:X)\in\mathcal{S}}{\Gamma\vdash_v \mathbf{k}:X}\,(\mathbf{k})\qquad \frac{\Gamma,x:A\vdash_t t:B}{\Gamma\vdash_v \lambda x.t:A\to B}\,(\to_r)\qquad \frac{(x:A)\in\Gamma}{\Gamma\vdash_V x:A}\,(x)\qquad \frac{\Gamma\vdash_v v:A}{\Gamma\vdash_V v:A}\,(\uparrow^V)$$

$$\frac{(\boldsymbol{\kappa}:A)\in\mathcal{S}}{\Gamma\vdash_F \boldsymbol{\kappa}:A^\perp}\,(\kappa)\qquad \frac{\Gamma\vdash_t t:A \quad \Gamma\vdash_E E:B^\perp}{\Gamma\vdash_F t\cdot E:(A\to B)^\perp}\,(\to_l)\qquad \frac{(\alpha:A)\in\Gamma}{\Gamma\vdash_E \alpha:A^\perp}\,(\alpha)$$

$$\frac{\Gamma\vdash_F F:A^\perp}{\Gamma\vdash_E F:A^\perp}\,(\uparrow^E)\qquad \frac{\Gamma\vdash_V V:A}{\Gamma\vdash_t V:A}\,(\uparrow^t)\qquad \frac{\Gamma,\alpha:A^\perp\vdash_c c}{\Gamma\vdash_t \mu\alpha.c:A}\,(\mu)\qquad \frac{\Gamma\vdash_E E:A^\perp}{\Gamma\vdash_e E:A^\perp}\,(\uparrow^e)$$

$$\frac{\Gamma,x:A\vdash_c c}{\Gamma\vdash_e \tilde{\mu}x.c:A^\perp}\,(\tilde{\mu})\qquad \frac{\Gamma,x:A,\Gamma'\vdash_F F:A^\perp \quad \Gamma\vdash_\tau \tau:\Gamma'}{\Gamma\vdash_E \tilde{\mu}[x].\langle x\|F\rangle\tau:A^\perp}\,(\tilde{\mu}^{[]})$$

$$\frac{\Gamma\vdash_t t:A \quad \Gamma\vdash_e e:A^\perp}{\Gamma\vdash_c \langle t\|e\rangle}\,(c)\qquad \frac{\Gamma,\Gamma'\vdash_c c \quad \Gamma\vdash_\tau \tau:\Gamma'}{\Gamma\vdash_l c\tau}\,(l)\qquad \frac{}{\Gamma\vdash_\tau \varepsilon:\varepsilon}\,(\varepsilon)$$

$$\frac{\Gamma\vdash_\tau \tau:\Gamma' \quad \Gamma,\Gamma'\vdash_t t:A}{\Gamma\vdash_\tau \tau[x:=t]:\Gamma',x:A}\,(\tau t)\qquad \frac{\Gamma\vdash_\tau \tau:\Gamma' \quad \Gamma,\Gamma'\vdash_E E:A^\perp}{\Gamma\vdash_\tau \tau[\alpha:=E]:\Gamma',\alpha:A^\perp}\,(\tau_E)$$

Fig. 2. Typing rules of the $\overline{\lambda}_{[lv\tau\star]}$-calculus

2.4 A Type System for the $\overline{\lambda}_{[lv\tau\star]}$-calculus

We have nine kinds of (one-sided) sequents, one for typing each of the nine syntactic categories. We write them with an annotation on the $\vdash$ sign, using one of the letters v, V, t, F, E, e, l, c, τ. Sequents typing values and terms are asserting a type, with the type written on the right; sequents typing contexts are expecting a type A with the type written $A^\perp$; sequents typing commands and closures are black boxes neither asserting nor expecting a type; sequents typing substitutions are instantiating a typing context. In other words, we have the following nine kinds of sequents:

$$\begin{array}{lll} \Gamma\vdash_l l & \Gamma\vdash_t t:A & \Gamma\vdash_e e:A^\perp \\ \Gamma\vdash_c c & \Gamma\vdash_V V:A & \Gamma\vdash_E E:A^\perp \\ \Gamma\vdash_\tau \tau:\Gamma' & \Gamma\vdash_v v:A & \Gamma\vdash_F F:A^\perp \end{array}$$

where types and typing contexts are defined by:

$$A,B ::= X \mid A\to B \qquad\qquad \Gamma ::= \varepsilon \mid \Gamma,x:A \mid \Gamma,\alpha:A^\perp$$

The typing rules are given on Fig. 2 where we assume that a variable x (resp. co-variable α) only occurs once in a context Γ (we implicitly assume the possibility of renaming variables by α-conversion). We also adopt the convention that constants $\mathbf{k}$ and co-constants $\boldsymbol{\kappa}$ come with a signature $\mathcal{S}$ which assigns them a type. This type system enjoys the property of subject reduction.

Theorem 1 (Subject reduction). *If $\Gamma\vdash_l c\tau$ and $c\tau\to c'\tau'$ then $\Gamma\vdash_l c'\tau'$.*

Proof. By induction on typing derivations. □

3 Normalization of the $\overline{\lambda}_{[lv\tau\star]}$-calculus

3.1 Normalization by Realizability

The proof of normalization for the $\overline{\lambda}_{[lv\tau\star]}$-calculus that we present in this section is inspired from techniques of Krivine's classical realizability [19], whose notations we borrow. Actually, it is also very close to a proof by reducibility[5]. In a nutshell, to each type A is associated a set $|A|_t$ of terms whose execution is guided by the structure of A. These terms are the ones usually called *realizers* in Krivine's classical realizability. Their definition is in fact indirect, and is done by orthogonality to a set of "correct" computations, called a *pole*. The choice of this set is central when studying models induced by classical realizability for second-order-logic, but in the present case we only pay attention to the particular pole of terminating computations. This is where lies one of the difference with usual proofs by reducibility, where everything is done with respect to SN, while our definition are parametric in the pole (which is chosen to be SN in the end). The adequacy lemma, which is the central piece, consists in proving that typed terms belong to the corresponding sets of realizers, and are thus normalizing.

More in details, our proof can be sketched as follows. First, we generalize the usual notion of closed term to the notion of closed *term-in-store*. Intuitively, this is due to the fact that we are no longer interested in closed terms and substitutions to close opened terms, but rather in terms that are closed when considered in the current store. This is based on the simple observation that a store is nothing more than a shared substitution whose content might evolve along the execution. Second, we define the notion of *pole* $\perp\!\!\!\perp$, which are sets of closures closed by anti-evaluation and store extension. In particular, the set of normalizing closures is a valid pole. This allows to relate terms and contexts thanks to a notion of orthogonality with respect to the pole. We then define for each formula A and typing level o (of e, t, E, V, F, v) a set $|A|_o$ (resp. $\|A\|_o$) of terms (resp. contexts) in the corresponding syntactic category. These sets correspond to reducibility candidates, or to what is usually called truth values and falsity values in Krivine realizability. Finally, the core of the proof consists in the adequacy lemma, which shows that any closed term of type A at level o is in the corresponding set $|A|_o$. This guarantees that any typed closure is in any pole, and in particular in the pole of normalizing closures. Technically, the proof of adequacy evaluates in each case a state of an abstract machine (in our case a closure), so that the proof also proceeds by evaluation. A more detailed explanation of this observation as well as a more introductory presentation of normalization proofs by classical realizability are given in an article by Dagand and Scherer [7].

3.2 Realizability Interpretation for the $\overline{\lambda}_{[lv\tau\star]}$-calculus

We begin by defining some key notions for stores that we shall need further in the proof.

[5] See for instance the proof of normalization for system D presented in [17, Sect. 3.2].

Definition 2 (Closed store). *We extend the notion of free variable to stores:*

$$\begin{array}{rcl} FV(\varepsilon) & \triangleq & \emptyset \\ FV(\tau[x := t]) & \triangleq & FV(\tau) \cup \{y \in FV(t) : y \notin \mathtt{dom}(\tau)\} \\ FV(\tau[\alpha := E]) & \triangleq & FV(\tau) \cup \{\beta \in FV(E) : \beta \notin \mathtt{dom}(\tau)\} \end{array}$$

so that we can define a closed store *to be a store τ such that $FV(\tau) = \emptyset$.*

Definition 3 (Compatible stores). *We say that two stores τ and τ' are* independent *and write $\tau \# \tau'$ when $\mathtt{dom}(\tau) \cap \mathtt{dom}(\tau') = \emptyset$. We say that they are* compatible *and write $\tau \diamond \tau'$ whenever for all variables x (resp. co-variables α) present in both stores: $x \in \mathtt{dom}(\tau) \cap \mathtt{dom}(\tau')$; the corresponding terms (resp. contexts) in τ and τ' coincide. Finally, we say that τ' is an* extension *of τ and write $\tau \lhd \tau'$ whenever $\mathtt{dom}(\tau) \subseteq \mathtt{dom}(\tau')$ and $\tau \diamond \tau'$.*

We denote by $\overline{\tau\tau'}$ the compatible union $\mathtt{join}(\tau\tau')$ of closed stores τ and τ', defined by:

$$\begin{array}{rcll} \mathtt{join}(\tau_0[x := t]\tau_1, \tau_0'[x := t]\tau_1') & \triangleq & \tau_0\tau_0'[x := t]\mathtt{join}(\tau_1, \tau_1') & (\text{if } \tau_0 \# \tau_0') \\ \mathtt{join}(\tau, \tau') & \triangleq & \tau\tau' & (\text{if } \tau \# \tau') \\ \mathtt{join}(\varepsilon, \tau) & \triangleq & \tau & \\ \mathtt{join}(\tau, \varepsilon) & \triangleq & \tau & \end{array}$$

The following lemma (which follows easily from the previous definition) states the main property we will use about union of compatible stores.

Lemma 4. *If τ and τ' are two compatible stores, then $\tau \lhd \overline{\tau\tau'}$ and $\tau' \lhd \overline{\tau\tau'}$. Besides, if τ is of the form $\tau_0[x := t]\tau_1$, then $\overline{\tau\tau'}$ is of the form $\tau_2[x := t]\tau_3$ with $\tau_0 \lhd \tau_2$ and $\tau_1 \lhd \tau_3$.*

Proof. This follows easily from the previous definition. □

As we explained in the introduction of this section, we will not consider closed terms in the usual sense. Indeed, while it is frequent in the proofs of normalization (*e.g.* by realizability or reducibility) of a calculus to consider only closed terms and to perform substitutions to maintain the closure of terms, this only makes sense if it corresponds to the computational behavior of the calculus. For instance, to prove the normalization of $\lambda x.t$ in typed call-by-name $\lambda\mu\tilde{\mu}$-calculus, one would consider a substitution ρ that is suitable for with respect to the typing context Γ, then a context $u \cdot e$ of type $A \to B$, and evaluates:

$$\langle \lambda x.t_\rho \| u \cdot e \rangle \quad \to \quad \langle t_\rho[u/x] \| e \rangle$$

Then we would observe that $t_\rho[u/x] = t_{\rho[x:=u]}$ and deduce that $\rho[x := u]$ is suitable for $\Gamma, x : A$, which would allow us to conclude by induction.

However, in the $\overline{\lambda}_{[lv\tau\star]}$-calculus we do not perform global substitution when reducing a command, but rather add a new binding $[x := u]$ in the store:

$$\langle \lambda x.t \| u \cdot E \rangle\tau \quad \to \quad \langle t \| E \rangle\tau[x := u]$$

Therefore, the natural notion of closed term invokes the closure under a store, which might evolve during the rest of the execution (this is to contrast with a substitution).

Definition 5 (Term-in-store). *We call* closed term-in-store *(resp.* closed context-in-store, closed closures*) the combination of a term t (resp. context e, command c) with a closed store τ such that $FV(t) \subseteq \mathtt{dom}(\tau)$. We use the notation $(t|\tau)$ (resp. $(e|\tau), (c|\tau)$) to denote such a pair.*

We should note that in particular, if t is a closed term, then $(t|\tau)$ is a term-in-store for any closed store τ. The notion of closed term-in-store is thus a generalization of the notion of closed terms, and we will (ab)use of this terminology in the sequel. We denote the sets of closed closures by $\mathcal{C}_0$, and will identify $(c|\tau)$ and the closure $c\tau$ when c is closed in τ. Observe that if $c\tau$ is a closure in $\mathcal{C}_0$ and τ' is a store extending τ, then $c\tau'$ is also in $\mathcal{C}_0$. We are now equipped to define the notion of pole, and verify that the set of normalizing closures is indeed a valid pole.

Definition 6 (Pole). *A subset $\perp\!\!\!\perp \subseteq \mathcal{C}_0$ is said to be* saturated *or* closed by anti-reduction *whenever for all $(c|\tau), (c'|\tau') \in \mathcal{C}_0$, if $c'\tau' \in \perp\!\!\!\perp$ and $c\tau \to c'\tau'$ then $c\tau \in \perp\!\!\!\perp$. It is said to be* closed by store extension *if whenever $c\tau \in \perp\!\!\!\perp$, for any store τ' extending τ: $\tau \lhd \tau'$, $c\tau' \in \perp\!\!\!\perp$. A* pole *is defined as any subset of $\mathcal{C}_0$ that is closed by anti-reduction and store extension.*

The following proposition is the one supporting the claim that our realizability proof is almost a reducibility proof whose definitions have been generalized with respect to a pole instead of the fixed set SN.

Proposition 7. *The set $\perp\!\!\!\perp_{\Downarrow} = \{c\tau \in \mathcal{C}_0 : \ c\tau \text{ normalizes } \}$ is a pole.*

Proof. As we only considered closures in $\mathcal{C}_0$, both conditions (closure by anti-reduction and store extension) are clearly satisfied:

- if $c\tau \to c'\tau'$ and $c'\tau'$ normalizes, then $c\tau$ normalizes too;
- if c is closed in τ and $c\tau$ normalizes, if $\tau \lhd \tau'$ then $c\tau'$ will reduce as $c\tau$ does (since c is closed under τ, it can only use terms in τ' that already were in τ) and thus will normalize. □

Definition 8 (Orthogonality). *Given a pole $\perp\!\!\!\perp$, we say that a term-in-store $(t|\tau)$ is* orthogonal *to a context-in-store $(e|\tau')$ and write $(t|\tau)\perp\!\!\!\perp(e|\tau')$ if τ and τ' are compatible and $\langle t\|e\rangle\overline{\tau\tau'} \in \perp\!\!\!\perp$.*

Remark 9. The reader familiar with Krivine's forcing machine [20] might recognize his definition of orthogonality between terms of the shape (t, p) and stacks of the shape (π, q), where p and q are forcing conditions[6]:

$$(t,p)\perp\!\!\!\perp(\pi,q) \Leftrightarrow (t \star \pi, p \wedge q) \in \perp\!\!\!\perp$$

[6] The meet of forcing conditions is indeed a refinement containing somewhat the "union" of information contained in each, just like the union of two compatible stores.

We can now relate closed terms and contexts by orthogonality with respect to a given pole. This allows us to define for any formula A the sets $|A|_v, |A|_V, |A|_t$ (resp. $\|A\|_F, \|A\|_E, \|A\|_e$) of realizers (or reducibility candidates) at level v, V, t (resp. F, E, e) for the formula A. It is to be observed that realizers are here closed terms-in-store.

Definition 10 (Realizers). *Given a fixed pole* $\bot\!\!\!\bot$, *we set:*

$$\begin{array}{ll}
|X|_v & = \{(\mathbf{k}|\tau) : \ \vdash \mathbf{k} : X\} \\
|A \to B|_v & = \{(\lambda x.t|\tau) : \forall u\tau', \tau \diamond \tau' \wedge (u|\tau') \in |A|_t \Rightarrow (t|\overline{\tau\tau'}[x := u]) \in |B|_t\} \\
\|A\|_F & = \{(F|\tau) : \forall v\tau', \tau \diamond \tau' \wedge (v|\tau') \in |A|_v \Rightarrow (v|\tau')\bot\!\!\!\bot(F|\tau)\} \\
|A|_V & = \{(V|\tau) : \forall F\tau', \tau \diamond \tau' \wedge (F|\tau') \in \|A\|_F \Rightarrow (V|\tau)\bot\!\!\!\bot(F|\tau')\} \\
\|A\|_E & = \{(E|\tau) : \forall V\tau', \tau \diamond \tau' \wedge (V|\tau') \in |A|_V \Rightarrow (V|\tau')\bot\!\!\!\bot(E|\tau)\} \\
|A|_t & = \{(t|\tau) : \forall E\tau', \tau \diamond \tau' \wedge (E|\tau') \in \|A\|_E \Rightarrow (t|\tau)\bot\!\!\!\bot(E|\tau')\} \\
\|A\|_e & = \{(e|\tau) : \forall t\tau', \tau \diamond \tau' \wedge (t|\tau') \in |A|_t \Rightarrow (t|\tau')\bot\!\!\!\bot(e|\tau)\}
\end{array}$$

Remark 11. We draw the reader attention to the fact that we should actually write $|A|_v^{\bot\!\!\!\bot}, \|A\|_F^{\bot\!\!\!\bot}$, etc. and $\tau \Vdash_{\bot\!\!\!\bot} \Gamma$, because the corresponding definitions are parameterized by a pole $\bot\!\!\!\bot$. As it is common in Krivine's classical realizability, we ease the notations by removing the annotation $\bot\!\!\!\bot$ whenever there is no ambiguity on the pole. Besides, it is worth noting that if co-constants do not occur directly in the definitions, they may still appear in the realizers by mean of the pole.

If the definition of the different sets might seem complex at first sight, we claim that they are quite natural in regards of the methodology of Danvy's semantics artifacts presented in [2]. Indeed, having an abstract machine in context-free form (the last step in this methodology before deriving the CPS) allows us to have both the term and the context (in a command) that behave independently of each other. Intuitively, a realizer at a given level is precisely a term which is going to behave well (be in the pole) in front of any opponent chosen in the previous level (in the hierarchy v, F, V, etc.). For instance, in a call-by-value setting, there are only three levels of definition (values, contexts and terms) in the interpretation, because the abstract machine in context-free form also has three. Here the ground level corresponds to strong values, and the other levels are somewhat defined as terms (or context) which are well-behaved in front of any opponent in the previous one. The definition of the different sets $|A|_v, \|A\|_F, |A|_V$, etc. directly stems from this intuition.

In comparison with the usual definition of Krivine's classical realizability, we only considered orthogonal sets restricted to some syntactical subcategories. However, the definition still satisfies the usual monotonicity properties of bi-orthogonal sets:

Proposition 12. *For any type A and any given pole* $\bot\!\!\!\bot$, *we have:*

1. $|A|_v \subseteq |A|_V \subseteq |A|_t$; *2.* $\|A\|_F \subseteq \|A\|_E \subseteq \|A\|_e$.

Proof. All the inclusions are proved in a similar way. We only give the proof for $|A|_v \subseteq |A|_V$. Let $\bot\!\!\!\bot$ be a pole and $(v|\tau)$ be in $|A|_v$. We want to show that $(v|\tau)$

is in $|A|_V$, that is to say that v is in the syntactic category V (which is true), and that for any $(F|\tau') \in \|A\|_F$ such that $\tau \diamond \tau'$, $(v|\tau) \perp\!\!\!\perp (F|\tau')$. The latter holds by definition of $(F|\tau') \in \|A\|_F$, since $(v|\tau) \in |A|_v$. □

We now extend the notion of realizers to stores, by stating that a store τ realizes a context Γ if it binds all the variables x and α in Γ to a realizer of the corresponding formula.

Definition 13. *Given a closed store τ and a fixed pole $\perp\!\!\!\perp$, we say that τ* realizes *Γ, which we write*[7] *$\tau \Vdash \Gamma$, if:*

1. *for any $(x : A) \in \Gamma$, $\tau \equiv \tau_0[x := t]\tau_1$ and $(t|\tau_0) \in |A|_t$*
2. *for any $(\alpha : A^{\perp}) \in \Gamma$, $\tau \equiv \tau_0[\alpha := E]\tau_1$ and $(E|\tau_0) \in \|A\|_E$*

In the same way than weakening rules (for the typing context) are admissible for each level of the typing system:

$$\frac{\Gamma \vdash_t t : A \quad \Gamma \subseteq \Gamma'}{\Gamma' \vdash_t t : A} \qquad \frac{\Gamma \vdash_e e : A^{\perp} \quad \Gamma \subseteq \Gamma'}{\Gamma' \vdash_e e : A^{\perp}} \qquad \cdots \qquad \frac{\Gamma \vdash_\tau \tau : \Gamma'' \quad \Gamma \subseteq \Gamma'}{\Gamma' \vdash_\tau \tau : \Gamma''}$$

the definition of realizers is compatible with a weakening of the store.

Lemma 14 (Store weakening). *Let τ and τ' be two stores such that $\tau \lhd \tau'$, let Γ be a typing context and let $\perp\!\!\!\perp$ be a pole. The following statements hold:*

1. *$\overline{\tau\tau'} = \tau'$*
2. *If $(t|\tau) \in |A|_t$ for some closed term $(t|\tau)$ and type A, then $(t|\tau') \in |A|_t$. The same holds for each level e, E, V, F, v of the typing rules.*
3. *If $\tau \Vdash \Gamma$ then $\tau' \Vdash \Gamma$.*

Proof. 1. Straightforward from the definition of $\overline{\tau\tau'}$.

2. This essentially amounts to the following observations. First, one remarks that if $(t|\tau)$ is a closed term, so then so is $(t|\overline{\tau\tau'})$ for any closed store τ' compatible with τ. Second, we observe that if we consider for instance a closed context $(E|\tau'') \in \|A\|_E$, then $\overline{\tau\tau'} \diamond \tau''$ implies $\tau \diamond \tau''$, thus $(t|\tau) \perp\!\!\!\perp (E|\tau'')$ and finally $(t|\overline{\tau\tau'}) \perp\!\!\!\perp (E|\tau'')$ by closure of the pole under store extension. We conclude that $(t|\tau') \perp\!\!\!\perp (E|\tau'')$ using the first statement.
3. By definition, for all $(x : A) \in \Gamma$, τ is of the form $\tau_0[x := t]\tau_1$ such that $(t|\tau_0) \in |A|_t$. As τ and τ' are compatible, we know by Lemma 4 that $\overline{\tau\tau'}$ is of the form $\tau_0'[x := t]\tau_1'$ with τ_0' an extension of τ_0, and using the first point we get that $(t|\tau_0') \in |A|_t$. □

Definition 15 (Adequacy). *Given a fixed pole $\perp\!\!\!\perp$, we say that:*

- *A typing judgment $\Gamma \vdash_t t : A$ is* adequate *(w.r.t. the pole $\perp\!\!\!\perp$) if for all stores $\tau \Vdash \Gamma$, we have $(t|\tau) \in |A|_t$.*

---

[7] Once again, we should formally write $\tau \Vdash_{\perp\!\!\!\perp} \Gamma$ but we will omit the annotation by $\perp\!\!\!\perp$ as often as possible.

– *More generally, we say that an inference rule*

$$\frac{J_1 \quad \cdots \quad J_n}{J_0}$$

is adequate (w.r.t. the pole $\perp\!\!\!\perp$*) if the adequacy of all typing judgments* $J_1, \ldots, J_n$ *implies the adequacy of the typing judgment* J_0.

Remark 16. From the latter definition, it is clear that a typing judgment that is derivable from a set of adequate inference rules is adequate too.

We will now show the main result of this section, namely that the typing rules of Fig. 2 for the $\overline{\lambda}_{[lv\tau\star]}$-calculus without co-constants are adequate with any pole. Observe that this result requires to consider the $\overline{\lambda}_{[lv\tau\star]}$-calculus without co-constants. Indeed, we consider co-constants as coming with their typing rules, potentially giving them any type (whereas constants can only be given an atomic type). Thus, there is *a priori* no reason[8] why their types should be adequate with any pole.

However, as observed in the previous remark, given a fixed pole it suffices to check whether the typing rules for a given co-constant are adequate with this pole. If they are, any judgment that is derivable using these rules will be adequate.

Theorem 17 (Adequacy). *If* Γ *is a typing context,* $\perp\!\!\!\perp$ *is a pole and* τ *is a store such that* $\tau \Vdash \Gamma$*, then the following holds in the* $\overline{\lambda}_{[lv\tau\star]}$*-calculus without co-constants:*

1. *If* v *is a strong value such that* $\Gamma \vdash_v v : A$*, then* $(v|\tau) \in |A|_v$.
2. *If* F *is a forcing context such that* $\Gamma \vdash_F F : A^\perp$*, then* $(F|\tau) \in \|A\|_F$.
3. *If* V *is a weak value such that* $\Gamma \vdash_V V : A$*, then* $(V|\tau) \in |A|_V$.
4. *If* E *is a catchable context such that* $\Gamma \vdash_E E : A^\perp$*, then* $(E|\tau) \in \|A\|_F$.
5. *If* t *is a term such that* $\Gamma \vdash_t t : A$*, then* $(t|\tau) \in |A|_t$.
6. *If* e *is a context such that* $\Gamma \vdash_e e : A^\perp$*, then* $(e|\tau) \in \|A\|_e$.
7. *If* c *is a command such that* $\Gamma \vdash_c c$*, then* $c\tau \in \perp\!\!\!\perp$.
8. *If* τ' *is a store such that* $\Gamma \vdash_\tau \tau' : \Gamma'$*, then* $\tau\tau' \Vdash \Gamma, \Gamma'$.

Proof. The different statements are proved by mutual induction over typing derivations. We only give the most important cases here.

Rule $(\to_l)$. Assume that

$$\frac{\Gamma \vdash_t u : A \quad \Gamma \vdash_E E : B^\perp}{\Gamma \vdash_F u \cdot E : (A \to B)^\perp}\ (\to_l)$$

and let $\perp\!\!\!\perp$ be a pole and τ a store such that $\tau \Vdash \Gamma$. Let $(\lambda x.t|\tau')$ be a closed term in the set $|A \to B|_v$ such that $\tau \diamond \tau'$, then we have:

$$\langle \lambda x.t \| u \cdot E \rangle \overline{\tau\tau'} \quad \to \quad \langle u \| \tilde{\mu}x.\langle t \| E \rangle \rangle \overline{\tau\tau'} \quad \to \quad \langle t \| E \rangle \overline{\tau\tau'}[x := u]$$

[8] Think for instance of a co-constant of type $(A \to B)^\perp$, there is no reason why it should be orthogonal to any function in $|A \to B|_v$.

By definition of $|A \to B|_v$, this closure is in the pole, and we can conclude by anti-reduction.

Rule (x). Assume that

$$\frac{(x:A) \in \Gamma}{\Gamma \vdash_V x : A}\ (x)$$

and let $\perp\!\!\!\perp$ be a pole and τ a store such that $\tau \Vdash \Gamma$. As $(x:A) \in \Gamma$, we know that τ is of the form $\tau_0[x := t]\tau_1$ with $(t|\tau_0) \in |A|_t$. Let $(F|\tau')$ be in $\|A\|_F$, with $\tau \diamond \tau'$. By Lemma 4, we know that $\overline{\tau\tau'}$ is of the form $\overline{\tau_0}[x := t]\overline{\tau_1}$. Hence we have:

$$\langle x\|F\rangle\overline{\tau_0}[x := t]\overline{\tau_1} \quad \to \quad \langle t\|\tilde{\mu}[x].\langle x\|F\rangle\overline{\tau_1}\rangle\overline{\tau_0}$$

and it suffices by anti-reduction to show that the last closure is in the pole $\perp\!\!\!\perp$. By induction hypothesis, we know that $(t|\tau_0) \in |A|_t$ thus we only need to show that it is in front of a catchable context in $\|A\|_E$. This corresponds exactly to the next case that we shall prove now.

Rule $(\tilde{\mu}^{[]})$. Assume that

$$\frac{\Gamma, x:A, \Gamma' \vdash_F F : A \quad \Gamma, x:A \vdash \tau' : \Gamma'}{\Gamma \vdash_E \tilde{\mu}[x].\langle x\|F\rangle\tau' : A}\ (\tilde{\mu}^{[]})$$

and let $\perp\!\!\!\perp$ be a pole and τ a store such that $\tau \Vdash \Gamma$. Let $(V|\tau_0)$ be a closed term in $|A|_V$ such that $\tau_0 \diamond \tau$. We have that:

$$\langle V\|\tilde{\mu}[x].\langle x\|F\rangle\overline{\tau'}\rangle\overline{\tau_0\tau} \quad \to \quad \langle V\|F\rangle\overline{\tau_0\tau}[x := V]\tau'$$

By induction hypothesis, we obtain $\tau[x := V]\tau' \Vdash \Gamma, x : A, \Gamma'$. Up to α-conversion in F and τ', so that the variables in τ' are disjoint from those in τ_0, we have that $\overline{\tau_0\tau} \Vdash \Gamma$ (by Lemma 14) and then $\tau'' \triangleq \overline{\tau_0\tau}[x := V]\tau' \Vdash \Gamma, x : A, \Gamma'$. By induction hypothesis again, we obtain that $(F|\tau'') \in \|A\|_F$ (this was an assumption in the previous case) and as $(V|\tau_0) \in |A|_V$, we finally get that $(V|\tau_0)\perp\!\!\!\perp(F|\tau'')$ and conclude again by anti-reduction. □

Corollary 18. *If $c\tau$ is a closure such that $\vdash_l c\tau$ is derivable, then for any pole $\perp\!\!\!\perp$ such that the typing rules for co-constants used in the derivation are adequate with $\perp\!\!\!\perp$, $c\tau \in \perp\!\!\!\perp$.*

We can now put our focus back on the normalization of typed closures. As we already saw in Proposition 7, the set $\perp\!\!\!\perp_{\Downarrow}$ of normalizing closure is a valid pole, so that it only remains to prove that any typing rule for co-constants is adequate with $\perp\!\!\!\perp_{\Downarrow}$.

Lemma 19. *Any typing rule for co-constants is adequate with the pole $\perp\!\!\!\perp_{\Downarrow}$,* i.e. *if Γ is a typing context, and τ is a store such that $\tau \Vdash \Gamma$, if $\boldsymbol{\kappa}$ is a co-constant such that $\Gamma \vdash_F \boldsymbol{\kappa} : A^{\perp}$, then $(\boldsymbol{\kappa}|\tau) \in \|A\|_F$.*

Proof. This lemma directly stems from the observation that for any store τ and any closed strong value $(v|\tau') \in |A|_v$, $\langle v\|\kappa\rangle\overline{\tau\tau'}$ does not reduce and thus belongs to the pole $\perp\!\!\!\perp_{\Downarrow}$.

As a consequence, we obtain the normalization of typed closures of the full calculus.

Theorem 20. *If $c\tau$ is a closure of the $\overline{\lambda}_{[lv\tau\star]}$-calculus such that $\vdash_l c\tau$ is derivable, then $c\tau$ normalizes.*

This is to be contrasted with Okasaki, Lee and Tarditi's semantics for the call-by-need λ-calculus, which is not normalizing in the simply-typed case, as shown in Ariola *et al.* [2].

3.3 Extension to 2nd-Order Type Systems

We focused in this article on simply-typed versions of the $\overline{\lambda}_{lv}$ and $\overline{\lambda}_{[lv\tau\star]}$ calculi. But as it is common in Krivine classical realizability, first and second-order quantifications (in Curry style) come for free through the interpretation. This means that we can for instance extend the language of types to first and second-order predicate logic:

$$\begin{aligned} e_1, e_2 &::= x \mid f(e_1, \ldots, e_k) \\ A, B &::= X(e_1, \ldots, e_k) \mid A \to B \mid \forall x.A \mid \forall X.A \end{aligned}$$

We can then define the following introduction rules for universal quantifications:

$$\frac{\Gamma \vdash_v v : A \quad x \notin FV(\Gamma)}{\Gamma \vdash_v v : \forall x.A}\ (\forall_r^1) \qquad \frac{\Gamma \vdash_v v : A \quad X \notin FV(\Gamma)}{\Gamma \vdash_v v : \forall X.A}\ (\forall_r^2)$$

Observe that these rules need to be restricted at the level of strong values, just as they are restricted to values in the case of call-by-value[9]. As for the left rules, they can be defined at any levels, let say the more general e:

$$\frac{\Gamma \vdash_e e : (A[n/x])^{\perp}}{\Gamma \vdash_e e : (\forall x.A)^{\perp}}\ (\forall_l^1) \qquad \frac{\Gamma \vdash_e e : (A[B/X])^{\perp}}{\Gamma \vdash_e e : (\forall X.A)^{\perp}}\ (\forall_l^2)$$

where n is any natural number and B any formula. The usual (call-by-value) interpretation of the quantification is defined as an intersection over all the possible instantiations of the variables within the model. We do not wish to enter into too many details[10] on this topic here, but first-order variable are to be instantiated by integers, while second order are to be instantiated by subset of terms at the lower level, *i.e.* closed strong-values in store (which we write $\mathcal{V}_0$):

$$|\forall x.A|_v = \bigcap_{n \in \mathbb{N}} |A[n/x]|_v \qquad |\forall X.A|_v = \bigcap_{S \in \mathbb{N}^k \to \mathcal{P}(\mathcal{V}_0)} |A[S/X]|_v$$

9 For further explanation on the need for a value restriction in Krivine realizability, we refer the reader to [29] or [25].

10 Once again, we advise the interested reader to refer to [29] or [25] for further details.

where the variable X is of arity k. It is then routine to check that the typing rules are adequate with the realizability interpretation.

4 Conclusion and Further Work

In this paper, we presented a system of simple types for a call-by-need calculus with control, which we proved to be safe in that it satisfies subject reduction (Theorem 1) and that typed terms are normalizing (Theorem 20). We proved the normalization by means of realizability-inspired interpretation of the $\overline{\lambda}_{[lv\tau\star]}$-calculus. Incidentally, this opens the doors to the computational analysis (in the spirit of Krivine realizability) of classical proofs using control, laziness and shared memory.

In further work, we intend to present two extensions of the present paper. First, following the definition of the realizability interpretation, we managed to type the continuation-and-store passing style translation for the $\overline{\lambda}_{[lv\tau\star]}$-calculus (see [2]). Interestingly, typing the translation emphasizes its computational content, and in particular, the store-passing part is reflected in a Kripke forcing-like manner of typing the extensibility of the store [28, Chap. 6].

Second, on a different aspect, the realizability interpretation we introduced could be a first step towards new ways of realizing axioms. In particular, the first author used in his Ph.D. thesis [28, Chap. 8] the techniques presented in this paper to give a normalization proof for dPA^ω, a proof system developed by the second author [15]. Indeed, this proof system allows to define a proof for the axiom of dependent choice thanks to the use of streams that are lazily evaluated, and was lacking a proper normalization proof.

Finally, to determine the range of our technique, it would be natural to investigate the relation between our framework and the many different presentations of call-by-need calculi (with or without control). Amongst other calculi, we could cite Chang-Felleisen presentation of call-by-need [4], Garcia *et al.* lazy calculus with delimited control [10] or Kesner's recent paper on normalizing by-need terms characterized by an intersection type system [16]. To this end, we might rely on Pédrot and Saurin's classical by-need [33]. They indeed relate (classical) call-by-need with linear head-reduction from a computational point of view, and draw the connections with the presentations of Ariola *et al.* [2] and Chang-Felleisen [4]. Ariola *et al.* $\overline{\lambda}_{lv}$-calculus being close to the $\overline{\lambda}_{[lv\tau\star]}$-calculus (see [2] for further details), our technique is likely to be adaptable to their framework, and thus to Pédrot and Saurin's system.

References

1. Ariola, Z., Felleisen, M.: The call-by-need lambda calculus. J. Funct. Program. **7**(3), 265–301 (1993)
2. Ariola, Z.M., Downen, P., Herbelin, H., Nakata, K., Saurin, A.: Classical call-by-need sequent calculi: the unity of semantic artifacts. In: Schrijvers, T., Thiemann, P. (eds.) FLOPS 2012. LNCS, vol. 7294, pp. 32–46. Springer, Heidelberg (2012). https://doi.org/10.1007/978-3-642-29822-6_6

3. Barbanera, F., Berardi, S.: A symmetric λ-calculus for classical program extraction. Inf. Comput. **125**(2), 103–117 (1996)
4. Chang, S., Felleisen, M.: The call-by-need lambda calculus, revisited. In: Seidl, H. (ed.) ESOP 2012. LNCS, vol. 7211, pp. 128–147. Springer, Heidelberg (2012). https://doi.org/10.1007/978-3-642-28869-2_7
5. Crolard, T.: A confluent lambda-calculus with a catch/throw mechanism. J. Funct. Program. **9**(6), 625–647 (1999)
6. Curien, P.-L., Herbelin, H.: The duality of computation. In: Proceedings of ICFP 2000, SIGPLAN Notices, vol. 35, no. 9, pp. 233–243. ACM (2000)
7. Dagand, P.-É., Scherer, G.: Normalization by realizability also evaluates. In: Baelde, D., Alglave, J. (eds.) Proceedings of JFLA 2015, Le Val d'Ajol, France, January 2015
8. Felleisen, M., Friedman, D.P., Kohlbecker, E.E., Duba, B.F.: Reasoning with continuations. In: Proceedings of LICS 1986, Cambridge, Massachusetts, USA, 16–18 June 1986, pp. 131–141. IEEE Computer Society (1986)
9. Gallier, J.: On girard's "candidats de reductibilité". In: Odifreddi, P. (ed.) Logic and Computer Science, pp. 123–203. Academic Press (1900)
10. Garcia, R., Lumsdaine, A., Sabry, A.: Lazy evaluation and delimited control. Log. Methods Comput. Sci. **6**(3) (2010)
11. Girard, J.-Y.: Une extension de L'interpretation de gödel a L'analyse, et son application a L'elimination des coupures dans L'analyse et la theorie des types. In: Fenstad, J.E., (ed.) Proceedings of the Second Scandinavian Logic Symposium. Studies in Logic and the Foundations of Mathematics, vol. 63, pp. 63–92. Elsevier (1971)
12. Guillermo, M., Miquel, A.: Specifying peirce's law in classical realizability. Math. Struct. Comput. Sci. **26**(7), 1269–1303 (2016)
13. Guillermo, M., Miquey, É.: Classical realizability and arithmetical formulæ. Math. Struct. Comput. Sci. 1–40 (2016)
14. Herbelin, H.: C'est maintenant qu'on calcule: au cœur de la dualité. Habilitation thesis, University Paris 11, December 2005
15. Herbelin, H.: A constructive proof of dependent choice, compatible with classical logic. In: Proceedings of the 27th Annual IEEE Symposium on Logic in Computer Science, LICS 2012, Dubrovnik, Croatia, 25–28 June 2012, pp. 365–374. IEEE Computer Society (2012)
16. Kesner, D.: Reasoning about call-by-need by means of types. In: Jacobs, B., Löding, C. (eds.) FoSSaCS 2016. LNCS, vol. 9634, pp. 424–441. Springer, Heidelberg (2016). https://doi.org/10.1007/978-3-662-49630-5_25
17. Krivine, J.-L.: Lambda-calculus, Types and Models. Ellis Horwood series in computers and their applications. Ellis Horwood, Masson (1993)
18. Krivine, J.-L.: Dependent choice, 'quote' and the clock. Theor. Comp. Sc. **308**, 259–276 (2003)
19. Krivine, J.-L.: Realizability in classical logic. Panoramas et synthèses **27**, 197–229 (2009). Interactive models of computation and program behaviour
20. Krivine, J.-L.: Realizability algebras: a program to well order r. Log. Methods Comput. Sci. **7**(3) (2011)
21. Krivine, J.-L.: Realizability algebras II: new models of ZF + DC. Log. Methods Comput. Sci. **8**(1:10), 1–28 (2012)
22. Krivine, J.-L.: On the structure of classical realizability models of ZF (2014)
23. Lafont, Y., Reus, B., Streicher, T.: Continuations semantics or expressing implication by negation. Technical report 9321, Ludwig-Maximilians-Universität, München (1993)

24. Lang, F.: Explaining the lazy Krivine machine using explicit substitution and addresses. High.-Order Symbolic Comput. **20**(3), 257–270 (2007)
25. Lepigre, R.: A classical realizability model for a semantical value restriction. In: Thiemann, P. (ed.) ESOP 2016. LNCS, vol. 9632, pp. 476–502. Springer, Heidelberg (2016). https://doi.org/10.1007/978-3-662-49498-1_19
26. Maraist, J., Odersky, M., Wadler, P.: The call-by-need lambda calculus. J. Funct. Program. **8**(3), 275–317 (1998)
27. Miquel, A.: Existential witness extraction in classical realizability and via a negative translation. Log. Methods Comput. Sci. **7**(2), 188–202 (2011)
28. Miquey, É.: Classical realizability and side-effects. Ph.D. thesis, Université Paris-Diderot, Universidad de la República (Uruguay) (2017)
29. Munch-Maccagnoni, G.: Focalisation and classical realisability. In: Grädel, E., Kahle, R. (eds.) CSL 2009. LNCS, vol. 5771, pp. 409–423. Springer, Heidelberg (2009). https://doi.org/10.1007/978-3-642-04027-6_30
30. Okasaki, C., Lee, P., Tarditi, D.: Call-by-need and continuation-passing style. Lisp Symbolic Comput. **7**(1), 57–82 (1994)
31. Parigot, M.: Free deduction: an analysis of "computations" in classical logic. In: Voronkov, A. (ed.) RCLP -1990. LNCS, vol. 592, pp. 361–380. Springer, Heidelberg (1992). https://doi.org/10.1007/3-540-55460-2_27
32. Parigot, M.: Strong normalization of second order symmetric λ-calculus. In: Kapoor, S., Prasad, S. (eds.) FSTTCS 2000. LNCS, vol. 1974, pp. 442–453. Springer, Heidelberg (2000). https://doi.org/10.1007/3-540-44450-5_36
33. Pédrot, P.-M., Saurin, A.: Classical by-need. In: Thiemann, P. (ed.) ESOP 2016. LNCS, vol. 9632, pp. 616–643. Springer, Heidelberg (2016). https://doi.org/10.1007/978-3-662-49498-1_24
34. Plotkin, G.D.: Call-by-name, call-by-value and the lambda-calculus. Theor. Comput. Sci. **1**(2), 125–159 (1975)
35. Polonovski, E.: Strong normalization of $\overline{\lambda}\mu\widetilde{\mu}$-calculus with explicit substitutions. In: Walukiewicz, I. (ed.) FoSSaCS 2004. LNCS, vol. 2987, pp. 423–437. Springer, Heidelberg (2004). https://doi.org/10.1007/978-3-540-24727-2_30
36. Tait, W.W.: Intensional interpretations of functionals of finite type I. J. Symbolic Log. **32**(2), 198–212 (1967)

Quotient Inductive-Inductive Types

Thorsten Altenkirch[1], Paolo Capriotti[1], Gabe Dijkstra[3], Nicolai Kraus[1(✉)], and Fredrik Nordvall Forsberg[2]

[1] University of Nottingham, Nottingham, UK
{thorsten.altenkirch,paolo.capriotti,nicolai.kraus}@nottingham.ac.uk
[2] University of Strathclyde, Glasgow, Scotland
fredrik.nordvall-forsberg@strath.ac.uk
[3] London, UK
gabe.dijkstra@gmail.com

Abstract. Higher inductive types (HITs) in Homotopy Type Theory allow the definition of datatypes which have constructors for equalities over the defined type. HITs generalise quotient types, and allow to define types with non-trivial higher equality types, such as spheres, suspensions and the torus. However, there are also interesting uses of HITs to define types satisfying uniqueness of equality proofs, such as the Cauchy reals, the partiality monad, and the well-typed syntax of type theory. In each of these examples we define several types that depend on each other mutually, i.e. they are inductive-inductive definitions. We call those HITs quotient inductive-inductive types (QIITs). Although there has been recent progress on a general theory of HITs, there is not yet a theoretical foundation for the combination of equality constructors and induction-induction, despite many interesting applications. In the present paper we present a first step towards a semantic definition of QIITs. In particular, we give an initial-algebra semantics. We further derive a *section induction principle*, stating that every algebra morphism into the algebra in question has a section, which is close to the intuitively expected elimination rules.

1 Introduction

This paper is about type theory in the sense of Martin-Löf [29], a theory which proof assistants such as Coq [7] and Lean [14] as well as programming languages such as Agda [31] and Idris [8] are based on. Recently, Homotopy type theory (HoTT) [34] has been introduced inspired by homotopy theoretic interpretations of type theory by Awodey and Warren [5] and Voevodsky [25,36].

A central concept in type theory is the concept of inductive definitions, which allows us to define inductive datatypes like the natural numbers, lists and trees just by presenting constructors with strictly positive occurrences of the inductive type being defined. Using the propositions as types explanation, we can use the same mechanism to inductively define predicates and relations, like an order on the natural numbers, or the derivability predicate for a logic defined by rules. Conceptually, HoTT changes what we mean by an inductive definition, because we view a type not only as given by its elements (points) but also by its equality proofs

C. Baier and U. Dal Lago (Eds.): FOSSACS 2018, LNCS 10803, pp. 293–310, 2018.
https://doi.org/10.1007/978-3-319-89366-2_16

(paths). Hence an inductive definition may not only feature constructors for elements but also for equalities. This concept of higher inductive types (HITs) has been used to represent the homotopical structure of geometric objects, like circles, spheres and tori, and gives rise to synthetic homotopy theory in HoTT [32].

However, as already noted in the HoTT Book [34], HITs have also more quotidian applications, such as a definition of the Cauchy reals for which the use of the axiom of choice can be avoided when proving e.g. Cauchy completeness. Instead of defining the real numbers as a quotient of sequences of rationals, a HIT is used to define them as the Cauchy completion of the rational numbers, with the quotienting happening simultaneously with the completion definition. Similarly, a definition of the partiality monad, which represents potentially diverging operations over a given type, was given using a HIT [2,13,35], again avoiding the axiom of choice when showing e.g. that the construction is a monad [12].

As we see from these examples, the idea of generating points and equalities of a type inductively is interesting, even if we do not care about the higher equality structure of types, or if we do not want it. For example: consider trees branching over an arbitrary type A, quotiented by arbitrary permutations of subtrees. We first define the type $T_0(A)$ of A-branching trees, given by the constructors

$$\begin{aligned} \mathsf{leaf}_0 &: T_0(A) \\ \mathsf{node}_0 &: (A \to T_0(A)) \to T_0(A). \end{aligned}$$

We then form the binary relation R on $T_0(A)$ that we want to quotient by as follows: R is the smallest relation such that for any auto-equivalence on A (i.e. any $e : A \to A$ which has an inverse) and $f : A \to T_0(A)$, we have a proof $p_{f,e} : R(\mathsf{node}_0(f), \mathsf{node}_0(f \circ e))$, and, secondly, for $g, h : A \to T_0(A)$ such that $(n : A) \to R(g(n), h(n))$, we have a proof $c_{f,g} : R(\mathsf{node}_0(g), \mathsf{node}_0(h))$. We can then form the quotient type $T_0(A)/R$, which is the type of unlabelled trees where each node has an A-indexed family of subtrees, and two trees which agree modulo the "order" of its subtrees are equal. For $A \equiv \mathbf{2}$, these are binary trees where the order of the two subtrees of each node does not matter.

Now, morally, from a family $A \to (T_0(A)/R)$, we should be able to construct an element of the quotient $T_0(A)/R$. This is indeed possible if A is $\mathbf{2}$ or another finite type, by applying the induction principle of the quotient type A times. However, it seems that, for a general type A, this would require the axiom of choice [34], which unfortunately is not a constructive principle [15]. But using a higher inductive type, we can give an alternative definition for the type of A-branching trees modulo permutation of subtrees.

Example 1. Given a type A, we define $T(A) : \mathsf{hSet}$ by

$$\begin{aligned} \mathsf{leaf} &: \; T(A) \\ \mathsf{node} &: \; (A \to T(A)) \to T(A) \\ \mathsf{mix} &: \; (f : A \to T) \to (e : A \cong A) \to \mathsf{node}(f) = \mathsf{node}(f \circ e). \end{aligned}$$

Note that the fact that $T(A)$ is a homotopy set (see *preliminaries* below) is implicitly included in the statement $T(A) : \mathsf{hSet}$. The construction we were looking for is now directly given by the constructor node. This demonstration of the usefulness of higher inductive constructions to increase the strength of quotients was first discussed in Altenkirch and Kaposi [1], where such set-truncated HITs are called *quotient inductive types* (QITs).

Another example of the use of higher inductive types is *type theory in type theory* [1], where the well-typed syntax of type theory is implemented as a higher *inductive-inductive* [30] type in type theory itself. A significantly simplified version of this will serve as a running example for us:

Example 2. We define the syntax of a (very basic) type theory by constructing types representing contexts and types as follows. A set $\mathsf{Con} : \mathsf{hSet}$ and a type family $\mathsf{Ty} : \mathsf{Con} \to \mathsf{hSet}$ are simultaneously defined by giving the constructors

$$\begin{aligned}
\varepsilon :\ & \mathsf{Con}\\
\mathsf{ext} :\ & (\Gamma : \mathsf{Con}) \to \mathsf{Ty}(\Gamma) \to \mathsf{Con}\\
\iota :\ & (\Gamma : \mathsf{Con}) \to \mathsf{Ty}(\Gamma)\\
\sigma :\ & (\Gamma : \mathsf{Con}) \to (A : \mathsf{Ty}(\Gamma)) \to \mathsf{Ty}(\mathsf{ext}\,\Gamma\,A) \to \mathsf{Ty}(\Gamma)\\
\sigma_{\mathsf{eq}} :\ & (\Gamma : \mathsf{Con}) \to (A : \mathsf{Ty}(\Gamma)) \to (B : \mathsf{Ty}(\mathsf{ext}\,\Gamma\,A))\\
& \qquad \to \mathsf{ext}\,(\mathsf{ext}\,\Gamma\,A)\,B =_{\mathsf{Con}} \mathsf{ext}\,\Gamma\,(\sigma\,\Gamma\,A\,B).
\end{aligned}$$

For simplicity, we do not consider terms. Contexts are either empty ε, or an extended context $\mathsf{ext}\,\Gamma\,A$ representing the context Γ extended by a fresh variable of type A. Types are either the base type ι (well-typed in any context), or Σ-types represented by $\sigma\,\Gamma\,A\,B$ (well-typed in context Γ if A is well-typed in context Γ, and B is well-typed in the extended context $\mathsf{ext}\,\Gamma\,A$). Type theory in type theory as in [1] has plenty of equality constructors, which play a role as soon as terms are introduced. To keep the example simple we instead use another equality, stating that extending a context by A followed by B is equal to extending it by $\sigma\,\Gamma\,A\,B$. This equality is given by σ_{eq}. Note that it is not possible to list the constructors of Con and Ty separately: due to the mutual dependency, the Ty-constructor σ has to be given in between of the two Con-constructors ext and σ_{eq}.

Despite a lot of work making use of concrete HITs [4,9–11,23,26,27], and despite the fact that it is usually — on some intuitive level — clear for the expert how the elimination principle for such a HIT can be derived, giving a general specification and a theoretical foundation for HITs has turned out to be a major difficulty. Several approaches have been proposed [6,18,28,33], and they do indeed give a satisfactory specification of HITs in the sense that they cover all HITs which have been used so far (see *related work* below). However, to the best of our knowledge, no approach covers *higher inductive-inductive* definitions such as Example 2. The purpose of the current paper is to remedy this. We restrict ourselves to sets, i.e. to *quotient inductive-inductive types* (QIITs). This is of course a serious restriction, since it means that we cannot capture many ordinary HITs such as e.g. the circle $\mathbb{S}^1$. At the same time, all higher inductive-inductive types that we know of are indeed sets — the Cauchy reals, the surreal numbers, the partiality monad, type theory in type theory, permutable trees — and will be instances of our framework, which allows arbitrarily complicated dependency structures. In particular, we allow intermixing of constructors as in Example 2.

Contributions. We give a formal specification of quotient inductive-inductive types with arbitrary dependency structure. This can be viewed as the generalisation of the usual semantics of inductive types as initial algebras of a functor to quotient inductive-inductive types. A QIIT is specified by (i) its *sorts*, which encode the types and type families that it consists of (Sect. 2), and (ii) by a sequence of *constructors*, that in turn are specified by *argument* and *target functors* (Sect. 3). This is a very general framework, covering in particular point (Sect. 3.2) and path constructors (Sect. 3.4). Each constructor specification gives rise to a category of algebras, and we establish conditions on the target functors that allow us to conclude that these categories of algebras are complete (Sect. 3.5). This is important, because it allows us to prove the equivalence of initiality and a principle that we call *section induction* (Sect. 4), stating that every algebra morphism into the algebra in question has a section; this principle is close to the intuitively expected elimination rules.

A full version of the paper, including all proofs, is available on the arXiv [3].

Related Work. Sojakova [33] shows the correspondence between initiality and induction (a variant of our Theorem 31) for W-suspensions, a restricted class of HITs. Basold, Geuvers and van der Weide [6] introduce a syntactic schema for HITs without higher path constructors, and derive their elimination rules. Dybjer and Moeneclaey [18] give a syntactic schema for finitary HITs with at most paths between paths, and give an interpretation in Hofmann and Streicher's groupoid model [22]. Finally, Lumsdaine and Shulman's work on the semantics of HITs in model categories [28] is similar to an external version of our approach.

Preliminaries. We work in a standard Martin-Löf style type theory and assume function extensionality. We do not assume univalence, but also do not contradict it; in particular, everything we do works in the type theory from the HoTT Book [34]. We write $\mathcal{U}$ for "the" universe of types, omitting universe indices in the *typical ambiguity* style [21]. A type is a set if all its equality proofs are equal, and hSet is defined as $\Sigma(A : \mathcal{U}).\mathsf{is\text{-}set}(A)$; we implicitly treat elements of hSet as their first projections — this allows us to view hSet as a universe. By a *category*, we mean a precategory [34, Definition 9.1.1] in the sense of the HoTT Book (all our categories become univalent categories if univalence is assumed). We write $\mathcal{C} \Rightarrow \mathcal{D}$ for functors and $X \to Y$ for functions between types. We denote the obvious category of sets and functions by hSet as well; consequently, $F : A \to \mathsf{hSet}$ denotes a type family, while $F : \mathcal{C} \Rightarrow \mathsf{hSet}$ denotes a functor. For such a functor $F : \mathcal{C} \Rightarrow \mathsf{hSet}$, we write $\int^{\mathcal{C}} F$ for the *category of elements* of F, whose objects are pairs (X, x) of an object X in $\mathcal{C}$ and an element $x : FX$. For a function $f : X \to Y$ and $z, w : X$, we write $\mathsf{ap}\, f\ : z = w \to f(z) = f(w)$ for the usual "action of a function to paths", ${}^{-1} : x = y \to y = x$ for "path reversal", and $\bullet : x = y \to y = z \to x = z$ for "path concatenation" [34, Lemmas 2.2.1, 2.1.1, 2.1.2].

2 Sorts

Single inductive (and quotient inductive) sets are simply elements of hSet. Inductive families [17] indexed over some fixed type A are families $A \to \mathsf{hSet}$. For the inductive-inductive definitions we are considering, the situation is more complicated, since we allow very general dependency structures. Our only requirement is that there is no looping dependency, since this is easily seen to lead to contradictions, e.g. we do not allow the definition of a family $A : B \to \mathsf{hSet}$ mutually with a family $B : A \to \mathsf{hSet}$ (whatever this would mean). Concretely, we will ensure that the collection of type formation rules (the type signatures) is given in a valid order, and we refer to the types used as family indices as the *sorts* of the definition. Hence our first step towards a specification of general QIITs is to explain what a valid specification of the sorts is.

Sorts do not only determine the formation rules of the inductive definitions, but also the types of the eliminators. To capture this, it is not enough to specify a type of sorts — in order to take the shape of the elimination rules into account, we need to specify a category.

Definition 3 (Sort specifications). *A specification of the* sorts *of a quotient inductive-inductive definition of n types is given by a list*

$$H_0, H_1, \ldots, H_{n-1},$$

where each H_i is a functor $H_i : \mathcal{C}_i \Rightarrow \mathsf{hSet}$. Here, $\mathcal{C}_0 :\equiv \mathbb{1}$ is the terminal category, and $\mathcal{C}_{i+1}$ is defined as follows:

- *objects are pairs (X, P), where X is an object in $\mathcal{C}_i$, and $P : H_i(X) \to \mathsf{hSet}$ is a family of sets;*
- *a morphism $(f, g) : (X, P) \to (Y, Q)$ consists of a morphism $f : X \to Y$ in $\mathcal{C}_i$, and a dependent function $g : (x : H_i(X)) \to P(x) \to Q(H_i(f)\, x)$ (in* hSet*).*

We say that $\mathcal{C}_n$ is the base category *for the sort signature $H_0, \ldots, H_{n-1}$.*

The following examples will hopefully make clear the connection between the specification in Definition 3 and common classes of data types.

Example 4 (Permutable trees). For a single inductive type such as the type of trees $T(A)$ in Example 1, the sorts are specified by a single functor $H_0 : \mathcal{C}_0 \Rightarrow \mathsf{hSet}$ which maps the single object $\star$ of $\mathcal{C}_0$ to the unit type $\mathbf{1}$. Objects in the base category $\mathcal{C}_1$ are thus pairs $(\star, W)$, where $W : \mathbf{1} \to \mathsf{hSet}$, and morphisms are given by $f : \star \to \star$ in $\mathbb{1}$ (necessarily the identity morphism), together with a dependent function $g : (x : \mathbf{1}) \to W(x) \to V(x)$. It is easy to see that this category $\mathcal{C}_1$ is equivalent to the category hSet.

Example 5 (The finite types). Consider the inductive family $\mathsf{Fin} : \mathbb{N} \to \mathsf{hSet}$ of finite types. Again, this is a single type family, i.e. we are in the case $n \equiv 1$. We have $H_0(\star) :\equiv \mathbb{N}$, and the base category $\mathcal{C}_1$ is equivalent to the category of $\mathbb{N}$-indexed families, where objects are families $X : \mathbb{N} \to \mathsf{hSet}$ and morphisms $\mathcal{C}_1(X, Y)$ are dependent functions $f : (n : \mathbb{N}) \to X(n) \to Y(n)$.

Example 6 (Contexts and types). Let us consider the QIIT (Con, Ty) from Example 2. Here, we need two functors H_0, H_1, the first corresponding to Con and the second to Ty. The first is given by $H_0(\star) :\equiv \mathbf{1}$ as in Example 4, since Con is a type on its own. Next, we need $H_1 : \mathcal{C}_1 \Rightarrow \mathsf{hSet}$. Applying the equivalence between $\mathcal{C}_1$ and hSet established in Example 4, we define H_1 to be the identity functor $H_1(A) :\equiv A$, since then $\mathsf{Ty} : H_1(\mathsf{Con}) \to \mathsf{hSet}$. The base category $\mathcal{C}_2$ is equivalent to the category Fam(hSet), whose objects are pairs (A, B) where $A : \mathsf{hSet}$ and $B : A \to \mathsf{hSet}$, and whose morphisms (A, B) to (A', B') consist of functions $f : A \to A'$ together with dependent functions $g : (x : A) \to B(x) \to B'(f\,x)$.

Example 7 (the Cauchy reals). Recall that the Cauchy reals in the HoTT book [34] are constructed by simultaneously defining $\mathbb{R} : \mathsf{hSet}$ and $\sim : \mathbb{R} \times \mathbb{R} \to \mathsf{hSet}$ (we ignore the fact that [34] uses $\mathcal{U}$ instead of hSet). This time the sorts H_0, H_1 are given by $H_0(\star) :\equiv \mathbf{1}$ and $H_1(A) :\equiv A \times A$, corresponding to the fact that $\sim$ is a binary relation on $\mathbb{R}$. The base category has (up to equivalence) pairs (X, Y) with $Y : X \times X \to \mathsf{hSet}$ as objects, and morphisms are defined accordingly.

Example 8 (The full syntax of type theory). Altenkirch and Kaposi [1] give the complete syntax of a basic type theory as a (at that point unspecified) QIIT. Although this construction is far too involved to be treated as an example in the rest of this paper (where we prefer to work with the simplified version of Example 2), we can give the sort signature H_0, H_1, H_2, H_3 of this QIIT. Apart from contexts Con and types Ty, this definition also involves context morphisms Tms and terms Tm:

$$\begin{aligned} &\mathsf{Con} : \mathsf{hSet} && \mathsf{Tms} : \mathsf{Con} \times \mathsf{Con} \to \mathsf{hSet} \\ &\mathsf{Ty} : \ \mathsf{Con} \to \mathsf{hSet} && \mathsf{Tm} : \ (\Sigma(\Gamma : \mathsf{Con}).\mathsf{Ty}(\Gamma)) \to \mathsf{hSet}. \end{aligned}$$

We have:

$$\begin{aligned} H_0(\star) &:\equiv \mathbf{1} && \mathcal{C}_1 \cong \mathsf{hSet} \text{ as in Example 4;} \\ H_1(A) &:\equiv A && \mathcal{C}_2 \cong \mathsf{Fam(hSet)} \text{ as in Example 6;} \\ H_2(A, B) &:\equiv A \times A && \mathcal{C}_3 \text{ has objects } (A, B, C), \text{ where } C : A \times A \to \mathsf{hSet}; \\ H_3(A, B, C) &:\equiv \Sigma\,A\,B && \mathcal{C}_4 \text{ has objects } (A, B, C, D), \text{ where } D : (\Sigma\,A\,B) \to \mathsf{hSet}. \end{aligned}$$

Remark 9. Although we work in type theory also in the meta-theory, we give the presentation informally in natural language. Formally, the specification of sorts and base categories of Definition 3 can be defined as an inductive-recursive definition [19] of the list $H_0, \ldots, H_n$ simultaneously with a function that turns such a list into a category. Details can be found in Dijkstra's thesis [16, Sect. 4.3].

The main result of this section states that base categories of sort signatures are complete, i.e. have all small limits. By a small limit, we mean a limit of a diagram $D : \mathcal{I} \to \mathcal{C}$, where the shape category $\mathcal{I}$ has a set of objects, and the collection of morphisms between any two objects is a set. This result will be needed later to show that categories of QIIT algebras are complete. Recall that hSet has all small limits by a standard construction.

Theorem 10 (Base categories are complete). *For any sort signature H_0, ..., H_{n-1}, the corresponding base category $\mathcal{C}_n$ has all small limits.*

Proof. All proofs can be found in the arXiv version of the paper [3]. □

3 Algebras

Once the sorts of an inductive definition have been established, the next step is to specify the *constructors*. In this section, we will give a very general definition of constructor specifications, although we will mainly focus on two specific kinds: *point constructors*, which can be thought of as the operations of an algebraic signature, and *path constructors*, which correspond to the axioms.

Similarly to how sorts are specified inductively in Sect. 2, we construct suitable categories of algebras by starting with a finitely complete category $\mathcal{C}$ such as the one obtained from a sort signature, specify a constructor on $\mathcal{C}$, and then extend $\mathcal{C}$ using this constructor specification to get a new finitely complete category $\mathcal{C}'$. This process is repeated until all constructors have been added, and we obtain the sought-after inductive type as the underlying set of an initial object of the category at the last stage, provided this initial object exists. In the case of the inductive definition of natural numbers, this process will turn out as follows:

- we start with hSet as our base category (only one trivial sort, as in Example 4);
- we add a point constructor for the constant corresponding to 0; the category of algebras at this stage is the category of pointed sets;
- we add a second point constructor for the operation corresponding to suc; the category of algebras at this stage is the category of sets equipped with a point and a unary operation;
- the set of natural numbers, together with its usual structure, can now be regarded as an initial object in the category of algebras just constructed. □

3.1 Relative Continuity and Constructor Specifications

Roughly speaking, constructors at each stage are given by pairs of hSet-valued functors F and G on $\mathcal{C}$, where G is continuous (i.e. preserves all small limits). The intuition is that F specifies the arguments of the constructor, while G determines its target. For instance, in the example of the natural numbers when specifying the constructor $\mathsf{suc} : \mathbb{N} \to \mathbb{N}$, $\mathcal{C}$ is the category of pointed sets, and both F and G are the forgetful functor to hSet. The continuity condition on G is needed for the corresponding category of algebras to be complete. Intuitively, this expresses that a constructor should only "construct" elements of one of the sorts, or equalities thereof.[1] In particular, a constant functor is usually not a valid choice for G.

Unfortunately, this simple description falls short of capturing many of the examples of QIITs mentioned in Sect. 1. The problem is that we want G to be able to depend on the elements of F. However, since F is assumed to be an arbitrary functor, its category of elements is not necessarily complete, and so we need to refine the notion of G being continuous to this case.

Definition 11 (Relative continuity). *Let $\mathcal{C}$ be a category, $\mathcal{C}_0$ a complete category, and $U : \mathcal{C} \Rightarrow \mathcal{C}_0$ a functor. If I is a small category, and $X : I \to \mathcal{C}$ is a diagram, we say that a cone $A \to X$ in $\mathcal{C}$ is a U-*limit cone*, or* limit cone relative to U*, if the induced cone $UA \to UX$ is a limit cone in $\mathcal{C}_0$. A functor $\mathcal{C} \Rightarrow \mathsf{hSet}$ is* continuous relative to U *if it maps U-limit cones to limit cones in* hSet.

In the special case $\mathcal{C}_0 \equiv \mathsf{hSet}$, the functor U in Definition 11 is continuous relative to itself. Also note that if $\mathcal{C}$ is complete and U creates limits, then relative continuity with respect to U reduces to ordinary continuity. If $\mathcal{C}$ is a complete category, and $F : \mathcal{C} \Rightarrow \mathsf{hSet}$ is an arbitrary functor, the category $\int^{\mathcal{C}} F$ of elements of F is equipped with a forgetful functor into $\mathcal{C}$. We will implicitly consider relative limit cones and relative continuity with respect to this forgetful functor, unless specified otherwise. Note that if $\mathcal{C}$ is complete and F is continuous, then $\int^{\mathcal{C}} F$ is also complete, and relative continuity of functors on $\int^{\mathcal{C}} F$ is the same as continuity, as observed above.

We can now give a precise definition of what is needed to specify a constructor:

Definition 12 (Constructor specifications). *A* constructor specification *on a complete category $\mathcal{C}$ is given by:*

- *a functor $F : \mathcal{C} \Rightarrow \mathsf{hSet}$, called the* argument *functor of the specification;*
- *a relatively continuous functor $G : \int^{\mathcal{C}} F \Rightarrow \mathsf{hSet}$, called the* target *functor.*

Given a constructor specification, we can define the corresponding category of algebras. In Theorem 25, we will see that the assumptions of Definition 12 guarantee that this category is complete.

Definition 13 (Category of algebras). *Let (F, G) be a constructor specification on a complete category $\mathcal{C}$. The* category of algebras *of (F, G) is denoted $\mathcal{C}.(F, G)$, and is defined as follows:*

---

[1] More concretely, elements of a sort correspond to representable functors for algebras over a single generator for that sort, while equalities correspond to algebras with no generators and the given equality as the only relation. Clearly, representable functors are continuous, and the converse holds for reasonable functors (e.g. accessible ones). However, we do not attempt to make this construction precise here, and the following results do not depend on it.

- *objects are pairs* (X, θ), *where* X *is an object of* $\mathcal{C}$, *and* $\theta : (x : FX) \to G(X, x)$ *is a dependent function (in* hSet*);*
- *morphisms* $(X, \theta) \to (Y, \psi)$ *are given by morphisms* $f : X \to Y$ *in* $\mathcal{C}$, *with the property that for all* $x : FX$,

$$\psi(F(f)\,x) = G(\overline{f})(\theta\,x),$$

where $\overline{f} : (X, x) \to (Y, F(f)\,x)$ *is the morphism in* $\int^{\mathcal{C}} F$ *determined by* f.

We think of $\mathcal{C}.(F, G)$ as a category of "dependent dialgebras" [20]. Note that there is an obvious forgetful functor $\mathcal{C}.(F, G) \to \mathcal{C}$.

Similarly to how we defined sort specifications (Definition 3), we now have all the necessary notions in place to be able to give the full definition of a QIIT.

Definition 14 (QIIT descriptions). *A* QIIT description *is given by*
- *a sort specification* $H_0, \ldots, H_{n-1}$*;*
- *a list of constructor specifications* $(F_0, G_0), \ldots, (F_{n-1}, G_{n-1})$ *on* $\mathcal{B}_0, \ldots, \mathcal{B}_{n-1}$ *respectively, where* $\mathcal{B}_0$ *is the base category of the given sort specification, and* $\mathcal{B}_{i+1}$ *is the category of algebras of* (F_i, G_i).

For Definition 14 to make sense, the categories $\mathcal{B}_i$ need to be complete, since constructor specifications are only defined on complete categories. This will follow from Theorem 25.

Example 15 (Permutable trees). The constructor $\mathsf{leaf} : T(A)$ from Example 1 can be specified by functors $F_0 : \mathsf{hSet} \Rightarrow \mathsf{hSet}$ and $G_0 : \int^{\mathsf{hSet}} F_0 \Rightarrow \mathsf{hSet}$, where $F_0(X) :\equiv \mathbf{1}$ and $G_0(X, l) :\equiv X$. Note how F_0 specifies the (trivial) arguments of leaf, and G_0 the target. Next the constructor $\mathsf{node} : (A \to T(A)) \to T(A)$ can be specified by functors $F_1 : \mathsf{hSet}_\bullet \Rightarrow \mathsf{hSet}$ and $G_1 : \int^{\mathsf{hSet}_\bullet} F_1 \Rightarrow \mathsf{hSet}$, where $\mathsf{hSet}_\bullet$ is the category of pointed sets (we think of the point as the previous constructor leaf): F_1 and G_1 are defined as $F_1(X, l) :\equiv A \to X$ and $G_1(X, l, f) :\equiv X$, so that

$$\mathsf{node} : (f : F_1(T(A), \mathsf{leaf})) \to G_1(T(A), \mathsf{leaf}, f).$$

Theorem 18 will show that G_0 and G_1 are relatively continuous.

The corresponding category of algebras for this constructor specification (F_1, G_1) for node is equivalent to the category whose objects are triples (X, l, n) where $X : \mathsf{hSet}$, $l : A$, and $n : (A \to X) \to X$. After specifying also the mix-constructor, the new category of algebras further contains a dependent function $p : (f : A \to X) \to (e : X \cong X) \to n(f) = n(f \circ e)$.

Example 16 (Contexts and types). The constructor σ_{eq} of type

$$(\Gamma : \mathsf{Con})(A : \mathsf{Ty}(\Gamma))(B : \mathsf{Ty}(\mathsf{ext}\,\Gamma\,A)) \to \mathsf{ext}\,(\mathsf{ext}\,\Gamma\,A)\,B =_{\mathsf{Con}} \mathsf{ext}\,\Gamma\,(\sigma\,\Gamma\,A\,B)$$

from Example 2 is specified in the context of the previous constructors ε, ext and σ by functors $F : \mathcal{C} \Rightarrow \mathsf{hSet}$ and $G : \int^{\mathcal{C}} F \Rightarrow \mathsf{hSet}$, where $\mathcal{C}$ is the category of algebras of the previous constructors, with

$$F(C, T, \varepsilon, \mathsf{ext}, \sigma) :\equiv \Sigma(\Gamma : C).\Sigma(A : T(\Gamma)).T(\mathsf{ext}, \Gamma\,A)$$

and

$$G(\mathcal{C}, T, \varepsilon, \mathsf{ext}, \sigma, \Gamma, A, B) :\equiv \mathsf{ext}\,(\mathsf{ext}\,\Gamma\,A)\,B =_{\mathcal{C}} \mathsf{ext}\,\Gamma\,(\sigma\,\Gamma\,A\,B).$$

Theorem 23 will show that G is relatively continuous. The corresponding category of algebras for this constructor specification has objects tuples $(C, T, e, c, b, s, s_{\mathrm{eq}})$ where (C, T, e, c, b, s) is an algebra for the previous constructors, and

$$s_{\mathrm{eq}} : (\Gamma : C) \to (A : T(\Gamma)) \to (B : T(c\,\Gamma\,A)) \to c\,(c\,\Gamma\,A)\,B =_C c\,\Gamma\,(s\,\Gamma\,A\,B).$$

3.2 Point Constructors

If $\mathcal{C}$ is the base category for a sort signature as in Definition 3, we can define specific target functors $\mathcal{C} \Rightarrow \mathsf{hSet}$ which are guaranteed to be relatively continuous. Constructors having those as targets are referred to as *point constructors*. Intuitively, a point constructor is an operation that returns an element (point) of one of the sorts. The corresponding target functor is the forgetful functor that projects out the chosen sort. However, sorts can be dependent, so such a projection needs to be defined on a category of elements.

Specifically, let $\mathcal{C}$ be a finitely complete category, $H : \mathcal{C} \Rightarrow \mathsf{hSet}$ a functor, and $\mathcal{C}'$ the extended base category with one more sort indexed over H. Recall from Definition 13 that the objects of $\mathcal{C}'$ are pairs (X, P), where X is an object of $\mathcal{C}$, and P is a family of sets indexed over HX. Let $V_H : \mathcal{C}' \Rightarrow \mathcal{C}$ be the forgetful functor. We define the *base target* functor corresponding to H to be the functor $U_H : \int^{\mathcal{C}'} (H \circ V_H) \Rightarrow \mathsf{hSet}$ given by

$$U_H(X, P, x) = P(x).$$

In other words, given an object X of $\mathcal{C}$, a family P over HX, and a point x in the base, the functor U_H returns the fibre of the family P over x. The action of U_H on morphisms is the obvious one.

Example 17 (Permutable trees). In Example 15, the functor $G_0 : \int^{\mathsf{hSet}} F_0 \Rightarrow \mathsf{hSet}$ specifying the target of leaf is the composition of the forgetful $\int^{\mathsf{hSet}} F_0 \Rightarrow \mathsf{hSet}$ with the base target functor for the only sort, in this case the identity $\mathsf{id} : \mathsf{hSet} \Rightarrow \mathsf{hSet}$.

Note that $U_H = \mathsf{id}$ in Example 17 is relatively continuous, as required by Definition 12. In the rest of this section, we will show that this is true in general. Given a category $\mathcal{C}$ and a functor $F : \mathcal{C} \Rightarrow \mathsf{hSet}$, it is well known that the slice category over F of the functor category $\mathcal{C} \Rightarrow \mathsf{hSet}$ is equivalent to the functor category $\int^{\mathcal{C}} F \Rightarrow \mathsf{hSet}$ (see for example [24, Proposition 1.1.7]). Given a functor $G : \mathcal{C} \Rightarrow \mathsf{hSet}$ and a natural transformation $\alpha : G \to F$, we will refer to the functor $\overline{G} : \int^{\mathcal{C}} F \Rightarrow \mathsf{hSet}$ corresponding to α as the *functor of fibres* of α. Concretely, $\overline{G}$ maps an object (X, x), where $x : FX$, to the fibre of α_X over x. The following theorem is proved by noting that U_H is a functor of fibres.

Theorem 18 (Base target functors are relatively continuous). *Let $\mathcal{C}$ be a complete category, $H : \mathcal{C} \Rightarrow \mathsf{hSet}$ any functor, and $\mathcal{C}'$ the extended base category corresponding to H. Then the base target functor U_H is relatively continuous.* □

3.3 Reindexing Target Functors

In many cases, we can obtain suitable target functors by composing the desired base target functor with the forgetful functor to the appropriate stage of the base category. When building constructors one at a time, it will follow from Theorems 25 and 10 applied to the previous steps that this forgetful functor is continuous, and the relative continuity of the target functor will follow. In more complicated examples, composing with a forgetful functor is not quite enough. We often want to "substitute into" or reindex a target functor to target a specific element. For example, in the context of Example 2, consider a hypothetical modified σ constructor of the form

$$\sigma' : (\Sigma(\Gamma : \mathsf{Con}).\Sigma(A : \mathsf{Ty}(\Gamma)).\mathsf{Ty}(\mathsf{ext}\, \Gamma\, A)) \to \mathsf{Ty}(\mathsf{ext}\, \Gamma\, A).$$

We want the target functor to return the set $\mathsf{Ty}(\mathsf{ext}\, \Gamma\, A)$, and not just $\mathsf{Ty}(x)$ for a new argument x, which is the result of the base target functor. We can obtain the desired target functor as a composition

$$\textstyle\int^{\mathcal{C}} F \xrightarrow{S} \int^{\mathsf{Fam}(\mathsf{hSet})} \pi_1 \xrightarrow{U_H} \mathsf{hSet}, \qquad (1)$$

where $\mathcal{C}$ is the category with objects tuples $(C, T, \varepsilon, \mathsf{ext})$, $F : \mathcal{C} \Rightarrow \mathsf{hSet}$ is the functor giving the arguments of the constructor σ', U_H is the base target functor corresponding to the second sort, and S is the functor defined by $S(C, T, \varepsilon, \mathsf{ext}, \Gamma, A, B) :\equiv (C, T, \mathsf{ext}\, \Gamma\, A)$.

Since the functors S that we compose with in order to "substitute" are of a special form, the resulting functor will still be relatively continuous when starting with a relatively continuous functor. This is made precise by the following result:

Lemma 19 (Preservation of relative limit cones). *Suppose given is a commutative diagram of categories and functors as shown on the right, where $\mathcal{C}_0$ and $\mathcal{D}_0$ are complete, and G maps U-limit cones to V-limit cones. Then F maps $(U \circ U')$-limit cones to $(V \circ V')$-limit cones. In particular, if $\mathcal{C}$ and $\mathcal{D}$ are complete and G is continuous, then F preserves relative limit cones.* □

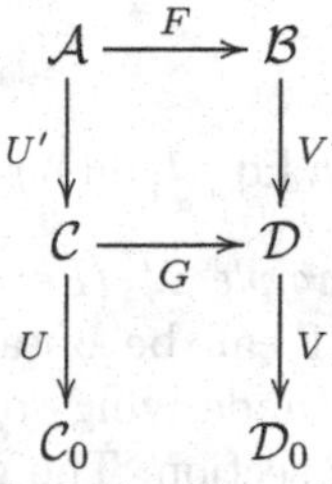

Example 20. Starting from the situation in (1) we can form the diagram shown on the left, where $V : \mathcal{C} \Rightarrow \mathsf{Fam}(\mathsf{hSet})$ is the forgetful functor and hence continuous. It follows from the second statement of Lemma 19 that S preserves relative limit cones, hence $G = U_H \circ S$ is relatively continuous by Theorem 18.

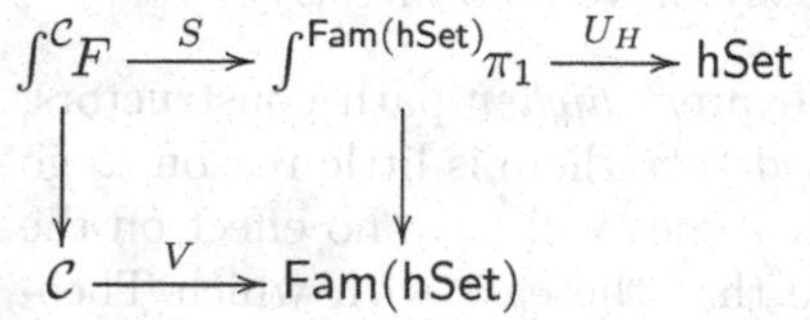

3.4 Path Constructors

Path constructors are constructors where the target functor G returns an *equality* type. They can e.g. be used to express laws when constructing an initial algebra of an algebraic theory as a QIT. We saw an example of this in Example 1, where we had a path constructor of the form

$$\mathsf{mix} : (f : A \to T) \to (e : A \cong A) \to \mathsf{node}(f) = \mathsf{node}(f \circ e).$$

The argument functor for mix is entirely unproblematic. However, it is perhaps not so clear that the target functor, which sends (X, l, n, f, e) to the equality type $n(f) =_X n(f \circ e)$, is relatively continuous. The aim of the current section is to show this for any functor of this form. We first observe that the prototypical such equality functor is relatively continuous, and then show that any other target functor for a path constructor can be obtained by substitution using Lemma 19.

Definition 21. *Let* $\mathsf{Eq} : \int^{\mathsf{hSet}}(\mathsf{id} \times \mathsf{id}) \Rightarrow \mathsf{hSet}$ *be the functor defined on objects by* $\mathsf{Eq}(X, x, y) :\equiv x =_X y$ *and on morphisms by* $\mathsf{Eq}(f, p_x, p_y) :\equiv p_x \cdot (\mathsf{ap}\ f\ -) \cdot p_y^{-1}$.

It is not hard to see that Eq is a functor. Furthermore, Eq is the functor of fibres of the obvious diagonal natural transformation $\Delta : \mathsf{id} \to \mathsf{id} \times \mathsf{id}$.

Lemma 22. *The standard equality functor* Eq *is relatively continuous.* □

The lemma we have just given is central to the observation that a large class of equality functors are suitable targets for constructors:

Theorem 23 (Equality functors are relatively continuous). *Let $\mathcal{C}$ be a complete category, $F : \mathcal{C} \Rightarrow \mathsf{hSet}$ any functor, and $G : \int^{\mathcal{C}} F \Rightarrow \mathsf{hSet}$ a relatively continuous functor. Suppose given two global elements l, r of G, i.e. natural transformations $l, r : \mathbf{1} \to G$. The map*

$$\mathsf{Eq}_G(l, r) : \int^{\mathcal{C}} F \to \mathsf{hSet}$$

with $\mathsf{Eq}_G(l, r)(Y) = (l_Y =_{G(Y)} r_Y)$ *extends to a relatively continuous functor.* □

Example 24 (Permutable trees). The target of the mix constructor from Example 1 can be obtained as an equality functor in this sense. We take G to be the underlying sort, which is relatively continuous by the results of the previous section. The global elements l and r are defined by $l_{(X,l,n,f,e)} :\equiv n(f)$ and $r_{(X,l,n,f,e)} :\equiv n(f \circ e)$. Their naturality can easily be verified directly.

Iterating equality functors, one can also express *higher* path constructors, but in our limited setting of inductively defined *sets*, there is little reason to go beyond one level of path constructors — higher ones will have no effect on the resulting inductive type. However, we believe that the ease with which Theorem 23 can be applied iteratively will be an important feature when generalising our technique to general higher inductive types. We discuss this further in Sect. 5.

3.5 Categories of Algebras are Complete

Recall from Definition 13 that the category of algebras $\mathcal{C}.(F, G)$ for a constructor specification (F, G) on a complete category $\mathcal{C}$ has "dependent (F, G)-dialgebras" as objects, and maps that commute with the dialgebra structure as morphisms. In this section, we will show that $\mathcal{C}.(F, G)$ is complete, and that its forgetful functor is continuous. The significance of this result is twofold: First of all, it enables the use of limits when reasoning about algebras; in particular, we will show in Sect. 4 how, using products and equalisers, one can extend the classical equivalence between initiality and induction for ordinary inductive types to our setting. Secondly, it goes a long way towards establishing existence of initial algebras; since a category of algebras over $n+1$ constructors is complete, and the forgetful functor to the category of algebras over the first n preserves limits, the adjoint functor theorem says that this functor has a left adjoint if and only if it satisfies the solution set condition. Applying this argument at every stage, we get a left adjoint for the forgetful functor down to hSet, and in particular an initial object. There is no reason to expect the solution set condition to hold at this generality, but we expect it to follow from appropriate "accessibility" conditions on the argument functors. This is discussed further in Sect. 5.

Theorem 25 (Categories of algebras are complete). *Let (F, G) be a constructor specification on a complete category $\mathcal{C}$. Then $\mathcal{C}.(F, G)$ is complete.* □

4 Elimination Principles

So far, we have given rules for specifying a QIIT by giving a sort signature and a list of constructors. As type-theoretical rules, these correspond to the formation and introduction rules for the QIIT. In this section, we introduce the corresponding elimination rules, stating that a QIIT is the smallest type closed under its constructors. We show that a categorical formulation of the elimination rules is equivalent to the universal property of initiality.

4.1 The Section Induction Principle

The elimination principle for an algebra X states that *every fibred algebra over X has a section*, where a fibred algebra over X is an algebra family "$Q : X \to \mathsf{hSet}$", and a section of it a dependent algebra morphism "$(x : X) \to Q(x)$".[2] The usual correspondence between type families and fibrations extends to algebras, and so we formulate the elimination rule for X as X being section inductive in the category of algebras in the following sense:

Definition 26 (Section inductive). *An object X of a category $\mathcal{C}$ is* section inductive *if for every object Y of $\mathcal{C}$ and morphism $p : Y \to X$, there exists $s : X \to Y$ such that $p \circ s = \mathsf{id}_X$.*

[2] See Dijkstra's thesis [16, Sect. 5.4] for the general definition of fibred algebras and their morphisms — here we restrict ourselves to examples only for space reasons.

For an algebra X, the existence of the underlying function(s) $X \to Y$ corresponds to the elimination rules, while the fact that they are algebra morphisms corresponds to the computation rules.

Example 27 (Permutable trees). Consider permutable-tree algebras, e.g. tuples (X, l, n, p) as in Example 15. A fibred permutable-tree algebra over (X, l, n, p) consists of $Q : X \to \mathsf{hSet}$ together with $m_l : Q(l)$ and

$$\begin{aligned} m_n :\ & (f : A \to X) \to (g : (a : A) \to Q(f\,a)) \to Q(n\,f) \\ m_p :\ & (f : A \to X) \to (g : (a : A) \to Q(f\,a)) \to (e : A \cong A) \\ & \qquad \to m_n\,f\,g =[\mathsf{ap}\ Q\ p]\ m_n\,(f \circ e)\,(g \circ e) \end{aligned}$$

Here the type $x = [p]\ y$ is the types of equalities between elements $x : A$ and $y : B$ in different types, themselves related by an equality proof $p : A = B$. This data can be arranged into an ordinary algebra $\Sigma(x : X).Q(x)$, together with an algebra morphism $\pi_1 : (\Sigma(x : X).Q(x)) \to X$. A section of π_1 is a dependent function $h : (x : X) \to Q(x)$. Since h comes from an algebra morphism, we further know e.g. $h(l) = m_l$ and $h(n\,f) = m_n\,f\,(h \circ f)$. Conversely, every algebra morphism $g : (X', l', n', p') \to (X, l, n, p)$ gives rise to a fibred algebra (Q, m_l, m_n, m_p) by considering the fibres $Q(x) = \Sigma(y : A').g(y) = x$ of p. The points m_l, m_n and the path m_p arise from the proof that g preserves l', n' and p'.

Example 28 (Contexts and types). For context-and-types algebras from Example 16, a fibred algebra over $(C, T, e, c, b, s, s_{\mathrm{eq}})$ consists of $Q : C \to \mathsf{hSet}$ and $R : (x : C) \to T(x) \to Q(x) \to \mathsf{hSet}$, together with $m_e : Q(e)$ and

$$\begin{aligned} m_c :\ & (\Gamma : C) \to (x : Q(\Gamma)) \to (A : T(\Gamma)) \to R(\Gamma, A, x) \to Q(c\,\Gamma\,A) \\ m_b :\ & (\Gamma : C) \to (x : Q(\Gamma)) \to R(\Gamma, b\,\Gamma, x) \\ m_s :\ & (\Gamma : C) \to (x : Q(\Gamma)) \to (A : T(\Gamma)) \to (y : R(\Gamma, A, x) \to (B : T(c\,\Gamma\,A)) \\ & \quad \to (z : R(c\,\Gamma\,A, B, m_c\,\Gamma\,x\,A\,y)) \to R(\Gamma, s\,\Gamma\,A\,B, x) \\ m_{s_{\mathrm{eq}}} :\ & (\Gamma : C) \to (x : Q(\Gamma)) \to (A : T(\Gamma)) \to (y : R(\Gamma, A, x)) \\ & \quad \to (B : T(c\,\Gamma\,A)) \to (z : R(c\,\Gamma\,A, B, m_c\,\Gamma\,x\,A\,y)) \\ & \quad \to m_c\,(c\,\Gamma\,A)\,(m_c\,\Gamma\,x\,A\,y)\,B\,z = [\mathsf{ap}\ Q\,(s_{\mathrm{eq}}\,\Gamma\,A\,B)] \\ & \qquad\qquad m_c\,\Gamma\,x\,(s\,\Gamma\,A\,B)\,(m_s\,\Gamma\,x\,A\,y\,B\,z) \end{aligned}$$

Again, this data can be arranged into an ordinary algebra with base $C' : \mathsf{hSet}$, $T' : C' \to \mathsf{hSet}$, where $C' = \Sigma(x : C).Q(x)$ and $T'(x, q) = \Sigma(y : T(x)).R(x, y, q)$, together with an algebra morphism $(\pi_1, \pi_1) : (C', T') \to (C, T)$. A section of this morphism gives functions $f : (x : C) \to Q(x)$ and $g : (x : C) \to (y : T(x)) \to R(x, y, f\,x)$ that preserve the algebra structure.

A general account of the equivalence between the usual formulation of the elimination rules and the section induction principle is in Dijkstra [16, Sect. 5.4].

4.2 Initiality, and its Relation to the Section Induction Principle

The section induction principle for an algebra X matches our intuitive understanding of the elimination rules for X quite well, but it is perhaps a priori not so clear that e.g. satisfying it defines an algebra uniquely up to equivalence. In this section, we show that this is the case by proving that the section induction principle is equivalent to the categorical property of initiality. Recall that a type is *contractible* if it is equivalent to the unit type [34, Definition 3.11.1].

Definition 29 (Initiality). *An object X of a category $\mathcal{C}$ is* (homotopy) initial *if for every object Y of $\mathcal{C}$, the set of morphisms $X \to Y$ is contractible.*

It is easy to see that initiality implies section induction, while the converse requires additional structure on $\mathcal{C}$:

Lemma 30. *If an object X in a category $\mathcal{C}$ is initial, then it is section inductive. If $\mathcal{C}$ has finite limits and X is section inductive, then X is initial.* □

From here, we can show the main theorem of the current section. The proof uses the fact that both statements involved are mere propositions, i.e. they have at most one proof.

Theorem 31 (Initiality ≅ section induction). *An object X in a in a category of algebras $\mathcal{C}.(F, G)$ being initial is equivalent to it being section inductive.* □

As an application, we can now reason about QIITs using their categories of algebras. For instance, we get a short proof of the following fact:

Corollary 32. *The interval is equivalent to the unit type.*

Proof. By Theorem 31, the interval is the initial object in the category with objects $\Sigma(X : \mathsf{hSet}).\Sigma(x : X).\Sigma(y : X).x =_X y$, while the unit type is the initial object in the category with objects $\Sigma(X : \mathsf{hSet}).X$. By contractibility of singleton types [34, Lemma 3.11.8], the former is equivalent to the latter, and since initiality is a universal property, the two initial objects coincide up to equivalence. □

5 Conclusions and Further Work

We have developed a semantic framework for QIITs: A QIIT description gives rise to a category of algebras, and the initial object of this category represent the types and constructors of the QIIT. This generalises the usual functorial semantics of inductive types to a more general setting. So far we have verified the appropriateness of this setting by means of examples. In future work, we would like to explicitly relate the syntax of QIITs to the corresponding semantics.

Our categories of algebras are complete. This is helpful for the metatheory of QIITs, as demonstrated by the proof of initiality being equivalent to section induction (Theorem 31), justifying elimination principles. Of course, completeness is not by itself sufficient to derive the existence of initial algebras, but it suggests that it should be possible to restrict the argument functors to guarantee this, possibly by reducing QIITs to a basic type former playing an analogous role to that of W-types for inductive types. We believe that completeness of the categories of algebras allows an existence proof using the adjoint functor theorem.

We have restricted our attention to QIITs, but we believe that our construction is applicable to general HITs (and even HIITs). While at first glance such an extension of our framework seems to require an internal theory of $(\infty, 1)$-categories, we believe that it is enough to keep track of only a very limited number of coherence conditions, making this extension possible even without solving the well-known problem of specifying an infinite tower of coherences in HoTT.

Other possible future directions include the combination of QIITs and induction-recursion, and the possibility of generalising coinductive types along similar lines. These generalisations should be driven by examples, similar to how the examples discussed in the current paper have motivated the need for a theory of QIITs.

Acknowledgements. We thank Ambrus Kaposi and Jakob von Raumer for many interesting discussions, and the anonymous referees for their valuable comments. This research was supported by EPSRC grants EP/M016994/1 and EP/K023837/1, as well as AFOSR award FA9550-16-1-0029.

References

1. Altenkirch, T., Kaposi, A.: Type theory in type theory using quotient inductive types. In: Principles of Programming Languages, pp. 18–29. ACM (2016)
2. Altenkirch, T., Danielsson, N.A., Kraus, N.: Partiality, revisited. In: Esparza, J., Murawski, A.S. (eds.) FoSSaCS 2017. LNCS, vol. 10203, pp. 534–549. Springer, Heidelberg (2017). https://doi.org/10.1007/978-3-662-54458-7_31
3. Altenkirch, T., Capriotti, P., Dijkstra, G., Kraus, N., Nordvall Forsberg, F.: Quotient inductive-inductive types (2018). arXiv:1612.02346
4. Angiuli, C., Morehouse, E., Licata, D.R., Harper, R.: Homotopical patch theory. In: International Conference on Functional Programming, pp. 243–256 (2014)
5. Awodey, S., Warren, M.A.: Homotopy theoretic models of identity types. Math. Proc. Camb. Philos. Soc. **146**(1), 45–55 (2009)
6. Basold, H., Geuvers, H., van der Weide, N.: Higher inductive types in programming. J. Univ. Comput. Sci. **23**(1), 63–88 (2016)
7. Bertot, Y., Castéran, P.: Interactive Theorem Proving and Program Development. Coq'Art: The Calculus of Inductive Constructions. Texts in Theoretical Computer Science. Springer, Heidelberg (2004). https://doi.org/10.1007/978-3-662-07964-5
8. Brady, E.: Idris, a general-purpose dependently typed programming language: design and implementation. J. Funct. Program. **23**(9), pp. 552–593 (2013)
9. Brunerie, G.: On the homotopy groups of spheres in homotopy type theory. Ph.D. thesis, Université de Nice (2016)
10. Buchholtz, U., Rijke, E.: The real projective spaces in homotopy type theory. In: Logic in Computer Science, pp. 1–8 (2017)

11. Cavallo, E.: Synthetic cohomology in Homotopy Type Theory. Master's thesis, Carnegie-Mellon University (2015)
12. Chapman, J., Uustalu, T., Veltri, N.: Quotienting the delay monad by weak bisimilarity. In: Leucker, M., Rueda, C., Valencia, F.D. (eds.) ICTAC 2015. LNCS, vol. 9399, pp. 110–125. Springer, Cham (2015). https://doi.org/10.1007/978-3-319-25150-9_8
13. Chapman, J., Uustalu, T., Veltri, N.: Quotienting the delay monad by weak bisimilarity. Math. Struct. Comput. Sci. 1–26 (2017)
14. de Moura, L., Kong, S., Avigad, J., van Doorn, F., von Raumer, J.: The Lean theorem prover. In: Conference on Automated Deduction (2015)
15. Diaconescu, R.: Axiom of choice and complementation. Proc. Am. Math. Soc. **51**(1), 176–178 (1975)
16. Dijkstra, G.: Quotient inductive-inductive types. Ph.D. thesis, University of Nottingham (2017)
17. Dybjer, P.: Inductive families. Formal Aspects Comput. **6**(4), 440–465 (1994)
18. Dybjer, P., Moeneclaey, H.: Finitary higher inductive types in the groupoid model. In: Silva, A. (ed.) Mathematical Foundations of Programming Semantics (2017)
19. Dybjer, P., Setzer, A.: A finite axiomatization of inductive-recursive definitions. In: Girard, J.-Y. (ed.) TLCA 1999. LNCS, vol. 1581, pp. 129–146. Springer, Heidelberg (1999). https://doi.org/10.1007/3-540-48959-2_11
20. Hagino, T.: A categorical programming language. Ph.D. thesis, University of Edinburgh (1987)
21. Harper, R., Pollack, R.: Type checking with universes. Theor. Comput. Sci. **89**(1), 107–136 (1991)
22. Hofmann, M., Streicher, T.: The groupoid interpretation of type theory. In: Twenty-Five Years of Constructive Type Theory of Oxford Logic Guides, vol. 36. Oxford University Press, New York, pp. 83–111 (1998)
23. Hou (Favonia), K.B., Finster, E., Licata, D.R., Lumsdaine, P.L.: A mechanization of the Blakers-Massey connectivity theorem in homotopy type theory. In: Logic in Computer Science (2016)
24. Johnstone, P.: Sketches of an Elephant: A Topos Theory Compendium. Oxford University Press, New York (2002)
25. Kapulkin, C., Lumsdaine, P.L.: The simplicial model of univalent foundations (after Voevodsky) (2016). arXiv:1211.2851
26. Licata, D.R., Finster, E.: Eilenberg-Maclane spaces in homotopy type theory. In: Logic in Computer Science, pp. 66:1–66:9 (2014)
27. Licata, D.R., Shulman, M.: Calculating the fundamental group of the circle in homotopy type theory. In: Logic in Computer Science, pp. 223–232 (2013)
28. Lumsdaine, P.L., Shulman, M.: Semantics of higher inductive types (2017). arXiv:1705.07088
29. Martin-Löf, P.: An intuitionistic theory of types. Published in Twenty-Five Years of Constructive Type Theory (1972)
30. Nordvall Forsberg, F.: Inductive-inductive definitions. Ph.D. thesis, Swansea University (2013)
31. Norell, U.: Towards a practical programming language based on dependent type theory. Ph.D. thesis, Chalmers University of Technology (2007)
32. Shulman, M.: Homotopy type theory: the logic of space. In: New Spaces for Mathematics and Physics (2017, to appear). arXiv:1703.03007
33. Sojakova, K.: Higher inductive types as homotopy-initial algebras. In: Principles of Programming Languages, pp. 31–42. ACM (2015)

34. The Univalent Foundations Program. Homotopy Type Theory: Univalent Foundations of Mathematics, Institute for Advanced Study (2013). https://homotopytypetheory.org/book
35. Veltri, N.: A type-theoretical study of nontermination. Ph.D. thesis, Tallinn University of Technology (2017)
36. Voevodsky, V.: The equivalence axiom and univalent models of type theory (talk at CMU on February 4, 2010). arXiv:1402.5556

Category Theory and Quantum Control

Guarded Traced Categories

Sergey Goncharov(✉) and Lutz Schröder(✉)

Friedrich-Alexander-Universität Erlangen-Nürnberg, Erlangen, Germany
{Sergey.Goncharov,Lutz.Schroeder}@fau.de

Abstract. Notions of guardedness serve to delineate the admissibility of cycles, e.g. in recursion, corecursion, iteration, or tracing. We introduce an abstract notion of guardedness structure on a symmetric monoidal category, along with a corresponding notion of guarded traces, which are defined only if the cycles they induce are guarded. We relate structural guardedness, determined by propagating guardedness along the operations of the category, to geometric guardedness phrased in terms of a diagrammatic language. In our setup, the Cartesian case (recursion) and the co-Cartesian case (iteration) become completely dual, and we show that in these cases, guarded tracedness is equivalent to presence of a guarded Conway operator, in analogy to an observation on total traces by Hasegawa and Hyland. Moreover, we relate guarded traces to unguarded categorical uniform fixpoint operators in the style of Simpson and Plotkin. Finally, we show that partial traces based on Hilbert-Schmidt operators in the category of Hilbert spaces are an instance of guarded traces.

1 Introduction

In models of computation, various notions of *guardedness* serve to control cyclic behaviour by allowing only guarded cycles, with the aim to ensure properties such as solvability of recursive equations or productivity. Typical examples are guarded process algebra specifications [6,29], coalgebraic guarded (co-)recursion [27,33], finite delay in online Turing machines [9], and productive definitions in intensional type theory [1,30], but also contractive maps in (ultra-)metric spaces [24].

A highly general model for unrestricted cyclic computations, on the other hand, are *traced monoidal categories* [22]; besides *recursion* and *iteration*, they cover further kinds of cyclic behaviour, e.g. in Girard's *Geometry of Interaction* [4,14] and quantum programming [3,34]. In the present paper we parametrize the framework of traced symmetric monoidal categories with a notion of guardedness, arriving at *(abstractly) guarded traced categories*, which effectively vary between two extreme cases: symmetric monoidal categories (nothing is guarded) and traced symmetric monoidal categories (everything is guarded). In terms of the standard diagrammatic language for traced monoidal categories, we decorate input and output gates of boxes to indicate guardedness; the diagram governing trace formation would then have the general form

C. Baier and U. Dal Lago (Eds.): FOSSACS 2018, LNCS 10803, pp. 313–330, 2018.
https://doi.org/10.1007/978-3-319-89366-2_17

depicted in Fig. 1 – that is, we can only form traces connecting guarded (black) output gates to input gates that are unguarded (black), i.e. not assumed to be already guarded.

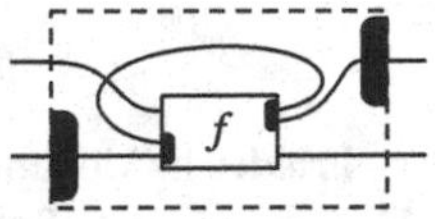

Fig. 1. Guarded trace

We provide basic structural results on our notion of abstract guardedness, and identify a wide array of examples. Specifically, we establish a geometric characterization of guardedness in terms of paths in diagrams; we identify a notion of *guarded ideal*, along with a construction of guardedness structures from guarded ideals and simplifications of this construction for the (co-)Cartesian and the Cartesian closed case; and we describe 'vacuous' guardedness structures where traces do not actually generate proper diagrammatic cycles. In terms of examples, we begin with the case where the monoidal structure is either product (Cartesian), corresponding to guarded recursion, or coproduct (co-Cartesian), for guarded iteration; the axioms for guardedness allow for a basic duality that indeed makes these two cases precisely dual. For total traces in Cartesian categories, Hasegawa and Hyland observed that trace operators are in one-to-one correspondence with *Conway fixpoint operators* [18,19]; we extend this correspondence to the guarded case, showing that guarded trace operators on a Cartesian category are in one-to-one correspondence with guarded Conway operators. In a more specific setting, we relate *guarded* traces in Cartesian categories to *unguarded* categorical uniform fixpoints as studied by Crole and Pitts [11] and by Simpson and Plotkin [37,38]. Concluding with a case where the monoidal structure is a proper tensor product, we show that the partial trace operation on (infinite-dimensional) Hilbert spaces is an instance of vacuous guardedness; this result relates to work by Abramsky, Blute, and Panangaden on traces over nuclear ideals, in this case over *Hilbert-Schmidt operators* [2].

Related Work. Abstract guardedness serves to determine definedness of a guarded trace operation, and thus relates to work on partial traces. We discuss work on nuclear ideals [2] in Sect. 6. In *partial traced categories* [17,26], traces are governed by a partial equational version (consisting of both strong and directed equations) of the Joyal-Street-Verity axioms; morphisms for which trace is defined are called *trace class*. A key difference to the approach via guardedness is that being trace class applies only to morphisms with inputs and outputs of matching types while guardedness applies to arbitrary morphisms, allowing for compositional propagation. Also, the axiomatizations are incomparable: Unlike for trace class morphisms [17, Remark 2.2], we require guardedness to be closed under composition with arbitrary morphisms (thus covering contractivity but not, e.g., monotonicity as in the modal μ-calculus); on the other hand, as noted by Jeffrey [21], guarded traces, e.g. of contractions, need not satisfy Vanishing II as a Kleene equality as assumed in partial traced categories. Some approaches treat traces as partial over objects [8,20]. In concrete algebraic categories, partial traces can be seen as induced by total traces in an ambient category of relations [5]. We discuss work on guardedness via endofunctors in Remark 23.

2 Preliminaries

We recall requisite categorical notions; see [25] for a comprehensive introduction.

Symmetric Monoidal Categories. A *symmetric monoidal category* $(\mathbf{C}, \otimes, I)$ consists of a category $\mathbf{C}$ (with object class $|\mathbf{C}|$), a bifunctor $\otimes$ (*tensor product*), and a *(tensor) unit* $I \in |\mathbf{C}|$, and coherent isomorphisms witnessing that $\otimes$ is, up to isomorphism, a commutative monoid structure with unit I. For the latter, we reserve the notation $\alpha_{A,B,C} : (A \otimes B) \otimes C \cong A \otimes (B \otimes C)$ (*associator*), $\gamma_{A,B} : A \otimes B \cong B \otimes A$ (*symmetry*), and $\upsilon_A : I \otimes A \cong A$ (*left unitor*); the *right unitor* $\hat{\upsilon}_A : A \otimes I \cong A$ is expressible via the symmetry. A symmetric monoidal category is *Cartesian* if the monoidal structure is finite product (i.e. $\otimes = \times$, and $I = 1$ is a terminal object), and, dually, *co-Cartesian* if the monoidal structure is finite coproduct (i.e. $\otimes = +$, and $I = \emptyset$ is an initial object). Coproduct injections are written $\mathsf{in}_i : X_i \to X_1 + X_2$ $(i = 1, 2)$, and product projections $\mathsf{pr}_i : X_1 \times X_2 \to X_i$. Various notions of algebraic tensor products also induce symmetric monoidal structures; see Sect. 6 for the case of Hilbert spaces. One has an obvious expression language for objects and morphisms in symmetric monoidal categories [36], the former obtained by postulating basic objects and closing under I and $\otimes$, and the latter by postulating basic morphisms of given profile and closing under $\otimes$, I, composition, identities, and the monoidal isomorphisms, subject to the evident notion of *well-typedness*. Morphism expressions are conveniently represented as *diagrams* consisting of boxes representing the basic morphisms, with input and output gates corresponding to the given profile. Tensoring is represented by putting boxes on top of each other, and composition by wires connecting outputs to inputs [36]. In a *traced symmetric monoidal category* one has an additional operation (*trace*) that essentially enables the formation of loops in diagrams, as in Fig. 1 (but without decorations).

Monads and (Co-)algebras. A(n F)-*coalgebra* for a functor $F : \mathbf{C} \to \mathbf{C}$ is a pair $(X, f : X \to FX)$ where $X \in |\mathbf{C}|$, thought of as modelling states and generalized transitions [33]. A *final coalgebra* is a final object in the category of coalgebras (with $\mathbf{C}$-morphisms $h : X \to Y$ such that $(Fh)f = gh$ as morphisms $(X, f) \to (Y, g)$), denoted $(\nu F, \mathsf{out} : \nu F \to F\nu F)$ if it exists. Dually, an F-*algebra* has the form $(X, f : FX \to X)$. A *monad* $\mathbb{T} = (T, \mu, \eta)$ on a category $\mathbf{C}$ consists of an endofunctor T on $\mathbf{C}$ and natural transformations $\eta : \mathrm{Id} \to T$ (*unit*) and $\mu : T^2 \to T$ (*multiplication*) subject to standard equations [25]. As observed by Moggi [31], monads can be seen as capturing *computational effects* of programs, with TX read as a type of computations with side effects from T and results in X. In this view, the *Kleisli category* $\mathbf{C}_{\mathbb{T}}$ of $\mathbb{T}$, which has the same objects as $\mathbf{C}$ and $\mathsf{Hom}_{\mathbf{C}_{\mathbb{T}}}(X, Y) = \mathsf{Hom}_{\mathbf{C}}(X, TY)$, is a category of side-effecting programs. A monad is *strong* if it is equipped with a *strength*, i.e. a natural transformation $X \times TY \to T(X \times Y)$ satisfying evident coherence conditions (e.g. [31]). A T-algebra (A, a) is an *(Eilenberg-Moore)* $\mathbb{T}$-*algebra* (for the *monad* $\mathbb{T}$) if additionally $a\eta = \mathsf{id}$ and $a(Ta) = a\mu_A$; the category of $\mathbb{T}$-algebras is denoted $\mathbf{C}^{\mathbb{T}}$.

3 Guarded Categories

We now introduce our notion of guarded structure. A standard example of guardedness are guarded definitions in process algebra. E.g. in the definition $P = a.P$, the right hand occurrence of P is guarded, ensuring unique solvability (by a process that keeps outputting a). A further example is contractivity of maps between complete metric spaces. We formulate abstract closure properties for *partial* guardedness where only some of the inputs and outputs of a morphism are guarded. Specifically, we distinguish *guarded outputs* and *guarded inputs* (D and B, respectively, in the following definition), with the intended reading that guarded outputs yield guarded data *provided* guarded data is already provided at guarded inputs, while unguarded inputs may be fed arbitrarily.

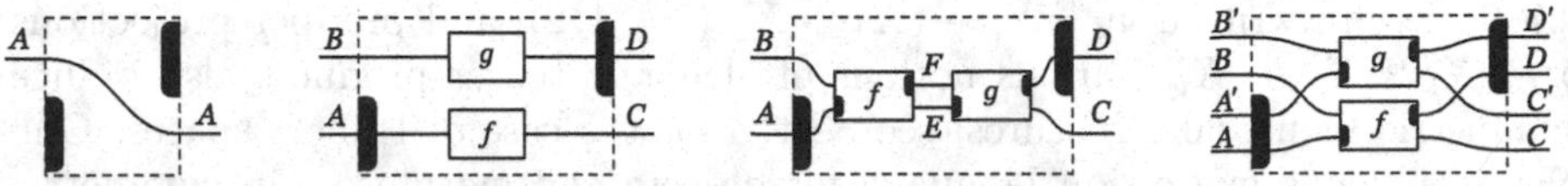

Fig. 2. Axioms of guarded categories

Definition 1 (Guarded category). An *(abstractly) guarded category* is a symmetric monoidal category $(\mathbf{C}, \otimes, I)$ equipped with distinguished subsets $\mathsf{Hom}^\bullet(A \otimes B, C \otimes D) \subseteq \mathsf{Hom}(A \otimes B, C \otimes D)$ of *partially guarded morphisms* for $A, B, C, D \in |\mathbf{C}|$, satisfying the following conditions:

(uni$_\otimes$) $\gamma_{I,A} \in \mathsf{Hom}^\bullet(I \otimes A, A \otimes I)$;
(vac$_\otimes$) $f \otimes g \in \mathsf{Hom}^\bullet(A \otimes B, C \otimes D)$ for all $f : A \to C$, $g : B \to D$;
(cmp$_\otimes$) $g \in \mathsf{Hom}^\bullet(A \otimes B, E \otimes F)$ and $f \in \mathsf{Hom}^\bullet(E \otimes F, C \otimes D)$ imply $fg \in \mathsf{Hom}^\bullet(A \otimes B, C \otimes D)$;
(par$_\otimes$) for $f \in \mathsf{Hom}^\bullet(A \otimes B, C \otimes D)$, $g \in \mathsf{Hom}^\bullet(A' \otimes B', C' \otimes D')$, the evident transpose of $f \otimes g$ is in $\mathsf{Hom}^\bullet((A \otimes A') \otimes (B \otimes B'), (C \otimes C') \otimes (D \otimes D'))$.

We emphasize that $\mathsf{Hom}^\bullet(A \otimes B, C \otimes D)$ is meant to depend individually on A, B, C, D and not just on $A \otimes B$ and $C \otimes D$.

One easily derives a *weakening* rule stating that if $f \in \mathsf{Hom}^\bullet((A \otimes A') \otimes B, C \otimes (D' \otimes D))$, then the obvious transpose of f is in $\mathsf{Hom}^\bullet(A \otimes (A' \otimes B), (C \otimes D') \otimes D)$.

We extend the standard diagram language for symmetric monoidal categories (Sect. 2), representing morphisms $f \in \mathsf{Hom}^\bullet(A \otimes B, C \otimes D)$ by *decorated boxes* as shown on the right,

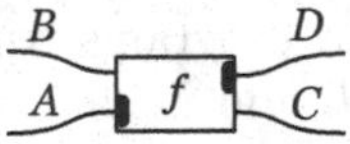

with black bars marking the *unguarded input* gates A and the *guarded output* gates D. Weakening then corresponds to shrinking the black bars of decorated boxes. Figure 2 depicts the above axioms in this language. Solid boxes represent the assumptions, while dashed boxes represent the conclusions. The latter only occur in the derivation process and do not form part of the actual diagrams representing concrete morphisms. We silently identify object expressions and sets of gates in diagrams. Given a (well-typed) morphism expression e, a judgement $e \in \mathsf{Hom}^\bullet(A \otimes B, C \otimes D)$, called a *guardedness typing* of e, is *derivable* if it can be derived from the assumed guardedness typing of the constituent basic boxes of e using the rules in Definition 1. We have an obvious notion of (directed) *paths* in diagrams; a path is *guarded* if it passes some basic box f through an unguarded input gate and a guarded output gate (intuitively, guardedness is then introduced along the path as the passage through f will guarantee guarded output without assuming guarded input). We then have the following geometric characterization of guardedness typing:

Theorem 2. *For a well-typed morphism expression* $e \in \mathsf{Hom}(A \otimes B, C \otimes D)$, *the guardedness typing* $e \in \mathsf{Hom}^\bullet(A \otimes B, C \otimes D)$ *is derivable iff in the diagram of* e*, every path from an input gate in* A *to an output gate in* D *is guarded.*

Every symmetric monoidal category has both a largest ($\mathsf{Hom}^\bullet(A \otimes B, C \otimes D) = \mathsf{Hom}(A \otimes B, C \otimes D)$) and a least guarded structure:

Lemma and Definition 3 (Vacuous guardedness). *Every symmetric monoidal category is guarded under taking* $f \in \mathsf{Hom}^\bullet(A \otimes B, C \otimes D)$ *iff* f *factors as*

$$A \otimes B \xrightarrow{\mathsf{id}_A \otimes g} A \otimes E \otimes D \xrightarrow{h \otimes \mathsf{id}_D} C \otimes D$$

(eliding associativity) with $g : B \to E \otimes D$, $h : A \otimes E \to C$. *This is the least guarded structure on* $\mathbf{C}$, *the* vacuous guarded structure.

E.g. the natural guarded structure on Hilbert spaces (Sect. 6) is vacuous.

Remark 4 (Duality). The rules and axioms in Fig. 2 are stable under 180°-rotation, that is, under reversing arrows and applying the monoidal symmetry on both sides (this motivates decorating the *unguarded* inputs). Consequently, if $\mathbf{C}$ is guarded, then so is the dual category $\mathbf{C}^{op}$, with guardedness given by $f \in \mathsf{Hom}^\bullet_{\mathbf{C}^{op}}(A \otimes B, C \otimes D)$ iff the obvious transpose of f is in $\mathsf{Hom}^\bullet_{\mathbf{C}}(D \otimes C, B \otimes A)$.

In case $\otimes$ is coproduct, we can simplify the description of partial guardedness:

Proposition 5. *Partial guardedness in a co-Cartesian category* $(\mathbf{C}, +, \emptyset)$ *is equivalently determined by distinguished subsets* $\mathsf{Hom}_\sigma(X, Y) \subseteq \mathsf{Hom}(X, Y)$ *with* σ *ranging over coproduct injections* $Y_2 \to Y_1 + Y_2 \cong Y$, *subject to the rules on the right hand side of Fig. 3, where* $f : X \to_\sigma Y$ *denotes* $f \in \mathsf{Hom}_\sigma(X, Y)$, *with* $f \in \mathsf{Hom}^\bullet(X_1 + X_2, Y_1 + Y_2)$ *iff* $(f \mathsf{in}_1) \in \mathsf{Hom}_{\mathsf{in}_2}(X_1, Y_1 + Y_2)$.

We have used the mentioned rules for $\to_\sigma$ in previous work on guarded iteration [16] (with **(vac$_\times$)** called **(trv)**, and together with weakening, which as indicated above turns out to be derivable). By duality (Remark 4), we immediately have a corresponding description for the Cartesian case:

Corollary 6. *Partial guardedness in a Cartesian category* $(\mathbf{C}, \times, 1)$ *is equivalently determined by distinguished subsets* $\mathsf{Hom}^\sigma(X, Y) \subseteq \mathsf{Hom}(X, Y)$ *with* σ *ranging over product projections* $X \cong X_1 \times X_2 \to X_1$, *subject to the rules on the left hand side of Fig. 3, where* $f : X \to^\sigma Y$ *denotes* $f \in \mathsf{Hom}^\sigma(X, Y)$, *with* $f \in \mathsf{Hom}^\bullet(X_1 \times X_2, Y_1 \times Y_2)$ *iff* $\mathsf{pr}_2 f \in \mathsf{Hom}^{\mathsf{pr}_1}(X_1 \times X_2, Y_2)$.

$$(\mathbf{vac}_\times)\ \frac{f : X \to Z}{f\,\mathsf{pr}_1 : X \times Y \to^{\mathsf{pr}_2} Z} \qquad (\mathbf{vac}_+)\ \frac{f : X \to Z}{\mathsf{in}_1 f : X \to_{\mathsf{in}_2} Z + Y}$$

$$(\mathbf{cmp}_\times)\ \frac{f : X \times Y \to^{\mathsf{pr}_2} Z \qquad g : V \to^\sigma X \qquad h : V \to Y}{f\langle g, h\rangle : V \to^\sigma Z} \qquad (\mathbf{cmp}_+)\ \frac{f : X \to_{\mathsf{in}_2} Y + Z \qquad g : Y \to_\sigma V \qquad h : Z \to V}{[g, h]\,f : X \to_\sigma V}$$

$$(\mathbf{par}_\times)\ \frac{f : X \to^\sigma Y \qquad g : X \to^\sigma Z}{\langle f, g\rangle : X \to^\sigma Y \times Z} \qquad (\mathbf{par}_+)\ \frac{f : X \to_\sigma Z \qquad f : Y \to_\sigma Z}{[f, g] : X + Y \to_\sigma Z}$$

Fig. 3. Axioms of Cartesian (left) and co-Cartesian (right) guarded categories

Remark 7. In a co-Cartesian category, vacuous guardedness (Lemma 3) can equivalently be described by $f \in \mathsf{Hom}^\bullet(A + B, C + D)$ iff f decomposes as $f = [\mathsf{in}_1 h, g]$ (uniquely provided that in_1 is monic), or in terms of the description from Proposition 5, $u \in \mathsf{Hom}_{\mathsf{in}_2}(X, Y + Z)$ iff u factors through in_1. Of course, the dual situation obtains in Cartesian categories.

Example 8 (Process algebra). Fix a monad $\mathbb{T}$ on $(\mathbf{C}, +, \emptyset)$ and an endofunctor $\Sigma : \mathbf{C} \to \mathbf{C}$ such that the generalized coalgebraic resumption transform $T_\Sigma = \nu\gamma.\, T(- + \Sigma\gamma)$ exists; we think of $T_\Sigma X$ as a type of processes that have side-effects in $\mathbb{T}$ and perform communication actions from Σ, seen as a generalized signature. The Kleisli category $\mathbf{C}_{\mathbb{T}_\Sigma}$ of $\mathbb{T}_\Sigma$ is again co-Cartesian. Putting

$$f : X \to_{\mathsf{in}_2} T_\Sigma(Y + Z) \iff \mathsf{out} f \in \{T(\mathsf{in}_1 + \mathsf{id})\,g \mid g : X \to T(Y + \Sigma T_\Sigma(Y + Z))\}$$

(cf. Sect. 2 for notation), we make $\mathbf{C}_{\mathbb{T}_\Sigma}$ into a guarded category [16]. The standard motivating example of finitely nondeterministic processes is obtained by taking $\mathbb{T} = \mathcal{P}_\omega$ (finite powerset monad) and $\Sigma = A \times -$ (action prefixing).

Example 9 (Metric spaces). Let $\mathbf{C}$ be the Cartesian category of metric spaces and non-expansive maps. Taking $f : X \times Y \to^{\mathsf{pr}_2} Z$ iff $\lambda y.\, f(x, y)$ is contractive for every $x \in X$ makes $\mathbf{C}$ into a guarded Cartesian category.

4 Guardedness via Guarded Ideals

Most of the time, the structure of a guarded category is determined by morphisms with only unguarded inputs and guarded outputs, which form an *ideal*:

Definition 10 (Guarded morphisms). A morphism $f : X \to Y$ in a guarded category is *guarded* (as opposed to only partially guarded) if $\upsilon_Y^{-1} f \hat{\upsilon}_X \in \mathsf{Hom}^\bullet(X \otimes I, I \otimes Y)$; we write $\mathsf{Hom}^\blacktriangleright(X, Y)$ for the set of guarded morphisms $f : X \to Y$.

Definition 11 (Guarded ideal). A family G of subsets $G(X, Y) \subseteq \mathsf{Hom}(X, Y)$ $(X, Y \in |\mathbf{C}|)$ in a monoidal category $(\mathbf{C}, \otimes, I)$ is a *guarded ideal* if it is closed under $\otimes$ and under composition with arbitrary $\mathbf{C}$-morphisms on both sides, and $G(I, I) = \mathsf{Hom}(I, I)$.

There is always a *least guarded ideal*, $G(X,Y) = \{gf \mid f : X \to I, g : I \to Y\}$. Moreover, as indicated above:

Lemma and Definition 12. *In a guarded category, the sets* $\mathsf{Hom}^{\blacktriangleright}(X,Y)$ *form a guarded ideal, the guarded ideal* induced *by the guarded structure.*

Conversely, it is clear that every guarded ideal *generates* a guarded structure by just closing under the rules of Definition 1.

Definition 13 (Ideally guarded category). A guarded category is *ideal* or *ideally guarded* (over G) if it is generated by some guarded ideal (G).

We give a more concrete description:

Theorem 14. *Let* $(\mathbf{C}, \otimes, I)$ *be ideally guarded over* G. *Then* $\mathsf{Hom}^{\bullet}(A \otimes B, C \otimes D)$ *consists of the morphisms of the form*

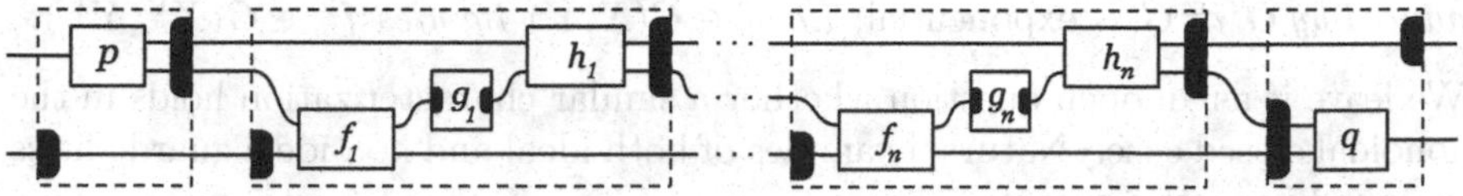

for g_i *in* G *and arbitrary* p, q, f_i, h_i.

The transitions between guarded ideals and guarded structures are not in general mutually inverse: The guarded structure generated the guarded ideal induced by a guarded structure may be smaller than the original one (Example 21), and the guarded ideal induced by the guarded structure generated by a guarded ideal G may be larger than G (Remark 16). We proceed to analyse details.

Proposition 15. *On every symmetric monoidal category, the least guarded structure (Lemma 3) is ideal.*

Remark 16. Vacuously guarded categories need not induce the least guarded ideal (although by the next results, this does hold in the Cartesian and the co-Cartesian case). In fact, by Lemma 3, the guarded ideal induced by the vacuous guarded structure consists of the morphisms of the form $(h \otimes \mathsf{id}_D)(\mathsf{id}_A \otimes g)$ (eliding associativity and the unitor) where $g : I \to E \otimes D$, $h : A \otimes E \to I$:

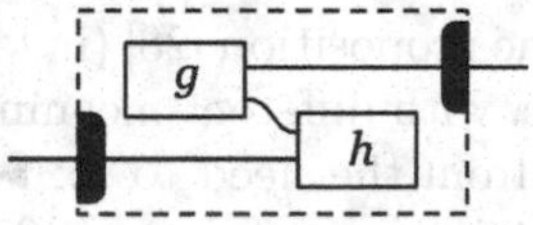

(1)

This ideal will resurface in the discussion of Hilbert spaces (Sect. 6).

The situation is simpler in the Cartesian and, dually, in the co-Cartesian case.

Lemma 17. *Let* $\mathbf{C}$ *be ideally guarded over* G, *and suppose that every* $f \in G(X \otimes Y, Z)$ *factors through* $\hat{f} \otimes \mathsf{id} : X \otimes Y \to V \otimes Y$ *for some* $\hat{f} \in G(X, V)$. *Then the guardedness structure of* $\mathbf{C}$ *induces* G.

If $\otimes = +$, the premise of the lemma is automatic, since $f \in G(X + Y, Z)$ can be represented as $[f\ \mathsf{in}_1, f\ \mathsf{in}_2] = [\mathsf{id}, f\ \mathsf{in}_2](f\ \mathsf{in}_1 + \mathsf{id})$ where $f\ \mathsf{in}_1 \in G(X, Z)$ by the closure properties of guarded ideals. Hence, we obtain

Theorem 18. *The guarded structure generated by a guarded ideal G on a co-Cartesian category is equivalently described by* $\mathsf{Hom}_{\mathsf{in}_2}(X, Y + Z) = \{[\mathsf{in}_1, g]h \mid g \in G(W, Y + Z), h : X \to Y + W\}$*, and hence induces G.*

Corollary 19. *The guarded structure generated by a guarded ideal G on a Cartesian category is equivalently described by* $\mathsf{Hom}^{\mathsf{pr}_1}(X \times Y, Z) = \{h\langle g, \mathsf{pr}_2\rangle \mid g \in G(X \times Y, W), h : W \times Y \to Z\}$*, and hence induces G.*

The description can be further simplified in the Cartesian closed case.

Corollary 20. *Given a guarded ideal G on a Cartesian closed category, put* $f : X \times Y \to^{\mathsf{pr}_1} Z$ *iff* $\mathsf{curry}\ f \in G(X, Z^Y)$*. This describes the guarded structure induced by G iff G is* exponential*, i.e.* $f \in G(X, Y)$ *implies* $f^V \in G(X^V, Y^V)$*.*

(We leave it as an open question whether a similar characterization holds in the monoidal closed case.) Natural examples of both ideal and non-ideal guardedness are found in metric spaces:

Example 21 (Metric spaces). The guarded structure on metric spaces from Example 9 fails to be ideal: It induces the guarded ideal of contractive maps, which however generates the (ideal) guarded structure described by $f : X \times Y \to^{\mathsf{pr}_2} Z$ iff $f(x, y)$ is *uniformly* contractive in y, i.e. there is $c < 1$ such that for every x, $\lambda y.\, f(x, y)$ is contractive with contraction factor c.

A large class of ideally guarded structures arises as follows.

Proposition 22. *Let* **C** *be a Cartesian category equipped with an endofunctor* $\blacktriangleright : \mathbf{C} \to \mathbf{C}$ *and a natural transformation* $\mathsf{next} : \mathrm{Id} \to \blacktriangleright$*. Then the following definition yields a guarded ideal in* **C**: $G(X, Y) = \{f\ \mathsf{next} \mid f : \blacktriangleright X \to Y\}$*. The arising guarded structure is* $\mathsf{Hom}^{\mathsf{pr}_1}(X \times Y, Z) = \{f\langle \mathsf{next}, \mathsf{pr}_2\rangle \mid f : \blacktriangleright(X \times Y) \times Y \to Z\}$*. If moreover* $\mathsf{next} : X \times Y \to \blacktriangleright(X \times Y)$ *factors through* $\mathsf{next} \times \mathsf{id} : X \times Y \to \blacktriangleright X \times Y$*, then* $\mathsf{Hom}^{\mathsf{pr}_1}(X \times Y, Z) = \{f(\mathsf{next} \times \mathsf{id}) \mid f : \blacktriangleright X \times Y \to Z\}$*.*

Remark 23. Proposition 22 connects our approach to previous work based precisely on the assumptions of the proposition [28] (in fact, the term guarded traced category is already used there, with different meaning). A limitation of the approach via a functor $\blacktriangleright$ arises from the need to fix $\blacktriangleright$ globally, so that, e.g., the ideal guarded structure on metric spaces (Example 21) is not covered – capturing contractivity via $\blacktriangleright$ requires fixing a single global contraction factor.

The following instance of Proposition 22 has received extensive recent interest in programming semantics:

Example 24 (Topos of Trees). Let **C** be the *topos of trees* [7], i.e. the presheaf category $\mathbf{Set}^{\omega^{op}}$ where ω is the preorder of natural numbers (starting from 1) ordered by inclusion. An object X of **C** is thus a family $(X(n))_{n=1,2...}$

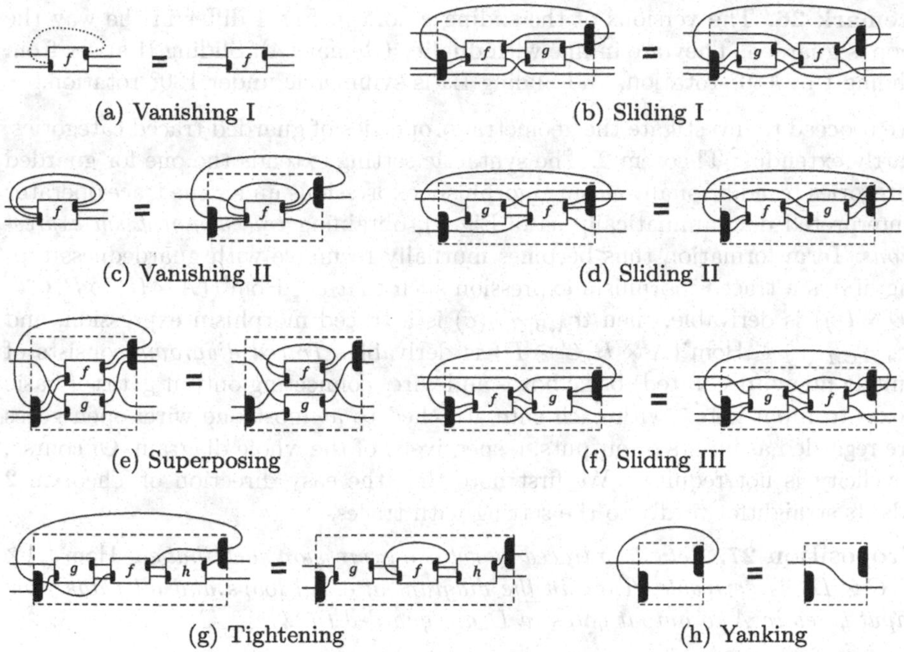

(a) Vanishing I (b) Sliding I (c) Vanishing II (d) Sliding II (e) Superposing (f) Sliding III (g) Tightening (h) Yanking

Fig. 4. Axioms of guarded traced categories

of sets with restriction maps $r_n : X(n+1) \to X(n)$. The *later*-endofunctor $\blacktriangleright : \mathbf{C} \to \mathbf{C}$ is defined by $\blacktriangleright X(1) = \{\star\}$ and $\blacktriangleright X(n+1) = X(n)$, and the natural transformation $\mathsf{next}_X : X \to \blacktriangleright X$ by $\mathsf{next}_X(1) = ! : X(1) \to \{\star\}$, $\mathsf{next}_X(n+1) = r_{n+1} : X(n+1) \to X(n)$. Guarded morphisms according to Proposition 22 are called *contractive*, generalizing the metric setup. Contractive morphisms form an exponential ideal, so partial guardedness is described as in Corollary 20, and hence agrees with contractivity in part of the input as in [7, Definition 2.2].

5 Guarded Traces

As indicated previously, the main purpose of our notion of abstract guardedness is to enable fine-grained control over the formation of feedback loops, viz, *traces.*

Definition 25 (Guarded traced category). We call a guarded category $(\mathbf{C}, \otimes, I)$ *guarded traced* if it is equipped with a *guarded trace operator*

$$\mathsf{tr}^U_{A,B,C,D} : \mathsf{Hom}^\bullet((A \otimes U) \otimes B, C \otimes (D \otimes U)) \to \mathsf{Hom}^\bullet(A \otimes B, C \otimes D),$$

visually corresponding to the diagram formation rule in Fig. 1, so that the adaptation of the Joyal-Street-Verity axiomatization of traced symmetric monoidal categories [22] shown in Fig. 4 is satisfied.

Remark 26. The versions of the sliding axiom in Fig. 4 differ in the way the loop is guarded. They are in line with duality (Remark 4): Sliding II arises from Sliding I by 180° rotation, and Sliding III is symmetric under 180° rotation.

We proceed to investigate the geometric properties of guarded traced categories, partly extending Theorem 2. The syntactic setting extends the one for guarded categories by additionally closing morphism expressions under the trace operator (interpreted diagrammatically as in Fig. 1), obtaining *traced morphism expressions.* Term formation thus becomes mutually recursive with guardedness typing: if e is a traced morphism expression such that $e \in \mathsf{Hom}^{\bullet}((A \otimes U) \otimes B, C \otimes (D \otimes U))$ is derivable, then $\mathsf{tr}_{A,B,C,D}(e)$ is a traced morphism expression, and $\mathsf{tr}_{A,B,C,D}(e) \in \mathsf{Hom}^{\bullet}(A \otimes B, C \otimes D)$ is derivable. *Traced diagrams* consists of finitely many (decorated) basic boxes and wires connecting output gates of basic boxes to input gates, with each gate attached to at most one wire; open gates are regarded as inputs or outputs, respectively, of the whole diagram. Of course, acyclicity is not required. We first note that the easy direction of Theorem 2 adapts straightforwardly to the setting with traces:

Proposition 27. *Let* e *be a traced morphism expression such that* $e \in \mathsf{Hom}^{\bullet}(A \otimes B, C \otimes D)$ *is derivable. Then in the diagram of* e*, all loops and all paths from input gates in* A *to output gates in* D *are guarded (p. 4).*

Remarkably, the converse of Proposition 27 in general fails in several ways:

Example 28. The left diagram below

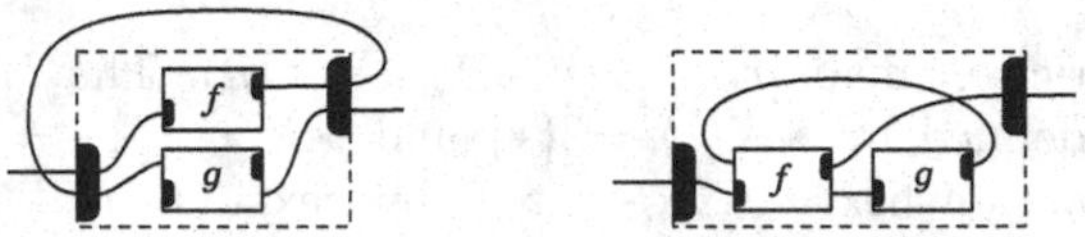

(2)

shows that guardedness typing is not closed under equality of traced morphism expressions: Write e for the expression inducing the dashed box. By Proposition 27, e, and hence $\mathsf{tr}(e)$, fail to type as indicated. However, $\mathsf{tr}(e) = gf$, for which the overall guardedness typing indicated is easily derivable.

Moreover, the diagram on the right above satisfies the necessary condition from Proposition 27 but is not induced by an expression for which the indicated guardedness typing is derivable, essentially because both ways of cutting the loop violate the necessary condition from Proposition 27.

However, if $\mathbf{C}$ is ideally guarded over a guarded ideal G, we do have a converse to Proposition 27: By Theorem 14, we can then restrict basic boxes in diagrams to be either *guarded*, i.e. have only black gates, or *unguarded*, i.e. have only white gates. We call the correspondingly restricted diagrams *ideally guarded.* (We emphasize that the guardedness typing of *composite* ideally guarded diagrams still needs to mix guarded and unguarded inputs and outputs.) A path in an ideally guarded diagram is guarded iff it passes through a guarded basic box.

The left-hand diagram in (2) is in fact ideally guarded, so guardedness typing fails to be closed under equality also in the ideally guarded case. However, for ideally guarded diagrams we have the following converse of Proposition 27.

Theorem 29. *Let Δ be an ideally guarded diagram, with sets of input and output gates disjointly decomposed as $A \uplus B$ and $C \uplus D$, respectively. If every loop in Δ and every path from a gate in A to a gate in D is guarded, then Δ is induced by a traced morphism expression e such that $e \in \mathsf{Hom}^{\bullet}(A \otimes B, C \otimes D)$ is derivable.*

We next take a look at the Cartesian and co-Cartesian cases. Recall that by Proposition 5, the definition of guarded category can be simplified if $\otimes = +$ (and dually if $\otimes = \times$). This simplification extends to guarded traced categories by generalizing Hyland-Hasegawa's equivalence between Cartesian trace operators and Conway fixpoint operators [18,19].

Definition 30 (Guarded Conway operators). Let $\mathbf{C}$ be a guarded co-Cartesian category. We call an operator $(-)^{\dagger}$ of profile

$$f \in \mathsf{Hom}_{\sigma+\mathsf{id}}(X, Y + X) \mapsto f^{\dagger} \in \mathsf{Hom}_{\sigma}(X, Y) \tag{3}$$

a *guarded iteration operator* if it satisfies

- *fixpoint:* $f^{\dagger} = [\mathsf{id}, f^{\dagger}]\, f$ for $f : X \to_{\mathsf{in}_2} Y + X$;

and a *Conway iteration operator* if it additionally satisfies

- *naturality:* $g f^{\dagger} = ((g + \mathsf{id})\, f)^{\dagger}$ for $f : X \to_{\mathsf{in}_2} Y + X$, $g : Y \to Z$;
- *dinaturality:* $([\mathsf{in}_1, h]\, g)^{\dagger} = [\mathsf{id}, ([\mathsf{in}_1, g]\, h)^{\dagger}]\, g$ for $g : X \to_{\mathsf{in}_2} Y + Z$ and $h : Z \to Y + X$ or $g : X \to Y + Z$ and $h : Z \to_{\mathsf{in}_2} Y + X$;
- *(co)diagonal:* $([\mathsf{id}, \mathsf{in}_2]\, f)^{\dagger} = f^{\dagger\dagger}$ for $f : X \to_{\mathsf{in}_2 + \mathsf{id}} (Y + X) + X$.

Furthermore, we distinguish the following principles:

- *squaring* [12]*:* $f^{\dagger} = ([\mathsf{in}_1, f]\, f)^{\dagger}$ for $f : X \to_{\mathsf{in}_2} Y + X$;
- *uniformity w.r.t. a subcategory* $\mathbf{S}$ *of* $\mathbf{C}$: $(\mathsf{id} + h)\, f = g\, h$ implies $f^{\dagger} = g^{\dagger} h$ for all $f : X \to_{\mathsf{in}_2} Z + X$, $g : Y \to_{\mathsf{in}_2} Z + Y$ and $h : Y \to X$ from $\mathbf{S}$;

and call $(-)^{\dagger}$ *squarable* or *uniform* if it satisfies squaring or uniformity, respectively.

Guarded (Conway) recursion operators $(-)_{\dagger}$ on guarded Cartesian categories are defined dually in a straightforward manner. We collect the following facts about guarded iteration operators for further reference.

Lemma 31. *Let $(-)^{\dagger}$ be a guarded iteration operator on $(\mathbf{C}, +, \emptyset)$.*

1. *If $(-)^{\dagger}$ is uniform w.r.t. some co-Cartesian subcategory of $\mathbf{C}$ and satisfies the codiagonal identity then it is squarable.*
2. *If $(-)^{\dagger}$ is squarable and uniform w.r.t. coproduct injections then it is dinatural.*
3. *If $(-)^{\dagger}$ is Conway then it is uniform w.r.t. coproduct injections.*

Proposition 32. *A guarded co-Cartesian category $\mathbf{C}$ is traced iff it is equipped with a guarded Conway iteration operator $(-)^{\dagger}$, with mutual conversions like in the total case [18,19].*

Example 33 (Guarded Conway operators). We list some examples of guarded Conway iteration/recursion operators. In all cases except 2, Conwayness follows from uniqueness of fixpoints [16, Theorem 17].

1. In a vacuously guarded co-Cartesian category (Remark 7), $f : X \to_{\mathsf{in}_2} Y + Z$ iff $f = \mathsf{in}_1 g$ for some $g : X \to Y$. If coproduct injections are monic, then g is uniquely determined, and $f^\dagger = g$ defines a guarded Conway operator.
2. Every Cartesian category $\mathbf{C}$ is guarded under $\mathsf{Hom}^\pi(X, Y) = \mathsf{Hom}(X, Y)$ (making every morphism guarded). Then $\mathbf{C}$ has a guarded Conway recursion operator iff $\mathbf{C}$ is a *Conway category* [13], i.e. models standard total recursion.
3. The guarded Cartesian category of complete metric spaces as in Example 9 is traced: For $f : X \times Y \to^{\mathsf{pr}_2} Y$, define $f^\dagger(x)$ as the unique fixpoint of $\lambda y.\, f(x, y)$ according to Banach's fixpoint theorem.
4. Similarly, the topos of trees, ideally guarded as in Example 24, has a guarded Conway recursion operator obtained by taking unique fixpoints [7, Theorem 2.4].
5. The guarded co-Cartesian category $\mathbf{C}_{\mathbb{T}_\Sigma}$ of side-effecting processes (Example 8) has a guarded Conway iteration operator obtained by taking unique fixpoints, thanks to the universal property of the final coalgebra $T_\Sigma X$ [32].

Guarded vs. Unguarded Recursion. We proceed to present a class of examples relating guarded and unguarded recursion. For motivation, consider the category $(\mathbf{Cpo}, \times, 1)$ of complete partial orders (cpos) and continuous maps. This category nearly supports recursion via least fixpoints, except that, e.g., $\mathsf{id} : X \to X$ only has a least fixpoint if X has a bottom. The following equivalent approaches involve the *lifting monad* $(-)_\bot$, which adjoins a fresh bottom $\bot$ to a given $X \in |\mathbf{Cpo}|$.

Classical approach [38,39]: Define a total recursion operator $(-)_\ddagger$ on the category $\mathbf{Cpo}_\bot$ of *pointed cpos* and continuous maps, using least fixpoints.

Guarded approach (cf. [28]): Extend $\mathbf{Cpo}$ to a guarded category: $f : X \times Y \to^{\mathsf{pr}_2} Z$ iff $f \in \{g(\mathsf{id} \times \eta) \mid g : X \times Y_\bot \to Z\}$ (see Proposition 22), and define a guarded recursion operator sending $f = g(\mathsf{id} \times \eta) : Y \times X \to^{\mathsf{pr}_2} X$ to $f_\dagger = g\langle \mathsf{id}, \hat{f}\rangle : Y \to X$ with $\hat{f}(y) \in X_\bot$ calculated as the least fixpoint of $\lambda z.\, \eta g(y, z)$.

Pointed cpos happen to be always of the form $X_\bot$ with $X \in |\mathbf{Cpo}|$, which indicates that $(-)_\ddagger$ is a special case of $(-)_\dagger$. This is no longer true in more general cases when the connection between $(-)_\ddagger$ and $(-)_\dagger$ is more intricate. We show that $(-)_\ddagger$ and $(-)_\dagger$ are nevertheless equivalent under reasonable assumptions.

Definition 34 ([11]). A *let-ccc with a fixpoint object* is a tuple $(\mathbf{C}, \mathbb{T}, \Omega, \omega)$, consisting of a Cartesian closed category $\mathbf{C}$, a strong monad $\mathbb{T}$ on it, an initial T-algebra (Ω, in) and an equalizer $\omega : 1 \to \Omega$ of $\mathsf{in}\,\eta : \Omega \to \Omega$ and $\mathsf{id} : \Omega \to \Omega$.

The key requirement is the last one, satisfied, e.g., for $\mathbf{Cpo}$ and the lifting monad. Given a monad $\mathbb{T}$ on $\mathbf{C}$, $\mathbf{C}^{\mathbb{T}}_\star$ denotes the category of $\mathbb{T}$-algebras and $\mathbf{C}$-morphisms (instead of $\mathbb{T}$-algebra homomorphisms).

Proposition 35 ([37, Theorem 4.6]). *Let $(\mathbf{C}, \mathbb{T}, \Omega, \omega)$ be a let-ccc with a fixpoint object. Then $\mathbf{C}^{\mathbb{T}}_{\star}$ has a unique $\mathbf{C}^{\mathbb{T}}$-uniform recursion operator $(-)_{\ddagger}$.*

By [38, Theorem 4], the operator $(-)_{\ddagger}$ in Proposition 35 is Conway, in particular, by Lemma 31, squarable, if $\mathbf{C}$ has a natural numbers object and $\mathbb{T}$ is an *equational lifting monad* [10], such as $(-)_{\bot}$. There are however further squarable operators obtained via Proposition 35, e.g. for the partial state monad $TX = (X \times S)^S_{\bot}$ [11]. By Lemma 31, the following result applies in particular in the setup of Proposition 35 under the additional assumption of squarability.

Theorem 36. *Let $\mathbb{T}$ be a strong monad on a Cartesian category $\mathbf{C}$. The following gives a bijective correspondence between squarable dinatural recursive operators $(-)_{\ddagger}$ on $\mathbf{C}^{\mathbb{T}}_{\star}$ and squarable dinatural guarded recursive operators $(-)_{\dagger}$ on $\mathbf{C}$ ideally guarded over $\mathsf{Hom}^{\blacktriangleright}(X,Y) = \{f\eta \mid f : TX \to Y\}$:*

$$(f : B \times A \to A)_{\ddagger} = a(\eta f(\mathsf{id} \times a))_{\dagger} \qquad \textit{for } (A,a) \in |\mathbf{C}^{\mathbb{T}}_{\star}| \tag{4}$$

$$(f = g(\mathsf{id} \times \eta) : Y \times X \to X)_{\dagger} = g\langle \mathsf{id}, (\eta g)_{\ddagger}\rangle \tag{5}$$

(in (5) we call on a slight extension of $(-)_{\ddagger}$; the right hand side of (4) is defined because $\eta f(\mathsf{id} \times a)$ factors as $\eta f(\mathsf{id} \times a(Ta)\eta)$). Moreover, $(-)_{\dagger}$ is Conway iff so is $(-)_{\ddagger}$.

6 Vacuous Guardedness and Nuclear Ideals

We proceed to discuss traces in vacuously guarded categories (Lemma 3), and show that the partial trace operation in the category of (possibly infinite-dimensional) Hilbert spaces [2] in fact lives over the vacuous guarded structure. We first note that vacuous guarded structures are traced as soon as a simple rewiring operation satisfies a suitable well-definedness condition (similar to one defining traced nuclear ideals [2, Definition 8.14]):

Proposition 37. *Let $(\mathbf{C}, \otimes, I)$ be vacuously guarded. If for $f \in \mathsf{Hom}^{\bullet}(A \otimes B, C \otimes D)$ with factorization $f = (h \otimes \mathsf{id}_{D\otimes U})(\mathsf{id}_{A\otimes U} \otimes g)$ (eliding associativity), $g : B \to E \otimes D \otimes U$, $h : A \otimes U \otimes E \to C$ as per Lemma 3, the composite*

$$A \otimes B \xrightarrow{\mathsf{id}_A \otimes g} A \otimes E \otimes D \otimes U \cong A \otimes U \otimes E \otimes D \xrightarrow{h \otimes \mathsf{id}_D} C \otimes D \tag{6}$$

depends only on f, then $\mathbf{C}$ is guarded traced, with $\mathsf{tr}^U_{A,B,C,D}(f)$ defined as (6).

Diagrammatically, the trace in a vacuously guarded category is thus given by

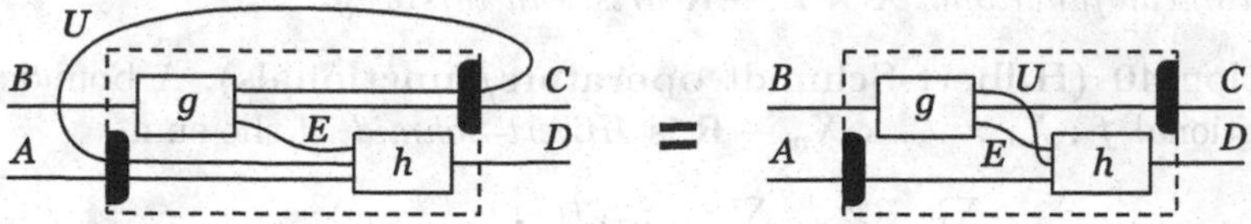

We proceed to instantiate the above to Hilbert spaces. On a more abstract level, a *dagger symmetric monoidal category* [35] (or *tensored $*$-category* [2]) is a symmetric monoidal category $(\mathbf{C}, \otimes, I)$ equipped with an identity-on-objects strictly involutive functor $(-)^\dagger : \mathbf{C} \to \mathbf{C}^{op}$ coherently preserving the symmetric monoidal structure. The main motivation for dagger symmetric monoidal categories is to capture categories that are similar to (dagger) compact closed categories in that they admit a canonical trace construction for certain morphisms, but fail to be closed, much less compact closed. The "compact closed part" of a dagger symmetric monoidal category is axiomatized as follows.

Definition 38 (Nuclear Ideal, [2]). A *nuclear ideal* N in a dagger symmetric monoidal category $(\mathbf{C}, \otimes, I, (-)^\dagger)$ is a family of subsets $\mathsf{N}(X, Y) \subseteq \mathsf{Hom}_{\mathbf{C}}(X, Y)$, $X, Y \in |\mathbf{C}|$, satisfying the following conditions:

1. N is closed under $\otimes$, $(-)^\dagger$, and composition with arbitrary morphisms on both sides;
2. There is a bijection $\theta : \mathsf{N}(X, Y) \to \mathsf{Hom}_{\mathbf{C}}(I, X^\dagger \otimes Y)$, natural in X and Y, coherently preserving the dagger symmetric monoidal structure.
3. (*Compactness*) For $f \in \mathsf{N}(B, A)$ and $g \in \mathsf{N}(B, C)$, the following diagram commutes:

$$\begin{array}{ccccc} A & \xrightarrow{\cong} & A \otimes I & \xrightarrow{\mathsf{id}_A \otimes \theta(g)} & A \otimes (B^\dagger \otimes C) \\ {\scriptstyle g f^\dagger}\downarrow & & & & \downarrow{\scriptstyle \cong} \\ C & \xleftarrow{\cong} & I \otimes C & \xleftarrow[(\theta(f))^\dagger \otimes \mathsf{id}_C]{} & (B^\dagger \otimes A) \otimes C \end{array}$$

The above definition is slightly simplified in that we elide a covariant involutive functor $\overline{(-)} : \mathbf{C} \to \mathbf{C}$, capturing, e.g. complex conjugation; i.e., we essentially restrict to spaces over the reals.

We proceed to present a representative example of a nuclear ideal in the category of Hilbert spaces. Recall that a *Hilbert space* [23] H over the field R of reals is a vector space with an *inner product* $\langle -, - \rangle : H \times H \to \mathsf{R}$ that is complete as a *normed space* under the induced *norm* $\|x\| = \sqrt{\langle x, x \rangle}$. Let **Hilb** be the category of Hilbert spaces and bounded linear operators.

Clearly, R itself is a Hilbert space; linear operators $X \to \mathsf{R}$ are conventionally called *functionals.* More generally, we consider *(multi-)linear* functionals $X_1 \times \ldots \times X_n \to \mathsf{R}$, i.e. maps that are linear in every argument. Such a functional is *bounded* if $|f(x_1, \ldots, x_n)| \leqslant c\|x_1\| \cdots \|x_n\|$ for some constant $c \in \mathsf{R}$. We can move between bounded linear operators and bounded linear functionals, similarly as we can move between relations and functions to the Booleans:

Proposition 39 ([23, Theorem 2.4.1])**.** *Given a bounded linear operator $f : X \to Y$, $f^\circ(x, y) = \langle fx, y \rangle$ defines a bounded linear functional f°, and every bounded linear functional $X \times Y \to \mathsf{R}$ arises in this way.*

Definition 40 (Hilbert-Schmidt operators/functionals). A bounded linear functional $f : X_1 \times \ldots \times X_n \to \mathsf{R}$ is *Hilbert-Schmidt* if the sum

$$\textstyle\sum_{x_1 \in B_1} \cdots \sum_{x_n \in B_n} (f(x_1, \ldots, x_n))^2$$

is finite for some, and then any, orthonormal bases $B_1, \ldots, B_n$ of $X_1, \ldots, X_n$, respectively. A bounded linear operator $f : X \to Y$ is *Hilbert-Schmidt* if the induced functional f° (Proposition 39) is Hilbert-Schmidt, equivalently if $\sum_{x \in B} \|fx\|^2$ is finite for some, and then any, orthonormal basis B of X. We denote by $\mathsf{HS}(X, Y)$ the space of all Hilbert-Schmidt operators from X to Y.

For $X, Y \in |\mathbf{Hilb}|$, the space of Hilbert-Schmidt functionals $X \times Y \to \mathbf{R}$ is itself a Hilbert space, denoted $X \otimes Y$, with the pointwise vector space structure and the inner product $\langle f, g \rangle = \sum_{x \in B} \sum_{y \in B'} f(x, y)\, g(x, y)$ where B and B' are orthonormal bases of X and Y, respectively. By virtue of the equivalence between f and f°, this induces a Hilbert space structure on $\mathsf{HS}(X, Y)$, with induced norm $\|f\|_2 = \sqrt{\sum_{x \in B} \|fx\|^2}$. The operator $\otimes$ forms part of a dagger symmetric monoidal structure on $\mathbf{Hilb}$, with unit $\mathbf{R}$. For a bounded linear operator $f : X \to Y$, $f^\dagger : Y \to X$ is the *adjoint operator* uniquely determined by equation $\langle x, f^\dagger y \rangle = \langle fx, y \rangle$. The tensor product of $f : A \to B$ and $g \colon C \to D$ is the functional sending $h : A \times C \to \mathbf{R}$ to $h(f^\dagger \times g^\dagger) : B \times D \to \mathbf{R}$. Given $a \in A$ and $c \in C$, let us denote by $a \otimes c \in A \otimes C$ the functional $(a', c') \mapsto \langle a, a' \rangle \langle c, c' \rangle$, and so, with the above f and g, $(f \otimes g)(a \otimes c) = f(a) \otimes g(c)$.

Proposition 41 ([2]). *The Hilbert-Schmidt operators form a nuclear ideal in* $\mathbf{Hilb}$ *with* $\theta : \mathsf{HS}(X, Y) \cong \mathsf{Hom}(\mathbf{R}, X^\dagger \otimes Y)$ *defined by*

$$\theta(f : X \to Y)(r : \mathbf{R})(x : X, y : Y) = r \langle fx, y \rangle.$$

A crucial fact underlying the proof of Proposition 41 is that $\mathsf{HS}(X, Y)$ is isomorphic to $X^\dagger \otimes Y$, naturally in X and Y. We emphasize that what makes the case of $\mathbf{Hilb}$ significant is that we do not restrict to finite-dimensional Hilbert spaces. In that case all bounded linear operators would be Hilbert-Schmidt and the corresponding category would be (dagger) compact closed [35]. In the infinite-dimensional case, identities need not be Hilbert-Schmidt, so HS is indeed only an ideal and not a subcategory.

Let $\mathsf{N}^2(X, Y) = \{g^\dagger h : X \to Y \mid h \in \mathsf{N}(X, Z), g \in \mathsf{N}(Y, Z)\}$ for any nuclear ideal N. The main theorem of the section now can be stated as follows.

Theorem 42. *1. The guarded ideal induced by the vacuous guarded structure on* $\mathbf{Hilb}$ *(see (1)) is precisely* HS^2*, and* $\mathbf{Hilb}$ *is guarded traced over* HS^2*.*
2. Guarded traces in $\mathbf{Hilb}$ *commute with* $(-)^\dagger$ *in the sense that if* $f \in \mathsf{Hom}^\bullet((A \otimes U) \otimes B, C \otimes (D \otimes U))$*, then* $\gamma_{B, A \otimes U} f^\dagger \gamma_{D \otimes U, C} \in \mathsf{Hom}^\bullet((D \otimes U) \otimes C, B \otimes (A \otimes U))$ *and* $\mathsf{tr}^U_{D,C,B,A}(\gamma_{B, A \otimes U} f^\dagger \gamma_{D \otimes U, C}) = \gamma_{A,B} (\mathsf{tr}^U_{A,B,C,D}(f))^\dagger \gamma_{C,D}$*.*

Clause 1 is a generalization of the result in [2, Theorem 8.16] to parametrized traces. Specifically, we obtain agreement with the conventional mathematical definition of trace: given $f \in \mathsf{HS}^2(X, X)$, $\mathsf{tr}(f) = \sum_i \langle f(e_i), e_i \rangle$ for any choice of an orthonormal basis $(e_i)_i$, and $\mathsf{HS}^2(X, X)$ contains precisely those f for which this sum is absolutely convergent independently of the basis.

7 Conclusions and Further Work

We have presented and investigated a notion of abstract *guardedness* and guarded *traces*, focusing on foundational results and important classes of examples. We have distinguished a more specific notion of *ideal guardedness*, which in many respects appears to be better behaved than the unrestricted one, in particular ensures closer agreement between structural and geometric guardedness. An unexpectedly prominent role is played by 'vacuous' guardedness, characterized by the absence of paths connecting unguarded inputs to guarded outputs; e.g., partial traces in Hilbert spaces [2] turn out to be based on this form of guardedness. Further research will concern a coherence theorem for guarded traced categories generalizing the well-known unguarded case [22,34], and a generalization of the Int-construction [22], which would relate guarded traced categories to a suitable guarded version of compact closed categories. Also, we plan to investigate guarded traced categories as a basis for generalized Hoare logics, extending and unifying previous work [5,15].

References

1. Abel, A., Pientka, B.: Wellfounded recursion with copatterns: a unified approach to termination and productivity. In: International Conference on Functional Programming, ICFP 2013, pp. 185–196. ACM (2013)
2. Abramsky, S., Blute, R., Panangaden, P.: Nuclear and trace ideals in tensored*-categories. J. Pure Appl. Algebra **143**, 3–47 (1999)
3. Abramsky, S., Coecke, B.: A categorical semantics of quantum protocols. In: Logic in Computer Science, LICS 2004, pp. 415–425. IEEE Computer Society (2004)
4. Abramsky, S., Haghverdi, E., Scott, P.: Geometry of interaction and linear combinatory algebras. Math. Struct. Comput. Sci. **12**(5), 625–665 (2002)
5. Arthan, R., Martin, U., Mathiesen, E., Oliva, P.: A general framework for sound and complete Floyd-Hoare logics. ACM Trans. Comput. Log. **11**, 7:1–7:31 (2009)
6. Baeten, J., Basten, T., Reniers, M.: Process Algebra: Equational Theories of Communicating Processes. Cambridge University Press, Cambridge (2010)
7. Birkedal, L., Møgelberg, R., Schwinghammer, J., Støvring, K.: First steps in synthetic guarded domain theory: step-indexing in the topos of trees. Log. Methods Comput. Sci. **8**(4:1), 1–45 (2012)
8. Blute, R., Cockett, R., Seely, R.: Feedback for linearly distributive categories: traces and fixpoints. J. Pure Appl. Algebra **154**, 27–69 (2000)
9. Book, R., Greibach, S.: Quasi-realtime languages. Math. Syst. Theory **4**(2), 97–111 (1970)
10. Bucalo, A., Führmann, C., Simpson, A.: An equational notion of lifting monad. Theoret. Comput. Sci. **294**, 31–60 (2003)
11. Crole, R., Pitts, A.: New foundations for fixpoint computations. In: Logic in Computer Science, LICS 1990, pp. 489–497. IEEE Computer Society (1990)
12. Ésik, Z.: Axiomatizing iteration categories. Acta Cybern. **14**(1), 65–82 (1999)
13. Ésik, Z.: Equational properties of fixed point operations in Cartesian categories: an overview. In: Italiano, G.F., Pighizzini, G., Sannella, D.T. (eds.) MFCS 2015. LNCS, vol. 9234, pp. 18–37. Springer, Heidelberg (2015). https://doi.org/10.1007/978-3-662-48057-1_2

14. Girard, J.-Y.: Towards a geometry of interaction. Contemp. Math. **92**(69–108), 6 (1989)
15. Goncharov, S., Schröder, L.: A relatively complete generic Hoare logic for order-enriched effects. In: Proceedings of 28th Annual Symposium on Logic in Computer Science (LICS 2013), pp. 273–282. IEEE (2013)
16. Goncharov, S., Schröder, L., Rauch, C., Piróg, M.: Unifying guarded and unguarded iteration. In: Esparza, J., Murawski, A.S. (eds.) FoSSaCS 2017. LNCS, vol. 10203, pp. 517–533. Springer, Heidelberg (2017). https://doi.org/10.1007/978-3-662-54458-7_30
17. Haghverdi, E., Scott, P.: Towards a typed geometry of interaction. Math. Struct. Comput. Sci. **20**, 473–521 (2010)
18. Hasegawa, M.: Recursion from cyclic sharing: traced monoidal categories and models of cyclic lambda calculi. In: de Groote, P., Roger Hindley, J. (eds.) TLCA 1997. LNCS, vol. 1210, pp. 196–213. Springer, Heidelberg (1997). https://doi.org/10.1007/3-540-62688-3_37
19. Hasegawa, M.: Models of Sharing Graphs: A Categorical Semantics of let and letrec. Distinguished Dissertations. Springer, London (1999). https://doi.org/10.1007/978-1-4471-0865-8
20. Jeffrey, A.: Premonoidal categories and flow graphs. In: Higher-Order Operational Techniques in Semantics, HOOTS 1997, vol. 10 of ENTCS, p. 51. Elsevier (1997)
21. Jeffrey, A.: LTL types FRP: linear-time temporal logic propositions as types, proofs as functional reactive programs. In: Programming Languages Meets Program Verification, PLPV 2012, pp. 49–60. ACM (2012)
22. Joyal, A., Street, R., Verity, D.: Traced monoidal categories. Math. Proc. Camb. Philos. Soc. **119**, 447–468 (1996)
23. Kadison, R., Ringrose, J.: Fundamentals of the Theory of Operator Algebras: Advanced Theory, vol. 2. AMS (1997)
24. Krishnaswami, N., Benton, N.: Ultrametric semantics of reactive programs. In: Logic in Computer Science, LICS 2011, pp. 257–266. IEEE Computer Society (2011)
25. MacLane, S.: Categories for the Working Mathematician. Springer, New York (1971). https://doi.org/10.1007/978-1-4612-9839-7
26. Malherbe, O., Scott, P.J., Selinger, P.: Partially traced categories. J. Pure Appl. Algebra **216**, 2563–2585 (2012)
27. Milius, S.: Completely iterative algebras and completely iterative monads. Inf. Comput. **196**, 1–41 (2005)
28. Milius, S., Litak, T.: Guard your daggers and traces: properties of guarded (co-)recursion. Fund. Inf. **150**, 407–449 (2017)
29. Milner, R.: Communication and Concurrency. Prentice-Hall Inc., Upper Saddle River (1989)
30. Møgelberg, R.: A type theory for productive coprogramming via guarded recursion. In: Computer Science Logic/Logic in Computer Science, CSL-LICS 2014, pp. 71:1–71:10. ACM (2014)
31. Moggi, E.: Notions of computation and monads. Inf. Comput. **93**, 55–92 (1991)
32. Piróg, M., Gibbons, J.: The coinductive resumption monad. In: Mathematical Foundations of Programming Semantics, MFPS 2014. ENTCS, vol. 308, pp. 273–288 (2014)
33. Rutten, J.: Universal coalgebra: a theory of systems. Theoret. Comput. Sci. **249**, 3–80 (2000)
34. Selinger, P.: Towards a quantum programming language. Math. Struct. Comput. Sci. **14**, 527–586 (2004)

35. Selinger, P.: Dagger compact closed categories and completely positive maps. In: Quantum Programming Languages, QPL 2005. ENTCS, vol. 170, pp. 139–163. Elsevier (2007)
36. Selinger, P.: A survey of graphical languages for monoidal categories. In: Coecke, B. (ed.) New Structures for Physics. Lecture Notes in Physics, vol. 813, pp. 289–355. Springer, Heidelberg (2010). https://doi.org/10.1007/978-3-642-12821-9_4
37. Simpson, A.: Recursive types in Kleisli categories. Technical report, University of Edinburgh (1992)
38. Simpson, A., Plotkin, G.: Complete axioms for categorical fixed-point operators. In: Logic in Computer Science, LICS 2000, pp. 30–41 (2000)
39. Winskel, G.: The Formal Semantics of Programming Languages. MIT Press, Cambridge (1993)

Proper Semirings and Proper Convex Functors

Ana Sokolova[1(✉)] and Harald Woracek[2]

[1] University of Salzburg, Salzburg, Austria
ana.sokolova@cs.uni-salzburg.at
[2] TU Vienna, Vienna, Austria
harald.woracek@tuwien.ac.at

Abstract. Esik and Maletti introduced the notion of a proper semiring and proved that some important (classes of) semirings – Noetherian semirings, natural numbers – are proper. Properness matters as the equivalence problem for weighted automata over a semiring which is proper and finitely and effectively presented is decidable. Milius generalised the notion of properness from a semiring to a functor. As a consequence, a semiring is proper if and only if its associated "cubic functor" is proper. Moreover, properness of a functor renders soundness and completeness proofs for axiomatizations of equivalent behaviour.

In this paper we provide a method for proving properness of functors, and instantiate it to cover both the known cases and several novel ones: (1) properness of the semirings of positive rationals and positive reals, via properness of the corresponding cubic functors; and (2) properness of two functors on (positive) convex algebras. The latter functors are important for axiomatizing trace equivalence of probabilistic transition systems. Our proofs rely on results that stretch all the way back to Hilbert and Minkowski.

Keywords: Proper semirings · Proper functors · Coalgebra
Weighted automata · Probabilistic transition systems

1 Introduction

In this paper we deal with algebraic categories and deterministic weighted automata functors on them. Such categories are the target of generalized determinization [10,22,23] and enable coalgebraic modelling beyond sets. For example, non-deterministic automata, weighted, or probabilistic ones are coalgebraically modelled over the categories of join-semilattices, semimodules for a semiring, and convex sets, respectively. Moreover, expressions for axiomatizing behavior semantics often live in algebraic categories.

In order to prove completeness of such axiomatizations, the common approach [4,21,23] is to prove finality of a certain object in a category of coalgebras

C. Baier and U. Dal Lago (Eds.): FOSSACS 2018, LNCS 10803, pp. 331–347, 2018.
https://doi.org/10.1007/978-3-319-89366-2_18

over an algebraic category. Proofs are significantly simplified if it suffices to verify finality only w.r.t. coalgebras carried by free finitely generated algebras, as those are the coalgebras that result from generalized determinization.

In recent work, Milius [16] proposed the notion of a proper functor on an algebraic category that provides a sufficient condition for this purpose. This notion is an extension of the notion of a proper semiring introduced by Esik and Maletti [8]: A semiring is proper if and only if its "cubic" functor is proper. A cubic functor is a functor $\mathbb{S} \times (-)^A$ where A is a finite alphabet and $\mathbb{S}$ is a free algebra with a single generator in the algebraic category. Cubic functors model deterministic weighted automata which are models of determinizations of non-deterministic and probabilistic transition systems.

Properness is the property that for any two states that are behaviourally equivalent in coalgebras with free finitely generated carriers, there is a zig-zag of homomorphisms (called a chain of simulations in the original works on weighted automata and proper semirings) that identifies the two states and whose nodes are all carried by free finitely generated algebras.

Even though the notion of properness is relatively new for a semiring and very new for a functor, results on properness of semirings can be found in more distant literature as well. Here is a brief history, to the best of our knowledge:

- The Boolean semiring was proven to be proper in [3].
- Finite commutative ordered semirings were proven to be proper in [7, Theorem 5.1]. Interestingly, the proof provides a zig-zag with at most seven intermediate nodes.
- Any euclidean domain and any skew field were proven proper in [1, Theorem 3]. In each case the zig-zag has two intermediate nodes.
- The semiring of natural numbers $\mathbb{N}$, the Boolean semiring $\mathbb{B}$, the ring of integers $\mathbb{Z}$ and any skew field were proven proper in [2, Theorem 1]. All zig-zags were spans, i.e., had a single intermediate node with outgoing arrows.
- Noetherian semirings were proven proper in [8, Theorem 4.2], commutative rings also in [8, Corollary 4.4], and finite semirings as well in [8, Corollary 4.5], all with a zig-zag being a span. Moreover, the tropical semiring is not proper, as proven in [8, Theorem 5.4].

Having properness of a semiring, together with the property of the semiring being finitely and effectively presentable, yields decidability of the equivalence problem (decidability of trace equivalence) for weighted automata.

In this paper, motivated by the wish to prove properness of a certain functor $\widehat{F}$ on convex algebras used for axiomatizing trace semantics of probabilistic systems in [23], as well as by the open questions stated in [16, Example 3.19], we provide a framework for proving properness. We instantiate this framework on known cases like Noetherian semirings and $\mathbb{N}$ (with a zig-zag that is a span), and further prove new results of properness:

- The semirings $\mathbb{Q}_+$ and $\mathbb{R}_+$ of non-negative rationals and reals, respectively, are proper. The shape of the zig-zag is a span as well.
- The functor $[0,1] \times (-)^A$ on $\mathtt{PCA}$ is proper, again the zig-zag being a span.

– The functor $\widehat{F}$ on PCA is proper. This proof is the most involved, and interestingly, provides the only case where the zig-zag is not a span: it contains three intermediate nodes of which the middle one forms a span.

Our framework requires a proof of so-called *extension* and *reduction lemmas* in each case. While the extension lemma is a generic result that covers all cubic functors of interest, the reduction lemma is in all cases a nontrivial property intrinsic to the algebras under consideration. For the semiring of natural numbers it is a consequence of a result that we trace back to Hilbert; for the case of convex algebra $[0, 1]$ the result is due to Minkowski. In the case of $\widehat{F}$, we use Kakutani's set-valued fixpoint theorem.

It is an interesting question for future work whether these new properness results may lead to new complete axiomatizations of expressions for certain weighted automata.

The organization of the rest of the paper is as follows. In Sect. 2 we give some basic definitions and introduce the semirings, the categories, and the functors of interest. Section 3 provides the general framework as well as proofs of properness of the cubic functors. Sections 4, 5 and 6 lead us to properness of $\widehat{F}$ on PCA. For space reasons, we present the ideas of proofs and constructions in the main paper and defer all detailed proofs to the arXiv-version [24].

2 Proper Functors

We start with a brief introduction of the basic notions from algebra and coalgebra needed in the rest of the paper, as well as the important definition of proper functors [16]. We refer the interested reader to [9,11,20] for more details. We assume basic knowledge of category theory, see e.g. [14] or [24, Appendix A].

Let C be a category and F a C-endofunctor. The category $\mathsf{Coalg}(F)$ of *F-coalgebras* is the category having as objects pairs (X, c) where X is an object of C and c is a C-morphism from X to FX, and as morphisms $f\colon (X, c) \to (Y, d)$ those C-morphisms from X to Y that make the diagram on the right commute.

$$\begin{array}{ccc} X & \xrightarrow{f} & Y \\ {\scriptstyle c}\downarrow & & \downarrow{\scriptstyle d} \\ FX & \xrightarrow{Ff} & FY \end{array}$$

All base categories C in this paper will be *algebraic categories*, i.e., categories $\mathtt{Set}^T$ of Eilenberg-Moore algebras of a finitary monad [1] in Set. Hence, all base categories are concrete with forgetful functor that is identity on morphisms.

In such categories behavioural equivalence [13,25,26] can be defined as follows. Let (X, c) and (Y, d) be F-coalgebras and let $x \in X$ and $y \in Y$. Then x and y are *behaviourally equivalent*, and we write $x \sim y$, if there exists an F-coalgebra (Z, e) and $\mathsf{Coalg}(F)$-morphisms $f\colon (X, c) \to (Z, e)$, $g\colon (Y, d) \to (Z, e)$, with $f(x) = g(y)$.

$$(X, c) \xrightarrow{\;f\;} \underset{f(x)=g(y)}{(Z, e)} \xleftarrow{\;g\;} (Y, d)$$

[1] The notions of monads and algebraic categories are central to this paper. We recall them in [24, Appendix A] to make the paper better accessible to all readers.

If there exists a final coalgebra in $\mathsf{Coalg}(F)$, and all functors considered in this paper will have this property, then two elements are behaviourally equivalent if and only if they have the same image in the final coalgebra. If we have a *zig-zag diagram* in $\mathsf{Coalg}(F)$

$$\begin{array}{ccccccccccccc} (X,c) & & & & (Z_2,e_2) & & & & \cdots & & & & (Y,d) \\ & \overset{f_1}{\searrow} & & \overset{f_2}{\swarrow} & & \overset{f_3}{\searrow} & & \overset{f_4}{\swarrow} & & \overset{f_{2n-1}}{\searrow} & & \overset{f_{2n}}{\swarrow} & \\ & & (Z_1,e_1) & & & & (Z_3,e_1) & & & & (Z_{2n-1},e_1) & & \end{array} \tag{1}$$

which relates x with y in the sense that there exist elements $z_{2k} \in Z_{2k}$, $k = 1,\ldots,n-1$, with (setting $z_0 = x$ and $z_{2n} = y$)

$$f_{2k}(z_{2k}) = f_{2k-1}(z_{2k-2}), \quad k = 1,\ldots,n,$$

then $x \sim y$.

We now recall the notion of a proper functor, introduced by Milius [16] which is central to this paper. It is very helpful for establishing completeness of regular expressions calculi, cf. [16, Corollary 3.17].

Definition 2.1. Let $T\colon \mathtt{Set} \to \mathtt{Set}$ be a finitary monad with unit η and multiplication μ. A $\mathtt{Set}^T$-endofunctor F is *proper*, if the following statement holds.

For each pair (TB_1, c_1) and (TB_2, c_2) of F-coalgebras with B_1 and B_2 finite sets, and each two elements $b_1 \in B_1$ and $b_2 \in B_2$ with $\eta_{B_1}(b_1) \sim \eta_{B_2}(b_2)$, there exists a zig-zag (1) in $\mathsf{Coalg}(F)$ which relates $\eta_{B_1}(b_1)$ with $\eta_{B_2}(b_2)$, and whose nodes (Z_j, e_j) all have free and finitely generated carrier.

This notion generalizes the notion of a proper semiring introduced by Esik and Maletti in [8, Definition 3.2], cf. [16, Remark 3.10].

Remark 2.2. In the definition of properness the condition that intermediate nodes have free *and* finitely generated carrier is necessary for nodes with incoming arrows (the nodes Z_{2k-1} in (1)). For the intermediate nodes with outgoing arrows (Z_{2k} in (1)), it is enough to require that their carrier is finitely generated. This follows since every F-coalgebra with finitely generated carrier is the image under an F-coalgebra morphism of an F-coalgebra with free and finitely generated carrier.

Moreover, note that zig-zags which start (or end) with incoming arrows instead of outgoing ones, can also be allowed since a zig-zag of this form can be turned into one of the form (1) by appending identity maps.

Some Concrete Monads and Functors

We deal with the following base categories.

- The category $\mathbb{S}$-$\mathtt{SMOD}$ of semimodules over a semiring $\mathbb{S}$ induced by the monad $T_{\mathbb{S}}$ of finitely supported maps into $\mathbb{S}$, see, e.g., [15, Example 4.2.5].
- The category $\mathtt{PCA}$ of positively convex algebras induced by the monad of finitely supported subprobability distributions, see, e.g., [5,6] and [17].

For $n \in \mathbb{N}$, the free algebra with n generators in $\mathbb{S}$-SMOD is the direct product $\mathbb{S}^n$, and in PCA it is the n-simplex $\Delta^n = \{(\xi_1, \dots, \xi_n) \mid \xi_j \geq 0, \sum_{j=1}^n \xi_j \leq 1\}$.

Concerning semimodule-categories, we mainly deal with the semirings $\mathbb{N}$, $\mathbb{Q}_+$, and $\mathbb{R}_+$, and their ring completions $\mathbb{Z}$, $\mathbb{Q}$, and $\mathbb{R}$. For these semirings the categories of $\mathbb{S}$-semimodules are

- CMON of commutative monoids for $\mathbb{N}$,
- AB of abelian groups for $\mathbb{Z}$,
- CONE of convex cones for $\mathbb{R}_+$,
- $\mathbb{Q}$-VEC and $\mathbb{R}$-VEC of vector spaces over the field of rational and real numbers, respectively, for $\mathbb{Q}$ and $\mathbb{R}$.

We consider the following functors, where A is a fixed finite alphabet. Recall that we use the term *cubic functor* for the functor $T1 \times (-)^A$ where T is a monad on Set. We chose the name since $T1 \times (-)^A$ assigns to objects X a full direct product, i.e., a full cube.

- The *cubic functor* $F_{\mathbb{S}}$ on $\mathbb{S}$-SMOD, i.e., the functor acting as

$$F_{\mathbb{S}}X = \mathbb{S} \times X^A \text{ for } X \text{ object of } \mathbb{S}\text{-SMOD},$$
$$F_{\mathbb{S}}f = \mathrm{id}_{\mathbb{S}} \times (f \circ -) \text{ for } f \colon X \to Y \text{ morphism of } \mathbb{S}\text{-SMOD}.$$

  The underlying Set functors of cubic functors are also sometimes called deterministic-automata functors, see e.g. [10], as their coalgebras are deterministic weighted automata with output in the semiring.
- The *cubic functor* $F_{[0,1]}$ on PCA, i.e., the functor $F_{[0,1]}X = [0,1] \times X^A$ and $F_{[0,1]}f = \mathrm{id}_{[0,1]} \times (f \circ -)$.
- A *subcubic convex functor* $\widehat{F}$ on PCA whose action will be introduced in Definition 4.1.[2] The name originates from the fact that $\widehat{F}X$ is a certain convex subset of $F_{[0,1]}X$ and that $\widehat{F}f = (F_{[0,1]}f)|_{\widehat{F}X}$ for $f \colon X \to Y$.

Cubic functors are liftings of Set-endofunctors, in particular, they preserve surjective algebra homomorphisms. It is easy to see that also the functor $\widehat{F}$ preserves surjectivity, cf. [24, Lemma D.1]. This property is needed to apply the work of Milius, cf. [16, Assumptions 3.1].

Remark 2.3. We can now formulate precisely the connection between proper semirings and proper functors mentioned after Definition 2.1. A semiring $\mathbb{S}$ is proper in the sense of [8], if and only if for every finite input alphabet A the cubic functor $F_{\mathbb{S}}$ on $\mathbb{S}$-SMOD is proper.

We shall interchangeably think of direct products as sets of functions or as sets of tuples. Taking the viewpoint of tuples, the definition of $F_{\mathbb{S}}f$ reads as

$$(F_{\mathbb{S}}f)\big((o, (x_a)_{a \in A})\big) = \big(o, (f(x_a))_{a \in A}\big), \quad o \in \mathbb{S},\ x_a \in X \text{ for } a \in A.$$

[2] This functor was denoted $\hat{G}$ in [23] where it was first studied in the context of axiomatization of trace semantics.

A coalgebra structure $c\colon X \to F_{\mathbb{S}}X$ writes as

$$c(x) = \big(c_{\mathrm{o}}(x), (c_a(x))_{a \in A}\big), \quad x \in X,$$

and we use $c_{\mathrm{o}} : X \to \mathbb{S}$ and $c_a : X \to X$ as generic notation for the components of the map c. More generally, we define $c_w\colon X \to X$ for any word $w \in A^*$ inductively as $c_\varepsilon = \mathrm{id}_X$ and $c_{wa} = c_a \circ c_w$, $w \in A^*, a \in A$.

The map from a coalgebra (X, c) into the final $F_{\mathbb{S}}$-coalgebra, the *trace map*, is then given as $\mathrm{tr}_c(x) = \big((c_{\mathrm{o}} \circ c_w)(x)\big)_{w \in A^*}$ for $x \in X$. Behavioural equivalence for cubic functors is the kernel of the trace map.

3 Properness of Cubic Functors

Our proofs of properness in this section and in Sect. 6 below start from the following idea. Let $\mathbb{S}$ be a semiring, and assume we are given two $F_{\mathbb{S}}$-coalgebras which have free finitely generated carrier, say $(\mathbb{S}^{n_1}, c_1)$ and $(\mathbb{S}^{n_2}, c_2)$. Moreover, assume $x_1 \in \mathbb{S}^{n_1}$ and $x_2 \in \mathbb{S}^{n_2}$ are two elements having the same trace. For $j = 1, 2$, let $d_j\colon \mathbb{S}^{n_1} \times \mathbb{S}^{n_2} \to F_{\mathbb{S}}(\mathbb{S}^{n_1} \times \mathbb{S}^{n_2})$ be given by

$$d_j(y_1, y_2) = \Big(c_{j_{\mathrm{o}}}(y_j), ((c_{1a}(y_1), c_{2a}(y_2)))_{a \in A}\Big).$$

Denoting by $\pi_j\colon \mathbb{S}^{n_1} \times \mathbb{S}^{n_2} \to \mathbb{S}^{n_j}$ the canonical projections, both sides of the following diagram separately commute.

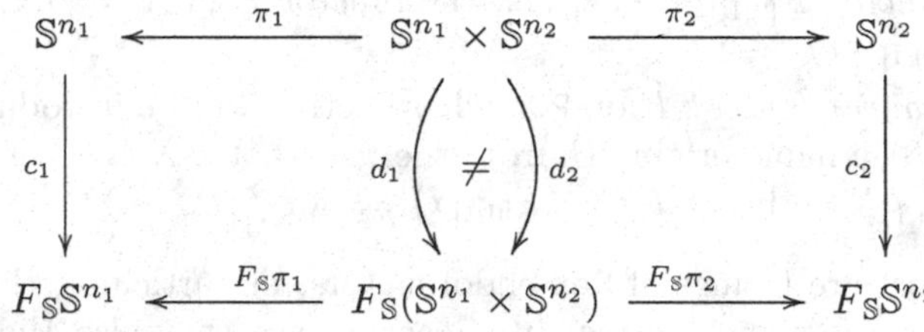

However, in general the maps d_1 and d_2 do not coincide.

The next lemma contains a simple observation: there exists a subsemimodule Z of $\mathbb{S}^{n_1} \times \mathbb{S}^{n_2}$, such that the restrictions of d_1 and d_2 to Z coincide and turn Z into an $F_{\mathbb{S}}$-coalgebra.

Lemma 3.1. *Let Z be the subsemimodule of $\mathbb{S}^{n_1} \times \mathbb{S}^{n_2}$ generated by the pairs $(c_{1w}(x_1), c_{2w}(x_2))$ for $w \in A^*$. Then $d_1|_Z = d_2|_Z$ and $d_j(Z) \subseteq F_{\mathbb{S}}(Z)$.*

The significance of Lemma 3.1 in the present context is that it leads to the diagram (we denote $d = d_j|_Z$)

$$\begin{array}{ccccc} & & \mathbb{S}^{n_1} \times \mathbb{S}^{n_2} & & \\ & & \cup\!| & & \\ \mathbb{S}^{n_1} & \xleftarrow{\pi_1} & Z & \xrightarrow{\pi_2} & \mathbb{S}^{n_2} \\ \downarrow{\scriptstyle c_1} & & \downarrow{\scriptstyle d} & & \downarrow{\scriptstyle c_2} \\ F_{\mathbb{S}}\mathbb{S}^{n_1} & \xleftarrow{F_{\mathbb{S}}\pi_1} & F_{\mathbb{S}}Z & \xrightarrow{F_{\mathbb{S}}\pi_2} & F_{\mathbb{S}}\mathbb{S}^{n_2} \\ & & \cap\!| & & \\ & & \mathbb{S} \times (\mathbb{S}^{n_1} \times \mathbb{S}^{n_2})^A & & \end{array}$$

In other words, it leads to the zig-zag in $\mathsf{Coalg}(F_{\mathbb{S}})$

$$(\mathbb{S}^{n_1}, c_1) \xleftarrow{\pi_1} (Z, d) \xrightarrow{\pi_2} (\mathbb{S}^{n_2}, c_2) \tag{2}$$

This zig-zag relates x_1 with x_2 since $(x_1, x_2) \in Z$. If it can be shown that Z is always finitely generated, it will follow that $F_{\mathbb{S}}$ is proper.

Let $\mathbb{S}$ be a Noetherian semiring, i.e., such that every $\mathbb{S}$-subsemimodule of some finitely generated $\mathbb{S}$-semimodule is itself finitely generated. Then Z is, as an $\mathbb{S}$-subsemimodule of $\mathbb{S}^{n_1} \times \mathbb{S}^{n_2}$, finitely generated. We reobtain [8, Theorem 4.2].

Corollary 3.2 (Esik–Maletti 2010). *Every Noetherian semiring is proper.*

Our first main result is Theorem 3.3 below, where we show properness of the cubic functors $F_{\mathbb{S}}$ on $\mathbb{S}$-`SMOD`, for $\mathbb{S}$ being one of the semirings $\mathbb{N}$, $\mathbb{Q}_+$, $\mathbb{R}_+$, and of the cubic functor $F_{[0,1]}$ on `PCA`. The case of $F_{\mathbb{N}}$ is known from [2, Theorem 4][3], the case of $F_{[0,1]}$ is stated as an open problem in [16, Example 3.19].

Theorem 3.3. *The cubic functors $F_{\mathbb{N}}$, $F_{\mathbb{Q}_+}$, $F_{\mathbb{R}_+}$, and $F_{[0,1]}$ are proper.*

In fact, for any two coalgebras with free finitely generated carrier and any two elements having the same trace, a zig-zag with free and finitely generated nodes relating those elements can be found, which is a span (has a single intermediate node with outgoing arrows).

The proof proceeds via relating to the Noetherian case. It always follows the same scheme, which we now outline. Observe that the ring completion of each of $\mathbb{N}$, $\mathbb{Q}_+$, $\mathbb{R}_+$, is Noetherian (for the last two it actually is a field), and that $[0, 1]$ is the positive part of the unit ball in $\mathbb{R}$.

Step 1. The extension lemma: We use an extension of scalars process to pass from the given category `C` to an associated category $\mathbb{E}$-`MOD` with a Noetherian ring $\mathbb{E}$. This is a general categorical argument.

[3] In [2] only a sketch of the proof is given, cf. [2, Sect. 3.3]. In this sketch one important point is not mentioned. Using the terminology of [2, Sect. 3.3]: it could a priori be possible that the size of the vectors in G and the size of G both oscillate.

To unify notation, we agree that $\mathbb{S}$ may also take the value $[0,1]$, and that $T_{[0,1]}$ is the monad of finitely supported subprobability distributions giving rise to the category $\mathtt{PCA}$.

$\mathbb{S}$	$\mathbb{N}$	$\mathbb{Q}_+$	$\mathbb{R}_+$	$[0,1]$
$\mathtt{C}$	$\mathbb{N}$-SMOD (CMON)	$\mathbb{Q}_+$-SMOD	$\mathbb{R}_+$-SMOD (CONE)	PCA
$\mathbb{E}$-MOD	$\mathbb{Z}$-MOD (AB)	$\mathbb{Q}$-MOD ($\mathbb{Q}$-VEC)	$\mathbb{R}$-MOD ($\mathbb{R}$-VEC)	$\mathbb{R}$-MOD ($\mathbb{R}$-VEC)

For the formulation of the extension lemma, recall that the starting category $\mathtt{C}$ is the Eilenberg-Moore category of the monad $T_\mathbb{S}$ and the target category $\mathbb{E}$-$\mathtt{MOD}$ is the Eilenberg-Moore category of $T_\mathbb{E}$. We write $\eta_\mathbb{S}$ and $\mu_\mathbb{S}$ for the unit and multiplication of $T_\mathbb{S}$ and analogously for $T_\mathbb{E}$. We have $T_\mathbb{S} \leq T_\mathbb{E}$, via the inclusion monad morphism $\iota\colon T_\mathbb{S} \Rightarrow T_\mathbb{E}$ given by $\iota_X(u) = u$, as $\eta_\mathbb{E} = \iota \circ \eta_\mathbb{S}$ and $\mu_\mathbb{E} \circ \iota\iota = \iota \circ \mu_\mathbb{S}$ where $\iota\iota \overset{\text{def}}{=} T_\mathbb{E}\iota \circ \iota \overset{\text{nat.}}{=} \iota \circ T_\mathbb{S}\iota$. Recall that a monad morphism $\iota\colon T_\mathbb{S} \to T_\mathbb{E}$ defines a functor $M_\iota\colon \mathbf{Set}^{T_\mathbb{E}} \to \mathbf{Set}^{T_\mathbb{S}}$ which maps a $T_\mathbb{E}$-algebra (X, α_X) to $(X, \iota_X \circ \alpha_X)$ and is identity on morphisms. Obviously, M_ι commutes with the forgetful functors $U_\mathbb{S} : \mathbf{Set}^{T_\mathbb{S}} \to \mathbf{Set}$ and $U_\mathbb{E} : \mathbf{Set}^{T_\mathbb{E}} \to \mathbf{Set}$, i.e., $U_\mathbb{S} \circ M_\iota = U_\mathbb{E}$.

Definition 3.4. Let $(X, \alpha_X) \in \mathbf{Set}^{T_\mathbb{S}}$ and $(Y, \alpha_Y) \in \mathbf{Set}^{T_\mathbb{E}}$ where $T_\mathbb{S}$ and $T_\mathbb{E}$ are monads with $T_\mathbb{S} \leq T_\mathbb{E}$ via $\iota\colon T_\mathbb{S} \Rightarrow T_\mathbb{E}$. A $\mathbf{Set}$-arrow $h\colon X \to Y$ is a $T_\mathbb{S} \leq T_\mathbb{E}$-homomorphism from (X, α_X) to (Y, α_Y) if and only if the following diagram commutes (in $\mathbf{Set}$)

$$\begin{array}{ccc} T_\mathbb{S}X & \xrightarrow{\ \iota h\ } & T_\mathbb{E}Y \\ {\scriptstyle \alpha_X}\downarrow & & \downarrow{\scriptstyle \alpha_Y} \\ X & \xrightarrow{\ h\ } & Y \end{array}$$

where ιh denotes the map $\iota h \overset{\text{def}}{=} T_\mathbb{E}h \circ \iota_X \overset{\text{nat.}}{=} \iota_Y \circ T_\mathbb{S}h$. In other words, a $T_\mathbb{S} \leq T_\mathbb{E}$-homomorphism from (X, α_X) to (Y, α_Y) is a morphism in $\mathbf{Set}^{T_\mathbb{S}}$ from (X, α_X) to $M(Y, \alpha_Y)$.

Now we can formulate the extension lemma.

Proposition 3.5 (Extension Lemma). *For every $F_\mathbb{S}$-coalgebra $T_\mathbb{S}B \xrightarrow{c} F_\mathbb{S}(T_\mathbb{S}B)$ with free finitely generated carrier $T_\mathbb{S}B$ for a finite set B, there exists an $F_\mathbb{E}$-coalgebra $T_\mathbb{E}B \xrightarrow{\tilde{c}} F_\mathbb{E}(T_\mathbb{E}B)$ with free finitely generated carrier $T_\mathbb{E}B$ such that*

$$\begin{array}{ccc} T_\mathbb{S}B & \xrightarrow{\ \iota_B\ } & T_\mathbb{E}B \\ {\scriptstyle c}\downarrow & & \downarrow{\scriptstyle \tilde{c}} \\ F_\mathbb{S}(T_\mathbb{S}B) & \xrightarrow{\ \iota_1 \times (\iota_B)^A\ } & F_\mathbb{E}(T_\mathbb{E}B) \end{array}$$

where the horizontal arrows (ι_B and $\iota_1 \times \iota_B^A$) are $T_\mathbb{S} \leq T_\mathbb{E}$-homomorphisms, and moreover they both amount to inclusion.

Step 2. The basic diagram: Let $n_1, n_2 \in \mathbb{N}$, let B_j be the n_j-element set consisting of the canonical basis vectors of $\mathbb{E}^{n_j}$, and set $X_j = T_{\mathbb{S}} B_j$. Assume we are given $F_{\mathbb{S}}$-coalgebras (X_1, c_1) and (X_2, c_2), and elements $x_j \in X_j$ with $\operatorname{tr}_{c_1} x_1 = \operatorname{tr}_{c_2} x_2$.

The extension lemma provides $F_{\mathbb{E}}$-coalgebras $(\mathbb{E}^{n_j}, \tilde{c}_j)$ with $\tilde{c}_j|_{X_j} = c_j$. Clearly, $\operatorname{tr}_{\tilde{c}_1} x_1 = \operatorname{tr}_{\tilde{c}_2} x_2$. Using the zig-zag diagram (2) in $\mathsf{Coalg}(F_{\mathbb{E}})$ and appending inclusion maps, we obtain what we call the *basic diagram*. In this diagram all solid arrows are arrows in $\mathbb{E}$-MOD, and all dotted arrows are arrows in C. The horizontal dotted arrows denote the inclusion maps, and π_j are the restrictions to Z of the canonical projections.

$$\mathbb{E}^{n_1} \times \mathbb{E}^{n_2}$$
$$\cup|$$

$$
\begin{array}{ccccccccc}
X_1 & \dashrightarrow & \mathbb{E}^{n_1} & \xleftarrow{\ \pi_1\ } & Z & \xrightarrow{\ \pi_2\ } & \mathbb{E}^{n_2} & \dashleftarrow & X_2 \\[2pt]
{\scriptstyle c_1}\big\downarrow & & {\scriptstyle \tilde{c}_1}\big\downarrow & & {\scriptstyle d}\big\downarrow & & {\scriptstyle \tilde{c}_2}\big\downarrow & & {\scriptstyle c_2}\big\downarrow \\[2pt]
F_{\mathbb{S}} X_1 & \dashrightarrow & F_{\mathbb{E}} \mathbb{E}^{n_1} & \xleftarrow{F_{\mathbb{E}} \pi_1} & F_{\mathbb{E}} Z & \xrightarrow{F_{\mathbb{E}} \pi_2} & F_{\mathbb{E}} \mathbb{E}^{n_2} & \dashleftarrow & F_{\mathbb{S}} X_2
\end{array}
$$
$$\cap|$$
$$\mathbb{E} \times (\mathbb{E}^{n_1} \times \mathbb{E}^{n_2})^A$$

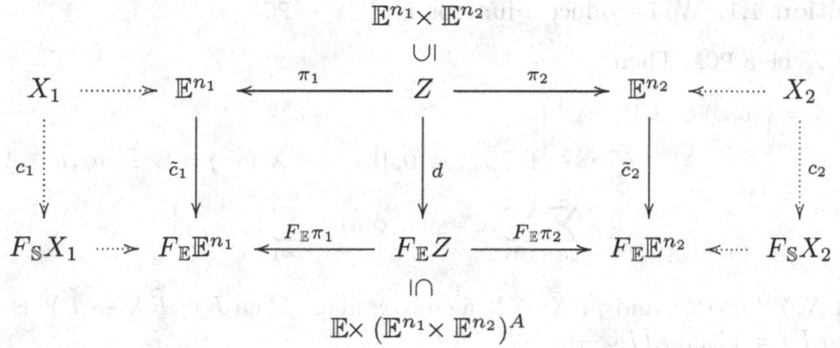

Commutativity of this diagram yields $d\big(\pi_j^{-1}(X_j)\big) \subseteq (F_{\mathbb{E}} \pi_j)^{-1}(F_{\mathbb{S}} X_j)$ for $j = 1, 2$. Now we observe the following properties of cubic functors.

Lemma 3.6. *We have* $F_{\mathbb{E}} X \cap F_{\mathbb{S}} Y = F_{\mathbb{S}}(X \cap Y)$. *Moreover, if* $Y_j \subseteq X_j$, *then* $(F_{\mathbb{E}} \pi_1)^{-1}(F_{\mathbb{S}} Y_1) \cap (F_{\mathbb{E}} \pi_2)^{-1}(F_{\mathbb{S}} Y_2) = F_{\mathbb{S}}(Y_1 \times Y_2)$.

Using this, yields

$$
\begin{aligned}
d\big(Z \cap (X_1 \times X_2)\big) &\subseteq F_{\mathbb{E}} Z \cap (F_{\mathbb{E}} \pi_1)^{-1}\big(F_{\mathbb{S}} X_1\big) \cap (F_{\mathbb{E}} \pi_2)^{-1}\big(F_{\mathbb{S}} X_2\big) \\
&= F_{\mathbb{E}} Z \cap F_{\mathbb{S}}(X_1 \times X_2) = F_{\mathbb{S}}\big(Z \cap (X_1 \times X_2)\big).
\end{aligned}
$$

This shows that $Z \cap (X_1 \times X_2)$ becomes an $F_{\mathbb{S}}$-coalgebra with the restriction $d|_{Z \cap (X_1 \times X_2)}$. Again referring to the basic diagram, we have the following zig-zag in $\mathsf{Coalg}(F_{\mathbb{S}})$ (to shorten notation, denote the restrictions of d, π_1, π_2 to $Z \cap (X_1 \times X_2)$ again as d, π_1, π_2):

$$(X_1, c_1) \xleftarrow{\ \pi_1\ } \big(Z \cap (X_1 \times X_2), d\big) \xrightarrow{\ \pi_2\ } (X_2, c_2) \tag{3}$$

This zig-zag relates x_1 with x_2 since $(x_1, x_2) \in Z \cap (X_1 \times X_2)$.

Step 3. The reduction lemma: In view of the zig-zag (3), the proof of Theorem 3.3 can be completed by showing that $Z \cap (X_1 \times X_2)$ is finitely generated as an algebra in C. Since Z is a submodule of the finitely generated module $\mathbb{E}^{n_1} \times \mathbb{E}^{n_2}$ over the Noetherian ring $\mathbb{E}$, it is finitely generated as an $\mathbb{E}$-module. The task thus is to show that being finitely generated is preserved when reducing scalars.

This is done by what we call the *reduction lemma*. Contrasting the extension lemma, the reduction lemma is not a general categorical fact, and requires specific proof in each situation.

Proposition 3.7 (Reduction Lemma). *Let $n_1, n_2 \in \mathbb{N}$, let B_j be the set consisting of the n_j canonical basis vectors of $\mathbb{E}^{n_j}$, and set $X_j = T_{\mathbb{S}}B_j$. Moreover, let Z be an $\mathbb{E}$-submodule of $\mathbb{E}^{n_1} \times \mathbb{E}^{n_2}$. Then $Z \cap (X_1 \times X_2)$ is finitely generated as an algebra in* $\mathtt{C}$.

4 A Subcubic Convex Functor

Recall the following definition from [23, p. 309].

Definition 4.1. We introduce a functor $\widehat{F}\colon \mathtt{PCA} \to \mathtt{PCA}$.

1. Let X be a $\mathtt{PCA}$. Then

$$\widehat{F}X = \Big\{(o, \phi) \in [0,1] \times X^A \mid \exists n_a \in \mathbb{N}.\ \exists p_{a,j} \in [0,1], x_{a,j} \in X \text{ for } j = 1, \ldots, n_a, a \in A.\ o + \sum_{a\in A}\sum_{j=1}^{n_a} p_{a,j} \le 1,\ \phi(a) = \sum_{j=1}^{n_a} p_{a,j}x_{a,j}\Big\}.$$

2. Let X, Y be $\mathtt{PCA}$s, and $f\colon X \to Y$ a convex map. Then $\widehat{F}f\colon \widehat{F}X \to \widehat{F}Y$ is the map $\widehat{F}f = \mathrm{id}_{[0,1]} \times (f \circ -)$.

For every X we have $\widehat{F}X \subseteq F_{[0,1]}X$, and for every $f\colon X \to Y$ we have $\widehat{F}f = (F_{[0,1]}f)|_{\widehat{F}X}$. For this reason, we think of $\widehat{F}$ as a *subcubic functor*.

The definition of $\widehat{F}$ can be simplified.

Lemma 4.2. *Let X be a* $\mathtt{PCA}$, *then*

$$\widehat{F}X = \Big\{(o, f) \in [0,1] \times X^A \mid \exists p_a \in [0,1], x_a \in X \text{ for } a \in A.\ o + \sum_{a\in A} p_a \le 1,\ f(a) = p_a x_a\Big\}.$$

From this representation it is obvious that $\widehat{F}$ is monotone in the sense that

- If $X_1 \subseteq X_2$, then $\widehat{F}X_1 \subseteq \widehat{F}X_2$.
- If $f_1\colon X_1 \to Y_1, f_2\colon X_2 \to Y_2$ with $X_1 \subseteq X_2, Y_1 \subseteq Y_2$ and $f_2|_{X_1} = f_1$, then $\widehat{F}f_2|_{\widehat{F}X_1} = \widehat{F}f_1$.

Note that $\widehat{F}$ does not preserve direct products.

For a $\mathtt{PCA}$ X whose carrier is a compact subset of a euclidean space, $\widehat{F}X$ can be described with help of a geometric notion, namely using the Minkowksi functional of X. Before we can state this fact, we have to make a brief digression to explain this notion and its properties.

Definition 4.3. Let $X \subseteq \mathbb{R}^n$ be a $\mathtt{PCA}$. The *Minkowski functional* of X is the map $\mu_X\colon \mathbb{R}^n \to [0,\infty]$ defined as $\mu_X(x) = \inf\{t > 0 \mid x \in tX\}$, where the infimum of the empty set is understood as ∞.

Minkowski functionals, sometimes also called *gauge*, are a central and exhaustively studied notion in convex geometry, see, e.g., [19, p. 34] or [18, p. 28].

We list some basic properties whose proof can be found in the mentioned textbooks.

1. $\mu_X(px) = p\mu_X(x)$ for $x \in \mathbb{R}^n, p \geq 0$,
2. $\mu_X(x+y) \leq \mu_X(x) + \mu_X(y)$ for $x, y \in \mathbb{R}^n$,
3. $\mu_{X\cap Y}(x) = \max\{\mu_X(x), \mu_Y(x)\}$ for $x \in \mathbb{R}^n$.
4. If X is bounded, then $\mu_X(x) = 0$ if and only if $x = 0$.

The set X can almost be recovered from μ_X.

5. $\{x \in \mathbb{R}^n \mid \mu_X(x) < 1\} \subseteq X \subseteq \{x \in \mathbb{R}^n \mid \mu_X(x) \leq 1\}$.
6. If X is closed, equality holds in the second inclusion of 5.
7. Let X, Y be closed. Then $X \subseteq Y$ if and only if $\mu_X \geq \mu_Y$.

Example 4.4. As two simple examples, consider the n-simplex $\Delta^n \subseteq \mathbb{R}^n$ and a convex cone $C \subseteq \mathbb{R}^n$. Then (here $\geq$ denotes the product order on $\mathbb{R}^n$)

$$\mu_{\Delta^n}(x) = \begin{cases} \sum_{j=1}^n \xi_j\,, & x = (\xi_1, \ldots, \xi_n) \geq 0, \\ \infty\,, & \text{otherwise.} \end{cases} \qquad \mu_C(x) = \begin{cases} 0\,, & x \in C, \\ \infty\,, & \text{otherwise.} \end{cases}$$

Observe that $\Delta^n = \{x \in \mathbb{R}^n \mid \mu_{\Delta^n}(x) \leq 1\}$.

Another illustrative example is given by general pyramids in a euclidean space. This example will play an important role later on.

Example 4.5. For $u \in \mathbb{R}^n$ consider the set

$$X = \{x \in \mathbb{R}^n \mid x \geq 0 \text{ and } (x, u) \leq 1\},$$

where $(\cdot, \cdot)$ denotes the euclidean scalar product on $\mathbb{R}^n$. The set X is intersection of the cone $\mathbb{R}^n_+$ with the half-space given by the inequality $(x, u) \leq 1$, hence it is convex and contains 0. Thus X is a PCA.

Let us first assume that u is strictly positive, i.e., $u \geq 0$ and no component of u equals zero. Then X is a pyramid (in 2-dimensional space, a triangle).

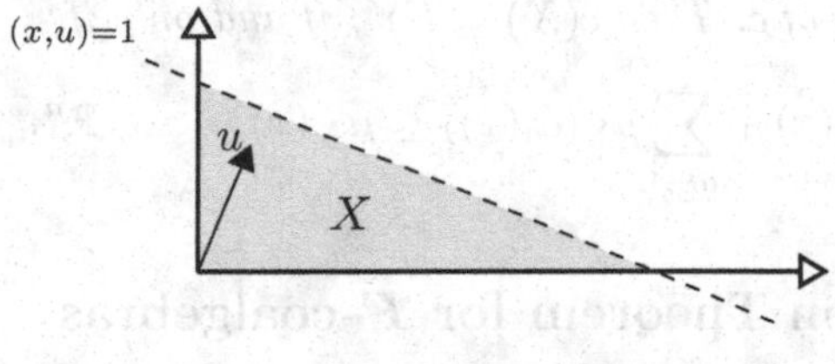

The n-simplex Δ^n is the pyramid obtained using $u = (1, \ldots, 1)$.

The Minkowski functional of the pyramid X associated with u is

$$\mu_X(x) = (x, u) \text{ if } x \geq 0, \quad \mu_X(x) = \infty \text{ otherwise.}$$

Write $u = \sum_{j=1}^{n} \alpha_j e_j$, where e_j is the j-th canonical basis vector, and set $y_j = \frac{1}{\alpha_j} e_j$. Clearly, $\{y_1, \ldots, y_n\}$ is linearly independent. Each vector $x = \sum_{j=1}^{n} \xi_j e_j$ can be written as $x = \sum_{j=1}^{n} (\xi_j \alpha_j) y_j$, and this is a subconvex combination if and only if $\xi_j \geq 0$ and $\sum_{j=1}^{n} \xi_j \alpha_j \leq 1$, i.e., if and only if $x \in X$. Thus X is generated by $\{y_1, \ldots, y_n\}$ as a PCA.

The linear map given by the diagonal matrix made up of the α_j's induces a bijection of X onto Δ^n, and maps the y_j's to the corner points of Δ^n. Hence, X is free with basis $\{y_1, \ldots, y_n\}$.

If u is not strictly positive, the situation changes drastically. Then X is not finitely generated as a PCA, because it is unbounded whereas the subconvex hull of a finite set is certainly bounded.

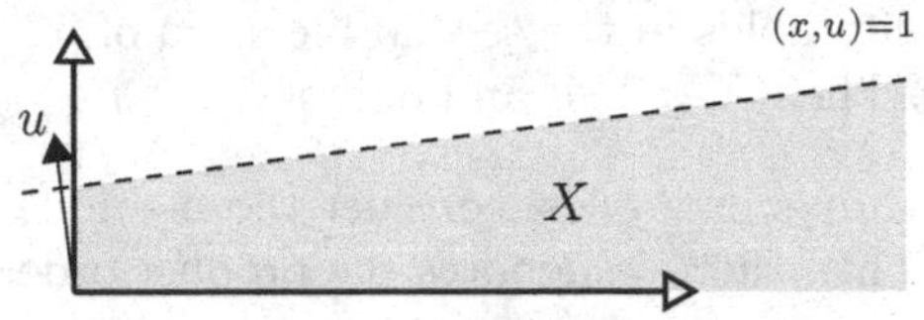

Now we return to the functor $\widehat{F}$.

Lemma 4.6. *Let $X \subseteq \mathbb{R}^n$ be a PCA, and assume that X is compact. Then*

$$\widehat{F}X = \Big\{(o, \phi) \in \mathbb{R} \times (\mathbb{R}^n)^A \mid o \geq 0,\ o + \sum_{a \in A} \mu_X(\phi(a)) \leq 1\Big\}.$$

In the following we use the elementary fact that every convex map has a linear extension.

Lemma 4.7. *Let V_1, V_2 be vector spaces, let $X \subseteq V_1$ be a PCA, and let $c\colon X \to V_2$ be a convex map. Then c has a linear extension $\tilde{c}\colon V_1 \to V_2$. If* span $X = V_1$, *this extension is unique.*

Rescaling in this representation of $\widehat{F}X$ leads to a characterisation of $\widehat{F}$-coalgebra maps. We give a slightly more general statement.

Corollary 4.8. *Let $X, Y \subseteq \mathbb{R}^n$ be PCA s, and assume that X and Y are compact. Further, let $c\colon X \to \mathbb{R}_+ \times (\mathbb{R}^n)^A$ be a convex map, and let $\tilde{c}\colon \mathbb{R}^n \to \mathbb{R} \times (\mathbb{R}^n)^A$ be a linear extension of c. Then $c(X) \subseteq \widehat{F}Y$, if and only if*

$$\tilde{c}_o(x) + \sum_{a \in A} \mu_Y(\tilde{c}_a(x)) \leq \mu_X(x), \quad x \in \mathbb{R}^n. \tag{4}$$

5 An Extension Theorem for $\widehat{F}$-coalgebras

In this section we establish an extension theorem for $\widehat{F}$-coalgebras. It states that an $\widehat{F}$-coalgebra, whose carrier has a particular geometric form, can, under a mild additional condition, be embedded into an $\widehat{F}$-coalgebra whose carrier is free and finitely generated.

Theorem 5.1. *Let (X, c) be an $\widehat{F}$-coalgebra whose carrier X is a compact subset of a euclidean space $\mathbb{R}^n$ with $\Delta^n \subseteq X \subseteq \mathbb{R}^n_+$. Assume that the output map c_o does not vanish on invariant coordinate hyperplanes in the sense that (e_j denotes again the j-th canonical basis vector in $\mathbb{R}^n$)*

$$\begin{aligned} &\nexists I \subseteq \{1, \ldots, n\}. \\ &I \neq \emptyset, \quad c_o(e_j) = 0, j \in I, \quad c_a(e_j) \subseteq \operatorname{span}\{e_i \mid i \in I\}, a \in A, j \in I. \end{aligned} \tag{5}$$

Then there exists an $\widehat{F}$-coalgebra (Y, d), such that $X \subseteq Y \subseteq \mathbb{R}^n_+$, the inclusion map $\iota \colon X \to Y$ is a $\mathsf{Coalg}(\widehat{F})$-morphism, and Y is the subconvex hull of n linearly independent vectors (in particular, Y is free with n generators).

The idea of the proof can be explained by geometric intuition. Say, we have an $\widehat{F}$-coalgebra (X, c) of the stated form, and let $\tilde{c} \colon \mathbb{R}^n \to \mathbb{R} \times (\mathbb{R}^n)^A$ be the linear extension of c to all of $\mathbb{R}^n$, cf. Lemma 4.7.

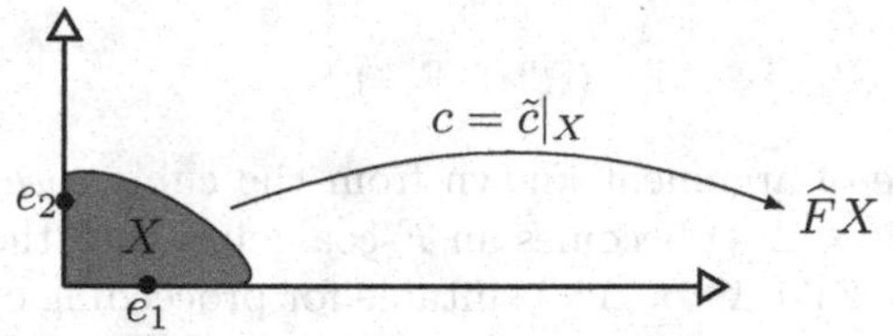

Remembering that pyramids are free and finitely generated, we will be done if we find a pyramid $Y \supseteq X$ which is mapped into $\widehat{F}Y$ by $\tilde{c}$:

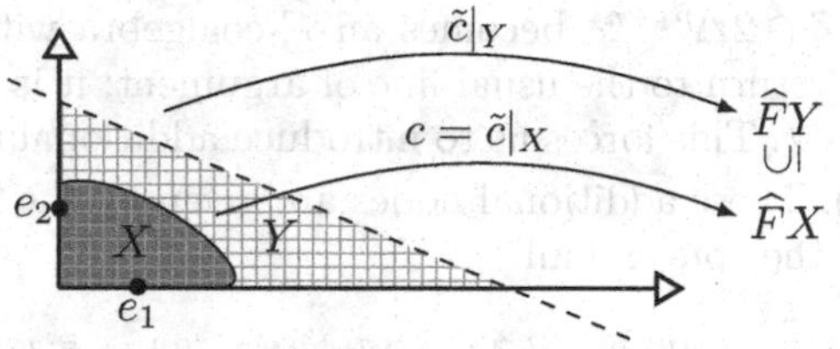

This task can be reformulated as follows: For each pyramid Y_1 containing X let $P(Y_1)$ be the set of all pyramids Y_2 containing X, such that $\tilde{c}(Y_2) \subseteq \widehat{F}Y_1$. If we find Y with $Y \in P(Y)$, we are done.

Existence of Y can be established by applying a fixed point principle for set-valued maps. The result sufficient for our present level of generality is Kakutani's generalisation [12, Corollary] of Brouwers fixed point theorem.

6 Properness of $\widehat{F}$

In this section we give the second main result of the paper.

Theorem 6.1. *The functor $\widehat{F}$ is proper.*

In fact, for each two given coalgebras with free finitely generated carrier and each two elements having the same trace, a zig-zag with free and finitely generated nodes relating those elements can be found, which has three intermediate nodes with the middle one forming a span.

We try to follow the proof scheme familiar from the cubic case. Assume we are given two $\widehat{F}$-coalgebras with free finitely generated carrier, say (Δ^{n_1}, c_1) and (Δ^{n_2}, c_2), and elements $x_1 \in \Delta^{n_1}$ and $x_2 \in \Delta^{n_2}$ having the same trace. Since $\widehat{F}\Delta^{n_j} \subseteq \mathbb{R} \times (\mathbb{R}^{n_j})^A$ we can apply Lemma 4.7 and obtain $F_{\mathbb{R}}$-coalgebras $(\mathbb{R}^{n_j}, \tilde{c}_j)$ with $\tilde{c}_j|_{\Delta^{n_j}} = c_j$. This leads to the basic diagram:

$$
\begin{array}{ccccccccc}
 & & & & \mathbb{R}^{n_1} \times \mathbb{R}^{n_2} & & & & \\
 & & & & \cup\!\!\mid & & & & \\
\Delta^{n_1} & \dashrightarrow & \mathbb{R}^{n_1} & \xleftarrow{\pi_1} & Z & \xrightarrow{\pi_2} & \mathbb{R}^{n_2} & \dashleftarrow & \Delta^{n_2} \\
{\scriptstyle c_1}\,\vdots & & \downarrow{\scriptstyle \tilde{c}_1} & & \downarrow{\scriptstyle d} & & \downarrow{\scriptstyle \tilde{c}_2} & & \vdots\,{\scriptstyle c_2} \\
\widehat{F}\Delta^{n_1} & \dashrightarrow & F_{\mathbb{R}}\mathbb{R}^{n_1} & \xleftarrow{F_{\mathbb{R}}\pi_1} & F_{\mathbb{R}}Z & \xrightarrow{F_{\mathbb{R}}\pi_2} & F_{\mathbb{R}}\mathbb{R}^{n_2} & \dashleftarrow & \widehat{F}\Delta^{n_2} \\
 & & & & \cap\!\!\mid & & & & \\
 & & & & \mathbb{R} \times (\mathbb{R}^{n_1} \times \mathbb{R}^{n_2})^A & & & &
\end{array}
$$

At this point the line of argument known from the cubic case breaks: it is *not* granted that $Z \cap (\Delta^{n_1} \times \Delta^{n_2})$ becomes an $\widehat{F}$-coalgebra with the restriction of d.

The substitute for $Z \cap (\Delta^{n_1} \times \Delta^{n_2})$ suitable for proceeding one step further is given by the following lemma, where we tacitly identify $\mathbb{R}^{n_1} \times \mathbb{R}^{n_2}$ with $\mathbb{R}^{n_1+n_2}$.

Lemma 6.2. *We have* $d(Z \cap 2\Delta^{n_1+n_2}) \subseteq \widehat{F}(Z \cap 2\Delta^{n_1+n_2})$.

This shows that $Z \cap 2\Delta^{n_1+n_2}$ becomes an $\widehat{F}$-coalgebra with the restriction of d. Still, we cannot return to the usual line of argument: it is *not* granted that $\pi_j(Z \cap 2\Delta^{n_1+n_2}) \subseteq \Delta^{n_j}$. This forces us to introduce additional nodes to produce a zig-zag in $\mathsf{Coalg}(\widehat{F})$. These additional nodes are given by the following lemma. There $\mathrm{co}(-)$ denotes the convex hull.

Lemma 6.3. *Set* $Y_j = \mathrm{co}(\Delta^{n_j} \cup \pi_j(Z \cap 2\Delta^{n_1+n_2}))$. *Then* $\tilde{c}_j(Y_j) \subseteq \widehat{F}Y_j$.

This shows that Y_j becomes an $\widehat{F}$-coalgebra with the restriction of $\tilde{c}_j$. We are led to a zig-zag in $\mathsf{Coalg}(\widehat{F})$:

$$
(\Delta^{n_1}, c_1) \xrightarrow{\subseteq} (Y_1, \tilde{c}_1) \xleftarrow{\pi_1} (Z \cap 2\Delta^{n_1+n_2}, d) \xrightarrow{\pi_2} (Y_2, \tilde{c}_2) \xleftarrow{\supseteq} (\Delta^{n_2}, c_2)
$$

This zig-zag relates x_1 and x_2 since $(x_1, x_2) \in Z \cap 2\Delta^{n_1+n_2}$.

Using Minkowski's Theorem and the argument from [24, Lemma B.8] shows that the middle node has finitely generated carrier. The two nodes with incoming arrows are, as convex hulls of two finitely generated `PCA`s, of course also finitely generated. But in general they will not be free (and this is essential, remember Remark 2.2). Now Theorem 5.1 comes into play.

Lemma 6.4. *Assume that each of (Δ^{n_1}, c_1) and (Δ^{n_2}, c_2) satisfies the following condition:*

$$\nexists I \subseteq \{1, \ldots, n\}.$$
$$I \neq \emptyset, \ c_{j_0}(e_k) = 0, k \in I, \ c_{j_a}(e_k) \subseteq \mathrm{co}(\{e_i \mid i \in I\} \cup \{0\}), a \in A, k \in I. \tag{6}$$

Then there exist free finitely generated PCAs U_j with $Y_j \subseteq U_j \subseteq \mathbb{R}_+^{n_j}$ which satisfy $\tilde{c}_j(U_j) \subseteq \widehat{F}U_j$.

This shows that U_j, under the additional assumption (6) on (Δ^{n_j}, c_j), becomes an $\widehat{F}$-coalgebra with the restriction of $\tilde{c}_j$. Thus we have a zig-zag in $\mathrm{Coalg}(\widehat{F})$ relating x_1 and x_2 whose nodes with incoming arrows are free and finitely generated, and whose node with outgoing arrows is finitely generated:

$$(\Delta^{n_1}, c_1) \overset{\subseteq}{\dashrightarrow} (Y_1, \tilde{c}_1) \overset{\pi_1}{\dashleftarrow} (Z \cap 2\Delta^{n_1+n_2}, d) \overset{\pi_2}{\dashrightarrow} (Y_2, \tilde{c}_2) \overset{\supseteq}{\dashleftarrow} (\Delta^{n_2}, c_2)$$

$$(U_1, \tilde{c}_1) \qquad\qquad (U_2, \tilde{c}_2)$$

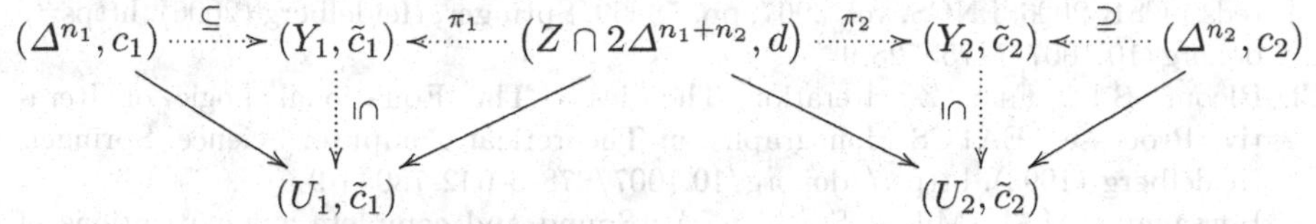

Removing the additional assumption on (Δ^{n_j}, c_j) is an easy exercise.

Lemma 6.5. *Let (Δ^n, c) be an $\widehat{F}$-coalgebra. Assume that I is a nonempty subset of $\{1, \ldots, n\}$ with*

$$c_0(e_k) = 0, \ k \in I \quad \text{and} \quad c_a(e_k) \in \mathrm{co}\left(\{e_i \mid i \in I\} \cup \{0\}\right), \ a \in A, k \in I. \tag{7}$$

Let X be the free PCA with basis $\{e_k \mid k \in \{1, \ldots, n\} \setminus I\}$, and let $f \colon \Delta^n \to X$ be the PCA-morphism with $f(e_k) = 0$ if $k \in I$ and $f(e_k) = e_k$ if $k \notin I$. Further, let $g \colon X \to [0,1] \times X^A$ be the PCA-morphism with

$$g(e_k) = \Big(c_0(e_k), \big(f(c_a(e_k))\big)_{a \in A}\Big), \quad k \in \{1, \ldots, n\} \setminus I.$$

Then (X, g) is an $\widehat{F}$-coalgebra, and f is an $\widehat{F}$-coalgebra morphism of (Δ^n, c) onto (X, g).

Corollary 6.6. *Let (Δ^n, c) be an $\widehat{F}$-coalgebra. Then there exists $k \leq n$, an $\widehat{F}$-coalgebra (Δ^k, g), such that (Δ^k, g) satisfies the assumption in Lemma 6.4 and such that there exists an $\widehat{F}$-coalgebra map f of (Δ^n, c) onto (Δ^k, g).*

The proof of Theorem 6.1 is now finished by putting together what we showed so far. Starting with $\widehat{F}$-coalgebras (Δ^{n_j}, c_j) without any additional assumptions, and elements $x_j \in \Delta^{n_j}$ having the same trace, we first reduce by means of Corollary 6.6 and then apply Lemma 6.4. This gives a zig-zag as required:

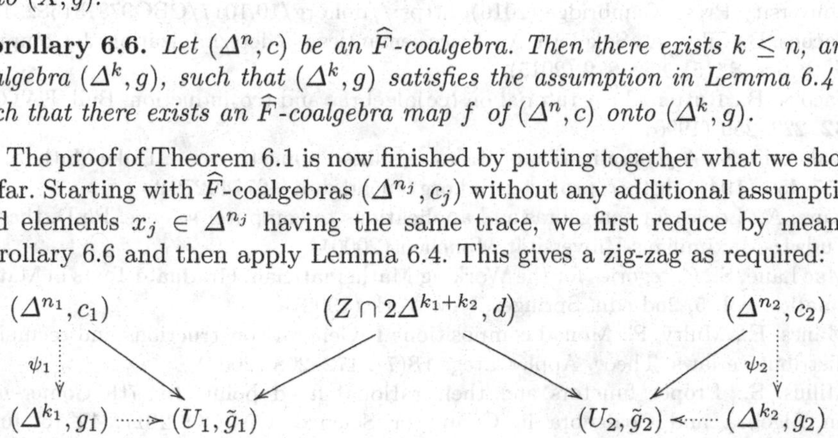

and completes the proof of properness of $\widehat{F}$.

Acknowledgements. We thank the anonymous reviewers for many valuable comments, in particular for reminding us of a categorical property that shortened the proof of the extension lemma.

References

1. Béal, M.-P., Lombardy, S., Sakarovitch, J.: On the equivalence of $\mathbb{Z}$-automata. In: Caires, L., Italiano, G.F., Monteiro, L., Palamidessi, C., Yung, M. (eds.) ICALP 2005. LNCS, vol. 3580, pp. 397–409. Springer, Heidelberg (2005). https://doi.org/10.1007/11523468_33
2. Béal, M.-P., Lombardy, S., Sakarovitch, J.: Conjugacy and equivalence of weighted automata and functional transducers. In: Grigoriev, D., Harrison, J., Hirsch, E.A. (eds.) CSR 2006. LNCS, vol. 3967, pp. 58–69. Springer, Heidelberg (2006). https://doi.org/10.1007/11753728_9
3. Bloom, S.L., Ésik, Z.: Iteration Theories - The Equational Logic of Iterative Processes. EATCS Monographs on Theoretical Computer Science. Springer, Heidelberg (1993). https://doi.org/10.1007/978-3-642-78034-9
4. Bonsangue, M.M., Milius, S., Silva, A.: Sound and complete axiomatizations of coalgebraic language equivalence. CoRR abs/1104.2803 (2011)
5. Doberkat, E.E.: Eilenberg-Moore algebras for stochastic relations. Inf. Comput. **204**(12), 1756–1781 (2006). https://doi.org/10.1016/j.ic.2006.09.001
6. Doberkat, E.E.: Erratum and addendum: Eilenberg-Moore algebras for stochastic relations. Inf. Comput. **206**(12), 1476–1484 (2008). https://doi.org/10.1016/j.ic.2008.08.002. [mr2277336]
7. Ésik, Z., Kuich, W.: A generation of Kozen's Axiomatization of the equational theory of the regular sets. In: Words, Semigroups, and Transductions - Festschrift in Honor of Gabriel Thierrin, pp. 99–114 (2001)
8. Ésik, Z., Maletti, A.: Simulation vs. equivalence. In: Proceedings of the 2010 International Conference on Foundations of Computer Science, FCS 2010, 12–15 July 2010, Las Vegas, Nevada, USA, pp. 119–124 (2010)
9. Jacobs, B.: Introduction to Coalgebra: Towards Mathematics of States and Observation. Cambridge Tracts in Theoretical Computer Science, vol. 59. Cambridge University Press, Cambridge (2016). https://doi.org/10.1017/CBO9781316823187
10. Jacobs, B., Silva, A., Sokolova, A.: Trace semantics via determinization. J. Comput. Syst. Sci. **81**(5), 859–879 (2015)
11. Jacobs, B., Rutten, J.: A tutorial on (co)algebras and (co)induction. Bull. EATCS **62**, 222–259 (1996)
12. Kakutani, S.: A generalization of Brouwer's fixed point theorem. Duke Math. J. **8**, 457–459 (1941). http://projecteuclid.org/euclid.dmj/1077492791
13. Kurz, A.: Logics for coalgebras and applications to computer science. Ph.D. thesis, Ludwig-Maximilians-Universität München (2000)
14. Mac Lane, S.: Categories for the Working Mathematician. Graduate Texts in Mathematics, vol. 5, 2nd edn. Springer, New York (1998)
15. Manes, E., Mulry, P.: Monad compositions I. General constructions and recursive distributive laws. Theor. Appl. Categ. **18**(7), 172–208 (2007)
16. Milius, S.: Proper functors and their rational fixed point. In: 7th Conference on Algebra and Coalgebra in Computer Science, CALCO 2017, 12–16 June 2017, Ljubljana, Slovenia, pp. 18:1–18:16 (2017). https://doi.org/10.4230/LIPIcs.CALCO.2017.18

17. Pumplün, D.: Regularly ordered Banach spaces and positively convex spaces. Res. Math. **7**(1), 85–112 (1984). https://doi.org/10.1007/BF03322493
18. Rockafellar, R.T.: Convex Analysis. Princeton Mathematical Series, vol. 28. Princeton University Press, Princeton (1970)
19. Rudin, W.: Functional Analysis. International Series in Pure and Applied Mathematics, 2nd edn. McGraw-Hill Inc., New York (1991)
20. Rutten, J.: Universal coalgebra: a theory of systems. Theor. Comput. Sci. **249**, 3–80 (2000)
21. Silva, A.: Kleene coalgebra. Ph.D. thesis, Radboud University Nijmegen (2010)
22. Silva, A., Bonchi, F., Bonsangue, M., Rutten, J.: Generalizing the powerset construction, coalgebraically. In: Proceedings of FSTTCS 2010. Leibniz International Proceedings in Informatics (LIPIcs), vol. 8, pp. 272–283 (2010)
23. Silva, A., Sokolova, A.: Sound and complete axiomatization of trace semantics for probabilistic systems. Electr. Notes Theor. Comput. Sci. **276**, 291–311 (2011). https://doi.org/10.1016/j.entcs.2011.09.027
24. Sokolova, A., Woracek, H.: Proper semirings and proper convex functors. arXiv 1802.07830 (2018). https://arxiv.org/abs/1802.07830
25. Staton, S.: Relating coalgebraic notions of bisimulation. Log. Methods Comput. Sci. **7**(1), 1–21 (2011)
26. Wolter, U.: On corelations, cokernels, and coequations. Electron. Notes Theor. Comput. Sci. **33**, 317–336 (2000)

From Symmetric Pattern-Matching to Quantum Control

Amr Sabry[1], Benoît Valiron[2](✉), and Juliana Kaizer Vizzotto[3]

[1] Indiana University, Bloomington, IN, USA
sabry@indiana.edu
[2] LRI, CentraleSupélec, Université Paris-Saclay, Orsay, France
benoit.valiron@lri.fr
[3] Universidade Federal de Santa Maria, Santa Maria, Brazil
juvizzotto@inf.ufsm.br

Abstract. One perspective on quantum algorithms is that they are classical algorithms having access to a special kind of memory with exotic properties. This perspective suggests that, even in the case of quantum algorithms, the control flow notions of sequencing, conditionals, loops, and recursion are entirely classical. There is however, another notion of control flow, that is itself quantum. The notion of quantum conditional expression is reasonably well-understood: the execution of the two expressions becomes itself a superposition of executions. The quantum counterpart of loops and recursion is however not believed to be meaningful in its most general form.

In this paper, we argue that, under the right circumstances, a reasonable notion of quantum loops and recursion is possible. To this aim, we first propose a classical, typed, reversible language with lists and fixpoints. We then extend this language to the *closed* quantum domain (without measurements) by allowing linear combinations of terms and restricting fixpoints to structurally recursive fixpoints whose termination proofs match the proofs of convergence of sequences in infinite-dimensional Hilbert spaces. We additionally give an operational semantics for the quantum language in the spirit of algebraic lambda-calculi and illustrate its expressiveness by modeling several common unitary operations.

1 Introduction

The control flow of a program describes how its elementary operations are organized along the execution. Usual primitive control mechanisms are sequences, tests, iteration and recursion. Elementary operations placed in sequence are executed in order. Tests allow conditionally executing a group of operations and changing the course of the execution of the program. Finally, iteration gives the

B. Valiron and J. K. Vizzotto—Partially funded by FoQCoss STIC AmSud project - STIC-AmSUD/Capes - Foundations of Quantum Computation: Syntax and Semantics.

C. Baier and U. Dal Lago (Eds.): FOSSACS 2018, LNCS 10803, pp. 348–364, 2018.
https://doi.org/10.1007/978-3-319-89366-2_19

possibility to iterate a process an arbitrary number of times and recursion generalizes iteration to automatically manage the history of the operations performed during iteration. The structure of control flow for conventional (classical) computation is well-understood. In the case of *quantum* computation, control flow is still subject to debate. This paper proposes a working notion of quantum control in closed quantum systems, shedding new light on the problem, and clarifying several of the previous concerns.

Quantum Computation. A good starting point for understanding quantum computation is to consider classical circuits over *bits* but replacing the bits with *qubits*, which are intuitively superpositions of bits weighed by complex number amplitudes. Computationally, a qubit is an abstract data type governed by the laws of quantum physics, whose values are normalized vectors of complex numbers in the Hilbert space $\mathbb{C}^2$ (modulo a global phase). By choosing an orthonormal basis, say the classical bits $\mathtt{tt}$ and $\mathtt{ff}$, a qubit can be regarded as a complex linear combination, $\alpha\ \mathtt{tt} + \beta\ \mathtt{ff}$, where α and β are complex numbers such that $|\alpha|^2 + |\beta|^2 = 1$. This generalizes naturally to multiple qubits: the state of a system of n qubits is a vector in the Hilbert space $(\mathbb{C}^2)^{\otimes n}$.

The operations one can perform on a quantum memory are of two kinds: quantum gates and measurements. Quantum gates are unitary operations that are "purely quantum" in the sense that they modify the quantum memory without giving any feedback to the outside world: the quantum memory is viewed as a *closed system*. A customary graphical representation for these operations is the *quantum circuit*, akin to conventional boolean circuits: wires represent qubits while boxes represents operations to perform on them. One of the peculiar aspects of quantum computation is that the state of a qubit is non-duplicable [1], a result known as the *no-cloning theorem*. A corollary is that a quantum circuit is a very simple kind of circuit: wires neither split nor merge.

Measurement is a fundamentally different kind of operation: it queries the state of the quantum memory and returns a classical result. Measuring the state of a quantum bit is a probabilistic and destructive operation: it produces a classical answer with a probability that depends on the amplitudes α, β in the state of the qubit while projecting this state onto $\mathtt{tt}$ or $\mathtt{ff}$, based on the result.

For a more detailed introduction to quantum computation, we refer the reader to recent textbooks (e.g., [2]).

Control Flow in Quantum Computation. In the context of quantum programming languages, there is a well-understood notion of control flow: the so-called *classical control flow*. A quantum program can be seen as the construction, manipulation and evaluation of quantum circuits [3,4]. In this setting, circuits are simply considered as special kinds of data without much computational content, and programs are ruled by regular classical control.

One can however consider the circuit being manipulated as a program in its own right: a particular sequence of execution on the quantum memory is then seen as a closed system. One can then try to derive a notion of *quantum control* [5], with "quantum tests" and "quantum loops". Quantum tests are a bit tricky to perform [5,6] but they essentially correspond to well-understood controlled operations. The situation with quantum loops is more subtle [6,7]. First, a hypothetical quantum loop *must* terminate. Indeed, a non-terminating quantum loop would entail an infinite quantum circuit, and this concept has so far no meaning. Second, the interaction of quantum loops with measurement is problematic: it is known that the canonical model of *open* quantum computation based on superoperators [8,9] is incompatible with such quantum control [6]. Finally, the mathematical operator corresponding to a quantum loop would need to act on an infinite-dimensional Hilbert space and the question of mixing programming languages with infinitary Hilbert spaces is still an unresolved issue.

Our Contribution. In this paper, we offer a novel solution to the question of quantum control: we define a purely quantum language, inspired by Theseus [10], featuring tests and fixpoints in the presence of lists. More precisely, we propose (1) a typed, reversible language, extensible to linear combinations of terms, with a reduction strategy akin to algebraic lambda-calculi [11–13]; (2) a model for the language based on unitary operators over infinite-dimensional Hilbert spaces, simplifying the Fock space model of Ying [7]. This model captures lists, tests, and structurally recursive fixpoints. We therefore settle two longstanding issues. (1) We offer a solution to the problem of quantum loops, with the use of *terminating, structurally recursive, purely quantum* fixpoints. We dodge previously noted concerns (e.g., [6]) by staying in the closed quantum setting and answer the problem of the external system of quantum "coins" [7] with the use of lists. (2) By using a linear language based on patterns and clauses, we give an extensible framework for reconciling algebraic calculi with quantum computation [11,12,16].

In the remainder of the paper, we first introduce the key idea underlying our classical reversible language in a simple first-order setting. We then generalize the setting to allow second-order functions, recursive types (e.g., lists), and fixpoints. After illustrating the expressiveness of this classical language, we adapt it to the quantum domain and give a semantics to the resulting quantum language in infinite-dimensional Hilbert spaces. Technical material that would interrupt the flow or that is somewhat complementary has been relegated to an extended version of the paper [17].

2 Pattern-Matching Isomorphisms

The most elementary control structure in a programming language is the ability to conditionally execute one of several possible code fragments. Expressing such an abstraction using predicates and nested **if**-expressions makes it difficult for both humans and compilers to reason about the control flow structure. Instead, in modern functional languages, this control flow paradigm is elegantly expressed using *pattern-matching*. This approach yields code that is not only more concise and readable but also enables the compiler to easily verify two crucial properties: (i) non-overlapping patterns and (ii) exhaustive coverage of a datatype using a collection of patterns. Indeed most compilers for functional languages perform these checks, warning the user when they are violated. At a more fundamental level, e.g., in type theories and proof assistants, these properties are actually necessary for correct reasoning about programs. Our first insight, explained in this section, is that these properties, perhaps surprisingly, are sufficient to produce a simple and intuitive first-order reversible programming language.

```
f :: Either Int Int -> a
f (Left 0)       = undefined
f (Left (n+1)) = undefined
f (Right n)      = undefined
```

Fig. 1. A skeleton

```
g :: (Bool,Int) -> a
g (False,n)  = undefined
g (True,0)   = undefined
g (True,n+1) = undefined
```

Fig. 2. Another skeleton

```
h :: Either Int Int <-> (Bool,Int)
h (Left 0)       = (True,0)
h (Left (n+1)) = (False,n)
h (Right n)      = (True,n+1)
```

Fig. 3. An isomorphism

2.1 An Example

We start with a small illustrative example, written in a Haskell-like syntax. Figure 1 gives the skeleton of a function `f` that accepts a value of type `Either Int Int`; the patterns on the left-hand side exhaustively cover every possible incoming value and are non-overlapping. Similarly, Fig. 2 gives the skeleton for a function `g` that accepts a value of type `(Bool,Int)`; again the patterns on the left-hand side exhaustively cover every possible incoming value and are non-overlapping. Now we claim that since the types `Either Int Int` and `(Bool,Int)` are isomorphic, we can combine the patterns of `f` and `g` into *symmetric pattern-matching clauses* to produce a reversible function between the types `Either Int Int` and `(Bool,Int)`. Figure 3 gives one such function; there, we suggestively use `<->` to indicate that the function can be executed in either direction. This reversible function is obtained by simply combining the non-overlapping exhaustive patterns on the two sides of a clause. In order to be well-formed in either direction, these clauses are subject to the constraint that each variable occurring on one side must occur exactly once on the other side (and with the same type). Thus it is acceptable to swap the second and third right-hand sides of `h` but not the first and second ones.

2.2 Terms and Types

We present a formalization of the ideas presented above using a simple typed first-order reversible language. The language is two-layered. The first layer contains values, which also play the role of patterns. These values are constructed from variables ranged over x and the introduction forms for the finite types a, b constructed from the unit type and sums and products of types. The second layer contains collections of pattern-matching clauses that denote isomorphisms of type $a \leftrightarrow b$. Computations are chained applications of isomorphisms to values:

$$\begin{array}{llcl}
\text{(Value types)} & a, b & ::= & \mathbb{1} \mid a \oplus b \mid a \otimes b \\
\text{(Iso types)} & T & ::= & a \leftrightarrow b \\
\text{(Values)} & v & ::= & () \mid x \mid \mathtt{inj}_l\, v \mid \mathtt{inj}_r\, v \mid \langle v_1, v_2 \rangle \\
\text{(Isos)} & \omega & ::= & \{ \mid v_1 \leftrightarrow v_1' \mid v_2 \leftrightarrow v_2' \ \dots \} \\
\text{(Terms)} & t & ::= & v \mid \omega\, t
\end{array}$$

The typing rules are defined using two judgments: $\Delta \vdash_v v : a$ for typing values (or *patterns*) and terms; and $\vdash_\omega \omega : a \leftrightarrow b$ for typing collections of pattern-matching clauses denoting an isomorphism. As it is customary, we write $a_1 \otimes a_2 \otimes \cdots \otimes a_n$ for $((a_1 \otimes a_2) \otimes \cdots \otimes a_n)$, and similarly $\langle x_1, x_2, \dots, x_n \rangle$ for $\langle \langle x_1, x_2 \rangle, \dots, x_n \rangle$.

The typing rules for values are the expected ones. The only subtlety is the fact that they are linear: because values act as patterns, we forbid the repetition of variables. A typing context Δ is a set of typed variables $x_1 : a_1, \dots, x_n : a_n$. A value typing judgment is valid if it can be derived from the following rules:

$$\frac{}{\vdash_v () : \mathbb{1}}, \qquad \frac{}{x : a \vdash_v x : a}, \qquad \frac{\Delta_1 \vdash_v v_1 : a \quad \Delta_2 \vdash_v v_2 : b}{\Delta_1, \Delta_2 \vdash_v \langle v_1, v_2 \rangle : a \otimes b}.$$

$$\frac{\Delta \vdash_v v : a}{\Delta \vdash_v \mathtt{inj}_l\, v : a \oplus b}, \qquad \frac{\Delta \vdash_v v : b}{\Delta \vdash_v \mathtt{inj}_r\, v : a \oplus b},$$

The typing rule for term construction is simple and forces the term to be closed:

$$\frac{\vdash_v t : a \quad \vdash_\omega \omega : a \leftrightarrow b}{\vdash_v \omega\, t : b}$$

The most interesting type rule is the one for isomorphisms. We present the rule and then explain it in detail:

$$\frac{\begin{array}{lcll} \Delta_1 \vdash_v v_1 : a & & \Delta_n \vdash_v v_n : a & \forall i \neq j, v_i \bot v_j \quad \dim(a) = n \\ \Delta_1 \vdash_v v_1' : b & \cdots & \Delta_n \vdash_v v_n' : b & \forall i \neq j, v_i' \bot v_j' \quad \dim(b) = n \end{array}}{\vdash_\omega \{ \mid v_1 \leftrightarrow v_1' \mid v_2 \leftrightarrow v_2' \ \dots \} : a \leftrightarrow b}, \tag{1}$$

The rule relies on two auxiliary conditions as motivated in the beginning of the section. These conditions are (i) the orthogonality judgment $v \bot v'$ that formalizes that patterns must be *non-overlapping* and (ii) the condition $\dim(a) = n$ which formalizes that patterns are *exhaustive*. The rules for deriving orthogonality of values or patterns are:

$$\frac{}{\mathtt{inj}_l\, v_1 \bot \mathtt{inj}_r\, v_2} \qquad \frac{}{\mathtt{inj}_r\, v_1 \bot \mathtt{inj}_l\, v_2}$$

$$\frac{v_1 \bot v_2}{\mathtt{inj}_l\, v_1 \bot \mathtt{inj}_l\, v_2} \quad \frac{v_1 \bot v_2}{\mathtt{inj}_r\, v_1 \bot \mathtt{inj}_r\, v_2} \quad \frac{v_1 \bot v_2}{\langle v, v_1 \rangle \bot \langle v', v_2 \rangle} \quad \frac{v_1 \bot v_2}{\langle v_1, v \rangle \bot \langle v_2, v' \rangle}$$

The idea is simply that the left and right injections are disjoint subspaces of values. To characterize that a set of patterns is exhaustive, we associate a *dimension* with each type. For finite types, this is just the number of elements in the type and is inductively defined as follows: $\dim(\mathbb{1}) = 1$; $\dim(a \oplus b) = \dim(a) + \dim(b)$;

and $\dim(a \otimes b) = \dim(a) \cdot \dim(b)$. For a given type a, if a set of non-overlapping clauses has cardinality $\dim(a)$, it is exhaustive. Conversely, any set of exhaustive clauses for a type a either has cardinality $\dim(a)$ or can be extended to an equivalent exhaustive set of clauses of cardinality $\dim(a)$.

2.3 Semantics

We equip our language with a simple operational semantics on terms, using the natural notion of matching. To formally define it, we first introduce the notion of variable assignation, or valuation, which is a partial map from a finite set of variables (the support) to a set of values. We denote the matching of a value w against a pattern v and its associated valuation σ as $\sigma[v] = w$ and define it as follows:

$$\frac{}{\sigma[()] = ()} \qquad \frac{\sigma = \{x \mapsto v\}}{\sigma[x] = v} \qquad \frac{\sigma[v] = w}{\sigma[\mathtt{inj}_l\ v] = \mathtt{inj}_l\ w} \qquad \frac{\sigma[v] = w}{\sigma[\mathtt{inj}_r\ v] = \mathtt{inj}_r\ w}$$

$$\frac{\sigma_2[v_1] = w_1 \quad \sigma_1[v_2] = w_2 \quad \mathrm{supp}(\sigma_1) \cap \mathrm{supp}(\sigma_2) = \emptyset \quad \sigma = \sigma_1 \cup \sigma_2}{\sigma[\langle v_1, v_2\rangle] = \langle w_1, w_2\rangle}$$

If σ is a valuation whose support contains the variables of v, we write $\sigma(v)$ for the value where the variables of v have been replaced with the corresponding values in σ.

Given these definitions, we can define the reduction relation on terms. The redex $\{ \mid v_1 \leftrightarrow v_1' \mid v_2 \leftrightarrow v_2' \ \dots \}\, v$ reduces to $\sigma(v_i')$ whenever $\sigma[v_i] = v_i'$. Because of the conditions on patterns, a matching pattern exists by exhaustivity of coverage, and this pattern is unique by the non-overlapping condition. Congruence holds: $\omega\, t \to \omega\, t'$ whenever $t \to t'$. As usual, we write $s \to t$ to say that s rewrites in one step to t and $s \to^* t$ to say that s rewrites to t in 0 or more steps.

Because of the conditions set on patterns, the rewrite system is deterministic. More interestingly, we can swap the two sides of all pattern-matching clauses in an isomorphism ω to get ω^{-1}. The execution of ω^{-1} is the reverse execution of ω in the sense that $\omega^{-1}(\omega\ t) \to^* t$ and $\omega(\omega^{-1}\ t') \to^* t'$.

3 Second-Order Functions, Lists, and Recursion

The first-order reversible language from the previous section embodies symmetric-pattern matching clauses as its core notion of control. Its expressiveness is limited, however. We now show that it is possible to extend it to have more in common with a conventional functional language. To that end, we extend the language with the ability to parametrically manipulate isomorphisms, with a recursive type (lists), and with recursion.

3.1 Terms and Types

Formally, the language is now defined as follows.

$$
\begin{array}{lrl}
\text{(Val \& term types)} & a, b ::= & \mathbb{1} \mid a \oplus b \mid a \otimes b \mid [a] \\
\text{(Iso types)} & T ::= & a \leftrightarrow b \mid (a \leftrightarrow b) \rightarrow T \\
\\
\text{(Values)} & v ::= & () \mid x \mid \mathtt{inj}_l\, v \mid \mathtt{inj}_r\, v \mid \langle v_1, v_2 \rangle \\
\text{(Products)} & p ::= & () \mid x \mid \langle p_1, p_2 \rangle \\
\text{(Extended Values)} & e ::= & v \mid \mathtt{let}\, p_1 = \omega\; p_2 \;\mathtt{in}\; e \\
\text{(Isos)} & \omega ::= & \{ \mid v_1 \leftrightarrow e_1 \mid v_2 \leftrightarrow e_2 \;\ldots\; \} \mid \lambda f.\omega \mid \\
 & & \mu f.\omega \mid f \mid \omega_1\, \omega_2 \\
\text{(Terms)} & t ::= & () \mid x \mid \mathtt{inj}_l\, t \mid \mathtt{inj}_r\, t \mid \langle t_1, t_2 \rangle \mid \\
 & & \omega\, t \mid \mathtt{let}\, p = t_1 \;\mathtt{in}\; t_2
\end{array}
$$

We use variables f to span a set of iso-variables and variables x to span a set of term-variables. We extend the layer of isos so that it can be parameterized by a fixed number of other isos, i.e., we now allow higher-order manipulation of isos using $\lambda f.\omega$, iso-variables, and applications. Isos can now be used inside the definition of other isos with a let-notation. These let-constructs are however restricted to products of term-variables: they essentially serve as syntactic sugar for composition of isos. An extended value is then a value where some of its free variables are substituted with the result of the application of one or several isos. Given an extended value e, we define its *bottom value*, denoted with $\mathrm{Val}(e)$ as the value "at the end" of the let-chain: $\mathrm{Val}(v) = v$, and $\mathrm{Val}(\mathtt{let}\, p = \omega p \;\mathtt{in}\; e) = \mathrm{Val}(e)$. The orthogonality of extended values is simply the orthogonality of their bottom value.

As usual, the type of lists $[a]$ of elements of type a is a recursive type and is equivalent to $\mathbb{1} \oplus (a \times [a])$. We build the value $[]$ (empty list) as $\mathtt{inj}_l\, ()$ and the term $t_1 : t_2$ (cons of t_1 and t_2) as $\mathtt{inj}_r\, \langle t_1, t_2 \rangle$. In addition, to take full advantage of recursive datatypes, it is natural to consider recursion. Modulo a termination guarantee it is possible to add a fixpoint to the language: we extend isos with the fixpoint constructor $\mu f.\omega$. Some reversible languages allow infinite loops and must work with partial isomorphisms instead. Since we plan on using our language as a foundation for a quantum language we insist of termination.

Since the language features two kinds of variables, there are typing contexts (written Δ) consisting of base-level typed variables of the form $x : a$, and typing context (written Ψ) consisting of typed iso-variables of the form $f : T$. As terms and values contain both base-level and iso-variables, one needs two typing contexts. Typing judgments are therefore written respectively as $\Delta; \Psi \vdash_v t : a$. The updated rules for $(\vdash_v)$ are found in Table 1. As the only possible free variables in isos are iso-variables, their typing judgments only need one context and are written as $\Psi \vdash_\omega \omega : T$.

The rules for typing derivations of isos are in Table 2. It is worthwhile mentioning that isos are treated in a usual, non-linear way: this is the purpose of the typing context separation. The intuition is that an iso is the description of a closed computation with respect to inputs: remark that isos cannot accept

Table 1. Typing rules for terms and values

$$\frac{}{\emptyset;\Psi\vdash_v ():\mathbb{1}} \qquad \frac{}{x:a;\Psi\vdash_v x:a}$$

$$\frac{\Delta;\Psi\vdash_v t:a}{\Delta;\Psi\vdash_v \mathtt{inj}_l\ t:a\oplus b} \quad \frac{\Delta;\Psi\vdash_v t:b}{\Delta;\Psi\vdash_v \mathtt{inj}_r\ t:a\oplus b} \quad \frac{\Delta_1;\Psi\vdash_v t_1:a \quad \Delta_2;\Psi\vdash_v t_2:b}{\Delta_1,\Delta_2;\Psi\vdash_v \langle t_1,t_2\rangle : a\otimes b}$$

$$\frac{\Psi\vdash_\omega \omega:a\leftrightarrow b \quad \Delta;\Psi\vdash_v t:a}{\Delta;\Psi\vdash_v \omega\ t:b} \quad \frac{\Delta;\Psi\vdash_v t_1:a\otimes b \quad \Delta,x:a,y:b;\Psi\vdash_v t_2:c}{\Delta;\Psi\vdash_v \mathtt{let}\ \langle x,y\rangle = t_1\ \mathtt{in}\ t_2 : c}$$

Table 2. Typing rules for isos

$$\frac{\begin{array}{lll}\Delta_1;\Psi\vdash_v v_1:a \quad \dots \quad \Delta_n;\Psi\vdash_v v_n:a & & \mathrm{OD}_a\{v_1,\dots,v_n\} \\ \Delta_1;\Psi\vdash_v e_1:b \quad \dots \quad \Delta_n;\Psi\vdash_v e_n:b & & \mathrm{OD}_b^{ext}\{e_1,\dots,e_n\}\end{array}}{\Psi\vdash_\omega \{\ \mid v_1\leftrightarrow e_1 \mid v_2\leftrightarrow e_2\ \dots\ \}:a\leftrightarrow b.}$$

$$\frac{\Psi,f:a\leftrightarrow b\vdash_\omega \omega:T}{\Psi\vdash_\omega \lambda f.\omega:(a\leftrightarrow b)\to T} \qquad \frac{}{\Psi,f:T\vdash_\omega f:T}$$

$$\frac{\Psi\vdash_\omega \omega_1:(a\leftrightarrow b)\to T \quad \Psi\vdash_\omega \omega_2:a\leftrightarrow b}{\Psi\vdash_\omega \omega_1\omega_2:T}$$

$$\frac{\begin{array}{c}\Psi,f:a\leftrightarrow b\vdash_\omega \omega:(a_1\leftrightarrow b_1)\to\dots\to(a_n\leftrightarrow b_n)\to(a\leftrightarrow b) \\ \mu f.\omega \text{ terminates in any finite context}\end{array}}{\Psi\vdash_\omega \mu f.\omega:(a_1\leftrightarrow b_1)\to\dots\to(a_n\leftrightarrow b_n)\to(a\leftrightarrow b)}$$

value-types. As computations, they can be erased or duplicated without issues. On the other hand, value-types still need to be treated linearly.

In the typing rule for recursion, the condition "$\mu f.\omega$ terminates in any finite context" formally refers to the following requirement. A well-typed fixpoint $\mu f.\omega$ of type $\Psi\vdash_\omega \mu f.\omega:(a_1\leftrightarrow b_1)\to\dots\to(a_n\leftrightarrow b_n)\to(a\leftrightarrow b)$ is *terminating in a* 0*-context* if for all closed isos $\omega_i:a_i\leftrightarrow b_i$ not using fixpoints and for every closed value v of type a, the term $((\mu f.\omega)\omega_1\dots\omega_n)v$ terminates. We say that the fixpoint is *terminating in an* $(n+1)$*-context* if for all closed isos $\omega_i:a_i\leftrightarrow b_i$ terminating in n-contexts, and for every closed value v of type a, the term $((\mu f.\omega)\omega_1\dots\omega_n)v$ terminates. Finally, we say that the fixpoint is *terminating in any finitary context* if for all n it is terminating in any n-context.

With the addition of lists, the non-overlapping and exhaustivity conditions need to be modified. The main problem is that we can no longer define the dimension of types using natural numbers: $[a]$ is in essence an infinite sum, and would have an "infinite" dimension. Instead, we combine the two conditions into the concept of *orthogonal decomposition.* Formally, given a type a, we say that a set S of patterns is an *orthogonal decomposition*, written $\mathrm{OD}_a(S)$, when these patterns are pairwise orthogonal and when they cover the whole type. We

Table 3. Reduction rules

$$\frac{t_1 \to t_2}{C[t_1] \to C[t_2]}\ \text{Cong} \qquad \frac{\sigma[p] = v_1}{\texttt{let}\, p = v_1\ \texttt{in}\ t_2 \to \sigma(t_2)}\ \text{LetE}$$

$$\frac{\sigma[v_i] = v}{\{\ |\ v_1 \leftrightarrow t_1\ |\ \dots\ |\ v_n \leftrightarrow t_n\ \}\ v \to \sigma(t_i)}\ \text{IsoApp} \qquad \frac{}{(\lambda f.\omega)\ \omega_2 \to \omega[\omega_2/f]}\ \text{HIsoApp}$$

$$\frac{\Psi, f : a \leftrightarrow b \vdash_\omega \omega : (a_1 \leftrightarrow b_1) \to \dots \to (a_n \leftrightarrow b_n) \to (a \leftrightarrow b)}{\mu f.\omega \to \lambda f_1 \dots f_n.(\omega[((\mu f.\omega) f_1 \dots f_n)/f]) f_1 \dots f_n}\ \text{IsoRec}$$

formally define $\mathrm{OD}_a(S)$ as follows. For all types a, $\mathrm{OD}_a\{x\}$ is valid. For the unit type, $\mathrm{OD}_{\mathbb{1}}\{()\}$ is valid. If $\mathrm{OD}_a(S)$ and $\mathrm{OD}_b(T)$, then

$$\begin{aligned} &\mathrm{OD}_{a\oplus b}(\{\texttt{inj}_l\ v \mid v \in S\} \cup \{\texttt{inj}_r\ v \mid v \in T\}) \\ \text{and}\quad &\mathrm{OD}_{a\otimes b}\{\langle v_1, v_2\rangle \mid v_1 \in S,\ v_2 \in T,\ \mathrm{FV}(v_1) \cap \mathrm{FV}(v_2) = \emptyset\}, \end{aligned}$$

where $\mathrm{FV}(t)$ stands for the set of free value-variables in t. We then extend the notion of orthogonal decomposition to extended values as follows. If S is a set of extended values, $\mathrm{OD}_a^{ext}(S)$ is true whenever $\mathrm{OD}_a\{\mathrm{Val}(e) \mid e \in S\}$. With this new characterization, the typing rule of iso in Eq. 1 still holds, and then can be re-written using this notion of orthogonal decomposition as shown in Table 2.

3.2 Semantics

In Table 3 we present the reduction rules for the reversible language. We assume that the reduction relation applies to well-typed terms. In the rules, the notation $C[-]$ stands for an *applicative context*, and is defined as: $C[-] ::= [-] \mid \texttt{inj}_l\ C[-] \mid \texttt{inj}_r\ C[-] \mid (C[-])\omega \mid \{\cdots\}\ (C[-]) \mid \texttt{let}\, p = C[-]\ \texttt{in}\ t_2 \mid \langle C[-], v\rangle \mid \langle v, C[-]\rangle$.

The inversion of isos is still possible but more subtle than in the first-order case. We define an inversion operation $(-)^{-1}$ on iso types with, $(a \leftrightarrow b)^{-1} := (b \leftrightarrow a)$, $((a \leftrightarrow b) \to T)^{-1} := ((b \leftrightarrow a) \to (T^{-1}))$. Inversion of isos is defined as follows. For fixpoints, $(\mu f.\omega)^{-1} = \mu f.(\omega^-1)$. For variables, $(f)^{-1} := f$. For applications, $(\omega_1\ \omega_2)^{-1} := (\omega_1)^{-1}\ (\omega_2)^{-1}$. For abstraction, $(\lambda f.\omega)^{-1} := \lambda f.(\omega^{-1})$. Finally, clauses are inverted as follows:

$$\begin{pmatrix} v_1 \leftrightarrow \texttt{let}\, p_1 = \omega_1\, p_1'\ \texttt{in} \\ \dots \\ \texttt{let}\, p_n = \omega_n\, p_n'\ \texttt{in}\ v_1' \end{pmatrix}^{-1} := \begin{pmatrix} v_1' \leftrightarrow \texttt{let}\, p_n' = \omega_n^{-1}\, p_n\ \texttt{in} \\ \dots \\ \texttt{let}\, p_1' = \omega_1^{-1}\, p_1\ \texttt{in}\ v_1 \end{pmatrix}.$$

Note that $(-)^{-1}$ only inverts first-order arrows $(\leftrightarrow)$, not second-order arrows $(\to)$. This is reflected by the fact that iso-variable are non-linear while value-variables are. This is due to the clear separation of the two layers of the language.

The rewriting system satisfies the usual properties for well-typed terms: it is terminating, well-typed closed terms have a unique normal value-form, and it preserves typing.

Theorem 1. *The inversion operation is well-typed, in the sense that if* $f_1 : a_1 \leftrightarrow b_1, \ldots, f_n : a_n \leftrightarrow b_n \vdash_\omega \omega : T$ *then we also have* $f_1 : b_1 \leftrightarrow a_1, \ldots, f_n : b_n \leftrightarrow a_n \vdash_\omega \omega^{-1} : T^{-1}$. □

Thanks to the fact that the language is terminating, we also recover the operational result of Sect. 2.3.

Theorem 2. *Consider a well-typed, closed iso* $\vdash_\omega \omega : a \leftrightarrow b$*, and suppose that* $\vdash_v v : a$ *and that* $\vdash_v w : b$*, then* $\omega^{-1}(\omega\ v) \to^* v$ *and* $\omega(\omega^{-1}\ w) \to^* w$. □

4 Examples

In the previous sections, we developed a novel classical reversible language with a familiar syntax based on pattern-matching. The language includes a limited notion of higher-order functions and (terminating) recursive functions. We illustrate the expressiveness of the language with a few examples and motivate the changes and extensions needed to adapt the language to the quantum domain.

We encode booleans as follows: $\mathbb{B} = \mathbb{1} \oplus \mathbb{1}$, $\mathtt{tt} = \mathtt{inj}_l\ ()$, and $\mathtt{ff} = \mathtt{inj}_r\ ()$. One of the easiest function to define is $\mathtt{not} : \mathbb{B} \leftrightarrow \mathbb{B}$ which flips a boolean. The controlled-not gate which flips the second bit when the first is true can also be expressed:

$$\mathtt{not} : \mathbb{B} \leftrightarrow \mathbb{B} = \begin{pmatrix} \mathtt{ff} & \leftrightarrow & \mathtt{tt} \\ \mathtt{tt} & \leftrightarrow & \mathtt{ff} \end{pmatrix}, \quad \mathtt{cnot} : \mathbb{B} \otimes \mathbb{B} \leftrightarrow \mathbb{B} \otimes \mathbb{B} = \begin{pmatrix} \langle \mathtt{ff}, x \rangle & \leftrightarrow & \langle \mathtt{ff}, x \rangle \\ \langle \mathtt{tt}, \mathtt{ff} \rangle & \leftrightarrow & \langle \mathtt{tt}, \mathtt{tt} \rangle \\ \langle \mathtt{tt}, \mathtt{tt} \rangle & \leftrightarrow & \langle \mathtt{tt}, \mathtt{ff} \rangle \end{pmatrix}.$$

All the patterns in the previous two functions are orthogonal decompositions which guarantee reversibility as desired.

By using the abstraction facilities in the language, we can define higher-order operations that build complex reversible functions from simpler ones. For example, we can define a conditional expression parameterized by the functions used in the two branches:

$$\mathtt{if} : (a \leftrightarrow b) \to (a \leftrightarrow b) \to (\mathbb{B} \otimes a \leftrightarrow \mathbb{B} \otimes b)$$
$$\mathtt{if} = \lambda g.\lambda h. \begin{pmatrix} \langle \mathtt{tt}, x \rangle & \leftrightarrow & \mathtt{let}\ y = g\ x\ \mathtt{in}\ \langle \mathtt{tt}, y \rangle \\ \langle \mathtt{ff}, x \rangle & \leftrightarrow & \mathtt{let}\ y = h\ x\ \mathtt{in}\ \langle \mathtt{ff}, y \rangle \end{pmatrix}$$

Using `if` and the obvious definition for the identity function `id`, we can define $\mathtt{ctrl} :: (a \leftrightarrow a) \to (\mathbb{B} \otimes a \leftrightarrow \mathbb{B} \otimes a)$ as $\mathtt{ctrl}\ f = \mathtt{if}\ f\ \mathtt{id}$ and recover an alternative definition of `cnot` as `ctrl not`. We can then define the controlled-controlled-not gate (aka the Toffoli gate) by writing `ctrl cnot`. We can even iterate this construction using fixpoints to produce an n-controlled-not function that takes a list of n control bits and a target bit and flips the target bit iff all the control bits are `tt`:

$$\texttt{cnot*} : ([\mathbb{B}] \otimes \mathbb{B}) \leftrightarrow ([\mathbb{B}] \otimes \mathbb{B})$$
$$\texttt{cnot*} = \mu f. \left(\begin{array}{rcl} \langle [], tb \rangle & \leftrightarrow & \texttt{let}\, tb' = \texttt{not}\ tb\ \texttt{in}\ \langle [], tb' \rangle \\ \langle \texttt{ff} : cbs, tb \rangle & \leftrightarrow & \langle \texttt{ff} : cbs, tb \rangle \\ \langle \texttt{tt} : cbs, tb \rangle & \leftrightarrow & \texttt{let}\ \langle cbs', tb' \rangle = f\ \langle cbs, tb \rangle\ \texttt{in}\ \langle \texttt{tt} : cbs', tb' \rangle \end{array} \right)$$

The language is also expressible enough to write conventional recursive (and higher-order) programs. We illustrate this expressiveness using the usual `map` operation and an accumulating variant `mapAccu`:

$$\texttt{map} : (a \leftrightarrow b) \to ([a] \leftrightarrow [b])$$
$$\lambda g.\mu f. \left(\begin{array}{rcl} [] & \leftrightarrow & [] \\ h : t & \leftrightarrow & \texttt{let}\, x = g\ h\ \texttt{in} \\ & & \texttt{let}\, y = f\ t\ \texttt{in}\ x : y \end{array} \right),$$

$$\texttt{mapAccu} : (a \otimes b \leftrightarrow a \otimes c) \to (a \otimes [b] \leftrightarrow a \otimes [c])$$
$$\lambda g.\mu f. \left(\begin{array}{rcl} \langle x, [] \rangle & \leftrightarrow & \langle x, [] \rangle \\ \langle x, (h : t) \rangle & \leftrightarrow & \texttt{let}\ \langle y, h' \rangle = g\ \langle x, h \rangle\ \texttt{in} \\ & & \texttt{let}\ \langle z, t' \rangle = f\ \langle y, t \rangle\ \texttt{in} \\ & & \langle z, (h' : t') \rangle \end{array} \right).$$

The three examples `cnot*`, `map` and `mapAccu` uses fixpoints which are clearly terminating in any finite context. Indeed, the functions are structurally recursive. A formal definition of this notion for the reversible language is as follows.

$$\begin{array}{c} \\ v_1' \\ v_2' \\ v_3' \end{array} \begin{array}{c} \begin{array}{ccc} v_1 & v_2 & v_3 \end{array} \\ \begin{pmatrix} 1 & 0 & 0 \\ 0 & 1 & 0 \\ 0 & 0 & 1 \end{pmatrix} \end{array}$$

Fig. 4. Classical iso

$$\begin{array}{c} \\ v_1' \\ v_2' \\ v_3' \end{array} \begin{array}{c} \begin{array}{ccc} v_1 & v_2 & v_3 \end{array} \\ \begin{pmatrix} a_{11} & a_{12} & a_{13} \\ a_{21} & a_{22} & a_{23} \\ a_{31} & a_{32} & a_{33} \end{pmatrix} \end{array}$$

Fig. 5. Quantum iso

$$\begin{array}{c} \\ \langle \texttt{tt}, x \rangle \\ \langle \texttt{ff}, x \rangle \end{array} \begin{array}{c} \begin{array}{cc} \langle \texttt{tt}, x \rangle & \langle \texttt{ff}, x \rangle \end{array} \\ \begin{pmatrix} \frac{1}{\sqrt{2}}\texttt{Had} & \frac{1}{\sqrt{2}}\texttt{Id} \\ \frac{1}{\sqrt{2}}\texttt{Had} & \frac{-1}{\sqrt{2}}\texttt{Id} \end{pmatrix} \end{array}$$

Fig. 6. Semantics of `Gate`

Definition 1. Define a *structurally recursive type* as a type of the form $[a] \otimes b_1 \otimes \ldots \otimes b_n$. Let $\omega = \{v_i \leftrightarrow e_i \mid i \in I\}$ be an iso such that $f : a \leftrightarrow b \vdash_\omega \omega : a \leftrightarrow c$ where a is a structurally recursive type. We say that $\mu f.\omega$ is *structurally recursive* provided that for each $i \in I$, the value v_i is either of the form $\langle [], p_1, \ldots p_n \rangle$ or of the form $\langle h : t, p_1, \ldots p_n \rangle$. In the former case, e_i does not contain f as a free variable. In the latter case, e_i is of the form $C[f \langle t, p_1', \ldots, p_n' \rangle]$ where C is a context of the form $C[-] ::= [-] \mid \texttt{let}\, p = C[-]\ \texttt{in}\, t \mid \texttt{let}\, p = t\ \texttt{in}\ C[-]$.

This definition will be critical for quantum loops in the next section.

5 From Reversible Isos to Quantum Control

In the language presented so far, an iso $\omega : a \leftrightarrow b$ describes a bijection between the set $\mathcal{B}_a$ of closed values of type a and the set $\mathcal{B}_b$ of closed values of type b. If one regards $\mathcal{B}_a$ and $\mathcal{B}_b$ as the basis elements of some vector space $[\![a]\!]$ and $[\![b]\!]$, the iso ω becomes a 0/1 matrix.

As an example, consider an iso ω defined using three clauses of the form $\{ \mid v_1 \leftrightarrow v_1' \mid v_2 \leftrightarrow v_2' \mid v_3 \leftrightarrow v_3' \}$. From the exhaustivity and non-overlapping conditions derives the fact that the space $[\![a]\!]$ can be split into the direct sum of the three subspaces $[\![a]\!]_{v_i}$ $(i = 1, 2, 3)$ generated by v_i. Similarly, $[\![b]\!]$ is split into the direct sum of the subspaces $[\![b]\!]_{v'_i}$ generated by v'_i. One can therefore represent ω as the matrix $[\![\omega]\!]$ in Fig. 4: The "1" in each column v_i indicates to which subspace $[\![b]\!]_{v'_j}$ an element of $[\![a]\!]_{v_i}$ is sent to.

In Sect. 2.2 we discussed the fact that $v_i \perp v_j$ when $i \neq j$. This notation hints at the fact that $[\![a]\!]$ and $[\![b]\!]$ could be seen as Hilbert spaces and the mapping $[\![\omega]\!]$ as a unitary map from $[\![a]\!]$ to $[\![b]\!]$. The purpose of this section is to extend and formalize precisely the correspondence between isos and unitary maps.

The definition of clauses is extended following this idea of seeing isos as unitaries, and not only bijections on basis elements of the input space. We therefore essentially propose to generalize the clauses to complex, linear combinations of values on the right-hand-side, such as shown on the left, with the side conditions on that the matrix of Fig. 5 is unitary. We define in Sect. 5.1 how this extends to second-order.

$$\left\{\begin{array}{l} \mid v_1 \leftrightarrow a_{11}v'_1 + a_{21}v'_2 + a_{31}v'_3 \\ \mid v_2 \leftrightarrow a_{12}v'_1 + a_{22}v'_2 + a_{23}v'_3 \\ \mid v_3 \leftrightarrow a_{31}v'_1 + a_{32}v'_2 + a_{33}v'_3 \end{array}\right\}$$

5.1 Extending the Language to Linear Combinations of Terms

The quantum unitary language extends the reversible language from the previous section by closing extended values and terms under complex, finite linear combinations. For example, if v_1 and v_2 are values and α and β are complex numbers, $\alpha \cdot v_1 + \beta \cdot v_2$ is now an extended value.

Several approaches exist for performing such an extension. One can update the reduction strategy to be able to reduce these sums and scalar multiplications to normal forms [12,18], or one can instead consider terms modulo the usual algebraic equalities [13,18]: this is the strategy we follow for this paper.

When extending a language to linear combination of terms in a naive way, this added structure might generate inconsistencies in the presence of unconstrained fixpoints [12,13,18]. The weak condition on termination we imposed on fixpoints in the classical language was enough to guarantee reversibility. With the presence of linear combinations, we want the much stronger guarantee of unitarity. For this reason, we instead impose fixpoints to be *structurally recursive*.

The quantum unitary language is defined by allowing sums of terms and values and multiplications by complex numbers: if t and t' are terms, so is $\alpha \cdot t + t'$. Terms and values are taken modulo the equational theory of modules. We furthermore consider the value and term constructs $\langle -, - \rangle$, $\texttt{let}\, p = -\ \texttt{in}\ -$, $\texttt{inj}_l\,(-)$, $\texttt{inj}_r\,(-)$ distributive over sum and scalar multiplication. We do *not* however take iso-constructions as distributive over sum and scalar multiplication: $\{\ \mid\ v_1 \leftrightarrow \alpha v_2 + \beta v_3\ \}$ is *not* the same thing as $\alpha\{\ \mid\ v_1 \leftrightarrow v_2\ \} + \beta\{\ \mid\ v_1 \leftrightarrow v_3\ \}$. This is in the spirit of Lineal [11,12].

The typing rules for terms and extended values are updated as follows. We only allow linear combinations of terms and values of the same type and of the same free variables. Fixpoints are now required to be *structurally recursive*, as introduced in Definition 1. Finally, an iso is now not only performing an "identity" as in Fig. 4 but a true unitary operation:

$$\frac{\begin{array}{lll}\Delta_1;\Psi\vdash_v v_1:a & \dots & \Delta_n;\Psi\vdash_v v_n:a \\ \Delta_1;\Psi\vdash_v e_1:b & \dots & \Delta_n;\Psi\vdash_v e_n:b \\ \mathrm{OD}_a\{v_1,\dots,v_n\} & & \mathrm{OD}_b^{ext}\{e_1,\dots,e_n\}\end{array} \quad \begin{pmatrix} a_{11} & \cdots & a_{1n} \\ \vdots & & \vdots \\ a_{n1} & \cdots & a_{nn}\end{pmatrix} \text{ is unitary}}{\Psi\vdash_\omega \left\{\begin{array}{l} v_1 \leftrightarrow a_{11}\cdot e_1 + \cdots + a_{1n}\cdot e_n \\ \quad\dots \\ v_n \leftrightarrow a_{n1}\cdot e_1 + \cdots + a_{nn}\cdot e_n\end{array}\right\} : a\leftrightarrow b.}$$

The reduction relation is updated in a way that it remains deterministic in this extended setting. It is split into two parts: the reduction of pure terms, i.e. non-extended terms or values, and linear combinations thereof. Pure terms and values reduce using the reduction rules found in Table 3. We do not extend applicative contexts to linear combinations. For linear combinations of pure terms, we simply ask that *all* pure terms that are not normal forms in the combination are reduced. This makes the extended reduction relation deterministic.

Example 1. This allows one to define an iso behaving as the Hadamard gate, or a slightly more complex iso conditionally applying another iso, whose behavior as a matrix is shown in Fig. 6.

$$\begin{array}{ll}\texttt{Had} : \mathbb{B}\leftrightarrow\mathbb{B} & \texttt{Gate} : \mathbb{B}\otimes\mathbb{B}\leftrightarrow\mathbb{B}\otimes\mathbb{B} \\ \begin{pmatrix}\texttt{tt} & \leftrightarrow & \frac{1}{\sqrt{2}}\texttt{tt}+\frac{1}{\sqrt{2}}\texttt{ff} \\ \texttt{ff} & \leftrightarrow & \frac{1}{\sqrt{2}}\texttt{tt}-\frac{1}{\sqrt{2}}\texttt{ff}\end{pmatrix}, & \begin{pmatrix}\langle\texttt{tt},x\rangle & \leftrightarrow & \texttt{let}\,y=\texttt{Had}\,x\ \texttt{in}\ \frac{1}{\sqrt{2}}\langle\texttt{tt},y\rangle+\frac{1}{\sqrt{2}}\langle\texttt{ff},y\rangle \\ \langle\texttt{ff},x\rangle & \leftrightarrow & \texttt{let}\,y=\texttt{Id}\,x\ \texttt{in}\ \frac{1}{\sqrt{2}}\langle\texttt{tt},y\rangle-\frac{1}{\sqrt{2}}\langle\texttt{ff},y\rangle\end{pmatrix}.\end{array}$$

With this extension to linear combinations of terms, one can characterize normal forms as follows.

Lemma 1 (Structure of the Normal Forms). *Let ω be such that $\vdash_\omega \omega : a\leftrightarrow b$. For all closed values v of type a, the term $\omega\, v$ rewrites to a normal form $\sum_{i=1}^{N}\alpha_i\cdot w_i$ where $N<\infty$, each w_i is a closed value of type b and $\sum_i|\alpha_i|=1$.*

Proof. The fact that $\omega\, v$ converges to a normal form is a corollary of the fact that we impose structural recursion on fixpoints. The property of the structure of the normal form is then proven by induction on the maximal number of steps it takes to reach it. It uses the restriction on the introduction of sums in the typing rule for clauses in isos and the determinism of the reduction. □

In the classical setting, isos describe bijections between sets of closed values: it was proven by considering the behavior of an iso against its inverse. In the presence of linear combinations of terms, we claim that isos describe more than bijections: they describe unitary maps. In the next section, we discuss how types can be understood as Hilbert spaces (Sect. 5.2) and isos as unitary maps (Sects. 5.3 and 5.4).

5.2 Modeling Types as Hilbert Spaces

By allowing complex linear combinations of terms, closed normal forms of finite types such as $\mathbb{B}$ or $\mathbb{B}\otimes\mathbb{B}$ can be regarded as complex vector spaces with basis consisting of closed values. For example, $\mathbb{B}$ is associated with $[\![\mathbb{B}]\!] = \{\alpha \cdot \texttt{tt} + \beta \cdot \texttt{ff} \mid \alpha, \beta \in \mathbb{C}\} \equiv \mathbb{C}^2$. We can consider this space as a complex Hilbert space where the scalar product is defined on basis elements in the obvious way: $\langle v|v\rangle = 1$ and $\langle v|w\rangle = 0$ if $v \neq w$. The map Had of Example 1 is then effectively a unitary map on the space $[\![\mathbb{B}]\!]$.

The problem comes from lists: the type $[\mathbb{1}]$ is inhabited by an infinite number of closed values: $[]$, $[()]$, $[(),()]$, $[(),(),()]$,... To account for this case, we need to consider infinitely dimensional complex Hilbert spaces. In general, a complex Hilbert space [19] is a complex vector space endowed with a scalar product that is complete with respect the distance induced by the scalar product. The completeness requirement implies for example that the infinite linear combination $[] + \frac{1}{2} \cdot [()] + \frac{1}{4}[(),()] + \frac{1}{8}[(),(),()] + \cdots$ needs to be an element of $[\![[\mathbb{B}]]\!]$. To account for these limit elements, we propose to use the standard [19] Hilbert space ℓ^2 of infinite sequences.

Definition 2. Let a be a value type. As before, we write $\mathcal{B}_a$ for the set of closed values of type a, that is, $\mathcal{B}_a = \{v \mid \vdash_v v : a\}$. The *span of* a is defined as the Hilbert space $[\![a]\!] = \ell^2(\mathcal{B}_a)$ consisting of sequences $(\phi_v)_{v\in\mathcal{B}_a}$ of complex numbers indexed by $\mathcal{B}_a$ such that $\sum_{v\in\mathcal{B}_a} |\phi_v|^2 < \infty$. The scalar product on this space is defined as $\langle(\phi_v)_{v\in\mathcal{B}_a}|(\psi_v)_{v\in\mathcal{B}_a}\rangle = \sum_{v\in\mathcal{B}_a} \overline{\phi_v}\psi_v$.

We shall use the following conventions. A closed value v of $[\![a]\!]$ is identified with the sequence $(\delta_{v,v'})_{v'\in\mathcal{B}_a}$ where $\delta_{v,v} = 1$ and $\delta_{v,v'} = 0$ if $v \neq v'$. An element $(\phi_v)_{v\in\mathcal{B}_a}$ of $[\![a]\!]$ is also written as the infinite, formal sum $\sum_{v\in\mathcal{B}_a} \phi_v \cdot v$.

5.3 Modeling Isos as Bounded Linear Maps

We can now define what is the linear map associated to an iso.

Definition 3. For each closed iso $\vdash_\omega \omega : a \leftrightarrow b$ we define $[\![\omega]\!]$ as the linear map from $[\![a]\!]$ to $[\![b]\!]$ sending the closed value $v : a$ to the normal form of $\omega\, v : b$ under the rewrite system.

In general, the fact that $[\![\omega]\!]$ is well-defined is not trivial. If it is formally stated in Theorem 3, we can first try to understand what could go wrong. The problem comes from the fact that the space $[\![a]\!]$ is not finite in general. Consider the iso `map Had` : $[\mathbb{B}] \leftrightarrow [\mathbb{B}]$. Any closed value $v : [\mathbb{B}]$ is a list and the term (`map Had`) v rewrites to a normal form consisting of a linear combination of lists. Denote the linear combination associated to v with L_v. An element of $[\![[\mathbb{B}]]\!]$ is a sequence $\phi = (\phi_v)_{v\in\mathcal{B}_{[\mathbb{B}]}}$. From Definition 3, the map $[\![\omega]\!]$ sends the element $\phi \in [\![[\mathbb{B}]]\!]$ to $\sum_{v\in\mathcal{B}_{[\mathbb{B}]}} \phi_v \cdot L_v$. This is an infinite sum of sums of complex numbers: we need to make sure that it is well-defined: this is the purpose of the next result. Because of the constraints on the language, we can even show that it is a *bounded* linear map.

In the case of the map `map Had`, we can understand why it works as follows. The space $[\![[\mathbb{B}]]\!]$ can be decomposed as the direct sum $\sum_{i=0}^{\infty} E_i$, where E_i is generated with all the lists in $\mathbb{B}$ of size i. The map `map Had` is acting locally on each finitely-dimensional subspace E_i. It is therefore well-defined. Because of the unitarity constraint on the linear combinations appearing in `Had`, the operation performed by `map Had` sends elements of norm 1 to elements of norm 1. This idea can be formalized and yield the following theorem.

Theorem 3. *For each closed iso $\vdash_\omega \omega : a \leftrightarrow b$ the linear map $[\![\omega]\!] : [\![a]\!] \to [\![b]\!]$ is well-defined and bounded.* □

5.4 Modeling Isos as Unitary Maps

In this section, we show that not only closed isos can be modeled as bounded linear maps, but that these linear maps are in fact unitary maps. The problem comes from fixpoints. We first consider the case of isos written without fixpoints, and then the case with fixpoints.

Without recursion. The case without recursion is relatively easy to treat, as the linear map modeling the iso can be compositionally constructed out of elementary unitary maps.

Theorem 4. *Given a closed iso $\vdash_\omega \omega : a \leftrightarrow b$ defined without the use of recursion, the linear map $[\![\pi]\!] : [\![a]\!] \to [\![b]\!]$ is unitary.* □

The proof of the theorem relies on the fact that to each closed iso $\vdash_\omega \omega : a \leftrightarrow b$ one can associate an operationally equivalent iso $\vdash_\omega \omega' : a \leftrightarrow b$ that does not use iso-variables nor lambda-abstractions. We can define a notion of *depth* of an iso as the number of nested isos. The proof is done by induction on this depth of the iso ω: it is possible to construct a unitary map for ω using the unitary maps for each ω_{ij} as elementary building blocks.

As an illustration, the semantics of `Gate` of Example 1 is given in Fig. 6.

Isos with structural recursion. When considering fixpoints, we cannot rely anymore on this finite compositional construction: the space $[\![a]\!]$ cannot anymore be regarded as a *finite* sum of subspaces described by each clause.

We therefore need to rely on the formal definition of unitary maps in general, infinite Hilbert spaces. On top of being bounded linear, a map $[\![\omega]\!] : [\![a]\!] \to [\![b]\!]$ is unitary if (1) it preserves the scalar product: $\langle [\![\omega]\!](e) | [\![\omega]\!](f) \rangle = \langle e | f \rangle$ for all e and f in $[\![a]\!]$ and (2) it is surjective.

Theorem 5. *Given a closed iso $\vdash_\omega \omega : a \leftrightarrow b$ that can use structural recursion, the linear map $[\![\pi]\!] : [\![a]\!] \to [\![b]\!]$ is unitary.* □

The proof uses the idea highlighted in Sect. 5.4: for a structurally recursive iso of type $[a] \otimes b \leftrightarrow c$, the Hilbert space $[\![[a] \otimes b]\!]$ can be split into a canonical decomposition $E_0 \oplus E_1 \oplus E_2 \oplus \cdots$, where E_i contains only the values of the form $\langle [x_1 \ldots x_i], y \rangle$, containing the lists of size i. On each E_i, the iso is equivalent to an iso without structural recursion.

6 Conclusion

In this paper, we proposed a reversible language amenable to quantum superpositions of values. The language features a weak form of higher-order that is nonetheless expressible enough to get interesting maps such as generalized Toffoli operators. We sketched how this language effectively encodes bijections in the classical case and unitary operations in the quantum case. It would be interesting to see how this relates to join inverse categories [14,15].

In the vectorial extension of the language we have the same control as in the classical, reversible language. Tests are captured by clauses, and naturally yield quantum tests: this is similar to what can be found in QML [5,6], yet more general since the QML approach is restricted to `if-then-else` constructs. The novel aspect of quantum control that we are able to capture here is a notion of *quantum loops*. These loops were believed to be hard, if not impossible. What makes it work in our approach is the fact that we are firmly within a closed quantum system, without measurements. This makes it possible to only consider unitary maps and frees us from the Löwer order on positive matrices [6]. As we restrict fixpoints to structural recursion, valid isos are regular enough to capture unitarity. Ying [7] also proposes a framework for quantum while-loops that is similar in spirit to our approach at the level of denotations: in his approach the control part of the loops is modeled using an external systems of "coins" which, in our case, correspond to conventional lists. Reducing the manipulation of this external coin system to iteration on lists allowed us to give a simple operational semantics for the language.

References

1. Wootters, W.K., Zurek, W.H.: A single quantum cannot be cloned. Nature **299**, 802–803 (1982)
2. Nielsen, M.A., Chuang, I.L.: Quantum Computation and Quantum Information. Cambridge University Press, Cambridge (2002)
3. Green, A.S., Lumsdaine, P.L., Ross, N.J., Selinger, P., Valiron, B.: Quipper: a scalable quantum programming language. In: Proceedings of PLDI 2013, pp. 333–342 (2013)
4. Paykin, J., Rand, R., Zdancewic, S.: QWIRE: A core language for quantum circuits. In: Proceedings of POPL 2017, pp. 846–858 (2017)
5. Altenkirch, T., Grattage, J.: A functional quantum programming language. In: Proceedings of LICS 2005, pp. 249–258 (2005)
6. Badescu, C., Panangaden, P.: Quantum alternation: Prospects and problems. In: Proceedings 12th International Workshop on Quantum Physics and Logic, QPL 2015, Oxford, UK, 15–17 July 2015, pp. 33–42 (2015)
7. Ying, M.: Foundations of Quantum Programming. Morgan Kaufmann, Cambridge (2016)
8. Selinger, P.: Towards a quantum programming language. Math. Struct. Comput. Sci. **14**(4), 527–586 (2004)
9. Vizzotto, J.K., Altenkirch, T., Sabry, A.: Structuring quantum effects: superoperators as arrows. Math. Struct. Comput. Sci. **16**(3), 453–468 (2006)

10. James, R.P., Sabry, A.: Theseus: a high-level language for reversible computation. In: Reversible Computation, Booklet of work-in-progress and short reports (2016)
11. Arrighi, P., Díaz-Caro, A., Valiron, B.: The vectorial λ-calculus. Inf. Comput. **254**(1), 105–139 (2017)
12. Arrighi, P., Dowek, G.: Lineal: a linear-algebraic lambda-calculus. Log. Methods Comput. Sci. **13**(1) (2013). https://doi.org/10.23638/LMCS-13(1:8)2017
13. Vaux, L.: The algebraic lambda calculus. Math. Struct. Comput. Sci. **19**(5), 1029–1059 (2009)
14. Glück, R., Kaarsgaard, R.: A categorical foundation for structured reversible flowchart languages: soundness and adequacy. arXiv:1710.03666 [cs.PL] (2017)
15. Kaarsgaard, R., Axelsen, H.B., Glück, R.: Join inverse categories and reversible recursion. J. Log. Algebraic Methods Program. **87**, 33–50 (2017)
16. van Tonder, A.: A lambda calculus for quantum computation. SIAM J. Comput. **33**(5), 1109–1135 (2004)
17. Sabry, A., Valiron, B., Vizzotto, J.K.: From symmetric pattern-matching to quantum control (extended version). In: FOSSACS 2018 (2018, to appear)
18. Assaf, A., Díaz-Caro, A., Perdrix, S., Tasson, C., Valiron, B.: Call-by-value, call-by-name and the vectorial behaviour of the algebraic λ-calculus. Log. Methods Comput. Sci. **10**(4:8) (2014)
19. Young, N.: An Introduction to Hilbert Space. Cambridge University Press, New York (1988)

Quantitative Models

The Complexity of Graph-Based Reductions for Reachability in Markov Decision Processes

Stéphane Le Roux[1](✉) and Guillermo A. Pérez[2]

[1] Department of Mathematics, Technische Universität Darmstadt, Darmstadt, Germany
leroux@mathematik.tu-darmstadt.de

[2] Departement d'Informatique, Université libre de Bruxelles, Brussels, Belgium
gperezme@ulb.ac.be

Abstract. We study the never-worse relation (NWR) for Markov decision processes with an infinite-horizon reachability objective. A state q is never worse than a state p if the maximal probability of reaching the target set of states from p is at most the same value from q, regardless of the probabilities labelling the transitions. Extremal-probability states, end components, and essential states are all special cases of the equivalence relation induced by the NWR. Using the NWR, states in the same equivalence class can be collapsed. Then, actions leading to sub-optimal states can be removed. We show that the natural decision problem associated to computing the NWR is coNP-complete. Finally, we extend a previously known incomplete polynomial-time iterative algorithm to under-approximate the NWR.

1 Introduction

Markov decision processes (MDPs) are a useful model for decision-making in the presence of a stochastic environment. They are used in several fields, including robotics, automated control, economics, manufacturing and in particular planning [20], model-based reinforcement learning [22], and formal verification [1]. We elaborate on the use of MDPs and the need for graph-based reductions thereof in verification and reinforcement learning applications below.

Several verification problems for MDPs reduce to reachability [1,5]. For instance, MDPs can be model checked against linear-time objectives (expressed in, say, LTL) by constructing an omega-automaton recognizing the set of runs that satisfy the objective and considering the product of the automaton with the original MDP [6]. In this product MDP, accepting end components—a generalization of strongly connected components—are identified and selected as target components. The question of maximizing the probability that the MDP behaviours satisfy the linear-time objective is thus reduced to maximizing the probability of reaching the target components.

The maximal reachability probability is computable in polynomial time by reduction to linear programming [1,6]. In practice, however, most model checkers

C. Baier and U. Dal Lago (Eds.): FOSSACS 2018, LNCS 10803, pp. 367–383, 2018.
https://doi.org/10.1007/978-3-319-89366-2_20

use value iteration to compute this value [9,17]. The worst-case time complexity of value iteration is pseudo-polynomial. Hence, when implementing model checkers it is usual for a graph-based pre-processing step to remove as many unnecessary states and transitions as possible while preserving the maximal reachability probability. Well-known reductions include the identification of extremal-probability states and maximal end components [1,5]. The intended outcome of this pre-processing step is a reduced amount of transition probability values that need to be considered when computing the number of iterations required by value iteration.

The main idea behind MDP reduction heuristics is to identify subsets of states from which the maximal probability of reaching the target set of states is the same. Such states are in fact redundant and can be "collapsed". Figure 1 depicts an MDP with actions and probabilities omitted for clarity. From p and q there are strategies to ensure that s is reached with probability 1. The same holds for t. For instance, from p, to get to t almost surely, one plays to go to the distribution directly below q; from q, to the distribution above q. Since from the state p, there is no strategy to ensure that q is reached with probability 1, p and q do not form an *end component*. In fact, to the best of our knowledge, no known MDP reduction heuristic captures this example (i.e., recognizes that p and q have the same maximal reachability probability for all possible values of the transition probabilities).

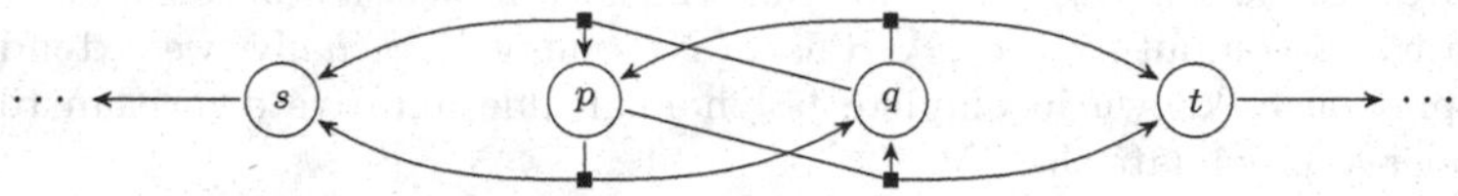

Fig. 1. An MDP with states depicted as circles and distributions as squares. The maximal reachability probability values from p and q are the same since, from both, one can enforce to reach s with probability 1, or t with probability 1, using different strategies.

In reinforcement learning the actual probabilities labelling the transitions of an MDP are not assumed to be known in advance. Thus, they have to be estimated by experimenting with different actions in different states and collecting statistics about the observed outcomes [14]. In order for the statistics to be good approximations, the number of experiments has to be high enough. In particular, when the approximations are required to be *probably approximately correct* [23], the necessary and sufficient number of experiments is pseudo-polynomial [13]. Furthermore, the expected number of steps before reaching a particular state even once may already be exponential (even if all the probabilities are fixed). The fact that an excessive amount of experiments is required is a known drawback of reinforcement learning [15,19].

A natural and key question to ask in this context is whether the maximal reachability probability does indeed depend on the actual value of the probability labelling a particular transition of the MDP. If this is not the case, then it need not be learnt. One natural way to remove transition probabilities which do not affect the maximal reachability value is to apply model checking MDP reduction techniques.

Contributions and Structure of the Paper. We view the directed graph underlying an MDP as a directed bipartite graph. Vertices in this graph are controlled by players *Protagonist* and *Nature*. Nature is only allowed to choose full-support probability distributions for each one of her vertices, thus instantiating an MDP from the graph; Protagonist has strategies just as he would in an MDP. Hence, we consider infinite families of MDPs with the same support. In the game played between Protagonist and Nature, and for vertices u and v, we are interested in knowing whether the maximal reachability probability from u is never (in any of the MDPs with the game as its underlying directed graph) worse than the same value from v.

In Sect. 2 we give the required definitions. We formalize the *never-worse relation* in Sect. 3. We also show that we can "collapse" sets of equivalent vertices with respect to the NWR (Theorem 1) and remove sub-optimal edges according to the NWR (Theorem 2). Finally, we also argue that the NWR generalizes most known heuristics to reduce MDP size before applying linear programming or value iteration. Then, in Sect. 4 we give a graph-based characterization of the relation (Theorem 3), which in turn gives us a coNP upper bound on its complexity. A matching lower bound is presented in Sect. 5 (Theorem 4). To conclude, we recall and extend an iterative algorithm to efficiently (in polynomial time) under-approximate the never-worse relation from [2].

Previous and Related Work. Reductions for MDP model checking were considered in [5,7]. From the reductions studied in both papers, extremal-probability states, essential states, and end components are computable using only graph-based algorithms. In [3], learning-based techniques are proposed to obtain approximations of the maximal reachability probability in MDPs. Their algorithms, however, do rely on the actual probability values of the MDP.

This work is also related to the widely studied model of interval MDPs, where the transition probabilities are given as intervals meant to model the uncertainty of the numerical values. Numberless MDPs [11] are a particular case of the latter in which values are only known to be zero or non-zero. In the context of numberless MDPs, a special case of the question we study can be simply rephrased as the comparison of the maximal reachability values of two given states.

In [2] a preliminary version of the iterative algorithm we give in Sect. 6 was described, implemented, and shown to be efficient in practice. Proposition 1 was first stated therein. In contrast with [2], we focus chiefly on characterizing the never-worse relation and determining its computational complexity.

2 Preliminaries

We use set-theoretic notation to indicate whether a letter $b \in \Sigma$ *occurs* in a word $\alpha = a_0 \dots a_k \in \Sigma^*$, i.e. $b \in \alpha$ if and only if $b = a_i$ for some $0 \le i \le k$.

Consider a directed graph $\mathcal{G} = (V, E)$ and a vertex $u \in V$. We write uE for the set of *successors* of u. That is to say, $uE := \{v \in V \mid (u, v) \in E\}$. We say that a path $\pi = u_0 \dots u_k \in V^*$ in $\mathcal{G}$ *visits* a vertex v if $v \in \pi$. We also say that π is a v–T path, for $T \subseteq V$, if $u_0 = v$ and $u_k \in T$.

2.1 Stochastic Models

Let S be a finite set. We denote by $\mathbb{D}(S)$ the set of all *(rational) probabilistic distributions* on S, i.e. the set of all functions $f : S \to \mathbb{Q}_{\ge 0}$ such that $\sum_{s \in S} f(s) = 1$. A probabilistic distribution $f \in \mathbb{D}(S)$ has *full support* if $f(s) > 0$ for all $s \in S$.

Definition 1 (Markov chains). *A* Markov chain $\mathcal{C}$ *is a tuple* (Q, δ) *where* Q *is a finite set of states and* δ *is a probabilistic transition function* $\delta : Q \to \mathbb{D}(Q)$.

A *run* of a Markov chain is a finite non-empty word $\varrho = p_0 \dots p_n$ over Q. We say ϱ *reaches* q if $q = p_i$ for some $0 \le i \le n$. The *probability of the run* is $\prod_{0 \le i < n} \delta(p_i, p_{i+1})$.

Let $T \subseteq Q$ be a set of states. The *probability of (eventually) reaching* T in $\mathcal{C}$ from q_0, which will be denoted by $\mathbb{P}^{q_0}_{\mathcal{C}}[\Diamond T]$, is the measure of the runs of $\mathcal{C}$ that start at q_0 and reach T. For convenience, let us first define the *probability of staying in states from* $S \subseteq Q$ *until* T *is reached*[1], written $\mathbb{P}^{q_0}_{\mathcal{C}}[S \cup T]$, as 1 if $q_0 \in T$ and otherwise

$$\sum \left\{ \prod_{0 \le i < n} \delta(q_i, q_{i+1}) \;\middle|\; q_0 \dots q_n \in (S \setminus T)^* T \text{ for } n \ge 1 \right\}.$$

We then define $\mathbb{P}^{q_0}_{\mathcal{C}}[\Diamond T] := \mathbb{P}^{q_0}_{\mathcal{C}}[Q \cup T]$.

When all runs from q_0 to T reach some set $U \subseteq Q$ before, the probability of reaching T can be decomposed into a finite sum as in the lemma below.

Lemma 1. *Consider a Markov chain* $\mathcal{C} = (Q, \delta)$*, sets of states* $U, T \subseteq Q$*, and a state* $q_0 \in Q \setminus U$*. If* $\mathbb{P}^{q_0}_{\mathcal{C}}[(Q \setminus U) \cup T] = 0$*, then*

$$\mathbb{P}^{q_0}_{\mathcal{C}}[\Diamond T] = \sum_{u \in U} \mathbb{P}^{q_0}_{\mathcal{C}}[(Q \setminus U) \cup u] \, \mathbb{P}^{u}_{\mathcal{C}}[\Diamond T].$$

Definition 2 (Markov decision processes). *A* (finite, discrete-time) Markov decision process $\mathcal{M}$*, MDP for short, is a tuple* (Q, A, δ, T) *where* Q *is a finite set of states,* A *a finite set of actions,* $\delta : Q \times A \to \mathbb{D}(Q)$ *a* probabilistic transition function, *and* $T \subseteq Q$ *a set of* target *states.*

For convenience, we write $\delta(q|p, a)$ instead of $\delta(p, a)(q)$.

[1] $S \cup T$ should be read as "S *until* T" and not understood as a set union.

Definition 3 (Strategies). *A (memoryless deterministic) strategy σ in an MDP $\mathcal{M} = (Q, A, \delta, T)$ is a function $\sigma : Q \to A$.*

Note that we have deliberately defined only memoryless deterministic strategies. This is at no loss of generality since, in this work, we focus on maximizing the probability of reaching a set of states. It is known that for this type of objective, memoryless deterministic strategies suffice [18].

From MDPs to Chains. An MDP $\mathcal{M} = (Q, A, \delta, T)$ and a strategy σ induce the Markov chain $\mathcal{M}^\sigma = (Q, \mu)$ where $\mu(q) = \delta(q, \sigma(q))$ for all $q \in Q$.

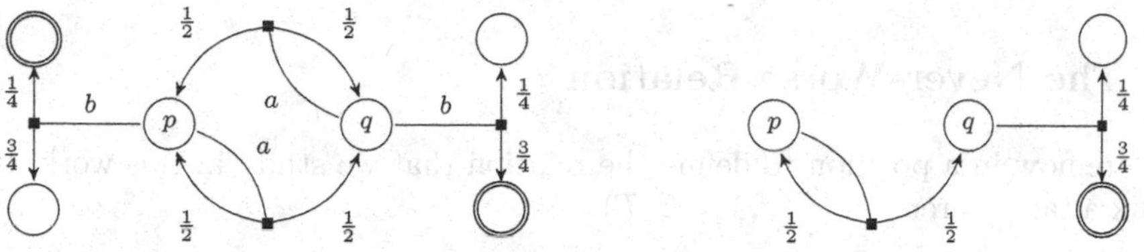

Fig. 2. On the left we have an MDP with actions $\{a, b\}$. On the right we have the Markov chain induced by the left MDP and the strategy $\{p \mapsto a, q \mapsto b\}$.

Example 1. Figure 2 depicts an MDP on the left. Circles represent states; double-circles, target states; and squares, distributions. The labels on arrows from states to distributions are actions; those on arrows from distributions to states, probabilities.

Consider the strategy σ that plays from p the action a and from q the action b, i.e. $\sigma(p) = a$ and $\sigma(q) = b$. The Markov chain on the right is the chain induced by σ and the MDP on the left. Note that we no longer have action labels.

The probability of reaching a target state from q under σ is easily seen to be 3/4. In other words, if we write $\mathcal{M}$ for the MDP and T for the set of target states then $\mathbb{P}^q_{\mathcal{M}^\sigma}[\Diamond T] = \frac{3}{4}$.

2.2 Reachability Games Against Nature

We will speak about families of MDPs whose probabilistic transition functions have the same support. To do so, we abstract away the probabilities and focus on a game played on a graph. That is, given an MDP $\mathcal{M} = (Q, A, \delta, T)$ we consider its *underlying directed graph* $\mathcal{G}_\mathcal{M} = (V, E)$ where $V := Q \cup (Q \times A)$ and $E := \{(q, \langle q, a\rangle) \in Q \times (Q \times A)\} \cup \{(\langle p, a\rangle, q) \mid \delta(q|p, a) > 0\}$. In $\mathcal{G}_\mathcal{M}$, *Nature* controls the vertices $Q \times A$. We formalize the game and the *arena* it is played on below.

Definition 4 (Target arena). *A target arena $\mathcal{A}$ is a tuple (V, V_P, E, T) such that $(V_P, V_N := V \setminus V_P, E)$ is a bipartite directed graph, $T \subseteq V_P$ is a set of target vertices, and $uE \neq \varnothing$ for all $u \in V_N$.*

Informally, there are two agents in a target arena: *Nature*, who controls the vertices in V_N, and *Protagonist*, who controls the vertices in V_P.

From Arenas to MDPs. A target arena $\mathcal{A} = (V, V_P, E, T)$ together with a family of probability distributions $\mu = (\mu_u \in \mathbb{D}(uE))_{u \in V_N}$ induce an MDP. Formally, let $\mathcal{A}_\mu$ be the MDP (Q, A, δ, T) where $Q = V_P \uplus \{\bot\}$, $A = V_N$, $\delta(q|p,a)$ is $\mu_a(q)$ if $(p,a),(a,q) \in E$ and 0 otherwise, for all $p \in V_P \cup \{\bot\}$ and $a \in A$ we have $\delta(\bot|p,a) = 1$ if $(p,a) \notin E$.

The Value of a Vertex. Consider a target arena $\mathcal{A} = (V, V_P, E, T)$ and a vertex $v \in V_P$. We define its *(maximal reachability probability) value* with respect to a family of full-support probability distributions μ as $\mathrm{Val}^\mu(v) := \max_\sigma \mathbb{P}^v_{\mathcal{A}^\sigma_\mu}[\Diamond T]$. For $u \in V_N$ we set $\mathrm{Val}^\mu(u) := \sum\{\mu_u(v)\mathrm{Val}^\mu(v) \mid v \in uE\}$.

3 The Never-Worse Relation

We are now in a position to define the relation that we study in this work. Let us fix a target arena $\mathcal{A} = (V, V_P, E, T)$.

Definition 5 (The never-worse relation (NWR)). *A subset $W \subseteq V$ of vertices is* never worse *than a vertex $v \in V$, written $v \trianglelefteq W$, if and only if*

$$\forall \mu = (\mu_u \in \mathbb{D}(uE))_{u \in V_N}, \exists w \in W : \mathrm{Val}^\mu(v) \leq \mathrm{Val}^\mu(w)$$

*where all the μ_u have **full support**. We write $v \sim w$ if $v \trianglelefteq \{w\}$ and $w \trianglelefteq \{v\}$.*

It should be clear from the definition that $\sim$ is an equivalence relation. For $u \in V$ let us denote by $\tilde{u}$ the set of vertices that are $\sim$-equivalent and belong to the same owner, i.e. $\tilde{u}$ is $\{v \in V_P \mid v \sim u\}$ if $u \in V_P$ and $\{v \in V_N \mid v \sim u\}$ otherwise.

Fig. 3. Two target arenas with $T = \{fin\}$ are shown. Round vertices are elements from V_P; square vertices, from V_N. In the left target arena we have that $p \trianglelefteq \{q\}$ and $q \trianglelefteq \{p\}$ since any path from either vertex visits t before T—see Lemma 1. In the right target arena we have that $t \trianglelefteq \{p\}$—see Proposition 1.

Example 2. Consider the left target arena depicted in Fig. 3. Using Lemma 1, it is easy to show that neither p nor q is ever worse than the other since t is visited before *fin* by all paths starting from p or q.

The literature contains various heuristics which consist in computing sets of states and "collapsing" them to reduce the size of the MDP without affecting the maximal reachability probability of the remaining states. We now show that we can collapse equivalence classes and, further, remove sub-optimal distributions using the NWR.

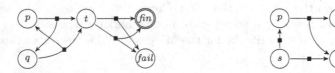

3.1 The Usefulness of the NWR

We will now formalize the idea of "collapsing" equivalent vertices with respect to the NWR. For convenience, we will also remove self-loops while doing so.

Consider a target arena $\mathcal{A} = (V, V_P, E, T)$. We denote by $\mathcal{A}_{/\sim}$ its $\sim$-*quotient*. That is, $\mathcal{A}_{/\sim}$ is the target arena (S, S_P, R, U) where $S_P = \{\tilde{v} \mid \exists v \in V_P\}$, $S = \{\tilde{v} \mid \exists v \in V_N\} \cup S_P$, $U = \{\tilde{t} \mid \exists t \in T\}$, and

$$R = \{(\tilde{u}, \tilde{v}) \mid \exists (u, v) \in (V_P \times V_N) \cap E : vE \setminus \tilde{u} \neq \varnothing\} \\ \cup \{(\tilde{u}, \tilde{v}) \mid \exists (u, v) \in (V_N \times V_P) \cap E\}.$$

For a family $\mu = (\mu_u \in \mathbb{D}(uE))_{u \in V_N}$ of full-support distributions we denote by $\mu_{/\sim}$ the family $\nu = (\nu_{\tilde{u}} \in \mathbb{D}(\tilde{u}R))_{\tilde{u} \in S_N}$ defined as follows. For all $\tilde{u} \in S_N$ and all $\tilde{v} \in \tilde{u}R$ we have $\nu_{\tilde{u}}(\tilde{v}) = \sum_{w \in \tilde{v}} \mu_u(w)$, where u is any element of $\tilde{u}$.

The following property of the $\sim$-quotient follows from the fact that all the vertices in $\tilde{v}$ have the same maximal probability of reaching the target vertices.

Theorem 1. *Consider a target arena $\mathcal{A} = (V, V_P, E, T)$. For all families $\mu = (\mu_u \in \mathbb{D}(uE))_{u \in V_N}$ of full-support probability distributions and all $v \in V_P$ we have*

$$\max_{\sigma} \mathbb{P}^{v}_{\mathcal{A}^{\sigma}_{\mu}}[\Diamond T] = \max_{\sigma'} \mathbb{P}^{\tilde{v}}_{\mathcal{B}^{\sigma'}_{\nu}}[\Diamond U],$$

where $\mathcal{B} = \mathcal{A}_{/\sim}$, $\nu = \mu_{/\sim}$, and $U = \{\tilde{t} \mid \exists t \in T\}$.

We can further remove edges that lead to sub-optimal Nature vertices. When this is done after $\sim$-quotienting the maximal reachability probabilities are preserved.

Theorem 2. *Consider a target arena $\mathcal{A} = (V, V_P, E, T)$ such that $\mathcal{A}_{/\sim} = \mathcal{A}$. For all families $\mu = (\mu_u \in \mathbb{D}(uE))_{u \in V_N}$ of full-support probability distributions, for all $(w, x) \in E \cap (V_P \times V_N)$ such that $x \trianglelefteq (wE \setminus \{x\})$, and all $v \in V_P$ we have*

$$\max_{\sigma} \mathbb{P}^{v}_{\mathcal{A}^{\sigma}_{\mu}}[\Diamond T] = \max_{\sigma'} \mathbb{P}^{v}_{\mathcal{B}^{\sigma'}_{\mu}}[\Diamond T],$$

where $\mathcal{B} = (V, V_P, E \setminus \{(w, x)\}, T)$.

3.2 Known Efficiently-Computable Special Cases

We now recall the definitions of the set of extremal-probability states, end components, and essential states. Then, we observe that for all these sets of states their maximal probability reachability coincide and their definitions are independent of the probabilities labelling the transitions of the MDP. Hence, they are subsets of the set of the equivalence classes induced by $\sim$.

Extremal-Probability States. The set of *extremal-probability states* of an MDP $\mathcal{M} = (Q, A, \delta, T)$ consists of the set of states with maximal probability reachability 0 and 1. Both sets can be computed in polynomial time [1,4]. We give below a game-based definition of both sets inspired by the classical polynomial-time algorithm to compute them (see, e.g., [1]). Let us fix a target arena $\mathcal{A} = (V, V_P, E, T)$ for the sequel.

For a set $T \subseteq V$, let us write $\mathbf{Z}_T := \{v \in V \mid T \text{ is not reachable from } v\}$.

(Almost-Surely Winning) Strategies. A strategy for Protagonist in a target arena is a function $\sigma : V_P \to V_N$. We then say that a path $v_0 \dots v_n \in V^*$ is *consistent with* σ if $v_i \in V_P \implies \sigma(v_i) = v_{i+1}$ for all $0 \leq i < n$. Let $\mathbf{Reach}(v_0, \sigma)$ denote the set of vertices reachable from v_0 under σ, i.e. $\mathbf{Reach}(v_0, \sigma) := \{v_k \mid v_0 \dots v_k \text{ is a path consistent with } \sigma\}$.

We say that a strategy σ for Protagonist is *almost-surely winning from* $u_0 \in V$ *to* $T \subseteq V_P$ if, after modifying the arena to make all $t \in T$ into sinks, for all $v_0 \in \mathbf{Reach}(u_0, \sigma)$ we have $\mathbf{Reach}(v_0, \sigma) \cap T \neq \varnothing$. We denote the set of all such strategies by $\mathbf{Win}_T^{v_0}$.

The following properties regarding almost-surely winning strategies in a target arena follow from the correctness of the graph-based algorithm used to compute extremal-probability states in an MDP [1, Lemma 10.108].

Lemma 2 (From [1]). *Consider a target arena* $\mathcal{A} = (V, V_P, E, T)$. *For all families* $\mu = (\mu_u \in \mathbb{D}(uE))_{u \in V_N}$ *of full-support probability distributions, for all* $v \in V_P$ *the following hold.*

(i) $\max_\sigma \mathbb{P}^v_{\mathcal{A}^\sigma_\mu}[\Diamond T] = 0 \iff v \in \mathbf{Z}_T$
(ii) $\forall \sigma : \sigma \in \mathbf{Win}_T^v \iff \mathbb{P}^v_{\mathcal{A}^\sigma_\mu}[\Diamond T] = 1$

End Components. Let us consider an MDP $\mathcal{M} = (Q, A, \delta, T)$. A set $S \subseteq Q$ of states is an *end component* in $\mathcal{M}$ if for all pairs of states $p, q \in S$ there exists a strategy σ such that $\mathbb{P}^p_{\mathcal{M}^\sigma}[S \mathsf{U} q] = 1$.

Example 3. Let us consider the MDP shown on the left in Fig. 2. The set $\{p, q\}$ is an end component since, by playing a from both states, one can ensure to reach either state from the other with probability 1.

It follows immediately from the definition of end component that the maximal probability of reaching T from states in the same end component is the same.

Lemma 3. *Let* $S \subseteq Q$ *be an end component in* $\mathcal{M}$. *For all* $p, q \in S$ *we have that* $\max_\sigma \mathbb{P}^p_{\mathcal{M}^\sigma}[\Diamond T] = \max_\sigma \mathbb{P}^q_{\mathcal{M}^\sigma}[\Diamond T]$.

We say an end component is *maximal* if it is maximal with respect to set inclusion. Furthermore, from the definition of end components in MDPs and Lemma 2 it follows that we can lift the notion of end component to target arenas. More precisely, a set $S \subseteq V_P$ is an end component in $\mathcal{A}$ if and only if for some family of full-support probability distributions μ we have that S is an end component in $\mathcal{A}_\mu$ (if and only if for all μ' the set S is an end component in $\mathcal{A}_{\mu'}$).

The set of all maximal end components of a target arena can be computed in polynomial time using an algorithm based on the strongly connected components of the graph [1,8].

Essential States. Consider a target arena $\mathcal{A} = (V, V_P, E, T)$ and let $\sqsubseteq$ be the smallest relation satisfying the following. For all $u \in V_P$ we have $u \sqsubseteq u$. For all $u_0, v \in V_P \setminus \mathbf{Z}_T$ such that $u_0 \neq v$ we have $u_0 \sqsubseteq v$ if for all paths $u_0 u_1 u_2$ we have that $u_2 \sqsubseteq v$ and there is at least one such path. Intuitively, $u \sqsubseteq v$ holds whenever all paths starting from u reach v. In [7], the maximal vertices according to $\sqsubseteq$ are called *essential states*[2].

Lemma 4 (From [7]). *Consider a target arena $\mathcal{A} = (V, V_P, E, T)$. For all families $\mu = (\mu_u \in \mathbb{D}(uE))_{u \in V_N}$ of full-support probability distributions, for all $v \in V_P$ and all essential states w, if $v \sqsubseteq w$ then* $\max_\sigma \mathbb{P}^v_{\mathcal{A}^\sigma_\mu}[\Diamond T] = \max_{\sigma'} \mathbb{P}^w_{\mathcal{A}^{\sigma'}_\mu}[\Diamond T]$.

Note that, in the left arena in Fig. 3, $p \sqsubseteq t$ does not hold since there is a cycle between p and q which does not visit t.

It was also shown in [7] that the $\sqsubseteq$ relation is computable in polynomial time.

4 Graph-Based Characterization of the NWR

In this section we give a characterization of the NWR that is reminiscent of the topological-based value iteration proposed in [5]. The main intuition behind our characterization is as follows. If $v \trianglelefteq W$ does not hold, then for all $0 < \varepsilon < 1$ there is some family μ of full-support distributions such that $\mathrm{Val}^\mu(v)$ is at least $1 - \varepsilon$, while $\mathrm{Val}^\mu(w)$ is at most ε for all $w \in W$. In turn, this must mean that there is a path from v to T which can be assigned a high probability by μ while, from W, all paths go with high probability to $\mathbf{Z}_T$.

We capture the idea of separating a "good" v–T path from all paths starting from W by using partitioning of V into layers $S_i \subseteq V$. Intuitively, we would like it to be easy to construct a family μ of probability distributions such that from all vertices in S_{i+1} all paths going to vertices outside of S_{i+1} end up, with high probability, in lower layers, i.e. some S_k with $k < i$. A formal definition follows.

Definition 6 (Drift partition and vertices). *Consider a target arena $\mathcal{A} = (V, V_P, E, T)$ and a partition $(S_i)_{0 \le i \le k}$ of V. For all $0 \le i \le k$, let $S_i^+ := \cup_{i<j} S_j$ and $S_i^- := \cup_{j<i} S_j$, and let $D_i := \{v \in S_i \cap V_N \mid vE \cap S_i^- \neq \varnothing\}$. We define the set $D := \cup_{0<i<k} D_i$ of* drift vertices. *The partition is called a* drift partition *if the following hold.*

- *For all $i \le k$ and all $v \in S_i \cap V_P$ we have $vE \cap S_i^+ = \varnothing$.*
- *For all $i \le k$ and all $v \in S_i \cap V_N$ we have $vE \cap S_i^+ \neq \varnothing \implies v \in D$.*

[2] This is not the usual notion of essential states from classical Markov chain theory.

Using drift partitions, we can now formalize our characterization of the negation of the NWR.

Theorem 3. *Consider a target arena $\mathcal{A} = (V, V_P, E, T)$, a non-empty set of vertices $W \subseteq V$, and a vertex $v \in V$. The following are equivalent*

(i) $\neg(v \trianglelefteq W)$
(ii) There exists a drift partition $(S_i)_{0 \le i \le k}$ and a simple path π starting in v and ending in T such that $\pi \subseteq S_k$ and $W \subseteq S_k^-$.

Before proving Theorem 3 we need an additional definition and two intermediate results.

Definition 7 (Value-monotone paths). *Let $\mathcal{A} = (V, V_P, E, T)$ be a target arena and consider a family of full-support probability distributions $\mu = (\mu_u \in \mathbb{D}(uE))_{u \in V_N}$. A path $v_0 \dots v_k$ is* μ-non-increasing *if and only if* $\mathrm{Val}^\mu(v_{i+1}) \le \mathrm{Val}^\mu(v_i)$ *for all* $0 \le i < k$*; it is* μ-non-decreasing *if and only if* $\mathrm{Val}^\mu(v_i) \le \mathrm{Val}^\mu(v_{i+1})$ *for all* $0 \le i < k$.

It can be shown that from any path in a target arena ending in T one can obtain a simple non-decreasing one.

Lemma 5. *Consider a target arena $\mathcal{A} = (V, V_P, E, T)$ and a family of full-support probability distributions $\mu = (\mu_u \in \mathbb{D}(uE))_{u \in V_N}$. If there is a path from some $v \in V$ to T, there is also a simple μ-non-decreasing one.*

Additionally, we will make use of the following properties regarding vertex-values. They formalize the relation between the value of a vertex, its owner, and the values of its successors.

Lemma 6. *Consider a target arena $\mathcal{A} = (V, V_P, E, T)$ and a family of full-support probability distributions $\mu = (\mu_u \in \mathbb{D}(uE))_{u \in V_N}$.*

(i) For all $u \in V_P$, for all successors $v \in uE$ it holds that $\mathrm{Val}^\mu(v) \le \mathrm{Val}^\mu(u)$.
(ii) For all $u \in V_N$ it holds that

$$(\exists v \in uE : \mathrm{Val}^\mu(u) < \mathrm{Val}^\mu(v)) \implies (\exists w \in uE : \mathrm{Val}^\mu(w) < \mathrm{Val}^\mu(u)).$$

Proof (of Theorem 3). Recall that, by definition, (i) holds if and only if there exists a family $\mu = (\mu_u \in \mathbb{D}(uE))_{u \in V_N}$ of full-support probability distributions such that $\forall w \in W : \mathrm{Val}^\mu(w) < \mathrm{Val}^\mu(v)$.

Let us prove (i) $\implies$ (ii). Let $x_0 < x_1 < \dots$ be the finitely many (i.e. at most $|V|$) values that occur in the MDP $\mathcal{A}_\mu$, and let k be such that $\mathrm{Val}^\mu(v) = x_k$. For all $0 \le i < k$ let $S_i := \{u \in V \mid \mathrm{Val}^\mu(u) = x_i\}$, and let $S_k := V \setminus \cup_{i<k} S_i$. Let us show below that the S_i form a drift partition.

- $\forall i \le k, \forall u \in S_i \cap S_P : uE \cap S_i^+ = \varnothing$ by Lemma 6(i) (for $i < k$) and since $S_k^+ = \varnothing$.
- $\forall i \le k, \forall u \in S_i \cap S_N : uE \cap S_i^+ \neq \varnothing \implies x \in D$ by Lemma 6(ii) (for $i < k$) and since $S_k^+ = \varnothing$.

We have that $\mathrm{Val}^\mu(w) < \mathrm{Val}^\mu(v) = x_k$ for all $w \in W$, by assumption, so $W \subseteq S_k^-$ by construction. By Lemma 5 there exists a simple μ-non-decreasing path π from v to T, so all the vertices occurring in π have values at least $\mathrm{Val}^\mu(v)$, so $\pi \subseteq S_k$.

We will prove (ii) $\implies$ (i) by defining some full-support distribution family μ. The definition will be partial only, first on $\pi \cap V_N$, and then on the drift vertices in $V \setminus S_k$. Let $0 < \varepsilon < 1$, which is meant to be small enough. Let us write $\pi = v_0 \dots v_n$ so that $v_0 = v$ and $v_n \in T$. Let us define μ on $\pi \cap V_N$ as follows: for all $i < n$, if $v_i \in V_N$ let $\mu_{v_i}(v_{i+1}) := 1 - \varepsilon$. Let σ be an arbitrary Protagonist strategy such that for all $i < n$, if $v_i \in V_P$ then $\sigma(v_i) := v_{i+1}$. Therefore

$$\begin{aligned}
(1-\varepsilon)^{|V|} &\leq (1-\varepsilon)^n && \text{since } \pi \text{ is simple} \\
&\leq \prod_{i<n, v_i \in S_N} \mu_{v_i}(v_{i+1}) && \text{by definition of } \mu \\
&\leq \mathbb{P}^v_{\mathcal{A}^\sigma_\mu}[\Diamond T] \\
&\leq \max_{\sigma'} \mathbb{P}^v_{\mathcal{A}^{\sigma'}_\mu}[\Diamond T] = \mathrm{Val}^\mu(v). && (1)
\end{aligned}$$

So, for $0 < \varepsilon < 1 - \frac{1}{\sqrt[|V|]{2}}$, we have $\frac{1}{2} < (1-\varepsilon)^{|V|} \leq \mathrm{Val}^\mu(v)$. Below we will further define μ such that $\mathrm{Val}^\mu(w) \leq 1 - (1-\varepsilon)^{|V|} < \frac{1}{2}$ for all $w \in W$ and all $0 < \varepsilon < 1 - \frac{1}{\sqrt[|V|]{2}}$, which will prove (ii) $\implies$ (i). However, the last part of the proof is more difficult.

For all $1 \leq i \leq k$, for all drift vertices $u \in S_i$, let $\varrho(u)$ be a successor of u in S_i^-. Such a $\varrho(u)$ exists by definition of the drift vertices. Then let $\mu_u(\varrho(u)) := 1 - \varepsilon$. We then claim that

$$\forall u \in D : (1-\varepsilon)(1 - \mathbb{P}^{\varrho(u)}_{\mathcal{A}^\sigma_\mu}[\Diamond T]) \leq 1 - \mathbb{P}^u_{\mathcal{A}^\sigma_\mu}[\Diamond T]. \tag{2}$$

Indeed, $1 - \mathbb{P}^u_{\mathcal{A}^\sigma_\mu}[\Diamond T]$ is the probability that, starting at u and following σ, T is never reached; and $(1-\varepsilon)(1 - \mathbb{P}^{\varrho(u)}_{\mathcal{A}^\sigma_\mu}[\Diamond T])$ is the probability that, starting at u and following σ, the second vertex is $\varrho(u)$ and T is never reached.

Now let σ be an arbitrary strategy, and let us prove the following by induction on j.

$$\forall 0 \leq j < k, \forall w \in S_j \cup S_j^- : \mathbb{P}^w_{\mathcal{A}^\sigma_\mu}[\Diamond T] \leq 1 - (1-\varepsilon)^j$$

Base case, $j = 0$: by assumption W is non-empty and included in S_k^-, so $0 < k$. Also by assumption $T \subseteq S_k$, so $T \cap S_0 = \varnothing$. By definition of a drift partition, there are no edges going out of S_0, regardless of whether the starting vertex is in V_P or V_N. So there is no path from w to T, which implies $\mathrm{Val}^\mu(w) = 0$ for all $w \in S_0$, and the claim holds for the base case. Inductive case, let $w \in S_j$, let $D' := D \cap (S_j \cup S_j^-)$ and let us argue that every path π from w to T must at some point leave $S_j \cup S_j^-$ to reach a vertex with higher index, i.e. there is some edge (π_i, π_{i+1}) from $\pi_i \in S_j \cup S_j^-$ to some $\pi_{i+1} \in S_\ell$ with $j < \ell$. By definition of a drift partition, π_i must also be a drift vertex, i.e. $\pi_i \in D'$. Thus, if we let $F := V_P \setminus D'$, Lemma 1 implies that $\mathbb{P}^w_{\mathcal{A}^\sigma_\mu}[\Diamond T] = \sum_{u \in D'} \mathbb{P}^w_{\mathcal{A}^\sigma_\mu}[F \cup u]\, \mathbb{P}^u_{\mathcal{A}^\sigma_\mu}[\Diamond T]$. Now, since

$$
\begin{aligned}
&\sum_{u \in D'} \mathbb{P}^u_{\mathcal{A}^\sigma_\mu}[\Diamond T] \\
&= \sum_{u \in D \cap S_j^-} \mathbb{P}^u_{\mathcal{A}^\sigma_\mu}[\Diamond T] + \sum_{u \in D_j} \mathbb{P}^u_{\mathcal{A}^\sigma_\mu}[\Diamond T] && \text{by splitting the sum} \\
&\leq \sum_{u \in D \cap S_j^-} \mathbb{P}^u_{\mathcal{A}^\sigma_\mu}[\Diamond T] + \sum_{u \in D_j} (1 - (1-\varepsilon)(1 - \mathbb{P}^{\varrho(u)}_{\mathcal{A}^\sigma_\mu}[\Diamond T])) && \text{by (2)} \\
&\leq \sum_{u \in D \cap S_j^-} (1 - (1-\varepsilon)^{j-1}) + && \text{by IH and since} \\
&\quad \sum_{u \in D_j} (1 - (1-\varepsilon)(1-\varepsilon)^{j-1}) && \forall x \in D_j : \varrho(x) \in S_j^- \\
&\leq \sum_{u \in D'} (1 - (1-\varepsilon)^j) && (1-\varepsilon)^j \leq (1-\varepsilon)^{j-1}
\end{aligned}
$$

and $\sum_{u \in D'} \mathbb{P}^w_{\mathcal{A}^\sigma_\mu}[F \cup u] \leq 1$, we have that $\mathbb{P}^w_{\mathcal{A}^\sigma_\mu}[\Diamond T] \leq 1 - (1-\varepsilon)^j$. The induction is thus complete. Since σ is arbitrary in the calculations above, and since $j < k \leq |V|$, we find that $\mathrm{Val}^\mu(w) \leq 1 - (1-\varepsilon)^{|V|}$ for all $w \in W \subseteq S_k^-$.

For $0 < \varepsilon < 1 - \frac{1}{\sqrt[|V|]{2}}$ we have $\frac{1}{2} < (1-\varepsilon)^{|V|}$, as mentioned after (1), so $\mathrm{Val}^\mu(w) \leq 1 - (1-\varepsilon)^{|V|} < \frac{1}{2}$. □

5 Intractability of the NWR

It follows from Theorem 3 that we can decide whether a vertex is sometimes worse than a set of vertices by guessing a partition of the vertices and verifying that it is a drift partition. The verification can clearly be done in polynomial time.

Corollary 1. *Given a target arena $\mathcal{A} = (V, V_P, E, T)$, a non-empty set $W \subseteq V$, and a vertex $v \in V$, determining whether $v \trianglelefteq W$ is decidable and in* coNP.

We will now show that the problem is in fact coNP-complete already for Markov chains.

Theorem 4. *Given a target arena $\mathcal{A} = (V, V_P, E, T)$, a non-empty vertex set $W \subseteq V$, and a vertex $v \in V$, determining whether $v \trianglelefteq W$ is* coNP*-complete even if $|uE| = 1$ for all $u \in V_P$.*

The idea is to reduce the 2-Disjoint Paths problem (2DP) to the existence of a drift partition witnessing that $v \trianglelefteq \{w\}$ does not hold, for some $v \in V$. Recall that 2DP asks, given a directed graph $\mathcal{G} = (V, E)$ and vertex pairs

$(s_1, t_1), (s_2, t_2) \in V \times V$, whether there exists an s_1–t_1 path π_1 and an s_2–t_2 path π_2 such that π_1 and π_2 are vertex disjoint, i.e. $\pi_1 \cap \pi_2 = \varnothing$. The problem is known to be NP-complete [10,12]. In the sequel, we assume without loss of generality that (a) t_1 and t_2 are reachable from all $s \in V \setminus \{t_1, t_2\}$; and (b) t_1 and t_2 are the only sinks $\mathcal{G}$.

Proof (of Theorem 4). From the 2DP input instance, we construct the target arena $\mathcal{A} = (S, S_P, R, T)$ with $S := V \cup E$, $R := \{(u, \langle u, v\rangle), (\langle u, v\rangle, v) \in S \times S \mid (u, v) \in E \text{ or } u = v \in \{t_1, t_2\}\}$, $S_P := V \times V$, and $T := \{\langle t_1, t_1\rangle\}$. We will show there are vertex-disjoint s_1–t_1 and s_2–t_2 paths in $\mathcal{G}$ if and only if there is a drift partition $(S_i)_{0 \le i \le k}$ and a simple s_1–t_1 path π such that $\pi \subseteq S_k$ and $s_2 \in S_k^-$. The result will then follow from Theorem 3.

Suppose we have a drift partition $(S_i)_{0 \le i \le k}$ with $s_2 \in S_k^-$ and a simple path $\pi = v_0\langle v_0, v_1\rangle \ldots \langle v_{n-1}, v_n\rangle v_n$ with $v_0 = s_1, v_n = t_1$. Since the set $\{t_2, \langle t_2, t_2\rangle\}$ is *trapping* in $\mathcal{A}$, i.e. all paths from vertices in the set visit only vertices from it, we can assume that $S_0 = \{t_2, \langle t_2, t_2\rangle\}$. (Indeed, for any drift partition, one can obtain a new drift partition by moving any trapping set to a new lowest layer.) Now, using the assumption that t_2 is reachable from all $s \in V \setminus \{t_1, t_2\}$ one can show by induction that for all $0 \le j < k$ and for all $\varrho = u_0 \in S_j$ there is a path $u_0 \ldots u_m$ in $\mathcal{G}$ with $u_m = t_2$ and $\varrho \subseteq S_{j+1}^-$. This implies that there is a s_2–t_2 path π_2 in $\mathcal{G}$ such that $\pi_2 \subseteq S_k^-$. It follows that π_2 is vertex disjoint with the s_1–t_1 path $v_0 \ldots v_n$ in $\mathcal{G}$.

Now, let us suppose that we have s_1–t_1 and s_2–t_2 vertex disjoint paths $\pi_1 = u_0 \ldots u_n$ and $\pi_2 = v_0 \ldots v_m$. Clearly, we can assume both π_1, π_2 are simple. We will construct a partition $(S_i)_{0 \le i \le m+1}$ and show that it is indeed a drift partition, that $u_0\langle u_0, u_1\rangle \ldots \langle u_{n-1}, u_n\rangle u_n \subseteq S_{m+1}$, and $s_2 = v_0 \in S_{m+1}^-$. Let us set $S_0 := \{\langle v_{m-1}, v_m\rangle, v_m, \langle t_2, t_2\rangle\}$, $S_i := \{\langle v_{m-i-1}, v_{m-i}\rangle, v_{m-i}\}$ for all $0 < i \le m$, and $S_{m+1} := S \setminus \cup_{0 \le i \le m} S_i$. Since π_2 is simple, $(S_i)_{0 \le i \le m+1}$ is a partition of V. Furthermore, we have that $s_2 = v_0 \in S_{m+1}^-$, and $u_0\langle u_0, u_1\rangle \ldots \langle u_{n-1}, u_n\rangle u_n \subseteq S_{m+1}$ since π_1 and π_2 are vertex disjoint. Thus, it only remains for us to argue that for all $0 \le i \le m+1$: for all $w \in S_i \cap S_N$ we have $wR \cap S_i^+ = \varnothing$, and for all $w \in S_i \cap V_N$ we have $wR \cap S_i^+ \neq \varnothing \implies wR \cap S_i^- \neq \varnothing$. By construction of the S_i, we have that $eR \subseteq S_i$ for all $0 \le i \le m$ and all $e \in S_i \cap S_P$. Furthermore, for all $0 < i \le m$, for all $x \in S_i \cap S_N = \{v_{m-i}\}$, there exists $y \in S_{i-1} \cap S_P = \{\langle v_{m-i}, v_{m-i+1}\rangle\}$ such that $(x, y) \in R$—induced by $(v_{m-i}, v_{m-1+1}) \in E$ from π_2. To conclude, we observe that since $S_0 = \{\langle v_{m-1}, v_m\rangle, v_m = t_2, \langle t_2, t_2\rangle\}$ and $\{t_2, \langle t_2, t_2\rangle\}$ is trapping in $\mathcal{A}$, the set t_2R is contained in S_0. □

6 Efficiently Under-Approximating the NWR

Although the full NWR cannot be efficiently computed for a given MDP, we can hope for "under-approximations" that are accurate and efficiently computable.

Definition 8 (Under-approximation of the NWR). *Let $\mathcal{A} = (V, V_P, E, T)$ be a target arena and consider a relation $\preceq : V \times \mathcal{P}(V)$. The relation $\preceq$ is an* under-approximation *of the NWR if and only if $\preceq \subseteq \trianglelefteq$.*

We denote by $\preceq^*$ the *pseudo transitive closure* of $\preceq$. That is, $\preceq^*$ is the smallest relation such that $\preceq \subseteq \preceq^*$ and for all $u \in V, X \subseteq V$ if there exists $W \subseteq V$ such that $u \preceq^* W$ and $w \preceq^* X$ for all $w \in W$, then $u \preceq^* X$.

Remark 1. The empty set is an under-approximation of the NWR. For all under-approximations $\preceq$ of the NWR, the pseudo transitive closure $\preceq^*$ of $\preceq$ is also an under-approximation of the NWR.

In [2], efficiently-decidable sufficient conditions for the NWR were given. In particular, those conditions suffice to infer relations such as those in the right MDP from Fig. 3. We recall (Proposition 1) and extend (Proposition 2) these conditions below.

Proposition 1 (From [2]**).** *Consider a target arena* $\mathcal{A} = (V, V_P, E, T)$ *and an under-approximation* $\preceq$ *of the NWR. For all vertices* $v_0 \in V$*, and sets* $W \subseteq V$ *the following hold.*

(i) If there exists $S \subseteq \{s \in V \mid s \preceq W\}$ *such that there exists no path* $v_0 \dots v_n \in (V \setminus S)^* T$*, then* $v_0 \trianglelefteq W$*.*
(ii) If $W = \{w\}$ *and there exists* $S \subseteq \{s \in V_P \mid w \preceq \{s\}\}$ *such that* $\mathbf{Win}^{v_0}_{S \cup T} \neq \varnothing$*, then* $w \trianglelefteq \{v_0\}$*.*

Proof (Sketch). The main idea of the proof of item (i) is to note that S is visited before T. The desired result then follows from Lemma 1. For item (ii), we intuitively have that there is a strategy to visit T with some probability or visit W, where the chances of visiting T are worse than before. We then show that it is never worse to start from v_0 to have better odds of visiting T. □

The above "rules" give an iterative algorithm to obtain increasingly better under-approximations of the NWR: from $\preceq_i$ apply the rules and obtain a new under-approximation $\preceq_{i+1}$ by adding the new pairs and taking the pseudo transitive closure; then repeat until convergence. Using the special cases from Sect. 3.2 we can obtain a nontrivial initial under-approximation $\preceq_0$ of the NWR in polynomial time.

The main problem is how to avoid testing all subsets $W \subseteq V$ in every iteration. One natural way to ensure we do not consider all subsets of vertices in every iteration is to apply the rules from Proposition 1 only on the successors of Protagonist vertices.

In the same spirit of the iterative algorithm described above, we now give two new rules to infer NWR pairs.

Proposition 2. *Consider a target arena* $\mathcal{A} = (V, V_P, E, T)$ *and* $\preceq$ *an under-approximation of the NWR.*

(i) For all $u \in V_N$*, if for all* $v, w \in uE$ *we have* $v \preceq \{w\}$ *and* $w \preceq \{v\}$*, then* $u \sim x$ *for all* $x \in uE$*.*
(i) For all $u, v \in V_P \setminus T$*, if for all* $w \in uE$ *such that* $w \preceq (uE \setminus \{w\})$ *does not hold we have that* $w \preceq vE$*, then* $u \trianglelefteq \{v\}$*.*

Proof (Sketch). Item (i) follows immediately from the definition of Val. For item (ii) one can use the Bellman optimality equations for infinite-horizon reachability in MDPs to show that since the successors of v are never worse than the non-dominated successors of u, we must have $u \trianglelefteq \{v\}$. □

Fig. 4. Two target arenas with $T = \{fin\}$ are shown. Using Propositions 1 and 2 one can conclude that $p \sim q$ in both target arenas.

The rules stated in Proposition 2 can be used to infer relations like those depicted in Fig. 4 and are clearly seen to be computable in polynomial time as they speak only of successors of vertices.

7 Conclusions

We have shown that the never-worse relation is, unfortunately, not computable in polynomial time. On the bright side, we have extended the iterative polynomial-time algorithm from [2] to under-approximate the relation. In that paper, a prototype implementation of the algorithm was used to empirically show that interesting MDPs (from the set of benchmarks included in PRISM [17]) can be drastically reduced.

As future work, we believe it would be interesting to implement an exact algorithm to compute the NWR using SMT solvers. Symbolic implementations of the iterative algorithms should also be tested in practice. In a more theoretical direction, we observe that the planning community has also studied maximizing the probability of reaching a target set of states under the name of MAXPROB (see, e.g., [16,21]). There, online approximations of the NWR would make more sense than the under-approximation we have proposed here. Finally, one could define a notion of never-worse for finite-horizon or quantitative objectives.

Acknowledgements. The research leading to these results was supported by the ERC Starting grant 279499: inVEST. Guillermo A. Pérez is an F.R.S.-FNRS Aspirant and FWA postdoc fellow.

We thank Nathanaël Fijalkow for pointing out the relation between this work and the study of interval MDPs and numberless MDPs. We also thank Shaull Almagor, Michaël Cadilhac, Filip Mazowiecki, and Jean-François Raskin for useful comments on earlier drafts of this paper.

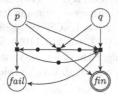

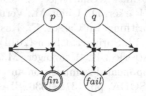

References

1. Baier, C., Katoen, J.-P.: Principles of Model Checking. MIT Press, New York (2008)
2. Bharadwaj, S., Le Roux, S., Pérez, G.A., Topcu, U.: Reduction techniques for model checking and learning in MDPs. In: IJCAI, pp. 4273–4279 (2017)
3. Brázdil, T., Chatterjee, K., Chmelík, M., Forejt, V., Křetínský, J., Kwiatkowska, M., Parker, D., Ujma, M.: Verification of markov decision processes using learning algorithms. In: Cassez, F., Raskin, J.-F. (eds.) ATVA 2014. LNCS, vol. 8837, pp. 98–114. Springer, Cham (2014). https://doi.org/10.1007/978-3-319-11936-6_8
4. Chatterjee, K., Henzinger, M.: Faster and dynamic algorithms for maximal end-component decomposition and related graph problems in probabilistic verification. In: SODA, pp. 1318–1336. SIAM (2011)
5. Ciesinski, F., Baier, C., Größer, M., Klein, J.: Reduction techniques for model checking Markov decision processes. In: QEST, pp. 45–54 (2008)
6. Courcoubetis, C., Yannakakis, M.: The complexity of probabilistic verification. J. ACM **42**(4), 857–907 (1995)
7. D'Argenio, P.R., Jeannet, B., Jensen, H.E., Larsen, K.G.: Reachability analysis of probabilistic systems by successive refinements. In: de Alfaro, L., Gilmore, S. (eds.) PAPM-PROBMIV 2001. LNCS, vol. 2165, pp. 39–56. Springer, Heidelberg (2001). https://doi.org/10.1007/3-540-44804-7_3
8. De Alfaro, L.: Formal verification of probabilistic systems. Ph.D. thesis, Stanford University (1997)
9. Dehnert, C., Junges, S., Katoen, J.-P., Volk, M.: A Storm is coming: a modern probabilistic model checker. In: Majumdar, R., Kunčak, V. (eds.) CAV 2017, Part II. LNCS, vol. 10427, pp. 592–600. Springer, Cham (2017). https://doi.org/10.1007/978-3-319-63390-9_31
10. Eilam-Tzoreff, T.: The disjoint shortest paths problem. Discret. Appl. Math. **85**(2), 113–138 (1998)
11. Fijalkow, N., Gimbert, H., Horn, F., Oualhadj, Y.: Two recursively inseparable problems for probabilistic automata. In: Csuhaj-Varjú, E., Dietzfelbinger, M., Ésik, Z. (eds.) MFCS 2014, Part I. LNCS, vol. 8634, pp. 267–278. Springer, Heidelberg (2014). https://doi.org/10.1007/978-3-662-44522-8_23
12. Fortune, S., Hopcroft, J.E., Wyllie, J.: The directed subgraph homeomorphism problem. Theor. Comput. Sci. **10**, 111–121 (1980)
13. Fu, J., Topcu, U.: Probably approximately correct MDP learning and control with temporal logic constraints. In: RSS (2014)
14. Kaelbling, L.P., Littman, M.L., Moore, A.W.: Reinforcement learning: a survey. JAIR **4**, 237–285 (1996)
15. Kawaguchi, K.: Bounded optimal exploration in MDP. In AAAI, pp. 1758–1764 (2016)
16. Kolobov, A., Mausam, M., Weld, D.S., Geffner, H.: Heuristic search for generalized stochastic shortest path MDPs. In: Bacchus, F., Domshlak, C., Edelkamp, S., Helmert, M. (eds.) ICAPS. AAAI (2011)
17. Kwiatkowska, M., Norman, G., Parker, D.: PRISM 4.0: verification of probabilistic real-time systems. In: Gopalakrishnan, G., Qadeer, S. (eds.) CAV 2011. LNCS, vol. 6806, pp. 585–591. Springer, Heidelberg (2011). https://doi.org/10.1007/978-3-642-22110-1_47
18. Puterman, M.L.: Markov Decision Processes. Wiley-Interscience, Hoboken (2005)

19. Russell, S.J., Dewey, D., Tegmark, M.: Research priorities for robust and beneficial artificial intelligence. AI Mag. **36**(4), 105–114 (2015)
20. Russell, S.J., Norvig, P.: Artificial Intelligence - A Modern Approach, 3rd Int. edn., Pearson Education, London (2010)
21. Steinmetz, M., Hoffmann, J., Buffet, O.: Goal probability analysis in probabilistic planning: exploring and enhancing the state of the art. JAIR **57**, 229–271 (2016)
22. Strehl, A.L., Li, L., Littman, M.L.: Reinforcement learning in finite MDPs: PAC analysis. J. Mach. Learn. Res. **10**, 2413–2444 (2009)
23. Valiant, L.: Probably Approximately Correct: Nature's Algorithms for Learning and Prospering in a Complex World. Basic Books, New York (2013)

A Hierarchy of Scheduler Classes for Stochastic Automata

Pedro R. D'Argenio[1,2,3], Marcus Gerhold[4], Arnd Hartmanns[4(✉)], and Sean Sedwards[5]

[1] Universidad Nacional de Córdoba, Córdoba, Argentina
dargenio@famaf.unc.edu.ar
[2] CONICET, Córdoba, Argentina
[3] Saarland University, Saarbrücken, Germany
[4] University of Twente, Enschede, The Netherlands
{m.gerhold,a.hartmanns}@utwente.nl
[5] University of Waterloo, Waterloo, Canada
sean.sedwards@uwaterloo.ca

Abstract. Stochastic automata are a formal compositional model for concurrent stochastic timed systems, with general distributions and nondeterministic choices. Measures of interest are defined over *schedulers* that resolve the nondeterminism. In this paper we investigate the power of various theoretically and practically motivated classes of schedulers, considering the classic complete-information view and a restriction to non-prophetic schedulers. We prove a hierarchy of scheduler classes w.r.t. unbounded probabilistic reachability. We find that, unlike Markovian formalisms, stochastic automata distinguish most classes even in this basic setting. Verification and strategy synthesis methods thus face a tradeoff between powerful and efficient classes. Using lightweight scheduler sampling, we explore this tradeoff and demonstrate the concept of a useful approximative verification technique for stochastic automata.

1 Introduction

The need to analyse continuous-time stochastic models arises in many practical contexts, including critical infrastructures [4], railway engineering [36], space mission planning [7], and security [28]. This has led to a number of discrete event simulation tools, such as those for networking [34,35,42], whose probabilistic semantics is founded on generalised semi-Markov processes (GSMP [21,33]). Nondeterminism arises through inherent concurrency of independent processes [11], but may also be deliberate underspecification. Modelling such uncertainty with probability is convenient for simulation, but not always adequate [3,29]. Various models and formalisms have thus been proposed to extend continuous-time

This work is supported by the 3TU.BSR, NWO BEAT (602.001.303) and JST ERATO HASUO Metamathematics for Systems Design (JPMJER1603) projects, by ERC grant 695614 (POWVER), and by SeCyT-UNC projects 05/BP12, 05/B497.

C. Baier and U. Dal Lago (Eds.): FOSSACS 2018, LNCS 10803, pp. 384–402, 2018.
https://doi.org/10.1007/978-3-319-89366-2_21

stochastic processes with nondeterminism [8,10,19,23,27,38]. It is then possible to *verify* such systems by considering the extremal probabilities of a property. These are the supremum and infimum of the probabilities of the property in the purely stochastic systems induced by classes of *schedulers* (also called *strategies*, *policies* or *adversaries*) that resolve all nondeterminism. If the nondeterminism is considered controllable, one may alternatively be interested in the *planning* problem of synthesising a scheduler that satisfies certain probability bounds.

We consider closed systems of stochastic automata (SA [16]), which extend GSMP and feature both generally distributed stochastic delays as well as discrete nondeterministic choices. The latter may arise from non-continuous distributions (e.g. deterministic delays), urgent edges, and edges waiting on multiple clocks. Numerical verification algorithms exist for very limited subclasses of SA only: Buchholz et al. [13] restrict to phase-type or matrix-exponential distributions, such that nondeterminism cannot arise (as each edge is guarded by a single clock). Bryans et al. [12] propose two algorithms that require an a priori fixed scheduler, continuous bounded distributions, and that all active clocks be reset when a location is entered. The latter forces regeneration on every edge, making it impossible to use clocks as memory between locations. Regeneration is central to the work of Ballarini et al. [6], however they again exclude nondeterminism. The only approach that handles nondeterminism is the region-based approximation scheme of Kwiatkowska et al. [30] for a model closely related to SA, but restricted to bounded continuous distributions. Without that restriction [22], error bounds and convergence guarantees are lost.

Evidently, the combination of nondeterminism and continuous probability distributions is a particularly challenging one. With this paper, we take on the underlying problem from a fundamental perspective: we investigate the power of, and relationships between, different classes of schedulers for SA. Our motivation is, on the one hand, that a clear understanding of scheduler classes is crucial to design verification algorithms. For example, Markov decision process (MDP) model checking works well because memoryless schedulers suffice for reachability, and the efficient time-bounded analysis of continuous-time MDP (CTMDP) exploits a relationship between two scheduler classes that are sufficiently simple, but on their own do not realise the desired extremal probabilities [14]. When it comes to planning problems, on the other hand, practitioners desire *simple* solutions, i.e. schedulers that need little information and limited memory, so as to be explainable and suitable for implementation on e.g. resource-constrained embedded systems. Understanding the capabilities of scheduler classes helps decide on the tradeoff between simplicity and the ability to attain optimal results.

We use two perspectives on schedulers from the literature: the classic complete-information *residual lifetimes* semantics [9], where optimality is defined via history-dependent schedulers that see the entire current state, and *non-prophetic* schedulers [25] that cannot observe the timing of *future* events. Within each perspective, we define classes of schedulers whose views of the state and history are variously restricted (Sect. 3). We prove their relative ordering w.r.t. achieving optimal reachability probabilities (Sect. 4). We find that SA distinguish most classes. In particular, memoryless schedulers suffice in the complete-information setting (as is implicit in the method of Kwiatkowska et al. [30]), but turn out to be suboptimal in the more realistic non-prophetic case. Considering only the relative order of clock expiration times, as suggested by the first algorithm of Bryans et al. [12], surprisingly leads to partly suboptimal, partly incomparable classes. Our distinguishing SA are small and employ a common nondeterministic gadget. They precisely pinpoint the crucial differences and how schedulers interact with the various features of SA, providing deep insights into the formalism itself.

Our study furthermore forms the basis for the application of *lightweight scheduler sampling* (LSS) to SA. LSS is a technique to use Monte Carlo simulation/statistical model checking with nondeterministic models. On every LSS simulation step, a pseudo-random number generator (PRNG) is re-seeded with a hash of the identifier of the current scheduler and the (restricted) information about the current state (and previous states, for history-dependent schedulers) that the scheduler's class may observe. The PRNG's first iterate then determines the scheduler's action deterministically. LSS has been successfully applied to MDP [18,31,32] and probabilistic timed automata [15,26]. Using only constant memory, LSS samples schedulers uniformly from a selected scheduler class to find "near-optimal" schedulers that conservatively approximate the true extremal probabilities. Its principal advantage is that it is largely indifferent to the size of the state space and of the scheduler space; in general, sampling efficiency depends only on the likelihood of selecting near-optimal schedulers. However, the mass of *near*-optimal schedulers in a scheduler class that also includes the optimal scheduler may be *less* than the mass in a class that does *not* include it. Given that the mass of optimal schedulers may be vanishingly small, it may be advantageous to sample from a class of less powerful schedulers. We explore these tradeoffs and demonstrate the concept of LSS for SA in Sect. 5.

Other Related Work. Alur et al. first mention nondeterministic stochastic systems similar to SA in [2]. Markov automata (MA [19]), interactive Markov chains (IMC [27]) and CTMDP are special cases of SA restricted to exponential distributions. Song et al. [37] look into partial information distributed schedulers for MA, combining earlier works of de Alfaro [1] and Giro and D'Argenio [20] for MDP. Their focus is on information flow and hiding in parallel specifications. Wolf et al. [39] investigate the power of classic (time-abstract, deterministic and memoryless) scheduler classes for IMC. They establish (non-strict) subset relationships for almost all classes w.r.t. trace distribution equivalence, a very strong measure. Wolovick and Johr [41] show that the class of measurable schedulers for CTMDP is complete and sufficient for reachability problems.

2 Preliminaries

For a given set S, its power set is $\mathcal{P}(S)$. We denote by $\mathbb{R}$, $\mathbb{R}^+$, and $\mathbb{R}_0^+$ the sets of real numbers, positive real numbers and non-negative real numbers, respectively. A (discrete) *probability distribution* over a set Ω is a function $\mu\colon \Omega \to [0,1]$, such that $\mathrm{support}(\mu) \stackrel{\text{def}}{=} \{\, \omega \in \Omega \mid \mu(\omega) > 0 \,\}$ is countable and $\sum_{\omega \in \mathrm{support}(\mu)} \mu(\omega) = 1$. $\mathrm{Dist}(\Omega)$ is the set of probability distributions over Ω. We write $\mathcal{D}(\omega)$ for the *Dirac* distribution for ω, defined by $\mathcal{D}(\omega)(\omega) = 1$. Ω is *measurable* if it is endowed with a σ-algebra $\sigma(\Omega)$: a collection of *measurable* subsets of Ω. A (continuous) *probability measure* over Ω is a function $\mu\colon \sigma(\Omega) \to [0,1]$, such that $\mu(\Omega) = 1$ and $\mu(\cup_{i \in I} B_i) = \sum_{i \in I} \mu(B_i)$ for any countable index set I and pairwise disjoint measurable sets $B_i \subseteq \Omega$. $\text{Prob}(\Omega)$ is the set of probability measures over Ω. Each $\mu \in \text{Dist}(\Omega)$ induces a probability measure. Given probability measures μ_1 and μ_2, we denote by $\mu_1 \otimes \mu_2$ the *product measure*: the unique probability measure such that $(\mu_1 \otimes \mu_2)(B_1 \times B_2) = \mu_1(B_1) \cdot \mu_2(B_2)$, for all measurable B_1 and B_2. For a collection of measures $(\mu_i)_{i \in I}$, we analogously denote the product measure by $\bigotimes_{i \in I} \mu_i$. Let $Val \stackrel{\text{def}}{=} V \to \mathbb{R}_0^+$ be the set of valuations for an (implicit) set V of (non-negative real-valued) variables. $\mathbf{0} \in Val$ assigns value zero to all variables. Given $X \subseteq V$ and $v \in Val$, we write $v[X]$ for the valuation defined by $v[X](x) = 0$ if $x \in X$ and $v[X](y) = v(y)$ otherwise. For $t \in \mathbb{R}_0^+$, $v + t$ is the valuation defined by $(v+t)(x) = v(x) + t$ for all $x \in V$.

Stochastic Automata [16] extend labelled transition systems with stochastic *clocks*: real-valued variables that increase synchronously with rate 1 over time and expire some random amount of time after having been *restarted*. Formally:

Definition 1. *A* stochastic automaton *(SA) is a tuple* $\langle Loc, \mathcal{C}, A, E, F, \ell_{init} \rangle$, *where Loc is a countable set of* locations, $\mathcal{C}$ *is a finite set of* clocks, A *is the finite* action alphabet, *and* $E\colon Loc \to \mathcal{P}(\mathcal{P}(\mathcal{C}) \times A \times \mathcal{P}(\mathcal{C}) \times \text{Dist}(Loc))$ *is the* edge function, *which maps each location to a finite set of edges that in turn consist of a* guard set *of clocks, a label, a* restart set *of clocks and a distribution over target locations.* $F\colon \mathcal{C} \to \text{Prob}(\mathbb{R}_0^+)$ *is the* delay measure function *that maps each clock to a probability measure, and* $\ell_{init} \in Loc$ *is the* initial location.

We also write $\ell \xrightarrow{G,a,R}_E \mu$ for $\langle G, a, R, \mu \rangle \in E(\ell)$. W.l.o.g. we restrict to SA where edges are fully characterised by source state and action label, i.e. whenever $\ell \xrightarrow{G_1,a,R_1}_E \mu_1$ and $\ell \xrightarrow{G_2,a,R_2}_E \mu_2$, then $G_1 = G_2$, $R_1 = R_2$ and $\mu_1 = \mu_2$.

Intuitively, an SA starts in ℓ_{init} with all clocks expired. An edge $\ell \xrightarrow{G,a,R}_E \mu$ may be taken only if all clocks in G are expired. If any edge is enabled, some edge must be taken (i.e. all actions are *urgent* and thus the SA is *closed*). When an edge is taken, its action is a, all clocks in R are restarted, other expired clocks remain expired, and we move to successor location ℓ' with probability $\mu(\ell')$. There, another edge may be taken immediately or we may need to wait until some further clocks expire, and so on. When a clock c is restarted, the time until it expires is chosen randomly according to the probability measure $F(c)$.

Example 1. We show an example SA, M_0, in Fig. 1. Its initial location is ℓ_0. It has two clocks, x and y, with $F(x)$ and $F(y)$ both being the continuous uniform distribution over the interval $[0,1]$. No time can pass in locations ℓ_0 and ℓ_1, since they have outgoing edges with empty guard sets. We omit action labels and assume every edge to have a unique label. On entering ℓ_1, both clocks are restarted. The choice of going to either ℓ_2 or ℓ_3 from ℓ_1 is nondeterministic, since

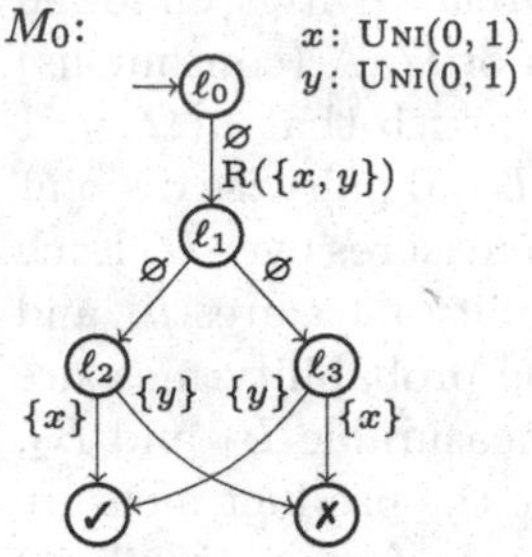

Fig. 1. Example SA M_0

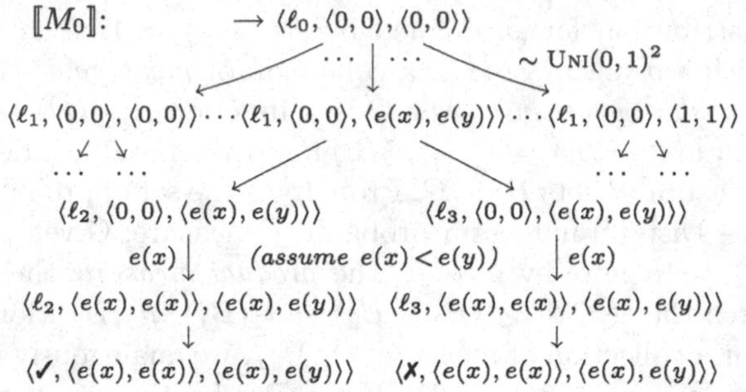

Fig. 2. Excerpt of the TPTS semantics of M_0

the two edges are always enabled at the same time. In ℓ_2, we have to wait until the first of the two clocks expires. If that is x, we have to move to location ✓; if it is y, we have to move to ✗. The probability that both expire at the same time is zero. Location ℓ_3 behaves analogously, but with the target states interchanged.

Timed Probabilistic Transition Systems form the semantics of SA. They are finitely-nondeterministic uncountable-state transition systems:

Definition 2. *A (finitely nondeterministic)* timed probabilistic transition system *(TPTS) is a tuple $\langle S, A', T, s_{init}\rangle$. S is a measurable set of states. $A' = \mathbb{R}^+ \uplus A$ is the* alphabet, *partitioned into* delays *in $\mathbb{R}^+$ and* jumps *in A. $T \colon S \to \mathcal{P}(A' \times \mathrm{Prob}(S))$ is the* transition function, *which maps each state to a finite set of transitions, each consisting of a label in A' and a measure over target states. The initial state is $s_{init} \in S$. For all $s \in S$, we require $|T(s)| = 1$ if $\exists \langle t, \mu\rangle \in T(s) \colon t \in \mathbb{R}^+$, i.e. states admitting delays are deterministic.*

We also write $s \xrightarrow{a}_T \mu$ for $\langle a, \mu\rangle \in T(s)$. A *run* is an infinite alternating sequence $s_0 a_0 s_1 a_1 \ldots \in (S \times A')^\omega$, with $s_0 = s_{init}$. A *history* is a finite prefix of a run ending in a state, i.e. an element of $(S \times A')^* \times S$. Runs resolve all nondeterministic and probabilistic choices. A *scheduler* resolves only the nondeterminism:

Definition 3. *A measurable function $\mathfrak{s} \colon (S \times A')^* \times S \to \mathrm{Dist}(A' \times \mathrm{Prob}(S))$ is a* scheduler *if, for all histories $h \in (S \times A')^* \times S$, $\langle a, \mu\rangle \in \mathrm{support}(\mathfrak{s}(h))$ implies $lst_h \xrightarrow{a}_T \mu$, where lst_h is the last state of h.*

Once a scheduler has chosen $s_i \xrightarrow{a}_T \mu$, the successor state s_{i+1} is picked randomly according to μ. Every scheduler $\mathfrak{s}$ defines a probability measure $\mathbb{P}_{\mathfrak{s}}$ on the space of all runs. For a formal definition, see [40]. As is usual, we restrict to *non-Zeno* schedulers that make time diverge with probability one: we require $\mathbb{P}_{\mathfrak{s}}(\Pi_\infty) = 1$, where Π_∞ is the set of runs where the sum of delays is ∞. In the remainder of this paper we consider extremal probabilities of reaching a set of goal locations G:

Definition 4. *For $G \subseteq Loc$, let $J_G \stackrel{\text{def}}{=} \{\langle \ell, v, e\rangle \in S \mid \ell \in G\}$. Let $\mathfrak{S}$ be a class of schedulers. Then $\mathrm{P}^{\mathfrak{S}}_{\min}(G)$ and $\mathrm{P}^{\mathfrak{S}}_{\max}(G)$ are the minimum and maximum* reachability probabilities *for G under $\mathfrak{S}$, defined as $\mathrm{P}^{\mathfrak{S}}_{\min}(G) = \inf_{\mathfrak{s} \in \mathfrak{S}} \mathbb{P}_{\mathfrak{s}}(\Pi_{J_G})$ and $\mathrm{P}^{\mathfrak{S}}_{\max}(G) = \sup_{\mathfrak{s} \in \mathfrak{S}} \mathbb{P}_{\mathfrak{s}}(\Pi_{J_G})$, respectively.*

Semantics of Stochastic Automata. We present here the residual lifetimes semantics of [9], simplified for closed SA: any delay step must be of the minimum delay that makes some edge become enabled.

Definition 5. *The semantics of an SA* $M = \langle Loc, \mathcal{C}, A, E, F, \ell_{init} \rangle$ *is the TPTS*

$$[\![M]\!] = \langle Loc \times Val \times Val, A \uplus \mathbb{R}^+, T_M, \langle \ell_{init}, \mathbf{0}, \mathbf{0} \rangle \rangle$$

where the states are triples $\langle \ell, v, e \rangle$ *of the current location* ℓ*, a valuation* v *assigning to each clock its current value, and a valuation* e *keeping track of all clocks' expiration times.* T_M *is the smallest transition function satisfying inference rules*

$$\frac{\ell \xrightarrow{G,a,R}_E \mu \quad \mathrm{En}(G, v, e)}{\langle \ell, v, e \rangle \xrightarrow{a}_{T_M} \mu \otimes \mathcal{D}(v[R]) \otimes \mathrm{Sample}_e^R}$$

$$\frac{t \in \mathbb{R}^+ \quad \exists \ell \xrightarrow{G,a,R}_E \mu \colon \mathrm{En}(G, v+t, e) \quad \forall t' \in [0, t), \ell \xrightarrow{G,a,R}_E \mu \colon \neg\, \mathrm{En}(G, v+t', e)}{\langle \ell, v, e \rangle \xrightarrow{t}_{T_M} \mathcal{D}(\langle \ell, v+t, e \rangle)}$$

with $\mathrm{En}(G, v, e) \stackrel{\text{def}}{=} \forall x \in G \colon v(x) \geq e(x)$ *characterising the enabled edges and*

$$\mathrm{Sample}_e^R \stackrel{\text{def}}{=} \bigotimes_{c \in \mathcal{C}} \begin{cases} F(c) & \text{if } c \in R \\ \mathcal{D}(e(c)) & \text{if } c \notin R. \end{cases}$$

The second rule creates *delay* steps of t time units if no edge is enabled from now until just before t time units have elapsed (third premise) but then, after exactly t time units, some edge becomes enabled (second premise). The first rule applies if an edge $\ell \xrightarrow{G,a,R}_E \mu$ is enabled: a transition is taken with the edge's label, the successor state's location is chosen by μ, v is updated by resetting the clocks in R to zero, and the expiration times for the restarted clocks are resampled. All other expiration times remain unchanged. Notice that $[\![M]\!]$ is also a nondeterministic labelled Markov process [40] (a proof can be found in [17]).

Example 2. Figure 2 outlines the semantics of M_0. The first step from ℓ_0 to all the states in ℓ_1 is a single transition. Its probability measure is the product of $F(x)$ and $F(y)$, sampling the expiration times of the two clocks. We exemplify the behaviour of all of these states by showing it for the case of expiration times $e(x)$ and $e(y)$, with $e(x) < e(y)$. In this case, to maximise the probability of reaching ✓, we should take the transition to the state in ℓ_2. If a scheduler $\mathfrak{s}$ can see the expiration times, noting that only their order matters here, it can always make the optimal choice and achieve $\mathrm{P}_{\max}^{\{\mathfrak{s}\}}(\{✓\}) = 1$.

3 Classes of Schedulers

We now define classes of schedulers for SA with restricted information, hiding in various combinations the history and parts of states such as clock values and expiration times. All definitions consider TPTS as in Definition 5 with states $\langle \ell, v, e \rangle$ and we require for all $\mathfrak{s}$ that $\langle a, \mu \rangle \in \mathrm{support}(\mathfrak{s}(h)) \Rightarrow lst_h \xrightarrow{a}_T \mu$, as in Definition 3.

3.1 Classic Schedulers

We first consider the "classic" complete-information setting where schedulers can in particular see expiration times. We start with restricted classes of history-dependent schedulers. Our first restriction hides the values of all clocks, only revealing the total time since the start of the history. This is inspired by the step-counting or time-tracking schedulers needed to obtain optimal step-bounded or time-bounded reachability probabilities on MDP or Markov automata:

Definition 6. *A classic history-dependent* global-time *scheduler is a measurable function* $\mathfrak{s}\colon (S|_{\ell,t,e} \times A')^* \times S|_{\ell,t,e} \to \text{Dist}(A' \times \text{Prob}(S))$, *where* $S|_{\ell,t,e} \stackrel{\text{def}}{=} Loc \times \mathbb{R}_0^+ \times Val$ *with the second component being the total time t elapsed since the start of the history. We write* $\mathfrak{S}^{hist}_{\ell,t,e}$ *for the set of all such schedulers.*

We next hide the values of all clocks, revealing only their expiration times:

Definition 7. *A classic history-dependent* location-based *scheduler is a measurable function* $\mathfrak{s}\colon (S|_{\ell,e} \times A')^* \times S|_{\ell,e} \to \text{Dist}(A' \times \text{Prob}(S))$, *where* $S|_{\ell,e} \stackrel{\text{def}}{=} Loc \times Val$, *with the second component being the clock expiration times e. We write* $\mathfrak{S}^{hist}_{\ell,e}$ *for the set of all such schedulers.*

Having defined three classes of classic history-dependent schedulers, $\mathfrak{S}^{hist}_{\ell,v,e}$, $\mathfrak{S}^{hist}_{\ell,t,e}$ and $\mathfrak{S}^{hist}_{\ell,e}$, noting that $\mathfrak{S}^{hist}_{\ell,v,e}$ denotes all schedulers of Definition 3, we also consider them with the restriction that they only see the relative order of clock expiration, instead of the exact expiration times: for each pair of clocks c_1, c_2, these schedulers see the relation $\sim\, \in \{<, =, >\}$ in $e(c_1) - v(c_1) \sim e(c_2) - v(c_2)$. E.g. in ℓ_1 of Example 2, the scheduler would not see $e(x)$ and $e(y)$, but only whether $e(x) < e(y)$ or vice-versa (since $v(x) = v(y) = 0$, and equality has probability 0 here). We consider this case because the expiration order is sufficient for the first algorithm of Bryans et al. [12], and would allow optimal decisions in M_0 of Fig. 1. We denote the relative order information by o, and the corresponding scheduler classes by $\mathfrak{S}^{hist}_{\ell,v,o}$, $\mathfrak{S}^{hist}_{\ell,t,o}$ and $\mathfrak{S}^{hist}_{\ell,o}$. We now define memoryless schedulers, which only see the current state and are at the core of e.g. MDP model checking. On most formalisms, they suffice to obtain optimal reachability probabilities.

Definition 8. *A classic* memoryless *scheduler is a measurable function* $\mathfrak{s}\colon S \to \text{Dist}(A' \times \text{Prob}(S))$. *We write* $\mathfrak{S}^{ml}_{\ell,v,e}$ *for the set of all such schedulers.*

We apply the same restrictions as for history-dependent schedulers:

Definition 9. *A classic memoryless global-time scheduler is a measurable function* $\mathfrak{s}\colon S|_{\ell,t,e} \to \text{Dist}(A' \times \text{Prob}(S))$, *with* $S|_{\ell,t,e}$ *as in Definition 6. We write* $\mathfrak{S}^{ml}_{\ell,t,e}$ *for the set of all such schedulers.*

Definition 10. *A classic memoryless location-based scheduler is a measurable function* $\mathfrak{s}\colon S|_{\ell,e} \to \text{Dist}(A' \times \text{Prob}(S))$, *with* $S|_{\ell,e}$ *as in Definition 7. We write* $\mathfrak{S}^{ml}_{\ell,e}$ *for the set of all such schedulers.*

Again, we also consider memoryless schedulers that only see the expiration order, so we have memoryless scheduler classes $\mathfrak{S}^{ml}_{\ell,v,e}$, $\mathfrak{S}^{ml}_{\ell,t,e}$, $\mathfrak{S}^{ml}_{\ell,e}$, $\mathfrak{S}^{ml}_{\ell,v,o}$, $\mathfrak{S}^{ml}_{\ell,t,o}$ and $\mathfrak{S}^{ml}_{\ell,o}$. Class $\mathfrak{S}^{ml}_{\ell,o}$ is particularly attractive because it has a compact finite domain.

3.2 Non-prophetic Schedulers

Consider the SA M_0 in Fig. 1. No matter which of the previously defined scheduler classes we choose, we always find a scheduler that achieves probability 1 to reach ✓, and a scheduler that achieves probability 0. This is because they can all see the expiration times or expiration order of x and y when in ℓ_1. When in ℓ_1, x and y have not yet expired—this will only happen later, in ℓ_2 or ℓ_3—yet the schedulers already know which clock will "win". The classic schedulers can thus be seen to make decisions based on the timing of *future* events. This *prophetic* scheduling has already been observed in [9], where a "fix" in the form of the *spent lifetimes* semantics was proposed. Hartmanns et al. [25] have shown that this not only still permits prophetic scheduling, but even admits *divine* scheduling, where a scheduler can *change* the future. The authors propose a complex *non-prophetic* semantics that provably removes all prophetic and divine behaviour.

Much of the complication of the non-prophetic semantics of [25] is due to it being specified for open SA that include delayable actions. For the closed SA setting of this paper, prophetic scheduling can be more easily excluded by hiding from the schedulers all information about what will happen in the future of the system's evolution. This information is only contained in the expiration times e or the expiration order o. We can thus keep the semantics of Sect. 2 and modify the definition of schedulers to exclude prophetic behaviour by construction.

In what follows, we thus also consider all scheduler classes of Sect. 3.1 with the added constraint that the expiration times, resp. the expiration order, are not visible, resulting in the *non-prophetic* classes $\mathfrak{S}^{hist}_{\ell,v}$, $\mathfrak{S}^{hist}_{\ell,t}$, $\mathfrak{S}^{hist}_{\ell}$, $\mathfrak{S}^{ml}_{\ell,v}$, $\mathfrak{S}^{ml}_{\ell,t}$ and $\mathfrak{S}^{ml}_{\ell}$. Any non-prophetic scheduler can only reach ✓ of M_0 with probability $\frac{1}{2}$.

4 The Power of Schedulers

Now that we have defined a number of classes of schedulers, we need to determine what the effect of the restrictions is on our ability to optimally control an SA. We thus evaluate the power of scheduler classes w.r.t. unbounded reachability probabilities (Definition 4) on the semantics of SA. We will see that this simple setting already suffices to reveal interesting differences between scheduler classes.

For two scheduler classes $\mathfrak{S}_1$ and $\mathfrak{S}_2$, we write $\mathfrak{S}_1 \succcurlyeq \mathfrak{S}_2$ if, for all SA and all sets of goal locations G, $\mathrm{P}^{\mathfrak{S}_1}_{\min}(G) \leq \mathrm{P}^{\mathfrak{S}_2}_{\min}(G)$ and $\mathrm{P}^{\mathfrak{S}_1}_{\max}(G) \geq \mathrm{P}^{\mathfrak{S}_2}_{\max}(G)$. We write $\mathfrak{S}_1 \succ \mathfrak{S}_2$ if additionally there exists at least one SA and set G' where $\mathrm{P}^{\mathfrak{S}_1}_{\min}(G') < \mathrm{P}^{\mathfrak{S}_2}_{\min}(G')$ or $\mathrm{P}^{\mathfrak{S}_1}_{\max}(G') > \mathrm{P}^{\mathfrak{S}_2}_{\max}(G')$. Finally, we write $\mathfrak{S}_1 \approx \mathfrak{S}_2$ for $\mathfrak{S}_1 \succcurlyeq \mathfrak{S}_2 \wedge \mathfrak{S}_2 \succcurlyeq \mathfrak{S}_1$, and $\mathfrak{S}_1 \not\approx \mathfrak{S}_2$, i.e. the classes are incomparable, for $\mathfrak{S}_1 \not\succcurlyeq \mathfrak{S}_2 \wedge \mathfrak{S}_2 \not\succcurlyeq \mathfrak{S}_1$. Unless noted otherwise, we omit proofs for $\mathfrak{S}_1 \succcurlyeq \mathfrak{S}_2$ when it is obvious that the information available to $\mathfrak{S}_1$ includes the information available to $\mathfrak{S}_2$. All our distinguishing examples are based on the resolution of a single nondeterministic choice between two actions to eventually reach one of two locations. We therefore prove only w.r.t. the maximum probability, $p_{\max}$, for these examples since the minimum probability is given by $1 - p_{\max}$ and an analogous proof for $p_{\min}$ can be made by relabelling locations. We may write $\mathrm{P}_{\max}(\mathfrak{S}^y_x)$ for $\mathrm{P}^{\mathfrak{S}^y_x}_{\max}(\{ ✓ \})$ to improve readability.

$$\begin{array}{ccccccc}
\mathfrak{S}^{ml}_{\ell,o} & \prec & \mathfrak{S}^{ml}_{\ell,t,o} & \prec & \mathfrak{S}^{ml}_{\ell,v,o} & \prec & \mathfrak{S}^{hist}_{\ell,o} \\
\rotatebox{90}{$\not\approx$} & & \rotatebox{90}{$\not\approx$} & & \rotatebox{90}{$\prec$} & & \rotatebox{90}{$\approx$} \\
\mathfrak{S}^{ml}_{\ell,e} & \prec & \mathfrak{S}^{ml}_{\ell,t,e} & \prec & \mathfrak{S}^{ml}_{\ell,v,e} & \succ & \mathfrak{S}^{hist}_{\ell,t,o} \\
\rotatebox{90}{$\prec$} & & \rotatebox{90}{$\prec$} & & \rotatebox{90}{$\approx$} & & \rotatebox{90}{$\approx$} \\
\mathfrak{S}^{hist}_{\ell,e} & \approx & \mathfrak{S}^{hist}_{\ell,t,e} & \approx & \mathfrak{S}^{hist}_{\ell,v,e} & \succ & \mathfrak{S}^{hist}_{\ell,v,o}
\end{array}$$

Fig. 3. Hierarchy of classic scheduler classes

$$\begin{array}{ccccc}
\mathfrak{S}^{ml}_{\ell} & \prec & \mathfrak{S}^{ml}_{\ell,t} & \prec & \mathfrak{S}^{ml}_{\ell,v} \\
\rotatebox{90}{$\prec$} & & \rotatebox{90}{$\prec$} & & \rotatebox{90}{$\prec$} \\
\mathfrak{S}^{hist}_{\ell} & \approx & \mathfrak{S}^{hist}_{\ell,t} & \approx & \mathfrak{S}^{hist}_{\ell,v}
\end{array}$$

Fig. 4. Non-prophetic classes

4.1 The Classic Hierarchy

We first establish that all classic history-dependent scheduler classes are equivalent:

Proposition 1. $\mathfrak{S}^{hist}_{\ell,v,e} \approx \mathfrak{S}^{hist}_{\ell,t,e} \approx \mathfrak{S}^{hist}_{\ell,e}$.

Proof. From the transition labels in $A' = A \uplus \mathbb{R}^+$ in the history $(S' \times A')^*$, with $S' \in \{ S, S|_{\ell,t,e}, S|_{\ell,e} \}$ depending on the scheduler class, we can reconstruct the total elapsed time as well as the values of all clocks: to obtain the total elapsed time, sum the labels in $\mathbb{R}^+$ up to each state; to obtain the values of all clocks, do the same per clock and perform the resets of the edges identified by the actions.

The same argument applies among the expiration-order history-dependent classes:

Proposition 2. $\mathfrak{S}^{hist}_{\ell,v,o} \approx \mathfrak{S}^{hist}_{\ell,t,o} \approx \mathfrak{S}^{hist}_{\ell,o}$.

However, the expiration-order history-dependent schedulers are strictly less powerful than the classic history-dependent ones:

Proposition 3. $\mathfrak{S}^{hist}_{\ell,v,e} \succ \mathfrak{S}^{hist}_{\ell,v,o}$.

Proof. Consider the SA M_1 in Fig. 5. Note that the history does not provide any information for making the choice in ℓ_1: we always arrive after having spent zero time in ℓ_0 and then having taken the single edge to ℓ_1. We can analytically determine that $\mathrm{P}_{\max}(\mathfrak{S}^{hist}_{\ell,v,e}) = \frac{3}{4}$ by going from ℓ_1 to ℓ_2 if $e(x) \leq \frac{1}{2}$ and to ℓ_3 otherwise. We would obtain a probability equal to $\frac{1}{2}$ by always going to either ℓ_2 or ℓ_3 or by picking either edge with equal probability. This is the best we can do if e is not visible, and thus $\mathrm{P}_{\max}(\mathfrak{S}^{hist}_{\ell,v,o}) = \frac{1}{2}$: in ℓ_1, $v(x) = v(y) = 0$ and the expiration order is always "y before x" because y has not yet been started.

Just like for MDP and unbounded reachability probabilities, the classic history-dependent and memoryless schedulers with complete information are equivalent:

Proposition 4. $\mathfrak{S}^{hist}_{\ell,v,e} \approx \mathfrak{S}^{ml}_{\ell,v,e}$.

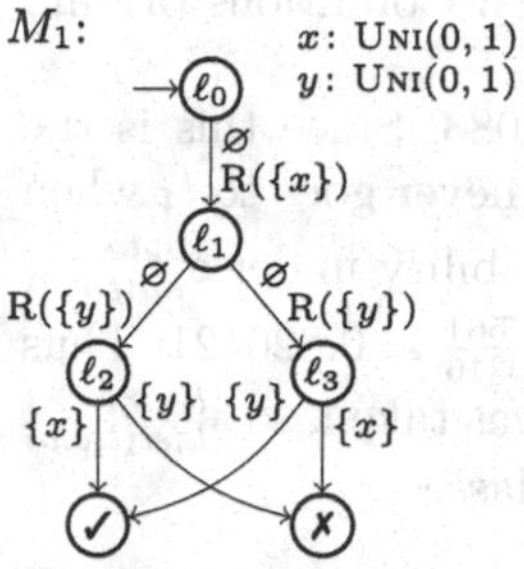

Fig. 5. SA M_1

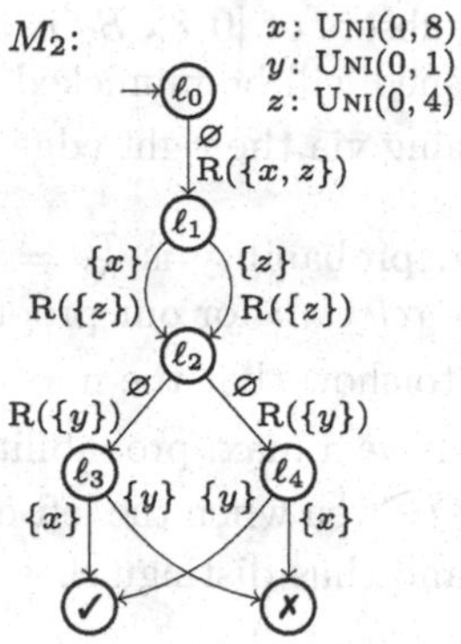

Fig. 6. SA M_2

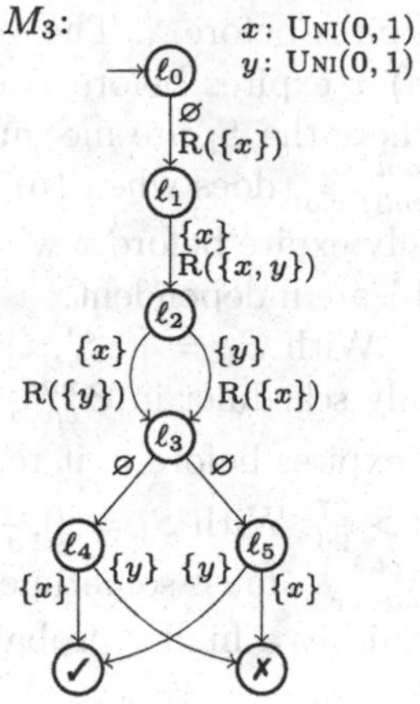

Fig. 7. SA M_3

Proof sketch. Our definition of TPTS only allows finite nondeterministic choices, i.e. we have a very restricted form of continuous-space MDP. We can thus adapt the argument of the corresponding proof for MDP [5, Lemma 10.102]: For each state (of possibly countably many), we construct a notional optimal memoryless (and deterministic) scheduler in the same way, replacing the summation by an integration for the continuous measures in the transition function. It remains to show that this scheduler is indeed measurable. For TPTS that are the semantics of SA, this follows from the way clock values are used in the guard sets so that optimal decisions are constant over intervals of clock values and expiration times (see e.g. the arguments in [12] or [30]).

On the other hand, when restricting schedulers to see the expiration order only, history-dependent and memoryless schedulers are no longer equivalent:

Proposition 5. $\mathfrak{S}^{hist}_{\ell,v,o} \succ \mathfrak{S}^{ml}_{\ell,v,o}$.

Proof. Consider the SA M_2 in Fig. 6. Let $\mathfrak{s}^{opt}_{ml(l,v,o)}$ be the (unknown) optimal scheduler in $\mathfrak{S}^{ml}_{\ell,v,o}$ w.r.t. the max. probability of reaching ✓. Define $\mathfrak{s}^{better}_{hist(l,v,o)} \in \mathfrak{S}^{hist}_{\ell,v,o}$ as: when in ℓ_2 and the last edge in the history is the left one (i.e. x is expired), go to ℓ_3; otherwise, behave like $\mathfrak{s}^{opt}_{ml(l,v,o)}$. This scheduler distinguishes $\mathfrak{S}^{hist}_{\ell,v,o}$ and $\mathfrak{S}^{ml}_{\ell,v,o}$ (by achieving a strictly higher max. probability than $\mathfrak{s}^{opt}_{ml(l,v,o)}$) if and only if there are some combinations of clock values (aspect v) and expiration orders (aspect o) in ℓ_2 that can be reached with positive probability via the left edge into ℓ_2, for which $\mathfrak{s}^{opt}_{ml(l,v,o)}$ must nevertheless decide to go to ℓ_4.

All possible clock valuations in ℓ_2 can be achieved via either the left or the right edge, but taking the left edge implies that x expires before z in ℓ_2. It is thus sufficient to show that $\mathfrak{s}^{opt}_{ml(l,v,o)}$ must go to ℓ_4 in *some* cases where x

expires before z. The general form of schedulers in $\mathfrak{S}^{ml}_{\ell,v,o}$ in ℓ_2 is "go to ℓ_3 iff (a) x expires before z and $v(x) \in S_1$ or (b) z expires before x and $v(x) \in S_2$" where the S_i are measurable subsets of $[0, 8]$. S_2 is in fact *irrelevant*: whatever $\mathfrak{s}^{opt}_{ml(l,v,o)}$ does when (b) is satisfied will be mimicked by $\mathfrak{s}^{better}_{hist(l,v,o)}$ because z can only expire before x when coming via the right edge into ℓ_2. Conditions (a) and (b) are independent.

With $S_1 = [0, 8]$, the max. probability is $\frac{77}{96} = 0.80208\overline{3}$. Since this is the only scheduler in $\mathfrak{S}^{ml}_{\ell,v,o}$ that is *relevant* for our proof and never goes to l_4 when x expires before z, it remains to show that the max. probability under $\mathfrak{s}^{opt}_{ml(l,v,o)}$ is $> \frac{77}{96}$. With $S_1 = [0, \frac{35}{12})$, we have a max. probability of $\frac{7561}{9216} \approx 0.820421$. Thus $\mathfrak{s}^{opt}_{ml(l,v,o)}$ must sometimes go to l_4 even when the left edge was taken, so $\mathfrak{s}^{better}_{hist(l,v,o)}$ achieves a higher probability and thus distinguishes the classes.

Knowing only the global elapsed time is less powerful than knowing the full history or the values of all clocks:

Proposition 6. $\mathfrak{S}^{hist}_{\ell,t,e} \succ \mathfrak{S}^{ml}_{\ell,t,e}$ *and* $\mathfrak{S}^{ml}_{\ell,v,e} \succ \mathfrak{S}^{ml}_{\ell,t,e}$.

Proof sketch. Consider the SA M_3 in Fig. 7. We have $\mathrm{P}_{\max}(\mathfrak{S}^{hist}_{\ell,t,e}) = 1$: when in ℓ_3, the scheduler sees from the history which of the two incoming edges was used, and thus knows whether x or y is already expired. It can then make the optimal choice: go to ℓ_4 if x is already expired, or to ℓ_5 otherwise. We also have $\mathrm{P}_{\max}(\mathfrak{S}^{ml}_{\ell,v,e}) = 1$: the scheduler sees that either $v(x) = 0$ or $v(y) = 0$, which implies that the other clock is already expired, and the argument above applies. However, $\mathrm{P}_{\max}(\mathfrak{S}^{ml}_{\ell,t,e}) < 1$: the distribution of elapsed time t on entering ℓ_3 is itself independent of which edge is taken. With probability $\frac{1}{4}$, exactly one of $e(x)$ and $e(y)$ is below t in ℓ_3, which implies that that clock has just expired and thus the scheduler can decide optimally. Yet with probability $\frac{3}{4}$, the expiration times are not useful: they are both positive and drawn from the same distribution, but one unknown clock is expired. The wait for x in ℓ_1 ensures that comparing t with the expiration times in e does not reveal further information in this case.

In the case of MDP, knowing the total elapsed time (i.e. steps) does not make a difference for unbounded reachability. Only for step-bounded properties is that extra knowledge necessary to achieve optimal probabilities. With SA, however, it makes a difference even in the unbounded case:

Proposition 7. $\mathfrak{S}^{ml}_{\ell,t,e} \succ \mathfrak{S}^{ml}_{\ell,e}$.

Proof. Consider SA M_4 in Fig. 8. We have $\mathrm{P}_{\max}(\mathfrak{S}^{ml}_{\ell,t,e}) = 1$: in ℓ_2, the remaining time until y expires is $e(y)$ and the remaining time until x expires is $e(x) - t$ for the global time value t as ℓ_2 is entered. The scheduler can observe all of these quantities and thus optimally go to ℓ_3 if x will expire first, or to ℓ_4 otherwise. However, $\mathrm{P}_{\max}(\mathfrak{S}^{ml}_{\ell,e}) < 1$: $e(x)$ only contains the absolute expiration time of x, but without knowing t or the expiration time of z in ℓ_1, and thus the current value $v(x)$, this scheduler cannot know with certainty which of the clocks will expire first and is therefore unable to make an optimal choice in ℓ_2.

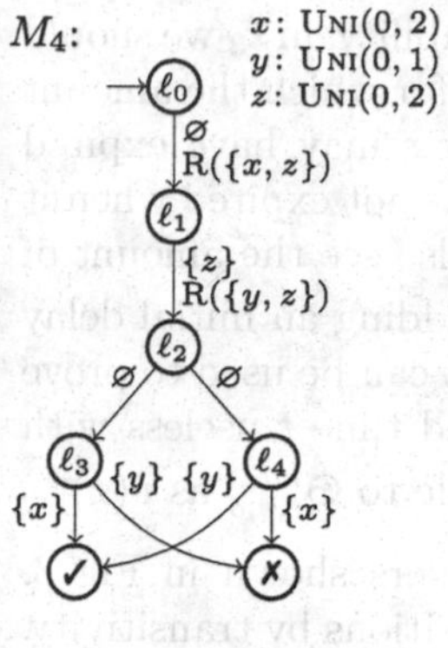

Fig. 8. SA M_4

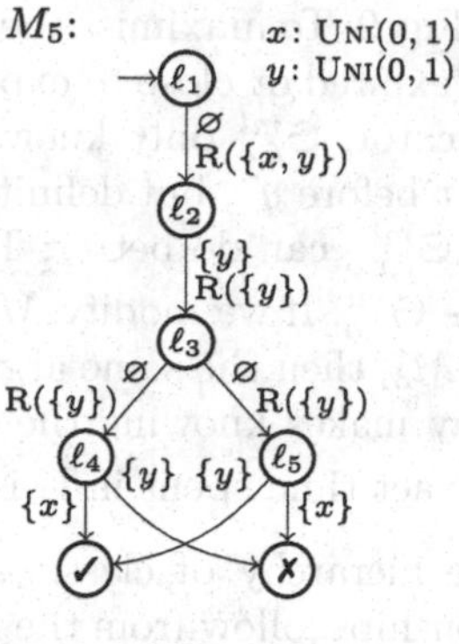

Fig. 9. SA M_5

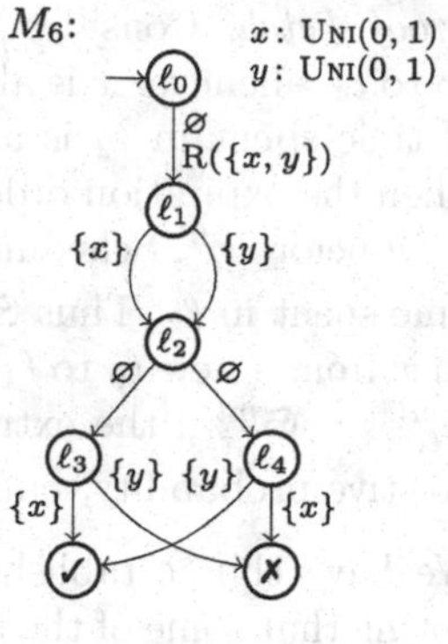

Fig. 10. SA M_6

Finally, we need to compare the memoryless schedulers that see the clock expiration times with memoryless schedulers that see the expiration order. As noted in Sect. 3.1, these two views of the current state are incomparable unless we also see the clock values:

Proposition 8. $\mathfrak{S}^{ml}_{\ell,v,e} \succ \mathfrak{S}^{ml}_{\ell,v,o}$.

Proof. $\mathfrak{S}^{ml}_{\ell,v,e} \not\preccurlyeq \mathfrak{S}^{ml}_{\ell,v,o}$ follows from the same argument as in the proof of Proposition 3. $\mathfrak{S}^{ml}_{\ell,v,e} \succcurlyeq \mathfrak{S}^{ml}_{\ell,v,o}$ is because knowing the current clock values v and the expiration times e is equivalent to knowing the expiration order, since that is precisely the order of the differences $e(c) - v(c)$ for all clocks c.

Proposition 9. $\mathfrak{S}^{ml}_{\ell,t,e} \approx\!\!\!\!/\; \mathfrak{S}^{ml}_{\ell,t,o}$.

Proof. $\mathfrak{S}^{ml}_{\ell,t,e} \not\preccurlyeq \mathfrak{S}^{ml}_{\ell,t,o}$ follows from the same argument as in the proof of Proposition 3. For $\mathfrak{S}^{ml}_{\ell,t,e} \not\succcurlyeq \mathfrak{S}^{ml}_{\ell,t,o}$, consider the SA M_3 of Fig. 7. We know from the proof of Proposition 6 that $\mathrm{P}_{\max}(\mathfrak{S}^{ml}_{\ell,t,e}) < 1$. However, if the scheduler knows the order in which the clocks will expire, it knows which one has already expired (the first one in the order), and can thus make the optimal choice in ℓ_3 to achieve $\mathrm{P}_{\max}(\mathfrak{S}^{ml}_{\ell,t,o}) = 1$.

Proposition 10. $\mathfrak{S}^{ml}_{\ell,e} \approx\!\!\!\!/\; \mathfrak{S}^{ml}_{\ell,o}$.

Proof. The argument of Proposition 9 applies by observing that, in M_3 of Fig. 7, we also have $\mathrm{P}_{\max}(\mathfrak{S}^{ml}_{\ell,e}) < 1$ via the same argument as for $\mathfrak{S}^{ml}_{\ell,t,e}$ in the proof of Proposition 6.

Among the expiration-order schedulers, the hierarchy is as expected:

Proposition 11. $\mathfrak{S}^{ml}_{\ell,v,o} \succ \mathfrak{S}^{ml}_{\ell,t,o} \succ \mathfrak{S}^{ml}_{\ell,o}$.

Proof sketch. Consider M_5 of Fig. 9. To maximise the probability, in ℓ_3 we should go to ℓ_4 whenever x is already expired or close to expiring, for which the amount of time spent in ℓ_2 is an indicator. $\mathfrak{S}^{ml}_{\ell,o}$ only knows that x may have expired when the expiration order is "x before y", but definitely has not expired when it is "y before x". Schedulers in $\mathfrak{S}^{ml}_{\ell,t,o}$ can do better: They also see the amount of time spent in ℓ_2. Thus $\mathfrak{S}^{ml}_{\ell,t,o} \succ \mathfrak{S}^{ml}_{\ell,o}$. If we modify M_5 by adding an initial delay on x from a new ℓ_0 to ℓ_1 as in M_3, then the same argument can be used to prove $\mathfrak{S}^{ml}_{\ell,v,o} \succ \mathfrak{S}^{ml}_{\ell,t,o}$: the extra delay makes knowing the elapsed time t useless with positive probability, but the exact time spent in l_2 is visible to $\mathfrak{S}^{ml}_{\ell,v,o}$ as $v(x)$.

We have thus established the hierarchy of classic schedulers shown in Fig. 3, noting that some of the relationships follow from the propositions by transitivity.

4.2 The Non-prophetic Hierarchy

Each non-prophetic scheduler class is clearly dominated by the classic and expiration-order scheduler classes that otherwise have the same information, for example $\mathfrak{S}^{hist}_{\ell,v,e} \succ \mathfrak{S}^{hist}_{\ell,v}$ (with very simple distinguishing SA). We show that the non-prophetic hierarchy follows the shape of the classic case, including the difference between global-time and pure memoryless schedulers, with the notable exception of memoryless schedulers being weaker than history-dependent ones.

Proposition 12. *$\mathfrak{S}^{hist}_{\ell,v} \approx \mathfrak{S}^{hist}_{\ell,t} \approx \mathfrak{S}^{hist}_{\ell}$.*

Proof. This follows from the argument of Proposition 1.

Proposition 13. *$\mathfrak{S}^{hist}_{\ell,v} \succ \mathfrak{S}^{ml}_{\ell,v}$.*

Proof. Consider the SA M_6 in Fig. 10. It is similar to M_4 of Fig. 8, and our arguments are thus similar to the proof of Proposition 7. On M_6, we have $P_{max}(\mathfrak{S}^{hist}_{\ell,v}) = 1$: in ℓ_2, the history reveals which of the two incoming edges was used, i.e. which clock is already expired, thus the scheduler can make the optimal choice. However, if neither the history nor e is available, we get $P_{max}(\mathfrak{S}^{ml}_{\ell,v}) = \frac{1}{2}$: the only information that can be used in ℓ_2 are the values of the clocks, but $v(x) = v(y)$, so there is no basis for an informed choice.

Proposition 14. *$\mathfrak{S}^{hist}_{\ell,t} \succ \mathfrak{S}^{ml}_{\ell,t}$ and $\mathfrak{S}^{ml}_{\ell,v} \succ \mathfrak{S}^{ml}_{\ell,t}$.*

Proof. Consider the SA M_3 in Fig. 7. We have $P_{max}(\mathfrak{S}^{hist}_{\ell,t}) = P_{max}(\mathfrak{S}^{ml}_{\ell,v}) = 1$, but $P_{max}(\mathfrak{S}^{ml}_{\ell,t}) = \frac{1}{2}$ by the same arguments as in the proof of Proposition 6.

Proposition 15. *$\mathfrak{S}^{ml}_{\ell,t} \succ \mathfrak{S}^{ml}_{\ell}$.*

Proof. Consider the SA M_4 in Fig. 8. The schedulers in $\mathfrak{S}^{ml}_{\ell}$ have no information but the current location, so they cannot make an informed choice in ℓ_2. This and the simple loop-free structure of M_4 make it possible to analytically calculate the resulting probability: $P_{max}(\mathfrak{S}^{ml}_{\ell}) = \frac{17}{24} = 0.708\overline{3}$. If information about the global elapsed time t in ℓ_2 is available, however, the value of x is revealed. This allows making a better choice, e.g. going to ℓ_3 when $t \leq \frac{1}{2}$ and to ℓ_4 otherwise, resulting in $P_{max}(\mathfrak{S}^{ml}_{\ell,t}) \approx 0.771$ (statistically estimated with high confidence).

We have thus established the hierarchy of non-prophetic schedulers shown in Fig. 4, where some relationships follow from the propositions by transitivity.

5 Experiments

We have built a prototype implementation of lightweight scheduler sampling for SA by extending the MODEST TOOLSET's [24] MODES simulator, which already supports deterministic stochastic timed automata (STA [8]). With some care, SA can be encoded into STA. Using the original algorithm for MDP of [18], our prototype works by providing to the schedulers a discretised view of the continuous components of the SA's semantics, which, we recall, is a continuous-space MDP. The currently implemented discretisation is simple: for each real-valued quantity (the value $v(c)$ of clock c, its expiration time $e(c)$, and the global elapsed time t), it identifies all values that lie within the same interval $[\frac{i}{n}, \frac{i+1}{n})$, for integers i, n. We note that better static discretisations are almost certainly possible, e.g. a region construction for the clock values as in [30].

We have modelled M_1 through M_6 as STA in MODEST. For each scheduler class and model in the proof of a proposition, and discretisation factors $n \in \{1, 2, 4\}$, we sampled 10 000 schedulers and performed statistical model checking for each of them in the lightweight manner. In Fig. 11 we report the min. and max. estimates, $(\hat{p}_{\min}, \hat{p}_{\max})_{\dots}$, over all sampled schedulers. Where different discretisations lead to different estimates, we report the most extremal values. The subscript denotes the discretisation factors that achieved the reported estimates. The analysis for each sampled scheduler was performed with a number of simulation runs sufficient for the overall max./min. estimates to be within $\pm\, 0.01$ of the true maxima/minima of the *sampled* set of schedulers with probability $\geq$0.95 [18]. Note that $\hat{p}_{\min}$ is an upper bound on the true minimum probability and $\hat{p}_{\max}$ is a lower bound on the true maximum probability.

Increasing the discretisation factor or increasing the scheduler power generally increases the number of decisions the schedulers *can* make. This may also increase the number of *critical* decisions a scheduler *must* make to achieve the extremal probability. Hence, the sets of discretisation factors associated to specific experiments may be informally interpreted in the following way:

- $\{1, 2, 4\}$: Fine discretisation is not important for optimality and optimal schedulers are not rare.
- $\{1, 2\}$: Fine discretisation is not important for optimality, but increases rarity of optimal schedulers.
- $\{2, 4\}$: Fine discretisation is important for optimality, optimal schedulers are not rare.
- $\{1\}$: Optimal schedulers are very rare.
- $\{2\}$: Fine discretisation is important for optimality, but increases rarity of schedulers.
- $\{4\}$: Fine discretisation is important for optimality and optimal schedulers are not rare.

M_1:
Proposition 3:
$\mathfrak{S}^{hist}_{\ell,v,e}$: $(0.24, 0.76)_{2,4}$
$\mathfrak{S}^{hist}_{\ell,v,o}$: $(0.49, 0.51)_{1,2,4}$
Proposition 8:
$\mathfrak{S}^{ml}_{\ell,v,e}$: $(0.24, 0.76)_{2,4}$
$\mathfrak{S}^{ml}_{\ell,v,o}$: $(0.49, 0.51)_{1,2,4}$
Proposition 9:
$\mathfrak{S}^{ml}_{\ell,t,e}$: $(0.24, 0.76)_{2,4}$
$\mathfrak{S}^{ml}_{\ell,t,o}$: $(0.49, 0.51)_{1,2,4}$
Proposition 10:
$\mathfrak{S}^{ml}_{\ell,e}$: $(0.24, 0.76)_{2,4}$
$\mathfrak{S}^{ml}_{\ell,o}$: $(0.49, 0.51)_{1,2,4}$

M_3:
Proposition 6:
$\mathfrak{S}^{hist}_{\ell,t,e}$: $(0.00, 1.00)_{1}$
$\mathfrak{S}^{ml}_{\ell,v,e}$: $(0.12, 0.86)_{4}$
$\mathfrak{S}^{ml}_{\ell,t,e}$: $(0.37, 0.65)_{2}$
Proposition 9:
$\mathfrak{S}^{ml}_{\ell,t,e}$: $(0.37, 0.65)_{2}$
$\mathfrak{S}^{ml}_{\ell,t,o}$: $(0.00, 1.00)_{1,2}$
Proposition 10:
$\mathfrak{S}^{ml}_{\ell,e}$: $(0.34, 0.67)_{4}$
$\mathfrak{S}^{ml}_{\ell,o}$: $(0.00, 1.00)_{1,2,4}$
Proposition 14:
$\mathfrak{S}^{hist}_{\ell,t}$: $(0.00, 1.00)_{1,2}$
$\mathfrak{S}^{ml}_{\ell,v}$: $(0.21, 0.78)_{4}$
$\mathfrak{S}^{ml}_{\ell,t}$: $(0.49, 0.51)_{1,2,4}$

M_2:
Proposition 5:
$\mathfrak{S}^{hist}_{\ell,o}$: $(0.06, 0.94)_{1,2,4}$
$\mathfrak{S}^{ml}_{\ell,v,o}$: $(0.18, 0.83)_{1}$

M_4:
Proposition 7:
$\mathfrak{S}^{ml}_{\ell,t,e}$: $(0.25, 0.79)_{1}$
$\mathfrak{S}^{ml}_{\ell,e}$: $(0.29, 0.71)_{1}$
Proposition 15:
$\mathfrak{S}^{ml}_{\ell,t}$: $(0.22, 0.78)_{2,4}$
$\mathfrak{S}^{ml}_{\ell}$: $(0.28, 0.72)_{1,2,4}$

M_5:
Proposition 11:
$\mathfrak{S}^{ml}_{\ell,t,o}$: $(0.15, 0.86)_{4}$
$\mathfrak{S}^{ml}_{\ell,o}$: $(0.16, 0.84)_{1,2,4}$

M_6:
Proposition 13:
$\mathfrak{S}^{hist}_{\ell,v}$: $(0.00, 1.00)_{1,2}$
$\mathfrak{S}^{ml}_{\ell,v}$: $(0.49, 0.51)_{1,2,4}$

Fig. 11. Results from the prototype of lightweight scheduler sampling for SA

The results in Fig. 11 respect and differentiate our hierarchy. In most cases, we found schedulers whose estimates were within the statistical error of calculated optima or of high confidence estimates achieved by alternative statistical techniques. The exceptions involve M_3 and M_4. We note that M_4 makes use of an additional clock, increasing the dimensionality of the problem and potentially making near-optimal schedulers rarer. The best result for M_3 and class $\mathfrak{S}^{ml}_{l,t,e}$ was obtained using discretisation factor $n = 2$: a compromise between nearness to optimality and rarity. A greater compromise was necessary for M_4 and classes $\mathfrak{S}^{ml}_{l,t,e}, \mathfrak{S}^{ml}_{l,e}$, where we found near-optimal schedulers to be very rare and achieved best results using discretisation factor $n = 1$.

The experiments demonstrate that lightweight scheduler sampling can produce useful and informative results with SA. The present theoretical results will allow us to develop better abstractions for SA and thus to construct a refinement algorithm for efficient lightweight verification of SA that will be applicable to realistically sized case studies. As is, they already demonstrate the importance of selecting a proper scheduler class for efficient verification, and that restricted classes are useful in planning scenarios.

6 Conclusion

We have shown that the various notions of information available to a scheduler class, such as history, clock order, expiration times or overall elapsed time, almost all make distinct contributions to the power of the class in SA. Our choice of notions was based on classic scheduler classes relevant for other stochastic models, previous literature on the character of nondeterminism in and verification of SA, and the need to synthesise simple schedulers in planning. Our distinguishing examples clearly expose how to exploit each notion to improve the probability of reaching a goal. For verification of SA, we have demonstrated the feasibility of lightweight scheduler sampling, where the different notions may be used to finely control the power of the lightweight schedulers. To solve stochastic timed planning problems defined via SA, our analysis helps in the case-by-case selection of an appropriate scheduler class that achieves the desired tradeoff between optimal probabilities and ease of implementation of the resulting plan.

We expect the arguments of this paper to extend to steady-state/frequency measures (by adding loops back from absorbing to initial states in our examples), and that our results for classic schedulers transfer to SA with delayable actions. We propose to use the results to develop better abstractions for SA, the next goal being a refinement algorithm for efficient lightweight verification of SA.

References

1. de Alfaro, L.: The verification of probabilistic systems under memoryless partial-information policies is hard. Technical report, DTIC Document (1999)
2. Alur, R., Courcoubetis, C., Dill, D.: Model-checking for probabilistic real-time systems. In: Albert, J.L., Monien, B., Artalejo, M.R. (eds.) ICALP 1991. LNCS, vol. 510, pp. 115–126. Springer, Heidelberg (1991). https://doi.org/10.1007/3-540-54233-7_128
3. Andel, T.R., Yasinsac, A.: On the credibility of MANET simulations. IEEE Comput. **39**(7), 48–54 (2006)
4. Avritzer, A., Carnevali, L., Ghasemieh, H., Happe, L., Haverkort, B.R., Koziolek, A., Menasché, D.S., Remke, A., Sarvestani, S.S., Vicario, E.: Survivability evaluation of gas, water and electricity infrastructures. Electr. Notes Theor. Comput. Sci. **310**, 5–25 (2015)
5. Baier, C., Katoen, J.P.: Principles of Model Checking. MIT Press, Cambridge (2008)
6. Ballarini, P., Bertrand, N., Horváth, A., Paolieri, M., Vicario, E.: Transient analysis of networks of stochastic timed automata using stochastic state classes. In: Joshi, K., Siegle, M., Stoelinga, M., D'Argenio, P.R. (eds.) QEST 2013. LNCS, vol. 8054, pp. 355–371. Springer, Heidelberg (2013). https://doi.org/10.1007/978-3-642-40196-1_30
7. Bisgaard, M., Gerhardt, D., Hermanns, H., Krčál, J., Nies, G., Stenger, M.: Battery-aware scheduling in low orbit: the GomX–3 case. In: Fitzgerald, J., Heitmeyer, C., Gnesi, S., Philippou, A. (eds.) FM 2016. LNCS, vol. 9995, pp. 559–576. Springer, Cham (2016). https://doi.org/10.1007/978-3-319-48989-6_34
8. Bohnenkamp, H.C., D'Argenio, P.R., Hermanns, H., Katoen, J.P.: MoDeST: a compositional modeling formalism for hard and softly timed systems. IEEE Trans. Softw. Eng. **32**(10), 812–830 (2006)
9. Bravetti, M., D'Argenio, P.R.: Tutte le algebre insieme: concepts, discussions and relations of stochastic process algebras with general distributions. In: Baier, C., Haverkort, B.R., Hermanns, H., Katoen, J.-P., Siegle, M. (eds.) Validation of Stochastic Systems. LNCS, vol. 2925, pp. 44–88. Springer, Heidelberg (2004). https://doi.org/10.1007/978-3-540-24611-4_2
10. Bravetti, M., Gorrieri, R.: The theory of interactive generalized semi-Markov processes. Theor. Comput. Sci. **282**(1), 5–32 (2002)

11. Brázdil, T., Krčál, J., Křetínský, J., Řehák, V.: Fixed-delay events in generalized semi-Markov processes revisited. In: Katoen, J.-P., König, B. (eds.) CONCUR 2011. LNCS, vol. 6901, pp. 140–155. Springer, Heidelberg (2011). https://doi.org/10.1007/978-3-642-23217-6_10
12. Bryans, J., Bowman, H., Derrick, J.: Model checking stochastic automata. ACM Trans. Comput. Log. **4**(4), 452–492 (2003)
13. Buchholz, P., Kriege, J., Scheftelowitsch, D.: Model checking stochastic automata for dependability and performance measures. In: DSN, pp. 503–514. IEEE Computer Society (2014)
14. Butkova, Y., Hatefi, H., Hermanns, H., Krčál, J.: Optimal continuous time Markov decisions. In: Finkbeiner, B., Pu, G., Zhang, L. (eds.) ATVA 2015. LNCS, vol. 9364, pp. 166–182. Springer, Cham (2015). https://doi.org/10.1007/978-3-319-24953-7_12
15. D'Argenio, P.R., Hartmanns, A., Legay, A., Sedwards, S.: Statistical approximation of optimal schedulers for probabilistic timed automata. In: Ábrahám, E., Huisman, M. (eds.) IFM 2016. LNCS, vol. 9681, pp. 99–114. Springer, Cham (2016). https://doi.org/10.1007/978-3-319-33693-0_7
16. D'Argenio, P.R., Katoen, J.P.: A theory of stochastic systems part I: stochastic automata. Inf. Comput. **203**(1), 1–38 (2005)
17. D'Argenio, P.R., Lee, M.D., Monti, R.E.: Input/output stochastic automata. In: Fränzle, M., Markey, N. (eds.) FORMATS 2016. LNCS, vol. 9884, pp. 53–68. Springer, Cham (2016). https://doi.org/10.1007/978-3-319-44878-7_4
18. D'Argenio, P.R., Legay, A., Sedwards, S., Traonouez, L.M.: Smart sampling for lightweight verification of Markov decision processes. STTT **17**(4), 469–484 (2015)
19. Eisentraut, C., Hermanns, H., Zhang, L.: On probabilistic automata in continuous time. In: LICS, pp. 342–351. IEEE Computer Society (2010)
20. Giro, S., D'Argenio, P.R.: Quantitative model checking revisited: neither decidable nor approximable. In: Raskin, J.-F., Thiagarajan, P.S. (eds.) FORMATS 2007. LNCS, vol. 4763, pp. 179–194. Springer, Heidelberg (2007). https://doi.org/10.1007/978-3-540-75454-1_14
21. Haas, P.J., Shedler, G.S.: Regenerative generalized semi-Markov processes. commun. stat. Stochast. Models **3**(3), 409–438 (1987)
22. Hahn, E.M., Hartmanns, A., Hermanns, H.: Reachability and reward checking for stochastic timed automata. In: Electronic Communications of the EASST, AVoCS 2014, vol. 70 (2014)
23. Harrison, P.G., Strulo, B.: SPADES - a process algebra for discrete event simulation. J. Log. Comput. **10**(1), 3–42 (2000)
24. Hartmanns, A., Hermanns, H.: The Modest Toolset: an integrated environment for quantitative modelling and verification. In: Ábrahám, E., Havelund, K. (eds.) TACAS 2014. LNCS, vol. 8413, pp. 593–598. Springer, Heidelberg (2014). https://doi.org/10.1007/978-3-642-54862-8_51
25. Hartmanns, A., Hermanns, H., Krčál, J.: Schedulers are no Prophets. In: Probst, C.W., Hankin, C., Hansen, R.R. (eds.) Semantics, Logics, and Calculi. LNCS, vol. 9560, pp. 214–235. Springer, Cham (2016). https://doi.org/10.1007/978-3-319-27810-0_11
26. Hartmanns, A., Sedwards, S., D'Argenio, P.: Efficient simulation-based verification of probabilistic timed automata. In: WSC. IEEE (2017). https://doi.org/10.1109/WSC.2017.8247885
27. Hermanns, H.: Interactive Markov Chains: The Quest for Quantified Quality. LNCS, vol. 2428. Springer, Heidelberg (2002). https://doi.org/10.1007/3-540-45804-2

28. Hermanns, H., Krämer, J., Krčál, J., Stoelinga, M.: The value of attack-defence diagrams. In: Piessens, F., Viganò, L. (eds.) POST 2016. LNCS, vol. 9635, pp. 163–185. Springer, Heidelberg (2016). https://doi.org/10.1007/978-3-662-49635-0_9
29. Kurkowski, S., Camp, T., Colagrosso, M.: MANET simulation studies: the incredibles. Mob. Comput. Commun. Rev. **9**(4), 50–61 (2005)
30. Kwiatkowska, M., Norman, G., Segala, R., Sproston, J.: Verifying quantitative properties of continuous probabilistic timed automata. In: Palamidessi, C. (ed.) CONCUR 2000. LNCS, vol. 1877, pp. 123–137. Springer, Heidelberg (2000). https://doi.org/10.1007/3-540-44618-4_11
31. Legay, A., Sedwards, S., Traonouez, L.M.: Estimating rewards & rare events in nondeterministic systems. In: Electronic Communications of the EASST, AVoCS 2015, vol. 72 (2015)
32. Legay, A., Sedwards, S., Traonouez, L.-M.: Scalable verification of Markov decision processes. In: Canal, C., Idani, A. (eds.) SEFM 2014. LNCS, vol. 8938, pp. 350–362. Springer, Cham (2015). https://doi.org/10.1007/978-3-319-15201-1_23
33. Matthes, K.: Zur Theorie der Bedienungsprozesse. In: 3rd Prague Conference on Information Theory, Stat. Dec. Fns. and Random Processes, pp. 513–528 (1962)
34. NS-3 Consortium: ns-3: A Discrete-event Network Simulator for Internet Systems. https://www.nsnam.org/
35. Pongor, G.: OMNeT: objective modular network testbed. In: MASCOTS, pp. 323–326. The Society for Computer Simulation (1993)
36. Ruijters, E., Stoelinga, M.: Better railway engineering through statistical model checking. In: Margaria, T., Steffen, B. (eds.) ISoLA 2016. LNCS, vol. 9952, pp. 151–165. Springer, Cham (2016). https://doi.org/10.1007/978-3-319-47166-2_10
37. Song, L., Zhang, L., Godskesen, J.C.: Late weak bisimulation for Markov automata. CoRR abs/1202.4116 (2012)
38. Strulo, B.: Process algebra for discrete event simulation. Ph.D. thesis, Imperial College of Science, Technology and Medicine. University of London, October 1993
39. Wolf, V., Baier, C., Majster-Cederbaum, M.E.: Trace semantics for stochastic systems with nondeterminism. Electr. Notes Theor. Comput. Sci. **164**(3), 187–204 (2006)
40. Wolovick, N.: Continuous probability and nondeterminism in labeled transition systems. Ph.D. thesis, Universidad Nacional de Córdoba, Córdoba, Argentina (2012)
41. Wolovick, N., Johr, S.: A characterization of meaningful schedulers for continuous-time Markov decision processes. In: Asarin, E., Bouyer, P. (eds.) FORMATS 2006. LNCS, vol. 4202, pp. 352–367. Springer, Heidelberg (2006). https://doi.org/10.1007/11867340_25
42. Zeng, X., Bagrodia, R.L., Gerla, M.: Glomosim: a library for parallel simulation of large-scale wireless networks. In: PADS, pp. 154–161. IEEE Computer Society (1998)

Symbolically Quantifying Response Time in Stochastic Models Using Moments and Semirings

Hugo Bazille[1], Eric Fabre[1], and Blaise Genest[2(✉)]

[1] Univ Rennes, Inria, SUMO Team, Rennes, France
[2] Univ Rennes, CNRS, IRISA, Rennes, France
bgenest@irisa.fr

Abstract. We study quantitative properties of the response time in stochastic models. For instance, we are interested in quantifying bounds such that a high percentage of the runs answers a query within these bounds. To study such problems, computing probabilities on a state-space blown-up by a factor depending on the bound could be used, but this solution is not satisfactory when the bound is large.

In this paper, we propose a new *symbolic* method to quantify bounds on the response time, using the moments of the distribution of simple stochastic systems. We prove that the distribution (and hence the bounds) is uniquely defined given its moments. We provide *optimal* bounds for the response time over all distributions having a pair of these moments. We explain how to *symbolically* compute in polynomial time any moment of the distribution of response times using adequately-defined semirings. This allows us to compute optimal bounds in parametric models and to reduce complexity for computing optimal bounds in hierarchical models.

1 Introduction

Response time has been considered lately as an important property of systems [8,15,21]. In this context, one does not simply want a query to be answered eventually, but to be answered in a reasonable amount of time. In the model-checking community, problems on response time have been studied mainly *qualitatively*, in the context of (pure, that is non stochastic) two-player games [8,21]. There, one looks for a strategy ensuring that the lim-sup of response time is finite. It ensures that under this strategy, there will be a bound on the response time to any query. This has been extended in [15] to a quantitative setting, where one wants to optimize the mean response time in a pure two-player game.

In this paper, we consider stochastic systems. In such systems, the response time is a random variable, unlikely to be bounded as even a single probabilistic loop on a reachable state will make the response time longer than T for a set of runs of small but positive probability, no matter T. Instead, we propose to quantify such response times. One way to do that is to obtain the distribution

C. Baier and U. Dal Lago (Eds.): FOSSACS 2018, LNCS 10803, pp. 403–419, 2018.
https://doi.org/10.1007/978-3-319-89366-2_22

of response times. Another way is to compute, for a probability $0 < p < 1$, the bound T that is satisfied (by a set of runs) with probability at least $1 - p$. In this paper, we tackle both problems. For that, we use the concept of *moments* of the distribution of response times, as described next.

The *moment of order* r of a probability distribution δ over $\mathbb{R}$ or $\mathbb{R}^+$ is defined as the integral of $x^r \delta(x)$ over the support of δ, when defined (that is if $x^r \delta(x)$ is measurable and the integral is defined). For instance, the moment of order 1 is the expected value of δ, while the moment of order 2 allows one to compute the standard deviation of δ. Inspired by the computation of entropy for automata [10] (see also [1] for the computation of entropy for (non-Zeno) timed-automata), we design new semirings in which each moment corresponds to the sum of weights of runs reaching a state. This construction can be applied to probabilistic automata (that is, labeled discrete time Markov chains), as well as labeled *continuous time Markov chains*, where time is continuous and is drawn according to some rate. Adapting the Floyd-Warshall algorithm provides a *symbolic* way to perform the computation of the n first moments in time cubic in the number of states of the Markov Chain, and quadratic in n. For any n, we can thus compute the value of the first n moments. In some sense, we extend the approach of [12,16] from computing probabilities to computing any moments. This allows us to evaluate the distribution of response times in two ways:

Firstly, thanks to the symbolic expression of moments, we prove that there is a unique distribution having the moments of a distribution of response times of a probabilistic automaton. We can then build a sequence of distributions matching the first n moments, for instance the maximal entropy one [11]. Here, maximal entropy means assuming the least information besides these moments. This sequence of distributions is then ensured to converge in law towards the distribution of response times.

Secondly, we study optimal symbolic bounds on the time to answer a high percentage of queries, obtained from moments. The Tchebychev inequality provides optimal symbolic bounds when considering the space of distributions having one given moment, of any order i. We obtain bounds optimal in the space of distributions having two given moments, of any orders i, j. We show how this improves Tchebychev bounds on some example. Having symbolic methods allows for instance to deal with parametric systems where the parameters represent uncertain probabilities. In this case, we can compute optimal bounds satisfying all valuations of parameters. For hierarchical systems [3], which are compact representations of large systems, our symbolic method allows to design a much more efficient algorithm (e.g. it does not consider twice the same component) to compute the moments, and thus the bounds. Missing proofs can be found in [5].

Related Work: Response times in stochastic systems have been studied for a long time by the perf.eval. community under the name "first passage times", e.g. in [22]. Techniques used in this community to compute moments of Markov chains are mostly based on numerical methods, e.g. [13]. While [13] has the same complexity as our symbolic technique, it is very efficient on explicit models. However, these numerical methods are less adaptable than our symbolic algorithm, in particular concerning parametric or hierarchical systems.

Concerning the determinacy of the distribution given moments, it is known [20] that phase-type distributions of order n are determined by their first $2n-1$ moments. First passage distribution time in Markov chains with n states are phase type distribution of order n. However, [20] does not help characterizing bounds as it does not ensure that a non-phase type distribution cannot have the exact same moments as a phase type distribution, unlike our result.

Bounding the response time has also been studied in the perf.eval. community. Again, methods used there are mostly numerical [6,19]. In [19] (pp. 68–69), a symbolic bound is also provided in the particular case of moments of order 1, 2 and 3. In [2], it is shown how to use the two first moments of response time across various components to compute general bounds, using techniques close to ours, but restricted to moments of order 1 and 2. In our paper, we provide *optimal* bounds for any order $(i, j) \in \mathbb{N}^2$. Taking into account moments of order $i, j > 3$ is important when the proportion of runs to answer is close to 1.

Last, computing moments find other applications. For instance, in [4,7,14], complex functions describing the evolution of molecular species are approximated using the first k moments, for some k.

2 Probabilistic Automata

We first introduce a simple class of models, namely *probabilistic automata* (also called *labeled discrete time Markov chains*), on which we can demonstrate our techniques. Later, we will extend our results to handle continuous time, considering Continuous-Time Markov Chains (CTMC), as well as parametric and hierarchical systems.

Definition 1. *A probabilistic automaton A over a finite alphabet Σ is a tuple (S, Pr, δ_0) where:*

- *S is a finite set of states,*
- *$Pr : S \times \Sigma \times S \to [0,1]$ is a stochastic transition function such that for all $s \in S$, $\sum_{a\in\Sigma, t\in S} Pr(s,a,t) = 1$: the weights of paths leaving s sum to 1,*
- *$\delta_0 : S \to [0,1]$ is the initial distribution over states such that $\sum_{s\in S} \delta_0(s) = 1$.*

Example 1. For instance, the model depicted on Fig. 1 is a probabilistic automaton with 3 states $\{1, 2, 3\}$. There is a transition between 1 and 2 labeled **query** with probability 1. From state 2, with probability .9 we stay in state 2 with a transition labeled **wait**, and with probability .1 we go to state 3 with a transition labeled **response**. We loop in state 3 with probability 1.

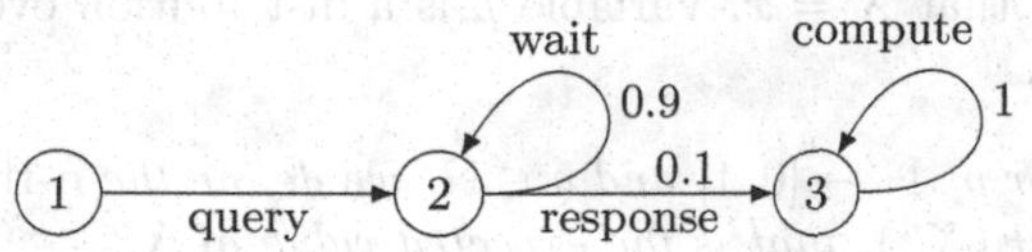

Fig. 1. A simple example of a query-response model

A finite sequence $\pi = s_0, a_1, s_1, \ldots, a_n, s_n \in (S\Sigma)^n S$ is called a *finite path* starting from s_0 and ending in s_n, and a transition $t \in \pi$ if $t = s_i a_{i+1} s_{i+1}$ for some i. We denote $|\pi| = n$ the length of the path π. For a path π_1 ending in s_n and a path π_2 starting from s_n, we can define the concatenated path $\pi_1 \cdot \pi_2$ where the last node of π_1 and the first node of π_2 are merged. A path π_1 is a *prefix* of π if there exists a path π_2 such that $\pi_1 \cdot \pi_2 = \pi$.

For a path π starting in a state s_0, we define $\mathbb{P}(\pi) = \prod_{t \in \pi} Pr(t)$ the probability that a path with prefix π is executed from s_0. A path π is realizable if $\mathbb{P}(\pi) > 0$.

Let s be a state, and Π be a set of finite paths starting from s such that no path in Π is a prefix of another path in Π. Then the probability that a path starting from s has a prefix in Π is $\mathbb{P}(\Pi) = \sum_{\rho \in \Pi} \mathbb{P}(\rho)$. We say that Π is *disjoint* if no path ρ of Π is a prefix of another path $\rho' \neq \rho$ of Π or similarly, $Cyl(\rho) \cap Cyl(\rho') = \emptyset$ with $Cyl(\rho) = \{\pi, \rho \text{ prefix of } \pi\}$.

Some labels of an automaton will be of particular interest concerning response time. Let $\Sigma_Q \subseteq \Sigma$ be a subset of labels standing for queries, and $\Sigma_R \subseteq \Sigma$ be a subset of labels standing for responses. For simplicity, we will assume that there is a unique query type $\Sigma_Q = \{q\}$ and a unique response type $\Sigma_R = \{r\}$, with $q \neq r$. We will also assume that there is no path with two (similar) queries q. To handle cases with several query/response types, it suffices for each type to consider only queries and answers of that type and disregard other types.

Problem Statement: We are interested in quantifying the time between queries and responses, called the *response time*, which is a random variable. A way to quantify it is to produce the distribution of response times, either for each transition labeled by a query, or averaged on these transitions, weighted by the probability to see each of these transitions. Another way is to answer model-checking questions such as: what is the smallest delay T such that the mass of paths unanswered after T units of time is smaller than some probability p?

To compute both the distribution and the delay T, we will use the so called *moments of the distribution of response times*. The moment of order 1 is the mean value, and the moment of order 2 allows to compute the standard deviation.

3 *Symbolically* Computing Moments Using Semirings

In this section, we define moments and explain how to compute them *symbolically* using appropriately-defined semirings.

Let X be the random variable of the response time. If all queries are answered, then X takes values in N_{max}, else X takes values in $N_{max} \cup \{\infty\}$. Let $p(x)$ be the probability that the response is obtained x units of time after the query, that is, the probability that $X = x$. Variable p is a distribution over response time, with $\sum_x p(x) = 1$.

Definition 2. *For* $p : \mathbb{N} \to [0, 1]$ *and* $n \in \mathbb{N}$, *we define the* n-th moment *of* p *by* $\sum_{x \in \mathbb{N}} p(x) \cdot x^n = E(X^n)$, *that is the expected value of* X^n.

3.1 Semirings Associated with Moments

We will compute moments of the distribution of response times by considering each query individually. We can then take e.g. the average over all queries (as we assumed that there are no two queries on the same path). Thus, we first fix a state q, target of a transition labeled by a query. State q symbolizes that a query has just been asked. We then let R be the set of target states of transitions labeled by a response. A state is in R if a response to this query has just been given. For instance, on Fig. 1, we have $q = 2$ and $R = \{3\}$.

We introduce a set of semirings that will allow us to compute symbolically the moment of order n of the distribution of response times to the query associated with state q, for all $n \in \mathbb{N}$. We will compute the moment inductively on a disjoint subset Π of paths of A from q to R. For an integer n, we denote $\mu_n(\Pi) = \sum_{\rho \in \Pi} \mathbb{P}(\rho)|\rho|^n$. Let $\mathbf{Path}_q^R$ be the set of paths in the automaton A between q and the first occurrence of R. Notice that $\mathbf{Path}_q^R$ is disjoint. Thus, we have that $\mu_n(\mathbf{Path}_q^R)$ is the moment of order n of the distribution of response times to the query associated with state q. To avoid some heavy notations, when R is reduced to one state t, let $\mu_n(\mathbf{Path}_s^t)$ be the set of paths between s to the first occurrence of t and we denote $\mu_n(s,t) = \mu_n(\mathbf{Path}_s^t)$.

We now give some properties of μ. Let Π_1 be a set of paths ending in some state s and let Π_2 be a set of paths starting from s. We denote by $\Pi_1 \cdot \Pi_2$ the set of paths $\rho_1\rho_2$ with $\rho_1 \in \Pi_1$ and $\rho_2 \in \Pi_2$.

Proposition 1. *For all n, we have $\mu_n(\Pi_1 \cdot \Pi_2) = \sum_{i=0}^{n} \binom{n}{i}\mu_i(\Pi_1) \cdot \mu_{n-i}(\Pi_2)$*

This property hints to a set of semirings $(\mathbb{R}, \oplus_n, \otimes_n, \overline{0}_n, \overline{1}_n)$ with good properties to compute moments. For $(n+1)$-tuples $(x_0, \ldots, x_n)$ and $(y_0, \ldots, y_n)$, we define operations $\oplus_n$ and $\otimes_n$:

- $(x_0, \ldots, x_n) \oplus_n (y_0, \ldots, y_n) = (x_0 + y_0, \ldots, x_n + y_n)$
- $(x_0, \ldots, x_n) \otimes_n (y_0, \ldots, y_n) = (z_0, \ldots, z_n)$ with $z_i = \sum_{j=0}^{i} \binom{i}{j} x_j y_{i-j}$

The neutral element for $\oplus_n$ is $\overline{0}_n = (0, \ldots, 0)$. $\overline{0}_n$ is an annihilator for $\otimes_n$. The neutral element for $\otimes_n$ is $\overline{1}_n = (1, 0, \ldots, 0)$. In the following, we will denote the different laws and elements by $\oplus$, $\otimes$, $\overline{0}$ and $\overline{1}$.

Proposition 2. *For $n \geq 0$, $(\mathbb{R}^{n+1}, \oplus, \otimes, \overline{0}, \overline{1})$ defines a commutative semiring.*

Notice that if for all $i \leq n$, we have $x_i = \mu_i(\Pi_1)$ and $y_i = \mu_i(\Pi_2)$, denoting $(z_0, \ldots, z_n) = (x_0, \ldots, x_n) \otimes_n (y_0, \ldots, y_n)$, we get $\mu_i(\Pi_1 \cdot \Pi_2) = z_i$. Further, if both Π_1, Π_2 are disjoint, and if no path of Π_1 (resp. Π_2) is a prefix of a path of Π_2 (resp. Π_1), then $\mu_i(\Pi_1 \cup \Pi_2) = x_i + y_i$.

3.2 Computations in a Semiring

Following the Floyd-Warshall algorithm to sum weights of paths reaching a state, we will decompose inductively $\mathbf{Path}_q^R$ using operations $\cup$ and $\cdot$. We will then use the semiring $(\mathbb{R}^{n+1}, \oplus, \otimes, \overline{0}, \overline{1})$ to perform these computations inductively. The induction will be over the number of states in S. Let G be a subset of S disjoint with R: $G \cap R = \emptyset$. For all state $s \in S \setminus R$, we define $\mathbf{Path}_s^t(G) = \{s_0 \cdots s_n \mid s_0 = s, s_n = t, \forall 1 \leq i \leq n-1, s_i \in G\}$ the set of paths from state s to state t using only states G, except for the initial state, which is s and for the last state which is t, even if $s, t \in R$ or $s, t \notin G$.

For a set of paths Π, we define $w_n(\Pi) = (\mathbb{P}(\Pi), \mu_1(\Pi), \ldots, \mu_n(\Pi))$. Let $g \in G$ be a state of G. A path ρ in $\mathbf{Path}_s^t(G)$ has two possibilities: either it does not use g, or it uses g one or several times. We deduce the inductive formula:

Proposition 3. $w_n(\mathbf{Path}_s^t(G)) = \quad w_n(\mathbf{Path}_s^t(G \setminus \{g\})) \oplus$
$w_n(\mathbf{Path}_s^g(G \setminus \{g\})) \otimes \big(\bigoplus_{k=1}^{\infty} w_n(\mathbf{Path}_g^g(G \setminus \{g\}))^{\otimes k}\big) \otimes w_n(\mathbf{Path}_g^t(G \setminus \{g\}))$

Proof (Sketch of). If ρ does not use g, we have ρ is in $\mathbf{Path}_s^t(G \setminus \{g\})$. Otherwise, ρ can be expressed as $\rho_0 \ldots \rho_k$ with:

- ρ_0 is in $\mathbf{Path}_s^g(G \setminus \{g\})$,
- ρ_k is in $\mathbf{Path}_g^t(G \setminus \{g\})$,
- and for all $0 < j < k$, $\rho_j \in \mathbf{Path}_g^g(G \setminus \{g\})$.

We can then write an inductive formula satisfied by $\mathbf{Path}_s^t(G)$:

$$\mathbf{Path}_s^t(\emptyset) = \{(s, a, t) \mid Pr(s, a, t) \neq 0\}$$

$$\mathbf{Path}_s^t(G) = \mathbf{Path}_s^t(G \setminus \{g\}) \cup \bigcup_{k=1}^{\infty} \{\rho_0 \ldots \rho_k \mid \rho_0 \in \mathbf{Path}_s^g(G \setminus \{g\}),$$
$$\rho_k \in \mathbf{Path}_g^t(G \setminus \{g\}), \forall j \in [1, k-1], \rho_j \in \mathbf{Path}_g^g(G \setminus \{g\})\} \qquad \square$$

In order to use this formula, we need to compute $\bigoplus_{k=1}^{\infty} w_n(\mathbf{Path}_g^g(G \setminus \{g\}))^{\otimes k} = w_n(\mathbf{Path}_g^g(G))$, which represents what happens along a cycle from g to g. Let (g, Π) a pair with g a state and Π a set of paths (cycles) using g exactly twice: the first state and the last states are g. The pair $(g, \mathbf{Path}_g^g(G \setminus \{g\}))$ satisfies this property. We define $w_n^*(\Pi) = \bigoplus_{k=1}^{\infty} w_n(\Pi)^{\otimes k}$. The restriction on (r, Π) ensures that $\bigcup_{k=1}^{\infty} \Pi^{\otimes k}$ is disjoint. We show that $w_n^*(\Pi)$ is defined in most cases, namely when $\mathbb{P}(\Pi) < 1$.

Proposition 4. *Let Π be a set of paths using state g exactly twice, as first and last state. If $\mathbb{P}(\Pi) < 1$, then*

$$w_n^*(\Pi)[0] = w_0^*(\Pi) = \mathbb{P}(\bigcup_{k=1}^{\infty} \Pi^{\otimes k}) = \frac{1}{1-\mathbb{P}(\Pi)}, \textit{ and for } i > 0$$

$$w_n^*(\Pi)[i] = \mu_i(\bigcup_{k=1}^{\infty} \Pi^{\otimes k}) = \frac{1}{1-\mathbb{P}(\Pi)} \sum_{j=0}^{i-1} \binom{i}{j} w_n(\Pi)[i-j] \times w_n^*(\Pi)[j]$$

Notice that $P(\Pi) = 1$ describes cases where s cannot reach t (as $t \notin G$, if $\mathbb{P}(w_n(\mathbf{Path}_g^g(G)) = 1$, it would mean that every path reaching g stays in G forever, and in particular never meets t). Thus, we first compute the set of states S_1 from which there exists a path to R. Notice that for each set Π of paths ending in $g \in S_1 \setminus R$, we have $\mathbb{P}(\Pi) < 1$, because there is a positive probability to reach R from g, which is not captured by paths in Π.

3.3 A Symbolic Algorithm

From the inductive formulae to compute set of paths from subsets of paths and to compute $w_n^*(\Pi)[i]$ from $w_n^*(\Pi)[j]$ for $j < i$, we deduce Algorithm 1, following the ideas of Floyd-Warshall, incrementally adding non response states from $S_1 \setminus R$, which can be used as intermediate states. Notice that states in $S \setminus S_1$ cannot reach R anyway. This algorithm is *symbolic* (or *algebraic*) in that every constant (e.g. $Pr(s, a, t)$) can be replaced by a variable (see e.g. Sect. 4.2).

Theorem 1. *Let $A = (S, \delta, \delta_0)$ be a probabilistic automaton. One can compute $\mu_i(s, t)$ for all $i \leq n$ and $s, t \in S$ in time $O(n^2 \times |S|^3)$.*

Proof. In Algorithm 1, after running the outer **for**-loop on $g_1, \ldots, g_j$, we have $w_n(s, t)[n] = \mu_n(\mathbf{Path}_s^t(\{g_1, \ldots, g_j\}))$. At the end of Algorithm 1, we obtain $w_n(s, t)[n] = \mu_n(\mathbf{Path}_s^t) = \mu_n(s, t)$.

Algorithm 1: Algorithm computing the moment of order n

```
for s ∈ S do
    for t ∈ S do
        %Initialization
        w := Σ_{a∈Σ} Pr(s, a, t)
        w_n(s, t) := (w, w, ..., w)
    end
end
for g ∈ S_1 \ R do
    for s ∈ S do
        for t ∈ S do
            w_n(s, t) := w_n(s, t) ⊕ w_n(s, g) ⊗ w_n^*(g, g) ⊗ w_n(g, t)
        end
    end
end
```

To obtain $\mu_i(s,t)$ for all $i \leq n$, it suffices to run Algorithm 1 inductively on moment of order $1,\ldots,n$. Computing $w_n^*[i](s,t)$ in the inner **for**-loop takes time $O(i)$ as $w_n[j](s,t) = w_j[j](s,t)$ has already been computed inductively for all $j < i$. This yields the complexity of $O(\sum_{j=1}^n i \times |S|^3) = O(n^2 \times |S|^3)$. □

Now, for each query q, we have $\mu_i(\mathbf{Path}_q^R) = \sum_{r\in R} \mu_i(q,r)$, as $\mathbf{Path}_q^{r_1}$ and $\mathbf{Path}_q^{r_2}$ have no path prefix of each other for $r_1 \neq r_2, r_1, r_2 \in R$. Now, the moment of order n of the distribution of response times of q is formally either ∞ if $\mu_0(\mathbf{Path}_q^R) < 1$ (there is positive probability to never answer q, that is have infinite response time), and $\mu_n(\mathbf{Path}_q^R)$ otherwise.

Example 2. For the example of Fig. 1, unfolding the algorithm for $n = 2$ (that is for probability, and moments of order 1 and 2) gives after initialization:

$w(1,2) = (1,1,1)$, $w(2,2) = (0.9, 0.9, 0.9)$, $w(2,3) = (0.1, 0.1, 0.1)$, and $w(1,3) = (0,0,0)$, as there is no direct transition from state 1 to state 3.

There are no paths with intermediary states 1 or 3, so $g = 1$ or $g = 3$ does not have any impact. For paths with intermediary states $g = 2$, the algorithm gives:

- $w(2,2) \leftarrow w(2,2) \oplus w(2,2) \otimes w(2,2)^* \otimes w(2,2) = w(2,2) \otimes w(2,2)^*$
- $w(2,3) \leftarrow w(2,3) \oplus w(2,2) \otimes w(2,2)^* \otimes w(2,3) = w(2,3) \otimes w(2,2)^*$
- $w(1,3) \leftarrow w(1,3) \oplus w(1,2) \otimes w(2,2)^* \otimes w(2,3)$

We have $w(2,2)^* = (\frac{1}{1-0.9}, \frac{0.9}{(1-0.9)^2}, \frac{0.9}{(1-0.9)^2} + \frac{2\times 0.9^2}{(1-0.9)^3}) = (10, 90, 1710)$

At the end of the algorithm, we obtain $\mu_i(2,3) = \mu_i(\mathbf{Path}_2^{\{2\}}) = w(2,3) = (0.1, 0.1, 0.1) \otimes (10, 90, 1710) = (1, 10, 190)$. Hence, in this probabilistic automata, the probability of responding to the query is 1, in a mean time of 10, with a standard deviation of $\sqrt{190 - 10^2} = 9.5$.

3.4 Extension to Continuous Time

We now extend the symbolic computation of moments to *continuous time Markov Chains (CTMCs)*. In order to be as close as possible to the setting of probabilistic automata, we use the sojourn time representation of CTMCs. This representation is fully equivalent with the more usual representation of CTMCs with transition rates, see Chap. 7.3 of [9].

Definition 3. *A CTMC is a tuple $(S, Pr, \delta_0, (\lambda_s)_{s\in S})$ with:*

- *(S, Pr, δ_0) is a probabilistic automata, and*
- *for all s, λ_s is the sojourn parameter associated with state s. That is, the PDF function of the sojourn time is $X_s(t) = \lambda_s e^{-\lambda_s \cdot t}$ and the probability to stay in s at least t units of time is $e^{-\lambda_s \cdot t}$.*

In this continuous context, we need integrals instead of sums to define the i-th moment of a variable X: $\mu_i(X) = \int_0^\infty X(t) t^i dt = 1$. For every state $s \in S$, let $X_s(t) = \lambda_s e^{-\lambda_s \cdot t}$. For all i, for all s, $\mu_i(X_s)$ is well defined and $\mu_i(X_s) = \frac{i!}{\lambda_s^i}$

We can easily extend the computation of moments for CTMCs. The inductive formulas for probabilities and moments of the reaching time distribution remain unchanged. We only need to change the definition of moments for every transition, which is input at the initialization phase of the Algorithm 1: for all $s, t \in S$, we set $w_n(s,t)$ to be $(w^0(s,t), w^1(s,t), \ldots, w^n(s,t))$, where $w^0(s,t) = \sum_{a \in \Sigma} Pr(s,a,t)$ and $w^i(s,t) = \sum_{a \in \Sigma} Pr(s,a,t) \frac{i!}{\lambda_s^i}$ for all $i \in [1,n]$.

Theorem 2. *Let $A = (S, Pr, \delta_0, (\lambda_s)_{s \in S})$ be a CTMC. One can compute $\mu_i(s,t)$ for all $i \leq n$ and $s, t \in S$ in time $O(n^2 \times |S|^3)$.*

4 Uniqueness of Distribution, Parameters and Hierarchy

In this section, we present cases where having a symbolic algorithm allows efficient techniques, compared to numerical methods. We start with hierarchical systems which are a way to compactly describe systems. Then, we present the possibility to work on systems with parameters. Finally, thanks to the symbolic expression of moments, we prove that there is a unique distribution having the moments of a distribution of reaching times of a (continuous-time) Markov chain.

4.1 Hierarchical Probabilistic Automata

We use notations mainly from [3] to describe hierarchical structures:

Definition 4. *A hierarchical probabilistic automaton (HPA) A over a finite alphabet Σ is a tuple of n modules $(S_i, Pr_i, \lambda_i, s_i^0, s_i^f)_{1 \leq i \leq n}$ where for all i,*

- *S_i is the finite set of states of module i,*
- *$s_i^0 \in S_i$ is the initial state of module i, and s_i^f the finial state of module i,*
- *$Pr_i : S_i \setminus \{s_i^f\} \times \Sigma \times S_i \to [0,1]$ is a stochastic transition function such that for all $s \in S_i \setminus \{s_f\}$ (resp. $s \in S_1$ for $i = 1$), $\sum_{a \in \Sigma, t \in S_i} Pr_i(s,a,t) = 1$,*
- *$\lambda_i : S_i \to \{i+1, \ldots, n\}$ is a* partial *mapping associating some states of S_i from module i to deeper modules.*

Intuitively, the system starts in module 1, in state s_1^0. Each time a state $s \in S_i$ associated with a module $j > i$, that is $\lambda_i(s) = j$, is entered by a

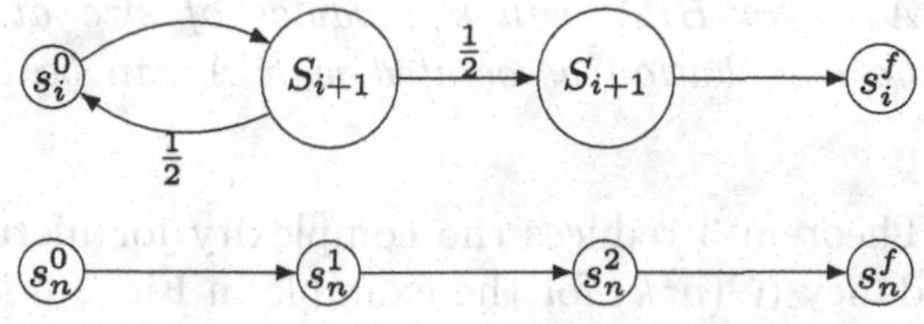

Fig. 2. An HPA with an exponential number of states.

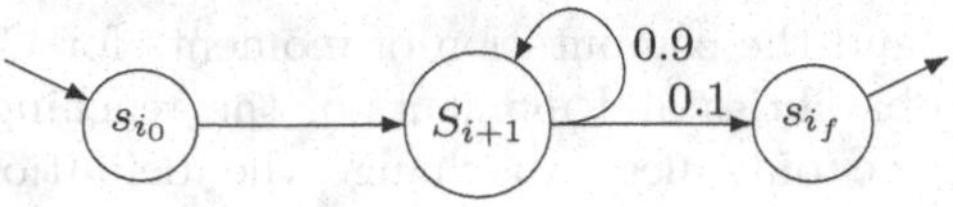

Fig. 3. An HPA without redundancy

transition $t \to s$, the system goes to state s_j^0 and stays in S_j till s_j^f is seen, in which case it comes back to state s and takes a transition $s \to t'$ (according to the probability distribution from s). This process can be repeated from any state in a module i to any module j as long as $j > i$.

To define the semantics of $(S_i, Pr_i, \lambda_i, s_i^0, s_i^f)_{1 \le i \le n}$ formally, we inductively replace states associated with the deepest module by their definition. Indeed, nodes from the deepest module are not associated with any module by definition. Once every module has been replaced, a (flat) probabilistic automaton is obtained with the intended semantics.

Hence, HPA have the same expressive power as probabilistic automata. Yet, they may be much more compact: we denote by $|A|$ the size of the description of the hierarchical automaton and by $||A||$ the size of the unfolded automaton. The interest of such a description is that it may be exponentially smaller than the size of the unfolded automaton, as depicted in Fig. 2: here, every module contains two copies of the next module, with the exception of the last one. While the number of states in the description is linear ($4n$), the number of states in the unfolded automaton is equal to $3 \cdot 2^n - 2$.

The symbolic Algorithm 1 is naturally modular, in that computations on a module used several times can be performed only once by considering states of the deepest module first. Indeed, one module can be summarized by three information items: the probability (and moments) to answer the query in this module, the probability (and moments) to leave this module without answering the query in the module and the probability to stay forever in this module without answering the query. Then the information can be used for shallower modules: every time a state s in a module i is associated with the deepest module, it can be replaced by this small set of states containing all the relevant information about the deepest module (and computed only once). Then, this process can be repeated to eliminate modules recursively. This leads to a complexity in the small size $|A|$ of the compact HPA representation rather than in the large size $||A||$ of the unfolded PA:

Theorem 3. *Let A be an HPA with k modules of size at most m. The n first moments of the distribution associated with A can be computed in time $O(n^2km^3)$.*

Not only does Theorem 3 reduces the complexity for hierarchical representations with redundancy ($O(n^2k)$ for the example in Fig. 2 instead of $O(n^2 2^{3k})$ when running the algorithm in [13] on the equivalent flat PA), it also gives a better complexity on structure without redundancy. Consider the example in

Fig. 3, without redundancy, with an unfolded PA with $3k + 1$ states. Theorem 3 takes time $O(n^2 k 3^3)$, while the algorithm in [13] on the equivalent flat PA would take time $O(n^2 (3k)^3)$.

4.2 Parametric Systems

Another case where having a symbolic algorithm is helpful is when the system has parameters standing for probability values (see for instance Fig. 4, where p is such a parameter). We illustrate two cases here.

The first case is when parameters help with redundancy. Often, stochastic systems reuse the same constructions, but with different probability values. This would be naturally encoded as a module M of a hierarchical system using a set of parameters P. This module M would be used several times, with different values of parameters specified in each module using it.

In this case, one can run Algorithm 1 on M, using the parameter values literally in the equations. This yields rational functions $f_n : [0,1]^P \to (0,1]$ of the parameters expressing the moments of order n for module M, for all n. For instance with the example of Fig. 4, the probability to reach state 4 from state 1 is equal to $\frac{2p+4}{5p+4}$, and the mean time is equal to $\frac{112+44p-12p^2}{(5p+4)(2p+4)}$. Each time module M is used, f_n can be evaluated using the value of the parameters P for this particular usage.

Another possible use of parameters is to model uncertainty of values. In the example of Fig. 4, we may not know exactly the value of parameter p, but only know that it is above 0.8. In this case, one may be interested of synthesizing the largest (resp. smallest) moment of order n which is smaller (resp. larger) than the moment of any system realizing the parametric system, that is where p is replaced by any value above 0.8. This will be particularly interesting in the next section discussing bounds. To do so, one can use the rational function f_n to compute its minimal and maximal values (e.g. deriving it and looking for 0 with Euler's method). In this way, we also obtain the best/worst value for p.

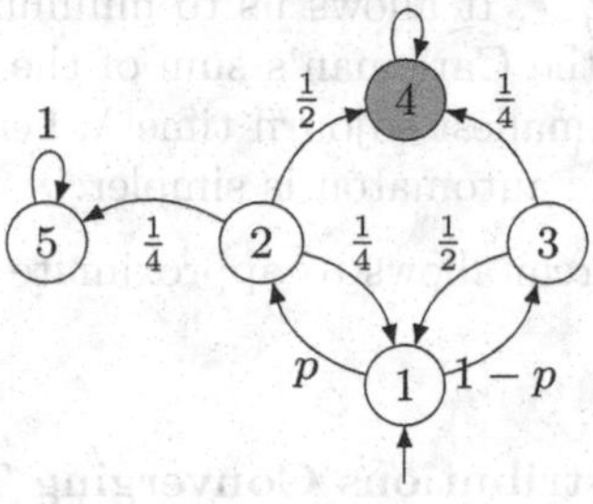

Fig. 4. Example of a parametric system with set of parameters $\{p\}$

4.3 Uniqueness of the Distribution

Last, we use the symbolic expression of moments obtained in Sect. 3 in order to prove the uniqueness of the distribution having moments of first passage times of (continuous-time) Markov chains. Thus this distribution is the distribution of response times of the system considered.

Notice that in general, there may be several distributions that correspond to a given sequence of moments $(\mu_n)_{n\in\mathbb{N}}$. This would compromise approximating the distribution using moments, as there would not be a unique such distribution.

Example 3. Let us consider a distribution δ on $\mathbb{R}^+$. If δ has the sequence of moments $\{\mu_n = n! \mid n \in \mathbb{N}\}$, then δ is the exponential distribution with parameter 1. Similarly, the sequence of moments $\{\mu_n = (2n)! \mid n \in \mathbb{N}\}$ for a distribution on $\mathbb{R}^+$ is characteristic of the square of the exponential distribution of parameter 1.

Now, consider the cube of the exponential distribution of parameter 1. Its sequence of moments is $\{\mu_n = (3n!) \mid n \in \mathbb{N}\}$. However, there exist an infinite number of distributions with this sequence of moments [18].

We now prove answer positively to the Stieljes moment problem for the case of the distribution of response time in a (continuous-time) Markov chain, that is its sequence of moments respects the Carleman's condition from year 1922, that guarantees the uniqueness of the distribution. The condition is that $\sum_{n\in\mathbb{N}} \mu_n(\delta)^{-\frac{1}{2n}} = \infty$.

Theorem 4. *Let A be a probabilistic automaton or a CTMC. For all $n \in \mathbb{N}$, let μ_n be the moment of order n of the times of first passage in a set of state R of A. Then there exists a unique distribution δ such that $\mu_n(\delta) = \mu_n$ for all $n \in \mathbb{N}$.*

Sketch of Proof: We first consider CTMC where all states have the same sojourn time λ. Then, a path that uses i transitions to answer a query will follow the gamma distribution with parameters (i, λ). We have a symbolic expression for moments of this distribution thanks to Sect. 3. This can be used to minimize $\sum_{n=0}^{\infty} \mu_n(\delta)^{-\frac{1}{2n}}$ by a diverging sum.

For general CTMCs, we use the fact that $\mathbb{E}(\Gamma(i, \lambda_1)^n) \leq \mathbb{E}((E(\lambda_1) + \cdots + E(\lambda_i))^n)$ iff $\lambda_1 = min(\lambda_j)_{j=1}^{i}$. It allows us to minimize the Carleman's sum of the CTMC considered by the Carleman's sum of the CTMC where all sojourn times are replaced by the smallest sojourn time λ, hence the divergence.

The case of probabilistic automaton is simpler. □

We show how this theorem allows to approximate distribution δ in the next subsection.

4.4 A Sequence of Distributions Converging Towards δ

Since we have unicity of the distribution corresponding to the sequence of moments of the distribution of response time of a probabilistic automaton, we obtain the following convergence in law:

Proposition 5 ([17]). *Let δ be the distribution of response times of a probabilistic automaton. Let $(\delta_i)_{i \in \mathbb{N}}$ be a sequence of distributions on $\mathbb{R}^+$ such that for all n, $\lim_{i \to \infty} \mu_n(\delta_i) = \mu_n(\delta)$. Then, if C_i is the cumulative distribution function of δ_i and C the cumulative distribution function of δ, then for all x $\lim_{i \to \infty} C_i(x) = C(x)$.*

Thus, C can be approximated by taking a sequence $(\delta_n)_{n \in \mathbb{N}}$ of distribution such that for all $i \leq n$, $\mu_i(\delta_n) = \mu_i(\delta)$. A reasonable choice for δ_n is to consider the distribution of maximal entropy corresponding to the moments $\mu_1, \ldots, \mu_n$, as presented in [11]. The distribution of maximal entropy can be understood as the distribution that assume the least information. It can be approximated as close as desired, for instance $\frac{1}{n}$ close to the distribution of maximal entropy having moments $(\mu_1(\delta), \ldots, \mu_n(\delta))$. Applying Proposition 5, we thus obtain that the cumulative distribution function associated with δ_i converges towards the cumulative distribution function associated with δ.

5 Bounding the Response Time

We now explain how to use moments in order to obtain optimal bounds on the response time. First, notice that as soon as there exists a loop between a query and a response (as in Fig. 1), then there will be runs with arbitrarily long response times, although there might be probability 1 to eventually answer every query (which is the case for Fig. 1). We thus turn to a more quantitative evaluation of the response time.

Let $0 < p < 1$. We are interested in a bound T on the delay between a query and a response such that more than $1 - p$ of the queries are answered before this bound. For a distribution $\delta : \mathbb{R}^+ \to \mathbb{R}^+$ of response times, we denote by $B(\delta, p)$ the lowest T such that the probability to have a response time above T is lower than p. Equivalently, we look for the highest T such that the probability of a response time above T is at least p.

We place ourselves in the general setting of continuous distributions, where Dirac delta functions are allowed for simplicity. Discrete distributions form a special case, with delta functions at integer values. One could get rid of Dirac delta functions by ϵ-approximating them without changing the moments, obtaining the same bounds as we prove here.

5.1 Tchebychev Bounds Associated with One Moment

Let $i \in \mathbb{N}$ and $\mu_i > 0$. We let Δ_{i,μ_i} be the set of distributions of response time which have μ_i as moment of order i. We are interested in bounding $B(\delta, p)$ for all $\delta \in \Delta_{i,\mu_i}$, that is for all distributions with μ_i as moment of order i. Such a bound is provided by *Tchebychev inequality*, and it is optimal:

Proposition 6. *Let $i \in \mathbb{N}$ and μ_i. Let $\alpha_i(\mu_i, p) = \sqrt[i]{\frac{\mu_i}{p}}$. Then for all $\delta \in \Delta_{i,\mu_i}$, we have $B(\delta, p) \leq \alpha_i(\mu_i, p)$. Further, $\exists \delta \in \Delta_{i,\mu_i}$ such that $B(\delta, p) = \alpha_i(\mu_i, p)$.*

Proof. It suffices to remark that $\mu_i > pb^i$ for b the bound we want to reach. Further, this bound is trivially optimal: it suffices to consider a distribution with a Dirac of mass $(1-p)$ at 0 and a Dirac of mass p at $\alpha_i(\mu_i, p)$. □

Given a probabilistic automaton, let δ be its associated distribution of response time. We can compute its associated moments μ_i using Algorithm 1, described in the previous section. We thus know that $\delta \in \Delta_{i,\mu_i}$. Given different values of i, one can compute the different moments and apply for each of the Tchebychev bound and use the minimal bound obtained.

Understanding the relationship between the α_i is thus important. For $i < j$, one can use Jensen's inequality for the convex function $f : x \to x^{\frac{j}{i}}$ over $\mathbb{R}^+$, and obtain: $(\mu_i)^j \leq (\mu_j)^i$. For instance, $\mu_1^2 < \mu_2$.

For $p = 1$, this gives $\alpha_i(p = 1) < \alpha_j(p = 1)$. On the other hand, for p sufficiently close to 0, we have $\alpha_j(p) < \alpha_i(p)$. That is, when p is very small, moments of high orders will give better bounds than moments of lower order. On the other hand, if p is not that small, moments of small order will suffice.

5.2 Optimal Bounds for a Pair of Moments

We now explain how to extend Tchebychev bounds to pairs of moments: We consider the set of distributions where two moments are fixed. Let $i < j$ be two orders of moments and $\mu_i, \mu_j > 0$. We denote by $\Delta_{i,\mu_i}^{j,\mu_j}$ the set of distributions with μ_i, μ_j as moments of order i, j respectively. As $\Delta_{i,\mu_i}^{j,\mu_j}$ is strictly included into Δ_{i,μ_i} and in Δ_{j,μ_j}, $\min(\alpha_i(p), \alpha_j(p))$ is a bound for any $\delta \in \Delta_{i,\mu_i}^{j,\mu_j}$. However, it may be the case that $\min(\alpha_i(p), \alpha_j(p))$ is not optimal. We now provide *optimal* bounds $\alpha_i^j(p)$ for any pair $i < j$ of order of moments and probability p:

Theorem 5. *Let $i < j$ be natural integers, $p \in (0,1)$, and let $\mu_i, \mu_j > 0$. Let $\alpha_i = (\frac{\mu_i}{p})^{\frac{1}{i}}$ and $\alpha_j = (\frac{\mu_j}{p})^{\frac{1}{j}}$. We define $\alpha_i^j(p)$ to be:*

- *α_i if $\alpha_i \leq \alpha_j$,*
- *$(\frac{\mu_j - M}{p})^{\frac{1}{j}}$ otherwise, where $0 \leq M \leq \mu_j$ is the smallest positive real root of:*

$$\mu_i = (1-p)^{\frac{j-i}{j}} M^{\frac{i}{j}} + p^{\frac{j-i}{j}} (\mu_j - M)^{\frac{i}{j}}.$$

For all $\delta \in \Delta_{i,\mu_i}^{j,\mu_j}$, we have $B(\delta, p) \leq \alpha_i^j$, and $\exists \delta \in \Delta_{i,\mu_i}^{j,\mu_j}$ with $B(\delta, p) = \alpha_i^j$

To obtain a value for M, one can use for instance Newton's method. For $i = 1, j = 2$, we can compute explicitly M and obtain:

$$\alpha_1^2 = \mu_1 + \sqrt{\frac{(1-p)}{p}(\mu_2 - \mu_1^2)}.$$

Example 4. Consider the distribution associated with the system of Fig. 1. We obtain the following bounds $\alpha_i(p), \alpha_i^{i-1}(p)$ considering different values of p and i:

i	μ_i	$\alpha_i(0.1)$	$\alpha_i^{i-1}(0.1)$	$\alpha_i(0.01)$	$\alpha_i^{i-1}(0.01)$
1	10	100	100	1000	1000
2	190	43.6	38.5	137.8	104.9
3	5410	37.8	**36.8**	81.5	73.9
4	205390	37.9	37.8	67.4	63.8
5	9747010	39.6	37.9	64.2	**61.43**
6	555066190	42.1	39.6	62.8	61.47

For $p = 0.1$, it is not useful to consider moments of order higher than 3. For $p = 0.01$, moment of order 5 provides better bounds than moment of lower orders.

For hierarchical systems, one can compute moments in an efficient way using Theorem 3, and then use Theorem 5 to obtain the associated optimal bounds. In order to handle parametric systems, we use the following result which allows to underapproximate the value of M, and thus overapproximate the optimal bound, by iterating the following operator f from $x = 0$:

$$f : x \mapsto \frac{(\mu_i - [\mu_j - x]^{\frac{i}{j}} p^{\frac{j-i}{j}})^{\frac{j}{i}}}{(1-p)^{\frac{j-i}{i}}}$$

Lemma 1. *$(f^n(0))_{n\in\mathbb{N}}$ is strictly increasing and converges towards M.*

We show how to ϵ-approximate the *optimal* bound B of a *parametric* probabilistic automaton A with set of parameters P, that is such that for all $val \in V^P$, the probabilistic automaton A with valuation val for parameter values has a bound $b(val) \leq B$ and there exists a $val \in V^P$ such that $b(val) = B$. First, we obtain the moments as symbolic functions of the parameters using Sect. 4.2. Then, we compute $M_1 = f(0)$ as a function of the parameters, using Lemma 1 and replacing μ_i, μ_j by their expression. One can then compute the minimum m_1 of function M_1 over all the parameters. We then proceed with $M_2 = f(m_1)$, and so on till obtaining a value m. This allows to obtain a lower bound m over values of M for all parameter values. Computing the largest μ_j over all parameters allows to obtain an upper bound B_{up}: $B \leq B_{up} = (\frac{\mu_j - m}{p})^{\frac{1}{j}}$. A lower bound B_{lw} is easily obtained by considering the value $\geq m$ of M for the parameters maximizing μ_j. If the distance between B_{up} and B_{lw} is larger than ϵ, one can partition the space of parameter values in zones and proceed in the same way on each zone, forgetting zones for which B_{up} is lower than the B_{lw} of another zone, till the distance between $\max(B_{lw})$ and $\max(B_{up})$ over zones is smaller than ϵ.

6 Conclusion

In this paper, we have shown how to compute moments symbolically for probabilistic automata and CTMCs, using adequately defined semirings. This method has the same complexity as the efficient numerical methods already known [13]. The proof of this symbolic computation allows proving that there is a unique distribution of response time corresponding to a probabilistic automaton or a CTMC. This allows obtaining simple approximated distributions scheme converging in law towards the distribution of response time. The symbolic computation of moments also allows computing moments in compact (hierarchical) models faster, as well as finding lowest/highest value of moments in parametric systems.

We also provide optimal bounds on the delay after which very few queries stay unanswered. It is optimal when considering distribution displaying a given pair of moments, and we showed on a simple example how this improves Tchebychev bounds. This can be used efficiently to obtain bounds for compact (hierarchical) models or to compute an optimal bound which fulfills the response of almost all queries even for systems where some parameter values are not known exactly.

References

1. Asarin, E., Basset, N., Degorre, A.: Entropy of regular timed languages. In: Information and Computation, vol. 241, pp. 142–176. Elsevier (2015)
2. Angrish, R., Chakraborty, S.: Probabilistic timing analysis of asynchronous systems with moments of delay. In: ASYNC 2002. IEEE (2002)
3. Alur, R.: Formal analysis of hierarchical state machines. In: Verification: Theory and Practice, pp. 42–66 (2002)
4. Backenköhler, M., Bortolussi, L., Wolf, V.: Generalized method of moments for stochastic reaction networks in equilibrium. In: Bartocci, E., Lio, P., Paoletti, N. (eds.) CMSB 2016. LNCS, vol. 9859, pp. 15–29. Springer, Cham (2016). https://doi.org/10.1007/978-3-319-45177-0_2
5. Bazille, H., Fabre, E., Genest, B.; Symbolically quantifying response time in stochastic models using moments and semirings. https://perso.crans.org/~genest/BFG18.pdf
6. Bradley, J., Dingle, N., Harder, U., Harrison, P., Knottenbelt, W.: Response time densities and quantiles in large Markov and semi-Markov Models. In: Performance Evaluation of Parallel, Distributed and Emergent Systems, vol. 1 (2006)
7. Bogomolov, S., Henzinger, T.A., Podelski, A., Ruess, J., Schilling, C.: Adaptive moment closure for parameter inference of biochemical reaction networks. In: Roux, O., Bourdon, J. (eds.) CMSB 2015. LNCS, vol. 9308, pp. 77–89. Springer, Cham (2015). https://doi.org/10.1007/978-3-319-23401-4_8
8. Chatterjee, K., Henzinger, T.A., Horn, F.: The complexity of request-response games. In: Dediu, A.-H., Inenaga, S., Martín-Vide, C. (eds.) LATA 2011. LNCS, vol. 6638, pp. 227–237. Springer, Heidelberg (2011). https://doi.org/10.1007/978-3-642-21254-3_17
9. Cassandras, C., Lafortune, S.: Introduction to Discrete Event Systems. Springer, Boston (2007). https://doi.org/10.1007/978-0-387-68612-7
10. Cortes, C., Mohri, M., Rastogi, A., Riley, M.: On the computation of the relative entropy of probabilistic automata. Int. J. Found. Comput. Sci. (IJFCS) **19**(1), 219–242 (2006)
11. Cover, T., Thomas, J.: Elements of Information Theory. Wiley, New York (2006)

12. Daws, C.: Symbolic and parametric model checking of discrete-time Markov chains. In: Liu, Z., Araki, K. (eds.) ICTAC 2004. LNCS, vol. 3407, pp. 280–294. Springer, Heidelberg (2005). https://doi.org/10.1007/978-3-540-31862-0_21
13. Dayar, T., Akar, N.: Computing moments of first passage times to a subset of states in Markov chains. SIAM J. Matrix Anal. Appl. **27**(2), 396–412 (2005)
14. Gonzalez, A.M., Uhlendorf, J., Schaul, J., Cinquemani, E., Batt, G., Ferrari-Trecate, G.: Identification of biological models from single-cell data: a comparison between mixed-effects and moment-based inference. In: ECC 2013, pp. 3652–3657. IEEE (2013)
15. Horn, F., Thomas, W., Wallmeier, N., Zimmerman, M.: Optimal strategy synthesis for request-response games. RAIRO **49**(3), 179–203 (2015)
16. Jansen, N., Corzilius, F., Volk, M., Wimmer, R., Ábrahám, E., Katoen, J.-P., Becker, B.: Accelerating parametric probabilistic verification. In: Norman, G., Sanders, W. (eds.) QEST 2014. LNCS, vol. 8657, pp. 404–420. Springer, Cham (2014). https://doi.org/10.1007/978-3-319-10696-0_31
17. Prohorov, Y., Rozanov, Y.: Probability Theory, Basic Concepts · Limit Theorems Random Processes. Springer, Heidelberg (1969). Translated from Russian
18. Stoyanov, J.: Determinacy of distributions by their moments. In: ICMSM 2006 (2006)
19. Tari, Á: Moments based bounds in stochastic models, Ph.D. Thesis. Budapesti Műszaki és Gazdaságtudományi Egyetem (2005)
20. Telek, M., Horvéth, G.: A minimal representation of Markov arrival processes and a moments matching method. Perform. Eval. **64**(9–12), 1153–1168 (2007)
21. Wallmeier, N., Hütten, P., Thomas, W.: Symbolic synthesis of finite-state controllers for Request-Response specifications. In: CIAA 2003 (2003)
22. Yao, D.: First-passage-time moments of Markov processes. J. Appl. Probab. **22**(4), 939–945 (1985)

Comparator Automata in Quantitative Verification

Suguman Bansal(✉), Swarat Chaudhuri(✉), and Moshe Y. Vardi(✉)

Rice University, Houston, TX 77005, USA
{suguman,swarat}@rice.edu, vardi@cs.rice.edu

Abstract. The notion of comparison between system runs is fundamental in formal verification. This concept is implicitly present in the verification of qualitative systems, and is more pronounced in the verification of quantitative systems. In this work, we identify a novel mode of comparison in quantitative systems: the online comparison of the aggregate values of two sequences of quantitative weights. This notion is embodied by *comparator automata* (*comparators*, in short), a new class of automata that read two infinite sequences of weights synchronously and relate their aggregate values.

We show that comparators that are finite-state and accept by the Büchi condition lead to generic algorithms for a number of well-studied problems, including the quantitative inclusion and winning strategies in quantitative graph games with incomplete information, as well as related non-decision problems, such as obtaining a finite representation of all counterexamples in the quantitative inclusion problem.

We study comparators for two aggregate functions: discounted-sum and limit-average. We prove that the discounted-sum comparator is ω-regular for all integral discount factors. Not every aggregate function, however, has an ω-regular comparator. Specifically, we show that the language of sequence-pairs for which limit-average aggregates exist is neither ω-regular nor ω-context-free. Given this result, we introduce the notion of *prefix-average* as a relaxation of limit-average aggregation, and show that it admits ω-context-free comparators.

1 Introduction

Many classic questions in formal methods can be seen as involving *comparisons* between different system runs or inputs. Consider the problem of verifying if a system S satisfies a linear-time temporal property P. Traditionally, this problem is phrased language-theoretically: S and P are interpreted as sets of (infinite) words, and S is determined to satisfy P if $S \subseteq P$. The problem, however, can also be framed in terms of a *comparison* between words in S and P. Suppose a word w is assigned a weight of 1 if it belongs to the language of the system

Electronic supplementary material The online version of this chapter (https://doi.org/10.1007/978-3-319-89366-2_23) contains supplementary material, which is available to authorized users.

C. Baier and U. Dal Lago (Eds.): FOSSACS 2018, LNCS 10803, pp. 420–437, 2018.
https://doi.org/10.1007/978-3-319-89366-2_23

or property, and 0 otherwise. Then determining if $S \subseteq P$ amounts to checking whether the weight of every word in S is less than or equal to its weight in P [5].

The need for such a formulation is clearer in quantitative systems, in which every run of a word is associated with a sequence of (rational-valued) weights. The weight of a run is given by *aggregate function* $f : \mathbb{Q}^\omega \rightarrow \mathbb{R}$, which returns the real-valued *aggregate value* of the run's weight sequence. The weight of a word is given by the supremum or infimum of the weight of all its runs. Common examples of aggregate functions include discounted-sum and limit-average.

In a well-studied class of problems involving quantitative systems, the objective is to check if the aggregate value of words of a system exceed a constant threshold value [14–16]. This is a natural generalization of emptiness problems in qualitative systems. Known solutions to the problem involve arithmetic reasoning via linear programming and graph algorithms such as negative-weight cycle detection, computation of maximum weight of cycles etc. [4,18].

A more general notion of comparison relates aggregate values of two weight sequences. Such a notion arises in the *quantitative inclusion problem* for weighted automata [1], where the goal is to determine whether the weight of words in one weighted automaton is less than that in another. Here it is necessary to compare the aggregate value along runs between the two automata. Approaches based on arithmetic reasoning do not, however, generalize to solving such problems. In fact, the known solution to discounted-sum inclusion with integer discount-factor combines linear programming with a *specialized* subset-construction-based determinization step, rendering an EXPTIME algorithm [4,6]. Yet, this approach does not match the PSPACE lower bound for discounted-sum inclusion.

In this paper, we present an automata-theoretic formulation of this form of comparison between weighted sequences. Specifically, we introduce *comparator automata* (*comparators*, in short), a class of automata that read pairs of infinite weight sequences synchronously, and compare their aggregate values in an online manner. While comparisons between weight sequences happen implicitly in prior approaches to quantitative systems, comparator automata make these comparisons explicit. We show that this has many benefits, including generic algorithms for a large class of quantitative reasoning problems, as well as a direct solution to the problem of discounted-sum inclusion that also closes its complexity gap.

A *comparator for aggregate function* f is an automaton that accepts a pair (A, B) of sequences of bounded rational numbers iff $f(A)\ R\ f(B)$, where R is an inequality relation ($>, <, \geq, \leq$) or the equality relation. A comparator could be finite-state or (pushdown) infinite-state. This paper studies such comparators.

A comparator is ω*-regular* if it is finite-state and accepts by the Büchi condition. We show that ω-regular comparators lead to generic algorithms for a number of well-studied problems including the quantitative inclusion problem, and in showing existence of winning strategies in incomplete-information quantitative games. Our algorithm yields PSPACE-completeness of quantitative inclusion when the ω-regular comparator is provided. The same algorithm extends to obtaining finite-state representations of counterexample words in inclusion.

Next, we show that the discounted-sum aggregation function admits an ω-regular comparator when the discount-factor $d > 1$ is an integer. Using properties of ω-regular comparators, we conclude that the discounted-sum inclusion is PSPACE-complete, hence resolving the complexity gap. Furthermore, we prove that the discounted-sum comparator for $1 < d < 2$ cannot be ω-regular. We suspect this result extends to non-integer discount-factors as well.

Finally, we investigate the limit-average comparator. Since limit-average is only defined for sequences in which the average of prefixes converge, limit-average comparison is not well-defined. We show that even a Büchi pushdown automaton cannot separate sequences for which limit-average exists from those for which it does not. Hence, we introduce the novel notion of *prefix-average comparison* as a relaxation of limit-average comparison. We show that the prefix-average comparator admits a comparator that is ω-context-free, i.e., given by a Büchi pushdown automaton, and we discuss the utility of this characterization.

This paper is organized as follows: Preliminaries are given in Sect. 2. Comparator automata is formally defined in Sect. 3. Generic algorithms for ω-regular comparators are discussed in Sects. 3.1 and 3.2. The construction and properties of discounted-sum comparator, and limit-average and prefix-average comparator are given in Sects. 4 and 5, respectively. We conclude with future directions in Sect. 6.

Related Work. The notion of comparison has been widely studied in quantitative settings. Here we mention only a few of them. Such aggregate-function based notions appear in weighted automata [1,17], quantitative games including mean-payoff and energy games [16], discounted-payoff games [3,4], in systems regulating cost, memory consumption, power consumption, verification of quantitative temporal properties [14,15], and others. Common solution approaches include graph algorithms such as weight of cycles or presence of cycle [18], linear-programming-based approaches, fixed-point-based approaches [8], and the like. The choice of approach for a problem typically depends on the underlying aggregate function. In contrast, in this work we present an automata-theoretic approach that unifies solution approaches to problems on different aggregate functions. We identify a class of aggregate functions, ones that have an ω-regular comparator, and present generic algorithms for some of these problems.

While work on finite-representations of counterexamples and witnesses in the qualitative setting is known [5], we are not aware of such work in the quantitative verification domain. This work can be interpreted as automata-theoretic arithmetic, which has been explored in regular real analysis [12].

2 Preliminaries

Definition 1 (Büchi automata [21]**).** *A (finite-state)* Büchi automaton *is a tuple* $\mathcal{A} = (S, \Sigma, \delta, Init, \mathcal{F})$*, where* S *is a finite set of* states*,* Σ *is a finite* input alphabet*,* $\delta \subseteq (S \times \Sigma \times S)$ *is the* transition relation*,* $Init \subseteq S$ *is the set of* initial states*, and* $\mathcal{F} \subseteq S$ *is the set of* accepting states.

A Büchi automaton is *deterministic* if for all states s and inputs a, $|\{s'|(s,a,s') \in \delta \text{ for some } s'\}| \leq 1$ and $|Init| = 1$. Otherwise, it is *nondeterministic*. For a word $w = w_0 w_1 \cdots \in \Sigma^\omega$, a *run* ρ of w is a sequence of states $s_0 s_1 \ldots$ s.t. $s_0 \in Init$, and $\tau_i = (s_i, w_i, s_{i+1}) \in \delta$ for all i. Let $inf(\rho)$ denote the set of states that occur infinitely often in run ρ. A run ρ is an *accepting run* if $inf(\rho) \cap \mathcal{F} \neq \emptyset$. A word w is an accepting word if it has an accepting run. Büchi automata are known to be closed under set-theoretic union, intersection, and complementation [21]. Languages accepted by these automata are called *ω-regular languages.*

Definition 2 (Weighted ω-automaton [10,20]). *A weighted ω-automaton over infinite words is a tuple $\mathcal{A} = (\mathcal{M}, \gamma)$, where $\mathcal{M} = (S, \Sigma, \delta, Init, S)$ is a Büchi automaton, and $\gamma : \delta \rightarrow \mathbb{Q}$ is a* weight function.

Words and *runs* in weighted ω-automata are defined as they are in Büchi automata. Note that all states are accepting states in this definition. The *weight sequence* of run $\rho = s_0 s_1 \ldots$ of word $w = w_0 w_1 \ldots$ is given by $wt_\rho = n_0 n_1 n_2 \ldots$ where $n_i = \gamma(s_i, w_i, s_{i+1})$ for all i. The *weight of a run* ρ is given by $f(wt_\rho)$, where $f : \mathbb{Q}^\omega \rightarrow \mathbb{R}$ is an *aggregate function*. We use $f(\rho)$ to denote $f(wt_\rho)$.

Here the *weight of a word* $w \in \Sigma^\omega$ in weighted ω-automata is defined as $wt_\mathcal{A}(w) = sup\{f(\rho)|\rho \text{ is a run of } w \text{ in } \mathcal{A}\}$. It can also be defined as the infimum of the weight of all its runs. By convention, if word $w \notin \mathcal{A}$, $wt_\mathcal{A}(w) = 0$ [10].

Definition 3 (Quantitative inclusion). *Given two weighted ω-automata P and Q with aggregate function f, the* quantitative-inclusion problem, *denoted by $P \subseteq_f Q$, asks whether for all words $w \in \Sigma^\omega$, $wt_P(w) \leq wt_Q(w)$.*

Quantitative inclusion is PSPACE-complete for limsup and liminf [10], and undecidable for limit-average [16]. For discounted-sum with integer discount-factor it is in EXPTIME [6,10], and decidability is unknown for rational discount-factors

Definition 4 (Incomplete-information quantitative games). *An* incomplete-information quantitative game *is a tuple $\mathcal{G} = (S, s_\mathcal{I}, O, \Sigma, \delta, \gamma, f)$, where S, O, Σ are sets of* states, observations, *and* actions, *respectively, $s_\mathcal{I} \in S$ is the* initial state, *$\delta \subseteq S \times \Sigma \times S$ is the* transition relation, *$\gamma : S \rightarrow \mathbb{N} \times \mathbb{N}$ is the* weight function, *and $f : \mathbb{N}^\omega \rightarrow \mathbb{R}$ is the* aggregate function.

The transition relation δ is *complete*, i.e., for all states p and actions a, there exists a state q s.t. $(p, a, q) \in \delta$. A *play* ρ is a sequence $s_0 a_0 s_1 a_1 \ldots$, where $\tau_i = (s_i, a_i, s_{i+1}) \in \delta$. The *observation of state* s is denoted by $O(s) \in O$. The *observed play* o_ρ of ρ is the sequence $o_0 a_0 o_1 a a_1 \ldots$, where $o_i = O(s_i)$. Player P_0 has incomplete information about the game $\mathcal{G}$; it only perceives the observation play o_ρ. Player P_1 receives full information and witnesses play ρ. Plays begin in the initial state $s_0 = s_\mathcal{I}$. For $i \geq 0$, Player P_0 selects action a_i. Next, player P_1 selects the state s_{i+1}, such that $(s_i, a_i, s_{i+1}) \in \delta$. The *weight of state* s is the pair of payoffs $\gamma(s) = (\gamma(s)_0, \gamma(s)_1)$. The *weight sequence* wt_i of player P_i along ρ is given by $\gamma(s_0)_i \gamma(s_1)_i \ldots$, and its payoff from ρ is given by $f(wt_i)$ for aggregate function f, denoted by $f(\rho_i)$, for simplicity. A play on which a player receives a greater payoff is said to be a *winning play* for the player. A strategy for player P_0 is given by a function $\alpha : O^* \to \Sigma$ since it only sees observations. Player P_0 follows strategy α if for all i, $a_i = \alpha(o_0 \ldots o_i)$. A strategy α is said to be a *winning strategy* for player P_0 if all plays following α are winning plays for P_0.

Definition 5 (Büchi pushdown automata [13]**).** *A* Büchi pushdown automaton (Büchi PDA) *is a tuple* $\mathcal{A} = (S, \Sigma, \Gamma, \delta, Init, Z_0, \mathcal{F})$, *where* S, Σ, Γ, *and* $\mathcal{F}$ *are finite sets of* states, input alphabet, pushdown alphabet *and* accepting states, *respectively.* $\delta \subseteq (S \times \Gamma \times (\Sigma \cup \{\epsilon\}) \times S \times \Gamma)$ *is the* transition relation, $Init \subseteq S$ *is a set of* initial states, $Z_0 \in \Gamma$ *is the* start symbol.

A *run* ρ on a word $w = w_0 w_1 \dots \in \Sigma^\omega$ of a Büchi PDA $\mathcal{A}$ is a sequence of configurations $(s_0, \gamma_0), (s_1, \gamma_1) \dots$ satisfying (1) $s_0 \in Init$, $\gamma_0 = Z_0$, and (2) $(s_i, \gamma_i, w_i, s_{i+1}, \gamma_{i+1}) \in \delta$ for all i. Büchi PDA consists of a *stack*, elements of which are the tokens Γ, and initial element Z_0. Transitions *push* or *pop* token(s) to/from the top of the stack. Let $inf(\rho)$ be the set of states that occur infinitely often in state sequence $s_0 s_1 \dots$ of run ρ. A run ρ is an *accepting run* in Büchi PDA if $inf(\rho) \cap \mathcal{F} \neq \emptyset$. A word w is an *accepting word* if it has an accepting run. Languages accepted by Büchi PDA are called *ω-context-free languages* (ω-CFL).

We introduce some notation. For an infinite sequence $A = (a_0, a_1, \dots)$, $A[i]$ denotes its i-th element. Abusing notation, we write $w \in \mathcal{A}$ and $\rho \in \mathcal{A}$ if w and ρ are an accepting word and an accepting run of $\mathcal{A}$ respectively.

For missing proofs and constructions, refer to the supplementary material.

3 Comparator Automata

Comparator automata (often abbreviated as *comparators*) are a class of automata that can read pairs of weight sequences synchronously and establish an equality or inequality relationship between these sequences. Formally, we define:

Definition 6 (Comparator automata). *Let* Σ *be a finite set of rational numbers, and* $f : \mathbb{Q}^\omega \to \mathbb{R}$ *denote an aggregate function. A* comparator automaton for aggregate function f *is an automaton over the alphabet* $\Sigma \times \Sigma$ *that accepts a pair* (A, B) *of (infinite) weight sequences iff* $f(A)\ R\ f(B)$, *where* R *is an inequality or the equality relation.*

From now on, unless mentioned otherwise, we assume that all weight sequences are bounded, natural number sequences. The boundedness assumption is justified since the set of weights forming the alphabet of a comparator is bounded. For all aggregate functions considered in this paper, the result of comparison of weight sequences is preserved by a uniform linear transformation that converts rational-valued weights into natural numbers; justifying the natural number assumption.

We explain comparators through an example. The *limit supremum* (limsup, in short) of a bounded, integer sequence A, denoted by $\mathsf{LimSup(A)}$, is the largest integer that appears infinitely often in A. The *limsup comparator* is a Büchi automaton that accepts the pair (A, B) of sequences iff $\mathsf{LimSup(A)} \geq \mathsf{LimSup(B)}$.

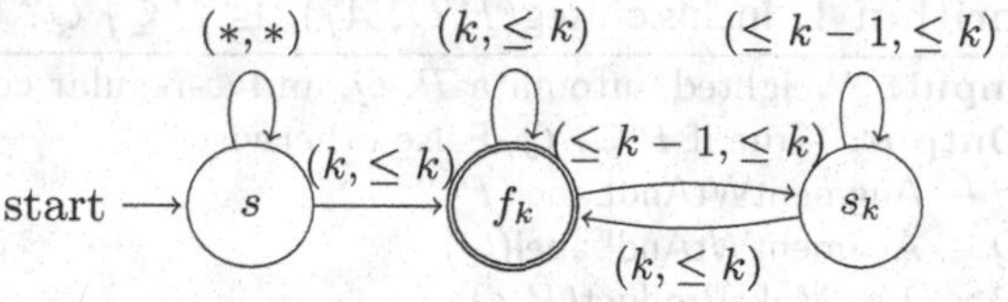

Fig. 1. State f_k is an accepting state. Automaton $\mathcal{A}_k$ accepts (A, B) iff $\mathsf{LimSup(A)} = k$, $\mathsf{LimSup(B)} \leq k$. $*$ denotes $\{0, 1 \ldots \mu\}$, $\leq m$ denotes $\{0, 1 \ldots, m\}$

The working of the limsup comparator is based on non-deterministically guessing the limsup of sequences A and B, and then verifying that $\mathsf{LimSup(A)} \geq \mathsf{LimSup(B)}$. Büchi automaton $\mathcal{A}_k$ (Fig. 1) illustrates the basic building block of the limsup comparator. Automaton $\mathcal{A}_k$ accepts pair (A, B) of number sequences iff $\mathsf{LimSup(A)} = k$, and $\mathsf{LimSup(B)} \leq k$, for integer k. To see why this is true, first note that all incoming edges to accepting state f_k occur on alphabet $(k, \leq k)$ while all transitions between states f_k and s_k occur on alphabet $(\leq k, \leq k)$, where $\leq k$ denotes the set $\{0, 1, \ldots k\}$. So, the integer k must appear infinitely often in A and all elements occurring infinitely often in A and B are less than or equal to k. Together these ensure that $\mathsf{LimSup(A)} = k$, and $\mathsf{LimSup(B)} \leq k$. The union of such automata $\mathcal{A}_k$ for $k \in \{0, 1, \ldots \mu\}$ for upper bound μ, results in the limsup comparator. The *limit infimum* (liminf, in short) of an integer sequence is the smallest integer that appears infinitely often in it; its comparator is similar.

When the comparator for an aggregate function is a Büchi automaton, we call it an *ω-regular comparator*. Likewise, when the comparator for an aggregate function is a Büchi pushdown automaton, we call it an *ω-context-free comparator*. As seen here, the limsup and liminf comparators are ω-regular. Later, we see that discounted-sum comparator and prefix-average comparator are ω-regular and ω-context-free respectively (Sects. 4 and 5). We call an aggregate function *ω-regular* when it has an ω-regular comparator for at least one inequality relation. Due to closure properties of Büchi automata, comparators for all inequality and equality relations of an ω-regular aggregate function are also ω-regular.

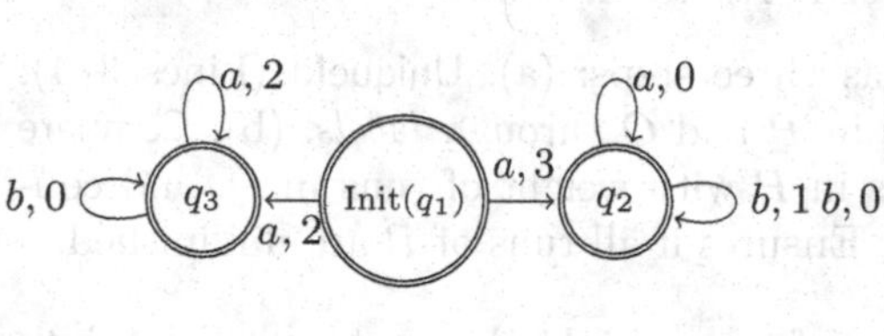

Fig. 2. Weighted automaton P

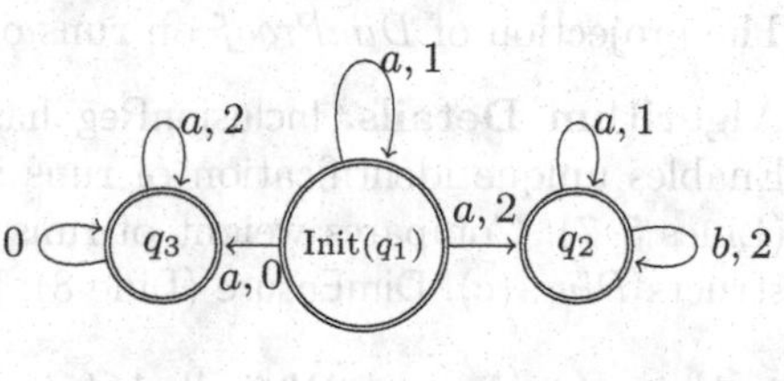

Fig. 3. Weighted automaton Q

Motivating Example. Let weighted ω-automata P and Q be as illustrated in Figs. 2 and 3. The word $w = a(ab)^\omega$ has two runs $\rho_1^P = q_1(q_2)^\omega$, $\rho_2^P = q_1(q_3)^\omega$

Algorithm 1. InclusionReg($P, Q, \mathcal{A}_f$), Is $P \subseteq_f Q$?

1: **Input:** Weighted automata P, Q, and ω-regular comparator $\mathcal{A}_f$ (Inequality $\leq$)
2: **Output:** True if $P \subseteq_f Q$, False otherwise
3: $\hat{P} \leftarrow$ AugmentWtAndLabel(P)
4: $\hat{Q} \leftarrow$ AugmentWtAndLabel(Q)
5: $\hat{P} \times \hat{Q} \leftarrow$ MakeProduct($\hat{P}, \hat{Q}$)
6: *DimProof* $\leftarrow$ Intersect($\hat{P} \times \hat{Q}, \mathcal{A}_{\succeq}$)
7: *Dim* $\leftarrow$ FirstProject(*DimProof*)
8: **return** $\hat{P} \equiv$ *Dim*

in P, and four runs $\rho_1^Q = q_1(q_2)^\omega$, $\rho_2^Q = q_1(q_3)^\omega$, $\rho_3^Q = q_1q_1(q_2)^\omega$ $\rho_4^Q = q_1q_1(q_3)^\omega$ in Q. Their weight-sequences are $wt_1^P = 3, (0,1)^\omega$, $wt_2^P = 2, (2,0)^\omega$ in P, and $wt_1^Q = (2,1)^\omega$, $wt_2^Q = (0,2)^\omega$, $wt_3^Q = 1, 2, (2,1)^\omega$, $wt_4^Q = 1, 0, (0,2)^\omega$ in Q.

To determine if w has greater weight in P or in Q, compare aggregate value of weight-sequences of runs in P and Q. Take the comparator for aggregate function f that accepts a pair (A, B) of weight-sequence iff $f(A) \leq f(B)$. For $wt_P(w) \leq wt_Q(w)$, for every run ρ_i^P in P, there exists a run ρ_j^Q in Q s.t. (ρ_i^P, ρ_j^Q) is accepted by the comparator. This forms the basis for quantitative inclusion.

3.1 Quantitative Inclusion

InclusionReg (Algorithm 1) is an algorithm for quantitative inclusion for ω-regular aggregate functions. For weighted ω-automata P, Q, and ω-regular comparator $\mathcal{A}_f$, InclusionReg returns True iff $P \subseteq_f Q$. We assume $P \subseteq Q$ (qualitative inclusion) to avoid trivial corner cases.

Key Ideas. $P \subseteq_f Q$ holds if for every run ρ_P in P on word w, there exists a run ρ_Q in Q on the same word w such that $f(\rho_P) \leq f(\rho_Q)$. We refer to such runs of P by *diminished run*. Hence, $P \subseteq_f Q$ iff all runs of P are diminished.

InclusionReg constructs Büchi automaton *Dim* that consists of exactly the diminished runs of P. It returns True iff *Dim* contains all runs of P. To obtain *Dim*, it constructs Büchi automaton *DimProof* that accepts word (ρ_P, ρ_Q) iff ρ_P and ρ_Q are runs of the same word in P and Q respectively, and $f(\rho_P) \leq f(\rho_Q)$. The ω-regular comparator for inequality $\leq$ for function f ensures $f(\rho_P) \leq f(\rho_Q)$. The projection of *DimProof* on runs of P results in *Dim*.

Algorithm Details. InclusionReg has three steps: (a). UniqueId (Lines 3–4): Enables unique identification of runs in P and Q through *labels*. (b). Compare (Lines 5–7): Compares weight of runs in P with weight of runs in Q, and constructs *Dim*. (c). DimEnsure (Line 8): Ensures if all runs of P are diminished.

1. UniqueId: AugmentWtAndLabel transforms weighted ω-automaton $\mathcal{A}$ into Büchi automaton $\hat{\mathcal{A}}$ by converting transition $\tau = (s, a, t)$ with weight $\gamma(\tau)$ in $\mathcal{A}$ to transition $\hat{\tau} = (s, (a, \gamma(\tau), l), t)$ in $\hat{\mathcal{A}}$, where l is a unique label assigned to transition τ. The word $\hat{\rho} = (a_0, n_0, l_0)(a_1, n_1, l_1) \cdots \in \hat{A}$ iff run $\rho \in \mathcal{A}$ on word $a_0a_1 \ldots$ with weight sequence $n_0n_1 \ldots$. Labels ensure bijection between

runs in $\mathcal{A}$ and words in $\hat{\mathcal{A}}$. Words of $\hat{\mathcal{A}}$ have a single run in $\hat{\mathcal{A}}$. Hence, transformation of weighted ω-automata P and Q to Büchi automata $\hat{P}$ and $\hat{Q}$ enables disambiguation between runs of P and Q (Line 3–4).
2. Compare: The output of this step is the Büchi automaton *Dim*, that contains the word $\hat{\rho} \in \hat{P}$ iff ρ is a diminished run in P (Lines 5–7). MakeProduct$(\hat{P}, \hat{Q})$ constructs $\hat{P} \times \hat{Q}$ s.t. word $(\hat{\rho_P}, \hat{\rho_Q}) \in \hat{P} \times \hat{Q}$ iff ρ_P and ρ_Q are runs of the same word in P and Q respectively (Line 5). Concretely, for transition $\hat{\tau_{\mathcal{A}}} = (s_{\mathcal{A}}, (a, n_{\mathcal{A}}, l_{\mathcal{A}}), t_{\mathcal{A}})$ in automaton $\mathcal{A}$, where $\mathcal{A} \in \{\hat{P}, \hat{Q}\}$, transition $\hat{\tau_P} \times \hat{\tau_Q} = ((s_P, s_Q), (a, n_P, l_P, n_Q, l_Q), (t_P, t_Q))$ is in $\hat{P} \times \hat{Q}$. Intersect intersects the weight components of $\hat{P} \times \hat{Q}$ with comparator $\mathcal{A}_f$ (Line 6). The resulting automaton *DimProof* accepts word $(\hat{\rho_P}, \hat{\rho_Q})$ iff $f(\rho_P) \leq f(\rho_Q)$, and ρ_P and ρ_Q are runs on the same word in P and Q respectively. The projection of *DimProof* on the words of $\hat{P}$ returns *Dim* which contains the word $\hat{\rho_P}$ iff ρ_P is a diminished run in P (Line 7).
3. DimEnsure: $P \subseteq_f Q$ iff $\hat{P} \equiv Dim$ (qualitative equivalence) since $\hat{P}$ consists of all runs of P and *Dim* consists of all diminished runs of P (Line 8).

Lemma 1. *Given weighted ω-automata P and Q with an ω-regular aggregate function f.* InclusionReg$(P, Q, \mathcal{A}_f)$ *returns* True *iff* $P \subseteq_f Q$.

Further, InclusionReg is adapted for quantitative *strict*-inclusion $P \subset_f Q$ i.e. for all words w, $wt_P(w) < wt_Q(w)$ by taking the ω-regular comparator $\mathcal{A}_f$ that accepts (A, B) iff $f(A) < f(B)$. Similarly for quantitative equivalence $P \equiv_f Q$.

Complexity Analysis. All operations in InclusionReg until Line 7 are polytime operations in the size of weighted ω-automata P, Q and comparator $\mathcal{A}_f$. Hence, *Dim* is polynomial in size of P, Q and $\mathcal{A}_f$. Line 8 solves a PSPACE-complete problem. Therefore, the quantitative inclusion for ω-regular aggregate function f is in PSPACE in size of the inputs P, Q, and $\mathcal{A}_f$.

The PSPACE-hardness of the quantitative inclusion is established via reduction from the *qualitative* inclusion problem, which is PSPACE-complete. The formal reduction is as follows: Let P and Q be Büchi automata (with all states as accepting states). Reduce P, Q to weighted automata $\overline{P}$, $\overline{Q}$ by assigning a weight of 1 to each transition. Since all runs in $\overline{P}$, $\overline{Q}$ have the same weight sequence, weight of all words in $\overline{P}$ and $\overline{Q}$ is the same for any function f. It is easy to see $P \subseteq Q$ (qualitative inclusion) iff $\overline{P} \subseteq_f \overline{Q}$ (quantitative inclusion).

Theorem 1. *Let P and Q be weighted ω-automata and $\mathcal{A}_f$ be an ω-regular comparator. The complexity of the quantitative inclusion problem, quantitative strict-inclusion problem, and quantitative equivalence problem for ω-regular aggregate function f is* PSPACE*-complete.*

Theorem 1 extends to weighted ω-automata when weight of words is the *infimum* of weight of runs. The key idea for $P \subseteq_f Q$ here is to ensure that for every run ρ_Q in Q there exists a run on the same word in ρ_P in P s.t. $f(\rho_P) \leq f(\rho_Q)$.

Representation of Counterexamples. When $P \not\subseteq_f Q$, there exists word(s) $w \in \Sigma^*$ s.t $wt_P(w) > wt_Q(w)$. Such a word w is said to be a *counterexample* *word.* Previously, finite-state representations of counterexamples have been useful in verification and synthesis in qualitative systems [5], and could be useful in quantitative settings as well. However, we are not aware of procedures for such representations in the quantitative settings. Here we show that a trivial extension of InclusionReg yields Büchi automata-representations for all counterexamples of the quantitative inclusion problem for ω-regular functions.

For word w to be a counterexample, it must contain a run in P that is not diminished. Clearly, all non-diminished runs of P are members of $\hat{P} \setminus \mathit{Dim}$. The counterexamples words can be obtained from $\hat{P} \setminus \mathit{Dim}$ by modifying its alphabet to the alphabet of P by dropping transition weights and their unique labels.

Theorem 2. *All counterexamples of the quantitative inclusion problem for an ω-regular aggregate function can be expressed by a Büchi automaton.*

3.2 Incomplete-Information Quantitative Games

Given an incomplete-information quantitative game $\mathcal{G} = (S, s_{\mathcal{I}}, O, \Sigma, \delta, \gamma, f)$, our objective is to determine if player P_0 has a winning strategy $\alpha : O^* \rightarrow \Sigma$ for ω-regular aggregate function f. We assume we are given the ω-regular comparator $\mathcal{A}_f$ for function f. Note that a function $A^* \rightarrow B$ can be treated like a B-labeled A-tree, and vice-versa. Hence, we proceed by finding a Σ-labeled O-tree – the *winning strategy tree*. Every branch of a winning strategy-tree is an observed play o_ρ of $\mathcal{G}$ for which every actual play ρ is a winning play for P_0.

We first consider all *game trees* of $\mathcal{G}$ by interpreting $\mathcal{G}$ as a tree-automaton over Σ-labeled S-trees. Nodes $n \in S^*$ of the game-tree correspond to states in S and labeled by actions in Σ taken by player P_0. Thus, the *root node* ε corresponds to $s_{\mathcal{I}}$, and a node $s_{i_0}, \dots, s_{i_k}$ corresponds to the state s_{i_k} reached via $s_{\mathcal{I}}, s_{i_0}, \dots, s_{i_{k-1}}$. Consider now a node x corresponding to state s and labeled by an action σ. Then x has children $xs_1, \dots xs_n$, for every $s_i \in S$. If $s_i \in \delta(s, \sigma)$, then we call xs_i a *valid* child, otherwise we call it an *invalid* child. Branches that contain invalid children correspond to invalid plays.

A game-tree τ is a *winning tree* for player P_0 if every branch of τ is either a winning play for P_0 or an invalid play of $\mathcal{G}$. One can check, using an automata, if a play is invalid by the presence of invalid children. Furthermore, the winning condition for P_0 can be expressed by the ω-regular comparator $\mathcal{A}_f$ that accepts (A, B) iff $f(A) > f(B)$. To use the comparator $\mathcal{A}_f$, it is determinized to parity automaton D_f. Thus, a product of game $\mathcal{G}$ with D_f is a deterministic parity tree-automaton accepting precisely winning-trees for player P_0.

Winning trees for player P_0 are Σ-labeled S-trees. We need to convert them to Σ-labeled O-trees. Recall that every state has a unique observation. We can simulate these Σ-labeled S-trees on strategy trees using the technique of *thinning* states S to observations O [19]. The resulting alternating parity tree automaton $\mathcal{M}$ will accept a Σ-labeled O-tree τ_o iff for all actual game-tree τ of τ_o, τ is a winning-tree for P_0 with respect to the strategy τ_o. The problem of existence of winning-strategy for P_0 is then reduced to non-emptiness checking of $\mathcal{M}$.

Theorem 3. *Given an incomplete-information quantitative game $\mathcal{G}$ and ω-regular comparator $\mathcal{A}_f$ for the aggregate function f, the complexity of determining whether P_0 has a winning strategy is exponential in $|\mathcal{G}| \cdot |D_f|$, where $|D_f| = |\mathcal{A}_f|^{O(|\mathcal{A}_f|)}$.*

Since, D_f is the deterministic parity automaton equivalent to $\mathcal{A}_f, |D_f| = |\mathcal{A}_f|^{O(|\mathcal{A}_f|)}$. The thinning operation is linear in size of $|\mathcal{G} \times D_f|$, therefore $|\mathcal{M}| = |\mathcal{G}| \cdot |D_f|$. Non-emptiness checking of alternating parity tree automata is exponential. Therefore, our procedure is doubly exponential in size of the comparator and exponential in size of the game. The question of tighter bounds is open.

4 Discounted-Sum Comparator

The discounted-sum of an infinite sequence A with discount-factor $d > 1$, denoted by $DS(A, d)$, is defined as $\Sigma_{i=0}^{\infty} A[i]/d^i$. The discounted-sum comparator (DS-comparator, in short) for discount-factor d, denoted by $\mathcal{A}_{\succ_{DS(d)}}$, accepts a pair (A, B) of weight sequences iff $DS(A, d) < DS(B, d)$. We investigate properties of the DS-comparator, and show that the DS-comparator is ω-regular for all integral discount-factors d, and cannot be ω-regular when $1 < d < 2$.

Theorem 4. *DS-comparator for rational discount-factor $1 < d < 2$ is not ω-regular.*

For discounted-sum automaton $\mathcal{A}$ with discount factor d, the *cut-point language* of $\mathcal{A}$ w.r.t. $r \in \mathbb{R}$ is defined as $L^{\geq r} = \{w \in L(\mathcal{A}) | DS(w, d) \geq r\}$. It is known that the cut-point language $L^{\geq 1}$ with discount-factor $1 < d < 2$ is not ω-regular [9]. One can show that if DS-comparator for discount-factor $1 < d < 2$ were ω-regular, then cut-point language $L^{\geq 1}$ is also ω-regular; thus proving Theorem 4.

We provide the construction of DS-comparator with integer discount-factor.

Key Ideas. The core intuition is that sequences bounded by μ can be converted to their value in base d via a finite-state transducer. Lexicographic comparison of the resulting sequences renders the desired result. Conversion of sequences to base d requires a certain amount of *book-keeping* by the transducer. Here we describe a direct method for book-keeping and lexicographic comparison.

For natural-number sequence A and integer discount-factor $d > 1$, $DS(A, d)$ can be interpreted as a value in base d i.e. $DS(A, d) = A[0] + \frac{A[1]}{d} + \frac{A[2]}{d^2} + \cdots = (A[0].A[1]A[2]\ldots)_d$ [12]. Unlike comparison of numbers in base d, the lexicographically larger sequence may not be larger in value. This occurs because (i) The elements of weight sequences may be larger in value than base d, and (ii) Every value has multiple infinite-sequence representations.

To overcome these challenges, we resort to arithmetic techniques in base d. Note that $DS(B, d) > DS(A, d)$ iff there exists a sequence C such that $DS(B, d) = DS(A, d) + DS(C, d)$, and $DS(C, d) > 0$. Therefore, to compare the discounted-sum of A and B, we obtain a sequence C. Arithmetic in base d also results in sequence X of carry elements. Then, we see:

Lemma 2. *Let A, B, C, X be number sequences, $d > 1$ be a positive integer such that following equations holds true:*

1. *When $i = 0$, $A[0] + C[0] + X[0] = B[0]$*
2. *When $i \geq 1$, $A[i] + C[i] + X[i] = B[i] + d \cdot X[i-1]$*

Then $DS(B, d) = DS(A, d) + DS(C, d)$.

Hence, to determine $DS(B, d) - DS(A, d)$, systematically guess sequences C and X using the equations, element-by-element beginning with the 0-th index and moving rightwards. There are two crucial observations here: (i) Computation of i-th element of C and X only depends on i-th and $(i-1)$-th elements of A and B. Therefore guessing $C[i]$ and $X[i]$ requires *finite memory* only. (ii) C refers to a representation of value $DS(B, d) - DS(A, d)$ in base d, and X is the carry-sequence. Hence if A and B are bounded-integer sequences, not only are X and C bounded sequences, they can be constructed from a *fixed finite set of integers*:

Lemma 3. *Let $d > 1$ be an integer discount-factor. Let A and B be nonnegative integer sequences bounded by μ s.t. $DS(A, d) < DS(B, d)$. Let C and X be as constructed in Lemma 2. There exists at least one pair of integer-sequences C and X that satisfy the following two equations*

1. *For all $i \geq 0$, $0 \leq C[i] \leq \mu \cdot \frac{d}{d-1}$. and*
2. *For all $i \geq 0$, $0 \leq |X[i]| \leq 1 + \frac{\mu}{d-1}$*

In Büchi automaton $\mathcal{A}_{\succ_{DS(d)}}$ (i) states are represented by (x, c) where x and c range over all possible elements of X and C, which are finite, (ii) a special start state s, (iii) transitions from the start state $(s, (a, b), (x, c))$ satisfy $a + c + x = b$ to replicate Eq. 1 (Lemma 2) at the 0-th index, (iv) all other transitions $((x_1, c_1), (a, b), (x_2, c_2))$ satisfy $a + c_2 + x_2 = b + d \cdot x_1$ to replicate Eq. 2 (Lemma 2) at indexes $i > 0$, and (v) all (x, c) states are accepting. Lemma 2 ensures that $\mathcal{A}_{\succ_{DS(d)}}$ accepts (A, B) iff $DS(B, d) = DS(A, d) + DS(C, d)$.

However, $\mathcal{A}_{\succ_{DS(d)}}$ is yet to guarantee $DS(C, d) > 0$. For this, we include non-accepting states $(x, \perp)$, where x ranges over all possible (finite) elements of X. Transitions into and out of states $(x, \perp)$ satisfy Eqs. 1 or 2 (depending on whether transition is from start state s) where $\perp$ is treated as $c = 0$. Transition from $(x, \perp)$-states to (x, c)-states occurs only if $c > 0$. Hence, any valid execution of (A, B) will be an accepting run only if the execution witnesses a non-zero value of c. Since C is a non-negative sequence, this ensures $DS(C, d) > 0$.

Construction. Let $\mu_C = \mu \cdot \frac{d}{d-1}$ and $\mu_X = 1 + \frac{\mu}{d-1}$. $\mathcal{A}_{\succ_{DS(d)}} = (S, \Sigma, \delta_d, \mathit{Init}, \mathcal{F})$

- $S = \mathit{Init} \cup \mathcal{F} \cup S_\perp$ where
  $\mathit{Init} = \{s\}$, $\mathcal{F} = \{(x, c) | |x| \leq \mu_X, 0 \leq c \leq \mu_C\}$, and
  $S_\perp = \{(x, \perp) | |x| \leq \mu_X\}$ where $\perp$ is a special character, and $c \in \mathbb{N}$, $x \in \mathbb{Z}$.
- $\Sigma = \{(a, b) : 0 \leq a, b \leq \mu\}$ where a and b are integers.
- $\delta_d \subset S \times \Sigma \times S$ is defined as follows:
  1. Transitions from start state s:
     i $(s, (a, b), (x, c))$ for all $(x, c) \in \mathcal{F}$ s.t. $a + x + c = b$ and $c \neq 0$

   ii $(s, (a, b), (x, \bot))$ for all $(x, \bot) \in S_\bot$ s.t. $a + x = b$
2. Transitions within $S_\bot$: $((x, \bot), (a, b), (x', \bot))$ for all $(x, \bot), (x', \bot) \in S_\bot$, if $a + x' = b + d \cdot x$
3. Transitions within $\mathcal{F}$: $((x, c), (a, b), (x', c'))$ for all $(x, c), (x', c') \in \mathcal{F}$ where $c' < d$, if $a + x' + c' = b + d \cdot x$
4. Transition between $S_\bot$ and $\mathcal{F}$: $((x, \bot), (a, b), (x', c'))$ for all $(x, \bot) \in S_\bot$, $(x', c') \in \mathcal{F}$ where $0 < c' < d$, if $a + x' + c' = b + d \cdot x$

Theorem 5. *The DS-comparator with maximum bound μ, is ω-regular for integer discount-factors $d > 1$. Size of the discounted-sum comparator is $\mathcal{O}(\frac{\mu^2}{d})$.*

DS-comparator with non-strict inequality $\leq$ and equality $=$ follow similarly. Consequently, properties of ω-regular comparators hold for DS-comparator with integer discount-factor. Specifically, DS-inclusion is PSPACE-complete in size of the input weighted automata and DS-comparator. Since, size of DS-comparator is polynomial w.r.t. to upper bound μ (in unary), DS-inclusion is PSPACE in size of input weighted automata and μ. Not only does this bound improve upon the previously known upper bound of EXPTIME but it also closes the gap between upper and lower bounds for DS-inclusion.

Corollary 1. *Given weighted automata P and Q, maximum weight on their transitions μ in unary form and integer discount-factor $d > 1$, the DS-inclusion, DS-strict-inclusion, and DS-equivalence problems are* PSPACE*-complete.*

As mentioned earlier, the known upper bound for discounted-sum inclusion with integer discount-factor is exponential [6,10]. This bound is based on an exponential determinization construction (subset construction) combined with arithmetical reasoning. We observe that the determinization construction can be performed on-the-fly in PSPACE. To perform, however, the arithmetical reasoning on-the-fly in PSPACE would require essentially using the same bit-level $((x, c)$-state) techniques that we have used to construct DS-comparator automata.

5 Limit-Average Comparator

The limit-average of an infinite sequence M is the point of convergence of the average of prefixes of M. Let $\mathsf{Sum}(M[0, n-1])$ denote the sum of the n-length prefix of sequence M. The *limit-average infimum*, denoted by $\mathsf{LimInfAvg}(M)$, is defined as $\lim\inf_{n\to\infty} \frac{1}{n} \cdot \mathsf{Sum}(M[0, n-1])$. Similarly, the *limit-average supremum*, denoted by $\mathsf{LimSupAvg}(M)$, is defined as $\lim\sup_{n\to\infty} \frac{1}{n} \cdot \mathsf{Sum}(M[0, n-1])$. The limit-average of sequence M, denoted by $\mathsf{LimAvg}(M)$, is defined *only if* the limit-average infimum and limit-average supremum coincide, and then $\mathsf{LimAvg}(M) = \mathsf{LimInfAvg}(M)$ $(= \mathsf{LimSupAvg}(M))$. Note that while limit-average infimum and supremum exist for all bounded sequences, the limit-average may not.

In existing work, limit-average is defined as the limit-average infimum (or limit-average supremum) to ensure that limit-average exists for all sequences [7,10,11,22]. While this definition is justified in context of the application, it may lead to a misleading comparison in some cases. For example, consider sequence A s.t. $\mathsf{LimSupAvg}(A) = 2$ and $\mathsf{LimInfAvg}(A) = 0$, and sequence B s.t. $\mathsf{LimAvg}(B) = 1$. Clearly, limit-average of A does not exist. Suppose, $\mathsf{LimAvg}(A) = \mathsf{LimInfAvg}(A) = 0$, then $\mathsf{LimAvg}(A) < \mathsf{LimAvg}(B)$, deluding that average of prefixes of A are always less than those of B in the limit. This is untrue since $\mathsf{LimSupAvg}(A) = 2$.

Such inaccuracies in limit-average comparison may occur when the limit-average of at least one sequence does not exist. However, it is not easy to distinguish sequences for which limit-average exists from those for which it doesn't.

We define *prefix-average comparison* as a relaxation of limit-average comparison. Prefix-average comparison coincides with limit-average comparison when limit-average exists for both sequences. Otherwise, it determines whether eventually the average of prefixes of one sequence are greater than those of the other. This comparison does not require the limit-average to exist to return intuitive results. Further, we show that the *prefix-average comparator* is ω-context-free.

5.1 Limit-Average Language and Comparison

Let $\Sigma = \{0, 1, \dots, \mu\}$ be a finite alphabet with $\mu > 0$. The *limit-average language* $\mathcal{L}_{LA}$ contains the sequence (word) $A \in \Sigma^\omega$ iff its limit-average exists. Suppose $\mathcal{L}_{LA}$ were ω-regular, then $\mathcal{L}_{LA} = \bigcup_{i=0}^{n} U_i \cdot V_i^\omega$, where $U_i, V_i \subseteq \Sigma^*$ are regular languages over *finite* words. The limit-average of sequences is determined by its behavior in the limit, so limit-average of sequences in V_i^ω exists. Additionally, the average of all (finite) words in V_i must be the same. If this were not the case, then two words in V_i with unequal averages l_1 and l_2, can generate a word $w \in V_i^\omega$ s.t the average of its prefixes oscillates between l_1 and l_2. This cannot occur, since limit-average of w exists. Let the average of sequences in V_i be a_i, then limit-average of sequences in V_i^ω and $U_i \cdot V_i^\omega$ is also a_i. This is contradictory since there are sequences with limit-average different from the a_i (see appendix). Similarly, since every ω-CFL is represented by $\bigcup_{i=1}^{n} U_i \cdot V_i^\omega$ for CFLs U_i, V_i over finite words [13], a similar argument proves that $\mathcal{L}_{LA}$ is not ω-context-free.

Quantifiers $\exists^\infty i$ and $\exists^f i$ denote the existence of *infinitely* many and *only finitely* many indices i, respectively.

Theorem 6. *$\mathcal{L}_{LA}$ is neither an ω-regular nor an ω-context-free language.*

In the next section, we will define *prefix-average comparison* as a relaxation of limit-average comparison. To show how prefix-average comparison relates to limit-average comparison, we will require the following two lemmas:

Lemma 4. *Let A and B be sequences s.t. their limit average exists. If $\exists^\infty i, \mathsf{Sum}(A[0, i-1]) \geq \mathsf{Sum}(B[0, i-1])$ then $\mathsf{LimAvg}(A) \geq \mathsf{LimAvg}(B)$.*

Lemma 5. *Let A, B be sequences s.t their limit-average exists. If $\mathsf{LimAvg}(A) > \mathsf{LimAvg}(B)$ then $\exists^f i, \mathsf{Sum}(B[0, i-1]) \geq \mathsf{Sum}(A[0, i-1])$ and $\exists^\infty i, \mathsf{Sum}(A[0, i-1]) > \mathsf{Sum}(B[0, i-1])$.*

5.2 Prefix-Average Comparison and Comparator

The previous section relates limit-average comparison with the sums of equal length prefixes of the sequences (Lemmas 4 and 5). The comparison criteria is based on the number of times sum of prefix of one sequence is greater than the other, which does not rely on the existence of limit-average. Unfortunately, this criteria cannot be used for limit-average comparison since it is incomplete (Lemma 5). Specifically, for sequences A and B with equal limit-average it is possible that $\exists^\infty i, \mathsf{Sum}(A[0,n-1]) > \mathsf{Sum}(B[0,n-1])$ and $\exists^\infty i, \mathsf{Sum}(B[0,n-1]) > \mathsf{Sum}(A[0,n-1])$. Instead, we use this criteria to define *prefix-average comparison*. In this section, we define prefix-average comparison and explain how it relaxes limit-average comparison. Lastly, we construct the prefix-average comparator, and prove that it is not ω-regular but is ω-context-free.

Definition 7 (Prefix-average comparison). *Let A and B be number sequences. We say* $\mathsf{PrefixAvg}(A) \geq \mathsf{PrefixAvg}(B)$ *if* $\exists^f i, \mathsf{Sum}(B[0,i-1]) \geq \mathsf{Sum}(A[0,i-1])$ *and* $\exists^\infty i, \mathsf{Sum}(A[0,i-1]) > \mathsf{Sum}(B[0,i-1])$.

Intuitively, prefix-average comparison states that $\mathsf{PrefixAvg}(A) \geq \mathsf{PrefixAvg}(B)$ if eventually the sum of prefixes of A are always greater than those of B. We use $\geq$ since the average of prefixes may be equal when the difference between the sum is small. It coincides with limit-average comparison when the limit-average exists for both sequences. Definition 7 and Lemmas 4, 5 relate limit-average comparison and prefix-average comparison:

Corollary 2. *When limit-average of A and B exists, then*

- $\mathsf{PrefixAvg}(A) \geq \mathsf{PrefixAvg}(B) \implies \mathsf{LimAvg}(A) \geq \mathsf{LimAvg}(B)$.
- $\mathsf{LimAvg}(A) > \mathsf{LimAvg}(B) \implies \mathsf{PrefixAvg}(A) \geq \mathsf{PrefixAvg}(B)$.

Therefore, limit-average comparison and prefix-average comparison return the same result on sequences for which limit-average exists. In addition, prefix-average returns intuitive results when even when limit-average may not exist. For example, suppose limit-average of A and B do not exist, but $\mathsf{LimInfAvg}(A) > \mathsf{LimSupAvg}(B)$, then $\mathsf{PrefixAvg}(A) \geq \mathsf{PrefixAvg}(B)$. Therefore, prefix-average comparison relaxes limit-average comparison.

The rest of this section describes *prefix-average comparator* $\mathcal{A}_{\succeq PA(\cdot)}$, an automaton that accepts the pair (A, B) of sequences iff $\mathsf{PrefixAvg}(A) \geq \mathsf{PrefixAvg}(B)$.

Lemma 6 ***(Pumping Lemma for ω-regular language [2]).*** *Let L be an ω-regular language. There exists $p \in \mathbb{N}$ such that, for each $w = u_1 w_1 u_2 w_2 \cdots \in L$ such that $|w_i| \geq p$ for all i, there are sequences of finite words $(x_i)_{i\in\mathbb{N}}$, $(y_i)_{i\in\mathbb{N}}$, $(z_i)_{i\in\mathbb{N}}$ s.t., for all i, $w_i = x_i y_i z_i$, $|x_i y_i| \leq p$ and $|y_i| > 0$ and for every sequence of pumping factors $(j_i)_{i\in\mathbb{N}} \in \mathbb{N}$, the pumped word $u_1 x_1 y_1^{j_1} z_1 u_2 x_2 y_2^{j_2} z_2 \cdots \in L$.*

Theorem 7. *The prefix-average comparator is not ω-regular.*

Proof (Proof Sketch). We use Lemma 6 to prove that $\mathcal{A}_{\succeq_{PA(\cdot)}}$ is not ω-regular. Suppose $\mathcal{A}_{\succeq_{PA(\cdot)}}$ were ω-regular. For $p > 0 \in \mathbb{N}$, let $w = (A, B) = ((0,1)^p(1,0)^{2p})^\omega$. The segment $(0,1)^*$ can be pumped s.t the resulting word is no longer in $\mathcal{L}_{\succeq_{PA(\cdot)}}$.

Concretely, $A = (0^p1^{2p})^\omega$, $B = (1^p0^{2p})^\omega$, $\mathsf{LimAvg}(A) = \frac{2}{3}$, $\mathsf{LimAvg}(B) = \frac{1}{3}$. So, $w = (A, B) \in \mathcal{A}_{\succeq_{PA(\cdot)}}$. Select as factor w_i (from Lemma 6) the sequence $(0,1)^p$. Pump each y_i enough times so that the resulting word is $\hat{w} = (\hat{A}, \hat{B}) = ((0,1)^{m_i}(1,0)^{2p})^\omega$ where $m_i > 4p$. It is easy to show that $\hat{w} = (\hat{A}, \hat{B}) \notin \mathcal{L}_{\succeq_{PA(\cdot)}}$.

We discuss key ideas and sketch the construction of the prefix average comparator. The term *prefix-sum difference at* i indicates $\mathsf{Sum}(A[0, i-1]) - \mathsf{Sum}(B[0, i-1])$, i.e. the difference between sum of i-length prefix of A and B.

Key Ideas. For sequences A and B to satisfy $\mathsf{PrefixAvg}(A) \geq \mathsf{PrefixAvg}(B)$, $\exists^f i, \mathsf{Sum}(B[0, i-1]) \geq \mathsf{Sum}(A[0, i-1])$ and $\exists^\infty i, \mathsf{Sum}(A[0, i-1]) > \mathsf{Sum}(B[0, i-1])$. This occurs iff there exists an index N s.t. for all indices $i > N$, $\mathsf{Sum}(A[0, i-1]) - \mathsf{Sum}(B[0, i-1]) > 0$. While reading a word, the prefix-sum difference is maintained by states and the stack of ω-PDA: states maintain whether it is negative or positive, while number of tokens in the stack equals its absolute value. The automaton non-deterministically guesses the aforementioned index N, beyond which the automaton ensure that prefix-sum difference remains positive.

Construction Sketch. The push-down comparator $\mathcal{A}_{\succeq_{PA(\cdot)}}$ consists of three states: (i) State s_P and (ii) State s_N that indicate that the prefix-sum difference is greater than zero and or not respectively, (iii) accepting state s_F. An execution of (A, B) begins in state s_N with an empty stack. On reading letter (a, b), the stack pops or pushes $|(a - b)|$ tokens from the stack depending on the current state of the execution. From state s_P, the stack pushes tokens if $(a - b) > 0$, and pops otherwise. The opposite occurs in state s_N. State transition between s_N and s_P occurs only if the stack action is to pop but the stack consists of $k < |a - b|$ tokens. In this case, stack is emptied, state transition is performed and $|a - b| - k$ tokens are pushed into the stack. For an execution of (A, B) to be an accepting run, the automaton non-deterministically transitions into state s_F. State s_F acts similar to state s_P except that execution is terminated if there aren't enough tokens to pop out of the stack. $\mathcal{A}_{\succeq_{PA(\cdot)}}$ accepts by accepting state.

To see why the construction is correct, it is sufficient to prove that at each index i, the number of tokens in the stack is equal to $|\mathsf{Sum}(A[0, i-1]) - \mathsf{Sum}(B[0, i-1])|$. Furthermore, in state s_N, $\mathsf{Sum}(A[0, i-1]) - \mathsf{Sum}(B[0, i-1]) \leq 0$, and in state s_P and s_F, $\mathsf{Sum}(A[0, i-1]) - \mathsf{Sum}(B[0, i-1]) > 0$. Next, the index at which the automaton transitions to the accepting state s_F coincides with index N. The execution is accepted if it has an infinite execution in state s_F, which allows transitions only if $\mathsf{Sum}(A[0, i-1]) - \mathsf{Sum}(B[0, i-1]) > 0$.

Theorem 8. *The prefix-average comparator is an* ω*-CFL.*

While ω-CFL can be easily expressed, they do not possess closure properties, and problems on ω-CFL are easily undecidable. Hence, the application of ω-context-free comparator will require further investigation.

6 Conclusion

In this paper, we identified a novel mode for comparison in quantitative systems: the online comparison of aggregate values of sequences of quantitative weights. This notion is embodied by comparators automata that read two infinite sequences of weights synchronously and relate their aggregate values. We showed that ω-regular comparators not only yield generic algorithms for problems including quantitative inclusion and winning strategies in incomplete-information quantitative games, they also result in algorithmic advances. We show that the discounted-sum inclusion problem is PSAPCE-complete for integer discount-factor, hence closing a complexity gap. We also studied the discounted-sum and prefix-average comparator, which are ω-regular and ω-context-free, respectively.

We believe comparators, especially ω-regular comparators, can be of significant utility in verification and synthesis of quantitative systems, as demonstrated by the existence of finite-representation of counterexamples of the quantitative inclusion problem. Another potential application is computing equilibria in quantitative games. Applications of the prefix-average comparator, in general ω-context-free comparators, is open to further investigation. Another direction to pursue is to study aggregate functions in more detail, and develop a clearer understanding of when aggregate functions are ω-regular.

Acknowledgements. We thank the anonymous reviewers for their comments. We thank K. Chatterjee, L. Doyen, G. A. Perez and J. F. Raskin for corrections to earlier drafts, and their contributions to this paper. We thank P. Ganty and R. Majumdar for preliminary discussions on the limit-average comparator. This work was partially supported by NSF Grant No. 1704883, "Formal Analysis and Synthesis of Multiagent Systems with Incentives".

References

1. Almagor, S., Boker, U., Kupferman, O.: What's decidable about weighted automata? In: Bultan, T., Hsiung, P.-A. (eds.) ATVA 2011. LNCS, vol. 6996, pp. 482–491. Springer, Heidelberg (2011). https://doi.org/10.1007/978-3-642-24372-1_37
2. Alur, R., Degorre, A., Maler, O., Weiss, G.: On omega-languages defined by mean-payoff conditions. In: de Alfaro, L. (ed.) FoSSaCS 2009. LNCS, vol. 5504, pp. 333–347. Springer, Heidelberg (2009). https://doi.org/10.1007/978-3-642-00596-1_24
3. Andersen, G., Conitzer, V.: Fast equilibrium computation for infinitely repeated games. In: Proceedings of AAAI, pp. 53–59 (2013)
4. Andersson, D.: An improved algorithm for discounted payoff games. In: ESSLLI Student Session, pp. 91–98 (2006)
5. Baier, C., Katoen, J.-P., et al.: Principles of Model Checking. MIT Press, Cambridge (2008)
6. Boker, U., Henzinger, T.A.: Exact and approximate determinization of discounted-sum automata. LMCS **10**(1) (2014)
7. Brim, L., Chaloupka, J., Doyen, L., Gentilini, R., Raskin, J.-F.: Faster algorithms for mean-payoff games. Formal Methods Syst. Des. **38**(2), 97–118 (2011)

8. Chatterjee, K., Doyen, L.: Energy parity games. In: Abramsky, S., Gavoille, C., Kirchner, C., Meyer auf der Heide, F., Spirakis, P.G. (eds.) ICALP 2010, Part II. LNCS, vol. 6199, pp. 599–610. Springer, Heidelberg (2010). https://doi.org/10.1007/978-3-642-14162-1_50
9. Chatterjee, K., Doyen, L., Henzinger, T.A.: Expressiveness and closure properties for quantitative languages. In: Proceedings of LICS, pp. 199–208. IEEE (2009)
10. Chatterjee, K., Doyen, L., Henzinger, T.A.: Quantitative languages. Trans. Comput. Log. **11**(4), 23 (2010)
11. Chatterjee, K., Henzinger, T.A., Jurdzinski, M.: Mean-payoff parity games. In: Proceedings of LICS, pp. 178–187. IEEE (2005)
12. Chaudhuri, S., Sankaranarayanan, S., Vardi, M.Y.: Regular real analysis. In: Proceedings of LICS, pp. 509–518 (2013)
13. Cohen, R.S., Gold, A.Y.: Theory of ω-languages: characterizations of ω-context-free languages. J. Comput. Syst. Sci. **15**(2), 169–184 (1977)
14. de Alfaro, L., Faella, M., Henzinger, T.A., Majumdar, R., Stoelinga, M.: Model checking discounted temporal properties. In: Jensen, K., Podelski, A. (eds.) TACAS 2004. LNCS, vol. 2988, pp. 77–92. Springer, Heidelberg (2004). https://doi.org/10.1007/978-3-540-24730-2_6
15. de Alfaro, L., Faella, M., Stoelinga, M.: Linear and branching metrics for quantitative transition systems. In: Díaz, J., Karhumäki, J., Lepistö, A., Sannella, D. (eds.) ICALP 2004. LNCS, vol. 3142, pp. 97–109. Springer, Heidelberg (2004). https://doi.org/10.1007/978-3-540-27836-8_11
16. Degorre, A., Doyen, L., Gentilini, R., Raskin, J.-F., Toruńczyk, S.: Energy and mean-payoff games with imperfect information. In: Dawar, A., Veith, H. (eds.) CSL 2010. LNCS, vol. 6247, pp. 260–274. Springer, Heidelberg (2010). https://doi.org/10.1007/978-3-642-15205-4_22
17. Droste, M., Kuich, W., Vogler, H.: Handbook of Weighted Automata. Springer, Heidelberg (2009). https://doi.org/10.1007/978-3-642-01492-5
18. Karp, R.M.: A characterization of the minimum cycle mean in a digraph. Discret. Math. **23**(3), 309–311 (1978)
19. Kupferman, O., Vardi, M.Y.: Synthesis with incomplete informatio. In: Barringer, H., Fisher, M., Gabbay, D., Gough, G. (eds.) Advances in Temporal Logic, pp. 109–127. Springer, Dordrecht (2000). https://doi.org/10.1007/978-94-015-9586-5_6
20. Mohri, M.: Weighted automata algorithms. In: Mohri, M. (ed.) Handbook of Weighted Automata, pp. 213–254. Springer, Heidelberg (2009). https://doi.org/10.1007/978-3-642-01492-5_6
21. Grädel, E., Thomas, W., Wilke, T. (eds.): Automata Logics, and Infinite Games: A Guide to Current Research. LNCS, vol. 2500. Springer, Heidelberg (2002). https://doi.org/10.1007/3-540-36387-4
22. Zwick, U., Paterson, M.: The complexity of mean payoff games on graphs. Theor. Comput. Sci. **158**(1), 343–359 (1996)

Logics and Equational Theories

Modular Tableaux Calculi for Separation Theories

Simon Docherty[1(✉)] and David Pym[1,2]

[1] University College London, London, UK
{simon.docherty.14,d.pym}@ucl.ac.uk
[2] The Alan Turing Institute, London, UK

Abstract. In recent years, the key principles behind Separation Logic have been generalized to generate formalisms for a number of verification tasks in program analysis via the formulation of 'non-standard' models utilizing notions of separation distinct from heap disjointness. These models can typically be characterized by a *separation theory*, a collection of first-order axioms in the signature of the model's underlying ordered monoid. While all separation theories are interpreted by models that instantiate a common mathematical structure, many are undefinable in Separation Logic and determine different classes of valid formulae, leading to incompleteness for existing proof systems. Generalizing systems utilized in the proof theory of bunched logics, we propose a framework of tableaux calculi that are generically extendable by rules that correspond to separation theories axiomatized by coherent formulas. This class covers all separation theories in the literature—for both classical and intuitionistic Separation Logic—as well as axioms for a number of related formalisms appropriate for reasoning about complex systems, security, and concurrency. Parametric soundness and completeness of the framework is proved by a novel representation of tableaux systems as coherent theories, suggesting a strategy for implementation and a tentative first step towards a new logical framework for non-classical logics.

Keywords: Bunched logic · Coherent logic · Kripke semantics
Proof theory · Separation logic · Separation theories
Substructural logic · Tableaux

1 Introduction

Separation Logic [39], introduced by Ishtiaq and O'Hearn [32], Reynolds [44], Yang and O'Hearn [50], is a Hoare-style program logic suitable for reasoning about programs that mutate data structures. In its original formulation, the assertion language of Separation Logic is based on a model of O'Hearn and Pym's logic of bunched implications [40] formulated by considering heaps as possible worlds with internal structure that allows their decomposition into separate pieces of memory. This decomposition is witnessed in the logic by the

C. Baier and U. Dal Lago (Eds.): FOSSACS 2018, LNCS 10803, pp. 441–458, 2018.
https://doi.org/10.1007/978-3-319-89366-2_24

separating conjunction $*$, with $\phi * \psi$ informally read as 'the heap can be split into *separate* parts; one satisfying ϕ and the other satisfying ψ'.

Calcagno et al. [13] abstract the details of the heap model to a structure called a *separation algebra*, a partial-deterministic and cancellative monoid model of the Boolean logic of bunched implications (BBI), which can be used to generate bespoke separation logics suitable for program analysis tasks beyond that of the original formalism. Conflicting definitions of separation algebra have since been given by adding/removing first-order properties or strengthening/weakening the monoid properties [10,14,21,24]. These mutually exclusive definitions can be encompassed in a framework of *separation theories* [10], collections of first-order axioms (*separation properties*) common to separation logic models which the definition of (B)BI model can be extended by. All separation logics in the literature can be seen to be models of separation theories, while the frameworks Views [21] and Iris [33] explicitly implement the idea of generating program logics parametrically by separation theory.

Recent work has revealed an expressivity gap between the logic of bunched implications and common separation theories in the literature, however. Brotherston and Villard [10], Larchey-Wendling and Galmiche [36] show that separation properties like indivisibility of units and partial deterministic composition determine distinct sets of valid BBI formulae, leading to the incompleteness of standard proof systems with respect to typical classes of memory models. To make matters worse, Brotherston and Villard additionally show that many separation properties (among them partial determinism) are undefinable in BBI, and thus cannot be axiomatized by the logic. These results also hold for BI, the intuitionistic logic of bunched implications. This is an increasingly relevant issue given the growing number of intuitionistic separation logics, most prominent amongst them Iris, a framework that utilizes a 'later' modality [37] that can only be nontrivially defined in intuitionistic systems.

This expressivity gap is a significant problem for Separation Logic. A theorem prover for deriving assertions satisfied by the underlying model is a necessary component of any implementation of a separation logic, with the deployable proof theory of the standard formalism crucial for its scalability to large code bases [12,50]. Standard implementations are model-specific, however, and only suitable for the heap model. In order to account for the large numbers of bespoke separation logics, as well as Views/Iris-style frameworks, we require tools that support parametrization by separation theory.

Technical Approach. The present work generalizes methods pioneered on tableaux systems for a range of logics including and related to BI and BBI [20,22,28,34] to specify modular tableaux calculi for the breadth of separation theories in the literature, proved sound and complete uniformly and parametrically in choice of separation theory. While previous systems implicitly implement a systematic method for constructing tableaux proof theory for bunched logics, subtle but significant changes must be made to additionally capture separation theories. Past systems can be formulated as particular instances of our framework, thus making the systematic method explicit.

First, we specify tableaux proof systems for BI and BBI, the propositional basis for Separation Logic. The key difference between our calculi and tableaux systems previously given in the literature is that we do not outsource any part of the derivation of proofs to an algebra of labels or auxilliary proof system for constraints. Instead, we utilize *frame expansion rules* that are of the same form as the standard *logical expansion rules* of the system. These rules capture the same structural properties (and more) but can also be added/removed in a modular fashion. Crucially, this ensures separation properties—for example, partial determinism—are not hard-coded into the basic systems via the structure of labels, and facilitates the parametricity of our completeness theorem.

We extend these systems with a rule schema for separation properties that are axiomatized by *coherent formulae*; a subset of first-order formulae with a special syntactic form. This set contains every separation property that can be found in the literature and is expressive enough to include virtually any axiom that might be utilized in future. The strength of this statement can be justified by a folklore result recently reconstructed by Dyckhoff and Negri [25] that shows that *every* first-order axiom can be reconstructed as an equivalent system of coherent formulae. We thus obtain a modular framework of $(B)BI + \Sigma$-tableaux systems, where Σ is an arbitrary collection of coherent axioms.

In order to prove soundness and completeness of the system, we utilize a novel representation of labelled tableaux systems as theories of coherent logic. The key insight here is that the translation of coherent formulae into tableaux rules is not one way: tableaux rules can naturally be seen as coherent formulae in a signature augmented with special predicate symbols. The parametric soundness and completeness of the framework can then be reduced to proving the soundness and completeness of Tarskian truth for coherent logic with respect to a meta-tableaux method, a problem positively resolved by Bezem and Coquand [4]. To our knowledge, the application of this technique to labelled tableaux is new, although, in the aforementioned work, Bezem and Coquand show how to encode the tableaux method for first-order classical logic as a coherent theory, and trace the idea of abbreviating formulae with predicate symbols to Skolem [47].

Contributions. We identify three principal contributions.

1. A sound and complete proof theory for the full breadth of separation theories in the literature. Notably, this includes the first proof theoretic treatment of separation theories for intuitionistic Separation Logic.
2. A new technique for constructing proof systems for essentially any logic interpreted on Kripke structures that are axiomatized by coherent theories.
3. The identification of tableaux systems with theories of coherent logic.

On points 2 and 3, we believe many tableaux systems in the literature are subsumed by this method, with their respective 'Hintikka set' completeness proofs actually localized instances of the parametric completeness theorem given here. This suggests the possibility of a logical framework for non-classical logics via the representation of tableaux systems as coherent theories. This may be related to Schmitt and Tishkovsky's [45] technique for automatically synthesising tableaux calculi for logics that can be presented as first-order theories in a particular form. We believe the "rule refinement" post-processing their tableau rules undergo after synthesis can be made redundant by instead synthesising from coherent theories, but we defer such an investigation to another occasion.

Related Work. While much work has been done on the proof theory of BI and BBI [9,28,29,41], as well as proof systems for the concrete heap model of Separation Logic [5,27,30], very little exists for separation theories. A key exception to this is Hóu et al.'s [31] labelled sequent calculi for propositional abstract separation logic. There, a labelled sequent calculus for BBI is extended with rules corresponding to the most common separation properties – *partial determinism, cancellativity, indivisible unit* and *disjointness* – and completeness and cut elimination is proved. In Hóu's PhD dissertation [29] the properties *cross-split* and *splittability* are additionally handled, although completeness for these new rules requires 'non-trivial changes' to the previous proofs.

The classes of model captured by our systems strictly extend those of Hóu et al. [31]—in particular, by additionally considering classes of BI models that are appropriate for intuitionistic separation logics—and our calculi are proved complete uniformly. Our systems are also generically extendable according to a rule schema, meaning the framework should be suitable for new separation theories devised in the future. A deficiency of our approach with respect to Hóu et al.'s is a lack of implementation, though we note that the representation of our systems as theories of coherent logic suggests off-the-shelf coherent logic provers (cf. [43]) could be used to give naive implementations of our framework.

Brotherston and Villard [10] deal with the undefinability of separation theories by defining a conservative extension of BBI called HyBBI, extending the syntax with nominals, satisfaction operators and binders. This extra expressivity leads to the axiomatizability of the undefinable separation properties. This work is not specifically concerned with proof theory, giving only a Hilbert-style system for HyBBI, and has the defect of requiring modifications to the syntax of Separation Logic. In addition, a significant theoretical reformulation would be required to capture intuitionistic separation theories this way. In contrast, in our work the necessary machinery is internalized within the proof system and both Boolean and intuitionistic cases are taken care of uniformly.

Finally, we connect our work to a line of research in proof theory investigating the generation of proof rules from coherent theories. Simpson [46] and Braüner [8] have used this technique to produce natural deduction rules, while Negri [38] has extensively developed it to generate (systems of) labelled sequent rules from frame conditions axiomatized by (generalized) coherent formulae. To our knowledge the present work is the first application of these ideas to the tableaux method. In addition, we believe the encoding of the proof systems themselves as coherent theories is novel.

2 Preliminaries

The Logics of Bunched Implications. We first recall O'Hearn and Pym's *logics of bunched implications* BI and BBI [40], the propositional basis of Separation Logic's assertion language. BI and BBI are archetypal examples of *bunched logics*; systems given by combining the standard *additives* of classical or intutionistic propositional logic with the *multiplicatives* of a substructural logic. This idea has been developed to give logics for reasoning about concurrency [23] and the layering structure of complex systems [17,18,22], Hennessey-Milner-style process logics for reasoning about security and systems modelling [1,19] and modal and epistemic systems for reasoning about reachability/knowledge subject to the availability of resources [20,26].

Let Prop be a set of atomic propositions, ranged over by p. The set of all formulae of (B)BI is generated by the following grammar:

$$\phi ::= \mathrm{p} \mid \top \mid \bot \mid \mathrm{I} \mid \phi \wedge \phi \mid \phi \vee \phi \mid \phi \rightarrow \phi \mid \phi * \phi \mid \phi \mathrel{-\!\!*} \phi.$$

For BI, the standard connectives are interpreted intuitionistically; in BBI, classically. Negation is defined by $\neg\phi := \phi \rightarrow \bot$. Figure 1 gives Hilbert rules for the multiplicative fragment of the logics.

$$\frac{\xi \vdash \phi \quad \eta \vdash \psi}{\xi * \eta \vdash \phi * \psi} \qquad \frac{\eta * \phi \vdash \psi}{\eta \vdash \phi \mathrel{-\!\!*} \psi} \qquad \frac{\xi \vdash \phi \mathrel{-\!\!*} \psi \quad \eta \vdash \phi}{\xi * \eta \vdash \psi}$$

$$\frac{}{(\phi * \psi) * \xi \vdash \phi * (\psi * \xi)} \qquad \frac{}{\phi * \psi \vdash \psi * \phi} \qquad \frac{}{\phi * \mathrm{I} \dashv\vdash \phi}$$

Fig. 1. Rules for the multiplicative fragment of (B)BI.

A *BI frame* is given by a tuple $\mathcal{X} = (X, \leq, \circ, E)$, where $(X, \leq)$ is a partial order, $\circ : X^2 \rightarrow \mathcal{P}(X)$ a binary composition (where $\mathcal{P}(X)$ denotes the power set of X) and $E \subseteq X$ a set of units for $\circ$. This structure must satisfy the following axioms, where the outermost universal quantification is left implicit:

(Comm) $z \in x \circ y \rightarrow z \in y \circ x$ (Up) $e \in E \wedge e \leq e' \rightarrow e' \in E$
(Unit 1) $\exists e \in E(x \in x \circ e)$ (Unit 2) $x \in y \circ e \wedge e \in E \rightarrow y \leq x$
(Assoc) $t' \geq t \in x \circ y \wedge w \in t' \circ z \rightarrow \exists s, s', w'(s' \geq s \in y \circ z \wedge w \geq w' \in x \circ s')$.

The axioms formalize intuitive ideas about the composition of generic resources; for example, that the composition satisfies a generalized associativity that is compatible with the comparison order. This analysis is known as *resource semantics*.

A sound interpretation of BI is given by extending the standard poset semantics for propositional intuitionistic logic. This requires a *persistent* valuation: a map $\mathcal{V} : \mathrm{Prop} \rightarrow \mathcal{P}(X)$ such that $x \in \mathcal{V}(\mathrm{p})$ and $x \leq y$ entail $y \in \mathcal{V}(\mathrm{p})$. We call a BI frame $\mathcal{X}$ together with a persistent valuation $\mathcal{V}$ a *Kripke BI model*. The satisfaction relation $\vDash_{\mathcal{V}}$ is given in Fig. 2. As is standard for intuitionistic logics, persistence extends to all formulae of BI. *Kripke BBI models* and their

$r \vDash \mathrm{p}$	iff $r \in \mathcal{V}(\mathrm{p})$	$r \vDash \top$			$r \nvDash \bot$
$r \vDash \phi \wedge \psi$	iff $r \vDash \phi$ and $r \vDash \psi$	$r \vDash \phi \vee \psi$	iff	$r \vDash \phi$ or $r \vDash \psi$	
$r \vDash \phi \rightarrow \psi$	iff for all $r' \geq r, r' \vDash \phi$ implies $r' \vDash \psi$;	$r \vDash \mathrm{I}$	iff	$r \in E$	
$r \vDash \phi * \psi$	iff there exists r', s, t such that $r \geq r' \in s \circ t$, $s \vDash \phi$ and $t \vDash \psi$				
$r \vDash \phi \mathrel{-\!\!*} \psi$	iff for all r', s, t: $r \leq r'$, $t \in r' \circ s$ and $s \vDash \phi$ implies $t \vDash \psi$				

Fig. 2. Satisfaction for **(B)BI**. **BBI** is the case where $\leq$ is substituted with $=$.

associated semantics are given by the special case of the definitions for BI when the partial order $\leq$ is equality.

Coherent Logic. Coherent logic is the fragment of first-order logic consisting of formulae of the form $A_1(\vec{x}) \wedge \cdots \wedge A_n(\vec{x}) \rightarrow \exists \vec{y_1} B_1(\vec{x}, \vec{y_1}) \vee \cdots \vee \exists \vec{y_m} B_m(\vec{x}, \vec{y_m})$, for $n, m \geq 0$, where each A_i is an atomic formula involving only variables from the vector $\vec{x}$, and each B_i is the conjunction of atomic formulae involving only variables from the vectors $\vec{x}$ and $\vec{y_i}$. In a coherent formula, the variables $\vec{x}$ are implicitly universally quantified (with scope the whole formula) and both $\vec{x}$ and $\vec{y_i}$ may be empty. The case $n = 0$ is a consequent that is always true—$\top \rightarrow \exists \vec{y_1} B_1(\vec{x}, \vec{y_1}) \vee \cdots \vee \exists \vec{y_m} B_m(\vec{x}, \vec{y_m})$—similarly, the case $m = 0$ is an antecedent that is always false: $A_1(\vec{x}) \wedge \cdots \wedge A_n(\vec{x}) \rightarrow \bot$.

This fragment of first-order logic is sometimes referred to as *geometric logic*; however, we reserve this name for the generalization of the definition given here that permits the consequent to be an *infinite* disjunction. In turn, coherent logic generalizes—via the case $m = 1$ with empty $\vec{y_1}$—the *Horn clause* fragment of first-order logic utilized in logic programming and first-order theorem provers based on the resolution method.

We call a set of coherent formulae Φ a *coherent theory.* Models of coherent theories are given in a way standard for first-order logic: a *Tarskian model of* Φ is a non-empty set X together with an interpretation $\mathcal{I}$, which assigns to every n-ary relation symbol R in the signature a set $R^{\mathcal{I}} \subseteq X^n$ such that for each coherent formulae in Φ, for all $\vec{x} \in X$, the consequent $\exists \vec{y_1} \in X(B^{\mathcal{I}}(\vec{x}, \vec{y_1})) \vee \cdots \vee \exists \vec{y_m} \in X(B^{\mathcal{I}}(\vec{x}, \vec{y_m}))$ is true whenever the antecedent $A_1^{\mathcal{I}}(\vec{x}) \wedge \cdots \wedge A_n^{\mathcal{I}}(\vec{x})$ is true.

Many common mathematical structures are axiomatized by coherent theories. For example, algebraic structures like groups, rings, lattices, and fields, as well as total, partial, and linear orders. Further examples are found in the theory of confluence for term rewriting systems [4,48]. Of interest for our purposes, (B)BI frames are axiomatized by coherent theories. As we will see, every known separation property is given directly as a coherent axiom, with the exception of Splittability, which can be rewritten as a coherent theory.

3 Modular Tableaux Calculi for Separation Theories

The Base Tableaux Systems. We begin with tableaux systems designed for the semantics of (B)BI as outlined in Sect. 2. As is standard for tableaux systems,

Logical expansion rules

$$\langle \mathbb{T}\wedge \rangle \quad \frac{\mathbb{T}\phi \wedge \psi : x \in \mathcal{F}}{\langle \{\mathbb{T}\phi : x, \mathbb{T}\psi : x\}, \emptyset \rangle} \qquad \langle \mathbb{F}\wedge \rangle \quad \frac{\mathbb{F}\phi \wedge \psi : x \in \mathcal{F}}{\langle \{\mathbb{F}\phi : x\}, \emptyset \rangle \mid \langle \{\mathbb{F}\psi : x\}, \emptyset \rangle}$$

$$\langle \mathbb{T}\vee \rangle \quad \frac{\mathbb{T}\phi \vee \psi : x \in \mathcal{F}}{\langle \{\mathbb{T}\phi : x\}, \emptyset \rangle \mid \langle \{\mathbb{T}\psi : x\}, \emptyset \rangle} \qquad \langle \mathbb{F}\vee \rangle \quad \frac{\mathbb{F}\phi \vee \psi : x \in \mathcal{F}}{\langle \{\mathbb{F}\phi : x, \mathbb{F}\psi : x\}, \emptyset \rangle}$$

$$\langle \mathbb{T}\mathrm{I} \rangle \quad \frac{\mathbb{T}\mathrm{I} : x \in \mathcal{F}}{\langle \emptyset, \{Ex\} \rangle}$$

Frame expansion rules

$$\langle \mathrm{Ref} \rangle \quad \frac{Expr(x) \in \mathcal{C} \cup \mathcal{F}}{\langle \emptyset, \{x \sim x\} \rangle} \qquad \langle \mathrm{Trans} \rangle \quad \frac{x \sim y, y \sim z \in \mathcal{C}}{\langle \emptyset, \{x \sim z\} \rangle}$$

$$\langle \mathrm{Cong} \rangle \quad \frac{x \sim y, y \sim x, Expr(x) \in \mathcal{C}}{\langle \emptyset, \{Expr(y/x)\} \rangle} \qquad \langle \mathrm{Comm} \rangle \quad \frac{R_* xyz \in \mathcal{C}}{\langle \emptyset, \{R_* yxz\} \rangle}$$

$$\langle \mathrm{Unit\ 1} \rangle \quad \frac{Expr(x) \in \mathcal{F} \cup \mathcal{C}}{\langle \emptyset, \{Ec_i, R_* x c_i x\} \rangle} \qquad \langle \mathrm{Unit\ 2} \rangle \quad \frac{R_* xyz, Ey \in \mathcal{C}}{\langle \emptyset, \{x \sim z\} \rangle}$$

with c_i a fresh label and $Expr(x)$ any expression in which x occurs.

Fig. 3. Shared rules for the tableaux systems.

derivations in our calculi are implicit attempts to construct a countermodel for the formula ϕ to be proved. This is done via the derivation of syntactic expressions that give partial specifications of a (B)BI model that can be realized as a real model if the formula is invalid. If every possible countermodel construction (i.e., every branch of a tableau) results in a contradiction, then we may conclude that no countermodel exists and call such a tableau a proof of ϕ.

The calculi work with two types of syntactic expression. First we have *labelled formulae* $\mathbb{S}\phi : x$, given by a sign $\mathbb{S} \in \{\mathbb{T}, \mathbb{F}\}$ together with a (B)BI formula ϕ and a *label* $x \in \{c_i \mid i \in \mathbb{N}\}$. A labelled formula states that a (B)BI formula ϕ is true ($\mathbb{T}$) or false ($\mathbb{F}$) at the state represented by the label x. The other type are called *constraints*, and encode a partial specification of the structure of a (B)BI frame. For labels $x, y, z \in \{c_i \mid i \in \mathbb{N}\}$, a constraint is an expression of the form $x \sim y$, $R_* xyz$ or Ex, corresponding to the state represented by x being $\leq$ that represented by y, the state represented by z being a composition of those represented by x and y, or the state represented by x being a unit, respectively.

Unlike other bunched logic tableaux systems, we only utilize atomic labels, as opposed to a monoidal algebra of labels that encodes properties of the multiplicative connectives. New constraints are derived only by *frame expansion rules* (which directly reflect the axioms that define (B)BI frames and equality), rather than through the properties of a label algebra and a separate proof system for constraints. A *constrained set of statements* (CSS) is a pair $\langle \mathcal{F}, \mathcal{C} \rangle$, where $\mathcal{F}$ is a set of labelled formulae and $\mathcal{C}$ is a set of constraints. It is finite if $\mathcal{F}$ and $\mathcal{C}$ are.

Informally, tableaux are trees annotated with finite CSSs. Each branch determines a CSS $\langle \mathcal{F}, \mathcal{C} \rangle$ where $\mathcal{F}$ (respectively $\mathcal{C}$) is the union of the formula (constraint) sets that occur on the branch. Figures 3 and 4 give rules dictating the expansion of tableaux: Fig. 3 gives rules shared by both the BI and BBI systems, while Fig. 4 gives rules exclusive to each system. While c_i, c_j, c_k denote concrete fresh labels, x, y, z etc. are *label variables*. An instance of a rule is triggered for a branch CSS when a concrete substitution instance of the premiss holds of it, and the same label substitutions carry through to the (branching) CSS(s) that the conclusion dictates are added to the tree. We now define (B)BI tableaux formally, with $\oplus$ giving concatenation of lists.

Logical expansion rules for BI

$$\langle \mathbb{T}\rightarrow\rangle\ \frac{\mathbb{T}\phi \rightarrow \psi : x \in \mathcal{F} \text{ and } x \sim y \in \mathcal{C}}{\langle\{\mathbb{F}\phi : y\}, \emptyset\rangle \mid \langle\{\mathbb{T}\psi : y\}, \emptyset\rangle} \qquad \langle \mathbb{F}\rightarrow\rangle\ \frac{\mathbb{F}\phi \rightarrow \psi : x \in \mathcal{F}}{\langle\{\mathbb{T}\phi : c_i, \mathbb{F}\psi : c_i\}, \{x \sim c_i\}\rangle}$$

$$\langle \mathbb{T}*\rangle\ \frac{\mathbb{T}\phi * \psi : x \in \mathcal{F}}{\langle\{\mathbb{T}\phi : c_i, \mathbb{T}\psi : c_j\}, \{R_* c_i c_j c_k, c_k \sim x\}\rangle} \qquad \langle \mathbb{F}*\rangle\ \frac{\mathbb{F}\phi * \psi : x \in \mathcal{F} \text{ and } R_* yzw, w \sim x \in \mathcal{C}}{\langle\{\mathbb{F}\phi : y\}, \emptyset\rangle \mid \langle\{\mathbb{F}\psi : z\}, \emptyset\rangle}$$

$$\langle \mathbb{T}{-\!*}\rangle\ \frac{\mathbb{T}\phi -\!* \psi : x \in \mathcal{F} \text{ and } x \sim w, R_* wyz \in \mathcal{C}}{\langle\{\mathbb{F}\phi : y\}, \emptyset\rangle \mid \langle\{\mathbb{T}\psi : z\}, \emptyset\rangle} \qquad \langle \mathbb{F}{-\!*}\rangle\ \frac{\mathbb{F}\phi -\!* \psi : x \in \mathcal{F}}{\langle\{\mathbb{T}\phi : c_j, \mathbb{F}\psi : c_k\}, \{x \sim c_i, R_* c_i c_j c_k\}\rangle}$$

Frame expansion rules for BI

$$\langle \text{Assoc}\rangle\ \frac{t \sim t', R_* xyt, R_* t'zw \in \mathcal{C}}{\langle\emptyset, \{c_i \sim c_j, c_k \sim w, R_* yzc_i, R_* xc_jc_k\}\rangle} \qquad \langle \text{Up}\rangle\ \frac{Ex, x \sim y \in \mathcal{C}}{\langle\emptyset, \{Ey\}\rangle}$$

Logical expansion rules for BBI

$$\langle \mathbb{T}\neg\rangle\ \frac{\mathbb{T}\neg\phi : x \in \mathcal{F}}{\langle\{\mathbb{F}\phi : x\}, \emptyset\rangle} \qquad \langle \mathbb{F}\neg\rangle\ \frac{\mathbb{F}\neg\phi : x \in \mathcal{F}}{\langle\{\mathbb{T}\phi : x\}, \emptyset\rangle}$$

$$\langle \mathbb{T}\rightarrow\rangle\ \frac{\mathbb{T}\phi \rightarrow \psi : x \in \mathcal{F}}{\langle\{\mathbb{F}\phi : x\}, \emptyset\rangle \mid \langle\{\mathbb{T}\psi : x\}, \emptyset\rangle} \qquad \langle \mathbb{F}\rightarrow\rangle\ \frac{\mathbb{F}\phi \rightarrow \psi : x \in \mathcal{F}}{\langle\{\mathbb{T}\phi : x, \mathbb{F}\psi : x\}, \emptyset\rangle}$$

$$\langle \mathbb{T}*\rangle\ \frac{\mathbb{T}\phi * \psi : x \in \mathcal{F}}{\langle\{\mathbb{T}\phi : c_i, \mathbb{T}\psi : c_j\}, \{R_* c_i c_j x\}\rangle} \qquad \langle \mathbb{F}*\rangle\ \frac{\mathbb{F}\phi * \psi : x \in \mathcal{F} \text{ and } R_* yzx \in \mathcal{C}}{\langle\{\mathbb{F}\phi : y\}, \emptyset\rangle \mid \langle\{\mathbb{F}\psi : z\}, \emptyset\rangle}$$

$$\langle \mathbb{T}{-\!*}\rangle\ \frac{\mathbb{T}\phi -\!* \psi : x \in \mathcal{F} \text{ and } R_* xyz \in \mathcal{C}}{\langle\{\mathbb{F}\phi : y\}, \emptyset\rangle \mid \langle\{\mathbb{T}\psi : z\}, \emptyset\rangle} \qquad \langle \mathbb{F}{-\!*}\rangle\ \frac{\mathbb{F}\phi -\!* \psi : x \in \mathcal{F}}{\langle\{\mathbb{T}\phi : c_i, \mathbb{F}\psi : c_j\}, \{R_* x c_i c_j\}\rangle}$$

Frame expansion rules for BBI

$$\langle \text{Assoc}\rangle\ \frac{R_* xyt, R_* tzw \in \mathcal{C}}{\langle\emptyset, \{R_* yzc_i, R_* xc_iw\}\rangle} \qquad \langle \text{Sym}\rangle\ \frac{x \sim y \in \mathcal{C}}{\langle\emptyset, \{y \sim x\}\rangle}$$

with c_i, c_j, c_k fresh labels, $Expr(x)$ any expression in which x occurs.

Fig. 4. Tableaux rules for (B)BI

Definition 1 (Tableau). *A* (B)BI tableau *for a finite CSS* $\langle\mathcal{F}_0, \mathcal{C}_0\rangle$ *is a list of CSSs, called* branches, *built inductively according to the following rules:*

1. *The one branch list* $[\langle\mathcal{F}_0, \mathcal{C}_0\rangle]$ *is a tableau for* $\langle\mathcal{F}_0, \mathcal{C}_0\rangle$;
2. *If the list* $\mathcal{T}_m \oplus [\langle\mathcal{F}, \mathcal{C}\rangle] \oplus \mathcal{T}_n$ *is a tableau for* $\langle\mathcal{F}_0, \mathcal{C}_0\rangle$ *and*

$$\frac{Premiss}{\langle\mathcal{F}_1, \mathcal{C}_1\rangle \mid \ldots \mid \langle\mathcal{F}_k, \mathcal{C}_k\rangle}$$

is a (B)BI expansion rule from Figs. 3 or 4 for which a concrete instance of Premiss is fulfilled by $\langle\mathcal{F}, \mathcal{C}\rangle$, *then the list* $\mathcal{T}_m \oplus [\langle\mathcal{F} \cup \mathcal{F}_1, \mathcal{C} \cup \mathcal{C}_1\rangle; \ldots; \langle\mathcal{F} \cup \mathcal{F}_k, \mathcal{C} \cup \mathcal{C}_k\rangle] \oplus \mathcal{T}_n$ *is a tableau for* $\langle\mathcal{F}_0, \mathcal{C}_0\rangle$.

A (B)BI tableau for ϕ *is a (B)BI tableau for* $\langle\{\mathbb{F}\phi : c_0\}, \emptyset\rangle$. □

Partial Determinism	$z \in x \circ y \wedge z' \in x \circ y \rightarrow z = z'$
Total	$\exists z(z \in x \circ y)$
Cancellativity	$z \in x \circ y \wedge z \in x \circ y' \rightarrow y = y'$
Single Unit	$x \in E \wedge x' \in E \rightarrow x = x'$
Indivisible Units	$x \in y \circ z \wedge x \in E \rightarrow y \in E$
Disjointness	$x \in y \circ y \rightarrow y \in E$
Splittability	$x \in \overline{E} \wedge x \in E \rightarrow \bot, x \in \overline{E} \vee x \in E, x \in \overline{E} \rightarrow \exists y, z(y \in \overline{E} \wedge z \in \overline{E} \wedge x \in y \circ z)$
Cross-Split	$x \in t \circ u \wedge x \in v \circ w \rightarrow \exists a, b, c, d(t \in a \circ b \wedge u \in c \circ d \wedge v \in a \circ c \wedge w \in b \circ d)$
Upwards-Closed	$z \in x \circ y \wedge z \leq z' \rightarrow \exists x', y'(z' \in x' \circ y' \wedge x \leq x' \wedge y \leq y')$
Downwards-Closed	$z \in x \circ y \wedge x' \leq x \wedge y' \leq y \rightarrow \exists z'(z' \in x' \circ y' \wedge z' \leq z)$
Non-Branching	$x \leq y \wedge x \leq y' \rightarrow y \leq y' \vee y' \leq y$
Always-Joins	$x \leq y \wedge x \leq y' \rightarrow \exists z(y \leq z \wedge y' \leq z)$
Increasing	$z \in x \circ y \rightarrow y \leq z$
Unit Self Joining	$Ex \rightarrow x \in x \circ x$
Normal Increasing	$z \in x \circ y \wedge Ez \rightarrow x \leq z$

Fig. 5. Separation properties.

Definition 2 (Closed Tableau/Proof). *A CSS $\langle \mathcal{F}, \mathcal{C} \rangle$ is* closed *if one of the following* closure conditions *holds: (1) $\mathbb{T}\phi : x \in \mathcal{F}$, $\mathbb{F}\phi : y \in \mathcal{F}$ and $x \sim y \in \mathcal{C}$; (2) $\mathbb{F}\top : x \in \mathcal{F}$; (3) $\mathbb{T}\bot : x \in \mathcal{F}$; (4) $\mathbb{F}\mathrm{I} : x \in \mathcal{F}$ and $Ex \in \mathcal{C}$. A CSS is* open *iff it is not closed. A tableau is closed iff all its branches are closed. A* proof *for a formula ϕ is a closed tableau for ϕ.* □

We note that we could simply add $\langle \mathbb{T}\neg \rangle$, $\langle \mathbb{F}\neg \rangle$, and $\langle \mathrm{Sym} \rangle$ to the BI system and obtain one for BBI. However, this causes a significant amount of redundancy in the production of labels and constraints while requiring many more derivation steps in proofs, something that does not arise with the BBI rules given.

Extension with Separation Theories. A *separation property* is a first-order axiom in the language of (B)BI Kripke frames. Figure 5 gives separation properties taken from across the Separation Logic literature [10,13,14,24], presented as coherent formulae. A *separation theory* is thus a collection Σ of axioms from Fig. 5. The syntactic form of coherent formulae enables a uniform translation of separation properties into tableaux expansion rules and closure conditions. First, each first-order atomic formula is translated into constraints: $Tr(z \in x \circ y) = R_* xyz$, $Tr(x \in E) = Ex$, $Tr(x \leq y) = x \sim y$ and $Tr(x = x') = x \sim x', x' \sim x$. Given $A_1(\vec{x}) \wedge \cdots \wedge A_n(\vec{x}) \rightarrow \exists \vec{y_1} B_1(\vec{x}, \vec{y_1}) \vee \cdots \vee \exists \vec{y_m} B_m(\vec{x}, \vec{y_m})$ with $n, m \neq 0$, we obtain the frame expansion rule

$$\frac{Tr(A_1(\vec{x})), \ldots, Tr(A_n(\vec{x})) \in \mathcal{C}}{\langle \emptyset, \mathcal{C}_1 \rangle \mid \ldots \mid \langle \emptyset, \mathcal{C}_m \rangle},$$

where each $\mathcal{C}_i$ is the set of constraints translated from the conjuncts of B_i, using fresh labels $\vec{c_i}$ in place of the previously quantified $\vec{y_i}$. For example, the separation properties Cross-Split and Non-Branching are translated to the rules

$$\frac{R_* tux, R_* vwx \in \mathcal{C}}{\langle \emptyset, \{R_* c_i c_j t, R_* c_k c_l u, R_* c_i c_k v, R_* c_j c_l w\} \rangle} \text{ and } \frac{x \sim y, x \sim y' \in \mathcal{C}}{\langle \emptyset, \{y \sim y'\} \rangle \mid \langle \emptyset, \{y' \sim y\} \rangle},$$

where c_i, c_j, c_k, c_l are fresh labels. The special case $n = 0$ gives a rule with premiss $Expr_1(x_1), \ldots, Expr_p(x_p) \in \mathcal{F} \cup \mathcal{C}$, where each $Expr_i(x_i)$ is *any* expression in which x_i occurs and the x_i are the universally quantified variables in the original formula. The case $m = 0$ gives a new closure condition consisting of the conjunction of constraints translated from the antecedent of the original formula.

Note that the property Splittability is defined by a *system* of coherent axioms. These axioms force the new predicate $\overline{E}$ to be interpreted as the complement of E. When translated into tableaux rules, $x \in \overline{E}$ gives a new constraint $\overline{E}x$.

Given a separation theory Σ, a *(B)BI + Σ-tableau/proof* is defined in the same way as Definitions 1 and 2, except that a tableau can also be expanded by translated Σ-rules, and any new closure properties obtained from Σ can factor into the closure of a tableau and thus into proofs.

We give an example of a tableau proof in Fig. 6. The formula $(\neg \mathrm{I} \mathrel{-\!\!*} \bot) \to \mathrm{I}$ is valid in BBI models satisfying Total, but not in all BBI models [35], and Fig. 6—written, for clarity, using the traditional representation of tableaux and using $\otimes$ to denote closed branches—shows that the tableaux system for BBI + Total proves it. The left-hand branch is closed because both $\mathbb{F}\mathrm{I} : c_0$, $\mathbb{T}\mathrm{I} : c_0$ and $c_0 \sim c_0$ occur, while the right is closed because $\mathbb{T}\bot : c_1$ occurs.

4 Applications to Separation Logics

A *separation logic* can be determined by an assertion logic to describe machine state—a theory of (B)BI generated by validity in a concrete model of (B)BI + Σ for some separation theory Σ—and a specification logic to describe changes to machine state following program execution—typically a logic of Hoare triples $\{\phi\}C\{\psi\}$, where ϕ and ψ are formulas of the assertion language and C is a program in some programming language. Soundness of the *frame rule*,

$$\frac{\{\phi\}C\{\psi\}}{\{\phi * \chi\}C\{\psi * \chi\}},$$

where χ does not include any free variables modified by the program C, witnesses the coherence of these different aspects, and facilitates Separation Logic's characteristic 'local reasoning', which allows conclusions about a program's effect on the global state to be derived from reasoning on just the resource it accesses.

(1)	$\langle\{\mathbb{F}(\neg \mathrm{I} \mathrel{-\!\!*} \bot) \to \mathrm{I} : c_0\}, \emptyset\rangle$		Premiss
(2)	$\langle\{\mathbb{T}\neg \mathrm{I} \mathrel{-\!\!*} \bot : c_0, \mathbb{F}\mathrm{I} : c_0\}, \emptyset\rangle$		$\langle\mathbb{F} \to\rangle$, from (1)
(3)	$\langle\emptyset, \{R_* c_0 c_0 c_1\}\rangle$		Total, from (1)
(4)	$\langle\{\mathbb{F}\neg \mathrm{I} : c_0\}, \emptyset\rangle$	$\langle\{\mathbb{T}\bot : c_1\}, \emptyset\rangle$	$\langle\mathbb{T}\mathrel{-\!\!*}\rangle$, from (2), (3)
(5)	$\langle\{\mathbb{T}\mathrm{I} : c_0\}, \emptyset\rangle$	$\otimes$	$\langle\mathbb{F}\neg\rangle$, from (4)
(6)	$\langle\emptyset, \{c_0 \sim c_0\}\rangle$		$\langle$Ref$\rangle$, from (5)
	$\otimes$		

Fig. 6. Tableau proof of $(\neg \mathrm{I} \mathrel{-\!\!*} \bot) \to \mathrm{I}$ in the BBI + Total system.

To demonstrate the wide applicability of our framework we now give a number of separation logics that are models of separation theories. We note that our systems can be incomplete with respect to a given concrete model, but this is as expected for any proof system: the benefit versus a standard (B)BI system—which will be incomplete with respect to the class of models of a given separation theory—is the capability to make inferences based on the additional structure the model carries. Because of space constraints this selection is demonstrative rather than exhaustive. Other examples include Petri nets [13]; step-indexed models for storable locks [11] and the Iris framework [33]; separation logics incorporating named [42] and fractional [7] permissions; and separation logics designed for message passing [49] and amortized resource analysis [3].

Heaps. Our first example is given by the standard memory models of Separation Logic [32]. A *heap* is a partial function $h : \mathbb{N} \rightharpoonup \mathbb{Z}$, representing an allocation of memory addresses to values. Given heaps h, h', $h \# h'$ denotes that $dom(h) \cap dom(h') = \emptyset$; $h \cdot h'$ denotes the union of functions with disjoint domains, which is defined iff $h \# h'$. The *empty heap*, $[]$, is defined nowhere.

Let H denote the set of all heaps. Then $\mathrm{Heap}_{\mathrm{BBI}} = (H, \cdot, \{[]\})$ is a BBI frame. Letting $h \sqsubseteq h'$ denote that h' extends h, $\mathrm{Heap}_{\mathrm{BI}} = (H, \sqsubseteq, \cdot, H)$ defines a BI frame. These frames generate the standard classical and intuitionistic models of Separation Logic. $\mathrm{Heap}_{\mathrm{BBI}}$ satisfies Partial Determinism, Cancellativity, Single Unit, Indivisible Units, Cross-Split and Unit Self Joining; $\mathrm{Heap}_{\mathrm{BI}}$ additionally satisfies Splittability, Upwards-Closed, Downwards-Closed, Increasing and Normal Increasing while dropping Single Unit and Unit Self Joining.

One property distinguishing the standard memory models is that $*$-elimination—$\phi * \psi \to \psi$, useful for reasoning about garbage-collected languages—is valid in the intuitionistic heap model but not the classical. Cao et al. [14] show that this corresponds to the separation property Increasing. Figure 7—written with a traditional tableau presentation—shows a single branch tableaux proof of $\phi * \psi \to \psi$ for BI + Increasing, closed because $\mathbb{T}\psi : c_4$, $\mathbb{F}\psi : c_1$ and $c_4 \sim c_1$ occur.

Permissions. Permissions are incorporated into variants of separation logics that are designed to reason about certain kinds of concurrent algorithms and more fine-grained notions of memory disjointness: for example, disjointness modulo shared read permission. Hóu [29] reports a schema of Clouston that encompasses many such models: we recall it, with two concrete instances.

Let V be a set of values and $\star : V^2 \rightharpoonup V$ an associative and commutative partial function. Denote by H_V the set of V-valued heaps $h : \mathbb{N} \rightharpoonup V$. Then $\mathrm{Heap}_V = (H_V, \circ_\star, \{[]\})$ is a BBI frame, where $\circ_\star$ is defined by

$$h_1 \circ_\star h_2(n) = \begin{cases} h_1(n) \star h_2(n) & \text{if } n \in dom(h_1) \cap dom(h_2) \text{ and } h_1(n) \star h_2(n) \downarrow \\ h_1(n) & \text{if } n \in dom(h_1) \setminus dom(h_2) \\ h_2(n) & \text{if } n \in dom(h_2) \setminus dom(h_1) \\ \text{undefined} & \text{otherwise.} \end{cases}$$

(1)	$\langle\{\mathbb{F}\phi * \psi \to \psi : c_0\}, \emptyset\rangle$	Premiss
(2)	$\langle\{\mathbb{T}\phi * \psi : c_1, \mathbb{F}\psi : c_1\}, \{c_0 \sim c_1\}\rangle$	$\langle \mathbb{F} \to\rangle$, from (1)
(3)	$\langle\{\mathbb{T}\phi : c_3, \mathbb{T}\psi : c_4\}, \{R_* c_3 c_4 c_2, c_2 \sim c_1\}\rangle$	$\langle \mathbb{T}*\rangle$, from (2)
(4)	$\langle\emptyset, \{c_4 \sim c_2\}\rangle$	Increasing, from (3)
(5)	$\langle\emptyset, \{c_4 \sim c_1\}\rangle$	$\langle$Trans$\rangle$, from (2), (3)
	$\otimes$	

Fig. 7. Tableau proof of $\phi * \psi \to \psi$ in the BI + Increasing system.

Hóu defines Bornat et al.'s [6] *counting permissions model* with $V = \mathbb{Z}^2$ and

$$(x, i) \star (y, j) = \begin{cases} (x, i + j) & \text{if } x = y, i < 0 \text{ and } j < 0 \\ (x, i + j) & \text{if } x = y, i + j \geq 0 \text{ and } (i < 0 \text{ or } j < 0) \\ \text{undefined} & \text{otherwise.} \end{cases}$$

This frame satisfies Partial Determinism, Cancellativity, Indivisible Units, Single Unit, Cross-Split and Unit Self Joining.

Hóu defines Dockins et al.'s [24] *binary tree model* by considering the set T of non-empty binary trees with leaves labelled $\top$ or $\bot$ that are quotiented by the smallest congruence that identifies any subtree in which all leaves have the same label with a single leaf carrying that label. Then $V = \mathbb{Z} \times T$, and $\star$ is defined, where $\vee$ ($\wedge$) denotes pointwise disjunction (conjunction) of equivalent trees, by

$$(x, [t]) \star (y, [t']) = \begin{cases} (x, [t \vee t']) & \text{if } x = y \text{ and } [t \wedge t'] = [\bot] \\ \text{undefined} & \text{otherwise.} \end{cases}$$

This frame satisfies Partial Determinism, Cancellativity, Single Unit, Indivisible Units, Disjointness, Splittability, Cross-Split and Unit Self Joining.

Crash Hoare Logic. Chen et al. [16] use a separation logic to verify that the FSCQ file system meets its specification and secures its data under any sequence of crashes. Cao et. al. [14] give the underlying model as the following BI frame. Let V^+ be the set of non-empty lists over a set V and ϵ the empty list. Buffer heaps are defined to be heaps $h : \mathbb{N} \to V^+$. Let H_{buff} be the set of all buffer heaps. Then $Heap_{\text{buff}} = (H_{\text{buff}}, \leq, \cdot, \{[]\})$ is a BI frame, where $\cdot$ is the usual heap composition, and $h_1 \leq h_2$ iff $dom(h_1) = dom(h_2)$ and $\forall x \in \mathbb{N}, \exists l \in V^+ \cup \{\epsilon\}$ such that $h_1(x) = l \oplus h_2(x)$. This frame satisfies Partial Determinism, Cancellativity, Single Unit, Indivisible Units, Cross-Split, Upwards-Closed, Downwards-Closed, Always-Joins, Non-Branching, Unit Self Joining, and Normal Increasing.

Typed Heaps. Cao et al. [14] give an example derived from the handling of multibyte locks in Appel's [2] Verified System Toolchain separation logic for CompCert C. Let a *typed heap* be a partial map $h : \mathbb{N} \to \{\text{char}, \text{short}_1, \text{short}_2\}$ such that $h(n) = \text{short}_1$ implies $h(n+1) = \text{short}_2$. Let H_{typ} denote the set of all typed heaps. Then $\text{Heap}_{\text{Typ}} = (H_{\text{typ}}, \leq, \circ, H_{\text{typ}})$ is a BI frame, where $h_1 \leq h_2$ iff, for all $n \in dom(h_1)$ either $n \in dom(h_2)$ and $h_1(n) = h_2(n)$ or $h_1(n) = \text{char}$, and $h \in h_1 \circ h_2$ iff $h_1 \cdot h_2 \leq h$. This frame satisfies Indivisible Units, Disjointness,

Splittability, Cross-Split, Upwards-Closed, Downwards-Closed, Non-Branching, Increasing, and Normal Increasing.

5 Metatheory

Tableaux Systems as Coherent Theories. Just as coherent formulae yield tableaux rules, tableaux rules yield coherent formulae, allowing a complete specification of our calculi as coherent theories. Our framework determines a first-order signature: for each formula ϕ of (B)BI, we have unary relation symbols $\mathbb{T}\phi$ and $\mathbb{F}\phi$, together with the unary relation symbol E, the binary relation symbol $\sim$ and the ternary relation symbol R_*.

Given a rule premiss '$\mathbb{S}\phi : x \in \mathcal{F}$ and $A_1 x_1^1 \ldots x_{k_1}^1, \ldots, A_m x_1^m \ldots x_{k_m}^m \in \mathcal{C}$' we obtain the coherent antecedent $C(\vec{x}) \equiv \mathbb{S}\phi(x) \wedge \bigwedge_i A_i x_1^i \ldots x_{k_i}^i$. For the $j-th$ conclusion $\langle \mathcal{F}_j, \mathcal{C}_j \rangle$ of the rule we obtain $\exists \vec{y_j} C_j(\vec{x}, \vec{y_j})$, where C_j is the conjunction of atomic formulae translated from the constraints in $\mathcal{F}_j \cup \mathcal{C}_j$, with any fresh labels $\vec{c}$ that occurred substituted with $\vec{y_j}$. The translated rule is thus $C(\vec{x}) \rightarrow \exists \vec{y_1} C_1(\vec{x}, \vec{y_1}) \vee \cdots \vee \exists \vec{y_n} C_n(\vec{x}, \vec{y_n})$. For example, the instance of the BI rule $\langle \mathbb{F}\text{-}\!*\rangle$ for $\phi \mathrel{-\!\!*} \psi$ becomes $\mathbb{F}\phi \mathrel{-\!\!*} \psi(x) \rightarrow \exists y_1, y_2, y_3 (\mathbb{T}\phi(y_2) \wedge \mathbb{F}\psi(y_3) \wedge x \sim y_1 \wedge R_* y_1 y_2 y_3)$.

There are some special cases to pay attention to. For tableaux rules with premiss $\text{Expr}(x) \in \mathcal{F} \cup \mathcal{C}$ the antecedent of the translated coherent formula is $\top$. This is not the case for rules with premiss $\text{Expr}(x) \in \mathcal{C}$: these must be translated into a separate rule for each of the finitely many ways x can occur in each constraint. Finally, each closure condition '$\mathbb{S}_1\phi_1 : x_1, \ldots, \mathbb{S}_n\phi_n : x_n$, $A_1 y_1^1 \ldots y_{k_1}^1, \ldots$, and $A_m y_1^m \ldots y_{k_m}^m$' gives $\bigwedge_i \mathbb{S}_i\phi_i(x_i) \wedge \bigwedge_i A_i y_1^i \ldots y_{k_i}^i \rightarrow \bot$.

Given a (B)BI formula ϕ, the finite coherent theory $\Phi_\phi^{(B)BI+\Sigma}$ is given by the translated $(B)BI + \Sigma$-frame expansion rules, the translated closure conditions and the instances of translated logical expansion rules for subformulae of ϕ. We note that we could specify the whole tableaux system for (B)BI + Σ as an infinite coherent theory (similar to the axiomatization of a Hintikka set in standard tableaux completeness proofs), but finiteness is required for our argument.

Soundness and Completeness. We now prove soundness and completeness of the tableaux method via an analogous result for the Tarskian semantics of coherent logic. First, we show that the existence of a Kripke $(B)BI + \Sigma$-model with a state that doesn't satisfy ϕ is equivalent to the existence of a Tarskian model of $\Phi_\phi^{(B)BI+\Sigma} \cup \{\exists x. \mathbb{F}\phi(x)\}$.

Definition 3 (Induced Kripke Model of $\mathcal{M}$). *Given a Tarskian model $\mathcal{M}$ of $\Phi_\phi^{(B)BI+\Sigma}$, define $[a] = \{b \mid a \sim^{\mathcal{I}} b, b \sim^{\mathcal{I}} a\}$ and $X_{\mathcal{M}} = \{[a] \mid a \in X\}$. Then $[a] \leq_{\mathcal{M}} [b]$ iff $a \sim^{\mathcal{I}} b$, $[c] \in [a] \circ_{\mathcal{M}} [b]$ iff $R_*^{\mathcal{I}} abc$, and $E_{\mathcal{M}} = \{[a] \mid E^{\mathcal{I}} a\}$. $\mathcal{V}_{\mathcal{M}}(\mathrm{p}) = \{[a] \mid \exists b (b \sim^{\mathcal{I}} a$ and $\mathbb{T}\mathrm{p}^{\mathcal{I}}(b))\}$.*

1. *If $\mathcal{M}$ is a model of $\Phi_\phi^{BI+\Sigma}$, the induced Kripke frame is given by $\mathcal{X}_{\mathcal{M}} = (X_{\mathcal{M}}, \leq_{\mathcal{M}}, \circ_{\mathcal{M}}, E_{\mathcal{M}})$; the induced Kripke model is given by $(\mathcal{X}_{\mathcal{M}}, \mathcal{V}_{\mathcal{M}})$.*
2. *If $\mathcal{M}$ is a model of $\Phi_\phi^{BBI+\Sigma}$, the induced Kripke frame is given by $\mathcal{X}_{\mathcal{M}} = (X_{\mathcal{M}}, \circ_{\mathcal{M}}, E_{\mathcal{M}})$; the induced Kripke model is given by $(\mathcal{X}_{\mathcal{M}}, \mathcal{V}_{\mathcal{M}})$.*

The induced Kripke frame is a well-defined structure because of the frame tableaux rules, with $[-]$ forming equivalence classes and $\leq_{\mathcal{M}}$, $\circ_{\mathcal{M}}$, and $E_{\mathcal{M}}$ independent from the choice of representatives due to $\langle$Cong$\rangle$. The $(B)BI + \Sigma$-frame properties for the induced frame follow from their correspondent rules in the tableaux and the valuation $\mathcal{V}_{\mathcal{M}}$ is independent of choice of representative and persistent for induced Kripke $BI + \Sigma$-models.

Lemma 1. *Given a Tarskian model $\mathcal{M}$ of $\Phi_{\phi}^{(B)BI+\Sigma}$, the induced Kripke model $\mathcal{X}_{\mathcal{M}}$ is a Kripke $(B)BI + \Sigma$-model.* □

The significance of this model is that satisfiability of subformulae ψ of ϕ is determined by the interpretation of the relation symbols $\mathbb{S}\psi$ in the original Tarskian model. A simple proof by induction yields the next lemma.

Lemma 2. *Let $\mathcal{M}$ be a Tarskian model of the coherent theory $\Phi_{\phi}^{(B)BI+\Sigma}$, ψ a subformula of ϕ and $a \in X$. 1. If $\mathbb{T}\psi^{\mathcal{I}}(a)$ holds in $\mathcal{M}$, then $[a] \vDash_{\mathcal{V}_{\mathcal{M}}} \psi$; 2. If $\mathbb{F}\psi^{\mathcal{I}}(a)$ holds in $\mathcal{M}$, then $[a] \nvDash_{\mathcal{V}_{\mathcal{M}}} \psi$.* □

We can also induce Tarskian models from Kripke models. Let $(\mathcal{X}, \mathcal{V})$ be a Kripke $(B)BI + \Sigma$-model. We define the induced Tarskian model by taking X to be the carrier, and defining the interpretation $\mathcal{I}$ by $\sim^{\mathcal{I}} = \ \leq$, $R_*^{\mathcal{I}} = \{(a, b, c) \mid c \in a \circ b\}$, $E^{\mathcal{I}} = E$, $\mathbb{T}\psi^{\mathcal{I}} = \{x \mid x \vDash_{\mathcal{V}} \psi\}$ and $\mathbb{F}\psi^{\mathcal{I}} = \{x \mid x \nvDash_{\mathcal{V}} \psi\}$.

Lemma 3. *Every Kripke $(B)BI$+Σ-model $(\mathcal{X}, \mathcal{V})$ with a state x (not) satisfying ϕ induces a model of $\Phi_{\phi}^{(B)BI+\Sigma} \cup \{\exists x.\mathbb{T}\phi(x)\}$ ($\Phi_{\phi}^{(B)BI+\Sigma} \cup \{\exists x.\mathbb{F}\phi(x)\}$).* □

We now connect the existence of a closed tableaux to Bezem and Coquand's [4] *breadth-first forward reasoning* proof system for coherent logic. In their system, judgments of the form $X \Vdash^{\Phi} D$ are derived, where X is a set of atomic first-order sentences, Φ a finite coherent theory and D a *closed coherent disjunction*; a first-order sentence with the same syntactic shape as the consequent of a coherent formula. The derivation of the judgment $X \Vdash^{\Phi} D$ is defined inductively:

1. (Base): $X \Vdash^{\Phi} D$ holds if for one of the disjuncts $\exists\vec{y}.C$ of D, there are constants $\vec{a}$ such that all conjuncts of $C[\vec{y} := \vec{a}]$ occur in X;
2. (Inductive Step): Consider all closed instances $C_i \to D_i$ of Φ-axioms such that the conjuncts of C_i occur in X but the conjuncts of no disjunct $C_{i,j}$ of D_i do. There exist finitely many, with their consequents thus enumerated $D_0, \ldots, D_n$. Let $\exists\overrightarrow{y_{i,j}}.C_{i,j}$ denote the j-th of the m_i disjuncts of D_i, and denote by $\overline{C_{i,j}}$ the substitution of $\overrightarrow{y_{i,j}}$ with fresh constants. Infer $X \Vdash^{\Phi} D$ from $\forall j_0 \in \{1, \ldots, m_0\}, \ldots, \forall j_n \in \{1, \ldots, m_n\}(X, \overline{C_{0,j_0}}, \ldots, \overline{C_{n,j_n}} \Vdash^{\Phi} D)$. Importantly, if a D_i is $\bot$, then $m_i = 0$, and $X \Vdash^{\Phi} D$ is trivially inferred.

A derivation can be seen as a kind of tableau, branching at each stage by adding every possible consequence of Φ obtainable from the atomic first-order sentences at the current node. A semi-decidable procedure is given to systematically search for a derivation of $X \Vdash^{\Phi} D$. First check the base case. If it doesn't hold, apply the inductive step to any Φ-axioms fireable from X. If there are none, X forms an Herbrand countermodel of Φ against D. If the inductive step can be applied, apply the search procedure recursively to all premisses. Bezem and Coquand show that successful termination corresponds to Tarskian truth.

Theorem 1 **([4]).** *$X \Vdash^{\Phi} D$ is derivable iff the search procedure successfully terminates for $X \Vdash^{\Phi} D$ iff D is true in all Tarskian models of $X \cup \Phi$.* □

It is straightforward that the search procedure for $\{\mathbb{F}\phi(a)\} \Vdash^{\Phi_{\phi}^{(B)BI+\Sigma}} \bot$ corresponds precisely to an exhaustive search for a closed tableau for ϕ.

Lemma 4. *There exists a closed $(B)BI + \Sigma$-tableaux for ϕ iff the search procedure for $\{\mathbb{F}\phi(a)\} \Vdash^{\Phi_{\phi}^{(B)BI+\Sigma}} \bot$ successfully terminates.* □

Hence if a closed $(B)BI + \Sigma$-tableaux does not exist for ϕ, there exists a Tarskian model $\mathcal{M}$ of $\Phi_{\phi}^{(B)BI+\Sigma} \cup \{\exists x.\mathbb{F}\phi(x)\}$. By Lemma 2, the induced Kripke model $\mathcal{X}_{\mathcal{M}}$ has a state $[a]$ such that $[a] \not\models_{\mathcal{V}_{\mathcal{M}}} \phi$, establishing that ϕ fails to be valid for Kripke $(B)BI + \Sigma$-models. Conversely, if a closed tableaux does exist, then there is no Tarskian model of $\mathcal{M}$ of $\Phi_{\phi}^{(B)BI+\Sigma} \cup \{\exists x.\mathbb{F}\phi(x)\}$. By Lemma 3, ϕ is valid in Kripke $(B)BI + \Sigma$-models, as otherwise any countermodel would generate a Tarskian model $\mathcal{M}$ of $\Phi_{\phi}^{(B)BI+\Sigma} \cup \{\exists x.\mathbb{F}\phi(x)\}$, a contradiction.

Theorem 2 (Soundness and Completeness for $(B)BI+\Sigma$-Tableaux). *ϕ is valid in Kripke $(B)BI + \Sigma$-models iff ϕ is provable in the $(B)BI + \Sigma$-tableaux system.* □

6 Conclusions and Further Work

We have given a framework of tableaux systems that exhaustively captures the breadth of separation theories in the literature. Our framework is proven sound and complete parametrically by a novel representation of tableaux systems as coherent theories that allows us to apply existing theory from coherent logic. This resolves the expressivity gap between the logics of bunched implications and the separation logics defined upon them, and provides proof theory for the assertion languages of a wide array of program logics.

The completeness of tableaux systems is usually proved by defining a notion of a *Hintikka set*: a saturated set of (labelled) formulae (and possibly constraints) that specifies a term model of the logic. The existence of a Hintikka set is then shown to follow from non-existence of a tableau proof. Our method is a generalization of this idea, implemented parametrically by choice of tableaux system. While we have focused on Separation Logic, this technique is adaptable to virtually any logic interpreted on relational structures, including the breadth of bunched and modal logics. This suggests the significance of the coherent logic fragment extends beyond the generation of proof rules for frame conditions.

The implementation of our systems is of principal importance for future work. Our tableaux representation suggests existing coherent logic provers (see [43] for a survey) may already be suitable, though tactics designed specifically for tableaux coherent theories may have to be developed to make this efficient. A closely related goal is the development of parametric Separation Logic implementations that utilize our systems as assertion language provers. Finally, our results suggest interesting theoretical work. Coherent logic has close connections to topos theory, and Caramello [15] has developed techniques to transfer results between mathematical fields via bridges between the classifying topoi of coherent theories. We wish to investigate if any results of logical interest can be found in this way by utilizing the representation of tableaux as coherent theories.

References

1. Anderson, G., Pym, D.: A calculus and logic of bunched resources and processes. Theoret. Comput. Sci. **614**, 63–96 (2016)
2. Appel, A.W.: Program Logics for Certified Compilers. CUP (2014)
3. Atkey, R.: Amortised resource analysis with separation logic. Log. Methods Comput. Sci. **2**(17), 1–33 (2011)
4. Bezem, M., Coquand, T.: Automating coherent logic. In: Sutcliffe, G., Voronkov, A. (eds.) LPAR 2005. LNCS (LNAI), vol. 3835, pp. 246–260. Springer, Heidelberg (2005). https://doi.org/10.1007/11591191_18
5. Berdine, J., Calcagno, C., O'Hearn, P.W.: Smallfoot: modular automatic assertion checking with separation logic. In: de Boer, F.S., Bonsangue, M.M., Graf, S., de Roever, W.-P. (eds.) FMCO 2005. LNCS, vol. 4111, pp. 115–137. Springer, Heidelberg (2006). https://doi.org/10.1007/11804192_6
6. Bornat, R., Calcagno, C., O'Hearn, P., Parkinson, M.: Permission accounting in separation logic. In: Proceedings of POPL 2005, pp. 259–270. ACM (2005)
7. Boyland, J.: Checking interference with fractional permissions. In: Cousot, R. (ed.) SAS 2003. LNCS, vol. 2694, pp. 55–72. Springer, Heidelberg (2003). https://doi.org/10.1007/3-540-44898-5_4
8. Braüner, T.: Hybrid Logic and Its Proof-Theory. Applied Logic Series, vol. 37. Springer, Dordrecht (2011)
9. Brotherston, J.: Bunched logics displayed. Stud. Logica. **100**(6), 1223–1254 (2012)
10. Brotherston, J., Villard, J.: Parametric completeness for separation theories. In: Proceedings of POPL 2014, pp. 453–464. ACM (2014)
11. Buisse, A., Birkedal, L., Støvring, K.: A step-indexed Kripke model of separation logic for storable locks. In: Proceedings of MFPS XXVII, ENTCS, vol. 276, pp. 121–143 (2011)
12. Calcagno, C., Distefano, D., O'Hearn, P., Yang, H.: Compositional shape analysis by means of bi-abduction. J. ACM **58**(6), 26 (2011). https://doi.org/10.1145/2049697.2049700
13. Calcagno, C., O'Hearn, P., Yang, H.: Local action and abstract separation logic. In: Proceedings of LICS 2007, pp. 366–378. IEEE (2007)
14. Cao, Q., Cuellar, S., Appel, A.W.: Bringing order to the separation logic jungle. In: Chang, B.-Y.E. (ed.) APLAS 2017. LNCS, vol. 10695, pp. 190–211. Springer, Cham (2017). https://doi.org/10.1007/978-3-319-71237-6_10
15. Caramello, O.: Theories, Sites, Toposes: Relating and Studying Mathematical Theories Through Topos-Theoretic 'Bridges'. OUP, Oxford (2017)
16. Chen, H., Ziegler, D., Chajed, T., Chlipala, A., Kaashoek, M.F., Zeldovich, N.: Using crash hoare logic for certifying the FSCQ file system. In: Proceedings of SOSP 2015, pp. 18–37. ACM (2015)

17. Collinson, M., McDonald, K., Pym, D.: A substructural logic for layered graphs. J. Log. Comput. **24**(4), 953–988 (2014)
18. Collinson, M., McDonald, K., Pym, D.: Layered graph logic as an assertion language for access control policy models. J. Log. Comput. **27**(1), 41–80 (2017)
19. Collinson, M., Pym, D.: Algebra and logic for resource-based systems modelling. Math. Struct. Comput. Sci. **19**, 959–1027 (2009)
20. Courtault, J.-R., Galmiche, D., Pym, D.: A logic of separating modalities. Theoret. Comput. Sci. **637**, 30–58 (2016)
21. Dinsdale-Young, T., Birkedal, L., Gardner, P., Parkinson, M., Yang, H.: Views: compositional reasoning for concurrent programs. In: Proceedings of POPL 2013, pp. 287–300 (2013)
22. Docherty, S., Pym, D.: Intuitionistic layered graph logic. In: Olivetti, N., Tiwari, A. (eds.) IJCAR 2016. LNCS (LNAI), vol. 9706, pp. 469–486. Springer, Cham (2016). https://doi.org/10.1007/978-3-319-40229-1_32
23. Docherty, S., Pym, D.: Stone-Type Dualities for Separation Logics (Submitted)
24. Dockins, R., Hobor, A., Appel, A.W.: A fresh look at separation algebras and share accounting. In: Hu, Z. (ed.) APLAS 2009. LNCS, vol. 5904, pp. 161–177. Springer, Heidelberg (2009). https://doi.org/10.1007/978-3-642-10672-9_13
25. Dyckhoff, R., Negri, S.: Geometrisation of first-order logic. Bull. Symb. Log. **21**(2), 123–163 (2015)
26. Galmiche, D., Kimmel, P., Pym, D.: A substructural epistemic resource logic. In: Ghosh, S., Prasad, S. (eds.) ICLA 2017. LNCS, vol. 10119, pp. 106–122. Springer, Heidelberg (2017). https://doi.org/10.1007/978-3-662-54069-5_9
27. Galmiche, D., Méry, D.: Tableaux and resource graphs for separation logic. J. Log. Comput. **20**(1), 189–231 (2007)
28. Galmiche, D., Méry, D., Pym, D.: The semantics of BI and resource tableaux. Math. Struct. Comput. Sci. **15**, 1033–1088 (2005)
29. Hóu, Z.: Labelled sequent calculi and automated reasoning for assertions in separation logic. Ph.D. thesis, The Australian National University (2015)
30. Hóu, Z., Goré, R., Tiu, A.: Automated theorem proving for assertions in separation logic with all connectives. In: Felty, A.P., Middeldorp, A. (eds.) CADE 2015. LNCS (LNAI), vol. 9195, pp. 501–516. Springer, Cham (2015). https://doi.org/10.1007/978-3-319-21401-6_34
31. Hóu, Z., Clouston, R., Tiu, A., Goré, R.: Proof search for propositional abstract separation logics via labelled sequents. In: Proceedings of POPL 2014, pp. 465–476. ACM (2014)
32. Ishtiaq, S., O'Hearn, P.: BI as an assertion language for mutable data structures. In: Proceedings of POPL 2001, 14–26. ACM (2001)
33. Jung, R., Krebbers, R., Jourdan, J.-H., Bizjak, A., Birkedal, L., Dreyer, D.: Iris from the ground up: a modular foundation for higher-order concurrent separation logic (2017). Under consideration for publication in Journal of Functional Programming
34. Larchey-Wendling, D.: The formal strong completeness of partial monoidal Boolean BI. J. Log. Comput. **26**(2), 605–640 (2016)
35. Larchey-Wendling, D., Galmiche, D.: The undecidability of Boolean BI through phase semantics. In: Proceedings of LICS 2010, pp. 140–149. IEEE Computer Society Press (2010)
36. Larchey-Wendling, D., Galmiche, D.: Looking at separation algebras with Boolean BI-eyes. In: Diaz, J., Lanese, I., Sangiorgi, D. (eds.) TCS 2014. LNCS, vol. 8705, pp. 326–340. Springer, Heidelberg (2014). https://doi.org/10.1007/978-3-662-44602-7_25

37. Nakano, H.: A modality for recursion. In: Proceedings of LICS 2000, pp. 255–266. IEEE (2000)
38. Negri, S.: Proof analysis beyond geometric theories: from rule systems to systems of rules. J. Log. Comput. **26**(2), 513–537 (2016)
39. O'Hearn, P.: A Primer on Separation Logic. Software Safety and Security. NATO Science for Peace and Security Series, vol. 33, pp. 286–318 (2012)
40. O'Hearn, P., Pym, D.: The logic of bunched implications. Bull. Symb. Log. **5**(2), 215–244 (1999)
41. Park, J., Seo, J., Park, S.: A theorem prover for BBI. In: Proceedings of POPL 2013, pp. 219–232. ACM (2013)
42. Parkinson, M.: Local reasoning for Java. Ph.D. thesis, University of Cambridge (2005)
43. Polonsky, A.: Proofs, Types and Lambda Calculus. Ph.D. thesis, University of Bergen (2012)
44. Reynolds, J.: Separation logic: a logic for shared mutable data structures. In: Proceedings of LICS 2002, pp. 55–74. IEEE Computer Society Press (2002)
45. Schmidt, R.A., Tishkovsky, D.: Automated synthesis of tableau calculi. In: Giese, M., Waaler, A. (eds.) TABLEAUX 2009. LNCS (LNAI), vol. 5607, pp. 310–324. Springer, Heidelberg (2009). https://doi.org/10.1007/978-3-642-02716-1_23
46. Simpson, A.: The proof theory and semantics of intuitionistic modal logic. Ph.D. thesis, University of Edinburgh (1994)
47. Skolem, T.: Logisch-kombinatorische Untersuchungen über die Erfüllbarkeit und Beweisbarkeit mathematischen Sätze nebst einem Theoreme über dichte Mengen, Skrifter I, vol. 4, pp. 1–36. Det Norske Videnskaps-Akademi, (1920)
48. Terese: Term Rewriting Systems. Cambridge University Press (2003)
49. Villard, J., Lozes, É., Calcagno, C.: Proving copyless message passing. In: Hu, Z. (ed.) APLAS 2009. LNCS, vol. 5904, pp. 194–209. Springer, Heidelberg (2009). https://doi.org/10.1007/978-3-642-10672-9_15
50. Yang, H., O'Hearn, P.: A semantic basis for local reasoning. In: Nielsen, M., Engberg, U. (eds.) FoSSaCS 2002. LNCS, vol. 2303, pp. 402–416. Springer, Heidelberg (2002). https://doi.org/10.1007/3-540-45931-6_28

Differential Calculus with Imprecise Input and Its Logical Framework

Abbas Edalat[1] and Mehrdad Maleki[2](✉)

[1] Department of Computing, Imperial College London, London SW7 2RH, UK
a.edalat@imperial.ac.uk
[2] Institute for Research in Fundamental Sciences (IPM), Niavaran, Tehran, Iran
m.maleki@ipm.ir

Abstract. We develop a domain-theoretic Differential Calculus for locally Lipschitz functions on finite dimensional real spaces with imprecise input/output. The inputs to these functions are hyper-rectangles and the outputs are compact real intervals. This extends the domain of application of Interval Analysis and exact arithmetic to the derivative. A new notion of a tie for these functions is introduced, which in one dimension represents a modification of the notion previously used in the one-dimensional framework. A Scott continuous sub-differential for these functions is then constructed, which satisfies a weaker form of calculus compared to that of the Clarke sub-gradient. We then adopt a Program Logic viewpoint using the equivalence of the category of stably locally compact spaces with that of semi-strong proximity lattices. We show that given a localic approximable mapping representing a locally Lipschitz map with imprecise input/output, a localic approximable mapping for its sub-differential can be constructed, which provides a logical formulation of the sub-differential operator.

Keywords: Imprecise input/output · Interval analysis
Exact computation · Lipschitz maps · Clarke gradient
Domain theory · Stone duality

1 Introduction

A well-known hurdle in numerical computation is caused by accumulation of round-off errors in floating point arithmetic, which can create havoc and lead to catastrophic errors in compound calculations. In safety and critical systems, where reliability of numerical computation is of utmost importance, one way to avoid the pitfalls of floating point arithmetic is to use interval analysis or exact arithmetic. In both interval analysis and exact arithmetic as well as in computable analysis, a real number is represented by a nested shrinking sequence of compact intervals whose intersections is the real number. Similarly, a real n-vector can be represented by a nested sequence of hyper-rectangles in $\mathbb{R}^n$. This

C. Baier and U. Dal Lago (Eds.): FOSSACS 2018, LNCS 10803, pp. 459–475, 2018.
https://doi.org/10.1007/978-3-319-89366-2_25

leads to a framework in numerical computation and a framework for computational geometry where the inputs of algorithms or programmes are imprecise real numbers or real n-vectors; see for example [3,5,6,9,10,14,15,17,21–23,27].

All frameworks for interval analysis and exact real computation are based on functions whose input and output are real intervals. When we compose two such functions, the output of the first function serves as the input to the second function. An implementation of these frameworks in a functional programming language follows this same pattern; see for example the lazy Haskell implementation of IC-Reals for Exact Real Computation [1], which uses linear fractional transformations as developed in [14,22].

An important feature of working with a calculus consisting of functions with interval or imprecise input/output is that even when we deal with elementary functions such as polynomials we cannot restrict ourselves to their canonical (maximal) extensions to intervals [21]. These canonical extensions take a compact interval to its forward image under the function. In fact, these extensions are not closed under, for example, multiplication. Thus, the real-valued map of a real variable $x \mapsto x^2$ when implemented with interval input by $x \mapsto x \times x$, using multiplication of two copies of the input interval, is not the canonical extension of the quadratic map of real numbers: it evaluates for example $[-1,1]^2$ to $[-1,1]$ rather than $[0,1]$, which is what the canonical extension of the quadratic map evaluates to. In general, we need to work with any Scott continuous map of type $\mathbf{I}\mathbb{R} \to \mathbf{I}\mathbb{R}$ or, in higher dimension, of type $\mathbf{I}\mathbb{R}^n \to \mathbf{I}\mathbb{R}$, where $\mathbf{I}\mathbb{R}^n$ denotes the domain of hyper-rectangles of $\mathbb{R}^n$.

In the past 60 years, interval analysis has grown as a distinct interdisciplinary subject to impact on nearly all areas of mathematical and numerical analysis including computer arithmetic, linear algebra, integration, solution of initial value problems and partial differential equations to correct solutions in mathematical optimisation and robotics; see [20]. It is natural to ask if the domain of application of interval analysis and exact computation can be extended to the derivative of functions, i.e., whether one can take a kind of derivative of a map which takes a compact interval or a compact hyper-rectangle as input.

In [11], the notion of a domain-theoretic sub-differentiation of maps which have non-empty and compact intervals as inputs and outputs was introduced. The restriction of these maps to real numbers turns out to be locally Lipschitz maps of type $\mathbb{R} \to \mathbb{R}$ and the sub-differential restricted to real numbers has been shown to be the same as the Clarke sub-gradient [8]. A major problem, however, is that the framework in [11], which only deals with one-dimensional maps of type $\mathbf{I}\mathbb{R} \to \mathbf{I}\mathbb{R}$ is not accompanied with a Stone duality framework and thus, even in dimension one, cannot be used in order to handle program logic and predicate transformers.

In [7], a typed lambda calculus in the framework of an extension of Real PCF [6,17,22] was introduced in which in particular continuously differentiable and more generally Lipschitz functions can be defined. Given an expression representing a real-valued function of a real variable in this language, one is able to evaluate the expression on an argument, representing an interval, but also evaluate the generalised derivative, i.e., the L-derivative, equivalently the Clarke gradient, of the expression on an interval. The operational semantics of the language, which is equipped with min and a weighed average, enjoys adequacy and a definability result proving that any computable Lipschitz map is definable in it. The denotational semantics is based on domain theory which in principle allows a program logic formulation of the computation, although this challenge has not been taken up yet.

In [13], a point free framework for sub-differentiation of real-valued locally Lipschitz functions on finite dimensional Euclidean spaces has been developed which provides a Stone duality for the Clarke gradient and thus enables a program logic view of differentiation. However, the induced logical framework cannot be employed for the class of functions with imprecise input/output used in exact computation since, as already pointed out, this class necessarily contains general extensions of real-valued locally Lipschitz maps of finite dimensional Euclidean spaces.

In this paper, we formulate a new notion of a tie of functions with imprecise input/output, which, in one dimension, represents a modification of the corresponding notion in [12]. This allows us to develop a Scott continuous sub-differential for functions with hyper-rectangles in $\mathbb{R}^n$ as inputs and compact intervals in $\mathbb{R}$ as output, which are used in exact computation. We show that a weaker calculus compared to that for the Clarke sub-gradient is satisfied in this interval framework. In addition we construct a logical framework for sub-differentiation of locally Lipschitz maps of type $\mathbf{I}\mathbb{R}^n \to \mathbf{I}\mathbb{R}$. The basic Stone duality results developed in [13] are then extended to sub-differentiation of such interval maps.

1.1 Background

We assume the reader is familiar with basic elements of topology and domain theory. Following the definition in [18], by a domain we mean a continuous dcpo (directed complete partial order). All the domains we use in this paper are bounded complete as well. By $\mathbf{C}(\mathbb{R}^n)$, we denote the domain of non-empty convex and compact subsets of $\mathbb{R}^n$ ordered with reverse inclusion and augmented with $\bot = \mathbb{R}^n$ as the bottom element. If $C_1, C_2 \in \mathbf{C}(\mathbb{R}^n)$ then the way-below relation is given by $C_1 \ll C_2$ iff $C_1^\circ \supset C_2$, where S° is the interior of the set S. By $\mathbf{I}\mathbb{R}^n$, we denote the sub-domain of non-empty compact hyper-rectangles with faces parallel to coordinate hyper-planes of $\mathbb{R}^n$. The Euclidean norm of $x \in \mathbb{R}^n$ is denoted by $\|x\|$.

The lattice of open subsets of a topological space X is denoted by $\Omega(X)$. The Scott topology of a domain D is, however, written as σ_D. The closure of $S \subset X$ is denoted by $\overline{S}$. The upper topology, equivalently the Scott-topology, of $\mathbf{C}(\mathbb{R}^n)$ has a basis of the form

$$\Box O = \{C \in \mathbf{C}(\mathbb{R}^n) : C \subset O\},$$

where O belongs to a basis of open and convex subsets of $\mathbb{R}^n$.

Given an open set $a \subset X$ of a topological space and an element $b \in D$ of a domain D, the single-step function $b\chi_a : X \to D$ is defined by $b\chi_a(x) = b$ if $x \in a$ and $\bot$ otherwise. A non-empty compact real interval x is written as $x = [x^-, x^+]$. For a map $f : X \to Y$ of topological spaces, $f[S]$ denotes the image of the set $S \subset X$.

The three operations of addition of two vectors, scalar multiplication of a vector and a real number, and the inner product of two vectors can be extended to $\mathbf{C}(\mathbb{R}^n)$ to obtain the following three Scott continuous maps:

(i) $- + - : \mathbf{C}(\mathbb{R}^n) \times \mathbf{C}(\mathbb{R}^n) \to \mathbf{C}(\mathbb{R}^n)$ with $A + B = \{a + b : a \in A, b \in B\}$,
(ii) $- \times - : \mathbb{R} \times \mathbf{C}(\mathbb{R}^n) \to \mathbf{C}(\mathbb{R}^n)$ with $rA = \{rx : x \in A\}$, and,
(iii) $- \cdot - : \mathbf{C}(\mathbb{R}^n) \times \mathbf{C}(\mathbb{R}^n) \to \mathbf{I}\mathbb{R}$ with $A \cdot B = \{a \cdot b : a \in A, b \in B\}$.

These three operations have well-defined restrictions to $\mathbf{I}\mathbb{R}^n$. In addition, in this paper, we will consider their higher order extension to sets of sets. For example, if $a_1, a_2 \in \Omega(\mathbb{R})$ are open subsets, then $\Box a_1, \Box a_2 \in \sigma_{\mathbf{C}(\mathbb{R}^n)}$ and we have:

$$(\Box a_1) \cdot (\Box a_2) := \{x_1 \cdot x_2 : x_1 \in \Box a_1, x_2 \in \Box a_2\}$$

Moreover:

Proposition 1. *(i) The modal operator $\Box : \Omega(\mathbb{R}^n) \to \sigma_{\mathbf{C}(\mathbb{R}^n)}$ preserves meets, i.e., $\Box O_1 \wedge \Box O_2 = \Box(O_1 \wedge O_2)$ for all $O_1, O_2 \in \Omega(\mathbb{R}^n)$.*
(ii) The way-below relation satisfies $O_1 \ll O_2$ if and only if $\Box O_1 \ll \Box O_2$ for all $O_1, O_2 \in \Omega(\mathbb{R}^n)$.
(iii) If $O_1, O_2 \subset \mathbb{R}^n$ are open hyper-rectangles, then $\Box(O_1 + O_2) = \Box O_1 + \Box O_2$.
(iv) If $O \subset \mathbb{R}^n$ is a convex open set and $a \subset \mathbb{R}^n$ is a hyper-rectangle, then $\Box(O \cdot a) = (\Box O) \cdot (\Box a)$.

Next, we present the notion of Clarke's sub-gradient [4]. Recall that a map $f : U \subset \mathbb{R}^n \to \mathbb{R}$, where U is an open set, is locally Lipschitz if all points in U have an open neighbourhood $O \subset U$ with a constant $k \geq 0$ such that $|f(x) - f(y)| \leq k\|x - y\|$ for all $x, y \in O$. The generalized directional derivative of a locally Lipschitz f at x in the direction of v is defined as follow:

$$f^\circ(x; v) = \limsup_{y \to x\ t \to 0^+} \frac{f(y + tv) - f(y)}{t}$$

The Clarke subgradient of f at x, denoted by $\partial f(x)$ is a convex and compact subset of $\mathbb{R}^n$ and is defined by:

$$\partial f(x) = \{w \in \mathbb{R}^n : f^\circ(x; v) \geq w \cdot v \text{ for all } v \in \mathbb{R}^n\} \tag{1}$$

The sub-gradient function $\partial f : U \subset \mathbb{R}^n \to \mathbf{C}(\mathbb{R}^n)$ is upper continuous, equivalently Scott continuous. Moreover, the Clarke sub-gradient satisfies a weak calculus. For locally Lipschitz maps $f, g : U \subseteq \mathbb{R}^n \to \mathbb{R}$,

(i) Sum: $\partial f(x) + \partial g(x) \supseteq \partial(f + g)(x)$.
(ii) Product: $(\partial f(x))g(x) + f(x)(\partial g(x)) \supseteq \partial(f \cdot g)(x)$

(iii) Chain rule: For $f, g : \mathbb{R} \to \mathbb{R}, \partial f(g(x)) \cdot \partial g(x) \supseteq \partial(f \circ g)(x)$.

The notion of the L-derivative, equivalent to the Clarke sub-gradient, for real-valued functions on finite dimensional Euclidean spaces has the following ingredients [8]. A function $f : U \subset \mathbb{R}^n \to \mathbb{R}$ has a non-empty generalized Lipschitz constant $b \in \mathbf{C}(\mathbb{R}^n)$ in a non-empty convex open set $a \subset \mathbb{R}^n$ if for all $x, y \in a$ we have $f(x) - f(y) \in b \cdot (x - y)$. The collection of all functions that have generalized Lipschitz constant b in a is denoted by $\delta(a, b)$, called the tie of a with b. The collection of all single-step functions $b\chi_a$ with $a \subset U$ and $f \in \delta(a, b)$ is bounded in $(U \to \mathbf{C}(\mathbb{R}^n))$ and thus the L-derivative of f defined as

$$\mathcal{L}f = \sup\{b\chi_a : f \in \delta(a, b)\}$$

is Scott-continuous function. Moreover, we have $\mathcal{L}f = \partial f$.

1.2 Stably Locally Compact Space and Semi-strong Proximity Lattice

We recall that in geometric logic one uses the open sets of a topological space as propositions or semi-decidable properties [25,26]. If X is a topological space and $\Omega(X)$ its lattice of open sets, a propositional geometric theory is constructed as follows: For every open set $a \in \Omega(X)$, define a proposition P_a, i.e., every open set of X provides a property or predicate. For open sets a and b with $a \subseteq b$ stipulate: (i) $P_a \vdash P_b$. For a family of open sets S, stipulate: (ii) $P_{\bigcup S} \vdash \bigvee_{a \in S} P_a$. For a finite family of open sets S, stipulate: (iii) $\bigwedge_{a \in S} P_a \vdash P_{\bigcap S}$. The converses of (ii) and (iii) follow from (i). The nullary disjunction in (ii) is interpreted as **false** and the nullary conjunction in the converse of (iii) is interpreted as **truth**, i.e., $P_\emptyset \vdash$ **false** and $P_X \vdash$ **truth**.

We regard $x \in X$ as a model of the theory in which P_a is interpreted as **true** iff $x \in a$, i.e., $x \models a$ iff $x \in a$, or, a point is a model of a proposition if it is in the open set representing the proposition. It is possible that different points give rise to the same model, i.e., satisfy the same open sets, and it is also possible that a model does not arise by points in X in this way. For so-called sober spaces, as we will define below, we do have a one-to-one correspondence between points and models.

A topological space X is called *stably locally compact* [2,18] if it is sober, locally compact and if the intersection of two compact saturated sets is compact. Recall that X is sober if its points are in bijection with the completely prime filters of its lattice of open sets. (A set is saturated if it is the intersection of its open neighbourhoods.) Equivalently, X is stably locally compact if and only if its lattice of open sets is a distributive continuous lattice which is also arithmetic, i.e., its way-below relation satisfies:

$$O \ll O_1, O_2 \Rightarrow O \ll O_1 \wedge O_2$$

The spaces $\mathbb{R}^n$, $\mathbf{I}\mathbb{R}^n$ and $\mathbf{C}(\mathbb{R}^n)$ are all stably locally compact spaces. The way-below relation for $\Omega(\mathbb{R}^n)$ is given by $O_1 \ll O_2$ iff $\overline{O_1}$ is compact and $\overline{O_1} \subset O_2$, whereas the way-below relation in $\mathbf{C}(\mathbb{R}^n)$, and thus $\mathbf{I}\mathbb{R}^n$, is given by Proposition 1. We can obtain a finitary representation of these spaces by a sub-lattice of open sets as we will now describe.

A *semi-strong proximity lattice* [13] consists of a tuple $(B; \vee, \wedge, 0, 1; \prec)$ in which $(B; \vee, \wedge, 0, 1)$ is a distributive lattice such that $\prec$ is a binary relation on B with $\prec = \prec \circ \prec$ satisfying:

1. $\forall a \in B\ M \subset_f B.\, M \prec a \Leftrightarrow \bigvee M \prec a.$
2. $\forall a \in B.\, a \neq 1 \Rightarrow a \prec 1.$
3. $\forall a, a_1, a_2 \in B.\, a \prec a_1, a_2 \Leftrightarrow a \prec a_1 \wedge a_2.$
4. $\forall a, x, y \in B.\, a \prec x \vee y \Rightarrow$
   $\exists x', y' \in B.\, x' \prec x \,\&\, y' \prec y \,\&\, a \prec x' \vee y'.$

Here, $M \subset_f B$ means that M is a finite (possibly empty) subset of B, and $M \prec a$ means that $\forall m \in M.\, m \prec a$.

The relation $R \subseteq B_1 \times B_2$, between two semi-strong proximity lattice, is a *localic approximable mapping* if it satisfies:

1. $R \circ \prec_2 = R$
2. $\prec_1 \circ R = R.$
3. $\forall M \subset_f B_1 \forall b \in B_2.\, M\, R\, b \iff \bigvee M\, R\, b.$
4. $\forall a \in B_1.\, a \neq 1 \Rightarrow a\, R\, 1.$
5. $\forall a \in B_1 \forall a_1, a_2 \in B_2.\, a\, R\, a_1 \,\&\, a\, R\, a_2 \Leftrightarrow a\, R\, a_1 \wedge a_2.$
6. $\forall a \in B_1 \forall M \subset_f B_2.\, a\, R \bigvee M \Rightarrow$
   $\exists N \subset_f B_1.\, a \prec_1 \bigvee N \,\&\, \forall n \in N \exists m \in M.\, n\, R\, m.$

The identity approximable mapping on B is $\prec_B$ and composition of approximable mappings is the usual composition of the relations in the same order as for functions.

Let **SL-Compact** denote the category of all stably locally compact spaces and continuous functions and let **Semi-Strong PL** denote the category of semi-strong proximity lattice and approximable mappings. The following functors between these categories establish an equivalence between them [13,19].

$$A : \textbf{SL-Compact} \to \textbf{Semi-Strong PL}$$
$$G : \textbf{Semi-Strong PL} \to \textbf{SL-Compact}$$

Given a stably locally compact space X, fix a basis B of its topology which is closed under finite intersections and let $A(X)$ be the semi-strong proximity lattice based on B. Given a continuous function $f : X_1 \to X_2$ between two stably locally compact spaces, we have a localic approximable mapping $A_f : A(X_1) \to A(X_2)$ given by $a\, A_f\, b$ iff $a \ll f^{-1}(b)$.

Given a semi-strong proximity lattice B, the spectrum $\mathsf{spec(B)}$ of B is the set of all prime filters of B. For $x \in B$ let $\mathcal{O}_x = \{F \in \mathsf{spec(B)} : x \in F\}$. The collection of $\mathcal{O}_x$'s, $x \in B$, is a base of a topology over $\mathsf{spec(B)}$. Put,

$$G(B) = spec(B)$$

Given a localic approximable mapping $R : B_1 \to B_2$ define,

$$G_R : \mathsf{spec}(\mathsf{B}_1) \to \mathsf{spec}(\mathsf{B}_2)$$

by $G_R(F) = \{b_2 \in B_2 : \exists b_1 \in F. b_1 \, R \, b_2\}$. We have, $A_{G_R} = R$ and $G_{A_f} = f$. Thus, the category of semi-strong proximity lattice with approximable mappings is equivalent to the category of stably locally compact spaces and continuous functions [13].

We now construct some canonical bases of $\mathbf{C}(\mathbb{R}^n)$ and $\mathbf{I}\mathbb{R}^n$, which provide us with the semi-strong proximity lattices these spaces can be represented by. Let $B^0_{\mathbb{R}^n}$, respectively B^0_U, for $U \subset \mathbb{R}^n$, be any basis of $\mathbb{R}^n$, respectively U, that consists of bounded convex open sets and is closed under finite intersections. We let $B_{\mathbb{R}^n}$, respectively B_U, denote the semi-strong proximity lattice generated by $B^0_{\mathbb{R}}$, respectively B^0_U. This means that every element of $B_{\mathbb{R}}$, respectively B_U, is a finite join of elements of $B^0_{\mathbb{R}^n}$, respectively B^0_U [13].

It now follows, by Proposition 1, that $B^0_{\mathbf{C}(\mathbb{R}^n)} = \{\Box a : a \in B^0_{\mathbb{R}^n}\}$ is a basis of the Scott topology $\sigma_{\mathbf{C}(\mathbb{R}^n)}$, which is closed under finite intersections. Let $B_{\mathbf{C}(\mathbb{R}^n)}$ be the semi-strong proximity lattice generated by $B^0_{\mathbf{C}(\mathbb{R}^n)}$. Thus, each element of the semi-strong proximity lattice $B_{\mathbf{C}(\mathbb{R}^n)}$ is the finite join of elements of $B^0_{\mathbf{C}(\mathbb{R}^n)}$.

Finally, let $\mathcal{T}(U)$ be a basis of $U \subset \mathbb{R}^n$ consisting of open hyper-rectangles in U with faces parallel to the coordinate planes and let $\mathcal{T} := \mathcal{T}(\mathbb{R}^n)$. Then $B^0_{\mathbf{I}\mathbb{R}^n} = \{\Box a : a \in \mathcal{T}\}$ is a basis for $\sigma_{\mathbf{I}\mathbb{R}^n}$. By using $\mathcal{T}(U)$, we similarly obtain a basis $B^0_{\mathbf{I}U}$ for $\mathbf{I}U \subset \mathbf{I}\mathbb{R}^n$. Again by Proposition 1(i) these bases are closed under finite intersections. We let $B_{\mathbf{I}\mathbb{R}^n}$, respectively, $B_{\mathbf{I}U}$ be the semi-strong proximity lattices generated by $B^0_{\mathbf{I}\mathbb{R}^n}$, respectively, $B^0_{\mathbf{I}U}$. Thus, each element of $B_{\mathbf{I}\mathbb{R}^n}$, respectively, $B_{\mathbf{I}U}$, is the finite join of elements of $B^0_{\mathbf{I}\mathbb{R}^n}$, respectively, $B^0_{\mathbf{I}U}$.

The functors A and G thus provide a bijection between the two hom-sets:

$$(\mathbf{I}U \to \mathbf{I}\mathbb{R}) \underset{\mathrm{A}}{\overset{\mathrm{G}}{\leftrightarrows}} (B_{\mathbf{I}U} \to B_{\mathbf{I}\mathbb{R}})$$

and between the two hom-sets:

$$(\mathbf{I}U \to \mathbf{C}(\mathbb{R}^n)) \underset{\mathrm{A}}{\overset{\mathrm{G}}{\leftrightarrows}} (B_{\mathbf{I}U} \to B_{\mathbf{C}(\mathbb{R}^n)})$$

These bijections are used later to deduce our Stone duality results.

1.3 Related Work

Differentiation in logical form for functions of type $U \subseteq \mathbb{R}^n \to \mathbb{R}$ was introduced in [13]. These maps were represented by localic approximable mappings of type $B_U \to B_{\mathbb{R}}$, and the localic approximable mapping of the L-derivative of these functions have the type $B_U \to B_{\mathbf{C}(\mathbb{R}^n)}$. The strong tie of a with b, denoted by $\delta_s(a, b)$, was defined as the collection of all functions $f : a \subseteq U \to \mathbb{R}$ such that there exists $a' \in B^0_{\mathbb{R}}$ and $b' \in \mathbf{C}(\mathbb{R}^n)$ with $a \ll a'$, $b \ll b'$ and $f \in \delta(a', b')$.

The approximable mappings $R : B_U \to B_{\mathbb{R}}$ has Lipschitz constant $O \in B_{\mathbf{C}(\mathbb{R}^n)}$ in $a \in B_U$, denoted by $R \in \Delta(a, O)$, if we have:

$$\forall a_1, a_2 \prec a, (a_1, a_2) \in \mathsf{Sep}, \exists a_1', a_2' \in B_{\mathbb{R}}.$$
$$a_1 \, R \, a_1', a_2 \, R \, a_2', a_1' - a_2' \prec O \cdot (a_1 - a_2)$$

where the separation predicate $\mathsf{Sep} \subset B_U \times B_U$ means $(a_1, a_2) \in \mathsf{Sep}$ if there exists a_1', a_2' such that $a_1 \prec a_1', a_2 \prec a_2'$ and $a_1' \wedge a_2' = 0$. The strong knot $\Delta_s(a, O)$ is defined as the set of approximable mappings $R : B_U \to B_{\mathbb{R}}$ such that there exists $a' \in B_U$, $O' \in B_{\mathbf{C}(\mathbb{R}^n)}$ with $a \prec a'$, $O' \prec O$ and $R \in \Delta(a', O')$.

The strong ties and strong knots are dual to each others, i.e., $R \in \Delta_s(a, O)$ iff $G_R \in \delta_s(a, \overline{O})$. The Lipschitzian derivative of $R : B_U \to B_{\mathbb{R}}$ is defined as the approximable mapping

$$\mathsf{L}(R) = \sup\{A_{\overline{O}\chi_a} : R \in \Delta_s(a, O)\}$$

It turns out that $\mathsf{L}(R) = A_{\mathcal{L}G_R}$ and we have a weak calculus which matches that for the Clarke sub-gradient stated after Eq. (1), i.e., $\mathsf{L}(R_1) + \mathsf{L}(R_2) \sqsubseteq \mathsf{L}(R_1 + R_2)$ and $R_1 \cdot \mathsf{L}(R_2) + R_2 \cdot \mathsf{L}(R_1) \sqsubseteq \mathsf{L}(R_1 \cdot R_2)$, and if at least one of R_1 and R_2 is a continuously differentiable approximable mapping then equality holds. A weak form of the chain rule also holds for composition of approximable mappings corresponding to that for the Clarke sub-gradient.

2 L-derivative with Imprecise Inputs

We start by defining a notion of tie for Scott continuous map of type $f : \mathbf{I}U \to \mathbf{I}\mathbb{R}$, for an open convex subset $U \subset \mathbb{R}^n$. From now on, in the rest of the paper, we assume $f : \mathbf{I}U \to \mathbf{I}\mathbb{R}$ is Scott-continuous.

Definition 1. *Let $f : \mathbf{I}U \subseteq \mathbf{I}\mathbb{R}^n \to \mathbf{I}\mathbb{R}$ where $U \subset \mathbb{R}^n$ is an open set, be Scott continuous and $a \in \mathcal{T}(U)$, an open hyper-rectangle in U, and $b \in \mathbf{C}(\mathbb{R}^n)$. We say f has a* generalized Lipschitz constant *b in $\Box a$ and write $\delta(\Box a, b)$ if we have:*

$$\forall x, y \in \Box a, x \cap y = \emptyset. \, f(x) - f(y) \sqsubseteq b \cdot (x - y)$$

In the one dimensional case, this new notion is a modification of that in [12] as we in Definition 1, require the hyper-rectangles x and y to be disjoint, i.e., inconsistent in $\mathbf{I}U$. Thus, the condition for membership of a tie is weaker. We will need this weaker condition in order to develop the Stone duality result later in the paper.

We show that despite this weaker notion, if $f \in \delta(\Box a, b)$ with $b \neq \bot$, then f preserves maximal elements and its restriction to maximal elements gives a Lipschitz map. In other words f is the extension of a classical Lipschitz function in $\mathbf{I}a$.

Proposition 2. *Let $f \in \delta(\Box a, b)$, where $a \subset \mathbb{R}^n$ is a open hyper-rectangle and $b \in \mathbf{C}(\mathbb{R}^n) \setminus \{\bot\}$, then for each $x \in a$, $f(\{x\}) \in \mathbf{I}\mathbb{R}$ is maximal and the induced function $\hat{f} : a \subset \mathbb{R}^n \to \mathbb{R}$ is Lipschitz and satisfies:*

$$\forall x_1, x_2 \in a. (b \cdot (x_1 - x_2))^- \leq \hat{f}(x_1) - \hat{f}(x_2) \leq (b \cdot (x_1 - x_2))^+ \tag{2}$$

$$\forall x_1, x_2 \in a. |\hat{f}(x_1) - \hat{f}(x_2)| \leq \|b\| \|x_1 - x_2\|, \tag{3}$$

where $\|b\| = \max\{\|L\| | L \in b\}$.

Corollary 1. *If $f \in \delta(\Box a, b)$ then $\hat{f} \in \delta(a, b)$.*

Definition 2. *We say a Scott continuous function of type $\mathbf{I}U \subset \mathbf{I}\mathbb{R}^n \to \mathbf{I}\mathbb{R}$ is* locally Lipschitz *in $\Box a$, for $a \in \mathcal{T}(U)$, if it belongs to a tie $\delta(\Box a, b)$ with $b \neq \bot$.*

Given a continuous function $f : U \subseteq \mathbb{R}^n \to \mathbb{R}$, its maximal extension to a Scott continuous function $\mathbf{I}U \subseteq \mathbf{I}\mathbb{R}^n \to \mathbf{I}\mathbb{R}$ is denoted by $\mathbf{I}f$ with $\mathbf{I}f(x) = f[x]$ for $x \in \mathbf{I}U$ when $x \neq \bot$ and $\mathbf{I}f(\bot) = \bot$.

Corollary 2. *$f \in \delta(a, b)$ iff $\mathbf{I}f \in \delta(\Box a, b)$.*

If $(A, \sqsubseteq)$ is a dcpo then the consistency predicate $\mathsf{Con}_{(A,\sqsubseteq)}$ and $\mathsf{Con}_{(A,\ll)}$ for a finite subset $\{a_i : i \in I\}$ with respect to $\sqsubseteq$ and $\ll$ are defined as follow:

$$\mathsf{Con}_{(A,\sqsubseteq)}\{a_i : i \in I\} \iff \exists a \in A, \forall i \in I. a_i \sqsubseteq a$$

and

$$\mathsf{Con}_{(A,\ll)}\{a_i : i \in I\} \iff \exists a \in A, \forall i \in I. a_i \ll a$$

For the collection $(b_i \chi_{a_i})_{i \in I}$ or $(b_i \chi_{\Box a_i})_{i \in I}$ for finite indexing set I where $a_i \in \Omega(\mathbb{R}^n)$ are open hyper-rectangles and $b_i \in (D, \sqsubseteq)$, the function space consistency predicate $\mathsf{Con}_{\mathbb{R}^n \to D}$ or $\mathsf{Con}_{\mathbf{I}\mathbb{R}^n \to D}$ is defined as follows:

$$\mathsf{Con}_{\mathbb{R}^n \to D}(b_i \chi_{a_i})_{i \in I} \iff \forall J \subseteq I. [\mathsf{Con}_{(\Omega(\mathbb{R}^n), \gg)}\{a_i : i \in J\} \Rightarrow \mathsf{Con}_{(D,\sqsubseteq)}\{b_i : i \in J\}]$$

$$\mathsf{Con}_{\mathbf{I}\mathbb{R}^n \to D}(b_i \chi_{\Box a_i})_{i \in I} \iff \forall J \subseteq I. [\mathsf{Con}_{(\Omega(\mathbf{I}\mathbb{R}^n), \gg)}\{\Box a_i : i \in J\} \Rightarrow \mathsf{Con}_{(D,\sqsubseteq)}\{b_i : i \in J\}].$$

It follows that the supremum $\sup_{i \in I} b_i \chi_{a_i}$ exists iff $\mathsf{Con}_{\mathbb{R}^n \to D}(b_i \chi_{a_i})_{i \in I}$ and $\sup_{i \in I} b_i \chi_{\Box a_i}$ exists iff $\mathsf{Con}_{\mathbf{I}\mathbb{R}^n \to D}(b_i \chi_{\Box a_i})_{i \in I}$.

Proposition 3. *For any indexing set J the family of step functions $(b_j \chi_{\Box a_j})_{j \in J}$ is consistent if $\bigcap_{j \in J} \delta(\Box a_j, b_j) \neq \emptyset$.*

Proof. Suppose $f \in \bigcap_{j \in J} \delta(\Box a_j, b_j)$ then $\hat{f} \in \bigcap_{j \in J} \delta(a_j, b_j)$, and hence $(b_j \chi_{a_j})_{j \in J}$ is consistent, which implies $(b_j \chi_{\Box a_j})_{j \in J}$ is consistent. ∎

Recall that a crescent in $\mathbb{R}^n$ is the intersection of a closed and an open set. Given two points $p, q \in \mathbb{R}^n$, we denote the closed, respectively open, line segment between them by $[p, q] = \{\lambda p + (1 - \lambda) q : 0 \leq \lambda \leq 1\}$, respectively $(p, q) = \{\lambda p + (1 - \lambda) q : 0 < \lambda < 1\}$.

Proposition 4. *We have* $\delta(\Box a, b) \supseteq \bigcap_{j\in J} \delta(\Box a_j, b_j)$ *if* $b\chi_{\Box a} \sqsubseteq \sup_{j\in J} b_j\chi_{\Box a_j}$.

Proof. Let $g := \sup_{j\in J} b_j\chi_{\Box a_j}$. Suppose $b\chi_{\Box a} \sqsubseteq \sup_{j\in J} b_j\chi_{\Box a_j}$, then $\Box a \subset \bigcup_{j\in J} \Box a_j$ and thus $a \subset \bigcup_{j\in J} a_j$. In addition, by considering the restriction of g to the maximal elements of $\mathbf{I}\mathbb{R}^n$, we find that a is partitioned by the open sets a_j, $j \in J$, into a finite number of disjoint crescents c_i, $i \in I$, with

$$g(\{r\}) = \sup_{c_i \subset a_j} b_j \sqsupseteq b$$

for $r \in c_i$. Let $f \in \bigcap_{j\in J} \delta(\Box a_j, b_j)$. We show that $f \in \delta(\Box a, b)$. Suppose we have two hyper-rectangles $x, y \in \Box a$ with $x \cap y = \emptyset$. Let the points $p \in x$ and $q \in y$ be such that $\|p - q\|$ is the minimum distance between x and y. Then $[p, q]$ is partitioned by the crescents c_i, $i \in I$, into a finite number of one-dimensional intervals such that the one-dimensional interior of each is contained in c_i for some $i \in I$. Let $r_0, r_1, \ldots, r_k \in \mathbb{R}^n$ be the boundary points of these intervals ordered from p to q. Then, using the continuity of $\hat{f}$, we have:

$$f(\{r_t\}) - f(\{r_{t+1}\}) \subseteq \sup_{(r_t, r_{t+1}) \subseteq c_j} b_j \cdot (\{r_t\} - \{r_{t+1}\}) \subseteq b \cdot (\{r_t\} - \{r_{t+1}\})$$

for $0 \leq t \leq n - 1$. Since $x \in \Box a$, there exists $j \in J$ with $x \in \Box a_j$. Moreover, $x \subseteq a_j$ iff $r_0 \in \overline{a_j}$. Similarly, $y \subseteq a_j$ iff $r_k \in \overline{a_j}$. From these relations, we obtain:

$$f(x) - f(\{r_0\}) \subseteq \sup_{x \subset a_j} b_j \cdot (x - \{r_0\}), \quad f(\{r_k\}) - f(y) \subseteq \sup_{y \subset a_j} b_j \cdot (\{r_k\} - y)$$

Thus,

$$\begin{aligned} f(x) - f(y) &= f(x) - f(\{r_0\}) + f(\{r_0\}) - \cdots - f(\{r_k\}) + f(\{r_k\}) - f(y) \\ &\subseteq b \cdot (x - \left(\sum_{t=0}^{k-1} f(\{r_t\}) - f(\{r_t\}) \right) - y) = b \cdot (x - y) \blacksquare \end{aligned}$$

Definition 3. *The derivative of a Scott continuous map* $f : \mathbf{I}U \subset \mathbf{I}\mathbb{R}^n \to \mathbf{I}\mathbb{R}$ *is the map:*

$$\mathcal{L}f = \sup_{f \in \delta(\Box a, b)} b\chi_{\Box a} : \mathbf{I}U \to \mathbf{C}(\mathbb{R}^n)$$

where U *is a convex open subset of* $\mathbb{R}^n$.

Theorem 1. *(i)* $\mathcal{L}f$ *is well-defined and Scott continuous.*
(ii) $f \in \delta(\Box a, b)$ *iff* $b\chi_{\Box a} \sqsubseteq \mathcal{L}f$.

Proof. (i) Let the indexing set J be defined by $j \in J \iff f \in \delta(\Box a_j, b_j)$, then $f \in \bigcap_{j\in J} \delta(\Box a_j, b_j)$. Thus, by Proposition 3 $(b_j\chi_{\Box a_j})_{j\in J}$ is consistent therefore, $\mathcal{L}f = \sup_{f\in\delta(\Box a,b)} b\chi_{\Box a}$ exists and is Scott continuous.

(ii) If $f \in \delta(\Box a, b)$ then clearly $b\chi_{\Box a} \sqsubseteq \mathcal{L}f$. Now take $a' \ll a$ and $b' \ll b$. Then $b'\chi_{\Box a'} \ll b\chi_{\Box a} \sqsubseteq \mathcal{L}f$ and there exists a finite indexing set J such that $b'\chi_{\Box a'} \sqsubseteq \sup_{j\in J} b_j\chi_{\Box a_j}$ and $f \in \delta(\Box a_j, b_j)$ for $j \in J$. Now by Proposition 4, we have $\bigcap_{j\in J} \delta(\Box a_j, b_j) \subseteq \delta(\Box a', b')$, and thus, $f \in \delta(\Box a', b')$. From this, it follows that $f \in \delta(\Box a, b)$. ■

If $f : U \subseteq \mathbb{R}^n \to \mathbb{R}$ is a locally Lipschitz map, then the Clarke sub-gradient $\mathcal{L}f : U \to \mathbf{C}(\mathbb{R}^n)$ extends, by Scott's extension theory for densely injective spaces [24], to a Scott continuous map $\mathbf{I}(\mathcal{L}f) : \mathbf{I}U \to \mathbf{C}(\mathbb{R}^n)$. We then have:

Proposition 5.

$$\mathcal{L}(\mathbf{I}f) = \mathbf{I}(\mathcal{L}f)$$

Proof. This follows from the relation:

$$f \in \delta(a, b) \iff \mathbf{I}f \in \delta(\Box a, b),$$

for all $a \in \Omega(U)$ and $b \in \mathbf{C}(\mathbb{R}^n)$. ■

The following example shows that in the context of the L-derivative of interval functions, Clarke's weak calculus no longer holds for Sum.

Example 1. Let $f, g : \mathbb{IR} \to \mathbb{IR}$ defined by $f(x) = x$ and $g(x) = -x$, then $\mathcal{L}f(x) = \{1\}$ and $\mathcal{L}g(x) = \{-1\}$ and thus $\mathcal{L}f(x) + \mathcal{L}g(x) = \{0\}$. On the other hand, $(f + g)(x) = f(x) + g(x) = x - x$ and it follows that $f + g \notin \delta(\Box a, \{0\})$, for any open set $a \subset \mathbb{R}$, and consequently $\mathcal{L}(f + g) \neq \{0\}$. Hence, $\mathcal{L}(f + g)(x) \not\subseteq \mathcal{L}f(x) + \mathcal{L}g(x)$.

We say an interval $[r^-, r^+]$ is *positive* , respectively *negative*, if $r^- > 0$, respectively $r^+ < 0$. The above counter-example is the consequence of the fact that in interval arithmetic, while the relation $(u + v)w \subseteq uw + vw$ always holds for $u, v, w \in \mathbb{IR}$, the converse relation $(u + v)w \supseteq uw + vw$ may fail. However, if u and v are both positive or both negative then the converse also holds [21, p. 13].

We can obtain a weak calculus for sum and product of two functions f and g if we first use an operation that is routinely performed in interval analysis, namely to approximate the values $\mathcal{L}f(x)$ and $\mathcal{L}g(x)$ with the smallest axes aligned hyper-rectangle containing it, and then assume that the two induced hyper-rectangles have the same sign in each of their components. We now formalise this procedure.

Let $H : \mathbf{C}(\mathbb{R}^n) \to \mathbb{IR}^n$ be the map that takes every convex compact set to the smallest axes aligned hyper-rectangle containing it. Then, it is easy to check that H is Scott continuous. Let $\pi_i : \mathbb{R}^n \to \mathbb{R}$ be the projection of the ith coordinate and extend it pointwise to its maximal extension $\mathbf{I}\pi_i : \mathbb{IR}^n \to \mathbb{IR}$. Define the predicate $\mathsf{Sgn} \subset (\mathbb{IR}^n)^2$ by $(x, y) \in \mathsf{Sgn}$ if for each $i = 1, \ldots, n$ the two intervals $\mathbf{I}\pi_i(x)$ and $\mathbf{I}\pi_i(y)$ are either both positive or both negative.

Suppose $x, y, z \in \mathbb{IR}^n$ and $(y, z) \in \mathsf{Sgn}$, then the interval $\mathbf{I}\pi_i(y)\mathbf{I}\pi_i(z)$ is positive for each $i = 1 \ldots, n$ and we have $x(y + z) = xy + xz$. In fact,

$$\mathbf{I}\pi_i(x)(\mathbf{I}\pi_i(y) + \mathbf{I}\pi_i(z)) = \mathbf{I}\pi_i(x)\mathbf{I}\pi_i(y) + \mathbf{I}\pi_i(x)\mathbf{I}\pi_i(z),$$

and hence:

$$x(y + z) = \sum_{i=1}^{n} \mathbf{I}\pi_i(x)(\mathbf{I}\pi_i(y) + \mathbf{I}\pi_i(z)) = \sum_{i=1}^{n} \mathbf{I}\pi_i(x)\mathbf{I}\pi_i(y) + \mathbf{I}\pi_i(x)\mathbf{I}\pi_i(z)$$

$$= \sum_{i=1}^{n} \mathbf{I}\pi_i(x)\mathbf{I}\pi_i(y) + \sum_{i=1}^{n} \mathbf{I}\pi_i(x)\mathbf{I}\pi_i(z) = xy + xz$$

Proposition 6. *Suppose $f, g : \mathbf{I}U \subseteq \mathbf{I}\mathbb{R}^n \to \mathbf{I}\mathbb{R}$ are locally Lipschitz functions and $x \in \mathbf{I}U$ is such that $(H(\mathcal{L}f(x)), H(\mathcal{L}g(x))) \in \mathsf{Sgn}$. Then:*

1.
$$H(\mathcal{L}f(x)) + H(\mathcal{L}g(x)) \supseteq H(\mathcal{L}(f+g)(x))$$

2. *If, in addition, $(f(x), g(x)) \in \mathsf{Sgn}$, then we also have:*
$$f(x)H(\mathcal{L}g(x)) + g(x)H(\mathcal{L}f(x)) \supseteq H(\mathcal{L}(fg)(x))$$

We will provide the proof for a weak form of the chain rule, which is more involved compared to sum and product. First consider the extended scalar multiplication $M : \mathbf{C}(\mathbb{R}^n) \times \mathbf{I}\mathbb{R}^+ \to \mathbf{C}(\mathbb{R}^n)$, where $\mathbb{R}^+$ is the set of non-negative reals, with $M(b, x) = \{ur : u \in b, r \in x\}$. Then, M is well-defined and Scott continuous. For ease of presentation, we write $M(b, x) = bx$.

Proposition 7. *If $g : \mathbf{I}U_1 \subseteq \mathbf{I}\mathbb{R}^n \to \mathbf{I}\mathbb{R}$ and $f : \mathbf{I}U_2 \subseteq \mathbf{I}\mathbb{R} \to \mathbf{I}\mathbb{R}$ and $\mathsf{Im}(g) \subset \mathbf{I}U_2$ with $(\mathcal{L}f)(g(x)) \in \mathbf{I}\mathbb{R}^+$, are Scott-continuous, then:*
$$((\mathcal{L}f) \circ g)(x)\mathcal{L}g(x) \supseteq \mathcal{L}(f \circ g)(x)$$

3 Lipschitzian Approximable Mapping

Recall that, since $\mathbf{I}\mathbb{R}^n$, $\mathbf{C}(\mathbb{R}^n)$ and $\mathbb{R}^n$ are stably locally compact space and the category of stably locally compact spaces with continuous functions and the category of semi-strong proximity lattice with approximable mappings are equivalent, any continuous function $f : \mathbf{I}U \subset \mathbf{I}\mathbb{R}^n \to \mathbf{I}\mathbb{R}$ defines an approximable mapping $A_f : B_{\mathbf{I}U} \to B_{\mathbf{I}\mathbb{R}}$ by $\Box a A_f \Box a' \iff \Box a \ll f^{-1}(\Box a')$. On the other hand any approximable mapping with type $R : B_{\mathbf{I}\mathbb{R}^n} \to B_D$, where D is either $\mathbf{I}\mathbb{R}$ or $\mathbf{I}\mathbb{R}^n$ or $\mathbf{C}(\mathbb{R}^n)$, gives us a continuous function $G_R : \mathbf{I}\mathbb{R}^n \to D$.

Lemma 1. *Let $f : \mathbf{I}U \subset \mathbf{I}\mathbb{R}^n \to \mathbf{I}\mathbb{R}$ be a Scott continuous function such that $f(\{x\})$ is singleton for all $x \in U$. Suppose a_1 is an open hyper-rectangle in U and a_2 is an open interval. If $\hat{f} : U \subset \mathbb{R}^n \to \mathbb{R}$ is the induced function with $f(\{x\}) = \{\hat{f}(x)\}$ then:*
$$\Box a_1 \ll f^{-1}(\Box a_2) \Rightarrow a_1 \ll \hat{f}^{-1}(a_2) \qquad \Box a_1 \; A_f \; \Box a_2 \Rightarrow a_1 \; A_{\hat{f}} \; a_2$$

Recall the definition of the predicate $\mathsf{Sep} \subset B_{\mathbb{R}} \times B_{\mathbb{R}}$ from Subsect. 1.3.

Definition 4. *We say an approximable mapping $R : B_{\mathbf{I}U} \to B_{\mathbf{I}\mathbb{R}}$, where $U \subset \mathbb{R}^n$ is a convex open set, has Lipschitzian constant O in $\Box a$, with $O \in B^0_{\mathbb{R}^n}$ and $a \in \mathcal{T}(U)$, if:*
$$\forall a_1, a_2 \in \mathcal{T}(U).\, a_1, a_2 \prec a \,\&\, (a_1, a_2) \in \mathsf{Sep}\; \exists a_1', a_2' \in B_{\mathbb{R}}.$$
$$\Box a_1 \, R \, \Box a_1', \Box a_2 \, R \, \Box a_2' \;\&\; a_1' - a_2' \prec O \cdot (a_1 - a_2),$$

and we say R is Lipschitzian *in $\Box a$. The set of all approximable mappings with the above property is denoted by $\Delta(\Box a, O)$, called the* knot *of $\Box a$ and O.*

Note that, by Proposition 1, the last formula in Definition 4 is equivalent to $\Box a'_1 - \Box a'_2 \prec \Box O \cdot (\Box a_1 - \Box a_2)$. Given this equivalence, it is simpler to use the formula without the modal operator $\Box$ as we have done in this definition. By Proposition 1 and Stone duality, we have:

Proposition 8. *Suppose $f : \mathbf{I}U \to \mathbf{I}\mathbb{R}$ is a Scott continuous function such that $f(\{x\})$ is singleton for every $x \in U$. Then we have: $A_{\hat{f}} \in \Delta(a, O)$ if $A_f \in \Delta(\Box a, O)$.*

From $\Delta(\Box a, O)$, a Lipschitz property of G_R can be deduced as follows.

Proposition 9. *If $R : B_{\mathbf{I}U} \to B_{\mathbf{I}\mathbb{R}}$ is an approximable mapping such that $R \in \Delta(\Box a, O)$ then:*

$$\forall x, y \in \Box a.\, x \cap y = \emptyset \Rightarrow G_R(x) - G_R(y) \subseteq \overline{O} \cdot (x - y)$$

Proof. Let $x, y \in \Box a$ and $x \cap y = \emptyset$, then consider $a_1, a_2 \in \mathcal{T}(U)$ such that $(a_1, a_2) \in \mathsf{Sep}$ and $x \in \Box a_1, y \in \Box a_2$. Hence, there exist $a'_1, a'_2 \in B_{\mathbb{R}}$ such that $\Box a_i\, R\, \Box a'_i, i = 1, 2$ and:

$$a'_1 - a'_2 \prec O \cdot (a_1 - a_2)$$

By Stone duality we have $R = R_{G_R}$. Hence $\Box a_i \prec G_R^{-1}(\Box a'_i), i = 1, 2$, and thus:,

$$G_R(x) - G_R(y) \subseteq O \cdot (a_1 - a_2).$$

Since this holds for all sufficiently small a_1 and a_2 that contain x and y respectively, we obtain: $G_R(x) - G_R(y) \subseteq \overline{O} \cdot (x - y)$. ■

Corollary 3. *If $R \in \Delta(\Box a, O)$ then $G_R \in \delta(\Box a, \overline{O})$.*

Thus, if A_f is a Lipschitzian approximable mapping of type $B_{\mathbf{I}U} \to B_{\mathbf{I}\mathbb{R}}$ then f is a Lipschitz function of type $\mathbf{I}U \to \mathbf{I}\mathbb{R}$ and hence $f(\{x\})$ is a singleton for every $x \in U$ and the induced function $\hat{f} : U \to \mathbb{R}$ is also Lipschitz.

Now we are in a position to obtain duality results similar to those in [13] for functions of type $\mathbf{I}U \subseteq \mathbf{I}\mathbb{R}^n \to \mathbf{I}\mathbb{R}$.

Proposition 10. *Let $f \in \delta(\Box a, b)$ then for every $a_0 \in \mathcal{T}$ such that $a_0 \prec a$ and every $O \in B^0_{\mathbb{R}^n}$ such that $b \subset O$ we have $A_f \in \Delta(\Box a_0, O)$.*

Proof. Suppose $a_0 \prec a$. Let $a_1, a_2 \in \mathcal{T}(U)$ with $(a_1, a_2) \in \mathsf{Sep}$ and $a_1, a_2 \prec a_0$. Then, since $\overline{a_1}, \overline{a_2} \in \mathbf{I}U$, from definition of the tie $\delta(\Box a, b)$, we have,

$$\begin{aligned} f(\overline{a_1}) - f(\overline{a_2}) &\subseteq b \cdot (\overline{a_1} - \overline{a_2}) \\ &\subseteq O \cdot (a_1 - a_2). \end{aligned}$$

Since $f(\overline{a_1}), f(\overline{a_2}) \in \mathbf{I}\mathbb{R}$ are compact, there exist open hyper-rectangles $a'_1, a'_2 \in B_{\mathbb{R}}$ such that $f(\overline{a_i}) \subseteq a'_i, i = 1, 2$, and $a'_1 - a'_2 \prec O \cdot (a_1 - a_2)$. This implies $A_f \in \Delta(\Box a_0, O)$. ■

Example 2. Let $f : \mathbf{I}\mathbb{R} \to \mathbf{I}\mathbb{R}$ be given by:

$$f([x_1, x_2]) = [x_1 - \delta(x_2 - x_1), x_2 + \delta(x_2 - x_1)]$$

for $\delta > 0$. The restriction $\hat{f}$ of f to the maximal elements of $\mathbf{I}\mathbb{R}$ is the identity function of type $\hat{f} = \mathrm{Id} : \mathbb{R} \to \mathbb{R}$. Since $\mathbf{I}\mathrm{Id} \neq f$, the map f is not the maximal extension of the identity map Id. On the other hand, $A_f : B_{\mathbf{I}\mathbb{R}} \to B_{\mathbf{I}\mathbb{R}}$ satisfies $A_f \in \Delta(\Box\mathbb{R}, O)$ iff $(1 - \delta, 1 + \delta) \subseteq O$. However, $A_{\hat{f}} \in \Delta(\mathbb{R}, O)$ iff $1 \in O$.

The following two propositions represent a domain isomorphism between the function space $(\mathbf{I}U \to \mathbf{C}(\mathbb{R}^n))$ and the domain of approximable mappings $(B_{\mathbf{I}U} \to B_{\mathbf{C}(\mathbb{R}^n)})$ ordered by inclusion.

Proposition 11. *1. For $f_1, f_2 : \mathbf{I}U \to \mathbf{C}(\mathbb{R}^n)$ we have:*

$$f_1 \sqsubseteq f_2 \Longleftrightarrow A_{f_1} \subseteq A_{f_2}$$

2. For $R_1, R_2 : B_{\mathbf{I}U} \to B_{\mathbf{C}(\mathbb{R}^n)}$ we have:

$$R_1 \subseteq R_2 \Longleftrightarrow G_{R_1} \sqsubseteq G_{R_2}$$

Proposition 12. *1. If $(f_i)_{i\in I}$ is a directed set in $\mathbf{I}U \to \mathbf{C}(\mathbb{R}^n)$, with supremum $f = \sup_{i\in I} f_i$, then $\bigcup_{i\in I} A_{f_i} = A_f$ in $\mathsf{App}(B_{\mathbf{I}U}, B_{\mathbf{C}(\mathbb{R}^n)})$.*
2. If $(R_i)_{i\in I}$ is a directed set in $\mathsf{App}(B_{\mathbf{I}U}, B_{\mathbf{C}(\mathbb{R}^n)})$ then $\sup_{i\in I} G_{R_i} = G_R$ in $(\mathbf{I}U \to \mathbf{C}(\mathbb{R}^n))$ where $R = \sup_{i\in I} R_i$.

Definition 5. *If a is an open hyper-rectangle and O is a basic convex open set then the single-step approximable mapping $\eta_{(\Box a, O)}$ is defined as $\eta_{(\Box a, O)} = A_{\overline{O}\chi_{\Box a}} : B_{\mathbf{I}U} \to B_{\mathbf{C}(\mathbb{R}^n)}$.*

For defining the Lipschitzian derivative of an approximable mapping we first need to define the notions of a strong tie and a strong knot.

Definition 6. *We say $f : \mathbf{I}U \to \mathbf{I}\mathbb{R}$ has a* strong set-valued Lipschitz constant *$b \in \mathbf{C}(\mathbb{R}^n)$ in $\Box a$, for $a \in \mathcal{T}(U)$, denoted by $f \in \delta_s(\Box a, b)$, if there exist $a' \prec a$ and $b' \in \mathbf{C}(\mathbb{R}^n)$ with $b \ll_{\mathbf{C}(\mathbb{R}^n)} b'$ such that $f \in \delta(\Box a', b')$. We call $\delta_s(\Box a, b)$ the strong single-tie of $\Box a$ with b.*

From general results about single-step functions, [16] we know that if $b\chi_{\Box a} \ll \mathcal{L}f$, then for every $x \in \Box a$ we have $b \ll \mathcal{L}f(x)$, and hence, $\mathcal{L}f(x) \in \Uparrow b$. This means $\mathcal{L}f(\Box a) \subseteq \Uparrow b$. Moreover $\Box a \ll (\mathcal{L}f)^{-1}(\Uparrow b)$.

Similar to Proposition VII.3 in [13] and its corollary, we have:

Proposition 13. *If $f : \mathbf{I}U \to \mathbf{I}\mathbb{R}$ is locally Lipschitz, then:*

$$f \in \delta_s(\Box a, b) \Longleftrightarrow b\chi_{\Box a} \ll \mathcal{L}f$$

$$\mathcal{L}f = \sup\{b\chi_{\Box a} : b\chi_{\Box a} \ll \mathcal{L}f\} = \sup\{b\chi_{\Box a} : f \in \delta_s(\Box a, b)\}$$

Definition 7. *We say an approximable mapping $R : B_{IU} \to B_{I\mathbb{R}}$ has* strong Lipschitz constant *O in $\Box a$, for $O \in B^0_{\mathbb{R}^n}$ and $a \in \mathcal{T}(U)$, denoted by $R \in \Delta_s(\Box a, O)$, if there exist $a' \in \mathcal{T}(U)$ with $a \prec a'$ and $O' \in B^0_{\mathbb{R}^n}$ with $O' \prec O$ such that $R \in \Delta(\Box a', O')$.*

Proposition 14. *1. If $f \in \delta_s(\Box a, b)$ then for all $O \in B^0_{\mathbb{R}^n}$ with $b \subset O$ we have $A_f \in \Delta_s(\Box a, O)$.*
2. If $A_f \in \Delta_s(\Box a, O)$ then there exists $b \subset O$ such that $f \in \delta_s(\Box a, b)$.

Proof. 1. Let $f \in \delta_s(\Box a, b)$ and $b \subset O$, then there exists $a' \in \mathcal{T}(U)$ with $a \prec a'$ and b' with $b \ll b'$ such that $f \in \delta(\Box a', b')$. By the interpolation property of $\prec$ there exists a_0 with $a \prec a_0 \prec a'$ and O_0 with $b \subset O_0 \prec O$. By Proposition 10 we have $A_f \in \Delta(\Box a_0, O_0)$ and thus $A_f \in \Delta_s(\Box a, O)$.
2. Let $A_f \in \Delta_s(\Box a, O)$ then by the definition of strong knot there exists a' with $a \prec a'$ and O' with $O' \prec O$ such that $A_f \in \Delta(\Box a', O')$. By Corollary 3, $f \in \delta(\Box a', \overline{O'})$. By the interpolation property, there exists O'' with, $O' \prec O'' \prec O$. Let $b' = \overline{O'}$ and $b = \overline{O''}$ then $b \prec b'$ and $f \in \delta(\Box a', b')$. Hence, $f \in \delta_s(\Box a, b)$. ■

Finally, we obtain the duality between strong ties and strong knots extending the main result in [13] to functions with interval input and output.

Corollary 4. *We have $R \in \Delta_s(\Box a, O)$ iff $G_R \in \delta_s(\Box a, \overline{O})$. Dually, we have $f \in \delta_s(\Box a, b)$ iff $A_f \in \Delta_s(\Box a, b^\circ)$.*

Definition 8. *Let $R : B_{IU} \to B_{I\mathbb{R}}$ be a Lipschitzian approximable mapping. The Lipschitzian derivative of R is defied as:*

$$L(R) = \sup\{\eta_{(\Box a, O)} : R \in \Delta_s(\Box a, O)\}$$

which is of type $B_{IU} \to B_{\mathbf{C}(\mathbb{R}^n)}$.

The following theorem extends Theorem VII.12 in [13] to functions with interval input and output.

Theorem 2. *The Lipschitzian derivative of a Lipschitzian approximable mapping $R : B_{IU} \to B_{I\mathbb{R}}$ is an approximable mapping and we have: $L(R) = A_{\mathcal{L}G_R}$.*

4 Conclusion

We have developed a notion of sub-differentiation for Scott continuous maps which take hyper-rectangles in a finite dimensional Euclidean spaces to compact real intervals and is itself a Scott continuous map. This extends the domain of application of Interval Analysis to the classical derivative. It also extends Clarke's theory and that of the L-derivative to functions with imprecise input/output as one encounters in interval analysis and exact real number computation. The classical Clarke operator commutes with the extension operator that extends a non-empty convex and compact valued map of a finite dimensional Euclidean spaces to the space of the hyper-rectangles of the Euclidean space. We have derived a calculus for sub-differentiation of interval maps which is weaker than the corresponding Clarkes calculus for point maps. A Stone duality framework for sub-differentiation of interval maps is also constructed which allows for a program logic view of sub-differentiation. We envisage several areas for immediate further work, namely an implementation of this work in Haskell, an implementation in a theorem prover such as Coq and a derivation of a weak calculus for constructors of approximable mappings which would match the calculus for the interval functions.

References

1. Haskell Implementation of IC-Reals for Exact Real Computation. Imperial College London. http://www.doc.ic.ac.uk/exact-computation/Haskell
2. Abramsky, S., Jung, A.: Domain theory. In: Abramsky, S., Gabbay, D.M., Maibaum, T.S.E. (eds.) Handbook of Logic in Computer Science, vol. 3. Clarendon, Oxford (1994)
3. Bauer, A., Escardó, M.H., Simpson, A.: Comparing functional paradigms for exact real-number computation. In: Widmayer, P., Eidenbenz, S., Triguero, F., Morales, R., Conejo, R., Hennessy, M. (eds.) ICALP 2002. LNCS, vol. 2380, pp. 488–500. Springer, Heidelberg (2002). https://doi.org/10.1007/3-540-45465-9_42
4. Clarke, F.H.: Optimization and Nonsmooth Analysis. Wiley, New York (1983)
5. Di Gianantonio, P.: A functional approach to real number computation. Ph.D. thesis, University of Pisa (1993)
6. Di Gianantonio, P.: Real number computability and domain theory. Inf. Comput. **127**(1), 11–25 (1996)
7. Di Gianantonio, P., Edalat, A.: A language for differentiable functions. In: Proceedings of the 16th International Conference on Foundations of Software Science and Computation Structures (FoSSaCS) (2013)
8. Edalat, A.: A continuous derivative for real-valued functions. In: Cooper, S.B., Löwe, B., Sorbi, A. (eds.) CiE 2007. LNCS, vol. 4497, pp. 248–257. Springer, Heidelberg (2007). https://doi.org/10.1007/978-3-540-73001-9_26
9. Edalat, A., Escardó, M.: Integration in real PCF. In: Eleventh Annual IEEE Symposium on Logic in Computer Science (LICS). IEEE (1996)
10. Edalat, A., Heckmann, R.: Computing with real numbers. In: Barthe, G., Dybjer, P., Pinto, L., Saraiva, J. (eds.) APPSEM 2000. LNCS, vol. 2395, pp. 193–267. Springer, Heidelberg (2002). https://doi.org/10.1007/3-540-45699-6_5
11. Edalat, A., Lieutier, A.: Domain theory and differential calculus (functions of one variable). In: LICS. IEEE (2002). www.doc.ic.ac.uk/~ae/papers/diffcal.ps. Full paper to appear in MSCS
12. Edalat, A., Lieutier, A.: Foundation of a computable solid modelling. Theoret. Comput. Sci. **284**(2), 319–345 (2002)
13. Edalat, A., Maleki, M.: Differentiation in logical form. In: Proceedings of 32th ACM/IEEE Symposium on Logic in Computer Science (LICS 2017). ACM/IEEE (2017)
14. Edalat, A., Potts, P.J.: A new representation for exact real numbers. In: Proceedings of Mathematical Foundations of Programming Semantics 13, Electronic Notes in Theoretical Computer Science, vol. 6. Elsevier Science B.V. (1997). www.elsevier.nl/locate/entcs/volume6.html

15. Edalat, A., Potts, P.J., Sünderhauf, P.: Lazy computation with exact real numbers. In: Proceedings of the Third ACM SIGPLAN International Conference on Functional Programming, pp. 185–194. ACM (1998)
16. Erker, T., Escardó, M., Keimel, K.: The way-below relation of function spaces over semantic domains. Topol. Appl. **89**(1–2), 61–74 (1998)
17. Escardó, M.H.: PCF extended with real numbers. Theor. Comput. Sci. **162**(1), 79–115 (1996)
18. Gierz, G., Hofmann, K.H., Keimel, K., Lawson, J.D., Mislove, M., Scott, D.S.: Continuous Lattices and Domains. Cambridge University Press, Cambridge (2003)
19. Jung, A., Sünderhauf, P.: On the duality of compact vs. open. Ann. New York Acad. Sci. **806**(1), 214–230 (1996)
20. Moore, R., Kearfott, R., Cloud, M.: Introduction to Interval Analysis. Society for Industrial and Applied Mathematics, Philadelphia (2009)
21. Moore, R.E.: Interval Analysis. Prentice-Hall, Englewood Cliffs (1966)
22. Potts, P.J., Edalat, A., Escardó, M.: Semantics of exact real arithmetic. In: Twelfth Annual IEEE Symposium on Logic in Computer Science. IEEE (1997)
23. Pour-El, M.B., Richards, J.I.: Computability in Analysis and Physics. Springer, New York (1988)
24. Scott, D.S.: Continuous lattices. In: Lawvere, F.W. (ed.) Toposes, Algebraic Geometry and Logic. LNM, vol. 274, pp. 97–136. Springer, Heidelberg (1972). https://doi.org/10.1007/BFb0073967
25. Smyth, M.B.: Effectively given domains. Theor. Comput. Sci. **5**, 257–274 (1977)
26. Vickers, S.J.: Geometric logic in computer science. In: Burn, G.L., Gay, S.J., Ryan, M.D. (eds.) Theory and Formal Methods, pp. 37–54. Springer, Heidelberg (1993). https://doi.org/10.1007/978-1-4471-3503-6_4
27. Weihrauch, K.: Computable Analysis (An Introduction). Springer, Heidelberg (2000). https://doi.org/10.1007/978-3-642-56999-9

The Effects of Adding Reachability Predicates in Propositional Separation Logic

Stéphane Demri[1], Étienne Lozes[2], and Alessio Mansutti[1](✉)

[1] LSV, CNRS, ENS Paris-Saclay, Université Paris-Saclay, Cachan, France
alessio.mansutti@lsv.fr
[2] I3S, Université Côte d'Azur, Nice, France

Abstract. The list segment predicate `ls` used in separation logic for verifying programs with pointers is well-suited to express properties on singly-linked lists. We study the effects of adding `ls` to the full propositional separation logic with the separating conjunction and implication, which is motivated by the recent design of new fragments in which all these ingredients are used indifferently and verification tools start to handle the magic wand connective. This is a very natural extension that has not been studied so far. We show that the restriction without the separating implication can be solved in polynomial space by using an appropriate abstraction for memory states whereas the full extension is shown undecidable by reduction from first-order separation logic. Many variants of the logic and fragments are also investigated from the computational point of view when `ls` is added, providing numerous results about adding reachability predicates to propositional separation logic.

1 Introduction

Separation logic [20,25,28] is a well-known assertion logic for reasoning about programs with dynamic data structures. Since the implementation of Smallfoot and the evidence that the method is scalable [3,33], many tools supporting separation logic as an assertion language have been developed [3,8,9,16,17,33]. Even though the first tools could handle relatively limited fragments of separation logic, like symbolic heaps, there is a growing interest and demand to consider extensions with richer expressive power. We can point out three particular extensions of symbolic heaps (without list predicates) that have been proved decidable.

– Symbolic heaps with generalised inductive predicates, adding a fixpoint combinator to the language, is a convenient logic for specifying data structures that are more advanced than lists or trees. The entailment problem is known to be decidable by means of tree automata techniques for the bounded tree-width fragment [1,19], whereas satisfiability is ExpTime-complete [6]. Other related results can be found in [21].

C. Baier and U. Dal Lago (Eds.): FOSSACS 2018, LNCS 10803, pp. 476–493, 2018.
https://doi.org/10.1007/978-3-319-89366-2_26

- List-free symbolic heaps with all classical Boolean connectives $\wedge$ and $\neg$ (and with the separating conjunction $*$), called herein SL($*$), is a convenient extension when combinations of results of various analysis need to be expressed, or when the analysis requires a complementation. This extension already is PSPACE-complete [11].
- Propositional separation logic with separating implication, a.k.a. magic wand ($-\!\!*$), is a convenient fragment (called herein SL($*, -\!\!*$)) in which can be solved two problems of frame inference and abduction, that play an important role in static analysers and provers built on top of separation logic. SL($*, -\!\!*$) can be decided in PSPACE thanks to a small model property [32].

A natural question is how to combine these extensions, and which separation logic fragment that allows Boolean connectives, magic wand and generalised recursive predicates can be decided with some adequate restrictions. As already advocated in [7,18,24,29,31], dealing with the separating implication $-\!\!*$ is a desirable feature for program verification and several semi-automated or automated verification tools support it in some way, see e.g. [18,24,29,31].

Our Contribution. In this paper, we address the question of combining magic wand and inductive predicates in the extremely limited case where the only inductive predicate is the gentle list segment predicate `ls`. So the starting point of this work is this puzzling question: what is the complexity/decidability status of propositional separation logic SL($*, -\!\!*$) enriched with the list segment predicate `ls` (herein called SL($*, -\!\!*$, `ls`))? More precisely, we study the decidability/complexity status of extensions of propositional separation logic SL($*, -\!\!*$) by adding one of the reachability predicates among `ls` (precise predicate as usual in separation logic), `reach` (existence of a path, possibly empty) and `reach`$^+$ (existence of a non-empty path).

First, we establish that the satisfiability problem for the propositional separation logic SL($*, -\!\!*$, `ls`) is undecidable. Our proof is by reduction from the undecidability of first-order separation logic [5,14], using an encoding of the variables as heap cells (see Theorem 1). As a consequence, we also establish that SL($*, -\!\!*$, `ls`) is not finitely axiomatisable. Moreover, our reduction requires a rather limited expressive power of the list segment predicate, and we can strengthen our undecidability results to some fragments of SL($*, -\!\!*$, `ls`). For instance, surprisingly, the extension of SL($*, -\!\!*$) with the atomic formulae of the form $\mathtt{reach}(\mathtt{x},\mathtt{y}) = 2$ and $\mathtt{reach}(\mathtt{x},\mathtt{y}) = 3$ (existence of a path between `x` and `y` of respective length 2 or 3) is already undecidable, whereas the satisfiability problem for SL($*, -\!\!*$, $\mathtt{reach}(\mathtt{x},\mathtt{y}) = 2$) is known to be in PSPACE [15].

Second, we show that the satisfiability problem for SL($*$, `reach`$^+$) is PSPACE-complete, extending the well-known result on SL($*$). The PSPACE upper bound relies on a small heap property based on the techniques of test formulae, see e.g. [4,15,22,23], and the PSPACE-hardness of SL($*$) is inherited from [11]. The PSPACE upper bound can be extended to the fragment of SL($*, -\!\!*$, `reach`$^+$) made of Boolean combinations of formulae from SL($*$, `reach`$^+$) $\cup$ SL($*, -\!\!*$) (see the developments in Sect. 4). Even better, we show that the fragment of

SL($*, \mathrel{-\!\!*}$, $\mathtt{reach}^+$) in which $\mathtt{reach}^+$ is not in the scope of $\mathrel{-\!\!*}$ is decidable. As far as we know, this is the largest fragment including full Boolean expressivity, $\mathrel{-\!\!*}$ and $\mathtt{ls}$ for which decidability is established.

2 Preliminaries

Let PVAR $= \{\mathtt{x}, \mathtt{y}, \ldots\}$ be a countably infinite set of *program variables* and LOC $= \{\ell_0, \ell_1, \ell_2, \ldots\}$ be a countable infinite set of *locations*. A *memory state* is a pair (s, h) such that s : PVAR $\rightarrow$ LOC is a variable valuation (known as the *store*) and h : LOC $\rightarrow_{\mathrm{fin}}$ LOC is a partial function with finite domain, known as the *heap*. We write $\mathrm{dom}(h)$ to denote its domain and $\mathrm{ran}(h)$ to denote its range. Given a heap h with $\mathrm{dom}(h) = \{\ell_1, \ldots, \ell_n\}$, we also write $\{\ell_1 \mapsto h(\ell_1), \ldots, \ell_n \mapsto h(\ell_n)\}$ to denote h. Each $\ell_i \mapsto h(\ell_i)$ is understood as a *memory cell* of h.

As usual, the heaps h_1 and h_2 are said to be *disjoint*, written $h_1 \perp h_2$, if $\mathrm{dom}(h_1) \cap \mathrm{dom}(h_2) = \emptyset$; when this holds, we write $h_1 + h_2$ to denote the heap corresponding to the disjoint union of the graphs of h_1 and h_2, hence $\mathrm{dom}(h_1 + h_2) = \mathrm{dom}(h_1) \uplus \mathrm{dom}(h_2)$. When the domains of h_1 and h_2 are not disjoint, the composition $h_1 + h_2$ is not defined. Moreover, we write $h' \sqsubseteq h$ to denote that $\mathrm{dom}(h') \subseteq \mathrm{dom}(h)$ and for all locations $\ell \in \mathrm{dom}(h')$, we have $h'(\ell) = h(\ell)$. The formulae φ of the separation logic SL($*, \mathrel{-\!\!*}, \mathtt{ls}$) and its atomic formulae π are built from $\pi ::= \mathtt{x} = \mathtt{y} \mid \mathtt{x} \hookrightarrow \mathtt{y} \mid \mathtt{ls}(\mathtt{x}, \mathtt{y}) \mid \mathtt{emp} \mid \top$ and $\varphi ::= \pi \mid \neg\varphi \mid \varphi \wedge \varphi \mid \varphi * \varphi \mid \varphi \mathrel{-\!\!*} \varphi$, where $\mathtt{x}, \mathtt{y} \in$ PVAR ($\Rightarrow$, $\Leftrightarrow$ and $\vee$ are defined as usually). Models of the logic SL($*, \mathrel{-\!\!*}, \mathtt{ls}$) are memory states and the satisfaction relation $\models$ is defined as follows (omitting standard clauses for $\neg, \wedge$):

$$
\begin{array}{lcl}
(s,h) \models \mathtt{x} = \mathtt{y} & \Longleftrightarrow & s(\mathtt{x}) = s(\mathtt{y}) \\
(s,h) \models \mathtt{emp} & \Longleftrightarrow & \mathrm{dom}(h) = \emptyset \\
(s,h) \models \mathtt{x} \hookrightarrow \mathtt{y} & \Longleftrightarrow & s(\mathtt{x}) \in \mathrm{dom}(h) \text{ and } h(s(\mathtt{x})) = s(\mathtt{y}) \\
(s,h) \models \mathtt{ls}(\mathtt{x},\mathtt{y}) & \Longleftrightarrow & \text{either } (\mathrm{dom}(h) = \emptyset \text{ and } s(\mathtt{x}) = s(\mathtt{y})) \text{ or} \\
 & & h = \{\ell_0 \mapsto \ell_1, \ell_1 \mapsto \ell_2, \ldots, \ell_{n-1} \mapsto \ell_n\} \text{ with } n \geq 1, \\
 & & \ell_0 = s(\mathtt{x}), \ell_n = s(\mathtt{y}) \text{ and for all } i \neq j \in [0,n], \ell_i \neq \ell_j \\
(s,h) \models \varphi_1 * \varphi_2 & \Longleftrightarrow & \text{there are } h_1 \text{ and } h_2 \text{ such that } (h_1 \perp h_2, (h_1 + h_2) = h, \\
 & & (s,h_1) \models \varphi_1 \text{ and } (s,h_2) \models \varphi_2) \\
(s,h) \models \varphi_1 \mathrel{-\!\!*} \varphi_2 & \Longleftrightarrow & \forall h_1 \text{ if } (h_1 \perp h \text{ and } (s,h_1) \models \varphi_1) \text{ then } (s, h + h_1) \models \varphi_2.
\end{array}
$$

Note that the semantics for $*$, $\mathrel{-\!\!*}$, $\hookrightarrow$, $\mathtt{ls}$ and for all other ingredients is the usual one in separation logic and $\mathtt{ls}$ is the *precise* list segment predicate. In the sequel, we use the following abbreviations: $\mathtt{size} \geq 0 \stackrel{\text{def}}{=} \top$ and for all $\beta \geq 0$, $\mathtt{size} \geq \beta{+}1 \stackrel{\text{def}}{=} (\mathtt{size} \geq \beta) * \neg\mathtt{emp}$, $\mathtt{size} \leq \beta \stackrel{\text{def}}{=} \neg(\mathtt{size} \geq \beta{+}1)$ and $\mathtt{size} = \beta \stackrel{\text{def}}{=} (\mathtt{size} \leq \beta) \wedge (\mathtt{size} \geq \beta)$. Moreover, $\varphi_1 \circledast \varphi_2 \stackrel{\text{def}}{=} \neg(\varphi_1 \mathrel{-\!\!*} \neg\varphi_2)$ (*septraction connective*), $\mathtt{alloc}(\mathtt{x}) \stackrel{\text{def}}{=} (\mathtt{x} \hookrightarrow \mathtt{x}) \mathrel{-\!\!*} \perp$ and $\mathtt{x} \mapsto \mathtt{y} \stackrel{\text{def}}{=} (\mathtt{x} \hookrightarrow \mathtt{y}) \wedge \mathtt{size} = 1$. W.l.o.g., we can assume that LOC $= \mathbb{N}$ since none of the developments depend on the elements of LOC as the only predicate involving locations is the equality. We write SL($*, \mathrel{-\!\!*}$) to denote the restriction of SL($*, \mathrel{-\!\!*}, \mathtt{ls}$) without $\mathtt{ls}$. Similarly, we write SL($*$) to denote the restriction of SL($*, -\!\!*$) without $-\!\!*$. Given two formulae φ, φ' (possibly from different logical languages), we write $\varphi \equiv \varphi'$ whenever for all (s,h), we have $(s,h) \models \varphi$ iff $(s,h) \models \varphi'$. When $\varphi \equiv \varphi'$, the formulae φ and φ' are said to be *equivalent*.

Variants with Other Reachability Predicates. We use two additional reachability predicates $\mathtt{reach}(\mathtt{x},\mathtt{y})$ and $\mathtt{reach}^+(\mathtt{x},\mathtt{y})$ and we write SL($*, -\!\!*, \mathtt{reach}$) (resp. SL($*, -\!\!*, \mathtt{reach}^+$)) to denote the variant of SL($*, -\!\!*, \mathtt{ls}$) in which $\mathtt{ls}$ is replaced by $\mathtt{reach}$ (resp. by $\mathtt{reach}^+$). The relation $\models$ is extended as follows: $(s,h) \models \mathtt{reach}(\mathtt{x},\mathtt{y})$ holds when there is $i \geq 0$ such that $h^i(s(\mathtt{x})) = s(\mathtt{y})$ (i functional composition(s) of h is denoted by h^i) and $(s,h) \models \mathtt{reach}^+(\mathtt{x},\mathtt{y})$ holds when there is $i \geq 1$ such that $h^i(s(\mathtt{x})) = s(\mathtt{y})$. As $\mathtt{ls}(\mathtt{x},\mathtt{y}) \equiv \mathtt{reach}(\mathtt{x},\mathtt{y}) \wedge \neg(\neg\mathtt{emp} * \mathtt{reach}(\mathtt{x},\mathtt{y}))$ and $\mathtt{reach}(\mathtt{x},\mathtt{y}) \equiv \top * \mathtt{ls}(\mathtt{x},\mathtt{y})$, the logics SL($*, -\!\!*, \mathtt{reach}$) and SL($*, -\!\!*, \mathtt{ls}$) have identical decidability status. As far as computational complexity is concerned, a similar analysis can be done as soon as $*$, $\neg$, $\wedge$ and $\mathtt{emp}$ are parts of the fragments (the details are omitted here). Similarly, we have the equivalences: $\mathtt{reach}(\mathtt{x},\mathtt{y}) \equiv \mathtt{x} = \mathtt{y} \vee \mathtt{reach}^+(\mathtt{x},\mathtt{y})$ and $\mathtt{ls}(\mathtt{x},\mathtt{y}) \equiv (\mathtt{x} = \mathtt{y} \wedge \mathtt{emp}) \vee (\mathtt{reach}^+(\mathtt{x},\mathtt{y}) \wedge \neg(\neg\mathtt{emp} * \mathtt{reach}^+(\mathtt{x},\mathtt{y})))$. So clearly, SL($*, \mathtt{reach}$) and SL($*, \mathtt{ls}$) can be viewed as fragments of SL($*, \mathtt{reach}^+$) and, SL($*, -\!\!*, \mathtt{ls}$) as a fragment of SL($*, -\!\!*, \mathtt{reach}^+$). It is therefore stronger to establish decidability or complexity upper bounds with $\mathtt{reach}^+$ and to show undecidability or complexity lower bounds with $\mathtt{ls}$ or $\mathtt{reach}$. Herein, we provide the optimal results.

Decision Problems. Let $\mathfrak{L}$ be a logic defined above. As usual, the *satisfiability problem for* $\mathfrak{L}$ takes as input a formula φ from $\mathfrak{L}$ and asks whether there is (s,h) such that $(s,h) \models \varphi$. The *validity problem* is also defined as usual. The *model-checking problem for* $\mathfrak{L}$ takes as input a formula φ from $\mathfrak{L}$, (s,h) and asks whether $(s,h) \models \varphi$ (s is restricted to the variables occurring in φ and h is encoded as a finite and functional graph). Unless otherwise specified, the *size* of a formula φ is understood as its tree size, i.e. approximately its number of symbols.

The main purpose of this paper is to study the decidability/complexity status of SL($*, -\!\!*, \mathtt{ls}$) and its fragments.

3 Undecidability of SL($*, -\!\!*, \mathtt{ls}$)

In this section, we show that SL($*, -\!\!*, \mathtt{ls}$) has an undecidable satisfiability problem even though it does not admit first-order quantification.

Let SL($\forall, -\!\!*$) be the first-order extension of SL($-\!\!*$) obtained by adding the universal quantifier $\forall$. The formulae φ of SL($\forall, -\!\!*$) are built from $\pi ::= \mathtt{x} = \mathtt{y} \mid \mathtt{x} \hookrightarrow \mathtt{y}$ and $\varphi ::= \pi \mid \neg\varphi \mid \varphi \vee \varphi \mid \varphi -\!\!* \varphi \mid \forall \mathtt{x}\ \varphi$, where $\mathtt{x}, \mathtt{y} \in$ PVAR. Note that $\mathtt{emp}$ can be easily defined by $\forall\ \mathtt{x}, \mathtt{x}'\ \neg(\mathtt{x} \hookrightarrow \mathtt{x}')$. Models of the logic SL($\forall, -\!\!*$) are memory states and the satisfaction relation $\models$ is defined as for SL($-\!\!*$) with the additional clause:

$$(s,h) \models \forall \mathtt{x}\ \varphi \iff \text{for all } \ell \in \text{LOC}, \text{ we have } (s[\mathtt{x} \leftarrow \ell], h) \models \varphi.$$

Without any loss of generality, we can assume that the satisfiability [resp. validity] problem for $\mathrm{SL}(\forall, -\!\!*)$ is defined by taking as inputs closed formulae (i.e. without free occurrences of the variables).

Proposition 1. *[5,14] The satisfiability problem for* $\mathrm{SL}(\forall, -\!\!*)$ *is undecidable and the set of valid formulae for* $\mathrm{SL}(\forall, -\!\!*)$ *is not recursively enumerable.*

In a nutshell, we establish the undecidability of $\mathrm{SL}(*, -\!\!*, \mathtt{ls})$ by reduction from the satisfiability problem for $\mathrm{SL}(\forall, -\!\!*)$. The reduction is nicely decomposed in two intermediate steps: (1) the undecidability of $\mathrm{SL}(*, -\!\!*)$ extended with a few atomic predicates, to be defined soon, and (2) a *tour de force* resulting in the encoding of these atomic predicates in $\mathrm{SL}(*, -\!\!*, \mathtt{ls})$.

3.1 Encoding Quantified Variables as Cells in the Heap

In this section, we assume for a moment that we can express three atomic predicates $\mathtt{alloc}^{-1}(\mathtt{x})$, $n(\mathtt{x}) = n(\mathtt{y})$ and $n(\mathtt{x}) \hookrightarrow n(\mathtt{y})$, that will be used in the translation and have the following semantics:

- $(s,h) \models \mathtt{alloc}^{-1}(\mathtt{x})$ holds whenever $s(\mathtt{x}) \in \mathrm{ran}(h)$,
- $(s,h) \models n(\mathtt{x}) = n(\mathtt{y})$ holds iff $\{s(\mathtt{x}), s(\mathtt{y})\} \subseteq \mathrm{dom}(h)$ and $h(s(\mathtt{x})) = h(s(\mathtt{y}))$,
- $(s,h) \models n(\mathtt{x}) \hookrightarrow n(\mathtt{y})$ holds iff $\{s(\mathtt{x}), s(\mathtt{y})\} \subseteq \mathrm{dom}(h)$ and $h^2(s(\mathtt{x})) = h(s(\mathtt{y}))$.

Let us first intuitively explain how the two last predicates will help encoding $\mathrm{SL}(\forall, -\!\!*)$. By definition, the satisfaction of the quantified formula $\forall \mathtt{x}\, \psi$ from $\mathrm{SL}(\forall, -\!\!*)$ requires the satisfaction of the formula ψ for all the values in LOC assigned to $\mathtt{x}$. The principle of the encoding is to use a set L of locations initially not in the domain or range of the heap to mimic the store by modifying how they are allocated. In this way, a variable will be interpreted by a location in the heap and, instead of checking whenever $\mathtt{x} \hookrightarrow \mathtt{y}$ (or $\mathtt{x} = \mathtt{y}$) holds, we will check if $n(\mathtt{x}) \hookrightarrow n(\mathtt{y})$ (or $n(\mathtt{x}) = n(\mathtt{y})$) holds, where $\mathtt{x}$ and $\mathtt{y}$ correspond, after the translation, to the locations in L that mimic the store for those variables. Let X be the set of variables needed for the translation. In order to properly encode the store, each location in L only mimics exactly one variable, i.e. there is a bijection between X and L, and cannot be reached by any location. As such, the formula $\forall \mathtt{x}\, \psi$ will be encoded by the formula $(\mathtt{alloc}(\mathtt{x}) \wedge \mathtt{size} = 1) -\!\!* (\mathrm{OK}(X) \Rightarrow \mathrm{T}(\psi))$, where $\mathrm{OK}(X)$ (formally defined below) checks whenever the locations in L still satisfy the auxiliary conditions just described, whereas $\mathrm{T}(\psi)$ is the translation of ψ.

Unfortunately, the formula $\psi_1 -\!\!* \psi_2$ cannot simply be translated into $\mathrm{T}(\psi_1) -\!\!* (\mathrm{OK}(X) \Rightarrow \mathrm{T}(\psi_2))$ because the evaluation of $\mathrm{T}(\psi_1)$ in a disjoint heap may need the values of free variables occurring in ψ_1 but our encoding of the variable valuations via the heap does not allow to preserve these values through disjoint heaps. In order to solve this problem, for each variable $\mathtt{x}$ in the formula, X will contain an auxiliary variable $\overline{\mathtt{x}}$, or alternatively we define on X an involution $\overline{(.)}$. If the translated formula has q variables then the set X of variables needed for the translation will have cardinality $2q$. In the translation of a formula whose outermost connective is the magic wand, the locations corresponding to variables of the form $\overline{\mathtt{x}}$ will be allocated on the left side of the magic wand, and checked to be equal to their non-bar versions on the right side of the magic wand. As such, the left side of the magic wand will be translated into

$$((\bigwedge_{\mathtt{z}\in Z}\mathtt{alloc}(\overline{\mathtt{z}})) \wedge (\bigwedge_{\mathtt{z}\in X\setminus Z}\neg\mathtt{alloc}(\mathtt{z})) \wedge \mathrm{OK}(Z) \wedge \mathrm{T}(\psi_1)[\mathtt{z} \leftarrow \overline{\mathtt{z}} \mid \mathtt{z} \in X]),$$

where Z is the set of free variables in ψ_1, whereas the right side will be

$$(((\bigwedge_{\mathtt{z}\in Z} n(\mathtt{z}) = n(\overline{\mathtt{z}})) \wedge \mathrm{OK}(X)) \Rightarrow ((\bigwedge_{\mathtt{z}\in Z}\mathtt{alloc}(\overline{\mathtt{z}}) \wedge \mathtt{size} = \mathrm{card}(Z)) * \mathrm{T}(\psi_2))).$$

The use of the separating conjunction before the formula $\mathrm{T}(\psi_2)$ separates the memory cells corresponding to $\overline{\mathtt{x}}$ from the rest of the heap. By doing this, we can reuse $\overline{\mathtt{x}}$ whenever a magic wand appears in $\mathrm{T}(\psi_2)$.

For technical convenience, we consider a slight alternative for the semantics of the logics $\mathrm{SL}(\forall, -\!\!*)$ and $\mathrm{SL}(*, -\!\!*, \mathtt{ls})$, which does not modify the notion of satisfiability/validity and such that the set of formulae and the definition of the satisfaction relation $\models$ remain unchanged. So far, the memory states are pairs of the form (s, h) with $s : \mathrm{PVAR} \to \mathrm{LOC}$ and $h : \mathrm{LOC} \to_{\mathrm{fin}} \mathrm{LOC}$ for a *fixed* countably infinite set of locations LOC, say $\mathrm{LOC} = \mathbb{N}$. Alternatively, the models for $\mathrm{SL}(\forall, -\!\!*)$ and $\mathrm{SL}(*, -\!\!*, \mathtt{ls})$ can be defined as triples $(\mathrm{LOC}_1, s_1, h_1)$ such that LOC_1 is a countable infinite set, $s_1 : \mathrm{PVAR} \to \mathrm{LOC}_1$ and $h_1 : \mathrm{LOC}_1 \to_{\mathrm{fin}} \mathrm{LOC}_1$. As shown below, this does not change the notion of satisfiability and validity, but this generalisation will be handy in a few places. Most of the time, a generalised memory state $(\mathrm{LOC}_1, s_1, h_1)$ shall be written (s_1, h_1) when no confusion is possible.

Given a bijection $\mathfrak{f} : \mathrm{LOC}_1 \to \mathrm{LOC}_2$ and a heap $h_1 : \mathrm{LOC}_1 \to_{\mathrm{fin}} \mathrm{LOC}_1$ equal to $\{\ell_1 \mapsto h_1(\ell_1), \ldots, \ell_n \mapsto h_1(\ell_n)\}$, we write $\mathfrak{f}(h_1)$ to denote the heap $h_2 : \mathrm{LOC}_2 \to_{\mathrm{fin}} \mathrm{LOC}_2$ with $h_2 = \{\mathfrak{f}(\ell_1) \mapsto \mathfrak{f}(h_1(\ell_1)), \ldots, \mathfrak{f}(\ell_n) \mapsto \mathfrak{f}(h_1(\ell_n))\}$.

Definition 1. *Let* $(\mathrm{LOC}_1, s_1, h_1)$ *and* $(\mathrm{LOC}_2, s_2, h_2)$ *be generalised memory states and* $X \subseteq \mathrm{PVAR}$. *A* partial isomorphism *with respect to* X *from* $(\mathrm{LOC}_1, s_1, h_1)$ *to* $(\mathrm{LOC}_2, s_2, h_2)$ *is a bijection* $\mathfrak{f} : \mathrm{LOC}_1 \to \mathrm{LOC}_2$ *such that* $h_2 = \mathfrak{f}(h_1)$ *and for all* $\mathtt{x} \in X$, $\mathfrak{f}(s_1(\mathtt{x})) = s_2(\mathtt{x})$ *(we write* $(\mathrm{LOC}_1, s_1, h_1) \approx_X (\mathrm{LOC}_2, s_2, h_2)$*).*

A folklore result states that isomorphic memory states satisfy the same formulae since the logics $\mathrm{SL}(\forall, -\!\!*)$, $\mathrm{SL}(*, -\!\!*, \mathtt{ls})$ can only perform equality tests.

Lemma 1. *Let* $(\mathrm{LOC}_1, s_1, h_1)$ *and* $(\mathrm{LOC}_2, s_2, h_2)$ *be two generalised memory states such that* $(\mathrm{LOC}_1, s_1, h_1) \approx_X (\mathrm{LOC}_2, s_2, h_2)$*, for some* $X \subseteq \mathrm{PVAR}$. *(I) For all formulae* φ *in* $\mathrm{SL}(\forall, -\!\!*)$ *whose free variables are among* X*, we have* $(\mathrm{LOC}_1, s_1, h_1) \models \varphi$ *iff* $(\mathrm{LOC}_2, s_2, h_2) \models \varphi$. *(II) For all formulae* φ *in* $\mathrm{SL}(*, -\!\!*, \mathtt{ls})$ *built on variables among* X*, we have* $(\mathrm{LOC}_1, s_1, h_1) \models \varphi$ *iff* $(\mathrm{LOC}_2, s_2, h_2) \models \varphi$.

As a direct consequence, satisfiability in $\mathrm{SL}(*, -\!\!*, \mathtt{ls})$ as defined in Sect. 2, is equivalent to satisfiability with generalised memory states, the same holds for $\mathrm{SL}(\forall, -\!\!*)$. Next, we define the encoding of a generalised memory state. This can be seen as the semantical counterpart of the syntactical translation process and, as such, formalise the intuition of using part of a heap to mimic the store.

Definition 2. *Let* $X = \{\mathtt{x}_1, \ldots, \mathtt{x}_{2q}\}$, $Y \subseteq \{\mathtt{x}_1, \ldots, \mathtt{x}_q\}$ *and,* $(\mathrm{LOC}_1, s_1, h_1)$ *and* $(\mathrm{LOC}_2, s_2, h_2)$ *be two (generalised) memory states. We say that* $(\mathrm{LOC}_1, s_1, h_1)$ *is encoded by* $(\mathrm{LOC}_2, s_2, h_2)$ *w.r.t.* X, Y, *written* $(\mathrm{LOC}_1, s_1, h_1) \rhd_q^Y (\mathrm{LOC}_2, s_2, h_2)$, *if the following conditions hold:*

- $\mathrm{LOC}_1 = \mathrm{LOC}_2 \setminus \{s_2(\mathtt{x}) \mid \mathtt{x} \in X\}$,
- *for all* $\mathtt{x} \neq \mathtt{y} \in X$, $s_2(\mathtt{x}) \neq s_2(\mathtt{y})$,
- $h_2 = h_1 + \{s_2(\mathtt{x}) \mapsto s_1(\mathtt{x}) \mid \mathtt{x} \in Y\}$.

Notice that h_2 is equal to h_1 plus the heap $\{s_2(\mathtt{x}) \mapsto s_1(\mathtt{x}) \mid \mathtt{x} \in Y\}$ that encodes the store s_1. The picture below presents a memory state (left) and its encoding (right), where $Y = \{\mathtt{x}_i, \mathtt{x}_j, \mathtt{x}_k\}$. From the encoding, we can retrieve the initial heap by removing the memory cells corresponding to $\mathtt{x}_i$, $\mathtt{x}_j$ and $\mathtt{x}_k$. By way of example, the memory state on the left satisfies the formulae $\mathtt{x}_i = \mathtt{x}_j$, $\mathtt{x}_i \hookrightarrow \mathtt{x}_k$ and $\mathtt{x}_k \hookrightarrow \mathtt{x}_k$ whereas its encoding satisfies the formulae $n(\mathtt{x}_i) = n(\mathtt{x}_j)$, $n(\mathtt{x}_i) \hookrightarrow n(\mathtt{x}_k)$ and $n(\mathtt{x}_k) \hookrightarrow n(\mathtt{x}_k)$.

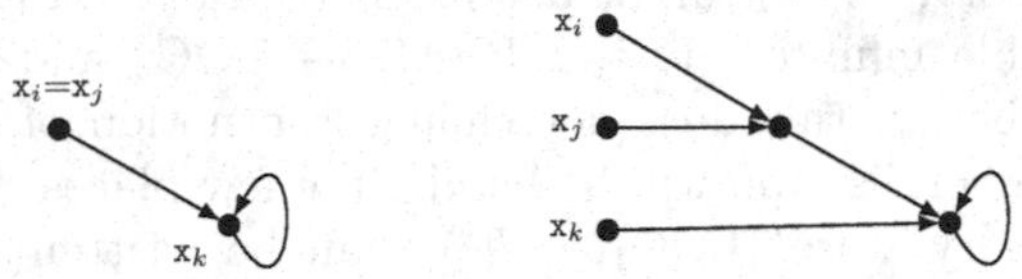

3.2 The Translation

We are now ready to define the translation of a first-order formula in propositional separation logic extended with the three predicates introduced at the beginning of the section. Let φ be a closed formula of $\mathrm{SL}(\forall, -\!\!*)$ with quantified variables $\{\mathtt{x}_1, \ldots, \mathtt{x}_q\}$. W.l.o.g., we can assume that distinct quantifications involve distinct variables. Moreover, let $X = \{\mathtt{x}_1, \ldots, \mathtt{x}_{2q}\}$ and $\overline{(.)}$ be the involution on X such that for all $i \in [1, q]$ $\overline{\mathtt{x}}_i \stackrel{\text{def}}{=} \mathtt{x}_{i+q}$.

We write $\mathrm{OK}(X)$ to denote the formula $(\bigwedge_{i \neq j} \mathtt{x}_i \neq \mathtt{x}_j) \wedge (\bigwedge_i \neg \mathtt{alloc}^{-1}(\mathtt{x}_i))$. The translation function T has two arguments: the formula in $\mathrm{SL}(\forall, -\!\!*)$ to be recursively translated and the total set of variables potentially appearing in the target formula (useful to check that $\mathrm{OK}(X)$ holds on every heap involved in the satisfaction of the translated formula). Let us come back to the definition of $\mathrm{T}(\psi, X)$ (homomorphic for Boolean connectives) with the assumption that the variables in ψ are among $\mathtt{x}_1, \ldots, \mathtt{x}_q$.

$$\mathrm{T}(\mathtt{x}_i = \mathtt{x}_j, X) \stackrel{\text{def}}{=} n(\mathtt{x}_i) = n(\mathtt{x}_j)$$
$$\mathrm{T}(\mathtt{x}_i \hookrightarrow \mathtt{x}_j, X) \stackrel{\text{def}}{=} n(\mathtt{x}_i) \hookrightarrow n(\mathtt{x}_j)$$
$$\mathrm{T}(\forall \mathtt{x}_i\ \psi, X) \stackrel{\text{def}}{=} (\mathtt{alloc}(\mathtt{x}_i) \wedge \mathtt{size} = 1) \mathrel{-\!\!*} (\mathrm{OK}(X) \Rightarrow \mathrm{T}(\psi, X))$$

Lastly, the translation $\mathrm{T}(\psi_1 \mathrel{-\!\!*} \psi_2, X)$ is defined as

$$((\bigwedge_{\mathtt{z} \in Z} \mathtt{alloc}(\bar{\mathtt{z}})) \wedge (\bigwedge_{\mathtt{z} \in X \setminus Z} \neg\mathtt{alloc}(\bar{\mathtt{z}})) \wedge \mathrm{OK}(X) \wedge \mathrm{T}(\psi_1, X)[\mathtt{x} \leftarrow \bar{\mathtt{x}}]) \mathrel{-\!\!*}$$

$$(((\bigwedge_{\mathtt{z} \in Z} n(\mathtt{z}) = n(\bar{\mathtt{z}})) \wedge \mathrm{OK}(X)) \Rightarrow ((\bigwedge_{\mathtt{z} \in Z} \mathtt{alloc}(\bar{\mathtt{z}}) \wedge \mathtt{size} = \mathrm{card}(Z)) * \mathrm{T}(\psi_2, X))),$$

where $Z \subseteq \{\mathtt{x}_1, \ldots, \mathtt{x}_q\}$ is the set of free variables in ψ_1.

Here is the main result of this section, which is essential for the correctness of $\mathcal{T}_{\mathrm{SAT}}(\varphi)$, defined below.

Lemma 2. *Let* $X = \{\mathtt{x}_1, \ldots, \mathtt{x}_{2q}\}$, $Y \subseteq \{\mathtt{x}_1, \ldots, \mathtt{x}_q\}$, ψ *be a formula in* SL($\forall, \mathrel{-\!\!*}$) *with free variables among* Y *that does not contain any bound variable of* ψ *and* $(\mathrm{LOC}_1, s_1, h_1) \rhd_q^Y (\mathrm{LOC}_2, s_2, h_2)$. *We have* $(s_1, h_1) \models \psi$ *iff* $(s_2, h_2) \models \mathrm{T}(\psi, X)$.

We define the translation $\mathcal{T}_{\mathrm{SAT}}(\varphi)$ in SL($*, \mathrel{-\!\!*}, \mathtt{ls}$) where $\mathrm{T}(\varphi, X)$ is defined recursively.

$$\mathcal{T}_{\mathrm{SAT}}(\varphi) \stackrel{\text{def}}{=} (\bigwedge_{i \in [1,2q]} \neg\mathtt{alloc}(\mathtt{x}_i)) \wedge \mathrm{OK}(X) \wedge \mathrm{T}(\varphi, X).$$

The first two conjuncts specify initial conditions, namely each variable y in X is interpreted by a location that is unallocated, it is not in the heap range and it is distinct from the interpretation of all other variables; in other words, the value for y is isolated. Similarly, let $\mathcal{T}_{\mathrm{VAL}}(\varphi)$ be the formula in SL($*, \mathrel{-\!\!*}, \mathtt{ls}$) defined by $((\bigwedge_{i \in [1,2q]} \neg\mathtt{alloc}(\mathtt{x}_i)) \wedge \mathrm{OK}(X)) \Rightarrow \mathrm{T}(\varphi, X)$. As a consequence of Lemma 2, φ and $\mathcal{T}_{\mathrm{SAT}}(\varphi)$ are shown equisatisfiable, whereas φ and $\mathcal{T}_{\mathrm{VAL}}(\varphi)$ are shown equivalid.

Corollary 1. *Let* φ *be a closed formula in* SL($\forall, \mathrel{-\!\!*}$) *using quantified variables among* $\{\mathtt{x}_1, \ldots, \mathtt{x}_q\}$. *(I)* φ *and* $\mathcal{T}_{\mathrm{SAT}}(\varphi)$ *are equisatisfiable. (II)* φ *and* $\mathcal{T}_{\mathrm{VAL}}(\varphi)$ *are equivalid.*

3.3 Expressing the Auxiliary Atomic Predicates

To complete the reduction, we briefly explain how to express the formulae $\mathtt{alloc}^{-1}(\mathtt{x})$, $n(\mathtt{x}) = n(\mathtt{y})$ and $n(\mathtt{x}) \hookrightarrow n(\mathtt{y})$ within SL($*, \mathrel{-\!\!*}, \mathtt{ls}$). Let us introduce a few macros that shall be helpful.

- Given φ in SL($*, -\!\!*, \mathtt{reach}^+$) and $\gamma \geq 0$, we write $[\varphi]_\gamma$ to denote the formula $(\mathtt{size} = \gamma \wedge \varphi) * \top$. It is easy to show that for any memory state (s, h), $(s, h) \models [\varphi]_\gamma$ iff there is $h' \sqsubseteq h$ such that $\mathrm{card}(\mathrm{dom}(h')) = \gamma$ and $(s, h') \models \varphi$.
- We write $\mathtt{reach}(\mathtt{x}, \mathtt{y}) = \gamma$ to denote the formula $[\mathtt{ls}(\mathtt{x}, \mathtt{y})]_\gamma$, which is satisfied in any memory state (s, h) where $h^\gamma(s(\mathtt{x})) = s(\mathtt{y})$. Lastly, we write $\mathtt{reach}(\mathtt{x}, \mathtt{y}) \leq \gamma$ to denote the formula $\bigvee_{0 \leq \gamma' \leq \gamma} \mathtt{reach}(\mathtt{x}, \mathtt{y}) = \gamma'$.

In order to define the existence of a predecessor (i.e. $\mathtt{alloc}^{-1}(\mathtt{x})$) in SL($*, -\!\!*, \mathtt{ls}$), we need to take advantage of an auxiliary variable y whose value is different from the one for x. Let $\mathtt{alloc}_{\mathtt{y}}^{-1}(\mathtt{x})$ be the formula

$$\mathtt{x} \hookrightarrow \mathtt{x} \vee \mathtt{y} \hookrightarrow \mathtt{x} \vee [(\mathtt{alloc}(\mathtt{y}) \wedge \neg(\mathtt{y} \hookrightarrow \mathtt{x}) \wedge \mathtt{size} = 1) \circledast\!\!\!- \mathtt{reach}(\mathtt{y}, \mathtt{x}) = 2]_1$$

Lemma 3. *Let* $\mathtt{x}, \mathtt{y} \in$ PVAR. *(I) For all memory states* (s, h) *such that* $s(\mathtt{x}) \neq s(\mathtt{y})$, *we have* $(s, h) \models \mathtt{alloc}_{\mathtt{y}}^{-1}(\mathtt{x})$ *iff* $s(\mathtt{x}) \in \mathrm{ran}(h)$. *(II) In the translation,* $\mathtt{alloc}^{-1}(\mathtt{x})$ *can be replaced with* $\mathtt{alloc}_{\overline{\mathtt{x}}}^{-1}(\mathtt{x})$.

As stated in Lemma 3(II), we can exploit the fact that in the translation of a formula with variables in $\{\mathtt{x}_1, \ldots, \mathtt{x}_q\}$, we use $2q$ variables that correspond to $2q$ distinguished locations in the heap in order to retain the soundness of the translation while using $\mathtt{alloc}_{\overline{\mathtt{x}}}^{-1}(\mathtt{x})$ as $\mathtt{alloc}^{-1}(\mathtt{x})$. Moreover, $\mathtt{alloc}_{\mathtt{y}}^{-1}(\mathtt{x})$ allows to express in SL($*, -\!\!*, \mathtt{ls}$) whenever a location corresponding to a program variable reaches itself in exactly two steps (we use this property in the definition of $n(\mathtt{x}) \hookrightarrow n(\mathtt{y})$). We write $\mathtt{x} \hookrightarrow_{\mathtt{y}}^2 \mathtt{x}$ to denote the formula $\neg(\mathtt{x} \hookrightarrow \mathtt{x}) \wedge (\mathtt{x} \hookrightarrow \mathtt{y} \Leftrightarrow \mathtt{y} \hookrightarrow \mathtt{x}) \wedge [\mathtt{alloc}(\mathtt{x}) \wedge \mathtt{alloc}_{\mathtt{y}}^{-1}(\mathtt{x}) \wedge (\top -\!\!* \neg\mathtt{reach}(\mathtt{x}, \mathtt{y}) = 2)]_2$. For any memory state (s, h) such that $s(\mathtt{x}) \neq s(\mathtt{y})$, we have $(s, h) \models \mathtt{x} \hookrightarrow_{\mathtt{y}}^2 \mathtt{x}$ if and only if $h^2(s(\mathtt{x})) = s(\mathtt{x})$ and $h(s(\mathtt{x})) \neq s(\mathtt{x})$.

The predicate $n(\mathtt{x}) = n(\mathtt{y})$ can be defined in SL($*, -\!\!*, \mathtt{ls}$) as

$$(\mathtt{x} \neq \mathtt{y} \Rightarrow [\mathtt{alloc}(\mathtt{x}) \wedge \mathtt{alloc}(\mathtt{y}) \wedge ((\mathtt{x} \hookrightarrow \mathtt{y} \wedge \mathtt{y} \hookrightarrow \mathtt{y}) \vee (\mathtt{y} \hookrightarrow \mathtt{x} \wedge \mathtt{x} \hookrightarrow \mathtt{x}) \vee \\ ((\bigwedge_{\mathtt{z}, \mathtt{z}' \in \{\mathtt{x}, \mathtt{y}\}} \neg(\mathtt{z} \hookrightarrow \mathtt{z}')) \wedge (\top -\!\!* \neg(\mathtt{reach}(\mathtt{x}, \mathtt{y}) = 2 \wedge \mathtt{reach}(\mathtt{y}, \mathtt{x}) = 2))))]_2) \wedge \mathtt{alloc}(\mathtt{x})$$

Lemma 4. *Let* $\mathtt{x}, \mathtt{y} \in$ PVAR. *For all memory states* (s, h), *we have* $(s, h) \models n(\mathtt{x}) = n(\mathtt{y})$ *iff* $h(s(\mathtt{x})) = h(s(\mathtt{y}))$.

Similarly to $\mathtt{alloc}^{-1}(\mathtt{x})$, we can show that $n(\mathtt{x}) \hookrightarrow n(\mathtt{y})$ is definable in SL($*, -\!\!*, \mathtt{ls}$) by using one additional variable z whose value is different from both x and y. Let $\varphi_{\hookrightarrow}(\mathtt{x}, \mathtt{y}, \mathtt{z})$ be $(n(\mathtt{x}) = n(\mathtt{y}) \wedge \varphi_{\hookrightarrow}^{=}(\mathtt{x}, \mathtt{y}, \mathtt{z})) \vee (n(\mathtt{x}) \neq n(\mathtt{y}) \wedge \varphi_{\hookrightarrow}^{\neq}(\mathtt{x}, \mathtt{y}))$ where $\varphi_{\hookrightarrow}^{=}(\mathtt{x}, \mathtt{y}, \mathtt{z})$ is defined as

$$(\mathtt{x} \hookrightarrow \mathtt{x} \wedge \mathtt{y} \hookrightarrow \mathtt{x}) \vee (\mathtt{y} \hookrightarrow \mathtt{y} \wedge \mathtt{x} \hookrightarrow \mathtt{y}) \vee (\mathtt{x} \hookrightarrow \mathtt{z} \wedge \mathtt{z} \hookrightarrow \mathtt{z}) \\ \vee [\mathtt{alloc}(\mathtt{x}) \wedge \neg\mathtt{alloc}_{\mathtt{z}}^{-1}(\mathtt{x}) \wedge (\top -\!\!* \neg\mathtt{reach}(\mathtt{x}, \mathtt{z}) \leq 3)]_2$$

whereas $\varphi^{\neq}_{\hookrightarrow}(\mathtt{x},\mathtt{y})$ is defined as

$$(\mathtt{x} \hookrightarrow \mathtt{y} \wedge \mathtt{alloc}(\mathtt{y})) \vee (\mathtt{y} \hookrightarrow \mathtt{y} \wedge \mathtt{reach}(\mathtt{x},\mathtt{y}) = 2) \vee (\mathtt{y} \hookrightarrow \mathtt{x} \wedge \mathtt{x} \hookrightarrow^2_{\mathtt{y}} \mathtt{x}) \vee$$
$$[\mathtt{alloc}(\mathtt{x}) \wedge \mathtt{alloc}(\mathtt{y}) \wedge (\textstyle\bigwedge_{\mathtt{z},\mathtt{z}' \in \{\mathtt{x},\mathtt{y}\}} \neg \mathtt{z} \hookrightarrow \mathtt{z}') \wedge \neg \mathtt{reach}(\mathtt{x},\mathtt{y}) \leq 3$$
$$\wedge ((\mathtt{size} = 1 \wedge \mathtt{alloc}^{-1}_{\mathtt{x}}(\mathtt{y})) \circledast (\mathtt{reach}(\mathtt{x},\mathtt{y}) = 3 \wedge \mathtt{y} \hookrightarrow^2_{\mathtt{x}} \mathtt{y}))]_3$$

Lemma 5. *Let* $\mathtt{x},\mathtt{y},\mathtt{z} \in \mathrm{PVAR}$. *(I) For all memory states* (s,h) *such that* $s(\mathtt{x}) \neq s(\mathtt{z})$ *and* $s(\mathtt{y}) \neq s(\mathtt{z})$, *we have* $(s,h) \models \varphi_{\hookrightarrow}(\mathtt{x},\mathtt{y},\mathtt{z})$ *iff* $\{s(\mathtt{x}), s(\mathtt{y})\} \subseteq \mathrm{dom}(h)$ *and* $h(h(s(\mathtt{x}))) = h(s(\mathtt{y}))$; *(II) In the translation,* $n(\mathtt{x}) \hookrightarrow n(\mathtt{y})$ *can be replaced by* $\varphi_{\hookrightarrow}(\mathtt{x},\mathtt{y},\overline{\mathtt{x}})$.

As for $\mathtt{alloc}^{-1}_{\mathtt{y}}(\mathtt{x})$, the properties of the translation imply the equivalence between $n(\mathtt{x}) \hookrightarrow n(\mathtt{y})$ and $\varphi_{\hookrightarrow}(\mathtt{x},\mathtt{y},\overline{\mathtt{x}})$ (as stated in Lemma 5(II)). By looking at the formulae herein defined, the predicate **reach** only appears bounded, i.e. in the form of $\mathtt{reach}(\mathtt{x},\mathtt{y}) = 2$ and $\mathtt{reach}(\mathtt{x},\mathtt{y}) = 3$. The three new predicates can therefore be defined in $\mathrm{SL}(*, \mathrel{-\!\!*})$ enriched with $\mathtt{reach}(\mathtt{x},\mathtt{y}) = 2$ and $\mathtt{reach}(\mathtt{x},\mathtt{y}) = 3$.

3.4 Undecidability Results and Non-finite Axiomatization

It is time to collect the fruits of all our efforts and to conclude this part about undecidability. As a direct consequence of Corollary 1 and the undecidability of $\mathrm{SL}(\forall, \mathrel{-\!\!*})$, here is one of the main results of the paper.

Theorem 1. *The satisfiability problem for* $\mathrm{SL}(*, \mathrel{-\!\!*}, \mathtt{ls})$ *is undecidable.*

As a by-product, the set of valid formulae for $\mathrm{SL}(*, \mathrel{-\!\!*}, \mathtt{ls})$ is not recursively enumerable. Indeed, suppose that the set of valid formulae for $\mathrm{SL}(*, \mathrel{-\!\!*}, \mathtt{ls})$ were r.e., then one can enumerate the valid formulae of the form $\mathcal{T}_{\mathrm{VAL}}(\varphi)$ as it is decidable in PTIME whether ψ in $\mathrm{SL}(*, \mathrel{-\!\!*}, \mathtt{ls})$ is syntactically equal to $\mathcal{T}_{\mathrm{VAL}}(\varphi)$ for some $\mathrm{SL}(\forall, \mathrel{-\!\!*})$ formula φ. This leads to a contradiction since this would allow the enumeration of valid formulae in $\mathrm{SL}(\forall, \mathrel{-\!\!*})$.

The essential ingredients to establish the undecidability of $\mathrm{SL}(*, \mathrel{-\!\!*}, \mathtt{ls})$ are the fact that the following properties $n(\mathtt{x}) = n(\mathtt{y})$, $n(\mathtt{x}) \hookrightarrow n(\mathtt{y})$ and $\mathtt{alloc}^{-1}(\mathtt{x})$ are expressible in the logic.

Corollary 2. $\mathrm{SL}(*, \mathrel{-\!\!*})$ *augmented with built-in formulae of the form* $n(\mathtt{x}) = n(\mathtt{y})$, $n(\mathtt{x}) \hookrightarrow n(\mathtt{y})$ *and* $\mathtt{alloc}^{-1}(\mathtt{x})$ *(resp. of the form* $\mathtt{reach}(\mathtt{x},\mathtt{y}) = 2$ *and* $\mathtt{reach}(\mathtt{x},\mathtt{y}) = 3$*) admits an undecidable satisfiability problem.*

This is the addition of $\mathtt{reach}(\mathtt{x},\mathtt{y}) = 3$ that is crucial for undecidability since the satisfiability problem for $\mathrm{SL}(*, \mathrel{-\!\!*}, \mathtt{reach}(\mathtt{x},\mathtt{y}) = 2)$ is in PSPACE [15]. Following a similar analysis, let $\mathrm{SL1}(\forall, *, \mathrel{-\!\!*})$ be the restriction of $\mathrm{SL}(\forall, *, \mathrel{-\!\!*})$ (i.e. $\mathrm{SL}(\forall, \mathrel{-\!\!*})$ plus $*$) to formulae of the form $\exists \mathtt{x}_1 \cdots \exists \mathtt{x}_q\ \varphi$, where $q \geq 1$, the variables in φ are among $\{\mathtt{x}_1, \ldots, \mathtt{x}_{q+1}\}$ and the only quantified variable in φ is $\mathtt{x}_{q+1}$. The satisfiability problem for $\mathrm{SL1}(\forall, *, \mathrel{-\!\!*})$ is PSPACE-complete [15]. Note that $\mathrm{SL1}(\forall, *, \mathrel{-\!\!*})$

can easily express $n(\mathtt{x}) = n(\mathtt{y})$ and $\mathtt{alloc}^{-1}(\mathtt{x})$. The distance between the decidability for SL1($\forall, *, -\!\!*$) and the undecidability for SL($*, -\!\!*, \mathtt{ls}$), is best witnessed by the corollary below, which solves an open problem [15, Sect. 6].

Corollary 3. *SL1($\forall, *, -\!\!*$) augmented with $n(\mathtt{x}) \hookrightarrow n(\mathtt{y})$ (resp. SL1($\forall, *, -\!\!*$) augmented with* ls*) admits an undecidable satisfiability problem.*

4 SL($*, \mathtt{reach}^+$) and Other PSPACE Variants

As already seen in Sect. 2, SL($*, \mathtt{ls}$) can be understood as a fragment of SL($*, \mathtt{reach}^+$). Below, we show that the satisfiability problem for SL($*, \mathtt{reach}^+$) can be solved in polynomial space. Refining the arguments used in our proof, we also show the decidability of the fragment of SL($*, -\!\!*, \mathtt{reach}^+$) where $\mathtt{reach}^+$ is constrained not to occur in the scope of $-\!\!*$, i.e. φ belongs to that fragment iff for any subformula ψ of φ of the form $\psi_1 -\!\!* \psi_2$, $\mathtt{reach}^+$ does not occur in ψ_1 and in ψ_2.

The proof relies on a small heap property: a formula φ is satisfiable if and only if it admits a model with a polynomial amount of memory cells. The PSPACE upper bound then follows by establishing that the model-checking problem for SL($*, \mathtt{reach}^+$) is in PSPACE too. To establish the small heap property, an equivalence relation on memory states with finite index is designed, following the standard approach in [10,32] and using test formulae as in [4,15,22,23].

4.1 Introduction to Test Formulae

Before presenting the test formulae for SL($*, \mathtt{reach}^+$), let us recall the standard result for SL($*, -\!\!*$) (that will be also used at some point later on).

Proposition 2. *[22, 32] Any formula φ in* SL($*, -\!\!*$) *built over variables in $\mathtt{x}_1, \ldots, \mathtt{x}_q$ is logically equivalent to a Boolean combination of formulae among $\mathtt{x}_i = \mathtt{x}_j$, $\mathtt{alloc}(\mathtt{x}_i)$, $\mathtt{x}_i \hookrightarrow \mathtt{x}_j$ and $\mathtt{size} \geq \beta$ ($i, j \in \{1, \ldots, q\}$, $\beta \in \mathbb{N}$).*

By way of example, $(\neg \mathtt{emp} * ((\mathtt{x}_1 \hookrightarrow \mathtt{x}_2) -\!\!* \perp))$ is equivalent to $\mathtt{size} \geq 2 \wedge \mathtt{alloc}(\mathtt{x}_1)$. As a corollary of the proof of Proposition 2, in $\mathtt{size} \geq \beta$ we can enforce that $\beta \leq 2 \times |\varphi|$ (rough upper bound) where $|\varphi|$ is the size of φ. Similar results will be shown for SL($*, \mathtt{reach}^+$) and for some of its extensions.

In order to define a set of test formulae that captures the expressive power of SL($*, \mathtt{reach}^+$), we need to study which basic properties on memory states can be expressed by SL($*, \mathtt{reach}^+$) formulae. For example, consider the memory states from Fig. 1.
The fragment memory states (s_1, h_1) and (s_2, h_2) can be distinguished by the formula $\top * (\mathtt{reach}(\mathtt{x}_i, \mathtt{x}_j) \wedge \mathtt{reach}(\mathtt{x}_j, \mathtt{x}_k) \wedge \neg\mathtt{reach}(\mathtt{x}_k, \mathtt{x}_i))$. Indeed, (s_1, h_1) satisfies this formula by considering a subheap that does not contain a path from $s(\mathtt{x}_k)$ to $s(\mathtt{x}_i)$, whereas it is impossible to find a subheap for (s_2, h_2) that retains the path from $s(\mathtt{x}_i)$ to $s(\mathtt{x}_j)$, the one from $s(\mathtt{x}_j)$ to $s(\mathtt{x}_k)$ but where the path from $s(\mathtt{x}_k)$ to $s(\mathtt{x}_i)$ is lost. This suggests that SL($*, \mathtt{reach}^+$) can express

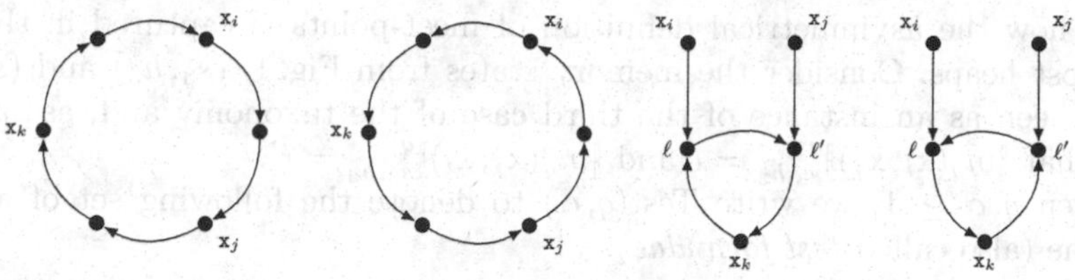

Fig. 1. Memory states (s_1, h_1), ..., (s_4, h_4) (from left to right)

whether, for example, any path from $s(\mathtt{x}_i)$ to $s(\mathtt{x}_j)$ also contains $s(\mathtt{x}_k)$. We will introduce the test formula $\mathtt{sees}_q(\mathtt{x}_i, \mathtt{x}_j) \geq \beta$ to capture this property.

Similarly, the memory states (s_3, h_3) and (s_4, h_4) can be distinguished by the formula $(\mathtt{size} = 1) * (\mathtt{reach}(\mathtt{x}_j, \mathtt{x}_k) \wedge \neg\mathtt{reach}(\mathtt{x}_i, \mathtt{x}_k) \wedge \neg\mathtt{reach}^+(\mathtt{x}_k, \mathtt{x}_k))$. The memory state (s_3, h_3) satisfies this formula by separating $\{\ell \mapsto \ell'\}$ from the rest of the heap, whereas the formula is not satisfied by (s_4, h_4). Indeed, there is no way to break the loop from $s(\mathtt{x}_k)$ to itself by removing just one location from the heap while retaining the path from $s(\mathtt{x}_j)$ to $s(\mathtt{x}_k)$ and loosing the path from $s(\mathtt{x}_i)$ to $s(\mathtt{x}_k)$. This suggests that the two locations ℓ and ℓ' are particularly interesting since they are reachable from several locations corresponding to program variables. Therefore by separating them from the rest of the heap, several paths are lost. In order to capture this, we introduce the notion of *meet-points*.

Let Terms_q be the set $\{\mathtt{x}_1, \ldots, \mathtt{x}_q\} \cup \{m_q(\mathtt{x}_i, \mathtt{x}_j) \mid i, j \in [1, q]\}$ understood as the set of *terms* that are either variables or expressions denoting a meet-point. We write $[\![\mathtt{x}_i]\!]^q_{s,h}$ to denote $s(\mathtt{x}_i)$ and $[\![m_q(\mathtt{x}_i, \mathtt{x}_j)]\!]^q_{s,h}$ to denote (if it exists) the first location reachable from $s(\mathtt{x}_i)$ that is also reachable from $s(\mathtt{x}_j)$. Moreover we require that this location can reach another location corresponding to a program variable. Formally, $[\![m_q(\mathtt{x}_i, \mathtt{x}_j)]\!]^q_{s,h}$ is defined as the unique location ℓ such that

- there are $L_1, L_2 \geq 0$ such that $h^{L_1}(s(\mathtt{x}_i)) = h^{L_2}(s(\mathtt{x}_j)) = \ell$, and
- for all $L_1' < L_1$ and for all $L_2' \geq 0$, $h^{L_1'}(s(\mathtt{x}_i)) \neq h^{L_2'}(s(\mathtt{x}_j))$, and
- there exist $k \in [1, q]$ and $L \geq 0$ such that $h^L(\ell) = s(\mathtt{x}_k)$.

These conditions hold for at most one location ℓ. One can easily show that the notion $[\![m_q(\mathtt{x}_i, \mathtt{x}_j)]\!]^q_{s,h}$ is well-defined. The picture below provides a taxonomy of meet-points, where arrows labelled by '+' represent paths of non-zero length and zig-zag arrows any path (possibly of zero length). Symmetrical cases, obtained by swapping $\mathtt{x}_i$ and $\mathtt{x}_j$, are omitted.

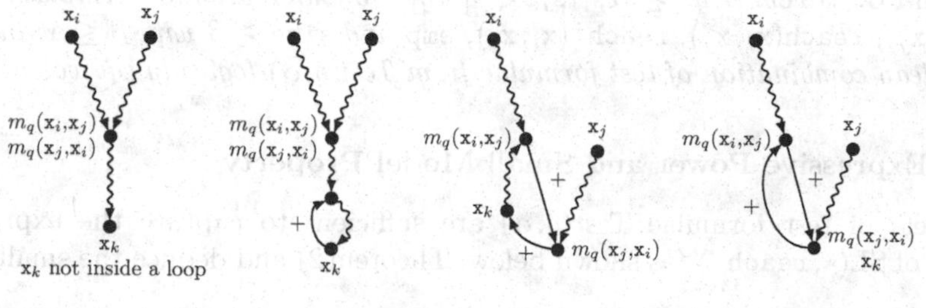

Notice how the asymmetrical definition of meet-points is captured in the two rightmost heaps. Consider the memory states from Fig. 1, (s_3, h_3) and (s_4, h_4) can be seen as an instance of the third case of the taxonomy and, as such, it holds that $[\![m_q(\mathtt{x}_i, \mathtt{x}_j)]\!]^q_{s_3,h_3} = \ell$ and $[\![m_q(\mathtt{x}_j, \mathtt{x}_i)]\!]^q_{s_3,h_3} = \ell'$.

Given $q, \alpha \geq 1$, we write $\mathrm{Test}(q, \alpha)$ to denote the following set of atomic formulae (also called *test formulae*):

$$v = v' \quad v \hookrightarrow v' \quad \mathtt{alloc}(v) \quad \mathtt{sees}_q(v, v') \geq \beta + 1 \quad \mathtt{sizeR}_q \geq \beta,$$

where $v, v' \in \mathrm{Terms}_q$ and $\beta \in [1, \alpha]$. It is worth noting that the $\mathtt{alloc}(v)$'s are not needed for the logic SL($*, \mathtt{reach}^+$) but it is required for extensions.

We identify as special locations the $s(\mathtt{x}_i)$'s and the meet-points of the form $[\![m_q(\mathtt{x}_i, \mathtt{x}_j)]\!]^q_{s,h}$ when it exists ($i, j \in [1, q]$). We call such locations, *labelled* locations, and the set of labelled locations is written $\mathrm{Labels}^q_{s,h}$. The formal semantics of the test formulae is provided below:

$$
\begin{array}{lcl}
(s,h) \models v = v' & \Longleftrightarrow & [\![v]\!]^q_{s,h}, [\![v']\!]^q_{s,h} \text{ are defined, } [\![v]\!]^q_{s,h} = [\![v']\!]^q_{s,h} \\
(s,h) \models \mathtt{alloc}(v) & \Longleftrightarrow & [\![v]\!]^q_{s,h} \text{ is defined and belongs to } \mathrm{dom}(h) \\
(s,h) \models v \hookrightarrow v' & \Longleftrightarrow & h([\![v]\!]^q_{s,h}) = [\![v']\!]^q_{s,h} \\
(s,h) \models \mathtt{sees}_q(v,v') \geq \beta+1 & \Longleftrightarrow & \exists L \geq \beta+1,\ h^L([\![v]\!]^q_{s,h}) = [\![v']\!]^q_{s,h} \text{ and} \\
 & & \forall\, 0 < L' < L,\ h^{L'}([\![v]\!]^q_{s,h}) \notin \mathrm{Labels}^q_{s,h} \\
(s,h) \models \mathtt{sizeR}_q \geq \beta & \Longleftrightarrow & \mathrm{card}(\mathrm{Rem}^q_{s,h}) \geq \beta
\end{array}
$$

where $\mathrm{Rem}^q_{s,h}$ is the set of locations that neither belong to a path between two locations interpreted by program variables nor are equal to program variable interpretations, i.e. $\mathrm{Rem}^q_{s,h} \stackrel{\text{def}}{=} \{\ell \in \mathrm{dom}(h) \mid \forall i \in [1,q],\ s(\mathtt{x}_i) \neq \ell$ and $\forall j \in [1,q]\ \nexists L, L' \geq 1, h^L(s(\mathtt{x}_i)) = \ell$ and $h^{L'}(\ell) = s(\mathtt{x}_j)\}$. There is no need for test formulae of the form $\mathtt{sees}_q(v, v') \geq 1$ since they are equivalent to $v \hookrightarrow v' \vee \mathtt{sees}_q(v, v') \geq 2$. One can check whether $[\![m_q(\mathtt{x}_i, \mathtt{x}_j)]\!]^q_{s,h}$ is defined thanks to the formula $m_q(\mathtt{x}_i, \mathtt{x}_j) = m_q(\mathtt{x}_i, \mathtt{x}_j)$. By contrast, $\mathtt{sizeR}_q \geq \beta$ states that the cardinality of the set $\mathrm{Rem}^q_{s,h}$ is at least β. Furthermore, $\mathtt{sees}_q(v, v') \geq \beta + 1$ states that there is a minimal path between v and v' of length at least $\beta + 1$ and strictly between v and v', there are no labelled locations. The satisfaction of $\mathtt{sees}_q(v, v') \geq \beta + 1$ entails the exclusion of labelled locations in the witness path, which is reminiscent to $T \xrightarrow{h \backslash T''} T'$ in the logic GRASS [26]. So, the test formulae are quite expressive since they capture the atomic formulae from SL($*, \mathtt{reach}^+$) and the test formulae for SL($*, -\!\!*$).

Lemma 6. *Given* $\alpha, q \geq 1$, $i, j \in [1, q]$, *for any atomic formula among* $\mathtt{ls}(\mathtt{x}_i, \mathtt{x}_j)$, $\mathtt{reach}(\mathtt{x}_i, \mathtt{x}_j)$, $\mathtt{reach}^+(\mathtt{x}_i, \mathtt{x}_j)$, $\mathtt{emp}$ *and* $\mathtt{size} \geq \beta$ *with* $\beta \leq \alpha$, *there is a Boolean combination of test formulae from* $\mathit{Test}(q, \alpha)$ *logically equivalent to it.*

4.2 Expressive Power and Small Model Property

The sets of test formulae $\mathrm{Test}(q, \alpha)$ are sufficient to capture the expressive power of SL($*, \mathtt{reach}^+$) (as shown below, Theorem 2) and deduce the small heap property of this logic (Theorem 3). We introduce an indistinguishability relation between memory states based on test formulae, see analogous relations in [13,15,22].

Definition 3. *Given $q, \alpha \geq 1$, we write $(s,h) \approx^q_\alpha (s',h') \overset{\text{def}}{\Leftrightarrow}$ for all $\psi \in$ Test(q,α), we have $(s,h) \models \psi$ iff $(s',h') \models \psi$.*

Theorem 2(I) states that if $(s,h) \approx^q_\alpha (s',h')$, then the two memory states cannot be distinguished by formulae whose syntactic resources are bounded in some way by q and α (details will follow, see the definition for $\mathtt{msize}(\varphi)$).

Below, we state the key intermediate result of the section that can be viewed as a distributivity lemma. The expressive power of the test formulae allows us to mimic the separation between two equivalent memory states with respect to the relation $\approx^q_\alpha$, which is essential in the proof of Theorem 2(I).

Lemma 7. *Let $q, \alpha, \alpha_1, \alpha_2 \geq 1$ with $\alpha = \alpha_1 + \alpha_2$ and (s,h), (s',h') be such that $(s,h) \approx^q_\alpha (s',h')$. For all heaps h_1, h_2 such that $h = h_1 + h_2$ there are heaps h'_1, h'_2 such that $h = h'_1 + h'_2$, $(s,h_1) \approx^q_{\alpha_1} (s',h'_1)$ and $(s,h_2) \approx^q_{\alpha_2} (s',h'_2)$.*

For each formula φ in $\mathrm{SL}(*, \mathtt{reach}^+)$, we define its *memory size* $\mathtt{msize}(\varphi)$ following the clauses below (see also [32]).

$$\mathtt{msize}(\pi) \overset{\text{def}}{=} 1 \quad \text{for any atomic formula } \pi$$
$$\mathtt{msize}(\psi * \psi') \overset{\text{def}}{=} \mathtt{msize}(\psi) + \mathtt{msize}(\psi')$$
$$\mathtt{msize}(\psi \wedge \psi') \overset{\text{def}}{=} \max(\mathtt{msize}(\psi), \mathtt{msize}(\psi'))$$
$$\mathtt{msize}(\neg\psi) \overset{\text{def}}{=} \mathtt{msize}(\psi).$$

We have $1 \leq \mathtt{msize}(\varphi) \leq |\varphi|$. Theorem 2 below establishes the properties that formulae in $\mathrm{SL}(*, \mathtt{reach}^+)$ can express.

Theorem 2. *Let φ be in $\mathrm{SL}(*, \mathtt{reach}^+)$ built over the variables in $\mathtt{x}_1, \ldots, \mathtt{x}_q$. (I) For all $\alpha \geq 1$ such that $\mathtt{msize}(\varphi) \leq \alpha$ and for all memory states (s,h), (s',h') such that $(s,h) \approx^q_\alpha (s',h')$, we have $(s,h) \models \varphi$ iff $(s',h') \models \varphi$. (II) φ is logically equivalent to a Boolean combination of test formulae from Test$(q, \mathtt{msize}(\varphi))$.*

The proof of Theorem 2(I) is by structural induction on φ. The basic cases for atomic formulae follow from Lemma 6 whereas the inductive cases for Boolean connectives are immediate. For the separating conjunction, suppose $(s,h) \models \varphi_1 * \varphi_2$ and $\mathtt{msize}(\varphi_1 * \varphi_2) \leq \alpha$. There are heaps h_1 and h_2 such that $h = h_1 + h_2$, $(s,h_1) \models \psi_1$ and $(s,h_2) \models \psi_2$. As $\alpha \geq \mathtt{msize}(\psi_1 * \psi_2) = \mathtt{msize}(\psi_1) + \mathtt{msize}(\psi_2)$, there exist α_1 and α_2 such that $\alpha = \alpha_1 + \alpha_2$, $\alpha_1 \geq \mathtt{msize}(\psi_1)$ and $\alpha_2 \geq \mathtt{msize}(\psi_2)$. By Lemma 7, there exist heaps h'_1 and h'_2 such that $h' = h'_1 + h'_2$, $(s,h_1) \approx^q_{\alpha_1} (s',h'_1)$ and $(s,h_2) \approx^q_{\alpha_2} (s',h'_2)$. By the induction hypothesis, we get $(s',h'_1) \models \psi_1$ and $(s',h'_2) \models \psi_2$. Consequently, we obtain $(s',h') \models \psi_1 * \psi_2$.

As an example, we can apply this result to the memory states from Fig. 1. We have already shown how we can distinguish (s_1,h_1) from (s_2,h_2) using a formula with only one separating conjunction. Theorem 2 ensures that these two memory states do not satisfy the same set of test formulae for $\alpha \geq 2$. Indeed, only (s_1, h_1) satisfies $\mathtt{sees}_q(\mathtt{x}_i, \mathtt{x}_j) \geq 2$. The same argument can be used with (s_3, h_3) and (s_4, h_4): only (s_3, h_3) satisfies the test formula $m_q(\mathtt{x}_i, \mathtt{x}_j) \hookrightarrow m_q(\mathtt{x}_j, \mathtt{x}_i)$. Clearly, Theorem 2(II) relates separation logic with classical logic as advocated also in the works [10,23]. Now, it is possible to establish a small heap property.

Theorem 3. *Let φ be a satisfiable* $\mathrm{SL}(*, \mathtt{reach}^+)$ *formula built over* $\mathtt{x}_1, \ldots, \mathtt{x}_q$. *There is (s, h) such that $(s, h) \models \varphi$ and* $\mathrm{card}(\mathrm{dom}(h)) \leq (q^2 + q) \cdot (|\varphi| + 1) + |\varphi|$.

The small heap property for $\mathrm{SL}(*, \mathtt{reach}^+)$ is inherited from the small heap property for the Boolean combinations of test formulae, which is analogous to the small model property for other theories of singly linked lists, see e.g. [13,27].

4.3 Complexity Upper Bounds

Let us draw some consequences of Theorem 3. First, for the logic $\mathrm{SL}(*, \mathtt{reach}^+)$, we get a PSPACE upper, which matches the lower bound for $\mathrm{SL}(*)$ [11].

Theorem 4. *The satisfiability problem for* $\mathrm{SL}(*, \mathtt{reach}^+)$ *is* PSPACE*-complete.*

Besides, we may consider restricting the usage of Boolean connectives. We note Bool(SHF) for the Boolean combinations of formulae from the symbolic heap fragment [2]. A PTIME upper bound for the entailment/satisfiability problem for the symbolic heap fragment is successfully solved in [12,17], whereas the satisfiability problem for a slight variant of Bool(SHF) is shown in NP in [26, Theorem 4]. Theorem 3 allows us to conclude this NP upper bound result as a by-product (we conjecture that our quadratic upper bound on the number of cells could be improved to a linear one in that case).

Corollary 4. *The satisfiability problem for* Bool(SHF) *is* NP*-complete.*

It is possible to push further the PSPACE upper bound by allowing occurrences of $-\!\!*$ in a controlled way. Let $\mathrm{SL}(*, \mathtt{reach}^+, \bigcup_{q,\alpha} \mathrm{Test}(q, \alpha))$ be the extension of $\mathrm{SL}(*, \mathtt{reach}^+)$ augmented with the test formulae. The memory size function is also extended: $\mathtt{msize}(v \hookrightarrow v') \stackrel{\text{def}}{=} 1$, $\mathtt{msize}(\mathtt{sees}_q(v, v') \geq \beta + 1) \stackrel{\text{def}}{=} \beta + 1$, $\mathtt{msize}(\mathtt{sizeR} \geq \beta) \stackrel{\text{def}}{=} \beta$ and $\mathtt{msize}(\mathtt{alloc}(v)) \stackrel{\text{def}}{=} 1$. When formulae are encoded as trees, we have $1 \leq \mathtt{msize}(\varphi) \leq |\varphi|\alpha_\varphi$ where α_φ is the maximal constant in φ. Theorem 2(I) admits a counterpart for $\mathrm{SL}(*, \mathtt{reach}^+, \bigcup_{q,\alpha} \mathrm{Test}(q, \alpha))$ and consequently, any formula built over $\mathtt{x}_1, \ldots, \mathtt{x}_q$ can be shown equivalent to a Boolean combination of test formulae from $\mathrm{Test}(q, |\varphi|\alpha_\varphi)$. By Theorem 3, any satisfiable formula has therefore a model with $\mathrm{card}(\mathrm{dom}(h)) \leq (q^2 + q) \cdot (|\varphi|\alpha_\varphi + 1) + |\varphi|\alpha_\varphi$. Hence, the satisfiability problem for $\mathrm{SL}(*, \mathtt{reach}^+, \bigcup_{q,\alpha} \mathrm{Test}(q, \alpha))$ is in PSPACE when the constants are encoded in unary. Now, we can state the new PSPACE upper bound for Boolean combinations of formulae from $\mathrm{SL}(*, -\!\!*) \cup \mathrm{SL}(*, \mathtt{reach}^+)$.

Theorem 5. *The satisfiability problem for Boolean combinations of formulae from* $\mathrm{SL}(*, -\!\!*) \cup \mathrm{SL}(*, \mathtt{reach}^+)$ *is* PSPACE*-complete.*

To conclude, let us introduce the largest fragment including $-\!\!*$ and $\texttt{ls}$ for which decidability can be established so far.

Theorem 6. *The satisfiability problem for the fragment of* $\mathrm{SL}(*, -\!\!*, \texttt{reach}^+)$ *in which* $\texttt{reach}^+$ *is not in the scope of* $-\!\!*$ *is decidable.*

5 Conclusion

We studied the effects of adding $\texttt{ls}$ to $\mathrm{SL}(*, -\!\!*)$ and variants. $\mathrm{SL}(*, -\!\!*, \texttt{ls})$ is shown undecidable (Theorem 1) and non-finitely axiomatisable, which remains quite unexpected since there are no first-order quantifications. This result is strengthened to even weaker extensions of $\mathrm{SL}(*, -\!\!*)$ such as the one augmented with $n(\texttt{x}) = n(\texttt{y})$, $n(\texttt{x}) \hookrightarrow n(\texttt{y})$ and $\texttt{alloc}^{-1}(\texttt{x})$, or the one augmented with $\texttt{reach}(\texttt{x},\texttt{y}) = 2$ and $\texttt{reach}(\texttt{x},\texttt{y}) = 3$. If the magic wand is discarded, we have established that the satisfiability problem for $\mathrm{SL}(*, \texttt{ls})$ is PSPACE-complete by introducing a class of test formulae that captures the expressive power of $\mathrm{SL}(*, \texttt{ls})$ and that leads to a small heap property. Such a logic contains the Boolean combinations of symbolic heaps and our proof technique allows us to get an NP upper bound for such formulae. Moreover, we show that the satisfiability problem for $\mathrm{SL}(*, -\!\!*, \texttt{reach}^+)$ restricted to formulae in which $\texttt{reach}^+$ is not in the scope of $-\!\!*$ is decidable, leading to the largest known decidable fragment for which $-\!\!*$ and $\texttt{reach}^+$ (or $\texttt{ls}$) cohabit. So, we have provided proof techniques to establish undecidability when $*$, $-\!\!*$ and $\texttt{ls}$ are present and to establish decidability based on test formulae. This paves the way to investigate the decidability status of $\mathrm{SL}(-\!\!*, \texttt{ls})$ as well as of the positive fragment of $\mathrm{SL}(*, -\!\!\circledast, \texttt{ls})$ from [30,31].

References

1. Antonopoulos, T., Gorogiannis, N., Haase, C., Kanovich, M., Ouaknine, J.: Foundations for decision problems in separation logic with general inductive predicates. In: Muscholl, A. (ed.) FoSSaCS 2014. LNCS, vol. 8412, pp. 411–425. Springer, Heidelberg (2014). https://doi.org/10.1007/978-3-642-54830-7_27
2. Berdine, J., Calcagno, C., O'Hearn, P.W.: A decidable fragment of separation logic. In: Lodaya, K., Mahajan, M. (eds.) FSTTCS 2004. LNCS, vol. 3328, pp. 97–109. Springer, Heidelberg (2004). https://doi.org/10.1007/978-3-540-30538-5_9
3. Berdine, J., Calcagno, C., O'Hearn, P.W.: Smallfoot: modular automatic assertion checking with separation logic. In: de Boer, F.S., Bonsangue, M.M., Graf, S., de Roever, W.-P. (eds.) FMCO 2005. LNCS, vol. 4111, pp. 115–137. Springer, Heidelberg (2006). https://doi.org/10.1007/11804192_6
4. Brochenin, R., Demri, S., Lozes, E.: Reasoning about sequences of memory states. APAL **161**(3), 305–323 (2009)
5. Brochenin, R., Demri, S., Lozes, E.: On the almighty wand. IC **211**, 106–137 (2012)
6. Brotherston, J., Fuhs, C., Gorogiannis, N., Navarro Perez, J.: A decision procedure for satisfiability in separation logic with inductive predicates. In: CSL-LICS 2014 (2014)

7. Brotherston, J., Villard, J.: Parametric completeness for separation theories. In: POPL 2014, pp. 453–464. ACM (2014)
8. Calcagno, C., Distefano, D.: Infer: an automatic program verifier for memory safety of C programs. In: Bobaru, M., Havelund, K., Holzmann, G.J., Joshi, R. (eds.) NFM 2011. LNCS, vol. 6617, pp. 459–465. Springer, Heidelberg (2011). https://doi.org/10.1007/978-3-642-20398-5_33
9. Calcagno, C., Distefano, D., O'Hearn, P., Yang, H.: Compositional shape analysis by means of bi-abduction. JACM **58**(6), 26:1–26:66 (2011)
10. Calcagno, C., Gardner, P., Hague, M.: From separation logic to first-order logic. In: Sassone, V. (ed.) FoSSaCS 2005. LNCS, vol. 3441, pp. 395–409. Springer, Heidelberg (2005). https://doi.org/10.1007/978-3-540-31982-5_25
11. Calcagno, C., Yang, H., O'Hearn, P.W.: Computability and complexity results for a spatial assertion language for data structures. In: Hariharan, R., Vinay, V., Mukund, M. (eds.) FSTTCS 2001. LNCS, vol. 2245, pp. 108–119. Springer, Heidelberg (2001). https://doi.org/10.1007/3-540-45294-X_10
12. Cook, B., Haase, C., Ouaknine, J., Parkinson, M., Worrell, J.: Tractable reasoning in a fragment of separation logic. In: Katoen, J.-P., König, B. (eds.) CONCUR 2011. LNCS, vol. 6901, pp. 235–249. Springer, Heidelberg (2011). https://doi.org/10.1007/978-3-642-23217-6_16
13. David, C., Kroening, D., Lewis, M.: Propositional reasoning about safety and termination of heap-manipulating programs. In: Vitek, J. (ed.) ESOP 2015. LNCS, vol. 9032, pp. 661–684. Springer, Heidelberg (2015). https://doi.org/10.1007/978-3-662-46669-8_27
14. Demri, S., Deters, M.: Expressive completeness of separation logic with two variables and no separating conjunction. ACM ToCL **17**(2), 12 (2016)
15. Demri, S., Galmiche, D., Larchey-Wendling, D., Mery, D.: Separation logic with one quantified variable. Theory Comput. Syst. **61**, 371–461 (2017)
16. Distefano, D., O'Hearn, P.W., Yang, H.: A local shape analysis based on separation logic. In: Hermanns, H., Palsberg, J. (eds.) TACAS 2006. LNCS, vol. 3920, pp. 287–302. Springer, Heidelberg (2006). https://doi.org/10.1007/11691372_19
17. Haase, C., Ishtiaq, S., Ouaknine, J., Parkinson, M.J.: SeLoger: a tool for graph-based reasoning in separation logic. In: Sharygina, N., Veith, H. (eds.) CAV 2013. LNCS, vol. 8044, pp. 790–795. Springer, Heidelberg (2013). https://doi.org/10.1007/978-3-642-39799-8_55
18. Hóu, Z., Goré, R., Tiu, A.: Automated theorem proving for assertions in separation logic with all connectives. In: Felty, A.P., Middeldorp, A. (eds.) CADE 2015. LNCS (LNAI), vol. 9195, pp. 501–516. Springer, Cham (2015). https://doi.org/10.1007/978-3-319-21401-6_34
19. Iosif, R., Rogalewicz, A., Simacek, J.: The tree width of separation logic with recursive definitions. In: Bonacina, M.P. (ed.) CADE 2013. LNCS (LNAI), vol. 7898, pp. 21–38. Springer, Heidelberg (2013). https://doi.org/10.1007/978-3-642-38574-2_2
20. Ishtiaq, S., O'Hearn, P.: BI as an assertion language for mutable data structures. In: POPL 2001, pp. 14–26. ACM (2001)
21. Le, Q.L., Tatsuta, M., Sun, J., Chin, W.-N.: A decidable fragment in separation logic with inductive predicates and arithmetic. In: Majumdar, R., Kunčak, V. (eds.) CAV 2017. LNCS, vol. 10427, pp. 495–517. Springer, Cham (2017). https://doi.org/10.1007/978-3-319-63390-9_26
22. Lozes, E.: Expressivité des Logiques Spatiales. Ph.D. thesis, ENS Lyon (2004)
23. Lozes, E.: Separation logic preserves the expressive power of classical logic. In: SPACE 2004 (2004)

24. Müller, P., Schwerhoff, M., Summers, A.J.: Viper: a verification infrastructure for permission-based reasoning. In: Jobstmann, B., Leino, K.R.M. (eds.) VMCAI 2016. LNCS, vol. 9583, pp. 41–62. Springer, Heidelberg (2016). https://doi.org/10.1007/978-3-662-49122-5_2
25. O'Hearn, P., Reynolds, J., Yang, H.: Local reasoning about programs that alter data structures. In: Fribourg, L. (ed.) CSL 2001. LNCS, vol. 2142, pp. 1–19. Springer, Heidelberg (2001). https://doi.org/10.1007/3-540-44802-0_1
26. Piskac, R., Wies, T., Zufferey, D.: Automating separation logic using SMT. In: Sharygina, N., Veith, H. (eds.) CAV 2013. LNCS, vol. 8044, pp. 773–789. Springer, Heidelberg (2013). https://doi.org/10.1007/978-3-642-39799-8_54
27. Ranise, S., Zarba, C.: A theory of singly-linked lists and its extensible decision procedure. In: SEFM 2006, pp. 206–215. IEEE (2006)
28. Reynolds, J.: Separation logic: a logic for shared mutable data structures. In: LICS 2002, pp. 55–74. IEEE (2002)
29. Schwerhoff, M., Summers, A.: Lightweight support for magic wands in an automatic verifier. In: ECOOP 2015, pp. 999–1023. Leibniz-Zentrum für Informatik, LIPICS (2015)
30. Thakur, A.: Symbolic Abstraction: Algorithms and Applications. Ph.D. thesis, University of Wisconsin-Madison (2014)
31. Thakur, A., Breck, J., Reps, T.: Satisfiability modulo abstraction for separation logic with linkedlists. In: SPIN 2014, pp. 58–67. ACM (2014)
32. Yang, H.: Local Reasoning for Stateful Programs. Ph.D. thesis, University of Illinois, Urbana-Champaign (2001)
33. Yang, H., Lee, O., Berdine, J., Calcagno, C., Cook, B., Distefano, D., O'Hearn, P.: Scalable shape analysis for systems code. In: Gupta, A., Malik, S. (eds.) CAV 2008. LNCS, vol. 5123, pp. 385–398. Springer, Heidelberg (2008). https://doi.org/10.1007/978-3-540-70545-1_36

The Equational Theory of the Natural Join and Inner Union is Decidable

Luigi Santocanale(✉)

LIS, CNRS UMR 7020, Aix-Marseille Université, Marseille, France
luigi.santocanale@lis-lab.fr

Abstract. The natural join and the inner union operations combine relations of a database. Tropashko and Spight [25] realized that these two operations are the meet and join operations in a class of lattices, known by now as the relational lattices. They proposed then lattice theory as an algebraic approach to the theory of databases, alternative to the relational algebra.

Previous works [17,23] proved that the quasiequational theory of these lattices—that is, the set of definite Horn sentences valid in all the relational lattices—is undecidable, even when the signature is restricted to the pure lattice signature.

We prove here that the equational theory of relational lattices is decidable. That, is we provide an algorithm to decide if two lattice theoretic terms t, s are made equal under all interpretations in some relational lattice. We achieve this goal by showing that if an inclusion $t \leq s$ fails in any of these lattices, then it fails in a relational lattice whose size is bound by a triple exponential function of the sizes of t and s.

1 Introduction

The natural join and the inner union operations combine relations (i.e. tables) of a database. SQL-like languages construct queries by making repeated use of the natural join and of the union. The inner union is a mathematically well behaved variant of the union—for example, it does not introduce empty cells. Tropashko and Spight realized [25,26] that these two operations are the meet and join operations in a class of lattices, known by now as the class of relational lattices. They proposed then lattice theory as an algebraic approach, alternative to Codd's relational algebra [4], to the theory of databases.

Roughly speaking, elements of the relational lattice $\mathsf{R}(D, A)$ are tables of a database, where A is a set of columns' names and D is the set of possible cells' values. Let us illustrate the two operations with examples. The natural join takes two tables and constructs a new one whose columns are indexed by the union of the headers, and whose rows are glueings of the rows along identical values in common columns:

Supported by the Project TICAMORE ANR-16-CE91-0002-01.

C. Baier and U. Dal Lago (Eds.): FOSSACS 2018, LNCS 10803, pp. 494–510, 2018.
https://doi.org/10.1007/978-3-319-89366-2_27

Author	Area
Santocanale	Logic
Santocanale	CS

$\bowtie$

Area	Reviewer
CS	Turing
Logic	Gödel

$=$

Author	Area	Reviewer
Santocanale	Logic	Gödel
Santocanale	CS	Turing

The inner union restricts two tables to the common columns and lists all the rows of the two tables. The following example suggests how to construct, using this operation, a table of users given two (or more) tables of people having different roles.

Author

Name	Surname	Conf
Luigi	Santocanale	FOSSACS

$\cup$

Reviewer

Name	Surname	Area
Alan	Turing	CS
Kurt	Gödel	Logic

$=$

User

Name	Surname
Luigi	Santocanale
Alan	Turing
Kurt	Gödel

Since we shall focus on lattice-theoretic considerations, we shall use the symbols $\wedge$ and $\vee$, in place of the symbols $\bowtie$ for $\cup$ used by database theorists.

A first important attempt to axiomatize these lattices was done by Litak et al. [17]. They proposed an axiomatization, comprising equations and quasiequations, in a signature that extends the pure lattice signature with a constant, the header constant. A main result of that paper is that the quasiequational theory of relational lattices is undecidable in this extended signature. Their proof mimics Maddux's proof that the equational theory of cylindric algebras of dimension $n \geq 3$ is undecidable [18].

Their result was further refined by us in [23]: the quasiequational theory of relational lattices is undecidable even when the signature considered is the least one, comprising only the meet (natural join) and the join operations (inner union). Our proof relied on a deeper algebraic insight: we proved that it is undecidable whether a finite subdirectly irreducible lattice can be embedded into a relational lattice—from this kind of result, undecidability of the quasiequational theory immediately follows. We proved the above statement by reducing to it an undecidable problem in modal logic, the coverability problem of a frame by a universal $\mathbf{S5}^3$-product frame [12]. In turn, this problem was shown to be undecidable by reducing it to the representability problem of finite simple relation algebras [11].

We prove here that the equational theory of relational lattices is decidable. That is, we prove that it is decidable whether two lattice terms t and s are such that $[\![t]\!]_v = [\![s]\!]_v$, for any valuation $v : \mathbb{X} \longrightarrow \mathsf{R}(D, A)$ of variables in a relational lattice $\mathsf{R}(D, A)$. We achieve this goal by showing that this theory has a kind of finite model property of bounded size. Out main result, Theorem 25, sounds as follows: *if an inclusion* $t \leq s$ *fails in a relational lattice* $\mathsf{R}(D, A)$, *then such inclusion fails in a finite lattice* $\mathsf{R}(E, B)$, *such that* B *is bound by an exponential function in the size of* t *and* s, *and* E *is linear in the size of* t. It follows that the size of $\mathsf{R}(E, B)$ can be bound by a triple exponential function in the size of t and s. In algebraic terms, our finite model theorem can be stated by saying that the variety generated by the relational lattices is actually generated by its finite generators, the relational lattices that are finite.

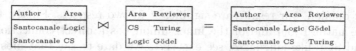

In our opinion, our results are significant in two respects. Firstly, the algebra of the natural join and of the inner union has a direct connection to the widespread SQL-like languages, see e.g. [17]. We dare to say that most of programmers that use a database—more or less explicitly, for example within server-side web programs—are using these operations. In view of the widespread use of these languages, the decidability status of this algebraic system deserved being settled. Moreover, we believe that the mathematical insights contained in our decidability proof shall contribute to understand further the algebraic system. For example, it is not known yet whether a complete finite axiomatic basis exists for relational lattices; finding it could eventually yield applications, e.g. on the side of automated optimization of queries.

Secondly, our work exhibits the equational theory of relational lattices as a decidable one within a long list of undecidable logical theories [11,12,17,18,23] that are used to model the constructions of relational algebra. We are exploring limits of decidability, a research direction widely explored in automata theoretic settings starting from [3]. We do this, within logic and with plenty of potential applications, coming from the undecidable side and crossing the border: after the quasiequational theory, undecidable, the next natural theory on the list, the equational theory of relational lattices, is decidable.

On the technical side, our work relies on [22] where the duality theory for finite lattices developed in [21] was used to investigate equational axiomatizations of relational lattices. A key insight from [22] is that relational lattices are, in some sense, duals of generalized ultrametric spaces over a powerset algebra. It is this perspective that made it possible to uncover the strong similarity between the lattice-theoretic methods and tools from modal logic—in particular the theory of combination of modal logics, see e.g. [15]. We exploit here this similarity to adapt filtrations techniques from modal logic [8] to lattice theory. Also, the notion of generalized ultrametric spaces over a powerset algebra and the characterization of injective objects in the category of these spaces have been fundamental tools to prove the undecidability of the quasiequational theory [23] as well as, in the present case, the decidability of the equational theory.

The paper is organised as follows. We recall in Sect. 2 some definitions and facts about lattices. The relational lattices $\mathsf{R}(D, A)$ are introduced in Sect. 3. In Sect. 4 we show how to construct a lattice $\mathsf{L}(X, \delta)$ from a generalized ultrametric space (X, δ). This construction generalizes the construction of the lattice $\mathsf{R}(D, A)$: if $X = D^A$ is the set of all functions from A to D and δ is as a sort of Hamming distance, then $\mathsf{L}(X, \delta) = \mathsf{R}(D, A)$. We use the functorial properties of L to argue that when a finite space (X, δ) has the property of being pairwise-complete, then $\mathsf{L}(X, \delta)$ belongs to the variety generated by the relational lattices. In Sect. 5 we show that if an inclusion $t \leq s$ fails in a lattice $\mathsf{R}(D, A)$, then we can construct a finite subset $T(f, t) \subseteq D^A$, a "tableau" witnessing the failure, such that if $T(f, t) \subseteq T$ and T is finite, then $t \leq s$ fails in a finite lattice of the form $\mathsf{L}(T, \delta_B)$, where the distance δ_B takes values in a finite powerset algebra $P(B)$. In Sect. 6, we show how to extend $T(f, t)$ to a finite bigger set G, so that (G, δ_B) as a space over the powerset algebra $P(B)$ is pairwise-complete. This lattice $\mathsf{L}(\mathsf{G}, \delta_B)$ fails the inclusion $t \leq s$; out of it, we build a lattice of the form $\mathsf{R}(E, B)$, which fails the same inclusion; the sizes of E and B can be bound by functions of the sizes of the terms t and s. Perspectives for future research directions appear in the last Sect. 7.

2 Elementary Notions on Orders and Lattices

We assume some basic knowledge of order and lattice theory as presented in standard monographs [5,9]. Most of the lattice theoretic tools we use originate from the monograph [7].

A *lattice* is a poset L such that every finite non-empty subset $X \subseteq L$ admits a smallest upper bound $\bigvee X$ and a greatest lower bound $\bigwedge X$. A lattice can also be understood as a structure $\mathfrak{A}$ for the functional signature $(\vee, \wedge)$, such that the interpretations of these two binary function symbols both give $\mathfrak{A}$ the structure of an idempotent commutative semigroup, the two semigroup structures being connected by the absorption laws $x \wedge (y \vee x) = x$ and $x \vee (y \wedge x) = x$. Once a lattice is presented as such structure, the order is recovered by stating that $x \leq y$ holds if and only if $x \wedge y = x$.

A lattice L is *complete* if any subset $X \subseteq L$ admits a smallest upper bound $\bigvee X$. It can be shown that this condition implies that any subset $X \subseteq L$ admits a greatest lower bound $\bigwedge X$. A lattice is *bounded* if it has a least element $\bot$ and a greatest element $\top$. A complete lattice (in particular, a finite lattice) is bounded, since $\bigvee \emptyset$ and $\bigwedge \emptyset$ are, respectively, the least and greatest elements of the lattice.

If P and Q are partially ordered sets, then a function $f : P \longrightarrow Q$ is *order-preserving* (or *monotone*) if $p \leq p'$ implies $f(p) \leq f(p')$. If L and M are lattices, then a function $f : L \longrightarrow M$ is a *lattice morphism* if it preserves the lattice operations $\vee$ and $\wedge$. A lattice morphism is always order-preserving. A lattice morphism $f : L \longrightarrow M$ between bounded lattices L and M is *bound-preserving* if $f(\bot) = \bot$ and $f(\top) = \top$. A function $f : P \longrightarrow Q$ is said to be *left adjoint* to an order-preserving $g : Q \longrightarrow P$ if $f(p) \leq q$ holds if and only if $p \leq g(q)$ holds, for every $p \in P$ and $q \in Q$; such a left adjoint, when it exists, is unique. Dually, a function $g : Q \longrightarrow P$ is said to be *right adjoint* to an order-preserving $f : P \longrightarrow Q$ if $f(p) \leq q$ holds if and only if $p \leq g(q)$ holds; clearly, f is left adjoint to g if and only if g is right adjoint to f, so we say that f and g form an adjoint pair. If P and Q are complete lattices, the property of being a left adjoint (resp., right adjoint) to some g (resp., to some f) is equivalent to preserving all (possibly infinite) joins (resp., all meets).

A *Moore family on* $P(U)$ is a collection $\mathcal{F}$ of subsets of U which is closed under arbitrary intersections. Given a Moore family $\mathcal{F}$ on $P(U)$, the correspondence sending $Z \subseteq U$ to $\overline{Z} := \bigcap\{Y \in \mathcal{F} \mid Z \subseteq Y\}$ is a *closure operator* on $P(U)$, that is, an order-preserving inflationary and idempotent endofunction of $P(U)$. The subsets in $\mathcal{F}$, called the *closed sets*, are exactly the fixpoints of this closure operator. A Moore family $\mathcal{F}$ has the structure of a complete lattice where

$$\bigwedge X := \bigcap X, \qquad \bigvee X := \overline{\bigcup X}. \tag{1}$$

The notion of Moore family can also be defined for an arbitrary complete lattice L. Moore families on L turns out to be in bijection with closure operators on L. We shall actually consider the dual notion: a *dual Moore family on a complete lattice* L is a subset $\mathcal{F} \subseteq L$ that is closed under arbitrary joins. Such an $\mathcal{F}$ determines an interior operator (an order-preserving decreasing and idempotent endofunction on L) by the formula $x^\circ = \bigvee\{ y \in \mathcal{F} \mid y \leq x \}$ and has the structure of a complete lattice, where $\bigvee_{\mathcal{F}} X := \bigvee_L X$ and $\bigwedge_{\mathcal{F}} X := (\bigwedge_L X)^\circ$. Dual Moore families on L are in bijection with interior operators on L. Finally, let us mention that closure (resp., interior) operators arise from adjoint pairs f and g (with f left adjoint to g) by the formula $\overline{x} = g(f(x))$ (resp., $x^\circ = f(g(x))$);

3 The Relational Lattices $\mathsf{R}(D, A)$

Throughout this paper we use the Y^X for the set of functions of domain Y and codomain X.

Let A be a collection of attributes (or column names) and let D be a set of cell values. A *relation* on A and D is a pair (α, T) where $\alpha \subseteq A$ and $T \subseteq D^\alpha$. Elements of the relational lattice[1] $\mathsf{R}(D, A)$ are relations on A and D. Informally, a relation (α, T) represents a table of a relational database, with α being the header, i.e. the collection of names of columns, while T is the collection of rows.

Before we define the natural join, the inner union operations, and the order on $\mathsf{R}(D, A)$, let us recall some key operations. If $\alpha \subseteq \beta \subseteq A$ and $f \in D^\beta$, then we shall use $f\restriction_\alpha \in D^\alpha$ for the restriction of f to α; if $T \subseteq D^\beta$, then $T\Restriction_\alpha$ shall denote projection to α, that is, the direct image of T along restriction, $T\Restriction_\alpha := \{ f\restriction_\alpha \mid f \in T \}$; if $T \subseteq D^\alpha$, then $i_\beta(T)$ shall denote cylindrification to β, that is, the inverse image of restriction, $i_\beta(T) := \{ f \in D^\beta \mid f\restriction_\alpha \in T \}$. Recall that i_β is right adjoint to $\Restriction_\alpha$. With this in mind, the natural join and the inner union of relations are respectively described by the following formulas:

$$
\begin{aligned}
(\alpha_1, T_1) \wedge (\alpha_2, T_2) &:= (\alpha_1 \cup \alpha_2, T) \\
\text{where } T &= \{ f \mid f\restriction_{\alpha_i} \in T_i, i = 1, 2 \} \\
&= i_{\alpha_1 \cup \alpha_2}(T_1) \cap i_{\alpha_1 \cup \alpha_2}(T_2), \\
(\alpha_1, T_1) \vee (\alpha_2, T_2) &:= (\alpha_1 \cap \alpha_2, T) \\
\text{where } T &= \{ f \mid \exists i \in \{1, 2\}, \exists g \in T_i \text{ s.t. } g\restriction_{\alpha_1 \cap \alpha_2} = f \} \\
&= T_1\Restriction_{\alpha_1 \cap \alpha_2} \cup T_2\Restriction_{\alpha_1 \cap \alpha_2}.
\end{aligned}
$$

The order is then given by $(\alpha_1, T_1) \leq (\alpha_2, T_2)$ iff $\alpha_2 \subseteq \alpha_1$ and $T_1\Restriction_{\alpha_2} \subseteq T_2$.

A convenient way of describing these lattices was introduced in [17, Lemma 2.1]. The authors showed that the relational lattices $\mathsf{R}(D, A)$ are isomorphic to the lattices of closed subsets of $A \cup D^A$, where $Z \subseteq A \cup D^A$ is said to be closed if it is a fixed-point of the closure operator $\overline{(-)}$ defined as

$$\overline{Z} := Z \cup \{ f \in D^A \mid A \setminus Z \subseteq Eq(f, g), \text{for some } g \in Z \},$$

[1] In [17] such a lattice is called *full* relational lattice. The wording "class of relational lattices" is used there for the class of lattices that have an embedding into some lattice of the form $\mathsf{R}(D, A)$.

where in the formula above $Eq(f,g)$ is the equalizer of f and g. Letting $\delta(f,g) := \{ x \in A \mid f(x) \neq g(x) \}$, the above definition of the closure operator is obviously equivalent to the following one:

$$\overline{Z} := \alpha \cup \{ f \in D^A \mid \delta(f,g) \subseteq \alpha, \text{ for some } g \in Z \cap D^A \}, \text{ with } \alpha = Z \cap A.$$

From now on, we rely on this representation of relational lattices.

4 Lattices from Metric Spaces

Generalized ultrametric spaces over a Boolean algebra $P(A)$ turn out to be a convenient tool for studying relational lattices [17,22]. Metrics are well known tools from graph theory, see e.g. [10]. Generalized ultrametric spaces over a Boolean algebra $P(A)$ were introduced in [20] to study equivalence relations.

Definition 1. *An* ultrametric space over $P(A)$ *(briefly, a* space*) is a pair* (X, δ), *with* $\delta : X \times X \to P(A)$ *such that, for every* $f, g, h \in X$,

$$\delta(f,f) \subseteq \emptyset, \qquad \delta(f,g) \subseteq \delta(f,h) \cup \delta(h,g), \tag{2}$$
$$\delta(f,g) = \emptyset \text{ implies } f = g, \qquad \delta(f,g) = \delta(g,f). \tag{3}$$

That is, we have defined an ultrametric space over $P(A)$ as a category (with a small set of objects) enriched over $(P(A)^{op}, \emptyset, \cup)$ (equation (2), see [16]) which moreover is *reduced* and *symmetric* (conditions (3)).

A *morphism* of spaces[2] $\psi : (X, \delta_X) \longrightarrow (Y, \delta_Y)$ is a function $\psi : X \to Y$ such that $\delta_Y(\psi(f), \psi(g)) \leq \delta_X(f,g)$, for each $f, g \in X$. Obviously, spaces and their morphisms form a category. If $\delta_Y(\psi(f), \psi(g)) = \delta_X(f,g)$, for each $f, g \in X$, then ψ is said to be an *isometry*. A space (X, δ) is said to be *pairwise-complete*, see [2], or *convex*, see [19], if, for each $f, g \in X$ and $\alpha, \beta \subseteq A$,

$$\delta(f,g) \subseteq \alpha \cup \beta \text{ implies } \delta(f,h) \subseteq \alpha \text{ and } \delta(h,g) \subseteq \beta, \quad \text{for some } h \in X.$$

Proposition 2 (see [2,20]). *If A is finite, then a space is injective in the category of spaces if and only if it is pairwise-complete.*

If (X, δ_X) is a space and $Y \subseteq X$, then the restriction of δ_X to Y induces a space (Y, δ_X); we say then that (Y, δ_X) is a *subspace* of X. Notice that the inclusion of Y into X yields an isometry of spaces.

Our main example of space over $P(A)$ is (D^A, δ), with D^A the set of functions from A to D and the distance defined by

$$\delta(f,g) := \{ a \in A \mid f(a) \neq g(a) \}. \tag{4}$$

A second example is a slight generalization of the previous one. Given a surjective function $\pi : D \to A$, let $\mathtt{Sec}_\pi$ denote the set of all the functions $f : A \to D$ such

[2] As $P(A)$ is not totally ordered, we avoid calling a morphism "*non-expanding map*" as it is often done in the literature.

that $\pi \circ f = id_A$. Then $\mathsf{Sec}_\pi \subseteq D^A$, so Sec_π with the distance inherited from (D^A, δ) can be made into a space. Considering the first projection $\pi_1 : A \times D \to A$, we see that (D^A, δ) is isomorphic to the space Sec_{π_1}. By identifying $f \in \mathsf{Sec}_\pi$ with a vector $\langle f(a) \in \pi^{-1}(a) \mid a \in A \rangle$, we see that

$$\mathsf{Sec}_\pi = \prod_{a \in A} D_a\,, \qquad \text{where } D_a := \pi^{-1}(a). \tag{5}$$

That is, the spaces of the form Sec_π are naturally related to Hamming graphs in combinatorics [13], dependent function types in type theory [6,14], universal $\mathbf{S5}^A$-product frames in modal logic [12].

Theorem 3 (see [23]). *Spaces of the form Sec_π are, up to isomorphism, exactly the injective objects in the category of spaces.*

4.1 The Lattice of a Space

The construction of the lattice $\mathsf{R}(D, A)$ can be carried out from any space. Namely, for a space (X, δ) over $P(A)$, say that $Z \subseteq X$ is *α-closed* if $g \in Z$ and $\delta(f, g) \subseteq \alpha$ implies $f \in Z$. Clearly, α-closed subsets of X form a Moore family so, for $Z \subseteq X$, we denote by $\overline{Z}^\alpha$ the least α-closed subset of X containing Z. Observe that $f \in \overline{Z}^\alpha$ if and only if $\delta(f, g) \subseteq \alpha$ for some $g \in Z$. Next and in the rest of the paper, we shall exploit the obvious isomorphism between $P(A) \times P(X)$ and $P(A \cup X)$ (where we suppose A and X disjoint) and notationally identify a pair $(\alpha, Z) \in P(A) \times P(X)$ with its image $\alpha \cup X \in P(A \cup X)$. Let us say then that (α, Z) is closed if Z is α-closed. Closed subsets of $P(A \cup X)$ form a Moore family, whence a complete lattice where the order is subset inclusion.

Definition 4. *For a space (X, δ), the lattice $\mathsf{L}(X, \delta)$ is the lattice of closed subsets of $P(A \cup X)$.*

Clearly, for the space (D^A, δ), we have $\mathsf{L}(D^A, \delta) = \mathsf{R}(D, A)$. Let us mention that meets and joins $\mathsf{L}(X, \delta)$ are computed using the formulas in (1). In particular, for joins,

$$(\alpha, Y) \vee (\beta, Z) = (\alpha \cup \beta, \overline{Y \cup Z}^{\alpha \cup \beta}).$$

The above formula yields that, for any $f \in X$, $f \in (\alpha, Y) \vee (\beta, Z)$ if and only if $\delta(f, g) \subseteq \alpha \cup \beta$, for some $g \in Y \cup Z$.

We argue next that the above construction is functorial. Below, for a function $\psi : X \to Y$, $\psi^{-1} : P(Y) \to P(X)$ is the inverse image of ψ, defined by $\psi^{-1}(Z) := \{x \in X \mid \psi(x) \in Z\}$.

Proposition 5. *If $\psi : (X, \delta_X) \to (Y, \delta_Y)$ is a space morphism and $(\alpha, Z) \in \mathsf{L}(Y, \delta_Y)$, then $(\alpha, \psi^{-1}(Z)) \in \mathsf{L}(X, \delta_X)$. Therefore, by defining $\mathsf{L}(\psi)(\alpha, Z) := (\alpha, \psi^{-1}(Z))$, the construction L lifts to a contravariant functor from the category of spaces to the category of complete meet-semilattices.*

Proof. Let $f \in X$ be such that, for some $g \in \psi^{-1}(Z)$ (i.e. $\psi(g) \in Z$), we have $\delta_X(f,g) \subseteq \alpha$. Then $\delta_Y(\psi(f),\psi(g)) \subseteq \delta_X(f,g) \subseteq \alpha$, so $\psi(f) \in Z$, since Z is α-closed, and $f \in \psi^{-1}(Z)$. In order to see that $\mathsf{L}(\psi)$ preserves arbitrary intersections, recall that ψ^{-1} does. □

Notice that $\mathsf{L}(\psi)$ might not preserve arbitrary joins.

Proposition 6. *The lattices $\mathsf{L}(\mathit{Sec}_\pi)$ generate the same lattice variety of the lattices $\mathsf{R}(D,A)$.*

That is, a lattice equation holds in all the lattices $\mathsf{L}(\mathtt{Sec}_\pi)$ if and only if it holds in all the relation lattices $\mathsf{R}(D,A)$.

Proof. Clearly, each lattice $\mathsf{R}(D,A)$ is of the form $\mathsf{L}(\mathtt{Sec}_\pi)$. Thus we only need to argue that every lattice of the form $\mathsf{L}(\mathtt{Sec}_\pi)$ belongs to the lattice variety generated by the $\mathsf{R}(D,A)$, that is, the least class of lattices containing the lattices $\mathsf{R}(D,A)$ and closed under products, sublattices, and homomorphic images. We argue as follows.

As every space $\mathtt{Sec}_\pi$ embeds into a space (D^A,δ) and a space $\mathtt{Sec}_\pi$ is injective, we have maps $\iota : \mathtt{Sec}_\pi \to (D^A,\delta)$ and $\psi : (D^A,\delta) \to \mathtt{Sec}_\pi$ such that $\psi \circ \iota = id_{\mathtt{Sec}_\pi}$. By functoriality, $\mathsf{L}(\iota) \circ \mathsf{L}(\psi) = id_{\mathsf{L}(\mathtt{Sec}_\pi)}$. Since $\mathsf{L}(\iota)$ preserves all meets, it has a left adjoint $\ell : \mathsf{L}(\mathtt{Sec}_\pi) \to \mathsf{L}(D^A,\delta) = \mathsf{R}(D,A)$. It is easy to see that $(\ell, \mathsf{L}(\psi))$ is an EA-duct in the sense of [24, Definition 9.1] and therefore $\mathsf{L}(\mathtt{Sec}_\pi)$ is a homomorphic image of a sublattice of $\mathsf{R}(D,A)$, by [24, Lemma 9.7]. □

Remark 7. For the statement of [24, Lemma 9.7] to hold, additional conditions are necessary on the domain and the codomain of an EA-duet. Yet the implication that derives being a homomorphic image of a sublattice from the existence of an EA-duet is still valid under the hypothesis that the two arrows of the EA-duet preserve one all joins and, the other, all meets.

4.2 Extension from a Boolean Subalgebra

We suppose that $P(B)$ is a Boolean subalgebra of $P(A)$ via an inclusion $i : P(B) \to P(A)$. If (X,δ_B) is a space over $P(B)$, then we can transform it into a space (X,δ_A) over $P(A)$ by setting $\delta_A(f,g) = i(\delta_B(f,g))$. We have therefore two lattices $\mathsf{L}(X,\delta_B)$ and $\mathsf{L}(X,\delta_A)$.

Proposition 8. *Let $\beta \subseteq B$ and $Y \subseteq X$. Then Y is β-closed if and only if it is $i(\beta)$-closed. Consequently the map i_*, sending $(\beta,Y) \in \mathsf{L}(X,\delta_B)$ to $i_*(\beta,Y) := (i(\beta),Y) \in \mathsf{L}(X,\delta_A)$, is a lattice embedding.*

Proof. Observe that $\delta_B(f,g) \subseteq \beta$ if and only if $\delta_A(f,g) = i(\delta_B(f,g)) \subseteq i(\beta)$. This immediately implies the first statement of the Lemma, but also that, for $Y \subseteq X$, $\overline{Y}^{\beta} = \overline{Y}^{i(\beta)}$. Using the fact that meets are computed as intersections and that i preserves intersections, it is easily seen that i_* preserves meets. For joins let us compute as follows:

$$\begin{aligned} i_*(\beta_1, Y_1) \vee i_*(\beta_2, Y_2) &= (i(\beta_1) \cup i(\beta_2), \overline{Y_1 \cup Y_2}^{i(\beta_1) \cup i(\beta_2)}) \\ &= (i(\beta_1 \cup \beta_2), \overline{Y_1 \cup Y_2}^{i(\beta_1 \cup \beta_2)}) = (i(\beta_1 \cup \beta_2), \overline{Y_1 \cup Y_2}^{\beta_1 \cup \beta_2}) \\ &= i_*(\beta_1 \cup \beta_2, \overline{Y_1 \cup Y_2}^{\beta_1 \cup \beta_2}) = i_*((\beta_1, Y_1) \vee (\beta_2, Y_2)). \end{aligned}$$ □

5 Failures from Big to Small Lattices

The set of lattice terms is generated by the following grammar:

$$t := x \mid \top \mid t \wedge t \mid \bot \mid t \vee t,$$

where x belongs to a set of variables $\mathbb{X}$. For lattice terms $t_1, \ldots, t_n$, we use $Vars(t_1, \ldots, t_n)$ to denote the set of variables (which is finite) occurring in any of these terms. The size of a term t is the number of nodes in the representation of t as a tree. If $v : \mathbb{X} \longrightarrow L$ is a valuation of variables into a lattice L, the value of a term t w.r.t. the valuation v is defined by induction in the obvious way; here we shall use $[\![t]\!]_v$ for it.

For t, s two lattice terms, the inclusion $t \leq s$ is the equation $t \vee s = s$. Any lattice-theoretic equation is equivalent to a pair of inclusions, so the problem of deciding the equational theory of a class of lattices reduces to the problem of decing inclusions. An inclusion $t \leq s$ is valid in a class of lattices $\mathcal{K}$ if, for any valuation $v : \mathbb{X} \longrightarrow L$ with $L \in \mathcal{K}$, $[\![v]\!]_v \leq [\![s]\!]_v$; it fails in $\mathcal{K}$ if for some $L \in \mathcal{K}$ and $v : \mathbb{X} \longrightarrow L$ we have $[\![t]\!]_v \not\leq [\![s]\!]_v$.

From now on, our goal shall be proving that if an inclusion $t \leq s$ fails in a lattice $\mathsf{R}(D, A)$, then it fails in a lattice $\mathsf{L}(\mathtt{Sec}_\pi)$, where $\mathtt{Sec}_\pi$ is a finite space over some finite Boolean algebra $P(B)$. The size of B and of the space $\mathtt{Sec}_\pi$, shall be inferred from of the sizes of t and s.

From now on, we us fix terms t and s, a lattice $\mathsf{R}(D, A)$, and a valuation $v : \mathbb{X} \longrightarrow \mathsf{R}(D, A)$ such that $[\![t]\!]_v \not\subseteq [\![s]\!]_v$.

Lemma 9. *If, for some $a \in A$, $a \in [\![t]\!]_v \setminus [\![s]\!]_v$, then the inclusion $t \leq s$ fails in the lattice $\mathsf{R}(E, B)$ with $B = \emptyset$ and E a singleton.*

Proof. The map sending $(\alpha, X) \in \mathsf{R}(D, A)$ to $\alpha \in P(A)$ is lattice morphism. Therefore if $t \leq s$ fails because of $a \in A$, then it already fails in the Boolean lattice $P(A)$. Since $P(A)$ is distributive, $t \leq s$ fails in the two elements lattice. Now, when $B = \emptyset$ and E is a singleton $\mathsf{R}(E, B)$ is (isomorphic to) the 2 elements lattice, so the same equation fails in $\mathsf{R}(E, B)$. □

Because of the Lemma, we shall focus on functions $f \in D^A$ such that $f \in [\![t]\!]_v \setminus [\![s]\!]_v$. In this case we shall say that f *witnesses the failure of* $t \leq s$ (in $\mathsf{R}(D, A)$, w.r.t. the valuation v).

5.1 The Lattices $\mathsf{R}(D, A)_T$

Let T be a subset of D^A and consider the subspace (T, δ) of D^A induced by the inclusion $i_T : T \subseteq D^A$. According to Proposition 5, the inclusion i_T induces a complete meet-semilattice homomorphism $\mathsf{L}(i_T) : \mathsf{R}(D, A) = \mathsf{L}(D^A, \delta) \to \mathsf{L}(T, \delta)$. Such a map has a right adjoint $j_T : \mathsf{L}(T, \delta) \to \mathsf{L}(D^A, \delta)$, which is a complete join-semilattice homomorphism; moreover j_T is injective, since $\mathsf{L}(i_T)$ is surjective.

Proposition 10. *For a subset $T \subseteq D^A$ and $(\alpha, X) \in \mathsf{R}(D, A)$, $(\alpha, \overline{X \cap T}^{\alpha}) = j_T(\mathsf{L}(i_T(\alpha, X))$. The set of elements of the form $(\alpha, \overline{X \cap T}^{\alpha})$, for $\alpha \subseteq A$ and $X \subseteq D^A$, is a complete sub-join-semilattice of* $\mathsf{R}(D, A)$.

Proof. It is easily seen that $\mathsf{L}(i_T)(\alpha, X) = (\alpha, X \cap T)$ and that, for $(\beta, Y) \in \mathsf{L}(T, \delta)$, $(\beta, Y) \subseteq (\alpha, X \cap T)$ if and only if $(\beta, \overline{Y}^{\beta}) \subseteq (\alpha, X)$, so $j_T(\beta, Y) = (\beta, \overline{Y}^{\beta})$.

It follows that the elements of the form $(\alpha, \overline{X \cap T}^{\alpha})$, where $(\alpha, X) \in \mathsf{R}(D, A)$, form a sub-complete join-semilattice of $\mathsf{R}(D, A)$: indeed, they are the image of lattice $\mathsf{L}(T, \delta)$ under the complete join-semilattice homomorphism j_T. We argue next that, for any pair (α, X) (we do not require that X is α-closed) there is a $Z \subseteq D^A$ which is α-closed and such that $\overline{X \cap T}^{\alpha} = \overline{Z \cap T}^{\alpha}$. Indeed, the equality

$$\overline{X \cap T}^{\alpha} = \overline{\overline{X \cap T}^{\alpha} \cap T}^{\alpha}$$

is easily verified, so we can let $Z = \overline{X \cap T}^{\alpha}$. □

Therefore, the set of pairs of the form $(\alpha, \overline{X \cap T}^{\alpha})$ is a dual Moore family and a complete lattice, where joins are computed as in $\mathsf{R}(D, A)$, and where meets are computed in a way that we shall make explicit. For the moment, let us fix the notation.

Definition 11. $\mathsf{R}(D, A)_T$ *is the lattice of elements of the form* $(\alpha, \overline{X \cap T}^{\alpha})$.

By the proof of Proposition 10, the lattice $\mathsf{R}(D, A)_T$ is isomorphic to the latttice $\mathsf{L}(T, \delta)$. We shall use the symbol $\bigwedge\!\!\!\bigwedge$ for meets in $\mathsf{R}(D, A)_T$; these are computed by the formula

$$\bigwedge\!\!\!\bigwedge_{i \in I}(\alpha_i, X_i) = \Big(\bigcap_{i \in I} \alpha_i, \bigcap_{i \in I} X_i\Big)^{\circ},$$

where, for each $(\alpha, X) \in \mathsf{R}(D, A)$, $(\alpha, X)^{\circ}$ is the greatest pair in $\mathsf{R}(D, A)_T$ that is below (α, X). Standard theory on adjoints yields

$$(\alpha, X)^{\circ} = (j_T \circ \mathsf{L}(i_T))(\alpha, X) = (\alpha, \overline{X \cap T}^{\alpha}).$$

We obtain in this way the explicit formula for the binary meet in $\mathsf{R}(D, A)_T$:

$$(\alpha, \overline{X \cap T}^{\alpha}) \wedge\!\!\!\wedge (\beta, \overline{Y \cap T}^{\beta}) = (\alpha \cap \beta, \overline{\overline{X \cap T}^{\alpha} \cap \overline{Y \cap T}^{\beta} \cap T}^{\alpha \cap \beta}).$$

Remark that we have

$$(\alpha, X) \wedge\!\!\!\wedge (\beta, Y) \subseteq (\alpha, X) \cap (\beta, Y)$$

whenever (α, X) and (β, Y) are in $\mathsf{R}(D, A)_T$.

Lemma 12. *Let $(\alpha, X), (\beta, Y) \in \mathsf{R}(D, A)_T$ and let $f \in T$. If $f \in (\alpha, X) \cap (\beta, Y)$, then $f \in (\alpha, X) \wedge\!\!\!\wedge (\beta, Y)$.*

Proof. This is immediate from the fact that

$$\overline{X \cap T}^{\alpha} \cap \overline{Y \cap T}^{\beta} \cap T \subseteq \overline{\overline{X \cap T}^{\alpha} \cap \overline{Y \cap T}^{\beta} \cap T}^{\alpha \cap \beta} .$$ □

5.2 Preservation of the Failure in the Lattices $\mathsf{R}(D, A)_T$

Recall that $v : \mathbb{X} \longrightarrow \mathsf{R}(D, A)$ is the valuation that we have fixed.

Definition 13. *For a susbset T of D^A, the valuation $v_T : \mathbb{X} \longrightarrow \mathsf{R}(D, A)_T$ is defined by the formula $v_T(x) = v(x)^{\circ}$, for each $x \in \mathbb{X}$.*

More explicitley, we have

$$v_T(x) := (\alpha, \overline{T \cap X}^{\alpha}), \quad \text{where } (\alpha, X) = v(x) .$$

The valuation v_T takes values in $\mathsf{R}(D, A)_T$, while v takes value in $\mathsf{R}(D, A)$. It is possible then to evaluate a lattice term t in $\mathsf{R}(D, A)_T$ using v_T and to evaluate it in $\mathsf{R}(D, A)$ using v. To improve readability, we shall use the notation $[\![t]\!]_T$ for the result of evaluating the term in $\mathsf{R}(D, A)_T$, and the notation $[\![t]\!]$ for the result of evaluating it in $\mathsf{R}(D, A)$. Since both $[\![t]\!]$ and $[\![t]\!]_T$ are subsets of $P(A \cup X)$, it is possible to compare them using inclusion.

Lemma 14. *The relation $[\![s]\!]_T \subseteq [\![s]\!]$ holds, for each $T \subseteq D^A$ and each lattice term s.*

Proof. The proof of the Lemma is a straightforward induction, considering that $v_T(x) \subseteq v(x)$ for all $x \in \mathbb{X}$. For example, using $[\![s_i]\!]_T \subseteq [\![s_i]\!]$, for $i = 1, 2$,

$$[\![s_1 \wedge s_2]\!]_T = [\![s_1]\!]_T \wedge\!\!\!\wedge [\![s_2]\!]_T \subseteq [\![s_1]\!]_T \cap [\![s_2]\!]_T \subseteq [\![s_1]\!] \cap [\![s_2]\!] = [\![s_1 \wedge s_2]\!] .$$ □

A straightforward induction also yields:

Lemma 15. *Let $T \subseteq D^A$ be a finite subset, let t be a lattice term and suppose that $[\![t]\!] = (\beta, Y)$. Then $[\![t]\!]_T$ is of the form (β, Y') for some $Y' \subseteq D^A$.*

Definition 16. *Let us define, for each term t and $f \in D^A$ such that $f \in [\![t]\!]$, a finite set $T(f, t) \subseteq D^A$ as follows:*

– *If t is the variable x, then we let $T(f,t) := \{ f \}$.*
– *If $t = s_1 \wedge s_2$, then $f \in [\![s_1]\!] \cap [\![s_2]\!]$, so we define $T(f,t) := T(f,s_1) \cup T(f,s_2)$.*
– *If $t = s_1 \vee s_2$ and $[\![s_i]\!] = (\alpha_i, X_i)$ for $i = 1,2$, then $f \in [\![s_1 \vee s_2]\!]$ gives that, for some $i \in \{1,2\}$ there exists $g \in X_i$ such that $\delta(f,g) \subseteq \alpha_1 \cup \alpha_2$. We set then $T(f,t) := \{ f \} \cup T(g,s_i)$.*

Obviously, we have:

Lemma 17. *For each lattice term t and $f \in D^A$ such that $f \in [\![t]\!]$, $f \in T(f,t)$.*

Proposition 18. *For each lattice term t and $f \in D^A$ such that $f \in [\![t]\!]$, if $T(f,t) \subseteq T$, then $f \in [\![t]\!]_T$.*

Proof. We prove the statement by induction on t.

– If t is the variable x and $f \in [\![x]\!] = v(x) = (\beta, Y)$, then $f \in Y$. We have $T(f,x) = \{ f \}$. Obviously, $f \in Y \cap \{ f \} = Y \cap T(f,t) \subseteq Y \cap T$, so $f \in (\beta, \overline{Y \cap T}^{\beta}) = v_T(x) = [\![t]\!]_T$.
– Suppose $t = s_1 \wedge s_2$ so $f \in [\![s_1 \wedge s_2]\!]$ yields $f \in [\![s_1]\!]$ and $f \in [\![s_2]\!]$. We have defined $T(f,t) = T(f,s_1) \cup T(f,s_2) \subseteq T$ and so, using $T(f,s_i) \subseteq T$ and the induction hypothesis, $f \in [\![s_i]\!]_T$ for $i = 1,2$. By Lemma 17 $f \in T$, so we can use Lemma 12 asserting that

$$f \in [\![s_1]\!]_T \wedge\!\!\!\wedge [\![s_2]\!]_T = [\![s_1 \wedge s_2]\!]_T \, .$$

– Suppose $t = s_1 \vee s_2$ and $f \in [\![s_1 \vee s_2]\!]$; let also $(\beta_i, Y_i) := [\![s_i]\!]$ for $i = 1,2$. We have defined $T(f,t) := \{ f \} \cup T(g,s_i)$ for some $i \in \{1,2\}$ and for some $g \in [\![s_i]\!]$ such that $\delta(f,g) \subseteq \beta_1 \cup \beta_2$. Now $g \in T(g,s_i) \subseteq T(f,t) \subseteq T$ so, by the induction hypothesis, $g \in [\![s_i]\!]_T$. According to Lemma 15, for each $i = 1,2$ $[\![s_i]\!]_T$ is of the form (β_i, Y_i'), for some subset $Y_i' \subseteq D^A$. Therefore $\delta(f,g) \subseteq \beta_1 \cup \beta_2$ and $g \in [\![s_i]\!]_T$ implies

$$f \in [\![s_1]\!]_T \vee [\![s_2]\!]_T = [\![s_1 \vee s_2]\!]_T \, .$$

□

Proposition 19. *Suppose f witnesses the failure of the inclusion $t \leq s$ in $\mathsf{R}(D,A)$ w.r.t. the valuation v. Then, for each subset $T \subseteq D^A$ such $T(f,t) \subseteq T$, f witnesses the failure of the inclusion $t \leq s$ in the lattice $\mathsf{R}(D,A)_T$ and w.r.t. valuation v_T.*

Proof. As f witnesses $t \not\leq s$ in $\mathsf{R}(D,A)$, $f \in [\![t]\!]$ and $f \notin [\![s]\!]$. By Lemma 18 $f \in [\![t]\!]_T$. If $f \in [\![s]\!]_T$, then $[\![s]\!]_T \subseteq [\![s]\!]$ (Lemma 14) implies $f \in [\![s]\!]$, a contradicition. Therefore $f \notin [\![s]\!]_T$, so f witnesses $t \not\leq s$ in $\mathsf{R}(D,A)_T$. □

5.3 Preservation of the Failure in a Finite Lattice $\mathsf{L}(X, \delta)$

From now on, we suppose that $T \subseteq D^A$ is finite and $T(f,t) \subseteq T$ with f witnessing the failure of $t \leq s$. Consider the sub-Boolean-algebra of $P(A)$ generated by the sets

$$\{ \delta(f,g) \mid f,g \in T \} \cup \{ A \cap v(x) \mid x \in Vars(t,s) \} \, . \qquad (6)$$

Let us call B this Boolean algebra (yet, notice the dependency of this definition on T, as well as on t, s and v). It is well known that a Boolean algebra generated by a finite set is finite.

Remark 20. If $n = \mathrm{card}(T)$ and $m = \mathrm{card}(Vars(t,s))$, then B can have at most $2^{\frac{n(n-1)}{2}+m}$ atoms. If we let k be the maximum of the sizes of t and s, then, for $T = T(f,t)$, both $n \leq k$ and $m \leq 2k$. We obtain in this case the over-approximation $2^{\frac{k^2+3k}{2}}$ on the number of atoms of B.

Let us also recall that B is isomorphic to the powerset $P(\mathsf{at}(\mathsf{B}))$, where $\mathsf{at}(\mathsf{B})$ is the set of atoms of B. Let $i : P(\mathsf{at}(\mathsf{B})) \longrightarrow P(A)$ be an injectve homomorphism of Boolean algebras whose image is B. Since $\delta(f,g) \in \mathsf{B}$ for every $f, g \in T$, we can transform the metric space (T, δ) induced from (D^A, δ) into a metric space $(T, \delta_{\mathsf{at}(\mathsf{B})})$ whose distance takes values in the powerset algebra $P(\mathsf{at}(\mathsf{B}))$:

$$\delta_{\mathsf{at}(\mathsf{B})}(f,g) = \beta \quad \text{if and only if} \quad \delta(f,g) = i(\beta)\,.$$

Recall from Proposition 8 that there is a lattice embedding $i_* : \mathsf{L}(T, \delta_{\mathsf{at}(\mathsf{B})}) \longrightarrow \mathsf{L}(T, \delta)$, defined in the obvious way: $i_*(\alpha, Y) = (i(\beta), Y)$.

Proposition 21. *If f witnesses the failure of the inclusion $t \leq s$ in $\mathsf{R}(D,A)$ w.r.t. the valuation v, then the same inclusion fails in all the lattices $\mathsf{L}(T, \delta_{\mathsf{at}(\mathsf{B})})$, where T is a finite set and $T(f,t) \subseteq T$.*

Proof. By Proposition 19 the inclusion $t \leq s$ fails in the lattice $\mathsf{R}(D,A)_T$. This lattice is isomorphic to the lattice $\mathsf{L}(T, \delta)$ via the map sending $(\alpha, X) \in \mathsf{R}(D,A)_T$ to $(\alpha, X \cap T)$. Up to this isomorphism, it is seen that the (restriction to the variables in t and s of) the valuation v_T takes values in the image of the lattice $\mathsf{L}(T, \delta_{\mathsf{at}(\mathsf{B})})$ via i_*, so $[\![t]\!]_T, [\![s]\!]_T$ belong to this sublattice and the inclusion fails in this lattice, and therefore also in $\mathsf{L}(T, \delta_{\mathsf{at}(\mathsf{B})})$. $\square$

6 Preservation of the Failure in a Finite Lattice $\mathsf{L}(\mathtt{Sec}_\pi)$

We have seen up to now that if $t \leq s$ fails in $\mathsf{R}(D,A)$, then it fails in many lattices of the form $\mathsf{L}(T, \delta_{\mathsf{at}(\mathsf{B})})$. Yet it is not obvious a priori that any of these lattices belongs to the variety generated by the relational lattices. We show in this section that we can extend any T to a finite set G while keeping B fixed, so that $(\mathsf{G}, \delta_{\mathsf{at}(\mathsf{B})})$ is a pairwise-complete space over $P(\mathsf{at}(\mathsf{B}))$. Thus, the inclusion $t \leq s$ fails in the finite lattice $\mathsf{L}(\mathsf{G}, \delta_{\mathsf{at}(\mathsf{B})})$. Since $(\mathsf{G}, \delta_{\mathsf{at}(\mathsf{B})})$ is isomorphic to a space of the form $\mathtt{Sec}_\pi$ with $\pi : E \longrightarrow \mathsf{at}(\mathsf{B})$, the inclusion $t \leq s$ fails in a lattice $\mathsf{L}(\mathtt{Sec}_\pi)$ which we have seen belongs to the variety generated by the relational lattices. This also leads to construct a finite relational lattice $\mathsf{R}(\mathsf{at}(\mathsf{B}), E)$ in which the equation $t \leq s$ fails. By following the chain of constructions, the sizes of $\mathsf{at}(\mathsf{B})$ and E can also be estimated, leading to decidability of the equational theory of relational lattices.

Definition 22. *A glue of* T *and* B *is a function* $g \in D^A$ *such that, for all* $\alpha \in \mathsf{at}(\mathsf{B})$*, there exists* $f \in T$ *with* $f{\restriction}\alpha = g$*. We denote by* G *the set of all functions that are glues of* T *and* B.

Observe that $T \subseteq \mathsf{G}$ and that G is finite, with

$$\mathrm{card}(\mathsf{G}) \leq \mathrm{card}(T)^{\mathrm{card}(\mathsf{at}(\mathsf{B}))}. \tag{7}$$

In order to prove the following Lemma, let, for each $\alpha \in \mathsf{at}(\mathsf{B})$ and $g \in \mathsf{G}$, $f(g,\alpha) \in T$ be such that $g{\restriction}\alpha = f(g,\alpha){\restriction}\alpha$.

Lemma 23. *If* $g_1, g_2 \in \mathsf{G}$*, then* $\delta(g_1, g_2) \in \mathsf{B}$.

Proof.

$$\delta(g_1,g_2) = \bigcup_{\alpha\in\mathsf{at}(\mathsf{B})} (\alpha \cap \delta(g_1,g_2)) = \bigcup_{\alpha\in\mathsf{at}(\mathsf{B})} (\alpha \cap \delta(f(g_1,\alpha), f(g_2,\alpha))).$$

Since $\delta(f(g_1,\alpha), f(g_2,\alpha)) \in \mathsf{B}$ and α is an atom of B, each expression of the form $\alpha \cap \delta(f(g_1,\alpha), f(g_2,\alpha))$ is either $\emptyset$ or α. It follows that $\delta(g_1,g_2) \in \mathsf{B}$. □

For a Boolean subalgebra B of $P(A)$, we say that a subset T of D^A is *pairwise-complete relative to* B if, for each $f, g \in T$,

1. $\delta(f,g) \in B$,
2. $\delta(f,g) \subseteq \beta \cup \gamma$, implies $\delta(f,h) \subseteq \beta$ and $\delta(h,g) \subseteq \gamma$ for some $h \in T$, for each $\beta, \gamma \in B$.

Lemma 24. *The set* G *is pairwise-complete relative to the Boolean algebra* B.

Proof. Let $f, g \in \mathsf{G}$ be such that $\delta(f,g) \subseteq \beta \cup \gamma$. Let $h \in D^A$ be defined so that, for each $\alpha \in \mathsf{at}(\mathsf{B})$, $h{\restriction}\alpha = f{\restriction}\alpha$ if $\alpha \not\subseteq \beta$ and $h{\restriction}\alpha = g{\restriction}\alpha$, otherwise. Obviously, $h \in \mathsf{G}$.

Observe that $\alpha \not\subseteq \beta$ if and only if $\alpha \subseteq \beta^{\mathsf{c}}$, for each $\alpha \in \mathsf{at}(\mathsf{B})$, since $\beta \in \mathsf{B}$. We deduce therefore $h{\restriction}\alpha = f{\restriction}\alpha$ if $\alpha \in \mathsf{at}(\mathsf{B})$ and $\alpha \subseteq \beta^{\mathsf{c}}$, so $f(a) = h(a)$ for each $a \in \beta^{\mathsf{c}}$. Consequently $\beta^{\mathsf{c}} \subseteq Eq(f,h)$ and $\delta(f,h) \subseteq \beta$.

We also have $h{\restriction}\alpha = g{\restriction}\alpha$ if $\alpha \in \mathsf{at}(\mathsf{B})$ and $\alpha \subseteq \gamma^{\mathsf{c}}$. As before, this implies $\delta(h,g) \subseteq \gamma$. Indeed, this is the case if $\alpha \subseteq \beta$, by definition of h. Suppose now that $\alpha \not\subseteq \beta$, so $\alpha \subseteq \beta^{\mathsf{c}} \cap \gamma^{\mathsf{c}} = (\beta \cup \gamma)^{\mathsf{c}}$. Since $\delta(f,g) \subseteq \beta \cup \gamma$, then $\alpha \subseteq \delta(f,g)^{\mathsf{c}} = Eq(f,g)$, i.e. $f{\restriction}\alpha = g{\restriction}\alpha$. Together with $h{\restriction}\alpha = f{\restriction}\alpha$ (by definition of h) we obtain $h{\restriction}\alpha = f{\restriction}\alpha$. □

We can finally bring together the observations developed so far and state our main results.

Theorem 25. *If an inclusion* $t \leq s$ *fails in all the lattices* $\mathsf{R}(D,A)$*, then it fails in a finite lattice* $\mathsf{R}(E,A')$*, where* $\mathrm{card}(A') \leq 2^{p(k)}$ *with* $k = \max(size(t), size(s))$*,* $p(k) = \frac{2^{k^2}+3k}{2}$*, and* $\mathrm{card}(E) \leq size(t)$.

Proof. By Proposition 19 the inclusion $t \leq s$ fails in all the lattices $\mathsf{R}(D, A)_T$ where $T(f,t) \subseteq T$. Once defined B as the Boolean subalgebra of $P(A)$ generated by the sets as in the display (6) (with $T = T(f,T)$) and G as the set of glues of $T(f,t)$ and B as in Definition 22, the inclusion fails in $\mathsf{R}(D, A)_{\mathsf{G}}$, since $T(f,T) \subseteq \mathsf{G}$, and then in $\mathsf{L}(\mathsf{G}, \delta_{\mathsf{at}(\mathsf{B})})$ by Proposition 21. The condition that G is pairwise-complete relative to B is equivalent to saying that the space $(\mathsf{G}, \delta_{\mathsf{at}(\mathsf{B})})$ is pairwise-complete. This space is therefore isomorphic to a space of the form $\mathtt{Sec}_\pi$ for some surjective $\pi : F \longrightarrow \mathsf{at}(\mathsf{B})$, and $t \leq s$ fails in $\mathsf{L}(\mathtt{Sec}_\pi)$.

Equation (7) shows that, for each $\alpha \in \mathsf{at}(\mathsf{B})$, $F_\alpha = \pi^{-1}(\alpha)$ has cardinality at most card$(T(f,t))$ and the size of t is an upper bound for card$(T(f,t))$. We can therefore embed the space $\mathtt{Sec}_\pi$ into a space of the form $(E^{\mathsf{at}(\mathsf{B})}, \delta)$ with the size of t an upper bound for card(E). The proof of Proposition 6 exhibits $\mathsf{L}(\mathtt{Sec}_\pi)$ as a homomorphic image of a sublattice of $\mathsf{L}(E^{\mathsf{at}(\mathsf{B})}, \delta)$ and therefore the inclusion $t \leq s$ also fails within $\mathsf{L}(E^{\mathsf{at}(\mathsf{B})}, \delta) = \mathsf{R}(E, \mathsf{at}(\mathsf{B}))$. The upper bound on the size of $\mathsf{at}(\mathsf{B})$ has been extimated in Remark 20. □

Remark 26. In the statement of the previous Theorem, the size of the lattice $\mathsf{R}(E, A')$ can be estimated out of the sizes of E and A' considering that

$$P(E^{A'}) \subseteq \mathsf{R}(E, A') \subseteq P(A' \cup E^{A'})\,.$$

An upper bound for card$(\mathsf{R}(E, A'))$ is therefore $2^{p(k)+k^{2^{p(k)}}}$ where $p(k)$ is the polynomial of degree 2 as in the statement of the Theorem and k is the maximum of $size(t), size(s)$.

A standard argument yields now:

Corollary 27. *The equational theory of the relational lattices is decidable.*

7 Conclusions

We argued that the equational theory of relational lattices is decidable. We achieved this goal by giving a finite (counter)model construction of bounded size.

Our result leaves open other questions that we might ask on relational lattices. We mentioned in the introduction the quest for a complete axiomatic base for this theory or, anyway, the need of a complete deductive system—so to develop automatic reasoning for the algebra of relational lattices. As part of future researches it is tempting to contribute achieving this goal using the mathematical insights contained in the decidability proof.

Our result also opens new research directions, in primis, the investigation of the complexity of deciding lattice-theoretic equations/inclusions on relational lattices. Of course, the obvious decision procedure arising from the finite model construction is not optimal; few algebraic considerations already suggest how the decision procedure can be improved.

Also, it would be desirable next to investigate decidability of equational theories in signatures extending of the pure lattice signature; many such extensions are proposed in [17]. It is not difficult to adapt the present decidability proof so to add to the signature the header constant.

A further interesting question is how this result translates back to the field of multidimensional modal logic [15]. We pointed out in [22] how the algebra of relational lattices can be encoded into multimodal framework; we conjecture that our decidability result yields the decidability of some positive fragments of well known undecidable logics, such as the products $\mathbf{S5}^n$ with $n \geq 3$. Moreover connections need to be established with other existing decidability results in modal logic and in database theory [1].

References

1. Abiteboul, S., Hull, R., Vianu, V.: Foundations of Databases: The Logical Level, 1st edn. Addison-Wesley Longman Publishing Co., Inc., Boston (1995)
2. Ackerman, N.: Completeness in generalized ultrametric spaces. p-Adic Numbers Ultrametric Anal. Appl. **5**(2), 89–105 (2013)
3. Caucal, D.: On infinite transition graphs having a decidable monadic theory. Theoret. Comput. Sci. **290**(1), 79–115 (2003)
4. Codd, E.F.: A relational model of data for large shared data banks. Commun. ACM **13**(6), 377–387 (1970)
5. Davey, B.A., Priestley, H.A.: Introduction to Lattices and Order. Cambridge University Press, New York (2002)
6. Dyckhoff, R., Tholen, W.: Exponentiable morphisms, partial products and pullback complements. J. Pure Appl. Algebra **49**(1–2), 103–116 (1987)
7. Freese, R., Ježek, J., Nation, J.: Free Lattices. American Mathematical Society, Providence (1995)
8. Gabbay, D.M.: Selective filtration in modal logic I. Semantic tableaux method. Theoria **36**, 323–330 (1970)
9. Grätzer, G.: General Lattice Theory. Birkhäuser Verlag, Basel (1998). New appendices by the author with B. A. Davey, R. Freese, B. Ganter, M. Greferath, P. Jipsen, H. A. Priestley, H. Rose, E. T. Schmidt, S. E. Schmidt, F. Wehrung and R. Wille
10. Hammack, R., Imrich, W., Klavzar, S.: Handbook of Product Graphs, 2nd edn. CRC Press Inc., Boca Raton (2011)
11. Hirsch, R., Hodkinson, I.: Representability is not decidable for finite relation algebras. Trans. Amer. Math. Soc. **353**, 1403–1425 (2001)
12. Hirsch, R., Hodkinson, I., Kurucz, A.: On modal logics between $\mathbf{K} \times \mathbf{K} \times \mathbf{K}$ and $\mathbf{S5} \times \mathbf{S5} \times \mathbf{S5}$. J. Symbolic Log. **67**(3), 221–234 (2002)
13. Imrich, W., Klavar, S.: Product Graphs. Wiley-Interscience Series in Discrete Mathematics and Optimization. Wiley-Interscience, New York (2000). Structure and recognition, With a foreword by Peter Winkler
14. Jacobs, B.: Categorical Logic and Type Theory. Studies in Logic and the Foundations of Mathematics, vol. 141. North-Holland Publishing Co., Amsterdam (1999)
15. Kurucz, A.: Combining modal logics. In: Patrick Blackburn, J.V.B., Wolter, F. (eds.) Handbook of Modal Logic, vol. 3 (Studies in Logic and Practical Reasoning), pp. 869–924. Elsevier (2007)

16. Lawvere, F.W.: Metric spaces, generalized logic and closed categories. Rendiconti del Seminario Matematico e Fisico di Milano XLIII, pp. 135–166 (1973)
17. Litak, T., Mikulás, S., Hidders, J.: Relational lattices: from databases to universal algebra. J. Logic. Algebraic Methods Program. **85**(4), 540–573 (2016)
18. Maddux, R.: The equational theory of CA_3 is undecidable. J. Symbolic Logic **45**(2), 311–316 (1980)
19. Pouzet, M.: Une approche métrique de la rétraction dans les ensembles ordonnés et les graphes. In: Proceedings of the Conference on Infinitistic Mathematics (Lyon, 1984), pp. 59–89. Publ. Dp. Math. Nouvelle Sér. B, 85–2, Univ. Claude-Bernard, Lyon (1985)
20. Priess-Crampe, S., Ribemboim, P.: Equivalence relations and spherically complete ultrametric spaces. C. R. Acad. Sci. Paris **320**(1), 1187–1192 (1995)
21. Santocanale, L.: A duality for finite lattices, September 2009. http://hal.archives-ouvertes.fr/hal-00432113
22. Santocanale, L.: Relational lattices via duality. In: Hasuo, I. (ed.) CMCS 2016. LNCS, vol. 9608, pp. 195–215. Springer, Cham (2016)
23. Santocanale, L.: Embeddability into relational lattices is undecidable. In: Höfner, P., Pous, D., Struth, G. (eds.) RAMICS 2017. LNCS, vol. 10226, pp. 258–273. Springer, Cham (2017). https://hal.archives-ouvertes.fr/hal-01474822
24. Santocanale, L., Wehrung, F.: The equational theory of the weak order on finite symmetric groups, P. 41, September 2014
25. Spight, M., Tropashko, V.: Relational lattice axioms (2008). http://arxiv.org/abs/0807.3795
26. Tropashko, V.: Relational algebra as non-distributive lattice (2006). http://arxiv.org/abs/cs/0501053

Graphs and Automata

Minimization of Graph Weighted Models over Circular Strings

Guillaume Rabusseau(✉)

Reasoning and Learning Lab, School of Computer Science,
McGill University, Montreal, Canada
guillaume.rabusseau@mcgill.ca

Abstract. Graph weighted models (GWMs) have recently been proposed as a natural generalization of weighted automata over strings, trees and 2-dimensional words to arbitrary families of labeled graphs (and hypergraphs). In this paper, we propose polynomial time algorithms for minimizing and deciding the equivalence of GWMs defined over the family of circular strings on a finite alphabet (GWM^{c}s). The study of GWM^{c}s is particularly relevant since circular strings can be seen as the simplest family of graphs with cycles. Despite the simplicity of this family and of the corresponding computational model, the minimization problem is considerably more challenging than in the case of weighted automata over strings and trees: while linear algebra tools are overall sufficient to tackle the minimization problem for classical weighted automata (defined over a field), the minimization of GWM^{c}s involves fundamental notions from the theory of finite dimensional algebra. We posit that the properties of GWM^{c}s unraveled in this paper willprove useful for the study of GWMs defined over richer families of graphs.

1 Introduction

Functions defined over syntactical structures such as strings, trees and graphs are ubiquitous in computer science. Automata models allow one to succinctly represent such functions. In particular, *weighted automata* can efficiently model functions mapping structured objects to values in a semi-ring. Weighted automata have been defined to handle functions whose domain are e.g. strings [9,26], trees [8,16] and 2-dimensional words [11]. More recently, Bailly et al. [2] proposed a computational model for functions mapping labeled graphs (or hypergraphs) to values in a field (see also [22, Chap. 2]): Graph Weighted Models (GWMs). GWMs extend the notion of *linear representation* of a function defined over strings and trees to functions defined over graphs labeled by symbols in a ranked alphabet: loosely speaking, while string weighted automata can be defined by associating each symbol in a finite alphabet to a linear map and tree weighted automata by associating each symbol in a ranked alphabet to a multilinear map, GWMs are defined by associating each arity k symbol from a ranked alphabet to a kth order tensor. The computation of a GWM boils down to mapping each vertex in a graph to the tensor associated to its label and performing contractions directed by the edges of the input graph to obtain a value in the supporting

C. Baier and U. Dal Lago (Eds.): FOSSACS 2018, LNCS 10803, pp. 513–529, 2018.
https://doi.org/10.1007/978-3-319-89366-2_28

field. When restricted to the families of strings, trees or 2-dimensional words, GWMs are expressively equivalent to the classical notions of weighted automata over these structures.

Weighted automata have recently received interest from the machine learning community due to their ability to represent functions defined over structured objects. Efficient (and often consistent) learning algorithms have been developed for such computational models defined over sequences [3,6,10,19] and trees [1,4,14]. Motivated by the relevance of learning functions defined over richer families of labeled graphs, our long term objective is to design efficient learning algorithms for GWMs. This is however a challenging task. Given the close relationship between minimization and learning for classical weighted automata (see e.g. [7,21,27]), we take a first step in this direction by tackling the problem of minimizing GWMs defined over the simple family of *circular strings*.

Circular strings are strings whose last symbol is connected to the first. A circular string can be seen as a directed graph where each vertex is labeled by a symbol from a finite alphabet and is connected to his unique successor (i.e. a labeled graph composed of a unique cycle). Circular strings are relevant in biology (see e.g. [20] and references therein) and have been studied from a formal language perspective in the non-quantitative setting in [24]. The study of GWMs defined over such graphs is particularly relevant since circular strings are in some sense the simplest family of graphs with cycles (and cycles can be seen as the key obstacle for going from strings and trees to general graphs). Moreover, GWMs defined over the family of circular strings—which we henceforth denote by GWM^cs to avoid confusions—take a simple form making them easily amenable to theoretical study: a GWM^c is given by a set of matrices $\mathbf{A}^\sigma$ for each symbol σ in a finite alphabet, and maps any circular string $\sigma_1\sigma_2\cdots\sigma_k$ to the trace of the products of the matrices associated with the letters in the string[1]. Despite the simplicity of this computational model and its strong connection with string weighted automata, the minimization problem is considerably more challenging than in the case of string or tree weighted automata. More precisely, while the minimization problem can easily be handled using notions from linear algebra for e.g. real-valued string weighted automata (see e.g. [7]), we show in this paper that the minimization of GWM^cs requires fundamental concepts from the theory of finite-dimensional algebras (such as the ones of radical and semi-simplicity).

Contributions. Throughout the paper, *we only consider automata defined over a field of characteristic* 0. After introducing notions on weighted automata, GWM^cs and finite-dimensional algebras in Sect. 2, we first tackle the problem of deciding the equivalence of GWM^cs in Sect. 3. The study of the equivalence problem is motivated by the simple observation that two minimal GWMs computing

[1] Note that this is a not a definition *per se* but rather a consequence of the definition of general GWMs (as introduced in [2,22]): when restricted to the family of circular strings, a GWM is given by a set of matrices and its computation can be succinctly expressed using the trace operator (whereas a general GWM is given by a set of *tensors* and its computation relies on *partial traces*).

the same function are not necessarily related by a change of basis, which is in contrast with a classical result stating that two minimal string weighted automata are equivalent if and only if they are related by a change of basis. Building from this observation, we unravel the fundamental notion of semi-simple GWM^c and we show that *any function recognizable by a GWM^c can be computed by a semi-simple* GWM^c (Corollary 1) and that *two semi-simple GWM^{c}s of equal dimensions computing the same function are necessarily related by a change of basis* (Corollary 2). These two results naturally give rise to *a polynomial time algorithm to decide whether two GWM^{c}s are equivalent.* We then move on to the minimization problem in Sect. 4, where we give *a polynomial time minimization algorithm for GWM^{c}s* which fundamentally relies on the notion of semi-simple GWM^c (Corollary 3). While the problem of minimizing a GWM defined over the simple family of circular strings is central to this paper, we see it as a test bed for developing the theory of general GWMs: beyond the minimization and equivalence algorithms we propose, we believe that one of our main contributions is to illustrate how the theory of GWMs will rely on advanced concepts from algebra theory and to unravel fundamental properties that will surely be central to the study of GWMs defined over more general families of graphs (such as the one of semi-simple GWM^c).

1.1 Notations

For any integer n we let $[n] = \{1, 2, \cdots, n\}$. We denote the set of integers by $\mathbb{N}$ and the fields of real and rational numbers by $\mathbb{R}$ and $\mathbb{Q}$ respectively. Let $\mathbb{F}$ be a field of characteristic 0, we denote by $\mathcal{M}_n(\mathbb{F}) = \mathbb{F}^{n\times n}$ the set of all $n \times n$ matrices over $\mathbb{F}$. We use lower case bold letters for vectors (e.g. $\mathbf{v} \in \mathbb{F}^{d_1}$) and upper case bold letters for matrices (e.g. $\mathbf{M} \in \mathbb{F}^{d_1\times d_2}$). We denote by $\mathbf{I}_n$ the $n \times n$ identity matrix (or simply $\mathbf{I}$ if the dimension is clear from context). Given a matrix $\mathbf{M} \in \mathbb{F}^{d_1\times d_2}$, we denote its entries by $\mathbf{M}_{i,j}$ and we use $\text{vec}(\mathbf{M}) \in \mathbb{F}^{d_1 d_2}$ to denote the column vector obtained by concatenating the columns of $\mathbf{M}$. We use $\ker(\mathbf{A})$ to denote the kernel (or null space) of a matrix $\mathbf{A}$. Given two matrices $\mathbf{A} \in \mathcal{M}_m(\mathbb{F})$ and $\mathbf{B} \in \mathcal{M}_n(\mathbb{F})$ we denote their Kronecker product by $\mathbf{A} \otimes \mathbf{B} \in \mathcal{M}_{mn}(\mathbb{F})$ and their direct sum by $\mathbf{A} \oplus \mathbf{B} \in \mathcal{M}_{m+n}(\mathbb{F})$: $\mathbf{A} \otimes \mathbf{B}$ is the block matrix with blocks $(\mathbf{A}_{i,j}\mathbf{B})_{i,j}$ and $\mathbf{A} \oplus \mathbf{B}$ is the block diagonal matrix with $\mathbf{A}$ in the upper diagonal block and $\mathbf{B}$ in the lower one. We denote by Σ^* the set of strings on a finite alphabet Σ and the empty string by λ. We denote by Σ^+ the set of non-empty strings and by Σ^k the set of all strings of length k.

2 Preliminaries

We first present notions on weighted automata, graph weighted models and finite dimensional algebras. The reader is referred to [9,16,25] for more details on weighted automata theory, to [2] and [22, Chap. 2] for an introduction to graph weighted models, and to [13,17] for a thorough introduction to finite dimensional algebras.

2.1 Weighted Automata and GWMs over Circular Strings

Let Σ be a finite alphabet. A *weighted finite automaton* (WFA) over a field $\mathbb{F}$ with n states is a tuple $M = (\boldsymbol{\alpha}, \{\mathbf{M}^\sigma\}_{\sigma\in\Sigma}, \boldsymbol{\omega})$ where $\boldsymbol{\alpha}, \boldsymbol{\omega} \in \mathbb{F}^n$ are the initial and final weight vectors respectively, and $\mathbf{M}^\sigma \in \mathcal{M}_n(\mathbb{F})$ is the transition matrix for each symbol $\sigma \in \Sigma$. A WFA computes a function $f_M\colon \Sigma^* \to \mathbb{F}$ defined for each word $x = x_1x_2\cdots x_k \in \Sigma^*$ by

$$f_M(x) = \boldsymbol{\alpha}^\top \mathbf{M}^{x_1}\mathbf{M}^{x_2}\cdots\mathbf{M}^{x_k}\boldsymbol{\omega}.$$

We will often use the shorthand notation $\mathbf{M}^x = \mathbf{M}^{x_1}\mathbf{M}^{x_2}\cdots\mathbf{M}^{x_k}$ for any word $x = x_1x_2\cdots x_k \in \Sigma^*$. A WFA M with n states is *minimal* if its number of states is minimal, i.e. any WFA M' such that $f_M = f_{M'}$ has at least n states. We say that a function $f\colon \Sigma^* \to \mathbb{R}$ is *WFA-recognizable* if there exists a WFA computing it.

Graph weighted models (GWMs) have been introduced as a computational model over arbitrary labeled graphs and hypergraphs in [2]. In this paper, we focus on the simple model of GWMs defined over the family of circular strings. A *circular string* is a string without a beginning or an end, one can think of it as a string closed onto itself (see Fig. 1).

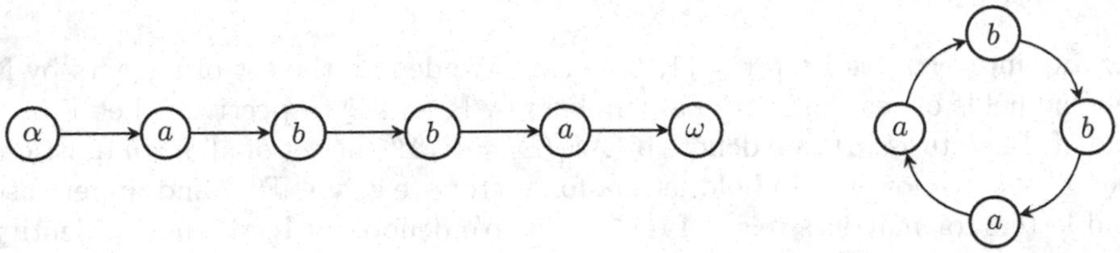

Fig. 1. (left) Graph representation of the string *abba* where the special vertices labeled with α and ω denote the beginning and end of the string respectively. (right) In contrast, the circular string *abba* has no beginning and no end, it is thus the same object as e.g. the circular string *baab*.

A *d-dimensional GWM A over circular strings* (GWM^c) on Σ is given by a set of matrices $\{\mathbf{A}^\sigma\}_{\sigma\in\Sigma} \subset \mathcal{M}_d(\mathbb{F})$. It computes a function $f_A\colon \Sigma^+ \to \mathbb{F}$ defined[2] for each word $x = x_1x_2\cdots x_k \in \Sigma^+$ by

$$f_A(x) = \mathrm{Tr}(\mathbf{A}^{x_1}\mathbf{A}^{x_2}\cdots\mathbf{A}^{x_k}) = \mathrm{Tr}(\mathbf{A}^x).$$

By invariance of the trace under cyclic permutation, we have $f_A(x_1x_2\cdots x_k) = f_A(x_2x_3\cdots x_kx_1) = f_A(x_3x_4\cdots x_kx_1x_2) = \cdots$. This is in accordance with the

[2] Observe that we exclude the empty string from the domain of f_A. This is on purpose since $f_A(\lambda)$ would be the dimension of A (using the convention $\mathbf{A}^\lambda = \mathbf{I}$): given two GWM^{c}s of different dimensions computing the same function on Σ^+, we want to consider them as equivalent even though they disagree on λ.

definition of a circular string: for any string x' obtained by cyclic permutation of the letters of a string x, both x and x' correspond to the same circular string. Similarly to WFAs, a GWM^c is *minimal* if its dimension is minimal and a function $f\colon \Sigma^+ \to \mathbb{F}$ is *GWM^c-recognizable* if it can be computed by a GWM^c.

It is immediate to see that there exist WFA-recognizable functions that are not GWM^c-recognizable, this is the case of any WFA-recognizable function that is not invariant under cyclic permutation of letters in a word[3]. In contrast, one can easily show that any GWM^c-recognizable function is WFA-recognizable. More precisely, we have the following result.

Proposition 1. *For any d-dimensional GWM^c $A = \{\mathbf{A}^\sigma\}_{\sigma\in\Sigma}$ on Σ, the WFA M with d^2 states $(\boldsymbol{\alpha}, \{\mathbf{M}^\sigma\}_{\sigma\in\Sigma}, \boldsymbol{\omega})$ where $\boldsymbol{\alpha} = \boldsymbol{\omega} = \mathrm{vec}(\mathbf{I}_d)$ and $\mathbf{M}^\sigma = \mathbf{I}_d \otimes \mathbf{A}^\sigma$ for each $\sigma \in \Sigma$, is such that $f_M(x) = f_A(x)$ for all $x \in \Sigma^*$.*

Proof. For any $w = w_1 \cdots w_n \in \Sigma^*$ we have $f_A(w) = \mathrm{Tr}(\mathbf{A}^w) = \sum_{i\in[d]} \mathbf{A}^w_{i,i} = \sum_{i\in[d]} \mathbf{e}_i^\top \mathbf{A}^w \mathbf{e}_i$ where $\mathbf{e}_i$ is the i-th vector of the canonical basis of $\mathbb{F}^d$. Since $\boldsymbol{\alpha} = \boldsymbol{\omega} = (\mathbf{e}_1^\top, \cdots, \mathbf{e}_d^\top)^\top$ and $\mathbf{M}^\sigma = \mathbf{I} \otimes \mathbf{A}^\sigma$ is the block-diagonal matrix with $\mathbf{A}^\sigma$ repeated d times on the diagonal, one can check that $f_M(w) = \boldsymbol{\alpha}^\top \mathbf{M}^w \boldsymbol{\omega} = \sum_{i\in[d]} \mathbf{e}_i^\top \mathbf{A}^w \mathbf{e}_i = f_A(w)$. □

It follows from this proposition that the learning and equivalence problems for GWM^{c}s could be handled by using the corresponding algorithms for WFAs. We will nonetheless study the equivalence problem in the next section[4] without falling back onto the theory of WFAs, which will allow us to unravel fundamental properties of GWMs that will be particularly relevant to further studies (moreover, the minimization problem obviously cannot be handled in such a way).

2.2 Finite-Dimensional Algebras

An *algebra* $\mathcal{A}$ over a field $\mathbb{F}$ (or $\mathbb{F}$-algebra) is a vector space over the field $\mathbb{F}$ equipped with a bilinear operation (called multiplication or product). An algebra is *associative* if its product is associative and it is *finite-dimensional* if it is of finite dimension as a vector space over $\mathbb{F}$. *In this paper, we will only consider finite-dimensional associative algebras.* A sub-algebra $\mathcal{B}$ of an algebra $\mathcal{A}$ is a linear subspace of $\mathcal{A}$ which is closed under product (i.e. $\mathcal{B}$ equipped with the operations of $\mathcal{A}$ is an algebra itself).

A classical example of finite-dimensional algebra is the set $\mathcal{L}(V)$ of linear operators on some finite-dimensional vector space V (where the product is composition). In this particular example, the algebra $\mathcal{L}(V)$ is isomorphic to the *full matrix algebra* $\mathcal{M}_d(\mathbb{F})$, where d is the dimension of V; we will mainly focus on matrix algebras in this paper, i.e. sub-algebras of the full matrix algebra $\mathcal{M}_d(\mathbb{F})$ for some d (an example of such an algebra is the set of $d \times d$ upper triangular matrices). In particular, we will often consider the *algebra generated by a finite*

[3] Note that this is not a necessary condition: the function f defined on $\{a, b\}^*$ by $f(x) = 1$ if $x = a$ and 0 otherwise is WFA-recognizable but not GWM^c-recognizable.

[4] The learning problem has been previously considered in [5,22].

set of matrices $\{\mathbf{A}^\sigma\}_{\sigma\in\Sigma} \subset \mathcal{M}_d(\mathbb{F})$ for some finite alphabet Σ, that is the set of all finite linear combinations of matrices of the form $\mathbf{A}^x = \mathbf{A}^{x_1}\mathbf{A}^{x_2}\cdots\mathbf{A}^{x_k}$ for $x = x_1x_2\cdots x_k \in \Sigma^*$. More formally, if we denote by $\mathcal{A}$ this algebra, we have

$$\mathcal{A} = \left\{\sum_{i=1}^n \alpha_i\mathbf{A}^{w_i} : n\in\mathbb{N},\ \alpha_1,\cdots,\alpha_n\in\mathbb{F},\ w_1,\cdots,w_n\in\Sigma^*\right\}.$$

Let $\mathcal{A}$ be a finite-dimensional algebra over $\mathbb{F}$. A sub-algebra $\mathcal{X}$ of $\mathcal{A}$ is called an *ideal of* $\mathcal{A}$ if both $xa\in\mathcal{X}$ and $ax\in\mathcal{X}$ for any $x\in\mathcal{X}$, $a\in\mathcal{A}$ (i.e. $\mathcal{X}$ is both left and right $\mathcal{A}$-invariant), which we will denote by $\mathcal{A}\mathcal{X} = \mathcal{X}\mathcal{A} = \mathcal{A}$. A sub-algebra $\mathcal{X}$ of $\mathcal{A}$ is *nilpotent* if there exists some integer k such that $\mathcal{X}^k = \{x_1x_2\cdots x_k : x_i\in\mathcal{X},\ i\in[k]\} = \{0\}$. The *factor algebra* $\mathcal{A}/\mathcal{X}$ of an algebra $\mathcal{A}$ by an ideal $\mathcal{X}$ is the algebra consisting of all cosets $a+\mathcal{X}$ for $a\in\mathcal{A}$, in other words $\mathcal{A}/\mathcal{X}$ is the quotient of $\mathcal{A}$ by the equivalence relation ($a\sim b$ if and only if $a-b\in\mathcal{X}$). The *radical*[5] *of* $\mathcal{A}$ is the maximal nilpotent ideal of $\mathcal{A}$ and will be denoted by $\mathrm{Rad}(\mathcal{A})$ (the existence of $\mathrm{Rad}(\mathcal{A})$ follows from the fact that $\mathcal{A}$ is of finite dimension). An algebra $\mathcal{A}$ is *semi-simple* if its radical is $\{0\}$.

Let us illustrate these definitions with a very simple example. Let $\mathcal{G}\subset\mathcal{M}_2(\mathbb{R})$ be the algebra generated by the matrix $\mathbf{G} = \begin{bmatrix}1&1\\0&1\end{bmatrix}$. One can easily check that $\mathcal{G} = \left\{\begin{bmatrix}\alpha&\beta\\0&\alpha\end{bmatrix} : \alpha,\ \beta\in\mathbb{R}\right\}$ and is thus of dimension 2. Consequently, both

$$\mathcal{G}_1 = \left\{\begin{bmatrix}\alpha&0\\0&\alpha\end{bmatrix} : \alpha\in\mathbb{R}\right\} \text{ and } \mathcal{G}_2 = \left\{\begin{bmatrix}0&\beta\\0&0\end{bmatrix} : \beta\in\mathbb{R}\right\} \tag{1}$$

are sub-algebras of $\mathcal{G}$. Moreover, $\mathcal{G}_2$ is a nilpotent ideal and one can check that it is maximal, i.e. $\mathrm{Rad}(\mathcal{G}) = \mathcal{G}_2$ and hence $\mathcal{G}$ is not semi-simple.

Intuitively, the radical of an algebra $\mathcal{A}$ contains its *bad elements* (in the sense that these elements annihilate all simple $\mathcal{A}$-modules). In our previous example, this *bad* property translates into the fact that the non-zero elements of $\mathcal{G}_2$ cannot be diagonalized. We will use two fundamental results from the theory of finite dimensional algebra. The first one is the Wedderburn-Malcev theorem which states that (under some conditions on the ground field $\mathbb{F}$) the elements of the radical can be *filtered out* from the algebra, i.e. one can find a sub-algebra of $\mathcal{A}$ that is isomorphic to $\mathcal{A}/\mathrm{Rad}(\mathcal{A})$ (see e.g. [17, Theorem 6.2.3]).

Theorem 1 (Wedderburn-Malcev Theorem). *Let $\mathcal{A}$ be a finite-dimensional algebra* over a field of characteristic0. *There exists a semi-simple subalgebra $\tilde{\mathcal{A}}$ of $\mathcal{A}$ which is isomorphic to $\mathcal{A}/\mathrm{Rad}(\mathcal{A})$ and such that $\mathcal{A} = \tilde{\mathcal{A}}\oplus\mathrm{Rad}(\mathcal{A})$ (direct sum of vector spaces).*

Going back to the example of the algebra $\mathcal{G}$ described above, we showed that it is not semi-simple, however one can easily check that $\mathcal{G}/\mathrm{Rad}(\mathcal{G})$ is isomorphic to the algebra $\mathcal{G}_1$ in Eq. (1) which is semi-simple, and furthermore that $\mathcal{G} = \mathcal{G}_1\oplus\mathrm{Rad}(\mathcal{G})$.

[5] Note that this definition is specific to the finite-dimensional case; for general rings, there exist distinct non-equivalent definitions of radicals, which all agree with the one given here in the case of finite-dimensional algebras.

The second fundamental result we will need is related to the notion of representation of an algebra. A *representation* of an $\mathbb{F}$-algebra $\mathcal{A}$ is a homomorphism of $\mathcal{A}$ into the algebra $\mathcal{L}(V)$ of the linear operators on some vector space V (over $\mathbb{F}$). Two representations $\rho\colon \mathcal{A} \to \mathcal{L}(V)$ and $\tau\colon \mathcal{A} \to \mathcal{L}(W)$ are *similar* if there exists an isomorphism $\phi\colon V \to W$ such that $\rho(a) = \phi^{-1}\tau(a)\phi$ for all $a \in \mathcal{A}$. For semi-simple algebras, the notion of similar representations is fundamentally related to the trace operator, which will be particularly relevant to the present study. Formally, we have the following theorem (see e.g. [17, Corollary 2.6.3]).

Theorem 2. *Let ρ and τ be two representations of a* semi-simple *algebra $\mathcal{A}$ over a* field of characteristic 0. *These representations are similar if and only if* $\mathrm{Tr}(\rho(a)) = \mathrm{Tr}(\tau(a))$ *for all $a \in \mathcal{A}$.*

3 Semi-Simple GWMs and the Equivalence Problem

In this section, we study the equivalence problem: given two GWMs over circular strings, how can we decide whether they compute the same function? In light of Proposition 1, one could solve this problem by simply *converting* the two GWM^{c}s into WFAs and checking whether these two WFAs compute the same function; indeed the equivalence problem for WFAs defined over a field is decidable in polynomial time [9]. Nonetheless, we will tackle this problem without relying on this proposition and, by doing so, we will unravel the notion of *semi-simple GWM^c* which will be relevant to the study of the minimization problem in the next section (and which should also be central to the study of GWMs defined over more general families of graphs).

3.1 Semi-Simplicity, Nilpotent Matrices and Traces

Let $\mathcal{A}$ be a finite dimensional matrix algebra. Recall that the radical of $\mathcal{A}$ is its maximal nilpotent ideal. A useful characterization of the elements of the radical relies on the notion of strongly nilpotent elements: $\mathbf{A} \in \mathcal{A}$ is *strongly nilpotent* if $\mathbf{AX}$ is nilpotent for any $\mathbf{X} \in \mathcal{A}$. It turns out that the radical of $\mathcal{A}$ is exactly the set of its strongly nilpotent elements [17, Corollary 3.1.10]. Since the computation of a GWM^c boils down to applying the trace operator, we will leverage this property to relate the notions of radical and semi-simplicity to simple properties of the elements of $\mathcal{A}$ with respect to the trace operator. We start with a simple lemma relating nilpotency and trace.

Lemma 1. *Let $\mathbb{F}$ be a field* of characteristic 0 *and let $\mathbf{A} \in \mathcal{M}_d(\mathbb{F})$. Then $\mathbf{A}$ is nilpotent if and only if* $\mathrm{Tr}(\mathbf{A}^n) = 0$ *for all $n \geq 1$.*

Proof. Let $\mathbf{A}$ be a nilpotent matrix and let k be such that $\mathbf{A}^k = 0$. Suppose $\mathbf{Av} = \gamma\mathbf{v}$ for some $\mathbf{v} \neq \mathbf{0}$ (where γ could belong to an algebraically closed field extension of $\mathbb{F}$). Then $\mathbf{A}^k\mathbf{v} = \gamma^k\mathbf{v} = 0$ hence $\gamma = 0$ since $\mathbb{F}$ is of characteristic 0, thus $\mathbf{A}$ has only 0 eigenvalues and $\mathrm{Tr}(\mathbf{A}^n) = 0$ for all $n \geq 1$.

Conversely, suppose that $\mathrm{Tr}(\mathbf{A}^n) = 0$ for all $n \geq 1$. Then, we have $\mathrm{Tr}(P(\mathbf{A})) = 0$ for any polynomial P with constant term 0. Suppose that $\mathbf{A}$

has a non-zero eigenvalue γ and let $m > 0$ be its multiplicity. Choose a polynomial P such that $P(\gamma) = 1$, $P(0) = 0$ and $P(\mu) = 0$ for any eigenvalue μ of $\mathbf{A}$ distinct from γ. We then have $0 = \mathrm{Tr}(P(\mathbf{A})) = m$, a contradiction. Hence $\mathbf{A}$ has only zero eigenvalues and is nilpotent. □

One can use the previous lemma to show that an element $\mathbf{A} \in \mathcal{A}$ is strongly nilpotent if and only if $\mathrm{Tr}(\mathbf{AX}) = 0$ for all $\mathbf{X} \in \mathcal{A}$, which leads to the following useful characterization of the semi-simplicity of an algebra.

Proposition 2. *Let $\mathcal{A} \subset \mathcal{M}_d(\mathbb{F})$ be a matrix algebra. We have*

$$\mathrm{Rad}(\mathcal{A}) = \{\mathbf{A} \in \mathcal{A} \,:\, \mathrm{Tr}(\mathbf{AX}) = 0 \text{ for all } \mathbf{X} \in \mathcal{A}\}.$$

Consequently, $\mathcal{A}$ is semi-simple if and only if for all $\mathbf{A} \in \mathcal{A}$ different from 0 there exists $\mathbf{X} \in \mathcal{A}$ such that $\mathrm{Tr}(\mathbf{AX}) \neq 0$.

Proof. We will show that $\mathbf{A} \in \mathcal{A}$ is strongly nilpotent if and only if $\mathrm{Tr}(\mathbf{AX}) = 0$ for all $\mathbf{X} \in \mathcal{A}$. The proposition will then directly follows from the fact that $\mathrm{Rad}(\mathcal{A})$ is the set of strongly nilpotent elements of $\mathcal{A}$ and from the fact that $\mathcal{A}$ is semi-simple if and only if $\mathrm{Rad}(\mathcal{A}) = \{0\}$.

Let $\mathbf{A} \in \mathcal{A}$ be such that $\mathrm{Tr}(\mathbf{AX}) = 0$ for all $\mathbf{X} \in \mathcal{A}$. Since $\mathbf{X}(\mathbf{AX})^{n-1} \in \mathcal{A}$ for all $n \geq 1$ and all $\mathbf{X} \in \mathcal{A}$ we have $\mathrm{Tr}((\mathbf{AX})^n) = 0$ for all $n \geq 1$ and all $\mathbf{X} \in \mathcal{A}$, hence $\mathbf{AX}$ is nilpotent for all $\mathbf{X} \in \mathcal{A}$ by Lemma 1, i.e. $\mathbf{A}$ is strongly nilpotent. Conversely, let $\mathbf{A}$ be a strongly nilpotent element of $\mathcal{A}$. By Lemma 1 we have $\mathrm{Tr}((\mathbf{AX})^n) = 0$ for all $\mathbf{X} \in \mathcal{A}$ and all $n \geq 1$, in particular $\mathrm{Tr}(\mathbf{AX}) = 0$. □

3.2 Equivalence of GWMs

We now consider the problem of deciding whether two GWM^{c}s are equivalent. Let us first briefly show how one can decide whether two real-valued WFAs compute the same function. One way to address this problem relies on the following result: two minimal real-valued WFAs computing the same function are related by a change of basis. Note that it is easy to check that WFAs are invariant under a change of basis of their weight vectors and transition matrices. The following proposition show that such a change of basis is actually the only way for two minimal WFAs to compute the same function [26] (see also [6, Corollary 4.2]).

Proposition 3. *If two WFAs $A = (\boldsymbol{\alpha}, \{\mathbf{A}^\sigma\}_{\sigma\in\Sigma}, \boldsymbol{\omega})$ and $\tilde{A} = (\tilde{\boldsymbol{\alpha}}, \{\tilde{\mathbf{A}}^\sigma\}_{\sigma\in\Sigma}, \tilde{\boldsymbol{\omega}})$ with d states taking their values in $\mathbb{R}$ are minimal and compute the same function, i.e. $f_A = f_{\tilde{A}}$, then there exists an invertible matrix $\mathbf{P} \in \mathcal{M}_d(\mathbb{R})$ such that*

$$\boldsymbol{\alpha}^\top = \tilde{\boldsymbol{\alpha}}^\top \mathbf{P}, \quad \boldsymbol{\omega} = \mathbf{P}^{-1}\tilde{\boldsymbol{\omega}} \text{ and } \mathbf{A}^\sigma = \mathbf{P}^{-1}\tilde{\mathbf{A}}^\sigma\mathbf{P} \text{ for each } \sigma \in \Sigma.$$

Hence, to decide whether two WFAs compute the same function one can simply minimize them and check whether the weight vectors and transition matrices obtained after minimization are related by a change of basis (which can both be done in polynomial time). In contrast, one can easily find an example of two minimal *GWM^{c}s* whose matrices are not related by a change of basis. Consider the constant function $f(x) = 2$ for all $x \in \Sigma^+$. One can check that the two GWM^{c}s G and $\tilde{G}$ with 2 states defined by the matrices

$$\mathbf{G} = \begin{bmatrix} 1 & 1 \\ 0 & 1 \end{bmatrix} \quad \text{and} \quad \tilde{\mathbf{G}} = \begin{bmatrix} 1 & 0 \\ 0 & 1 \end{bmatrix}$$

respectively are minimal and compute f, however $\mathbf{G}$ and $\tilde{\mathbf{G}}$ are not similar.

Let us now introduce the notion of *semi-simple*GWM^c. We say that a GWM^c A defined by a set of matrices $\{\mathbf{A}^\sigma\}_{\sigma\in\Sigma} \subset \mathcal{M}_d(\mathbb{F})$ is *semi-simple* if the algebra $\mathcal{A}$ generated by the matrices $\{\mathbf{A}^\sigma\}_{\sigma\in\Sigma}$ is semi-simple. It follows from the example presented in Sect. 2.2 that G is not semi-simple while $\tilde{G}$ is a semi-simple GWM^c computing the GWM^c*-recognizable function* f. We will now show that this simple example can be generalized: *any* GWM^c*-recognizable function can be computed by a semi-simple* GWM^c. This non-trivial result relies on the following theorem which is a direct consequence of the Wedderburn-Malcev theorem.

Theorem 3. *Let $\mathcal{A} \subset \mathcal{M}_d(\mathbb{F})$ be a matrix algebra over a field of characteristic 0. Then there exist a* semi-simple *sub-algebra $\tilde{\mathcal{A}}$ of $\mathcal{A}$ and a surjective homomorphism $\pi\colon \mathcal{A} \to \tilde{\mathcal{A}}$ such that* $\mathrm{Tr}(\mathbf{A}) = \mathrm{Tr}(\pi(\mathbf{A}))$ *for all* $\mathbf{A} \in \mathcal{A}$.

Proof. By Theorem 1 there exists a semi-simple sub-algebra $\tilde{\mathcal{A}}$ of $\mathcal{A}$ which is isomorphic to $\mathcal{A}/\mathrm{Rad}(\mathcal{A})$ and such that $\mathcal{A} = \tilde{\mathcal{A}} \oplus \mathrm{Rad}(\mathcal{A})$ (direct sum of vector spaces). Let $\pi\colon \mathcal{A} \to \tilde{\mathcal{A}}$ be the projection associated with this direct sum. Then for any $\mathbf{A} \in \mathcal{A}$ we have

$$\mathrm{Tr}(\mathbf{A}) = \mathrm{Tr}(\pi(\mathbf{A}) + (1-\pi)(\mathbf{A})) = \mathrm{Tr}(\pi(\mathbf{A})) + \mathrm{Tr}((1-\pi)(\mathbf{A})) = \mathrm{Tr}(\pi(\mathbf{A})).$$

Indeed, since $(1-\pi)(\mathbf{A}) \in \mathrm{Rad}(\mathcal{A})$, it is nilpotent, hence its trace is zero. □

Using the notations from Theorem 3, it follows that for any d-dimensional GWM^c A given by a set of matrices $\{\mathbf{A}^\sigma\}_{\sigma\in\Sigma} \subset \mathcal{M}_d(\mathbb{F})$ generating the algebra $\mathcal{A}$, the d-dimensional $\tilde{A}$ given by the matrices $\{\tilde{\mathbf{A}}^\sigma = \pi(\mathbf{A}^\sigma)\}_{\sigma\in\Sigma}$ is a semi-simple GWM^c computing the function f_A, hence the following corollary.

Corollary 1. *Any function that can be computed by a GWM^c can be computed by a semi-simple GWM^c of the same dimension.*

Given a finite dimensional algebra $\mathcal{A}$, one can compute the surjective homomorphism π from Theorem 3 in polynomial time when $\mathbb{F}$ allows efficient arithmetic computations (e.g. $\mathbb{F} = \mathbb{Q}$) [12,15]. The algorithm takes as input a basis $a_1, \cdots, a_n$ of $\mathcal{A}$ (as a vector space) and the structure coefficients of the algebra (which are the scalars $c^k_{i,j} \in \mathbb{F}$ satisfying $a_i a_j = \sum_k c^k_{i,j} a_k$). Since one can easily compute a basis and the structure coefficients of a matrix algebra $\mathcal{A}$ given a set of generators $\{\mathbf{A}^\sigma\}_{\sigma\in\Sigma}$ in polynomial time, it follows that any GWM^c can be transformed in polynomial time into a semi-simple GWM^c (of the same dimension) computing the same function.

We now show that a result similar to Proposition 3 holds for semi-simple GWM^{c}s: *two semi-simple d-dimensional* GWM^c*s* are equivalent if and only if they are related by a change of basis. This result relies on the following theorem.

Theorem 4. *Let Σ be a finite alphabet and let $\mathcal{A}, \mathcal{B} \subset \mathcal{M}_d(\mathbb{F})$ be the algebras generated by the sets of matrices $\{\mathbf{A}^\sigma\}_{\sigma\in\Sigma}$ and $\{\mathbf{B}^\sigma\}_{\sigma\in\Sigma}$ respectively.*

If $\mathcal{A}$ and $\mathcal{B}$ are semi-simple and $\mathrm{Tr}(\mathbf{A}^w) = \mathrm{Tr}(\mathbf{B}^w)$ for all $w \in \Sigma^*$ then $\mathcal{A}$ is isomorphic to $\mathcal{B}$. Moreover, the mapping $\tilde{\phi}\colon \mathcal{A} \to \mathcal{B}$ defined by extending the mapping

$$\phi\colon \mathbf{A}^x \mapsto \mathbf{B}^x \quad \text{for all } x \in \Sigma^*$$

by linearity is well-defined and is an isomorphism.

Proof. The mapping ϕ is by construction a trace-preserving surjective semi-group homomorphism. We first show[6] that ϕ can be extended to a homomorphism $\tilde{\phi}\colon \mathcal{A} \to \mathcal{B}$. By definition, any $\mathbf{A} \in \mathcal{A}$ can be written as $\mathbf{A} = \sum_{i=1}^n \alpha_i \mathbf{A}^{x_i}$ for some $n \in \mathbb{N}, \alpha_1, \cdots, \alpha_n \in \mathbb{F}, x_1, \cdots, x_n \in \Sigma^*$. We will show that the mapping

$$\tilde{\phi}\colon \sum_{i=1}^n \alpha_i \mathbf{A}^{x_i} \longmapsto \sum_{i=1}^n \alpha_i \phi(\mathbf{A}^{x_i})$$

is well-defined. By construction of $\tilde{\phi}$, it suffices to show that if $\sum_{i=1}^n \alpha_i \mathbf{A}^{x_i} = 0$ for some $\alpha_i \in \mathbb{F}$, $x_i \in \Sigma^*$, then $\tilde{\phi}(\sum_{i=1}^n \alpha_i \mathbf{A}^{x_i}) = 0$. Suppose $\sum_{i=1}^n \alpha_i \mathbf{A}^{x_i} = 0$, then $\sum_{i=1}^n \alpha_i \mathbf{A}^{x_i} \mathbf{A}^x = 0$ for any $x \in \Sigma^*$. By linearity of the trace and since ϕ is a trace-preserving morphism, it follows that

$$\begin{aligned} 0 &= \sum_{i=1}^n \alpha_i \mathrm{Tr}\left[\mathbf{A}^{x_i}\mathbf{A}^x\right] = \sum_{i=1}^n \alpha_i \mathrm{Tr}\left[\phi(\mathbf{A}^{x_i}\mathbf{A}^x)\right] = \sum_{i=1}^n \alpha_i \mathrm{Tr}\left[\phi(\mathbf{A}^{x_i})\phi(\mathbf{A}^x)\right] \\ &= \mathrm{Tr}\left[\left(\sum_{i=1}^n \alpha_i \phi(\mathbf{A}^{x_i})\right)\phi(\mathbf{A}^x)\right] = \mathrm{Tr}\left[\tilde{\phi}\left(\sum_{i=1}^n \alpha_i \mathbf{A}^{x_i}\right)\phi(\mathbf{A}^x)\right] \end{aligned}$$

for all $x \in \Sigma^*$. By linearity of the trace and since ϕ is surjective, we thus have $\mathrm{Tr}\left[\tilde{\phi}(\sum_{i=1}^n \alpha_i \mathbf{A}^{x_i})\mathbf{B}\right] = 0$ for any $\mathbf{B} \in \mathcal{B}$, hence $\tilde{\phi}(\sum_{i=1}^n \alpha_i \mathbf{A}^{x_i})$ belongs to $\mathrm{Rad}(\mathcal{B})$ by Proposition 2 and must be 0 since $\mathcal{B}$ is semi-simple.

One can easily check that $\tilde{\phi}$ is trace-preserving, is surjective and is a homomorphism. It remains to show that $\tilde{\phi}$ is injective. Let $\mathbf{A} \in \mathcal{A}$ be such that $\tilde{\phi}(\mathbf{A}) = 0$. Since $\tilde{\phi}$ is a homomorphism we have $\tilde{\phi}(\mathbf{AX}) = 0$ for any $\mathbf{X} \in \mathcal{A}$, and thus $0 = \mathrm{Tr}(\tilde{\phi}(\mathbf{AX})) = \mathrm{Tr}(\mathbf{AX})$ for all $\mathbf{X} \in \mathcal{A}$. Hence $\mathbf{A} \in \mathrm{Rad}(\mathcal{A})$ by Proposition 2 and must be 0 since $\mathcal{A}$ is semi-simple. □

The previous theorem can be leveraged to show that if two semi-simple GWM^{c}s of the same dimension compute the same function, then they are related by a change of basis (note that the converse of this statement is immediate since the trace is a basis independent operator). Let A and B be two d-dimensional semi-simple GWM^{c}s computing the same function and let $\mathcal{A}, \mathcal{B} \subset \mathcal{M}_d$ be the algebras generated by their respective sets of matrices $\{\mathbf{A}^\sigma\}_{\sigma\in\Sigma}$ and $\{\mathbf{B}^\sigma\}_{\sigma\in\Sigma}$. First observe that the identity mapping $\rho\colon \mathcal{A} \to \mathcal{L}(\mathbb{F}^d)$ defined by $\rho(\mathbf{A}) = \mathbf{A}$ for all $\mathbf{A} \in \mathcal{A}$ is (trivially) a representation of the algebra $\mathcal{A}$. Now, since A and B

[6] This part of the proof is adapted from the proof of Proposition 3.1 in [18].

compute the same function and are semi-simple, we have $\mathrm{Tr}(\mathbf{A}^w) = \mathrm{Tr}(\mathbf{B}^w)$ for all $w \in \Sigma^*$ and it follows from Theorem 4 that $\mathcal{A}$ is isomorphic to $\mathcal{B}$; let $\tilde{\phi}\colon \mathcal{A} \to \mathcal{B}$ be the isomorphism defined in this theorem. Then, the mapping $\tau\colon \mathcal{A} \to \mathcal{L}(\mathbb{F}^d)$ defined by $\tau(\mathbf{A}) = \tilde{\phi}(\mathbf{A})$ for all $\mathbf{A} \in \mathcal{A}$ is also a representation of $\mathcal{A}$, and since $\mathcal{A}$ is semi-simple it follows from Theorem 2 that ρ and τ are similar. That is, there exists an invertible matrix $\mathbf{P} \in \mathcal{M}_d(\mathbb{F})$ such that $\rho(\mathbf{A}) = \mathbf{P}^{-1}\tau(\mathbf{A})\mathbf{P}$ for all $\mathbf{A} \in \mathcal{A}$. In particular we have

$$\mathbf{A}^\sigma = \rho(\mathbf{A}^\sigma) = \mathbf{P}^{-1}\tau(\mathbf{A}^\sigma)\mathbf{P} = \mathbf{P}^{-1}\tilde{\phi}(\mathbf{A}^\sigma)\mathbf{P} = \mathbf{P}^{-1}\mathbf{B}^\sigma\mathbf{P}$$

for all $\sigma \in \Sigma$, hence the following corollary.

Corollary 2. *Two d-dimensional semi-simple GWM^{c}s A and B compute the same function if and only if they are related by a change of basis, i.e. there exists an invertible matrix* $\mathbf{P} \in \mathcal{M}_d(\mathbb{F})$ *such that* $\mathbf{A}^\sigma = \mathbf{P}^{-1}\mathbf{B}^\sigma\mathbf{P}$ *for all* $\sigma \in \Sigma$.

In the case where $\mathbb{F}$ allows for efficient arithmetic computations (e.g. $\mathbb{F} = \mathbb{Q}$), it follows that the equivalence of GWM^{c}s can be decided in polynomial time. Indeed, given two GWM^{c}s A and B of the same dimension defined by the matrices $\{\mathbf{A}^\sigma\}_{\sigma\in\Sigma}$ and $\{\mathbf{B}^\sigma\}_{\sigma\in\Sigma}$ respectively, one can first transform them into semi-simple GWM^{c}s using Theorem 3 and the algorithm in [12,15], and then check whether the resulting matrices are related by a change of basis. The case where the two GWM^{c}s are not of the same dimension can be easily handled. Without loss of generality, suppose that A and B are semi-simple GWM^{c}s of dimension d and d' respectively with $d' < d$. One can construct a d-dimensional GWM^c $\tilde{B}$ computing the same function as B by considering the block-diagonal matrices $\tilde{\mathbf{B}}^\sigma = \mathbf{B}^\sigma \oplus \mathbf{0}$ for each $\sigma \in \Sigma$ (where $\mathbf{0}$ is the $(d-d') \times (d-d')$ matrix with all entries equal to 0). It is easy to check that $\tilde{B}$ is semi-simple if B is semi-simple, hence one can decide if A is equivalent to B by checking whether the matrices $\mathbf{A}^\sigma$ and $\tilde{\mathbf{B}}^\sigma$ are related by a change of basis.

4 Minimization of GWMs over Circular Strings

We now consider the minimization problem: given a GWM^c A, can we find a minimal GWM^c computing f_A? We will show that the answer is in the positive and that such a minimal GWM^c can be computed in polynomial time. We start with a technical lemma that generalizes the classical result stating that for any $d \times d$ matrix $\mathbf{A}$, the kernel of $\mathbf{A}^d$ is equal to the kernel of $\mathbf{A}^{d+k}$ for any $k \geq 0$.

Lemma 2. *Let* $\{\mathbf{A}^\sigma\}_{\sigma\in\Sigma} \subset \mathcal{M}_d(\mathbb{F})$ *be a finite set of matrices. Then for all* $k \geq 0$ *we have*

$$\bigcap_{x\in\Sigma^d} \ker(\mathbf{A}^x) = \bigcap_{y\in\Sigma^{d+k}} \ker(\mathbf{A}^y).$$

Proof. For any integer i, let $E_i = \bigcap_{x\in\Sigma^i} \ker(\mathbf{A}^x)$. We start by showing that if $E_i = E_{i+1}$ for some i then $E_{i+1} = E_{i+2}$. The inclusion $E_{i+1} \subseteq E_{i+2}$ is immediate. Suppose $E_i = E_{i+1}$ for some integer i. If $\mathbf{v} \in E_{i+2}$ then $\mathbf{A}^\sigma\mathbf{v} \in$

$\ker(\mathbf{A}^x)$ for all $x \in \Sigma^{i+1}$ and all $\sigma \in \Sigma$, i.e. $\mathbf{A}^\sigma \mathbf{v} \in E_{i+1} = E_i$ for all $\sigma \in \Sigma$, which implies $\mathbf{A}^\sigma \mathbf{v} \in \ker(\mathbf{A}^y)$ for all $y \in \Sigma^i$ and all $\sigma \in \Sigma$ from which $\mathbf{v} \in E_{i+1}$ follows directly. To conclude, since each E_i is a linear subspace of $\mathbb{F}^d$, $E_i \subsetneq E_{i+1}$ implies $\dim E_i < \dim E_{i+1}$, hence there must exist an i for which $E_i = E_{i+1}$ and this i cannot be greater than d. □

We show in the following theorem that the linear space $E = \bigcap_{x\in\Sigma^d} \ker(\mathbf{A}^x)$ is not relevant to the computation of a GWM^c A with matrices $\{\mathbf{A}^\sigma\}_{\sigma\in\Sigma}$, i.e. one can project each matrix $\mathbf{A}^x$ onto the orthogonal complement of E without changing the function computed by A.

Theorem 5. *Let A be a GWM^c given by the set of matrices $\{\mathbf{A}^\sigma\}_{\sigma\in\Sigma} \subset \mathcal{M}_d(\mathbb{F})$. Consider the linear space*

$$E = \bigcap_{x\in\Sigma^d} \ker(\mathbf{A}^x) = \{\mathbf{v} \in \mathbb{F}^d : \mathbf{A}^x\mathbf{v} = \mathbf{0} \textit{ for all } x \in \Sigma^d\}$$

and let $\mathbf{\Pi} \in \mathbb{F}^{d\times d}$ be the matrix of the orthogonal projection onto E.

Then, the GWM^c $\hat{A}$ given by the matrices $\hat{\mathbf{A}}^\sigma = \mathbf{A}^\sigma(\mathbf{I} - \mathbf{\Pi})$ for each $\sigma \in \Sigma$ is such that $f_A = f_{\hat{A}}$.

Proof. Let $\mathcal{A}$ be the algebra generated by the matrices $\{\mathbf{A}^\sigma\}_{\sigma\in\Sigma}$. Let us first observe that E is $\mathcal{A}$-invariant, which follows from Lemma 2. Indeed, if $\mathbf{v} \in E$ and $y \in \Sigma^*$ we have $\mathbf{A}^x\mathbf{A}^y\mathbf{v} = \mathbf{0}$ for any $x \in \Sigma^d$ (since $|xy| \geq d$), hence $\mathbf{A}^y\mathbf{v} \in E$; the extension to an arbitrary element of $\mathbf{A}$ is immediate by linearity. This implies that for any $\mathbf{A} \in \mathcal{A}$, we have

$$\mathbf{\Pi}\mathbf{A}\mathbf{\Pi} = \mathbf{A}\mathbf{\Pi} \quad \text{and} \quad (\mathbf{I} - \mathbf{\Pi})\mathbf{A}\mathbf{\Pi} = \mathbf{0}. \tag{2}$$

Now, let $k \geq 1$, let $x = x_1x_2\cdots x_k \in \Sigma^k$ and let $\mathbf{P}_1 = \mathbf{\Pi}$ and $\mathbf{P}_2 = \mathbf{I} - \mathbf{\Pi}$. We can decompose $\mathbf{A}^x$ into

$$\begin{aligned}\mathbf{A}^x = \prod_{i=1}^k \mathbf{A}^{x_i} = \prod_{i=1}^k \mathbf{A}^{x_i}(\mathbf{P}_1 + \mathbf{P}_2) &= \sum_{j_1,\cdots,j_k\in\{1,2\}} \mathbf{A}^{x_1}\mathbf{P}_{j_1}\mathbf{A}^{x_2}\mathbf{P}_{j_2}\cdots\mathbf{A}^{x_k}\mathbf{P}_{j_k} \\ = \hat{\mathbf{A}}^x + \mathbf{A}^{x_1}\mathbf{\Pi}\mathbf{A}^{x_2}\mathbf{\Pi}\cdots\mathbf{A}^{x_k}\mathbf{\Pi} &+ \sum_{\substack{j_1,\cdots,j_k\in\{1,2\} \text{ s.t.} \\ \exists r,r' : j_r \neq j_{r'}}} \mathbf{A}^{x_1}\mathbf{P}_{j_1}\mathbf{A}^{x_2}\mathbf{P}_{j_2}\cdots\mathbf{A}^{x_k}\mathbf{P}_{j_k}.\end{aligned}$$

We will show that the traces of all the summands in this last expression, except for the first one, are equal to 0. First, using Eq. (2) we have $\mathbf{A}^{x_1}\mathbf{\Pi}\mathbf{A}^{x_2}\mathbf{\Pi}\cdots\mathbf{A}^{x_k}\mathbf{\Pi} = \mathbf{A}^x\mathbf{\Pi}$. Moreover, for any integer s such that $sk \geq d$ we have $(\mathbf{A}^x\mathbf{\Pi})^s = \mathbf{A}^{x^s}\mathbf{\Pi} = \mathbf{0}$ by definition of E and by Lemma 2, thus $\mathbf{A}^x\mathbf{\Pi}$ is nilpotent and its trace is 0 by Lemma 1. For the remaining terms, let $j_1, \cdots, j_k \in \{1, 2\}$ not all equal. Let $l \in [k]$ be an index such that $j_l = 2$ and $j_{\overline{l+1}} = 1$ where $\overline{l+1} = l+1$ if $l < k$ and 1 otherwise. Using the invariance of the trace under cyclic permutations of a matrix product, we obtain

$$\begin{aligned}\mathrm{Tr}(\mathbf{A}^{x_1}\mathbf{P}_{j_1}\mathbf{A}^{x_2}\mathbf{P}_{j_2}\cdots\mathbf{A}^{x_k}\mathbf{P}_{j_k}) &= \mathrm{Tr}(\mathbf{A}^{x_l}\mathbf{P}_{j_l}\mathbf{A}^{x_{\overline{l+1}}}\mathbf{P}_{j_{\overline{l+1}}}\cdots) \\ &= \mathrm{Tr}(\mathbf{A}^{x_l}(\mathbf{I} - \mathbf{\Pi})\mathbf{A}^{x_{\overline{l+1}}}\mathbf{\Pi}\cdots) = 0\end{aligned}$$

where we used Eq. 2 again for the last equality. To conclude, we have shown that $\mathrm{Tr}(\mathbf{A}^x) = \mathrm{Tr}(\hat{\mathbf{A}}^x)$ for all $x \in \Sigma^*$, hence A and $\hat{A}$ compute the same function on circular strings. □

Moreover, we now show that the subspace E from the previous theorem can be used to obtain a characterization of the minimality of a GWM^c.

Theorem 6. *Let A be a GWM^c given by the set of matrices $\{\mathbf{A}^\sigma\}_{\sigma\in\Sigma} \subset \mathcal{M}_d(\mathbb{F})$. Then, A is minimal if and only if the linear space*

$$E = \bigcap_{x\in\Sigma^d} \ker(\mathbf{A}^x) = \{\mathbf{v} \in \mathbb{F}^d : \mathbf{A}^x\mathbf{v} = \mathbf{0} \text{ for all } x \in \Sigma^d\}$$

is trivial, i.e. $E = \{\mathbf{0}\}$.

Proof. Suppose that E is not trivial and let $\mathbf{\Pi}$ be the matrix of the orthogonal projection onto E. Then, the rank R of $\mathbf{I} - \mathbf{\Pi}$ is strictly less than d and there exists an orthogonal matrix $\mathbf{U} \in \mathbb{R}^{d\times R}$ such that $\mathbf{I} - \mathbf{\Pi} = \mathbf{U}\mathbf{U}^\top$. It follows from the previous proposition that, for any non-empty word $x = x_1 \cdots x_k$, we have

$$\begin{aligned}\mathrm{Tr}(\mathbf{A}^x) &= \mathrm{Tr}(\mathbf{A}^{x_1}(\mathbf{I}-\mathbf{\Pi})\mathbf{A}^{x_2}(\mathbf{I}-\mathbf{\Pi})\cdots\mathbf{A}^{x_k}(\mathbf{I}-\mathbf{\Pi}))\\ &= \mathrm{Tr}(\mathbf{A}^{x_1}\mathbf{U}\mathbf{U}^\top\mathbf{A}^{x_2}\mathbf{U}\mathbf{U}^\top\cdots\mathbf{A}^{x_k}\mathbf{U}\mathbf{U}^\top) = \mathrm{Tr}((\mathbf{U}^\top\mathbf{A}^{x_1}\mathbf{U})(\mathbf{U}^\top\mathbf{A}^{x_2}\mathbf{U})\cdots(\mathbf{U}^\top\mathbf{A}^{x_k}\mathbf{U})).\end{aligned}$$

Hence, the R-dimensional GWM^c given by the matrices $\hat{\mathbf{A}}^\sigma = \mathbf{U}^\top\mathbf{A}^\sigma\mathbf{U}$ computes the same function as A, showing that A is not minimal.

Suppose now that A is not minimal. Let B be a GWM^c of dimension $d' < d$, given by the matrices $\{\mathbf{B}^\sigma\}_{\sigma\in\Sigma}$, such that $f_B = f_A$. Let $\mathcal{A}$ (resp. $\mathcal{B}$) be the algebra generated by the matrices $\{\mathbf{A}^\sigma\}_{\sigma\in\Sigma}$ (resp. $\{\mathbf{B}^\sigma\}_{\sigma\in\Sigma}$). By Corollary 1, we can assume that both A and B are semi-simple GWM^{c}s, i.e. that the algebras $\mathcal{A}$ and $\mathcal{B}$ are semi-simple. For each $\sigma \in \Sigma$, let $\hat{\mathbf{B}}^\sigma = \mathbf{B}^\sigma \oplus \mathbf{0} \in \mathbb{R}^{d\times d}$ be the block diagonal matrix having $\mathbf{B}^\sigma$ in the upper diagonal block and 0's elsewhere. Let $\hat{\mathcal{B}}$ be the algebras generated by the matrices $\{\hat{\mathbf{B}}^\sigma\}_{\sigma\in\Sigma} \subset \mathcal{M}_d(\mathbb{F})$. It is easy to check that the GWM^c $\hat{B}$ computes the same function as A and B and that the algebra $\hat{\mathcal{B}}$ is semi-simple (it is indeed isomorphic to the semi-simple algebra $\mathcal{B}$). It then follows from Corollary 2 that there exists an invertible matrix $\mathbf{P} \in \mathcal{M}_d(\mathbb{F})$ such that $\mathbf{A}^\sigma = \mathbf{P}\hat{\mathbf{B}}^\sigma\mathbf{P}^{-1}$ for all $\sigma \in \Sigma$. Let $\mathbf{e}_d$ be the dth vector of the canonical basis of $\mathbb{F}^d$, by definition of $\hat{\mathbf{B}}^\sigma$ we have $\hat{\mathbf{B}}^\sigma\mathbf{e}_d = \mathbf{0}$ for any $\sigma \in \Sigma$, and consequently $\mathbf{A}^\sigma\mathbf{P}\mathbf{e}_d = \mathbf{0}$ for any symbol σ, showing that $\mathbf{P}\mathbf{e}_d \in E$ and $E \neq \{\mathbf{0}\}$. □

It follows from the two previous theorems that by restricting the linear operators $\mathbf{A}^\sigma$ of a GWM^c A to the subspace $E^\perp$, one can obtain a minimal GWM^c computing f_A. We formally state this result in the following corollary.

Corollary 3. *Let A be a GWM^c given by the matrices $\{\mathbf{A}^\sigma\}_{\sigma\in\Sigma} \subset \mathcal{M}_d(\mathbb{F})$ and let $\mathbf{\Pi}$ be the matrix of the orthogonal projection onto the space $E = \bigcap_{x\in\Sigma^d}\ker(\mathbf{A}^x)$. For any orthogonal matrix $\mathbf{U} \in \mathbb{F}^{d\times R}$ such that $\mathbf{I} - \mathbf{\Pi} = \mathbf{U}\mathbf{U}^\top$ (where R is the dimension of $E^\perp$), the R-dimensional GWM^c $\hat{A}$ given by the matrices $\hat{\mathbf{A}}^\sigma = \mathbf{U}^\top\mathbf{A}^\sigma\mathbf{U}$ is a minimal GWM^c computing f_A.*

Proof. Using the invariance of the trace under cyclic permutations of a matrix product, it directly follows from Theorem 5 that $f_{\hat{A}} = f_A$. Moreover, one can check that $\hat{E} = \bigcap_{x\in\Sigma^d} \ker(\hat{\mathbf{A}}^x) = \{\mathbf{0}\}$ by construction of the matrices $\hat{\mathbf{A}}^\sigma$, hence $\hat{A}$ is minimal by Theorem 6. □

We showed that a GWM$^\mathrm{c}$ can be minimized by restricting its matrices to the subspace $E^\perp$. In order to do so, one needs to compute a basis of $E = \bigcap_{x\in\Sigma^d} \ker(\mathbf{A}^x)$. This can naively be done by first computing $\ker(\mathbf{A}^x)$ for each $x \in \Sigma^d$ and then computing a basis for the intersection of these linear subspaces, however the complexity of this approach is exponential in the dimension d. We show in the following proposition that for semi-simple GWM$^\mathrm{c}$s, one simply needs to compute a basis of the space $\bigcap_{\sigma\in\Sigma} \ker(\mathbf{A}^\sigma)$, which can be done in polynomial time (provided that the field $\mathbb{F}$ admits efficient symbolic arithmetic, e.g. $\mathbb{F} = \mathbb{Q}$).

Proposition 4. *Let $\mathcal{A} \subset \mathcal{M}_d(\mathbb{F})$ be the finite dimensional algebra generated by the set of matrices $\{\mathbf{A}^\sigma\}_{\sigma\in\Sigma}$. Then if $\mathcal{A}$ is semi-simple we have*

$$\bigcap_{x\in\Sigma^d} \ker(\mathbf{A}^x) = \bigcap_{\sigma\in\Sigma} \ker(\mathbf{A}^\sigma).$$

Proof. For any integer $i \geq 1$, let $E_i = \bigcap_{x\in\Sigma^i} \ker(\mathbf{A}^x)$. Recall from the proof of Lemma 2 that $E_i \subset E_{i+1}$ for all i and that $E_i = E_{i+1}$ implies $E_i = E_{i+k}$ for any integer $k \geq 0$, hence it will be sufficient to show that $E_1 = E_2$. One can check that each E_i is $\mathcal{A}$-invariant, i.e. each E_i is an $\mathcal{A}$-module. Since $\mathcal{A}$ is semi-simple, any $\mathcal{A}$-module is semi-simple [17, Theorem 2.6.2], which implies that if M is an $\mathcal{A}$-module, every submodule U of M has a complement [17, Proposition 2.2.1], i.e. there exists an $\mathcal{A}$-module V such that $M = U \oplus V$. Now since E_1 is a submodule of the $\mathcal{A}$-module E_2, E_1 has a complement U in E_2, i.e. U is $\mathcal{A}$-invariant and $E_2 = E_1 \oplus U$. Let $\mathbf{v} \in U$. We show $\mathbf{v} = \mathbf{0}$. Since $\mathbf{v} \in E_2$, we have $\mathbf{A}^{\sigma_1}\mathbf{A}^{\sigma_2}\mathbf{v} = \mathbf{0}$ for all $\sigma_1, \sigma_2 \in \Sigma$, hence $\mathbf{A}^\sigma\mathbf{v} \in E_1$ for all $\sigma \in \Sigma$. Moreover, we have $\mathbf{A}^\sigma\mathbf{v} \in U$ for all $\sigma \in \Sigma$ since U is $\mathcal{A}$-invariant. It follows that $\mathbf{A}^\sigma\mathbf{v} \in E_1 \cap U = \{\mathbf{0}\}$ and $\mathbf{A}^\sigma\mathbf{v} = \mathbf{0}$ for all $\sigma \in \Sigma$, hence $\mathbf{v} \in E_1$ and since $\mathbf{v} \in U$ we have $\mathbf{v} = \mathbf{0}$. To conclude, we have $U = \{\mathbf{0}\}$, hence $E_1 = E_2$. □

Since a GWM$^\mathrm{c}$ can be transformed into an equivalent semi-simple GWM$^\mathrm{c}$ in polynomial time (see Corollary 1 and the following discussion), the minimization of a GWM$^\mathrm{c}$ defined over circular strings can be achieved in polynomial time by first converting it to a semi-simple GWM$^\mathrm{c}$ and then applying Corollary 3 with Proposition 4. The overall minimization algorithm is summarized in Algorithm 1.

Algorithm 1. Minimization of a GWM defined over circular strings

Input: A d-dimensional GWM^c A given by a set of matrices $\{\mathbf{A}^\sigma\}_{\sigma\in\Sigma} \subset \mathcal{M}_d(\mathbb{F})$.
Output: A minimal GWM^c $\hat{A}$ computing f_A.

1: Let $\mathcal{A}$ be the algebra generated by the matrices $\{\mathbf{A}^\sigma\}_{\sigma\in\Sigma}$.
2: Compute a basis $(\mathbf{A}_1, \cdots, \mathbf{A}_n)$ of $\mathcal{A}$ (as an $\mathbb{F}$-vector space) and the structure coefficients $c_{i,j}^k \in \mathbb{F}$ for $i, j, k \in [n]$ satisfying $\mathbf{A}_i\mathbf{A}_j = \sum_{k=1}^n c_{i,j}^k \mathbf{A}_k$.
3: Compute the sub-algebra $\tilde{\mathcal{A}}$ and the corresponding surjective homomorphism $\pi\colon \mathcal{A} \to \tilde{\mathcal{A}}$ satisfying $\mathcal{A} = \mathrm{Rad}(\mathcal{A}) \oplus \tilde{\mathcal{A}}$ and $\tilde{\mathcal{A}} \cong \mathcal{A}/\mathrm{Rad}(\mathcal{A})$ (using the algorithm from [15], see Theorem 3).
4: Let $\tilde{A}$ be the semi-simple GWM given by the set of matrices $\{\tilde{\mathbf{A}}^\sigma = \pi(\mathbf{A}^\sigma)\}_{\sigma\in\Sigma}$.
5: Compute a basis of $E_1 = \{\mathbf{v} \in \mathbb{F}^d : \tilde{\mathbf{A}}^\sigma\mathbf{v} = \mathbf{0} \text{ for all } \sigma \in \Sigma\} = \bigcap_{\sigma\in\Sigma} \ker(\tilde{\mathbf{A}}^\sigma)$.
6: Let $\mathbf{\Pi} \in \mathbb{F}^{d\times d}$ be the matrix of the orthogonal projection onto E_1.
7: Let R be the rank of $\mathbf{I} - \mathbf{\Pi}$ and let $\mathbf{U} \in \mathbb{F}^{d\times R}$ be an orthogonal matrix such that $\mathbf{I} - \mathbf{\Pi} = \mathbf{U}\mathbf{U}^\top$.
8: **return** The R-dimensional GWM^c given by the matrices $\{\hat{\mathbf{A}}^\sigma = \mathbf{U}^\top\tilde{\mathbf{A}}^\sigma\mathbf{U}\}_{\sigma\in\Sigma}$.

5 Conclusion

We proposed polynomial time algorithms to handle both the minimization and the equivalence problems for GWMs defined over circular strings. By doing so, we unraveled fundamental notions from algebra theory that will be central to the study of GWMs. In particular, the notion of *semi-simple* GWM^c was paramount to our analysis. Intuitively, semi-simplicity can be thought of as a weak form of minimality: components from the radical do not contribute to the final computation of a GWM^c (semi-simplification thus corresponds to annihilating these irrelevant components from the algebra, i.e. from the GWM^c's dynamics).

The next step is of course to try to extend the results obtained in this paper to GWMs defined over more general families of graphs. One promising direction we are currently investigating relies on extending the central notion of semi-simple GWM^c to GWMs defined over arbitrary families of labeled graphs: by opening any edge e in a graph G one obtains a graph G_e with two *free ports* (i.e. edges having one end that is not connected to any vertex) which would be mapped by a d-dimensional GWM A to a matrix $\mathbf{A}^{G_e} \in \mathcal{M}_d(\mathbb{F})$ (indeed, a GWM naturally maps any graph with k free ports to a kth order tensor; see [22, Sect. 2.2.3] for more details). For circular strings, opening an edge corresponds to choosing a particular position in the circular string leading to an actual string $x \in \Sigma^*$ which is mapped to $\mathbf{A}^x$ by the GWM. For arbitrary labeled graphs, we have $f_A(G) = \mathrm{Tr}(\mathbf{A}^{G_e})$ similarly to the case of circular strings. One can then consider the algebra $\mathcal{A}$ generated by the matrices $\mathbf{A}^{G_e}$ for any graph G in some family of graphs and any edge e in G, and define a semi-simple GWM as a GWM for which this algebra $\mathcal{A}$ is semi-simple (note that one exactly recovers the notion of semi-simple GWM introduced here in the special case of circular strings). Hence, the fundamental results from algebra theory we leveraged in this paper should be directly relevant to the study of general GWMs. Beyond minimization, we intend to study the problem of approximate minimization (such as the ones considered in [7,23] for string and tree weighted automata) along with the closely related problem of learning GWMs defined over richer families of graphs than the one of circular strings.

Acknowledgements. The author acknowledges support of an IVADO postdoctoral fellowship and would like to thank the reviewers for their helpful comments as well as Philip Amortila, François Denis, Clara Lacroce, Prakash Panangaden and Joelle Pineau for fruitful discussions.

References

1. Bailly, R., Carreras Pérez, X., Luque, F.M., Quattoni, A.J.: Unsupervised spectral learning of WCFG as low-rank matrix completion. In: Proceedings of the 2013 Conference on Empirical Methods in Natural Language Processing, pp. 624–635. ACL (2013)
2. Bailly, R., Denis, F., Rabusseau, G.: Recognizable series on hypergraphs. In: Proceedings of the 9th International Conference on Language and Automata Theory and Applications, pp. 639–651 (2015)
3. Bailly, R., Denis, F., Ralaivola, L.: Grammatical inference as a principal component analysis problem. In: Proceedings of the 26th Annual International Conference on Machine Learning, pp. 33–40. ACM (2009)
4. Bailly, R., Habrard, A., Denis, F.: A spectral approach for probabilistic grammatical inference on trees. In: Hutter, M., Stephan, F., Vovk, V., Zeugmann, T. (eds.) ALT 2010. LNCS (LNAI), vol. 6331, pp. 74–88. Springer, Heidelberg (2010). https://doi.org/10.1007/978-3-642-16108-7_10
5. Bailly, R., Rabusseau, G., Denis, F.: Recognizable series on graphs and hypergraphs. J. Comput. Syst. Sci. (2017, in press)
6. Balle, B., Carreras, X., Luque, F.M., Quattoni, A.: Spectral learning of weighted automata. Mach. Learn. **96**(1–2), 33–63 (2014)
7. Balle, B., Panangaden, P., Precup, D.: A canonical form for weighted automata and applications to approximate minimization. In: 30th Annual Symposium on Logic in Computer Science, pp. 701–712. IEEE (2015)
8. Berstel, J., Reutenauer, C.: Recognizable formal power series on trees. Theor. Comput. Sci. **18**(2), 115–148 (1982)
9. Berstel, J., Reutenauer, C.: Rational Series and Their Languages. Springer, Heidelberg (1988)
10. Boots, B., Siddiqi, S.M., Gordon, G.J.: Closing the learning-planning loop with predictive state representations. Int. J. Robot. Res. **30**(7), 954–966 (2011)
11. Bozapalidis, S., Grammatikopoulou, A.: Recognizable picture series. J. Automata Lang. Comb. **10**(2/3), 159–183 (2005)
12. Bremner, M.R.: How to compute the Wedderburn decomposition of a finite-dimensional associative algebra. Groups Complex. Cryptol. **3**(1), 47–66 (2011)
13. Brešar, M.: Introduction to Noncommutative Algebra. Springer, Cham (2014). https://doi.org/10.1007/978-3-319-08693-4
14. Cohen, S.B., Stratos, K., Collins, M., Foster, D.P., Ungar, L.H.: Spectral learning of latent-variable PCFGs: algorithms and sample complexity. J. Mach. Learn. Res. **15**(1), 2399–2449 (2014)
15. de Graaf, W.A., Ivanyos, G., Küronya, K., Rónyai, L.: Computing Levi decompositions in lie algebras. Appl. Algebra Eng. Commun. Comput. **8**(4), 291–303 (1997)

16. Droste, M., Kuich, W., Vogler, H. (eds.): Handbook of Weighted Automata. Springer Science & Business Media, Heidelberg (2009). https://doi.org/10.1007/978-3-642-01492-5
17. Drozd, Y.A., Kirichenko, V.V.: Finite Dimensional Algebras. Springer Science & Business Media, Heidelberg (2012). https://doi.org/10.1007/978-3-642-76244-4
18. Hladnik, M., Omladic, M., Radjavi, H.: Trace-preserving homomorphisms of semigroups. J. Funct. Anal. **204**(2), 269–292 (2003)
19. Hsu, D.J., Kakade, S.M., Zhang, T.: A spectral algorithm for learning Hidden Markov Models. In: Proceedings of the Conference on Learning Theory (2009)
20. Lee, T., Na, J.C., Park, H., Park, K., Sim, J.S.: Finding consensus and optimal alignment of circular strings. Theoret. Comput. Sci. **468**, 92–101 (2013)
21. Marusic, I., Worrell, J.: Complexity of equivalence and learning for multiplicity tree automata. J. Mach. Learn. Res. **16**, 2465–2500 (2015)
22. Rabusseau, G.: A Tensor Perspective on Weighted Automata, Low-Rank Regression and Algebraic Mixtures. PhD thesis, Aix-Marseille Université (2016)
23. Rabusseau, G., Balle, B., Cohen, S.: Low-rank approximation of weighted tree automata. In: Proceedings of the 19th International Conference on Artificial Intelligence and Statistics, pp. 839–847 (2016)
24. Rittaud, B., Vivier, L.: Circular words and applications. In: WORDS, pp. 31–36 (2011)
25. Sakarovitch, J.: Elements of Automata Theory. Cambridge University Press, Cambridge (2009)
26. Tzeng, W.-G.: On the definition of a family of automata. Inf. Control **4**(2–3), 245–270 (1961)
27. Tzeng, W.-G.: A polynomial-time algorithm for the equivalence of probabilistic automata. SIAM J. Comput. **21**(2), 216–227 (1992)

Games on Graphs with a Public Signal Monitoring

Patricia Bouyer(✉)

LSV, CNRS, ENS Paris-Saclay, Université Paris-Saclay, Cachan, France
bouyer@lsv.fr

Abstract. We study pure Nash equilibria in games on graphs with an imperfect monitoring based on a public signal. In such games, deviations and players responsible for those deviations can be hard to detect and track. We propose a generic epistemic game abstraction, which conveniently allows to represent the knowledge of the players about these deviations, and give a characterization of Nash equilibria in terms of winning strategies in the abstraction. We then use the abstraction to develop algorithms for some payoff functions.

1 Introduction

Multiplayer concurrent games over graphs allow to model rich interactions between players. Those games are played as follows. In a state, each player chooses privately and independently an action, defining globally a move (one action per player); the next state of the game is then defined as the successor (on the graph) of the current state using that move; players continue playing from that new state, and form (an infinite) play. Each player then gets a reward given by a payoff function (one function per player). In particular, objectives of the players may not be contradictory: those games are non-zero-sum games, contrary to two-player games used for controller or reactive synthesis [23,30].

The problem of distributed synthesis [25] can be formulated using multiplayer concurrent games. In this setting, there is a global objective Φ, and one particular player called Nature. The question then is whether the so-called grand coalition (all players except Nature) can enforce Φ, whatever Nature does. While the players (except Nature) cooperate (and can initially coordinate), their choice of actions (or strategy) can only depend on what they see from the play so far. When modelling distributed synthesis as concurrent games, information players receive is given via a partial observation function of the states of the game. When the players have perfect monitoring of the play, the distributed synthesis problem reduces to a standard two-player zero-sum game. Distributed synthesis is a fairly hot topic, both using the formalization via concurrent games we have already described and using the formalization via an architecture of processes [26]. The most general decidability results in the concurrent game setting are under the

This work has been supported by ERC project EQualIS (FP7-308087).

C. Baier and U. Dal Lago (Eds.): FOSSACS 2018, LNCS 10803, pp. 530–547, 2018.
https://doi.org/10.1007/978-3-319-89366-2_29

assumption of hierarchical observation [6,36] (information received by the players is ordered) or more recently under recurring common knowledge [5].

While distributed synthesis involves several players, this remains nevertheless a zero-sum question. Using solution concepts borrowed from game theory, one can go a bit further in describing the interactions between the players, and in particular in describing rational behaviours of selfish players. One of the most basic solution concepts is that of Nash equilibria [24]. A Nash equilibrium is a strategy profile where no player can improve her payoff by unilaterally changing her strategy. The outcome of a Nash equilibrium can therefore be seen as a rational behaviour of the system. While very much studied by game theoretists (e.g. over matrix games), such a concept (and variants thereof) has been only rather recently studied over games on graphs. Probably the first works in that direction are [15,17,32,33]. Several series of works have followed. To roughly give an idea of the existing results, pure Nash equilibria always exist in turn-based games for ω-regular objectives [35] but not in concurrent games; they can nevertheless be computed for large classes of objectives [9,11,35]. The problem becomes harder with mixed (that is, stochastic) Nash equilibria, for which we often cannot decide the existence [10,34].

Computing Nash equilibria requires to (i) find a good behaviour of the system; (ii) detect deviations from that behaviour, and identify deviating players (called deviators); (iii) punish them. This simple characterization of Nash equilibria is made explicit in [18]. Variants of Nash equilibria require slightly different ingredients, but they are mostly of a similar vein.

In (almost) all these works though, perfect monitoring is implicitly assumed: in all cases, players get full information on the states which are visited; a slight imperfect monitoring is assumed in some works on concurrent games (like [9]), where actions which have been selected are not made available to all the players (we speak of hidden actions). This can yield some uncertainties for detecting deviators but not on states the game can be in, which is rather limited and can actually be handled.

In this work, we integrate imperfect monitoring into the problem of deciding the existence of pure Nash equilibria and computing witnesses. We choose to model imperfect monitoring via the notion of signal, which, given a joint decision of the players together with the next state the play will be in, gives some information to the players. To take further decisions, players get information from the signals they received, and have perfect recall about the past (their own actions and the signals they received). We believe this is a meaningful framework. Let us give an example of a wireless network in which several devices try to send data: each device can modulate its transmission power, in order to maximise its bandwidth and reduce energy consumption as much as possible. However there might be a degradation of the bandwidth due to other devices, and the satisfaction of each device is measured as a compromise between energy consumption and allocated bandwidth, and is given by a quantitative payoff function.[1] In such a problem, it is natural to assume that a device only gets a global information about the load of the network, and not about each other device which is connected to the network. This can be expressed using imperfect monitoring via public signals.

Following [31] in the framework of repeated matrix games, we put forward a notion of *public signal* (inspired by [31]). A signal will be said public whenever it is common to all players. That is, after each move, all the players get the same information (their own action remains of course private). We will also distinguish several kinds of payoff functions, depending on whether they are publicly visible (they only depend on the public signal), or privately visible (they depend on the public signal and on private actions: the corresponding player knows his payoff!), or invisible (players may not even be sure of their payoff).

The payoff functions we will focus on in this paper are Boolean ω-regular payoff functions and mean payoff functions. Some of the decidability results can be extended in various directions, which we will mention along the way.

As initial contributions of the paper, we show some undecidability results, and in particular that the hypothesis of public signal solely is not sufficient to enjoy all nice decidability results: for mean payoff functions, which are privately visible, one cannot decide the constrained existence of a Nash equilibrium. Constrained existence of a Nash equilibrium asks for the existence of a Nash equilibrium whose payoff satisfies some given constraint.

The main contribution of the paper is the construction of a so-called *epistemic game abstraction*. This abstraction is a two-player turn-based game in which we show that winning strategies of one of the players (Eve) actually correspond to Nash equilibria in the original game. The winning condition for Eve is rather complex, but can be simplified in the case of publicly visible payoff functions. The epistemic game abstraction is inspired by both the epistemic unfolding of [4] used for distributed synthesis, and the suspect game abstraction of [9] used to compute Nash equilibria in concurrent games with hidden actions. In our abstraction, we nevertheless not fully formalize epistemic unfoldings, and concentrate on the structure of the knowledge which is useful under the assumption of public signals; we show that several subset constructions (as done initially in [27], and since then used in various occasions, see e.g. [14,19,20,22]) made in parallel, are sufficient to represent the knowledge of all the players. The framework of [9] happens to be a special case of the public signal monitoring framework of the current paper. This construction can therefore be seen as an extension of the suspect game abstraction.

This generic construction can be applied to several frameworks with publicly visible payoff functions. We give two such applications, one with Boolean ω-regular payoff functions and one with mean payoff functions.

---

[1] This can be expressed by $\mathsf{payoff}_{\text{player } i} = \frac{R}{\mathsf{power}_i}\left(1 - e^{-0.5\gamma_i}\right)^L$ where γ_i is the signal-to-interference-and-noise ratio for player i, R is the rate at which the wireless system transmits the information and L is the size of the packets [29].

Further Related Works. We have already discussed several kinds of related works. Let us give some final remarks on related works.

We have mentioned earlier that one of the problems for computing Nash equilibria is to detect deviations and players who deviated. Somehow, the epistemic game abstraction tracks the potential deviators, and even though players might not know who exactly is responsible for the deviation (there might be several suspects), they can try to punish all potential suspects. And that what we do here. Very recently, [7] discusses the detection of deviators, and give some conditions for them to become common knowledge of the other players. In our framework, even though deviators may not become fully common knowledge, we can design mechanisms to punish the relevant ones.

Recently imperfect information has also been introduced in the setting of multi-agent temporal logics [2,3,20,21], and the main decidability results assume hierarchical information. However, while those logics allow to express rich interactions, it can somehow only consider qualitative properties. Furthermore, no tight complexity bounds are provided.

In [11], a deviator game abstraction is proposed. It twists the suspect game abstraction [9] to allow for more general solution concepts (so-called robust equilibria), but it assumes visibility of actions (hence remove any kind of uncertainties). Relying on results of [13], this deviator game abstraction allows to compute equilibria with mean payoff functions. Our algorithms for mean payoff functions will also rely on the polyhedron problem of [13].

A full version of this paper will all proofs is available as [8]. In this extended abstract, we made the choice to focus on the construction of the epistemic game abstraction and to be more sketchy on algorithms to compute Nash equilibria. We indeed believe the structure of the knowledge represented by the abstraction is the most important contribution, and that algorithms are more standard. However we believe it is important to be able to apply the abstract construction for algorithmics purpose.

2 Definitions

Throughout the paper, if $\mathbb{S} \subseteq \mathbb{R}$, we write $\overline{\mathbb{S}}$ for $\mathbb{S} \cup \{-\infty, +\infty\}$.

2.1 Concurrent Multiplayer Games with Signals

We consider the model of concurrent multi-player games, based on the two-player model of [1]. This model of games was used for instance in [9]. We equip games with *signals*, which will give information to the players.

Definition 1. *A* concurrent game with signals *is a tuple*

$$\mathcal{G} = \langle V, v_{\mathsf{init}}, \mathcal{P}, \mathsf{Act}, \Sigma, \mathsf{Allow}, \mathsf{Tab}, (\ell_A)_{A \in \mathcal{P}}, (\mathsf{payoff}_A)_{A \in \mathcal{P}} \rangle$$

where V is a finite set of vertices, $v_{\mathsf{init}} \in V$ is the initial vertex, $\mathcal{P}$ is a finite set of players, Act *is a finite set of actions, Σ is a finite alphabet,* Allow: $V \times \mathcal{P} \to$

$2^{\mathsf{Act}} \setminus \{\emptyset\}$ *is a mapping indicating the actions available to a given player in a given vertex,* $\mathsf{Tab}\colon V \times \mathsf{Act}^{\mathcal{P}} \to V$ *associates, with a given vertex and a given action tuple the target vertex, for every* $A \in \mathcal{P}$, $\ell_A\colon (\mathsf{Act}^{\mathcal{P}} \times V) \to \Sigma$ *is a signal, and* $\mathsf{payoff}_A\colon V \cdot (\mathsf{Act}^{\mathcal{P}} \cdot V)^\omega \to \mathbb{D}$ *is a payoff function with values in a domain* $\mathbb{D} \subseteq \overline{\mathbb{R}}$. *We say that the game has* public signal *if there is* $\ell\colon (\mathsf{Act}^{\mathcal{P}} \times V) \to \Sigma$ *such that for every* $A \in \mathcal{P}$, $\ell_A = \ell$.

The signals will help the players monitor the game: for taking decisions, a player will have the information given by her signal and the action she played earlier. A public signal will be a common information given to all the players. Our notion of public signal is inspired by [31] and encompasses the model of [9] where only action names were hidden to the players. Note that monitoring by public signal does not mean that all the players have the same information: they have private information implied by their own actions.

An element of $\mathsf{Act}^{\mathcal{P}}$ is called a move. When an explicit order is given on the set of players $\mathcal{P} = \{A_1, \ldots, A_{|\mathcal{P}|}\}$, we will write a move $m = (m_A)_{A \in \mathcal{P}}$ as $\langle m_{A_1}, \ldots, m_{A_{|\mathcal{P}|}} \rangle$. If $m \in \mathsf{Act}^{\mathcal{P}}$ and $A \in \mathcal{P}$, we write $m(A)$ for the A-component of m and $m(-A)$ for all but the A components of m. In particular, we write $m(-A) = m'(-A)$ whenever $m(B) = m'(B)$ for every $B \in \mathcal{P} \setminus \{A\}$.

A *full history* h in $\mathcal{G}$ is a finite sequence

$$v_0 \cdot m_0 \cdot v_1 \cdot m_1 \ldots m_{k-1} \cdot v_k \in V \cdot \left(\mathsf{Act}^{\mathcal{P}} \cdot V\right)^*$$

such that for every $0 \leq i < k$, $m_i \in \mathsf{Allow}(v_i)$ and $v_{i+1} = \mathsf{Tab}(v_i, m_i)$. For readability we will also write h as $v_0 \xrightarrow{m_0} v_1 \xrightarrow{m_1} \ldots \xrightarrow{m_{k-1}} v_k$.

We write *last*(h) for the last vertex of h (i.e., v_k). If $i \leq k$, we also write $h_{\leq i}$ for the prefix $v_0 \cdot m_0 \cdot v_1 \cdot m_1 \ldots m_{i-1} \cdot v_i$. We write $\mathsf{Hist}_{\mathcal{G}}(v_0)$ (or simply $\mathsf{Hist}(v_0)$ if $\mathcal{G}$ is clear in the context) for the set of full histories in $\mathcal{G}$ that start at v_0.

Let $A \in \mathcal{P}$. The projection of h for A is denoted $\pi_A(h)$ and is defined as:

$$v_0 \cdot (m_0(A), \ell_A(m_0, v_1)) \ldots (m_{k-1}(A), \ell_A(m_{k-1}, v_k)) \in V \cdot (\mathsf{Act} \times \Sigma)^*$$

This will be the information available to player A: it contains both the actions she played so far and the signal she received. Note that we assume perfect recall, that is, while playing, A will remember all her past knowledge, that is, all of $\pi_A(h)$ if h has been played so far. We define the *undistinguishability relation* $\sim_A$ as the equivalence relation over full histories induced by π_A: for two histories h and h', $h \sim_A h'$ iff $\pi_A(h) = \pi_A(h')$. While playing, if $h \sim_A h'$, A will not be able to know whether h or h' has been played. We also define the A-label of h as $\ell_A(h) = \ell_A(m_0, v_1) \cdot \ell_A(m_1, v_2) \ldots \ell_A(m_{k-1}, v_k)$.

We extend all the above notions to infinite sequences in a straightforward way and to the notion of *full play*. We write $\mathsf{Plays}_{\mathcal{G}}(v_0)$ (or simply $\mathsf{Plays}(v_0)$ if $\mathcal{G}$ is clear in the context) for the set of full plays in $\mathcal{G}$ that start at v_0.

We will say that the game $\mathcal{G}$ has *publicly (resp. privately) visible payoffs* if for every $A \in \mathcal{P}$, for every $v_0 \in V$, for every $\rho, \rho' \in \mathsf{Plays}(v_0)$, $\ell_A(\rho) = \ell_A(\rho')$ (resp. $\rho \sim_A \rho'$) implies $\mathsf{payoff}_A(\rho) = \mathsf{payoff}_A(\rho')$. Otherwise they are said *invisible*.

Private visibility of payoffs, while not always assumed (see for instance [3,19]), are reasonable assumptions: using only her knowledge, a player knows her payoff. Public visibility is more restrictive, but will be required for some of the results.

Let $A \in \mathcal{P}$ be a player. A *strategy* for player A from v_0 is a mapping $\sigma_A\colon \mathsf{Hist}(v_0) \to \mathsf{Act}$ such that for every history $h \in \mathsf{Hist}(v_0)$, $\sigma(h) \in \mathsf{Allow}(last(h))$. It is said ℓ_A-*compatible* whenever furthermore, for all histories $h, h' \in \mathsf{Hist}(v_0)$, $h \sim_A h'$ implies $\sigma_A(h) = \sigma_A(h')$. An *outcome* of σ_A is a(n infinite) play $\rho = v_0 \cdot m_0 \cdot v_1 \cdot m_1 \ldots$ such that for every $i \geq 0$, $\sigma_A(\rho_{\leq i}) = m_i(A)$. We write $\mathsf{out}(\sigma_A, v_0)$ for the set of outcomes of σ_A from v_0.

A *strategy profile* is a tuple $\sigma_{\mathcal{P}} = (\sigma_A)_{A \in \mathcal{P}}$, where, for every player $A \in \mathcal{P}$, σ_A is a strategy for player A. The strategy profile is said *info-compatible* whenever each σ_A is ℓ_A-compatible. We write $\mathsf{out}(\sigma_{\mathcal{P}}, v_0)$ for the unique full play from v_0, which is an outcome of all strategies part of $\sigma_{\mathcal{P}}$.

When $\sigma_{\mathcal{P}}$ is a strategy profile and σ'_A a player-A strategy, we write $\sigma_{\mathcal{P}}[A/\sigma'_A]$ for the profile where A plays according to σ'_A, and each other player B plays according to σ_B. The strategy σ'_A is a *deviation* of player A, or an A-*deviation.*

Definition 2. *A* Nash equilibrium *from v_0 is an info-compatible strategy profile σ such that for every $A \in \mathcal{P}$, for every player-A ℓ_A-compatible strategy σ'_A,* $\mathsf{payoff}_A\Big(\mathsf{out}(\sigma, v_0)\Big) \geq \mathsf{payoff}_A\Big(\mathsf{out}(\sigma[A/\sigma'_A], v_0)\Big)$.

In this definition, deviation σ'_A needs not be ℓ_A-compatible, since the only meaningful part of σ'_A is along $\mathsf{out}(\sigma[A/\sigma'_A], v_0)$, where there are no $\sim_A$-equivalent histories: any deviation can be made ℓ_A-compatible without affecting the profitability of the resulting outcome. Note also that there might be an A-deviation σ'_A which is not observable by another player B ($\mathsf{out}(\sigma, v_0) \sim_B \mathsf{out}(\sigma[A/\sigma'_A], v_0)$), and there might be two deviations σ'_B (by player B) and σ'_C (by player C) that cannot be distinguished by player A ($\mathsf{out}(\sigma[B/\sigma'_B], v_0) \sim_A \mathsf{out}(\sigma[C/\sigma'_C], v_0)$). Tracking such deviations will be the core of the abstraction we will develop.

Payoff Functions. In the following we will consider various payoff functions. Let Φ be an ω-regular property over some alphabet Γ. The function $pay_\Phi\colon \Gamma^\omega \to \{0, 1\}$ is defined by, for every $\mathbf{a} \in \Gamma^\omega$, $pay_\Phi(\mathbf{a}) = 1$ if and only if $\mathbf{a} \models \Phi$. A publicly (resp. privately) visible payoff function payoff_A for player A is said associated with Φ over Σ (resp. $\mathsf{Act} \times \Sigma$) whenever it is defined by $\mathsf{payoff}_A(\rho) = pay_\Phi(\ell_A(\rho))$ (resp. $\mathsf{payoff}_A(\rho) = pay_\Phi(\pi_A(\rho)_{-v_0})$, where $\pi_A(\rho)_{-v_0}$ crops the first v_0). Such a payoff function is called a Boolean ω-regular payoff function.

Let Γ be a finite alphabet and $w\colon \Gamma \to \mathbb{Z}$ be a weight assigning a value to every letter of that alphabet. We define two payoff functions over Γ^ω by, for every $\mathbf{a} = (a_i)_{i \geq 1} \in \Gamma^\omega$, $pay_{\underline{\mathrm{MP}}_w}(\mathbf{a}) = \liminf_{n \to \infty} \frac{1}{n} \sum_{i=1}^n w(a_i)$ and $pay_{\overline{\mathrm{MP}}_w}(\mathbf{a}) = \limsup_{n \to \infty} \frac{1}{n} \sum_{i=1}^n w(a_i)$. A publicly visible payoff function payoff_A for player A is said associated with the liminf (resp. limsup) mean payoff of w whenener it is defined by $\mathsf{payoff}_A(\rho) = pay_{\underline{\mathrm{MP}}_w}(\ell_A(\rho))$ (resp. $pay_{\overline{\mathrm{MP}}_w}(\ell_A(\rho))$). A privately visible payoff function payoff_A for player A is said associated with the liminf (resp. limsup) mean payoff of w whenener it is defined by $\mathsf{payoff}_A(\rho) = pay_{\underline{\mathrm{MP}}_w}(\pi_A(\rho)_{-v_0})$ (resp. $pay_{\overline{\mathrm{MP}}_w}(\pi_A(\rho)_{-v_0})$).

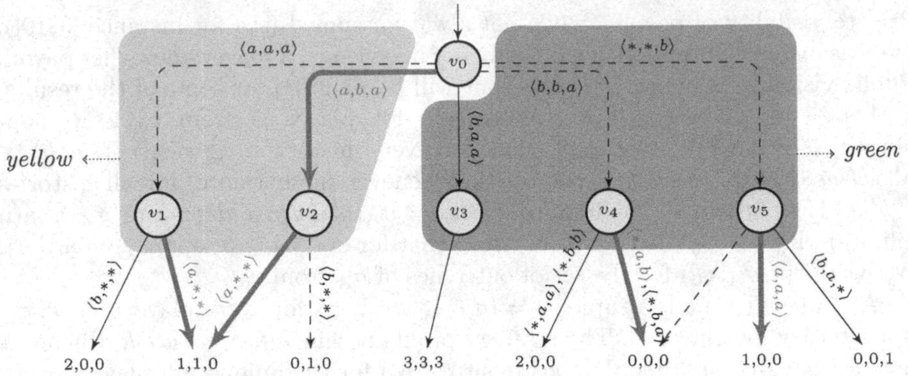

Fig. 1. An example of a concurrent game with public signal (yellow and green: public signal). Edges in red and bold are part of the strategy profile. Dashed edges are the possible deviations. One can notice that none of the deviations is profitable to the deviator, hence the strategy profile is a Nash equilibrium. Convention in the drawing: edges with no label are for complementary labels (for instance the edge from v_5 to $0,0,0$ is labelled by all $\langle a_1, a_2, a_3\rangle$ not in the set $\{\langle a,a,a\rangle, \langle b,a,a\rangle, \langle b,a,b\rangle\}$ (Color figure online))

Example 1. We now illustrate most notions on the game of Fig. 1. This is a game with three players A_1, A_2 and A_3, and which is played basically in two steps, starting at v_0. Graphically an edge labelled $\langle a_1, a_2, a_3\rangle$ between two vertices v and v' represents the fact that $a_i \in \mathsf{Allow}(v, A_i)$ for every $i \in \{1,2,3\}$ and that $v' = \mathsf{Tab}(v, \langle a_1, a_2, a_3\rangle)$. As a convention, $*$ stands for both a and b. For readability, bottom vertices explicitly indicate the payoffs of the three players (same order as for actions) if the game ends in that vertex.

After the first step of the game, signal yellow or green is sent to all the players. Histories $v_0 \cdot \langle a,b,a\rangle \cdot v_2$ and $v_0 \cdot \langle a,a,a\rangle \cdot v_1$ are undistinguishable by A_1 and A_3 (same action, same signal), but they can be distinguished by A_2 because of different actions (even if the same signal is observed).

In bold red, we have depicted a strategy profile, which is actually a Nash equilibrium. We analyze the possible deviations in this game to argue for this.

- First there is an A_2-deviation to v_1. This deviation is invisible to both players A_1 and A_3. For this reason, the strategy out of v_1 for A_1 is to play a (same as out of v_2). On the other hand, even though this would be profitable to her, A_1 cannot deviate from v_1, since we are in a branch where A_2 has already deviated, and at most one player is allowed to deviate at a time (and anyway A_1 does not know that they are in state v_1).
- There is an A_1-deviation from v_2 to $0,1,0$, which is not profitable to A_1.
- On the other hand, there is no possible deviation to v_3, since this would require two players to change their actions simultaneously (A_1 and A_2).
- Then, there is an A_1-deviation to v_4 and another A_3-deviation to v_5; both activate the green signal. A_2 knows there has been a deviation (because of the green signal), but she doesn't know who has deviated and whether the

game proceeds to v_4 or v_5 (but she knows that if A_1 has deviated, then we are in v_4, and if A_3 has deviated, we are in v_5). Then, A_2 has to find a way to punish both players, to be safe. On the other hand, both players A_1 and A_3 precisely know what has happened: in case she didn't deviate herself, she knows the other one deviated! And she knows in which state the game is in. Hence in state v_4, A_3 can help player A_2 punishing A_1, whereas in state v_5, A_1 can help player A_2 punishing A_3. Examples of punishing moves are therefore those depicted in red and bold; and they are part of the global strategy profile. Note that the action of A_2 out of v_5 has to be the same as the one out of v_4: this is required given the imperfect knowledge of A_2. On the other hand, the action of A_3 can be different out of v_4 and out of v_5 (which is the case in the given example profile).

Two-Player Turn-Based Game Structures. They are specific cases of the previous model, where at each vertex, at most one player has more than one action in her set of allowed actions. But for convenience, we will give a simplified definition, with only objects that will be useful. A two-player turn-based game structure is a tuple $G = \langle S, S_{\mathtt{Eve}}, S_{\mathtt{Adam}}, s_{\mathsf{init}}, A, \mathsf{Allow}, \mathsf{Tab}\rangle$, where $S = S_{\mathtt{Eve}} \sqcup S_{\mathtt{Adam}}$ is a finite set of states (states in $S_{\mathtt{Eve}}$ belong to player Eve whereas states in $S_{\mathtt{Adam}}$ belong to player Adam), $s_{\mathsf{init}} \in S$ is the initial state, A is a finite alphabet, $\mathsf{Allow}\colon S \to 2^A \setminus \{\emptyset\}$ gives the set of available actions, and $\mathsf{Tab}\colon S \times A \to S$ is the next-state function. If $s \in S_{\mathtt{Eve}}$ (resp. $S_{\mathtt{Adam}}$), $\mathsf{Allow}(s)$ is the set of actions allowed to Eve (resp. Adam) in state s.

In this context, strategies will see sequences of states and actions, with full information. Note that we do not include any winning condition or payoff function in the tuple, hence the name structure.

2.2 The Problem

We are interested in the constrained existence of a Nash equilibrium. For simplicity, we define constraints using non-strict thresholds constraints, but could well impose more involved constraints.

Problem 1 (Constrained existence problem). Given a game with signals $\mathcal{G} = \langle V, v_{\mathsf{init}}, \mathcal{P}, \mathsf{Act}, \Sigma, \mathsf{Allow}, \mathsf{Tab}, (\ell_A)_{A\in\mathcal{P}}, (\mathsf{payoff}_A)_{A\in\mathcal{P}}\rangle$ and threshold vectors $(\nu_A)_{A\in\mathcal{P}}, (\nu'_A)_{A\in\mathcal{P}} \in \overline{\mathbb{Q}}^{\mathcal{P}}$, can we decide whether there exists a Nash equilibrium $\sigma_{\mathcal{P}}$ from v_{init} such that for every $A \in \mathcal{P}$, $\nu_A \le \mathsf{payoff}_A(\mathsf{out}(\sigma_{\mathcal{P}}, v_{\mathsf{init}})) \le \nu'_A$? If so, compute one. If the constraints on the payoff are trivial (that is, $\nu_A = -\infty$ and $\nu'_A = +\infty$ for every $A \in \mathcal{P}$), we simply speak of the existence problem.

2.3 First Undecidability Results

In this section we state two preliminary undecidability results.

Theorem 1. – *The existence problem in games with signals is undecidable with three players and publicly visible Boolean ω-regular payoff functions.*

- *The constrained existence problem in games with a public signal is undecidable with two players and privately visible mean payoff functions.*

Proofs of these results rely on the distributed synthesis problem [26] for the first one, and on blind two-player mean-payoff games [19] for the second one. While there is no real surprise in the first result since we know that arbitrary partial information yields intrinsic difficulties, the second one suggests restrictions both to public signals and to publicly visible payoff functions.

In the following we will focus on public signals and develop an epistemic game abstraction, which will record and track possible deviations in the game. This will then be applied to get decidability results in two frameworks assuming publicly visible payoff functions.

3 The Epistemic Game Abstraction

Building over [4,9], we construct an epistemic game, which will record possible behaviours of the system, together with possible unilateral deviations. In [4], notions of epistemic Kripke structures are used to really track the precise knowledge of the players. These are mostly useful since undistinguishable states (expressed using signals here) are assumed arbitrary (no hierarchical structure). We could do the same here, but we think that would be overly complex and hide the real structure of knowledge in the framework of public signals. We therefore prefer to stick to simpler subset constructions, which are more commonly used (see e.g. [27] or later [14,19,22]), though it has to be a bit more involved here since also deviations have to be tracked.

Let $\mathcal{G} = \langle V, v_{\mathsf{init}}, \mathcal{P}, \mathsf{Act}, \Sigma, \mathsf{Allow}, \mathsf{Tab}, \ell, (\mathsf{payoff}_A)_{A\in\mathcal{P}}\rangle$ be a concurrent game with public signal. We will first define the epistemic abstraction as a two-player game structure $\mathcal{E}_\mathcal{G} = \langle S_{\mathtt{Eve}}, S_{\mathtt{Adam}}, s_{\mathsf{init}}, \Sigma', \mathsf{Allow}', \mathsf{Tab}'\rangle$, and then state the correspondence between $\mathcal{G}$ and $\mathcal{E}_\mathcal{G}$. The epistemic abstraction will later be used for decidability and algorithmics purposes. For clarity, we use the terminology "vertices" in $\mathcal{G}$ and "states" (or "epistemic states") in $\mathcal{E}_\mathcal{G}$.

3.1 Construction of the Game Structure $\mathcal{E}_\mathcal{G}$

The game $\mathcal{E}_\mathcal{G}$ will be played between two players, Eve and Adam. The aim of Eve is to build a suitable Nash equilibrium, whereas the aim of Adam is to prove that it is not an equilibrium; in particular, Adam will try to find a profitable deviation (to disprove the claim of Eve that she is building a Nash equilibrium). Choices available to Eve and Adam in the abstract game have to reflect partial knowledge of the players in the original game $\mathcal{G}$. States in the abstract game will therefore store information, which will be sufficient to infer the undistinguishability relation of all the players in the original game. Thanks to the public signal assumption, this information will be simple enough to have a simple structure.

In the following, we set $\mathcal{P}^\perp = \mathcal{P} \cup \{\perp\}$, where $\perp$ is a fresh symbol. For convenience, if $m \in \mathsf{Act}^\mathcal{P}$, we extend the notation $m(-A)$ when $A \in \mathcal{P}$ to $\mathcal{P}^\perp$ by setting $m(-\perp) = m$. We now describe all the components of $\mathcal{E}_\mathcal{G}$.

A state of Eve will store a set of vertices of the original game one can be in, together with possible deviators. More precisely, states of Eve are defined as $S_{\mathtt{Eve}} = \{s \colon \mathcal{P}^{\perp} \to 2^V \mid |s(\perp)| \leq 1\}$. Let $s \in S_{\mathtt{Eve}}$. If $A \in \mathcal{P}$, vertices of $s(A)$ are those where the game can be in, assuming one has followed the suggestions of Eve so far, up to an A-deviation; on the other hand, if $s(\perp) \neq \emptyset$, the single vertex $v \in s(\perp)$ is the one the game is in, assuming one has followed all suggestions by Eve so far (in particular, if Eve is building a Nash equilibrium, then this vertex belongs to the main outcome of the equilibrium). We define $\mathsf{sit}(s) = \{(v, A) \in V \times \mathcal{P}^{\perp} \mid v \in s(A)\}$ for the set of *situations* the game can be in at s:

(a) $(v, \perp) \in \mathsf{sit}(s)$ is the situation where the game has proceeded to vertex v without any deviation;
(b) $(v, A) \in \mathsf{sit}(s)$ with $A \in \mathcal{P}$ is the situation where the game has proceeded to vertex v benefitting, from an A-deviation.

Structure of state s will allow to infer the undistinguishability relation of all the players in game $\mathcal{G}$: basically (and we will formalize this later), if she is not responsible for a deviation, player $A \in \mathcal{P}$ will not know in which of the situations of $\mathsf{sit}(s) \setminus V \times \{A\}$ the game has proceeded; if she is responsible for a deviation, player A will know exactly in which vertex $v \in s(A)$ the game has proceeded.

Let $s \in S_{\mathtt{Eve}}$. From state s, Eve will suggest a tuple of moves M, one for each possible situation $(v, A) \in \mathsf{sit}(s)$. This tuple of moves has to satisfy the undistinguishability relation: if a player does not distinguish between two situations, her action should be the same in these two situations:

$$\mathsf{Allow}'(s) = \Big\{M \in \prod_{(v,A)\in\mathsf{sit}(s)} \mathsf{Allow}(v) \mid \forall (v_B, B), (v_C, C) \in \mathsf{sit}(s),\\ \forall A \in \mathcal{P} \setminus \{B, C\},\ M(v_B, B)(A) = M(v_C, C)(A)\Big\}$$

In the above set, the constraint $M(v_B, B)(A) = M(v_C, C)(A)$ expresses the fact that player A should play the same action in the two situations (v_B, B) and (v_C, C), since she does not distinguish between them. Obviously, we assume Σ' contains all elements of $\mathsf{Allow}'(s)$ above.

States of Adam are then copies of states of Eve with suggestions given by Eve, that is: $S_{\mathtt{Adam}} = \{(s, M) \mid s \in S_{\mathtt{Eve}} \times \mathsf{Allow}'(s)\}$. And naturally, we define $\mathsf{Tab}'(s, M) = (s, M)$ if $M \in \mathsf{Allow}'(s)$.

Let $(s, M) \in S_{\mathtt{Adam}}$. From state (s, M), Adam will choose a signal value which can be activated from some situation allowed in s, after no deviation or a single-player deviation w.r.t. M. From a situation $(v, A) \in \mathsf{sit}(s)$ with $A \in \mathcal{P}$, only A-deviations can be allowed (since we look for unilateral deviations), hence any signal activated by an A-deviation (w.r.t. $M(v, A)$) from v should be allowed. From the situation $(v, \perp) \in \mathsf{sit}(s)$ (if there is one), one can continue without any deviation, or any kind of single-player deviation should be allowed, hence the signal activated by $M(v, \perp)$ from v should be allowed, and any signal activated by some A-deviation (w.r.t. $M(v, \perp)$) from v should be allowed as well. Formally:

$$\mathsf{Allow}'(s,M) = \left\{ \beta \in \Sigma \;\middle|\; \begin{array}{l} \exists A \in \mathcal{P} \\ \exists v \in s(A) \\ \exists m \in \mathsf{Act}^{\mathcal{P}} \end{array} \text{ s.t. } \begin{array}{l} \text{(i) } m(-A) = M(v,A)(-A) \\ \text{(ii) } \ell(m, \mathsf{Tab}(v,m)) = \beta \end{array} \right\}$$

$$\cup \left\{ \beta \in \Sigma \;\middle|\; \begin{array}{l} \exists v \in s(\bot) \\ \exists m \in \mathsf{Act}^{\mathcal{P}} \\ \exists A \in \mathcal{P} \end{array} \text{ s.t. } \begin{array}{l} \text{(i) } m(-A) = M(v,\bot)(-A) \\ \text{(ii) } \ell(m, \mathsf{Tab}(v,m)) = \beta \end{array} \right\}$$

Note that we implicitly assume that Σ' contains Σ.

It remains to explain how one can compute the next state of some $(s, M) \in S_{\mathtt{Adam}}$ after some signal value $\beta \in \mathsf{Allow}'(s, M)$. The new state has to represent the new knowledge of the players in the original game when they have seen signal β; this has to take into account all possible deviations that we have already discussed which activate the signal value β. The new state is the result of several simultaneous subset constructions, which we formalize as follows: $s' = \mathsf{Tab}'((s, M), \beta)$, where for every $A \in \mathcal{P}^{\bot}$, $v' \in s'(A)$ if and only if there is $m \in \mathsf{Act}^{\mathcal{P}}$ such that $\beta = \ell(m, v')$, and

1. either there is $v \in s(A)$ such that $m(-A) = M(v, A)(-A)$ and $v' = \mathsf{Tab}(v, m)$;
2. or there is $v \in s(\bot)$ such that $m(-A) = M(v, \bot)(-A)$ and $v' = \mathsf{Tab}(v, m)$.

Note that in case $A = \bot$, the two above cases are redundant.

Before stating properties of $\mathcal{E}_{\mathcal{G}}$, we illustrate the construction.

Example 2. We consider again the example of Fig. 1, and we assume that the public signal when reaching the leaves of the game is uniformly orange. We depict (part of) the epistemic game abstraction of the game on Fig. 2. One can notice that from Eve-states s_1 and s_2, moves are multi-dimensional, in the sense that there is one move per vertex appearing in the state. There are nevertheless compatibility conditions which should be satisfied (expressed in condition Allow'); for instance, from s_2, player A_2 does not distinguish between the two options (i) A_1 has deviated and the game is in v_4, and (ii) A_3 has deviated and the game is in v_5, hence the action of player A_2 should be the same in the two moves (a in the depicted example, written in red).

3.2 Interpretation of this Abstraction

While we gave an intuitive meaning to the (epistemic) states of $\mathcal{E}_{\mathcal{G}}$, we now need to formalize this. And to do that, we need to explain how full histories and plays in $\mathcal{E}_{\mathcal{G}}$ can be interpreted as full histories and plays in $\mathcal{G}$.

Let $v_0 \in V$, and define $s_0 \colon \mathcal{P}^{\bot} \to 2^V \in S_{\mathtt{Eve}}$ such that $s_0(\bot) = \{v_0\}$ and $s_0(A) = \emptyset$ for every $A \in \mathcal{P}$. In the following, when $M \in \mathsf{Allow}'(s)$ for some $s \in S_{\mathtt{Eve}}$, if we speak of some $M(v, A)$, we implicitly assume that $(v, A) \in \mathsf{sit}(s)$. Given a full history $H = s_0 \xrightarrow{M_0} (s_0, M_0) \xrightarrow{\beta_0} s_1 \xrightarrow{M_1} (s_1, M_1) \xrightarrow{\beta_1} s_2 \ldots (s_{k-1}, M_{k-1}) \xrightarrow{\beta_{k-1}} s_k$ in $\mathcal{E}_{\mathcal{G}}$, we write *concrete*$(H)$ for the set of full histories in the original game, which correspond to H, up to a single deviation, that is: $v_0 \xrightarrow{m_0} v_1 \xrightarrow{m_1} v_2 \ldots v_{k-1} \xrightarrow{m_{k-1}} v_k \in$ *concrete*(H) whenever for every $0 \le i \le k-1$, $v_{i+1} = \mathsf{Tab}(v_i, m_i)$ and $\beta_i = \ell(m_i, v_{i+1})$, and:

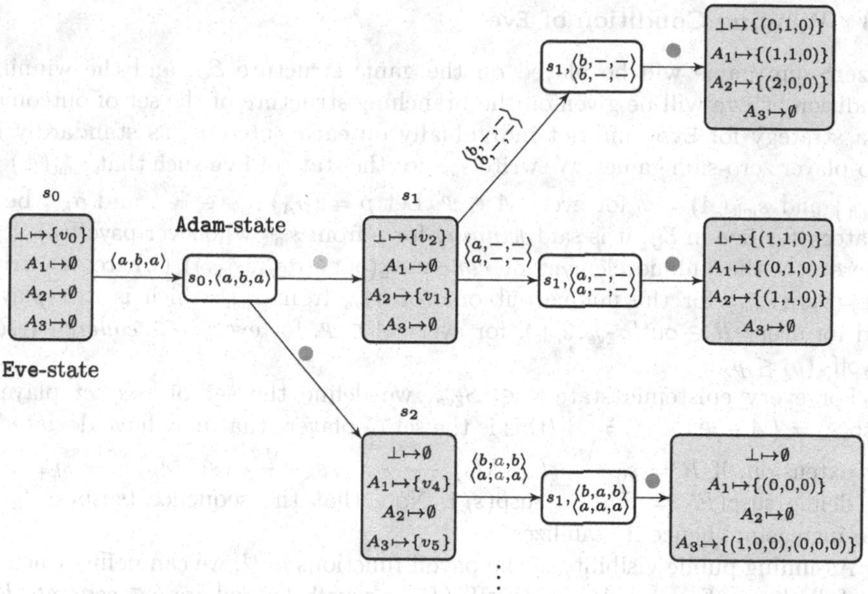

Fig. 2. Part of the epistemic game corresponding to the game of Fig. 1. For clarity, symbol $-$ is for any choice a or b (the precise choice is meaningless). (Color figure online)

(a) either $m_i = M_i(v_i, \bot)$ for every $0 \le i \le k-1$;
(b) or there exist $A \in \mathcal{P}$ and $0 \le i_0 \le k-1$ such that
  (i) for every $0 \le i < i_0$, $m_i = M_i(v_i, \bot)$;
  (ii) $m_{i_0} \ne M_{i_0}(v_{i_0}, \bot)$, but $m_{i_0}(-A) = M_{i_0}(v_{i_0}, \bot)(-A)$;
  (iii) for every $i_0 < i \le k-1$, $m_i(-A) = M_i(v_i, A)(-A)$.

Case (a) corresponds to a concrete history with no deviation (all moves suggested by Eve have been followed). Case (b) corresponds to a deviation by player A, and i_0 is the position at which player A has started deviating.

We write $concrete_\bot(H)$ for the set of histories of type (a); there is at most one such history, which is the real concrete history suggested by Eve. And we write $concrete_A(H)$ for the set of histories of the type (b) with deviator A. The correctness of the approach is obtained thanks to the following characterization of the undistinguishability relations along H: for every $A \in \mathcal{P}$, for every $h_1 \ne h_2 \in concrete(H)$,

$$h_1 \sim_A h_2 \text{ iff } h_1, h_2 \notin concrete_A(H).$$

In particular, a player may not distinguish between deviations by other players, or between a deviation by another player and the real concrete history suggested by Eve. But of course, in any case, a player will know that she has deviated!

We extend all these notions to full plays. A full play visiting only Eve-states s such that $s(\bot) \ne \emptyset$ is called a $\bot$-play.

3.3 Winning Condition of Eve

A zero-sum game will be played on the game structure $\mathcal{E}_\mathcal{G}$, and the winning condition of Eve will be given on the branching structure of the set of outcomes of a strategy for Eve, and not individually on each outcome, as standardly in two-player zero-sum games. We write s_{init} for the state of Eve such that $s_{\mathsf{init}}(\bot) = \{v_{\mathsf{init}}\}$ and $s_{\mathsf{init}}(A) = \emptyset$ for every $A \in \mathcal{P}$. Let $p = (p_A)_{A \in \mathcal{P}} \in \overline{\mathbb{R}}^{\mathcal{P}}$, and σ_{Eve} be a strategy for Eve in $\mathcal{E}_\mathcal{G}$; it is said *winning* for p from s_{init} whenever $\mathsf{payoff}(\rho) = p$, where ρ is the unique element of $\mathit{concrete}_\bot(\mathsf{out}_\bot(\sigma_{\mathsf{Eve}}, s_{\mathsf{init}}))$ (where we write $\mathsf{out}_\bot(\sigma_{\mathsf{Eve}}, s_{\mathsf{init}})$ for the unique outcome of σ_{Eve} from s_{init} which is a $\bot$-play), and for every $R \in \mathsf{out}(\sigma_{\mathsf{Eve}}, s_{\mathsf{init}})$, for every $A \in \mathcal{P}$, for every $\rho \in \mathit{concrete}_A(R)$, $\mathsf{payoff}_A(\rho) \le p_A$.

For every epistemic state $s \in S_{\mathsf{Eve}}$, we define the set of *suspect* players $\mathsf{susp}(s) = \{A \in \mathcal{P} \mid s(A) \neq \emptyset\}$ (this is the set of players that may have deviated). By extension, if $R = s_0 \xrightarrow{M_0} (s_0, M_0) \xrightarrow{\beta_0} s_1 \ldots s_k \xrightarrow{M_k} (s_k, M_k) \xrightarrow{\beta_k} s_{k+1} \cdots$, we define $\mathsf{susp}(R) = \lim_{k \to \infty} \mathsf{susp}(s_k)$. Note that the sequence $(\mathsf{susp}(s_k))_k$ is non-increasing, hence it stabilizes.

Assuming public visibility of the payoff functions in $\mathcal{G}$, we can define when R is a full play in $\mathcal{E}_\mathcal{G}$, and $A \in \mathcal{P}$, $\mathsf{payoff}'_A(R) = \mathsf{payoff}_A(\rho)$, where $\rho \in \mathit{concrete}(R)$. It is easy to show that payoff'_A is well-defined for every $A \in \mathcal{P}$. Under this assumption, the winning condition of Eve can be rewritten as: σ_{Eve} is winning for p from s_{init} whenever $\mathsf{payoff}'(\mathsf{out}_\bot(\sigma_{\mathsf{Eve}}, s_{\mathsf{init}})) = p$, and for every $R \in \mathsf{out}(\sigma_{\mathsf{Eve}}, s_{\mathsf{init}})$, for every $A \in \mathsf{susp}(R)$, $\mathsf{payoff}'_A(R) \le p_A$.

3.4 Correction of the Epistemic Abstraction

The epistemic abstraction tracks everything that is required to detect Nash equilibria in the original game, which we make explicit in the next result. Note that this theorem does not require public visibility of the payoff functions.

Theorem 2. *Let $\mathcal{G}$ be a concurrent game with public signal, and $p \in \overline{\mathbb{R}}^{\mathcal{P}}$. There is a Nash equilibrium in $\mathcal{G}$ with payoff p from v_{init} if and only if* Eve *has a winning strategy for p in $\mathcal{E}_\mathcal{G}$ from s_{init}.*

The proof of this theorem highlights a correspondence between Nash equilibria in $\mathcal{G}$ and winning strategies of Eve in $\mathcal{E}_\mathcal{G}$. In this correspondence, the main outcome of the equilibrium in $\mathcal{G}$ is the unique $\bot$-concretisation of the unique $\bot$-play generated by the winning strategy of Eve.

3.5 Remarks on the Construction

We did not formalize the epistemic unfolding as it is made in [4]. We believe we do not really learn anything for public signal using it. And the above extended subset construction can much better be understood.

One could argue that this epistemic game gives more information to the players, since Eve explicitely gives to everyone the move that should be played.

But in the real game, the players also have that information, which is obtained by an initial coordination of the players (this is required to achieve equilibria).

Finally, notice that the espitemic game constructed here generalizes the suspect game construction of [9], where all players have perfect information on the states of the game, but cannot see the actions that are precisely played. Somehow, games in [9] have a public signal telling the state the game is in (that is, $\ell(m, v) = v$). So, in the suspect game of [9], the sole uncertainty is in the players that may have deviated, not in the set of states that are visited.

Remark 1. Let us analyze the size of the epistemic game abstraction. The size of the alphabet is bounded by $|\Sigma| + |\mathsf{Act}|^{|\mathscr{P}|\cdot|V|\cdot(1+|\mathscr{P}|)}$. Furthermore, $|\Sigma|$ is bounded by $|V| \cdot |\mathsf{Act}|^{|\mathscr{P}|}$. The number of states is therefore in $O(2^{|\mathscr{P}|\cdot|V|} \cdot |\mathsf{Act}|^{|\mathscr{P}|^2\cdot|V|})$. The epistemic game is therefore of exponential size w.r.t. the initial game. Note that we could reduce the bounds by using tricks like those in [9, Proposition 4.8], but this would not avoid an exponential blowup.

4 Two Applications with Publicly Visible Payoffs

While the construction of the epistemic game has transformed the computation of Nash equilibria in a concurrent game with public signal to the computation of winning strategies in a two-player zero-sum turn-based game, we cannot apply standard algorithms out-of-the-box, because the winning condition is rather complex. In the following, we present two applications of that approach in the context of publicly visible payoffs, one with Boolean payoff functions, and another with mean payoff functions. Remember that in the latter case, public visibility is required to have decidability (Theorem 1).

The epistemic game has a specific structure, which can be used for algorithmics purpose. The main outcome of a potential Nash equilibrium is given by a $\perp$-play, that is, a play visiting only epistemic states s with $s(\perp) \neq \emptyset$. There are now two types of deviations:

(i) those that are invisible to all players (except the deviator): they are tracked along the main $\perp$-play. Assuming public visibility of the payoff functions, such a deviation cannot be profitable to any of the players (the payoff of all concrete plays along that $\perp$-play coincides with the payoff of the main outcome), hence no specific punishing strategy has to be played.
(ii) those that leave the main $\perp$-play at some point, and visit only epistemic states s such that $s(\perp) = \emptyset$ from that point on: those are the deviations that need to be punished. Note nevertheless that the deviator may not precisely be known by all the players, hence punishing strategies need to take this into account. However, the set of potential deviators along a deviating play is non-increasing, and we can solve subgames with specific subsets of potential deviators separately (e.g. in a bottom-up approach). The winning objectives in those subgames will depend on the payoff functions (and will mostly be conjunctions of constraints on those functions), and also on the value of those payoff functions along the main outcome.

Using such an approach and results of [16] on generalized parity games, we obtain the following result for Boolean ω-regular payoff functions:

Theorem 3. *The constrained existence problem is in* EXPSPACE *and* EXPTIME*-hard for concurrent games with public signal and publicly visible Boolean payoff functions associated with parity conditions. The lower bound holds even for Büchi conditions and two players.*

The same approach could be used for the ordered objectives of [9], which are finite preference relations over sets of ω-regular properties. Also, we believe we can enrich the epistemic game construction and provide an algorithm to decide the constrained existence problem for Boolean ω-regular invisible payoff functions.

We have also investigated publicly visible mean payoff functions. While we could have used the same bottom-up approach as above and applied results from [12,13], we adopt an approach similar to that of [11], which consists in transforming the winning condition of Eve in $\mathcal{E}_\mathcal{G}$ into a so-called *polyhedron query* in a multi-dimensional mean-payoff game. Given such a game, a polyhedron query asks whether there exists a strategy for Eve which achieves a payoff belonging to some given polyhedron. Using this approach, we get the following result:

Theorem 4. *The constrained existence problem is in* $\mathsf{NP}^{\mathsf{NEXPTIME}}$ *(hence in* EXPSPACE*) and* EXPTIME*-hard for concurrent games with public signal and publicly visible mean payoff functions.*

5 Conclusion

In this paper, we have studied concurrent games with imperfect monitoring modelled using signals. We have given some undecidability results, even in the case of public signals, when the payoff functions are not publicly visible. We have then proposed a construction to capture single-player deviations in games with public signals, and reduced the search of Nash equilibria to the synthesis of winning strategies in a two-player turn-based games (with a rather complex winning condition though). We have applied this general framework to two classes of payoff functions, and obtained decidability results.

As further work we wish to understand better if there could be richer communication patterns which would allow representable knowledge structures for Nash equilibria and thereby the synthesis of Nash equilibria under imperfect monitoring. A source of inspiration for further work will be [28].

References

1. Alur, R., Henzinger, T., Kupferman, O.: Alternating-time temporal logic. J. ACM **49**, 672–713 (2002)
2. Berthon, R., Maubert, B., Murano, A.: Decidability results for ATL* with imperfect information and perfect recall. In: Proceedings of 16th Conference Autonomous Agents and Multiagent Systems (AAMAS 2017), pp. 1250–1258. ACM (2017)
3. Berthon, R., Maubert, B., Murano, A., Rubin, S., Vardi, M.Y..: Strategy logic with imperfect information. In: Proceedings of 32nd Annual Symposium Logic in Computer Science (LICS 2017), pp. 1–12. IEEE Computer Society Press (2017)
4. Berwanger, D., Kaiser, Ł., Puchala, B.: Perfect-information construction for coordination in games. In: Proceedings of 30th Conference on Foundations of Software Technology and Theoretical Computer Science (FSTTCS 2011), LIPIcs, vol. 13, pp. 387–398. LZI (2011)
5. Berwanger, D., Mathew, A.B.: Infinite games with finite knowledge gaps. Inf. Comput. **254**, 217–237 (2017)
6. Berwanger, D., Mathew, A.B., Van den Bogaard, M.: Hierarchical information and the synthesis of distributed strategies. Acta Informatica, 1–33. Springer, Heidelberg (2017). https://doi.org/10.1007/s00236-017-0306-5
7. Berwanger, D., Ramanujam, R.: Deviator detection under imperfect monitoring. In: Proceedings of 5th International Workshop Strategic Reasoning (SR 2017) (2017)
8. Bouyer, P.: Games on graphs with a public signal monitoring. Research report, arXiv https://arxiv.org/abs/1710.07163 (2017)
9. Bouyer, P., Brenguier, R., Markey, N., Ummels, M.: Pure Nash equilibria in concurrent games. Log. Methods Comput.. Sci. **11**(2:9) (2015). https://doi.org/10.2168/LMCS-11(2:9)2015
10. Bouyer, P., Markey, N., Stan, D.: Mixed Nash equilibria in concurrent games. In: Proceedings of 33rd Conference on Foundations of Software Technology and Theoretical Computer Science (FSTTCS 2014), LIPIcs, vol. 29, pp. 351–363. LZI (2014)
11. Brenguier, R.: Robust equilibria in mean-payoff games. In: Jacobs, B., Löding, C. (eds.) FoSSaCS 2016. LNCS, vol. 9634, pp. 217–233. Springer, Heidelberg (2016). https://doi.org/10.1007/978-3-662-49630-5_13
12. Brenguier, R., Raskin, J.-F.: Optimal values of multidimensional mean-payoff games. Research report hal-00977352, Université Libre de Bruxelles, Belgium (2014). https://hal.archives-ouvertes.fr/hal-00977352
13. Brenguier, R., Raskin, J.-F.: Pareto curves of multidimensional mean-payoff games. In: Kroening, D., Păsăreanu, C.S. (eds.) CAV 2015, Part II. LNCS, vol. 9207, pp. 251–267. Springer, Cham (2015). https://doi.org/10.1007/978-3-319-21668-3_15
14. Chatterjee, K., Doyen, L., Henzinger, T., Raskin, J.-F.: Algorithms for ω-regular games with imperfect information. Log. Methods Comput. Sci. **3**(3) (2007). https://doi.org/10.2168/LMCS-3(3:4)2007
15. Chatterjee, K., Henzinger, T., Jurdziński, M.: Games with secure equilibria. Theor. Comput. Sci. **365**(1–2), 67–82 (2006)
16. Chatterjee, K., Henzinger, T.A., Piterman, N.: Generalized parity games. In: Seidl, H. (ed.) FoSSaCS 2007. LNCS, vol. 4423, pp. 153–167. Springer, Heidelberg (2007). https://doi.org/10.1007/978-3-540-71389-0_12
17. Chatterjee, K., Majumdar, R., Jurdziński, M.: On Nash equilibria in stochastic games. In: Marcinkowski, J., Tarlecki, A. (eds.) CSL 2004. LNCS, vol. 3210, pp. 26–40. Springer, Heidelberg (2004). https://doi.org/10.1007/978-3-540-30124-0_6

18. Condurache, R., Filiot, E., Gentilini, R., Raskin, J.-F.: The complexity of rational synthesis. In: Proceedings of 43rd International Colloquium on Automata, Languages and Programming (ICALP 2016), LIPIcs, vol. 55, pp. 121:1–121:15. Leibniz-Zentrum für Informatik (2016)
19. Degorre, A., Doyen, L., Gentilini, R., Raskin, J.-F., Toruńczyk, S.: Energy and mean-payoff games with imperfect information. In: Dawar, A., Veith, H. (eds.) CSL 2010. LNCS, vol. 6247, pp. 260–274. Springer, Heidelberg (2010). https://doi.org/10.1007/978-3-642-15205-4_22
20. Dima, C., Enea, C., Guelev, D.P.: Model-checking an alternating-time temporal logic with knowledge, imperfect information, perfect recall and communicating coalitions. In: Proceedings of 1st International Symposium Games, Automata, Logics and Formal Verification (GandALF 2010), Electronic Proceedings in Theoretical Computer Science, vol. 25, pp. 103–117 (2010)
21. Dima, C., Tiplea, F.L.: Model-checking ATL under imperfect information and perfect recall semantics is undecidable. Research report arXiV, http://arxiv.org/abs/1102.4225 (2011)
22. Doyen, L., Raskin, J.-F.: Games with imperfect information: theory and algorithms. Lectures in Game Theory for Computer Scientists, pp. 185–212. Cambridge University Press (2011)
23. Henzinger, T.: Games in system design and verification. In: Proceedings of 10th Conference Theoretical Aspects of Rationality and Knowledge (TARK 2005), pp. 1–4 (2005)
24. Nash, J.F.: Equilibrium points in n-person games. Proc. Nat. Acad. Sci. U.S.A. **36**(1), 48–49 (1950)
25. Peterson, G.L., Reif, J.H.: Multiple-person alternation. In: Proceedings of 20th Annual Symposium on Foundations of Computer Science (FOCS 1979), pp. 348–363. IEEE Computer Society Press (1979)
26. Pnueli, A., Rosner, R.: Distributed reactive systems are hard to synthesize. In: Proceedings of 31st Annual Symposium on Foundations of Computer Science (FOCS 1990), pp. 746–757. IEEE Computer Society Press (1990)
27. Reif, J.H.: The complexity of two-player games of incomplete information. J. Comput. System Sciences **29**(2), 274–301 (1984)
28. Renault, J., Tomala, T.: Repeated proximity games. Int. J. Game Theory **27**(4), 539–559 (1998)
29. Saraydar, C.U., Mandayam, N.B., Goodman, D.J.: Pareto efficiency of pricing-based power control in wireless data networks. In: Proceedings of IEEE Wireless Comm. and Networking Conference (WCNC 1999), pp. 231–235. IEEE Computer Society Press (1999)
30. Thomas, W.: Infinite games and verification. In: Brinksma, E., Larsen, K.G. (eds.) CAV 2002. LNCS, vol. 2404, pp. 58–65. Springer, Heidelberg (2002). https://doi.org/10.1007/3-540-45657-0_5
31. Tomala, T.: Pure equilibria of repeated games with public observation. Int. J. Game Theory **27**(1), 93–109 (1998)
32. Ummels, M.: Rational behaviour and strategy construction in infinite multiplayer games. In: Arun-Kumar, S., Garg, N. (eds.) FSTTCS 2006. LNCS, vol. 4337, pp. 212–223. Springer, Heidelberg (2006). https://doi.org/10.1007/11944836_21
33. Ummels, M.: The complexity of Nash equilibria in infinite multiplayer games. In: Amadio, R. (ed.) FoSSaCS 2008. LNCS, vol. 4962, pp. 20–34. Springer, Heidelberg (2008). https://doi.org/10.1007/978-3-540-78499-9_3

34. Ummels, M., Wojtczak, D.: The complexity of Nash equilibria in limit-average games. In: Katoen, J.-P., König, B. (eds.) CONCUR 2011. LNCS, vol. 6901, pp. 482–496. Springer, Heidelberg (2011). https://doi.org/10.1007/978-3-642-23217-6_32
35. Ummels, M., Wojtczak, D.: The complexity of Nash equilibria in stochastic multiplayer games. Log. Methods Comput. Sci. **7**(3) (2011). https://doi.org/10.2168/LMCS-7(3:20)2011
36. van der Meyden, R., Wilke, T.: Synthesis of distributed systems from knowledge-based specifications. In: Abadi, M., de Alfaro, L. (eds.) CONCUR 2005. LNCS, vol. 3653, pp. 562–576. Springer, Heidelberg (2005). https://doi.org/10.1007/11539452_42

WQO Dichotomy for 3-Graphs

Sławomir Lasota(✉) and Radosław Piórkowski

Institute of Informatics, University of Warsaw, Warsaw, Poland
sl@mimuw.edu.pl

Abstract. We investigate data-enriched models, like Petri nets with data, where executability of a transition is conditioned by a relation between data values involved. Decidability status of various decision problems in such models may depend on the structure of data domain. According to the WQO Dichotomy Conjecture, if a data domain is homogeneous then it either exhibits a well quasi-order (in which case decidability follows by standard arguments), or essentially all the decision problems are undecidable for Petri nets over that data domain.

We confirm the conjecture for data domains being 3-graphs (graphs with 2-colored edges). On the technical level, this results is a significant step beyond known classification results for homogeneous structures.

1 Introduction

In Petri nets with data, tokens carry values from some data domain, and executability of transitions is conditioned by a relation between data values involved. One can consider *unordered data*, like in [25], i.e., an infinite data domain with the equality as the only relation; or *ordered data*, like in [21], i.e., an infinite densely totally ordered data domain; or timed data, like in timed Petri nets [1] and timed-arc Petri nets [15]. In [19] an abstract setting of Petri nets with an arbitrary fixed data domain $\mathbb{A}$ has been introduced, parametric in a relational structure $\mathbb{A}$. The setting uniformly subsumes unordered, ordered and timed data (represented by $\mathbb{A} = (\mathbb{N}, =)$, $\mathbb{A} = (\mathbb{Q}, \leq)$ and $\mathbb{A} = (\mathbb{Q}, \leq, +1)$, respectively).

Following [19], in order to enable finite presentation of Petri nets with data, and in particular to consider such models as input to algorithms, we restrict to relational structures $\mathbb{A}$ that are *homogeneous* [23] and *effective* (the formal definitions are given in Sect. 2). Certain standard decision problems (like the termination problem, the boundedness problem, or the coverability problem, jointly called from now on *standard problems*) are all decidable for Petri nets with ordered data [21] (and in consequence also for Petri nets with unordered data), as the model fits into the framework of well-structured transition systems of [11].

S. Lasota—Partially supported by the European Research Council (ERC) project Lipa under the EU Horizon 2020 research and innovation programme (grant agreement No. 683080).

R. Piórkowski—Partially supported by the Polish NCN grant 2016/21/B/ST6/01505.

C. Baier and U. Dal Lago (Eds.): FOSSACS 2018, LNCS 10803, pp. 548–564, 2018.
https://doi.org/10.1007/978-3-319-89366-2_30

Most importantly, the structure $\mathbb{A} = (\mathbb{Q}, \leq)$ of ordered data *admits well quasi-order* (WQO) in the following sense: for any WQO X, the set of finite induced substructures of $(\mathbb{Q}, \leq)$ (i.e., finite total orders) labeled by elements of X, ordered naturally by embedding, is a WQO (this is exactly Higman's lemma). Moreover, essentially the same argument can be used for any other homogeneous effective data domain which admits WQO (see [19] for details). On the other hand, for certain homogeneous effective data domains $\mathbb{A}$ the standard problems become all undecidable. In the quest for understanding the decidability borderline, the following hypothesis has been formulated in [19]:

*Conjecture 1 (*WQO *Dichotomy Coinjecture* [19]*).* For an effective homogeneous structure $\mathbb{A}$, either $\mathbb{A}$ admits WQO (in which case the standard problems are decidable for Petri nets with data $\mathbb{A}$), or all the standard problems are undecidable for Petri nets with data $\mathbb{A}$.

According to [19], the conjecture could have been equivalently stated for another data-enriched models, e.g., for finite automata with one register [2]. In this paper we consider, for the sake of presentation, only Petri nets with data. WQO Dichotomy Conjecture holds in special cases when data domains $\mathbb{A}$ are undirected or directed graphs, due to the known classifications of homogeneous graphs [6,18].

Contributions. We confirm the WQO Dichotomy Conjecture for data domains $\mathbb{A}$ being *strongly*[1] homogeneous *3-graphs.* A 3-graph is a logical structure with three irreflexive symmetric binary relations such that every pair of elements of $\mathbb{A}$ belongs to exactly one of the relations (essentially, a clique with 3-colored edges).

Our main technical contribution is a complex analysis of possible shapes of strongly homogeneous 3-graphs, constituting the heart of the proof. We believe that this is a significant step towards full classification of homogeneous 3-graphs. The classification of homogeneous structures is a well-known challenge in model theory, and has been only solved in some cases by now: for undirected graphs [18], directed graphs (the proof of Cherlin spans a book [6]), multi-partite graphs [16], and few others (the survey [23] is an excellent overview of homogeneous structures). Although the full classification of homogeneous 3-graphs was not our primary objective, we believe that our analysis significantly improves our understanding of these structures and can be helpful for classification.

Our result does not fully settle the status of the WQO Dichotomy Conjecture. Dropping the (mild) strong homogeneity assumption, as well as extending the proof to arbitrarily many symmetric binary relations, is left for future work.

Related Research. Net models similar to Petri nets with data have been continuously proposed since the 80s, including, among the others, high-level Petri nets [13], colored Petri nets [17], unordered and ordered data nets [21], ν-Petri nets [25],

[1] Strong homogeneity is a mild strengthening of homogeneity.

and constraint multiset rewriting [5,8,9]. Petri nets with data can be also considered as a reinterpretation of the classical definition of Petri nets in sets with atoms [3,4], where one allows for *orbit-finite* sets of places and transitions instead of just finite ones. The decidability and complexity of standard problems for Petri nets over various data domains has attracted a lot of attention recently, see for instance [14,21,22,24,25].

WQOs are important for their wide applicability in many areas. Studies of WQOs similar to ours, in case of graphs, have been conducted by Ding [10] and Cherlin [7]; their framework is different though, as they concentrate on subgraph ordering while we investigate *induced* subgraph (or substructure) ordering.

2 Petri Nets with Homogeneous Data

In this section we provide all necessary preliminaries. Our setting follows [19] and is parametric in the underlying logical structure $\mathbb{A}$, which constitutes a *data domain.* Here are some example data domains:

- *Equality data domain*: natural numbers with equality $\mathbb{A}_= = (\mathbb{N}, =)$. Note that any other countably infinite set could be used instead of natural numbers, as the only available relation is equality.
- *Total order data domain*: rational numbers with the standard order $\mathbb{A}_\leq = (\mathbb{Q}, \leq)$. Again, any other countably infinite dense total order without extremal elements could be used instead.
- *Nested equality data domain*: $\mathbb{A}_1 = (\mathbb{N}^2, =_1, =)$ where $=_1$ is equality on the first component: $(n, m) =_1 (n', m')$ if $n = n'$ and $m \neq m'$. Essentially, $\mathbb{A}$ is an equivalence relation with infinitely many infinite equivalence classes.

Note that two latter structures essentially extend the first one: in each case the equality is either present explicitly, or is definable. From now on, we always assume a fixed countably infinite relational structure $\mathbb{A}$ with equality over a finite vocabulary (signature) Σ.

Petri Nets with Data. Petri nets with data are exactly like classical place/transition Petri nets, except that tokens carry data values and these data values must satisfy a prescribed constraint when a transition is executed. Formally, a *Petri net with data* $\mathbb{A}$ consists of two disjoint finite sets P (places) and T (transitions), the arcs $A \subseteq P \times T \cup T \times P$, and two labelings:

- arcs are labelled by pairwise disjoint finite nonempty sets of variables;
- transitions are labelled by first-order formulas over the vocabulary Σ of $\mathbb{A}$, such that free variables of the formula labeling a transition t belong to the union of labels of the arcs incident to t.

Example 1. For illustration consider a Petri net with equality data $\mathbb{A}_=$, with two places p_1, p_2 and two transitions t_1, t_2 depicted on Fig. 1. Transition t_1 outputs two tokens with arbitrary but distinct data values onto place p_1. Transition t_2

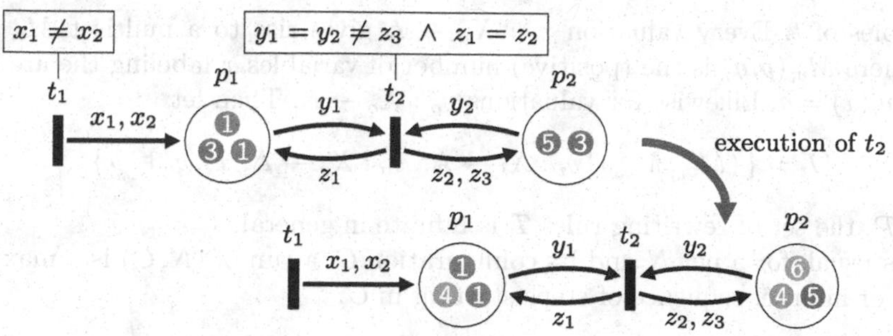

Fig. 1. A Petri net with equality data, with places $P = \{p_1, p_2\}$ and transitions $T = \{t_1, t_2\}$. In the shown configuration, t_2 can be fired: consume two tokens carrying 3, and put, e.g., token carrying 4 on p_1 and tokens carrying 4, 6 on p_2.

inputs two tokens with the same data value, say a, one from p_1 and one from p_2, and outputs 3 tokens: two tokens with arbitrary but equal data values, say b, one onto p_1 and the other onto p_2; and one token with a data value $c \neq a$ onto p_2. Note that the transition t_2 does not specify whether $b = a$, or $b = c$, or $b \neq a, c$, and therefore all three options are allowed. Variables y_1, y_2 can be considered as input variables of t_2, while variables z_1, z_2, z_3 can be considered as output ones; analogously, t_1 has no input variables, and two output ones x_1, x_2.

The formal semantics of Petri nets with data is given by translation to multiset rewriting. Given a set X, finite or infinite, a finite multiset over X is a finite (possibly empty) partial function from X to positive integers. In the sequel let $\mathcal{M}(X)$ stand for the set of all finite multisets over X. A *multiset rewriting system* $(\mathcal{P}, \mathcal{T})$ consists of a set $\mathcal{P}$ together with a set of rewriting rules:

$$\mathcal{T} \subseteq \mathcal{M}(\mathcal{P}) \times \mathcal{M}(\mathcal{P}).$$

Configurations $C \in \mathcal{M}(\mathcal{P})$ are finite multisets over $\mathcal{P}$, and the step relation $\longrightarrow$ between configurations is defined as follows: for every $(I, O) \in \mathcal{T}$ and every $M \in \mathcal{M}(\mathcal{P})$, there is the step (+ stands for multiset union)

$$M + I \longrightarrow M + O.$$

For instance, a classical Petri net induces a multiset rewriting system where $\mathcal{P}$ is the set of places, and $\mathcal{T}$ is essentially the set of transitions, both $\mathcal{P}$ and $\mathcal{T}$ being finite. Configurations correspond to markings.

A Petri net with data $\mathbb{A}$ induces a multiset rewriting system $(\mathcal{P}, \mathcal{T})$, where $\mathcal{P} = P \times \mathbb{A}$ and thus is infinite. Configurations are finite multisets over $P \times \mathbb{A}$ (cf. a configuration depicted in Fig. 1). The rewriting rules $\mathcal{T}$ are defined as

$$\mathcal{T} = \bigcup_{t \in T} \mathcal{T}_t,$$

where the relation $\mathcal{T}_t \subseteq \mathcal{M}(\mathcal{P}) \times \mathcal{M}(\mathcal{P})$ is defined as follows: Let ϕ denote the formula labeling the transition t, and let X_i, X_o be the sets of input and output variables of t. Every valuation $v_i : X_i \to \mathbb{A}$ gives rise to a multiset M_{v_i} over $\mathcal{P}$, where $M_{v_i}(p, a)$ is the (positive) number of variables x labeling the arc (p, t) with $v_i(x) = a$. Likewise for valuations $v_o : X_o \to \mathbb{A}$. Then let

$$\mathcal{T}_t = \{ (M_{v_i}, M_{v_o}) \mid v_i : X_i \to \mathbb{A},\ v_o : X_o \to \mathbb{A},\ v_i, v_o \models \phi \} .$$

Like $\mathcal{P}$, the set of rewriting rules $\mathcal{T}$ is infinite in general.

As usual, for a net N and its configuration C, a run of (N, C) is a maximal, finite or infinite, sequence of steps starting in C.

Remark 1. As for classical Petri nets, an essentially equivalent definition can be given in terms of vector addition systems (such a variant has been used in [14] for equality data). Petri nets with equality data are equivalent to (even if defined differently than) unordered data Petri nets of [21], and Petri nets with total ordered data are equivalent to ordered data Petri nets of [21].

Effective Homogeneous Structures. For two relational Σ-structures $\mathcal{A}$ and $\mathcal{B}$ we say that $\mathcal{A}$ *embeds* in $\mathcal{B}$, written $\mathcal{A} \trianglelefteq \mathcal{B}$, if $\mathcal{A}$ is isomorphic to an induced substructure of $\mathcal{B}$, i.e., to a structure obtained by restricting $\mathcal{B}$ to a subset of its domain. This is witnessed by an injective function[2] $h : \mathcal{A} \to \mathcal{B}$, which we call *embedding*. We write $\text{AGE}(\mathbb{A}) = \{ \mathcal{A} \text{ a finite structure} \mid \mathcal{A} \trianglelefteq \mathbb{A} \}$ for the class of all finite structures that embed into $\mathbb{A}$, and call it *the age of* $\mathbb{A}$.

Homogeneous structures are defined through their automorphisms: $\mathbb{A}$ is homogeneous if every isomorphism of two its finite induced substructures extends to an automorphism of $\mathbb{A}$. In the sequel we will also need an equivalent definition using amalgamation. An *amalgamation instance* consists of three structures $\mathcal{A}, \mathcal{B}_1, \mathcal{B}_2 \in \text{AGE}(\mathbb{A})$ and two embeddings $h_1 : \mathcal{A} \to \mathcal{B}_1$ and $h_2 : \mathcal{A} \to \mathcal{B}_2$. A solution of such instance is a structure $\mathcal{C} \in \text{AGE}(\mathbb{A})$ and two embeddings $g_1 : \mathcal{B}_1 \to \mathcal{C}$ and $g_2 : \mathcal{B}_2 \to \mathcal{C}$ such that $g_1 \circ h_1 = g_2 \circ h_2$ (we refer the reader to [12] for further details). Intuitively, $\mathcal{C}$ represents 'gluing' of $\mathcal{B}_1$ and $\mathcal{B}_2$ along the partial bijection $h_2 \circ ({h_1}^{-1})$. In this paper we will restrict ourselves to *singleton* amalgamation instances, where only one element of $\mathcal{B}_1$ is outside of $h_1(\mathcal{A})$, and likewise for $\mathcal{B}_2$.

An example singleton amalgamation instance is shown on the right, where the graph $\mathcal{A}$ consists of the single edge connecting two middle black nodes, $\mathcal{B}_1$ is the left triangle, and $\mathcal{B}_2$ the right one. The dashed line represents an edge that may (but does not have to) appear in a solution. $\mathbb{A}$ is homogeneous if, and only if every amalgamation instance has a solution; in such case we say that $\text{AGE}(\mathbb{A})$ has the *amalgamation property*. See [23] for further details.

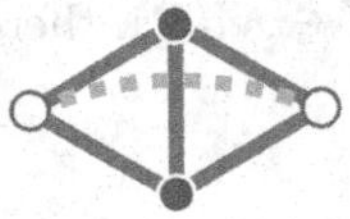

A solution $\mathcal{C}$ necessarily satisfies $g_1(h_1(\mathcal{A})) = g_2(h_2(\mathcal{A})) \subseteq g_1(\mathcal{B}_1) \cap g_2(\mathcal{B}_2)$; a solution is *strong* if $g_1(h_1(\mathcal{A})) = g_1(\mathcal{B}_1) \cap g_2(\mathcal{B}_2)$. Intuitively, this forbids additional gluing of $\mathcal{B}_1$ and $\mathcal{B}_2$ not specified by the partial bijection $h_2 \circ ({h_1}^{-1})$. If every amalgamation instance has a strong solution we call $\mathbb{A}$ *strongly homogeneous*. This is a mild restriction, as homogeneous structures are typically strongly homogeneous.

[2] We deliberately do not distinguish a structure $\mathcal{A}$ from its domain set.

The equality, nested equality, and total order data domains are strongly homogeneous structures. For instance, in the latter case finite induced substructures are just finite total orders, which satisfy the strong amalgamation property. Many other natural classes of structures have the amalgamation property: finite graphs, finite directed graphs, finite partial orders, finite tournaments, etc. Each of these classes is the age of a strongly homogeneous relational structure, namely the *universal graph* (called also random graph), the universal directed graph, the universal partial order, the universal tournament, respectively. Examples of homogeneous structures abound [23].

Homogeneous structures admit quantifier elimination: every first-order formula is equivalent to (i.e., defines the same set as) a quantifier-free one [23]. Thus it is safe to assume that formulas labeling transitions are quantifier-free.

Admitting WQO. A *well quasi-order* (WQO) is a well-founded quasi-order with no infinite antichains. For instance, finite multisets $\mathcal{M}(P)$ over a finite set P, ordered by multiset inclusion $\sqsubseteq$, are a WQO. Another example is the embedding quasi-order $\trianglelefteq$ in $\text{AGE}(\mathbb{A}_{\leq})$ (= all finite total orders) isomorphic to the ordering of natural numbers. Finally, the embedding quasi-order in $\text{AGE}(\mathbb{A})$ can be lifted from finite structures to finite structures *labeled* by elements of some ordered set $(X, \leq)$: for two such labeled structures $a : \mathcal{A} \to X$ and $b : \mathcal{B} \to X$ we define $a \trianglelefteq_X b$ if some embedding $h : \mathcal{A} \to \mathcal{B}$ satisfies $a(x) \leq b(h(x))$ for every $x \in \mathcal{A}$. We say that $\mathbb{A}$ *admits* WQO when for every WQO $(X, \leq)$, the lifted embedding order $\trianglelefteq_X$ is a WQO too. For instance, $\mathbb{A}_{\leq}$ admits WQO by Higman's lemma. The WQO Dichotomy Conjecture for homogeneous undirected (and also directed) graphs is easily shown by inspection of the classifications thereof [6,18]:

Theorem 1. *A homogeneous graph $\mathbb{A}$ either admits* WQO, *or all standard problems are undecidable for Petri nets with data $\mathbb{A}$.*

Note the natural correspondence between configurations of a Petri net with data $\mathbb{A}$, and structures $\mathcal{A} \in \text{AGE}(\mathbb{A})$ labeled by finite multisets over the set P of places:

$$\mathcal{M}(P \times \mathbb{A}) \quad \equiv \quad \{ m : \mathcal{A} \to \mathcal{M}(P) \mid \mathcal{A} \in \text{AGE}(\mathbb{A}) \} .$$

Thus the lifted embedding quasi-order $\trianglelefteq_{\mathcal{M}(P)}$ is an order on configurations.

Standard Decision Problems. A Petri net with data N can be finitely represented by finite sets P, T, A and appropriate labelings with variables and formulas. Due to the homogeneity of $\mathbb{A}$, a configuration C can be represented (up to automorphism of $\mathbb{A}$) by a structure $\mathcal{A} \in \text{AGE}(A)$ labeled by $\mathcal{M}(P)$. We can thus consider the classical decision problems that input Petri nets with data $\mathbb{A}$, like the *termination problem*: does a given (N, C) have only finite runs? The data domain is considered as a parameter, and hence itself does not constitute part of input. Another classical problem is the *place non-emptiness problem* (markability): given (N, C) and a place p of N, does (N, C) admit a run that puts at least one token on place p? One can also define the appropriate variants of the coverability problem (equivalent to the place non-emptiness problem), the boundedness problem, the evitability problem, etc. (see [19] for details). All the decision problems mentioned above we jointly call *standard problems.*

A Σ-structure $\mathbb{A}$ is called *effective* if the following *age problem* for $\mathbb{A}$ is decidable: given a finite Σ-structure $\mathcal{A}$, decide whether $\mathcal{A} \trianglelefteq \mathbb{A}$. If $\mathbb{A}$ admits WQO then application of the framework of well-structured transition systems [11] to the lifted embedding order $\trianglelefteq_{\mathcal{M}(P)}$ yields:

Theorem 2 ([19]). *If an effective homogeneous structure* $\mathbb{A}$ *admits* WQO *then all the standard problems are decidable for Petri nets with data* $\mathbb{A}$.

3 Results

A 3-graph $\mathbb{G} = (V, C_1, C_2, C_3)$ consists of a set V and three irreflexive symmetric binary relations $C_1, C_2, C_3 \subseteq V^2$ such that every pair of distinct elements of V belongs to exactly one of the three relations. In the sequel we treat a 3-graph as a clique with 3-colored edges. Any graph, including $\mathbb{A}_=$ and $\mathbb{A}_1$, can be seen as a 3-graph. Our main result confirms the WQO Dichotomy Conjecture for strongly homogeneous 3-graphs:

Theorem 3. *An effective strongly homogeneous 3-graph* $\mathbb{G}$ *either admits* WQO, *or all standard problems are undecidable for Petri nets with data* $\mathbb{G}$.

The core technical result of the paper is Theorem 4 below. A *path* is a finite graph with nodes $\{v_1, \ldots, v_n\}$ whose only edges are pairs $\{v_i, v_{i+1}\}$. The nodes v_1, v_n are *ends* of the path, and n is its length.

Theorem 4. *A strongly homogeneous 3-graph* $\mathbb{G}$ *either admits* WQO, *or for some* $i, j \in \{1, 2, 3\}$ *(not necessarily distinct) the graph* $(V, C_i \cup C_j)$ *contains arbitrarily long paths as induced subgraphs.*

In the rest of the paper we concentrate solely on (parts of) the proof of Theorem 4. The omitted parts, and well as the proof that Theorem 4 implies Theorem 3, are to be found in the full version of this paper [20].

Example 2. For a quasi-order $(X, \leq)$, the multiset inclusion is defined as follows for $m, m' \in \mathcal{M}(X)$: m' is included in m if m' is obtained from m by a sequence of operations, where each operation either removes some element, or replaces some element by a smaller one wrt. $\leq$. The structure $\mathbb{A}_= = (\mathbb{N}, =)$ admits WQO. Indeed, AGE($\mathbb{A}_=$) contains just finite pure sets, thus $\trianglelefteq_X$ is quasi-order-isomorphic to the multiset inclusion on $\mathcal{M}(X)$, and is therefore a WQO whenever the underlying quasi-order $(X, \leq)$ is. Similarly, $\mathbb{A}_1 = (\mathbb{N}^2, =_1, =)$ also admits WQO, as $\trianglelefteq_X$ is quasi-order-isomorphic to the multiset inclusion on $\mathcal{M}(\mathcal{M}(X))$.

On the other hand, consider a 3-graph $(\mathbb{N}^2, =_1, =_2, \neq_{12})$ where $=_2$ is symmetric to $=_1$ and $(n, m) \neq_{12} (n', m')$ if $n \neq n'$ and $m \neq m'$. It refines $\mathbb{A}_1$ and does not admit WQO. Indeed, in agreement with Theorem 4, the graph $(\mathbb{N}^2, =_1 \cup =_2)$ contains arbitrarily long paths of the shape presented on the right, where the two colors depict $=_1$ and $=_2$, respectively, and lack of color corresponds to $\neq_{12}$. Note that $(\mathbb{N}^2, =_1, =_2, \neq_{12})$ is homogeneous but not strongly so.

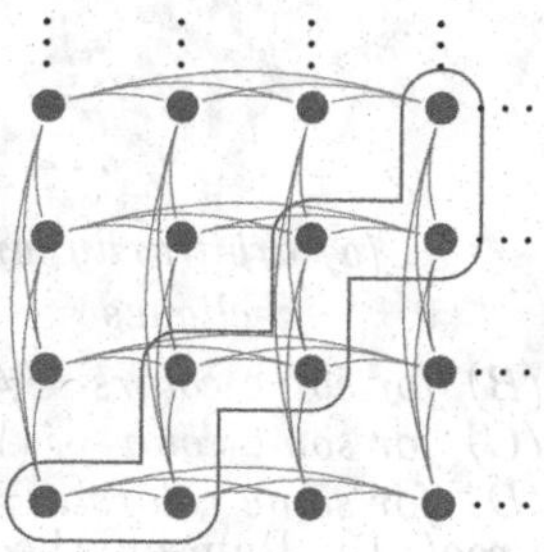

4 Proof of Theorem 4

From now on we consider a fixed 3-graph $\mathbb{G} = (V, C_1, C_2, C_3)$ as data domain, assuming $\mathbb{G}$ to be countably infinite and strongly homogeneous. We treat $\mathbb{G}$ as a clique with 3-colored edges: we call C_1, C_2 and C_3 *colors* and put *Colors* $= \{C_1, C_2, C_3\} \subset \mathcal{P}(V \times V)$. To denote individual colors from this set, we will use variables $\mathrm{a}, \mathrm{b}, \mathrm{c}$ and $\mathrm{x}, \mathrm{y}, \mathrm{z}$. A path in the graph $(V, \mathrm{a} \cup \mathrm{b})$ we call ab*-path* (ab $\in$ *Colors*); for simplicity, we will write a*-path* instead of aa-path. Likewise we speak of ab-cliques, a-cliques, ab-cycles, etc. A *triangle* $\triangle\mathrm{abc}$ is a 3-clique with edges colored by a, b, c. (Note that $\triangle\mathrm{abc} = \triangle\mathrm{bca} = \triangle\mathrm{cba}$).

Sketch of the Proof. The Lemma 1 below states that any 3-graph $\mathbb{G}$ has to meet one of the four listed cases. It splits the proof into four separate paths:

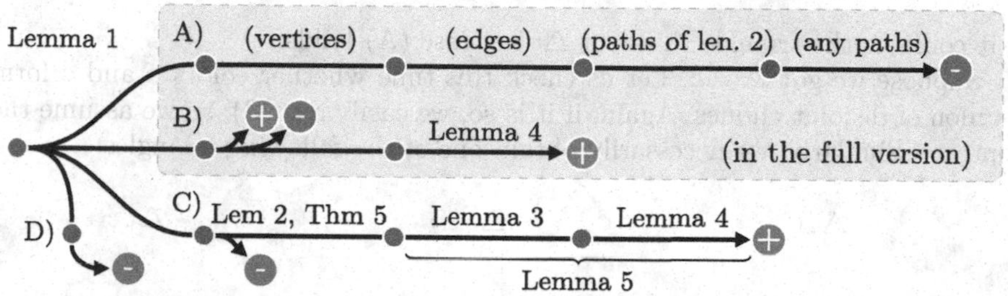

– $\mathbb{G}$ embeds arbitrarily long paths – $\mathbb{G}$ admits WQO

We present in detail only one of the three nontrivial paths – one corresponding to case (C). Cases (A) and (B) are treated in the full version [20]. Case (A) constitutes the most difficult part of the proof and involves a complex and delicate analysis of consequences of the amalgamation property. It consists of four step that deduce extension of the assumed induced substructures by individual vertices, individual edges, paths of length 2, resp., culminating in derivation of arbitrarily long paths. Thus in case (A) only the second condition of Theorem 4 is possible, while in the other two cases both conditions of Theorem 4 may hold true.

Lemma 1. *Every homogeneous 3-graph* $\mathbb{G} = (V, C_1, C_2, C_3)$ *satisfies one of the following conditions:*

(A) for some color $\mathrm{c} \in$ *Colors*, $\mathbb{G}$ *contains the following induced substructures:*

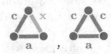

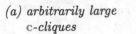

(a) arbitrarily large c*-cliques* *(b) two triangles:* $\triangle$axc *and* $\triangle$acc *for some colors* a, x *different than* c

(B) *for some colors* $x \neq y$, $(V, x \cup y)$ *is a union of disjoint cliques,*
(C) *for some color* x, (V, x) *is a union of finitely many disjoint infinite cliques,*
(D) *for some colors* $x \neq y$, $(V, x \cup y)$ *contains arbitrarily long paths.*

Proof. By Ramsey theorem, $\mathbb{G}$ contains an arbitrarily large monochromatic cliques. Let us state a bit stronger requirement:

Condition ♠: For some $a, c \in$ *Colors*, $\mathbb{G}$ contains arbitrarily large c-cliques and a triangle $\triangle$acc with exactly two c-edges ($a \neq c$).

Consider two cases, depending on whether the condition ♠ is satisfied or not.

Case 1°**.** Assume that $\mathbb{G}$ contains both arbitrarily large c-cliques and a triangle $\triangle$acc for some $a, c \in$ *Colors*. Let b be the third, remaining color. Our goal will be to show that either (A) or (B) holds.

If the graph $(V, a \cup b)$ is a disjoint sum of cliques, we immediately obtain (B). Suppose the contrary. We get that $\mathbb{G}$ has to contain one of the three possible counterexamples for transitivity of relation $a \cup b$:

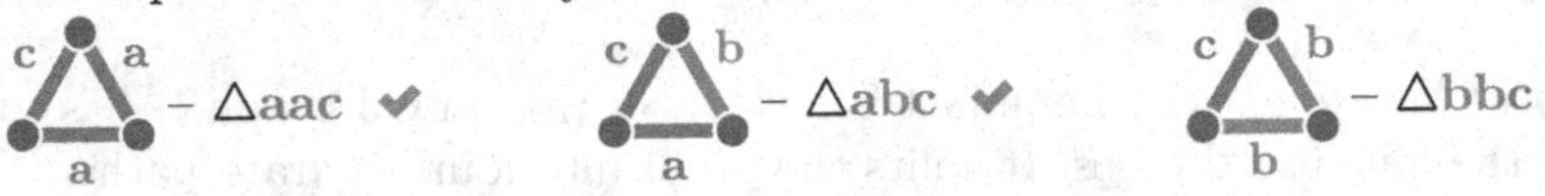

If it contains the triangle $\triangle$aac or $\triangle$abc, case (A) holds.

Suppose we got $\triangle$bbc. Let us check this time whether colors a and c form a union of disjoint cliques. Again, if it is so, we easily get (B), so we assume the contrary. Similarly, we necessarily obtain one of the following triangles:

This time case (A) also holds for two out of the three triangles above:

- for $\triangle$acb, because together with subgraphs resulting from assumption ♠ (i.e. with triangle $\triangle$acc and the c-cliques) we get all graphs required by (A).
- for $\triangle$ccb paired with the triangle $\triangle$bbc we just obtained, using color b appearing in those triangles in place of a in condition (A).

It only remains to consider the situation when we got $\triangle$aab. We use it together with previously obtained triangle $\triangle$bbc to build the following instance of singleton amalgamation:

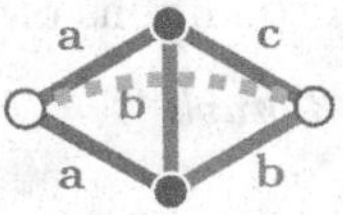

Depending on the color of the dashed edge, in the solution we get one of the following triangles:

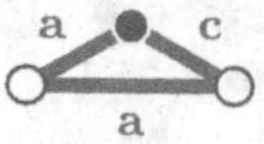

– Δaac

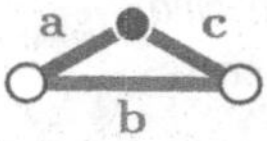

– Δabc

– Δabc

and each one alone completes the requirements of (A). This closes case 1°.

Case 2°**.** Suppose condition ♠ is false. Remind that $\mathbb{G}$ contains arbitrarily large c-cliques for some $c \in \mathbb{G}$. Since ♠ does not hold, the graph does not contain a triangle Δcca – in other words, the color c appears only within cliques. We conclude that (V, c) is a union of disjoint cliques. Clearly at least one of such cliques has to be infinite. By homogeneity we get that all the cliques in (V, c) have to be infinite. Now our target is to show that either (C) or (D) holds.

The case (C) is fulfilled when there are only finitely many c-cliques. Let us assume the contrary. In each of the c-cliques we chose one vertex. Edges between the chosen vertices form an infinite ab-clique K. Using Ramsey theorem again, we conclude that in K one of the colors a, b forms arbitrarily large monochromatic cliques. W.l.o.g. suppose that this is color b.

If the graph $\mathbb{G}$ contained Δybb for some $y \neq b$, then the assumptions of ♠ would be met, leading to a contradiction. Therefore we conclude that (V, b) is a union of disjoint infinite b-cliques.

When there are only finitely many b-cliques, condition (C) is fulfilled. Otherwise we know that $\mathbb{G}$ is a union of infinitely many x-cliques for both $x = c$ and $x = b$. Using homogenity, it is easy to show that then every pair of differently colored cliques has *exactly one* common vertex, so the graph $\mathbb{G}$ takes the form as depicted in Example 2. A graph of such form contains arbitrarily long bc-path, so the requirements of (D) are met. □

4.1 Case (C)

Let c be the color that satisfies condition (C), and a, b — the remaining two colors. In this section we often treat $\mathbb{G}$ as the k-partite graph $(V, a \cup b)$ (for some $k \in \mathbb{N}$): k cliques of color c allow to distinguish k groups of vertices $V_1 \cup V_2 \cup \dots \cup V_k = V$ (from now on we will refer to them as layers). The remaining two colors can be interpreted as existence (a) and nonexistence (b) of edges between these groups.

Remark ★*:* We observe that the special color c between vertices within each layer V_i ensures that the automorphisms of $\mathbb{G}$ will not 'mix' those layers: when two vertices u, v belong to a common layer V_i, then their images $f(u), f(v)$ will also belong to some common layer V_j, no matter what automorphism $f \in \mathrm{Aut}(\mathbb{G})$ we choose. Obviously, the automorphisms can switch positions of whole layers, e.g. move all vertices from V_i to some V_j and vice versa—in this respect the layers are undistinguishable.

Lemma 2. *For every $i, j \in \{1, 2, \ldots, k\}$ and $a \in Colors$ $(a \neq c)$ the bipartite graph $\mathbb{G}_{i,j} = (V_i \cup V_j, a \cap (V_i \cup V_j)^2, V_i, V_j)$ (with two distinguishable sides V_i, V_j) is homogeneous.*

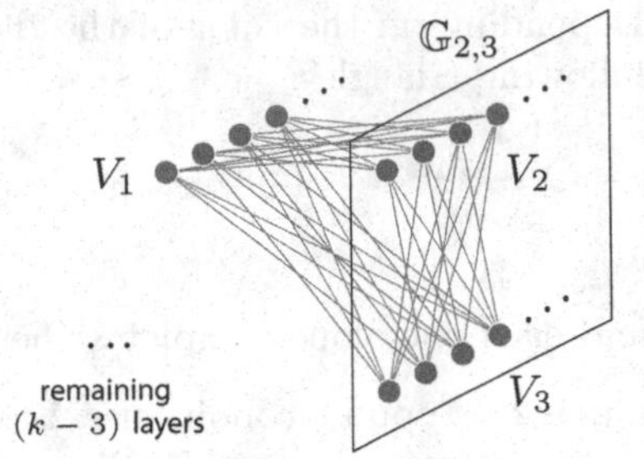

The vertex sets V_i and V_j are used here as unary relations that allow to tell the two layers of $\mathbb{G}_{i,j}$ (sides of $\mathbb{G}_{i,j}$) apart. An example is shown on the right, with three layers V_1, V_2 and V_3, and three bipartite graphs $\mathbb{G}_{1,2}$, $\mathbb{G}_{2,3}$ and $\mathbb{G}_{1,3}$.

Proof. Fix $\mathbb{G}_{i,j}$ a bipartite graph. To prove its homogeneity we have to show that each isomorphism of two of its finite induced subgraphs may be extended to some automorphism of $\mathbb{G}_{i,j}$. Let us then take some given automorphism $f : G_1 \to G_2$ for some finite induced subgraphs G_1, G_2 of $\mathbb{G}_{i,j}$. It is easy to extend it to a full automorphism when it 'touches' both layers of $G_{i,j}$, i.e.:

$$V(G_1) \cap V_i \neq \emptyset \ \wedge \ V(G_1) \cap V_j \neq \emptyset$$

where $V(G_1)$ is the set of vertices of G_1. In this case, by homogeneity of $\mathbb{G}$, we construct a full automorphism $f' : \mathbb{G} \to \mathbb{G}$, which extends f. It is easy to see that in this case f' has to fix the layers V_i and V_j, and hence f' restricted to the graph $\mathbb{G}_{i,j}$ is a correct automorphism of this graph.

Things get more complicated when f operates only on some single layer of $\mathbb{G}_{i,j}$. W.l.o.g. suppose that it 'touches' only V_i, so $V(G_1) \cap V_j = \emptyset$. Now the above construction will not work out of the box—if we were unlucky, the automorphism of $\mathbb{G}$ we get by homogeneity moves the whole layer V_j to some V_n located 'outside' the graph $\mathbb{G}_{i,j}$ $(n \notin \{i, j\})$.

It will be handy to make the following observation: when f 'touches' only V_i we may assume that $V(G_1) \cap V(G_2) = \emptyset$. Indeed, every function $g : G_1 \to G_2$ that violates this condition may be decomposed as $g = f_2 \circ f_1$ for some f_1, f_2:

$$G_1 \xrightarrow{f_1} H \xrightarrow{f_2} G_2$$

such that H is disjoint both with G_1 and with G_2.

Now, let $N = |V(G_1)| = |V(G_2)|$ be the size of the domain of isomorphism f. Let us take an arbitrary infinite family $(S_n)_{n \in \mathbb{N}}$ of subgraphs of $\mathbb{G}$ with disjoint vertex sets, such that the following conditions are met:

- $|V(S_n) \cap V_m| = 1$ for $m \neq i$ (and this single vertex will be denoted as $v_m^{(n)}$),
- $|V(S_n) \cap V_i| = N$ (denote these vertices as $s_1^{(n)}, s_2^{(n)}, s_3^{(n)}, \ldots, s_N^{(n)}$).

We define a *connection type* of a layer V_i with V_m in the graph S_n as the N-element sequence of colors of edges from the list bellow:

$$(\{s_1^{(n)}, v_m^{(n)}\}, \{s_2^{(n)}, v_m^{(n)}\}, \ldots, \{s_N^{(n)}, v_m^{(n)}\})$$

E.g. in the graph bellow, the connection type of layer $V_i = V_3$ with V_1 is abba, and with V_2 — aaba (remembering that b is treated as lack of an edge):

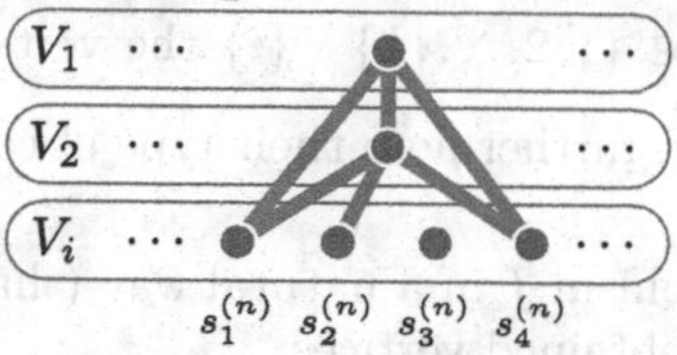

Furthermore, we define the type of graph S_n to be the sequence of types arising between V_i and other layers plus the list of edge-colors between all pairs of vertices $v_\bullet^{(n)}$ (enumerated in some consistent way). As there are only finitely many such types, by pigeonhole principle there exists a pair of graphs S_a and S_b with the same type.

Let us fix some order on vertices of G_1: $V(G_1) = \{g_1, g_2, \ldots, g_N\}$. Let h be the partial isomorphism that moves the vertices as follows:

$$s_1^{(a)} \to g_1 \qquad s_1^{(b)} \to f(g_1)$$
$$\cdots \qquad \cdots$$
$$s_N^{(a)} \to g_N \qquad s_N^{(b)} \to f(g_N)$$

By homogeneity, it has to extend to a full automorphism $h' \in \mathrm{Aut}(\mathbb{G})$. In particular, in the neighbourhood of G_1 and G_2 there will be images of all vertices $v_\bullet^{(\alpha)}$ of graphs S_a and S_b:

$$h'\left(v_1^{(\alpha)}\right), h'\left(v_2^{(\alpha)}\right), \ldots, h'\left(v_{i-1}^{(\alpha)}\right), h'\left(v_{i+1}^{(\alpha)}\right), \ldots, h'\left(v_k^{(\alpha)}\right)$$

(for α in $\{a, b\}$). What follows is that G_1 with added vertices $h'(v_\bullet^{(a)})$ has the same type as G_2 with $h'(v_\bullet^{(b)})$ respectively (that type may differ from the type of S_a and S_b though!). It is best illustrated on a picture:

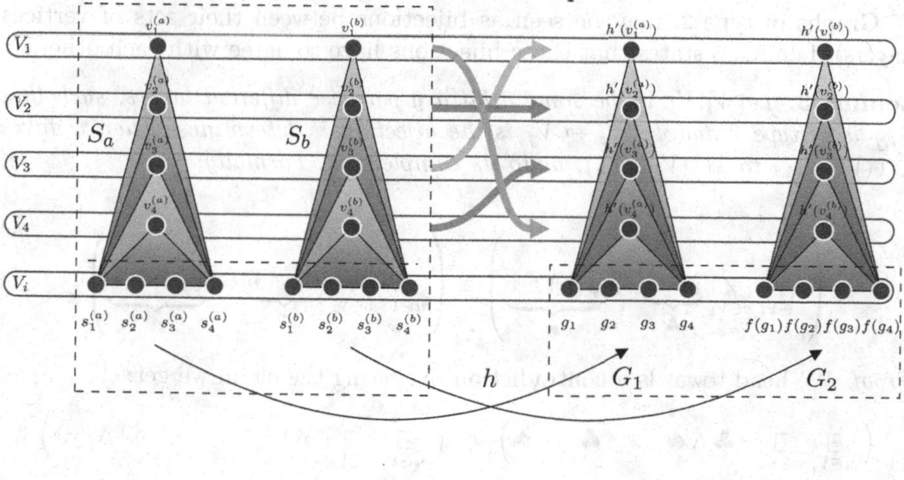

Above, the colored triangles represent the types of connections. The order of those types may get permuted when applying h', but still—in line with the remark ★ — for each $\beta \in \{1, 2, \ldots, k\} \setminus \{i\}$ the vertex $h'\left(v_\beta^{(a)}\right)$ must stay in the same layer as $h'\left(v_\beta^{(b)}\right)$, furthermore their type of connection with layer V_i is preserved.

Extending the isomorphism f in a natural way (thanks to the compatibility of types) on those newly obtained vertices:

$$h'\left(v_\bullet^{(a)}\right) \xrightarrow{\ f\ } h'\left(v_\bullet^{(b)}\right)$$

we get an isomorphism that this time 'operates' on all layers $V_\bullet$. If we now extend it to an automorphism of the whole $\mathbb{G}$, we will get a function that fixes all layers $V_\bullet$. This function may be safely restricted to $V_i \cup V_j$, staying a correct automorphism of our initial bipartite graph $\mathbb{G}_{i,j}$, which completes the proof. □

We are going to apply to graphs $\mathbb{G}_{i,j}$ the following classification result:

Theorem 5 ([16]). *A countably infinite homogeneous bipartite graph (with distinguishable sides) is either empty, or full, or a perfect matching, or the complement of a perfect matching, or a* universal *graph.*

From our point of view, all we need to know about the universal graph is that it contains arbitrarily long paths which – translated to our notation – would mean that $\mathbb{G}_{i,j}$ contains arbitrarily long a-paths. Therefore in our further considerations we assume that $\mathbb{G}_{i,j}$ is not universal which, in our notation, leaves two types of $\mathbb{G}_{i,j}$:

1. all edges of $\mathbb{G}_{i,j}$ have the same color $\mathrm{x} \in \{\mathrm{a}, \mathrm{b}\}$, i.e. $\mathbb{G}_{i,j}$ is a full or empty bipartite graph,
2. one of the colors $\mathrm{x} \in \{\mathrm{a}, \mathrm{b}\}$ forms a perfect matching in $\mathbb{G}_{i,j}$, the second one ($\mathrm{y} \neq \mathrm{x}$) is then the complement of this matching.

Graphs of type 2. may be seen as bijections between their sets of vertices (layers). Lemma 3 states that those bijections have to agree with each other.

Lemma 3. *Let V_i, V_j, V_k be some arbitrary pairwise different layers, such that $\mathbb{G}_{i,j}$ is of type 2 and $\psi : V_i \to V_j$ is the bijection it determines. Then ψ takes $\mathrm{a} \cap (V_i \cup V_k)$ to $\mathrm{a} \cap (V_j \cup V_k)$, or to its complement. Formally:*

$$\left(\underset{u \in V_i}{\forall} \ \underset{v \in V_k}{\forall} \ \underbrace{u \mathrel{\mathrm{a}} v}_{\clubsuit} \Leftrightarrow \underbrace{\psi(u) \mathrel{\mathrm{a}} v}_{\spadesuit} \right) \vee \left(\underset{u \in V_i}{\forall} \ \underset{v \in V_k}{\forall} \ \underbrace{\neg u \mathrel{\mathrm{a}} v}_{\heartsuit} \Leftrightarrow \underbrace{\psi(u) \mathrel{\mathrm{a}} v}_{\diamondsuit} \right)$$

Proof. We head towards a contradiction. Negating the claim we get:

$$\left(\underset{u \in V_i}{\exists} \ \underset{v \in V_k}{\exists} \ \neg\clubsuit \wedge \spadesuit \ \vee \ \clubsuit \wedge \neg\spadesuit \right) \wedge \left(\underset{u \in V_i}{\exists} \ \underset{v \in V_k}{\exists} \ \neg\heartsuit \wedge \diamondsuit \ \vee \ \heartsuit \wedge \neg\diamondsuit \right)$$

which leads to four cases with similar proofs. We will consider one of them (corresponding to $\neg\heartsuit \wedge \diamondsuit$ and $\clubsuit \wedge \neg\spadesuit$) and omit the other. Let us then assume that there exist $x, x' \in V_i$ and $y, y' \in V_k$ such that:

$$x \, \mathrm{a} \, y \;\wedge\; x' \, \mathrm{a} \, y' \;\wedge\; \psi(x) \, \mathrm{a} \, y \;\wedge\; \neg\psi(x') \, \mathrm{a} \, y'.$$

Let g be a partial isomorphism of the form $g = \{x \to x', y \to y'\}$. By homogeneity of $\mathbb{G}$, there is some full automorphism $g' \in \mathrm{Aut}(\mathbb{G})$ extending g. If additionally we were able to force g to fix the layer V_j, we would be almost done. Let us try to achieve that property.

For that purpose, in V_j we choose a vertex v such that:

I. $v \notin \psi(\{x, x'\})$,
II. if $\mathbb{G}_{j,k}$ is a graph of type 2. defining a bijection $\phi : V_k \to V_j$, then also $v \notin \phi(\{y, y'\})$.

Clearly such vertex must exist – two above conditions exclude at most 4 different vertices from the infinite set of candidates. The function g extended with $v \xrightarrow{g} v$ stays a correct isomorphism, because:

- in $\mathbb{G}_{i,j}$ by definition of isomorphism we need the edges $\{x, v\}$ and $\{g(x), g(v)\}$ to be equally colored, and, in fact, they are. We get this thanks to the condition I.: x is connected with all vertices from $V_j \setminus \{\psi(x)\}$ by x-edges, $\mathrm{x} \in \{\mathrm{a}, \mathrm{b}\}$. We similarly handle x'.
- in turn in $\mathbb{G}_{j,k}$ — if it is a graph of type 1, the needed equality of colors of edges $\{y, v\}$ and $\{g(y), g(v)\}$ trivially holds. If it is a graph of type 2, the equality of colors is derived similarly as in $\mathbb{G}_{i,j}$, using the condition II.

Presence of the vertex v ensures that layer V_j is preserved by the full automorphism $g' \in \mathrm{Aut}(\mathbb{G})$ we get by homogeneity.

Since $\mathbb{G}_{i,j}$ is of type 2, the vertex $\psi(x')$ is the only possible choice for the image of $\psi(x)$ under g' — this is the only vertex x' is connected to by an appropriately colored edge. Because g' is an automorphism, we get that $\psi(x')$ a y', which leads us to the contradiction. □

From the lemma we have just proved one easily derives the following corollary:

Corollary 1. *The following relation $\equiv$ on layers is transitive:*

$$V_i \equiv V_j \Leftrightarrow \text{the graph } \mathbb{G}_{i,j} \text{ is of type 2.}$$

Furthermore, if $V_i \equiv V_j$ and $V_j \equiv V_k$ then $f_{j,k} \circ f_{i,j} = f_{i,k}$, where $f_{i,j}, f_{i,k}, f_{j,k}$ are the bijections determined by graphs $\mathbb{G}_{i,j}, \mathbb{G}_{i,k}$ and $\mathbb{G}_{j,k}$.

In Lemma 5 below, which is the last step of the proof of case (C), we will apply the following fact:

Lemma 4. *Consider a homogeneous 3-graph $\mathbb{G}$ and a partition of its vertex set $V = \bigcup_{n \in \mathbb{N}} U_n$ into sets $U_\bullet$ of equal finite cardinality. Suppose further that for every $n \in \mathbb{N}$, there is an automorphism π_n of $\mathbb{G}$ that swaps U_0 with U_n and is identity elsewhere. Then $\mathbb{G}$ admits* WQO.

Proof. Let $\mathbb{G} = (V, \mathrm{a}, \mathrm{b}, \mathrm{c})$ be a 3-graph. Define for $u \in U_0$ the sets $V_u \subseteq V$, which we call *layers*:

$$V_u = \{ \pi_n(u) \mid n \in \mathbb{N} \}.$$

We will prove that the structure $\mathbb{G}' = (V, \mathrm{a}, \mathrm{b}, \mathrm{c}, (V_u)_{u \in U_0})$ admits WQO. This will imply that $\mathbb{G}$ admits WQO as well; indeed, compared to $\mathbb{G}$, structure $\mathbb{G}'$ is equipped with additional unary relations $V_\bullet$, which only makes the order $\trianglelefteq$ in $\mathrm{AGE}(\mathbb{G}')$ finer than the analogous order in $\mathrm{AGE}(\mathbb{G})$.

Let G_n denote the induced substructure of $\mathbb{G}'$ on vertex set U_n. By the assumptions, for every $n, m \in \mathbb{N}$ there is a swap of U_n and U_m that, extended with identity elsewhere, is an automorphism of $\mathbb{G}'$. In consequence, all structures $G_\bullet$ are isomorphic, and the embedding order $\trianglelefteq$ of induced substructures of $\mathbb{G}'$ is isomorphic to finite multisets over $\mathrm{AGE}(G_0)$, ordered by multiset inclusion. Thus $(\mathrm{AGE}(\mathbb{G}'), \trianglelefteq)$ is isomorphic to the multiset inclusion in $\mathcal{M}(\mathrm{AGE}(G_0))$, which is a WQO as U_0 is finite. For any WQO $(X, \leq)$, analogous isomorphism holds between the lifted embedding order $(\mathrm{AGE}(\mathbb{G}'), \trianglelefteq_X)$ and the multiset inclusion in multisets over induced substructures of G_0 labeled by elements of X, and again the latter order is a WQO. Thus $\mathbb{G}'$ admits WQO. □

Lemma 5. *The 3-graph* $\mathbb{G}$ *admits* WQO.

Proof. We are going to prepare the ground for the use of Lemma 4. By Corollary 1. the vertex set V partitions into $V = \bigcup_{n \in \mathbb{N}} U_n$ so that

(a) every layer V_i shares with every set U_n exactly one vertex: $U_n \cap V_i = \{v_i^{(n)}\}$,
(b) if $f_{i,j}$ is the bijection determined by $\mathbb{G}_{i,j}$ (a graph of type 2.), then $f_{i,j}(v_i^{(n)}) \in U_n$, so all the bijections preserve every set $U_\bullet$.

Intuitively, $\mathbb{G}$ can by cut into thin 'slices' perpendicular to the layers $V_\bullet$. By thin we mean that the slices have exactly one vertex in each layer. The cut is made along the bijections dictated by the graphs of type 2. as in the picture bellow:

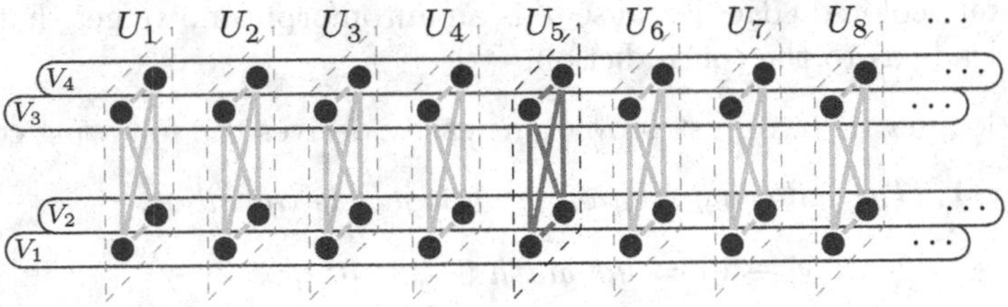

We observe that for every n, the bijection $h_n : V \to V$ that swaps U_1 and U_n along the only bijection $U_1 \to U_n$ that preserves layers, and is identity elsewhere, is an automorphism of $\mathbb{G}$. Indeed, for any three slices U_a, U_b, U_c we have that:

$$v_i^{(a)} \mathrm{\ a\ } v_j^{(c)} \Leftrightarrow v_i^{(b)} \mathrm{\ a\ } v_j^{(c)}$$

so the edges $\left\{v_i^{(a)}, v_j^{(c)}\right\}$ and $\left\{v_i^{(b)}, v_j^{(c)}\right\}$ are colored the same way. The above equivalence is obvious in case when $\mathbb{G}_{i,j}$ is a graph of type 1. In the case of graph of type 2, the vertex $v_i^{(c)}$ is connected with all vertices from V_j but one by x-edges for some $\mathrm{x} \in \{\mathrm{a}, \mathrm{b}\}$. However, the special vertex $f_{i,j}(v_i^{(c)})$ that is not connected by a x-edge, by the condition (b), also belongs to U_c, so it does not interfere with above equivalence.

By Lemma 4 we deduce that $\mathbb{G}$ admits WQO, which completes the proof. □

References

1. Abdulla, P.A., Nylén, A.: Timed Petri nets and BQOs. In: Colom, J.-M., Koutny, M. (eds.) ICATPN 2001. LNCS, vol. 2075, pp. 53–70. Springer, Heidelberg (2001). https://doi.org/10.1007/3-540-45740-2_5
2. Bojańczyk, M., Braud, L., Klin, B., Lasota, S.: Towards nominal computation. Proc. POPL **2012**, 401–412 (2012)
3. Bojańczyk, M., Klin, B., Lasota, S.: Automata theory in nominal sets. Logical Methods Comput. Sci. **10**(3:4) (2014). Paper 4
4. Bojańczyk, M., Klin, B., Lasota, S., Toruńczyk, S.: Turing machines with atoms. LICS **2013**, 183–192 (2013)
5. Cervesato, I., Durgin, N.A., Lincoln, P., Mitchell, J.C., Scedrov, A.: A meta-notation for protocol analysis. In: Proceedings of CSFW 1999, pp. 55–69 (1999)
6. Cherlin, G.: The classification of countable homogeneous directed graphs and countable homogeneous n-tournaments. Mem. Am. Math. Soc. **131**(621), xiv+161 (1998)
7. Cherlin, G.: Forbidden substructures and combinatorial dichotomies: WQO and universality. Discrete Math. **311**(15), 1543–1584 (2011)
8. Delzanno, G.: An overview of MSR(C): a CLP-based framework for the symbolic verification of parameterized concurrent systems. Electr. Notes Theor. Comput. Sci. **76**, 65–82 (2002)
9. Delzanno, G.: Constraint multiset rewriting. Technical report DISI-TR-05-08, DISI, Universitá di Genova (2005)
10. Ding, G.: Subgraphs and well-quasi-ordering. J. Graph Theor. **16**(5), 489–502 (1992)
11. Finkel, A., Schnoebelen, P.: Well-structured transition systems everywhere! Theor. Comput. Sci. **256**(1–2), 63–92 (2001)
12. Fraïssé, R.: Theory of Relations. North-Holland, Amsterdam (1953)
13. Genrich, H.J., Lautenbach, K.: System modelling with high-level Petri nets. Theor. Comput. Sci. **13**, 109–136 (1981)
14. Hofman, P., Lasota, S., Lazić, R., Leroux, J., Schmitz, S., Totzke, P.: Coverability trees for Petri nets with unordered data. In: Jacobs, B., Löding, C. (eds.) FoSSaCS 2016. LNCS, vol. 9634, pp. 445–461. Springer, Heidelberg (2016). https://doi.org/10.1007/978-3-662-49630-5_26
15. Jacobsen, L., Jacobsen, M., Møller, M.H., Srba, J.: Verification of timed-arc Petri nets. In: vCerná, I., Gyimóthy, T., Hromkovivc, J., Jefferey, K., Královič, R., Vukolić, M., Wolf, S. (eds.) SOFSEM 2011. LNCS, vol. 6543, pp. 46–72. Springer, Heidelberg (2011). https://doi.org/10.1007/978-3-642-18381-2_4
16. Jenkinson, T., Truss, J.K., Seidel, D.: Countable homogeneous multipartite graphs. Eur. J. Comb. **33**(1), 82–109 (2012)
17. Jensen, K.: Coloured Petri nets and the invariant-method. Theor. Comput. Sci. **14**, 317–336 (1981)

18. Lachlan, A.H., Woodrow, R.E.: Countable ultrahomogeneous undirected graphs. Trans. Amer. Math. Soc. **262**(1), 51–94 (1980)
19. Lasota, S.: Decidability border for Petri nets with data: WQO dichotomy conjecture. In: Kordon, F., Moldt, D. (eds.) PETRI NETS 2016. LNCS, vol. 9698, pp. 20–36. Springer, Cham (2016). https://doi.org/10.1007/978-3-319-39086-4_3
20. Lasota, S., Piórkowski, R.: WQO dichotomy for 3-graphs. CoRR, arXiv:1802.07612 (2018)
21. Lazić, R., Newcomb, T., Ouaknine, J., Roscoe, A.W., Worrell, J.: Nets with tokens which carry data. In: Kleijn, J., Yakovlev, A. (eds.) ICATPN 2007. LNCS, vol. 4546, pp. 301–320. Springer, Heidelberg (2007). https://doi.org/10.1007/978-3-540-73094-1_19
22. Lazic, R., Schmitz, S.: The complexity of coverability in ν-Petri nets. In: Proceedings of LICS 2016, pp. 467–476 (2016)
23. Macpherson, D.: A survey of homogeneous structures. Discrete Math. **311**(15), 1599–1634 (2011)
24. Rosa-Velardo, F.: Ordinal recursive complexity of unordered data nets. Inf. Comput. **254**, 41–58 (2017)
25. Rosa-Velardo, F., de Frutos-Escrig, D.: Decidability and complexity of Petri nets with unordered data. Theor. Comput. Sci. **412**(34), 4439–4451 (2011)

Verifying Higher-Order Functions with Tree Automata

Thomas Genet, Timothée Haudebourg(✉), and Thomas Jensen

Univ. Rennes, Inria, IRISA, Rennes, France
timothee.haudebourg@irisa.fr

Abstract. This paper describes a fully automatic technique for verifying safety properties of higher-order functional programs. Tree automata are used to represent sets of reachable states and functional programs are modeled using term rewriting systems. From a tree automaton representing the initial state, a completion algorithm iteratively computes an automaton which over-approximates the output set of the program to verify. We identify a subclass of higher-order functional programs for which the completion is guaranteed to terminate. Precision and termination are obtained conjointly by a careful choice of equations between terms. The verification objective can be used to generate sets of equations automatically. Our experiments show that tree automata are sufficiently expressive to prove intricate safety properties and sufficiently simple for the verification result to be certified in Coq.

1 Introduction

Higher-order functions are an integral feature of modern programming languages such as Java, Scala or JavaScript, not to mention Haskell and Caml. Higher-order functions are useful for program structuring but pose a challenge when it comes to reasoning about the correctness of programs that employ them. To this end, the correctness-minded software engineer can opt for proving properties interactively with the help of a proof assistant such as Coq [13] or Isabelle/HOL [30], or write a specification in a formalism such as Liquid Types [31] or Bounded Refinement Types [33,34] and ask an SMT solver whether it can prove the verification conditions generated from this specification. This approach requires expertise of the formal method used, and both the proof construction and the annotation phase can be time consuming.

Another approach is based on *fully automated* verification tools, where the proof is carried out automatically without annotations or intermediate lemmas. This approach is accessible to a larger class of programmers but applies to a more restricted class of program properties. The flow analysis of higher-order functions was studied by Jones [21] who proposed to model higher-order functions as term rewriting systems and use regular grammars to approximate the result. More recently, the breakthrough results of Ong [29] and Kobayashi [23,24,26] show that combining abstraction with model checking techniques can be used with

C. Baier and U. Dal Lago (Eds.): FOSSACS 2018, LNCS 10803, pp. 565–582, 2018.
https://doi.org/10.1007/978-3-319-89366-2_31

success to analyse higher-order functions automatically. Their approach relies on abstraction for computing over-approximations of the set of reachable states, on which safety properties can then be verified.

In this paper, we pursue the goals of higher-order functional verification using an approach based on the original term rewriting models of Jones. We present a formal verification technique based on Tree Automata Completion (TAC) [20], capable of checking a class of properties, called *regular properties*, of higher-order programs in a fully automatic manner. In our approach, a program is represented as a term rewriting system $\mathcal{R}$ and the set of (possibly infinite) inputs to this program as a tree automaton $\mathcal{A}$. The TAC algorithm computes a new automaton $\mathcal{A}^*$, by *completing* $\mathcal{A}$ with all terms reachable from $\mathcal{A}$ by $\mathcal{R}$-rewriting. This automaton representation of the *reachable terms* contains all intermediate states as well as the final output of the program. Checking correctness properties of the program is then reduced to checking properties of the computed automaton. Moreover, our completion-based approach permits to *certify* automatically $\mathcal{A}^*$ in Coq [6], i.e. given $\mathcal{A}$, $\mathcal{R}$ and $\mathcal{A}^*$, obtain the formal proof that $\mathcal{A}^*$ recognizes all terms reachable from $\mathcal{A}$ by $\mathcal{R}$-rewriting.

Example 1. The following term rewriting system $\mathcal{R}$ defines the *filter* function along with the two predicates *even* and *odd* on Peano's natural numbers.

$$@(@(\mathit{filter}, \underline{p}), \mathit{cons}(\underline{x}, \underline{l})) \rightarrow \textbf{if } @(\underline{p}, \underline{x}) \textbf{ then } \mathit{cons}(\underline{x}, @(@(\mathit{filter}, \underline{p}), \underline{l})) \textbf{ else } @(@(\mathit{filter}, \underline{p}), \underline{l})$$

$$@(@(\mathit{filter}, \underline{p}), \mathit{nil}) \rightarrow \mathit{nil}$$

$$@(\mathit{even}, 0) \rightarrow \mathit{true} \qquad @(\mathit{even}, s(\underline{x})) \rightarrow @(\mathit{odd}, \underline{x})$$

$$@(\mathit{odd}, 0) \rightarrow \mathit{false} \qquad @(\mathit{odd}, s(\underline{x})) \rightarrow @(\mathit{even}, \underline{x})$$

This function returns the input list where all elements not satisfying the input boolean function p are filtered out. Variables are underlined and the special symbol @ denotes function application where $@(f, x)$ means "x applied to f".

We want to check that for all lists l of natural numbers, $@(@(\mathit{filter}, \mathit{odd}), l)$ filters out all even numbers. One way to do this is to write a higher-order predicate, *exists*, and check that there exists no even number in the resulting list, i.e. that $@(@(\mathit{exists}, \mathit{even}), @(@(\mathit{filter}, \mathit{odd}), l))$ always rewrites to *false*. Let $\mathcal{A}$ be the tree automaton recognising terms of form $@(@(\mathit{exists}, \mathit{even}), @(@(\mathit{filter}, \mathit{odd}), l))$ where l is any list of natural numbers. The completion algorithm computes an automaton $\mathcal{A}^*$ recognising every term reachable from $L(\mathcal{A})$ (the set of terms recognised by A) using $\mathcal{R}$ with the definition of the *exists* function. Formally,

$$L(\mathcal{A}^*) = \mathcal{R}^*(L(\mathcal{A})) = \{t \mid \exists s \in L(\mathcal{A}), s \rightarrow^*_{\mathcal{R}} t\}$$

To prove the expected property, it suffices to check that *true* is not reachable, i.e. *true* does not belong to the regular set $L(\mathcal{A}^*)$. We denote by *regular properties* the family of properties characterised by a regular set. In particular, regular properties do not count symbols in terms, nor relate subterm heights (a property comparing the length of the list before and after *filter* is not regular)

Termination of the tree automata completion algorithm is not ensured in general [19]. For instance, if $\mathcal{R}^*(L(\mathcal{A}))$ is not regular, it cannot be represented as a tree automaton. In this case, the user can provide a set of *equations* that will force termination by introducing an approximation based on *equational abstraction* [27]: $L(\mathcal{A}^*) \supseteq \mathcal{R}^*(L(\mathcal{A}))$. Equations make TAC powerful enough to verify first-order functional programs [19]. However, state-of-the-art TAC has two short-comings. (i) Equations must be given by the user, which goes against full automation, and (ii) even with equations, termination is not guaranteed in the case of *higher-order programs.* In this paper we propose a solution to these short-comings with the following contributions:

- We state and prove a general termination theorem for the Tree Automata Completion algorithm (Sect. 3);
- From the conditions of the theorem we characterise a class of higher-order functional programs for which the completion algorithm terminates (Sect. 4). This class covers common usage of higher-order features in functional programming languages.
- We define an algorithm that is able to automatically generate equations for enforcing convergence, thus avoiding any user intervention (Sect. 5).

All proofs missing in this paper can be found in the accompanying technical report [17]. The paper is organised as follow: We describe the completion algorithm and how to use equations to ensure termination in Sect. 2.1. The technical contributions as described above are developed in Sects. 3 to 5. In Sect. 6, we present a series of experiments validating our verification technique, and discuss the certification of results in Coq. We present related work in Sect. 7. Section 8 concludes the paper.

2 Background

This section introduces basic concepts used throughout the paper. We recall the usual definitions of term rewriting systems and tree automata, and present the completion algorithm which forms the basis of our verification technique.

2.1 Term Rewriting and Tree Automata

Terms. An alphabet $\mathcal{F}$ is a finite set of symbols, with an arity function $ar : \mathcal{F} \to \mathbb{N}$. Symbols represent constructors such as *nil* or *cons*, or functions such as *filter*, etc. For simplicity, we also write $f \in \mathcal{F}^n$ when $f \in \mathcal{F}$ and $ar(f) = n$. For instance, $cons \in \mathcal{F}^2$ and $nil \in \mathcal{F}^0$. An alphabet $\mathcal{F}$ and finite set of variables $\mathcal{X}$ induces a set of *terms* $\mathcal{T}(\mathcal{F}, \mathcal{X})$ such that:

$$\underline{x} \in \mathcal{T}(\mathcal{F}, \mathcal{X}) \Leftarrow \underline{x} \in \mathcal{X}$$
$$f(t_1, \ldots, t_n) \in \mathcal{T}(\mathcal{F}, \mathcal{X}) \Leftarrow f \in \mathcal{F}^n \text{ and } t_1, \ldots, t_n \in \mathcal{T}(\mathcal{F}, \mathcal{X})$$

A *language* is a set of terms. A term t is *linear* if the multiplicity of each variable in t is at most 1, and *closed* if it contains no variables. The set of closed terms is written $\mathcal{T}(\mathcal{F})$. A *position* in a term t is a word over $\mathbb{N}$ pointing to a *subterm* of t. $Pos(t)$ is the set of positions in t, one for each subterm of t. It is defined by:

$$Pos(\underline{x}) = \{\lambda\}$$
$$Pos(f(t_1, \ldots, t_n)) = \{\lambda\} \cup \{i.p \mid 1 \leq i \leq n \wedge p \in Pos(t_i)\}$$

where λ is the empty word and "." in $i.p$ is the *concatenation* operator. For $p \in Pos(t)$, we write $t|_p$ for the subterm of t at position p, and $t[s]_p$ for the term t where the subterm at position p has been replaced by s. We write $s \trianglerighteq t$ if t is a subterm of s and $s \triangleright t$ if it is a subterm and $s \neq t$. If $\mathcal{L} \subseteq \mathcal{T}(\mathcal{F})$, we write $\mathcal{L}_{\trianglerighteq}$ for the language $\mathcal{L}$ and all its subterms. A *substitution* σ is an application of $\mathcal{X} \mapsto \mathcal{T}(\mathcal{F}, \mathcal{X})$, mapping variables to terms. We tacitly extend it to the endomorphism $\sigma : \mathcal{T}(\mathcal{F}, \mathcal{X}) \mapsto \mathcal{T}(\mathcal{F}, \mathcal{X})$ where $t\sigma$ is the result of the application of the term t to the substitution σ.

Term Rewriting Systems. [1] provide a flexible way of defining functional programs and their semantics. A rewriting system is a pair $\langle \mathcal{F}, \mathcal{R} \rangle$, where $\mathcal{F}$ is an alphabet and $\mathcal{R}$ a set of rewriting rules of the form $l \to r$, where $l, r \in \mathcal{T}(\mathcal{F}, \mathcal{X})$, $l \notin \mathcal{X}$ and $Var(r) \subseteq Var(l)$. A TRS can be seen as a set of rules, each of them defining one step of computation. We write $\mathcal{R}$ a rewriting system $\langle \mathcal{F}, \mathcal{R} \rangle$ if there is no ambiguity on $\mathcal{F}$. A rewriting rule $l \to r$ is said to be left-linear if the term l is linear. Example 1 shows a TRS representing a functional program, where each rule is left-linear. In that case we say that the TRS $\mathcal{R}$ is left-linear.

A rewriting system $\mathcal{R}$ induces a rewriting relation $\to_{\mathcal{R}}$ where for alls $s, t \in \mathcal{T}(\mathcal{F}, \mathcal{X})$, $s \to_{\mathcal{R}} t$ if it exists a rule $l \to r \in \mathcal{R}$, a position $p \in Pos(s)$ and a substitution σ such that $l\sigma = s|_p$ and $t = s[r\sigma]_p$. The reflexive-transitive closure of $\to_{\mathcal{R}}$ is written $\to^*_{\mathcal{R}}$. The rewriting system introduced in the previous example also derives a rewriting relation $\to_{\mathcal{R}}$ where

$$@(@(\mathit{filter}, \mathit{odd}), \mathit{cons}(0, \mathit{cons}(s(0), \mathit{nil}))) \to^*_{\mathcal{R}} \mathit{cons}(s(0), \mathit{nil})$$

The term $\mathit{cons}(s(0), \mathit{nil})$ is *irreducible* (no rule applies to it) and hence the result of the function call. We write $\mathit{IRR}(\mathcal{R})$ for the set of irreducible terms of $\mathcal{R}$.

Tree Automata. [12] are a convenient way to represent regular sets of terms. A tree automaton is a quadruple $\langle \mathcal{F}, \mathcal{Q}, \mathcal{Q}_f, \Delta \rangle$ where $\mathcal{F}$ is an alphabet, $\mathcal{Q}$ a finite set of states, $\mathcal{Q}_f$ the set of *final states*, and Δ a rewriting system on $\mathcal{F} \cup \mathcal{Q}$. Rules in Δ, called *transitions*, are of the form $l \to q$ where $q \in \mathcal{Q}$ and l is either a state ($\in \mathcal{Q}$), or a *configuration* of the form $f(q_1, \ldots, q_n)$ with $f \in \mathcal{F}, q_1 \ldots q_n \in \mathcal{Q}$. A term t is *recognised* by a state $q \in \mathcal{Q}$ if $t \to^*_{\Delta} q$, which we also write $t \to^*_{\mathcal{A}} q$. We write $L(\mathcal{A}, q)$ for the language of all terms recognised by q. A term t is recognised by $\mathcal{A}$ if there exists $q \in \mathcal{Q}_f$ s.t. $t \in L(\mathcal{A}, q)$. In that case we write $t \in L(\mathcal{A})$. *E.g.*, the tree automaton $\mathcal{A} = \langle \mathcal{F}, \mathcal{Q}, \mathcal{Q}_f, \Delta \rangle$ with $\mathcal{F} = \{0 : 0, s : 1\}$, $\mathcal{Q}_f = \{q_{pair}\}$ and $\Delta = \{0 \to q_{pair}, s(q_{odd}) \to q_{pair}, s(q_{pair}) \to q_{odd}, \mathit{nil} \to q_{list}, \mathit{cons}(q_{pair}, q_{list}) \to q_{list}\}$ recognises all lists of even natural numbers.

An *ϵ-transition* is a transition $q \to q'$ where $q \in \mathcal{Q}$. A tree automaton $\mathcal{A}$ is *ϵ-free* if it contains no ϵ-transitions. $\mathcal{A}$ is *deterministic* if for all terms t there is at most one state q such that $t \to_{\Delta}^{*} q$. $\mathcal{A}$ is *reduced* if for all q there is at least one term t such that $t \to_{\Delta}^{*} q$.

2.2 Tree Automata Completion Algorithm

The verification algorithm is based on **tree automata completion**. Given a program represented as a rewriting system $\mathcal{R}$, and its input represented as a tree automaton $\mathcal{A}^0$, the *tree automata completion algorithm* computes a new tree automaton $\mathcal{A}^*$ recognising the set of all *reachable terms* starting from a term in $L(\mathcal{A})$. For a given $\mathcal{R}$, we write this set $\mathcal{R}^*(L(\mathcal{A})) = \{t \mid \exists s \in L(\mathcal{A}), s \to_{\mathcal{R}}^{*} t\}$. It includes all intermediate computations and, in particular, the *output* of the functional program. The algorithm proceeds by computing iteratively $\mathcal{A}^1, \mathcal{A}^2, \ldots$ such that $\mathcal{A}^{i+1} = \mathcal{C}_{\mathcal{R}}(\mathcal{A}^i)$ until it reaches a fix-point, $\mathcal{A}^*$. Here, $\mathcal{C}_{\mathcal{R}}(\mathcal{A}^i)$ represents *one step* of completion and is performed by searching and *completing* the *critical pairs* of $\mathcal{A}^i$.

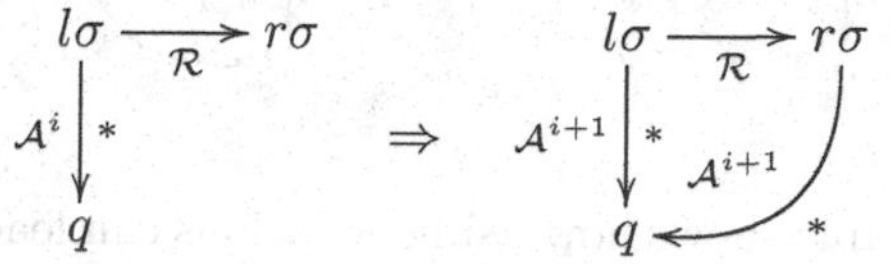

Definition 1 (Critical pair). *A critical pair is a triple $\langle l \to r, \sigma, q\rangle$ where $l \to r \in \mathcal{R}$, σ is a substitution, and $q \in \mathcal{Q}$ such that $l\sigma \to_{\mathcal{A}^i}^{*} q$ and $r\sigma \not\to_{\mathcal{A}^i}^{*} q$.*

Completing a critical pair consists in adding the necessary transitions in $\mathcal{A}^{i+1}$ to have $r\sigma \to_{\mathcal{A}^{i+1}}^{*} q$, and hence $r\sigma \in L(\mathcal{A}^{i+1}, q)$.

Example 2. Let $\mathcal{A}^0$ be the previously defined tree automaton recognising all lists of even natural numbers. Let $\mathcal{R} = \{s(s(x)) \to s(x)\}$. $\mathcal{A}^0$ has a critical pair $\langle s(s(x)) \to s(x), \sigma, q_{pair}\rangle$ with $\sigma(x) = q_{pair}$. To *complete* the automaton, we need to add transition such that $s(q_{pair}) \to_{\mathcal{A}^1}^{*} q_{pair}$. Since we already have the state q_{odd} recognising $s(q_{pair})$, we only add the transition $q_{odd} \to q_{pair}$. The formal definition of the completion step, including the procedure of choosing which new transition to introduce, can be found in [17].

Every completion step has the following property:

$$L(\mathcal{A}^i) \subseteq L(\mathcal{A}^{i+1}) \quad \text{and}$$
$$s \in L(\mathcal{A}^i) \Rightarrow s \to_{\mathcal{R}} t \Rightarrow t \in L(\mathcal{A}^{i+1})$$

It implies that, if a fix-point $\mathcal{A}^*$ then it recognises every term of $\mathcal{R}^*(L(\mathcal{A}))$. However it is in general impossible to compute a tree automaton recognising $\mathcal{R}^*(L(\mathcal{A}))$ exactly, and this may cause the completion algorithm to diverge. Instead we shall over-approximate it by an automaton $\mathcal{A}^*$ such that $L(\mathcal{A}^*) \supseteq$

$\mathcal{R}^*(L(\mathcal{A}))$. The approximation is performed by introducing a set E of *equations* of the form $l = r$ where $l, r \in \mathcal{T}(\mathcal{F}, \mathcal{X})$. From E we derive the relation $=_E$, the *smallest congruence* such that for all equation $l = r$ and substitution σ we have $l\sigma =_E r\sigma$. In this paper we also write $\vec{E}$ for the TRS $\{l \to r \mid l = r \in E\}$. At each completion step, the algorithm *simplifies* the automaton by merging states together according to E.

Definition 2 (Simplification Relation). *Let $\mathcal{A} = \langle \mathcal{F}, \mathcal{Q}, \mathcal{Q}_f, \Delta \rangle$ be a tree automaton and E be a set of equations. If $s = t \in E, \sigma : \mathcal{X} \mapsto \mathcal{Q}, q, q' \in \mathcal{Q}$ such that $s\sigma \to^*_{\mathcal{A}} q, t\sigma \to^*_{\mathcal{A}} q'$ and $q \neq q'$ then $\mathcal{A}$ can be simplified into $\mathcal{A}' = \mathcal{A}\{q' \mapsto q\}$ (where q' has been substitued by q), denoted by $\mathcal{A} \rightsquigarrow_E \mathcal{A}'$.*

We write $\mathcal{S}_E(\mathcal{A})$ for the unique automaton (up to renaming) $\mathcal{A}'$ such that $\mathcal{A} \rightsquigarrow^*_E \mathcal{A}'$ and $\mathcal{A}'$ is irreducible by $\rightsquigarrow_E$. One completion step is now defined by $\mathcal{A}^{i+1} = \mathcal{S}_E(\mathcal{C}_{\mathcal{R}}(\mathcal{A}^i))$.

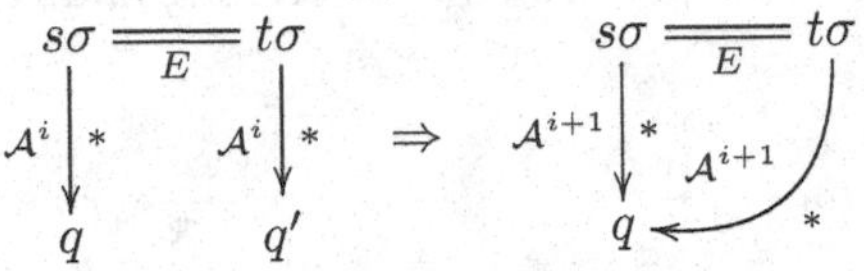

Example 3. This example shows how using equations can lead to approximations in tree automata. Let $\mathcal{A}$ be the tree automaton defined by the set of transitions $\Delta = \{0 \to q_0, s(q_0) \to q_1\}$. This automaton recognises the two terms 0 in q_0 and $s(0)$ (also known as 1) in q_1. Let $E = \{s(\underline{x}) = \underline{x}\}$ containing the equation that equates a number and its successor. For $\sigma = \{\underline{x} \mapsto 0\}$ we have $s(\underline{x})\sigma \to_{\mathcal{A}} q_1$, $\underline{x}\sigma \to_{\mathcal{A}} q_0$ and $s(\underline{x})\sigma =_E \underline{x}\sigma$. Then in $\mathcal{S}_E(\mathcal{A})$, q_0 and q_1 are merged. The resulting automaton has transitions $\{0 \to q_0, s(q_0) \to q_0\}$, which recognises $\mathbb{N}$ in q_0.

The idea behind the simplification is to overapproximate $\mathcal{R}^*(L(\mathcal{A}))$ when it is *not regular*. It has been shown in [19] that it is possible to tune the precision of the approximation. For a given TRS $\mathcal{R}$, initial state automaton $\mathcal{A}$ and set of equations E, the termination of the completion algorithm is undecidable in general, even with the use of equations. Our contribution in this paper consists in finding a class of TRS/programs and equations E for which the completion algorithm with equations terminates.

3 Termination of Tree Automata Completion

In this section, we show that termination of the completion algorithm with a set of equations E is ensured under the following conditions: if (i) $\mathcal{A}^k$ is reduced ϵ-free and deterministic (written **REFD** in the rest of the paper) for all k; (ii) every term of $\mathcal{A}^k$ can be rewritten into a term of a given language $\mathcal{L} \subseteq \mathcal{T}(\mathcal{F})$ using $\mathcal{R}$ (for instance if $\mathcal{R}$ is terminating); (iii) $\mathcal{L}$ has a finite number of equivalence classes w.r.t E. Completion is known to preserve $\not\epsilon$-reduceness and $\not\epsilon$-determinism if $E \supseteq E_r \cup E_{\mathcal{R}}$ [19] where $E_{\mathcal{R}} = \{s = t \mid s \to t \in \mathcal{R}\}$ and $E_r = \{f(x_1, \ldots, x_n) = f(x_1, \ldots, x_n) \mid f \in \mathcal{F}^n\}$. Condition (i) is ensured by showing that, in our verification setting, completion preserve REFD. The last condition is ensured by having $E \supseteq E^c_{\mathcal{L}}$ where $E^c_{\mathcal{L}}$ is a set of *contracting equations*.

Definition 3 (Contracting Equations). *Let $\mathcal{L} \subseteq \mathcal{T}(\mathcal{F})$. A set of equations is contracting for $\mathcal{L}$, denoted by $E^c_{\mathcal{L}}$, if all equations of $E^c_{\mathcal{L}}$ are of the form $u = u|_p$ with u a linear term of $\mathcal{T}(\mathcal{F}, \mathcal{X})$, $p \neq \lambda$ and if the set of normal forms of $\mathcal{L}$ w.r.t the TRS $\vec{E^c_{\mathcal{L}}} = \{u \to u|_p \mid u = u|_p \in E^c_{\mathcal{L}}\}$ is finite.*

Example 4. Assume that $\mathcal{F} = \{0 : 0, s : 1\}$. The set $E^c_{\mathcal{L}} = \{s(x) = x\}$ is contracting for $\mathcal{L} = \mathcal{T}(\mathcal{F})$ because the set of normal forms of $\mathcal{T}(\mathcal{F})$ with respect to $\vec{E^c_{\mathcal{L}}} = \{s(x) \to x\}$ is the (finite) set $\{0\}$. The set $E^c_{\mathcal{L}} = \{s(s(x)) = x\}$ is contracting because the normal forms of $\{s(s(x)) \to x\}$ are $\{0, s(0)\}$.

The contracting equations ensure that the completion algorithm will merge enough states during the simplification steps to terminate. Note that $E^c_{\mathcal{L}}$ cannot be empty, unless $\mathcal{L}$ is finite. To prove termination of completion, we first prove that it is possible to bound the number of states needed in $\mathcal{A}^*$ to recognise a language $\mathcal{L}$ by the number of normal forms of $\mathcal{L}$ with respect to $\vec{E^c_{\mathcal{L}}}$. In our case $\mathcal{L}$ will be the set of output terms of the program. Since $\mathcal{A}^*$ does not only recognises the output terms, we need additional states to recognise intermediate computation terms. In the proof of Theorem 1 we show that with $E_{\mathcal{R}}$, the simplification steps will merge the states recognising the intermediate computation with the states recognising the outputs. If the latter set of states is finite then we can show that $\mathcal{A}^*$ is finite.

Theorem 1. *Let $\mathcal{A}$ be an REFD tree automaton, $\mathcal{R}$ a left-linear TRS, E a set of equations and $\mathcal{L}$ a language closed by subterms such that for all $k \in \mathbb{N}$ and for all $s \in L_{\trianglerighteq}(\mathcal{A}^k)$, there exists $t \in \mathcal{L}$ s.t. $s \to^*_{\mathcal{R}} t$. If $E \supseteq E^r \cup E^c_{\mathcal{L}} \cup E_{\mathcal{R}}$ then the completion of $\mathcal{A}$ by R and E terminates with a REFD $\mathcal{A}^*$.*

4 A Class of Analysable Programs

The next step is to identify a class of functional programs and a language $\mathcal{L}$ for which Theorem 1 applies. By choosing $\mathcal{L} = \mathcal{T}(\mathcal{F})$ and providing a set of contracting equations $E^c_{\mathcal{T}(\mathcal{F})}$, the termination theorem above proves that the completion algorithm terminates on any functional program $\mathcal{R}$. If this works in theory, in practice we want to avoid introducing equations over the application symbol (such as $@(x, y) = y$). Contracting equations on applications makes sense in certain cases, *e.g.*, with idempotent functions ($@(sort, @(sort, x)) = @(sort, x)$), but in most cases, such equations dramatically lower the precision of the completion algorithm. Hence, we want to identify a language $\mathcal{L}$ with no contracting equations over @ in $E^c_{\mathcal{L}}$. Since such a language $\mathcal{L}$ still has to have a finite number of normal forms w.r.t. $\vec{E^c_{\mathcal{L}}}$ (Theorem 1), it cannot include terms containing an un-bounded *stack* of applications. For instance, $\mathcal{L}$ cannot contain all the terms of the form $@(f,x), @(f,@(f,x)), @(f,@(f,@(f,x)))$, etc. The @ stack must be bounded, even if the applications symbols are interleaved with other symbols (e.g. $@(f,s(@(f,s(@(f,s(x)))))))$. To do that we (i) define a set $\mathcal{B}^d$ of all terms where such stack size is bounded by $d \in \mathbb{N}$; (ii) define a set $\mathcal{K}^n$ and a class of TRS called $\mathcal{K}$-TRS such that for any TRS $\mathcal{R}$ in this class, $\mathcal{K}^n$ is closed by $\mathcal{R}$ and $\mathcal{K}^n \cap IRR(\mathcal{R}) \subseteq \mathcal{B}^{\phi(n)}$ for some function ϕ. This is done by first introducing a type system over the terms; (iii) finally define $\mathcal{L} = \mathcal{B}^{\phi(n)} \cap IRR(\mathcal{R})$ that can be used to instantiate Theorem 1.

Definition 4. *For a given alphabet $\mathcal{F} = \mathcal{C} \cup \{@\}$, $\mathcal{B}^d$ is the set of terms where every application depth is bounded by d. It is the smallest set defined by:*

$$\begin{aligned} f \in \mathcal{B}^0 &\Leftarrow f \in \mathcal{C}^0 \\ f(t_1,\ldots,t_n) \in \mathcal{B}^i &\Leftarrow f \in \mathcal{C}^n \wedge t_1 \ldots t_n \in \mathcal{B}^i \\ @(t_1,t_2) \in \mathcal{B}^{i+1} &\Leftarrow t_1, t_2 \in \mathcal{B}^i \\ t \in \mathcal{B}^{i+1} &\Leftarrow t \in \mathcal{B}^i \end{aligned}$$

In Sect. 5, we show how to produce E^c such that $\mathcal{B}^d \cap IRR(\mathcal{R})$ has a finite number of normal forms w.r.t. $\vec{E^c}$ with no equations on @. However we don't have for all k, for all term $t \in L_{\trianglerighteq}(\mathcal{A}^k)$ a term $s \in \mathcal{B}^d \cap IRR(\mathcal{R})$ s.t. $t \rightarrow_{\mathcal{R}}^* s$ in general. Theorem 1 cannot be instantiated with $\mathcal{L} = \mathcal{B}^d \cap IRR(\mathcal{R})$. Instead we define (i) a set $\mathcal{K}^n \subseteq \mathcal{T}(\mathcal{F})$ and ϕ such that $\mathcal{K}^n \cap IRR(\mathcal{R}) \subseteq \mathcal{B}^{\phi(d)}$ and (ii) a class of TRS, called $\mathcal{K}$-TRS for which $L_{\trianglerighteq}(\mathcal{A}^k) \subseteq \mathcal{K}^n_{\trianglerighteq}$. In $\mathcal{K}$-TRS, the right hand sides of TRS rules are contained in a set $\mathcal{K}$ whose purpose is to forbid the construction of unbounded partial applications during rewriting. If the initial automaton satisfies $L_{\trianglerighteq}(\mathcal{A}) \subseteq \mathcal{K}^n_{\trianglerighteq}$ then we can instantiate Theorem 1 with $\mathcal{L} = \mathcal{K}^n_{\trianglerighteq} \cap IRR(\mathcal{R})$ and prove termination.

4.1 Types

In order to define $\mathcal{K}$ and $\mathcal{K}^n$ we require the TRS to be well-typed. Our definition of types is inspired by [1]. Let $\mathscr{A}$ be a non-empty set of *algebraic types.* The set of *types* $\mathscr{T}$ is inductively defined as the least set containing $\mathscr{A}$ and all function types, *i.e.* $A \rightarrow B \in \mathscr{T} \Leftarrow A, B \in \mathscr{T}$. The function type constructor $\rightarrow$ is assumed to be right-associative. The *arity* of a type A is inductively defined on the structure of A by:

$$\begin{aligned} ar(A) &= 0 && \Leftarrow A \in \mathscr{A} \\ ar(A \rightarrow B) &= 1 + ar(B) && \Leftarrow A \rightarrow B \in \mathscr{T} \end{aligned}$$

Instead of using alphabets, in a typed terms environment we use *signatures* $\mathcal{F} = \mathcal{C} \cup \{@\}$ where $\mathcal{C}$ is a set of *constructor* symbols associated to a unique type and @ the application symbol (with no type). We also assign a type to every variable. We write $f : A$ if the symbol f has type A and $t : A$ a term $t \in \mathcal{T}(\mathcal{F},\mathcal{X})$

of type A. We write $\mathcal{W}(\mathcal{F}, \mathcal{X})$ for the set of all *well typed terms* using the usual definition. We extend the definition of term rewriting systems to typed TRS. A TRS is well typed if all rules are of the form $l : A \to r : A$ (type is preserved). In the same way, an equation $s = t$ is well typed if both s and t have the same type. In the rest of this paper we only consider well typed equations and TRSs.

Definition 5 (Functional TRS). *A higher-order functional TRS is composed of rules of the form*

$$@(\ldots @(f, t_1 : A_1) \ldots, t_n : A_n) : A \to r : A$$

where $f : A_1 \to \ldots \to A_n \to A \in \mathcal{C}^n$, $t_1 \ldots t_n \in \mathcal{W}(\mathcal{C}, \mathcal{X})$ *and* $r \in \mathcal{W}(\mathcal{F}, \mathcal{X})$. *A functional TRS is* complete *if for all term* $t = @(t_1, t_2) : A$ *such that* $ar(A) = 0$, *it is possible to rewrite t using* $\mathcal{R}$. *In other words, all defined functions are total.*

Types provides information about how a term can be rewritten. For instance we expect the term $@(f : A \to B, x : A) : B$ to be rewritten by every *complete* (no partial function) TRS $\mathcal{R}$ if $ar(A \to B) = 1$. Furthermore, for certain types, we can guarantee the absence of partial applications in the result of a computation using the type's *order*. For a given signature $\mathcal{F}$, the *order* of a type A, written $ord(A)$, is inductively defined on the structure of A by:

$$ord(A) = max\{ord(f) \mid f : \ldots \to A \in \mathcal{C}^n\}$$
$$ord(A \to B) = max\{ord(A) + 1, ord(B)\}$$

where $ord(f : A_1 \to \ldots \to A_n \to A) = max\{ord(A_1), \ldots, ord(A_n)\}$ (with, for $A_i = A$, $ord(A_i) = 0$). For instance $ord(int) = 0$ and $ord(int \to int) = 1$.

Example 5. Define two different types of lists *list* and *list'*. The first defines lists of *int* with the constructor $consA : int \to list \to list \in \mathcal{C}$, while the second defines lists of functions with the constructor $consB : (int \to int) \to list' \to list' \in \mathcal{C}$. The importance of order becomes manifest here: in the first case a fully reduced term of type *list* cannot contain any @ whereas in the second case it can. $ord(list) = 0$ and $ord(list') = 1$.

Lemma 1. *If* $\mathcal{R}$ *is a complete functional TRS and A a type such that* $ord(A) = 0$, *then all* closed *terms t of type A are rewritten into an irreducible term with no partial application:*

$$\forall s \in IRR(\mathcal{R}), \quad t \to_{\mathcal{R}}^{*} s \Rightarrow s \in \mathcal{B}^0.$$

4.2 The Class $\mathcal{K}$-TRS

Recall that we want to define (i) a set $\mathcal{K}^n \subseteq \mathcal{T}(\mathcal{F})$ and ϕ such that $\mathcal{K}^n_{\trianglerighteq} \cap IRR(\mathcal{R}) \subseteq \mathcal{B}^{\phi(n)}$ and (ii) a class of TRS $\mathcal{K}$-TRS for which $L_{\trianglerighteq}(\mathcal{A}^k) \subseteq \mathcal{K}^n_{\trianglerighteq}$. Assuming that $L_{\trianglerighteq}(\mathcal{A}) \subseteq \mathcal{K}^n_{\trianglerighteq}$ we instantiate Theorem 1 with $\mathcal{L} = \mathcal{K}^n_{\trianglerighteq} \cap IRR(\mathcal{R})$ and prove termination.

Definition 6 ($\mathcal{K}$-TRS). *A TRS $\mathcal{R}$ is part of $\mathcal{K}$-TRS if for all rules $l \rightarrow r \in \mathcal{R}$, $r \in \mathcal{K}$ where $\mathcal{K}$ is inductively defined by:*

$$\underline{x} : A \in \mathcal{K} \Leftarrow \underline{x} : A \in \mathcal{X}$$
$$f(t_1, \ldots, t_n) : A \in \mathcal{K} \Leftarrow f \in \mathcal{C}^n \wedge t_1, \ldots, t_n \in \mathcal{K}$$
$$@(t_1 : A \rightarrow B, t_2 : A) : B \in \mathcal{K} \Leftarrow t_1 \in \mathcal{Z}, t_2 \in \mathcal{K} \wedge B \in \mathscr{A} \quad (1)$$
$$@(t_1 : A \rightarrow B, t_2 : A) : B \in \mathcal{K} \Leftarrow t_1, t_2 \in \mathcal{K} \wedge ord(A) = 0 \quad (2)$$

with $\mathcal{Z}$ defined by:

$$t \in \mathcal{Z} \Leftarrow t \in \mathcal{K}$$
$$@(t_1, t_2) \in \mathcal{Z} \Leftarrow t_1 \in \mathcal{Z}, t_2 \in \mathcal{K}$$

By constraining the form of the right hand side of each rule of $\mathcal{R}$, $\mathcal{K}$ defines a set of TRS that cannot construct unbounded partial applications during rewriting. The definition of $\mathcal{K}$ takes advantage of the type structure and Lemma 1. The rules (1) and (2) ensure that an application $@(t_1, t_2)$ is either: (1) a total application, and the whole term can be rewritten; or (2) a partial application where t_2 can be rewritten into a term of $\mathcal{B}^0$ (Lemma 1). In (1), $\mathcal{Z}$ allows partial applications inside the total application of a multi-parameter function.

Example 6. Consider the classical *map* function. A typical call to this function is $@(@(map, f), l)$ of type *list*, where f is a mapping function, and l a list. The whole term belongs to $\mathcal{K}$ because of rule (1): *list* is an algebraic type and its subterm $@(map, f) : list \rightarrow list$ belongs to $\mathcal{Z}$. This subterm is a partial application, but there is no risk of stacking partial applications as it is part of a complete call (to the *map* function).

Example 7. Consider the function *stack* defined by:

$$@(@(stack, \underline{x}), 0) \rightarrow \underline{x}$$
$$@(@(stack, \underline{x}), S(\underline{n})) \rightarrow @(@(stack, @(g, \underline{x})), \underline{n})$$

Here g is a function of type $(A \rightarrow A) \rightarrow A \rightarrow A$. The *stack* function returns a stack of partial applications whose height is equal to the input parameter:

$$@(@(stack, f), \underbrace{S(S(S \ldots S}_{k}(0) \ldots))) \rightarrow^*_{\mathcal{R}} \underbrace{@(g, @(g, @(g, \ldots @(g}_{k}, f) \ldots)))$$

The depth of partial applications stacks in the output language is not bounded. With no equations on the @ symbol, the completion algorithm may not terminate. Notice that $\underline{x}$ is a function and $@(g, \underline{x})$ a partial application. Hence the term $@(@(stack, @(g, \underline{x})), \underline{n})$ is not in $\mathcal{K}$, so the TRS does not belong to the $\mathcal{K}$-TRS class.

We define $\mathcal{K}^n$ as $\{t\sigma \mid t \in \mathcal{K}, \quad \sigma : \mathcal{X} \mapsto \mathcal{B}^n \cap IRR(\mathcal{R})\}$ and claim that if for all rule $l \rightarrow r$ of the functional TRS $\mathcal{R}$, $r \in \mathcal{K}$ and if $L(\mathcal{A}) \subseteq \mathcal{K}^n$ then with Theorem 1 we can prove that the completion of $\mathcal{A}$ with $\mathcal{R}$ terminates. The idea is the following:

- Prove that if $\mathcal{A}$ recognises terms of $\mathcal{K}^n_{\trianglerighteq}$, then it is preserved by completion using the notion of $\mathcal{K}^n$-coherence of $\mathcal{A}$.
- Prove that $\mathcal{K}^n_{\trianglerighteq} \cap IRR(\mathcal{R}) \subseteq \mathcal{B}^{n+2B} \cap IRR(\mathcal{R})$ where $B \in \mathbb{N}$ is a fixed upper bound of the arity of all the types of the program.
- Prove that there is a finite number of normal form of $\mathcal{B}^{n+2B} \cap IRR(\mathcal{R})$ w.r.t $\vec{E^c_{\mathcal{L}}}$.
- Finally, we use those three properties combined, and instantiate Theorem 1 with $\mathcal{L} = \mathcal{B}^{n+2B} \cap IRR(\mathcal{R})$ to prove Theorem 2, defined as follows.

Theorem 2. *Let $\mathcal{A}$ be a $\mathcal{K}^n$-coherent REFD tree automaton, $\mathcal{R}$ a terminating functional TRS such that for all rule $l \to r \in \mathcal{R}, r \in \mathcal{K}$ and E a set of equations. Let $\mathcal{L} = \mathcal{B}^{n+2B} \cap IRR(\mathcal{R})$. If $E = E^r \cup E^c_{\mathcal{L}} \cup E_{\mathcal{R}}$ then the completion of $\mathcal{A}$ by $\mathcal{R}$ and E terminates.*

To prove that after each step of completion, the recognised language stays in $\mathcal{K}^n$, we require the considered automaton to be $\mathcal{K}^n$*-coherent*.

Definition 7 ($\mathcal{K}^n$-coherence). *Let $\mathcal{L} \subseteq \mathcal{W}(\mathcal{F})$ and $n \in \mathbb{N}$. $\mathcal{L}$ is $\mathcal{K}^n$-coherent if*

$$\mathcal{L} \subseteq \mathcal{K}^n \lor \mathcal{L} \subseteq \mathcal{Z}^n \setminus \mathcal{K}^n$$

By extension we say that a tree-automaton $\mathcal{A} = \langle \mathcal{F}, \mathcal{Q}, \mathcal{Q}_f, \Delta \rangle$ is $\mathcal{K}^n$-coherent if the language recognised by all states $q \in \mathcal{Q}$ is $\mathcal{K}^n$-coherent.

If $\mathcal{K}^n$-coherence is not preserved during completion, then some states in the completed automaton may recognise terms outside of $\mathcal{K}^n_{\trianglerighteq}$. Our goal is to show that it is preserved by $\mathcal{C}_{\mathcal{R}}(\cdot)$ (Lemma 2) then by $\mathcal{S}_E(\cdot)$ (Lemma 3).

Lemma 2 ($\mathcal{C}_{\mathcal{R}}(\mathcal{A})$ preserves $\mathcal{K}^n$-coherence). *Let $\mathcal{A}$ be a REFD tree automaton. If $\mathcal{A}$ is $\mathcal{K}^n$-coherent, then $\mathcal{C}_{\mathcal{R}}(\mathcal{A})$ is $\mathcal{K}^n$-coherent.*

Lemma 3 ($\mathcal{S}_E(\mathcal{A})$ preserves $\mathcal{K}^n$-coherence). *Let $\mathcal{A}$ be a REFD tree automaton, $\mathcal{R}$ a functional TRS and E a set of equations such that $E = E^r \cup E^c_{\mathcal{L}} \cup E_{\mathcal{R}}$ with $\mathcal{L} = \mathcal{B}^{n+2B} \cap IRR(\mathcal{R})$. If $\mathcal{A}$ is $\mathcal{K}^n$-coherent then $\mathcal{S}_E(\mathcal{A})$ is $\mathcal{K}^n$-coherent.*

By using Lemmas 2 and 3, we can prove that the completion algorithm, which is a composition of $C_{\mathcal{R}}(\mathcal{A})$ and $\mathcal{S}_E(\mathcal{A})$, preserves $\mathcal{K}^n$-coherence. The proofs of these two lemmas are based on a detailed analysis of the completion algorithm itself. The complete proofs are provided in [17].

Lemma 4 (Completion preserves $\mathcal{K}^n$-coherence). *Let $\mathcal{A} = \langle \mathcal{F}, \mathcal{Q}, \mathcal{Q}_f, \Delta \rangle$ be a tree automaton, $\mathcal{R}$ a functional TRS and E a set of equations. If $E = E^r \cup E^c_{\mathcal{L}} \cup E_{\mathcal{R}}$ with $\mathcal{L} = \mathcal{B}^{n+2B} \cap IRR(\mathcal{R})$ and $\mathcal{A}$ is $\mathcal{K}^n$-coherent then for all $k \in \mathbb{N}$, $\mathcal{A}^k$ is $\mathcal{K}^n$-coherent. In particular, $\mathcal{A}^*$ is $\mathcal{K}^n$-coherent.*

By construction we can prove that the depth of irreducible $\mathcal{K}^n_{\trianglerighteq}$ terms is bounded, which correspond to the following lemma.

Lemma 5. *For all $t : T \in \mathcal{K}^n_{\trianglerighteq}$, $t : T \in IRR(\mathcal{R}) \Rightarrow t : T \in \mathcal{B}^{n+2B-arity(T)}$.*

4.3 Proof of Theorem 2

Proof. According to Lemma 4, for all $k \in \mathbb{N}$, the completed automaton $\mathcal{A}^k$ is $\mathcal{K}^n$-coherent. By definition this implies that $\mathcal{L}_{\trianglerighteq}(\mathcal{A}^k) \subseteq \mathcal{K}^n_{\trianglerighteq}$. Moreover, we know that $IRR(\mathcal{R}) \cap \mathcal{K}^n_{\trianglerighteq} \subseteq \mathcal{B}^{n+2B}$ (Lemma 5). Let $\mathcal{L} = \mathcal{B}^{n+2B} \cap IRR(\mathcal{R})$. $\mathcal{R}$ is terminating, so for every term $s \in \mathcal{L}_{\trianglerighteq}(\mathcal{A}^k)$ there exists $t \in \mathcal{L}$ such that $s \rightarrow^*_{\mathcal{R}} t$. Since the number of normal form of $\mathcal{L}$ is finite w.r.t $\vec{E}$, Theorem 1 implies that the completion of $\mathcal{A}$ by $\mathcal{R}$ and E terminates.

5 Equation Generation

Theorem 2 states a number of hypotheses that must be satisfied in order to guarantee termination of the completion algorithm:

- The initial automaton $\mathcal{A}$ must be $\mathcal{K}^n$-coherent and REFD.
- $\mathcal{R}$ must be terminating.
- All left-hand sides of rules of $\mathcal{R}$ are in the set of terms $\mathcal{K}$. This is a straightforward syntactic check. If it is not verified, we can reject the TRS before starting the completion.
- The set of equations E must be of the form $E^r \cup E^c_{\mathcal{L}} \cup E_{\mathcal{R}}$. The equation sets E^r and $E_{\mathcal{R}}$ are determined directly from the syntactic structure of $\mathcal{R}$. However, there is no unique suitable set of contracting equations $E^c_{\mathcal{L}}$. This set must be generated carefully, because a bad choice of contracting equations (*i.e.*, equations that equate too many terms) will have a severe negative impact on the precision of the analysis result.

In this section, we describe a method for generating all possible sets of contracting equations $E^c_{\mathcal{L}}$. To simplify the presentation, we only present the case where $\mathcal{L} = \mathcal{W}(\mathcal{C})$ and $IRR(\mathcal{R}) \subseteq \mathcal{W}(\mathcal{C})$ (*i.e.*, all results are first-order terms). Our approach looks for contracting equations for the set of closed terms $\mathcal{W}(\mathcal{C})$ instead of the set $\mathcal{B}^{n+2B}$ mentioned in Theorem 2. More precisely, we generate the set of equations iteratively, as a series of equation sets $\mathbb{E}^k_c$ where the equations only equate terms of depth at most k. Recall that a contracting equation is of the form $u = u|_p$ with $p \neq \lambda$, *i.e.*, it equates a term with a strict subterm of the same type. A set of contracting equations over the set $\mathcal{W}(\mathcal{C})$ is then generated as follows: (i) generate the set of left-hand side of equations as a *covering set of terms* [25], so that for each term $t \in \mathcal{W}(\mathcal{C})$ there exists a left-hand side u of an equation and a substitution σ such that $t = u\sigma$. (ii) for each left-hand side, generate all possible equations of the form $u = u|_p$, satisfying that both sides have the same type. (iii) from all those equations, we build all possible $E^c_{\mathcal{L}}$ (with $\mathcal{L} = \mathcal{W}(\mathcal{C})$) such that the set of normal forms of $\mathcal{W}(\mathcal{C})$ w.r.t. $\vec{E^c_{\mathcal{L}}}$ is finite. Since $\vec{E^c_{\mathcal{L}}}$ is left-linear and $\mathcal{L} = \mathcal{W}(\mathcal{C})$, this can be decided efficiently [11].

Example 8. Assume that $\mathcal{C} = \{0 : 0, s : 1\}$. For $k = 1$, the covering set is $\{s(x), 0\}$ and $\mathbb{E}^1_c = \{\{s(x) = x\}\}$. For depth 2, the covering set is $\{s(s(x)), s(0), 0\}$ and $\mathbb{E}^2_c = \mathbb{E}^1_c \cup \{\{s(s(x)) = x\}, \{s(s(x)) = s(x)\}, \{s(0) = 0\}, \{s(0) = 0, s(s(x)) = x\}, \{s(0) = 0, s(s(x)) = s(x)\}\}$. All equation sets of $\mathbb{E}^1_c$ and $\mathbb{E}^2_c$ satisfy Definition 3 and lead to different approximations.

To verify a property φ on a program, we use completion and equation generation as follows. The program is represented by a TRS $\mathcal{R}$ and function calls are represented by an initial tree automaton $\mathcal{A}$. Both have to respect the hypothesis of Theorem 2. The algorithm searches for a set of contracting equations E_c such that verification succeeds, *i.e.* $\mathcal{L}(\mathcal{A}^*)$ satisfy φ. Starting from $k = 1$, we apply the following algorithm:

1. We first complete the tree automaton $\mathcal{A}_k$ recognising the *finite* subset of $\mathcal{L}(\mathcal{A})$ of terms of maximum depth k. Since $\mathcal{L}(\mathcal{A}_k)$ is finite and $\mathcal{R}$ is terminating, the set of reachable terms is finite, completion terminates without equations and computes an automaton $\mathcal{A}_k^*$ recognising exactly the set $\mathcal{R}^*(L(\mathcal{A}_k))$ [20].
2. If $\mathcal{L}(\mathcal{A}_k^*)$ does not satisfy φ then verification fails: a counterexample is found.
3. Otherwise, we search for a suitable set E_c. All E_c of $\mathbb{E}_c^k$ that introduce a counterexample in the completion of $\mathcal{A}_k$ with $\mathcal{R}$ and E_c are filtered out.
4. Then for all remaining E_c, we try to complete $\mathcal{A}$ with $\mathcal{R}$ and $E = E_r \cup E_{\mathcal{R}} \cup E_c$ and check φ on the completed automaton. If φ is true on $\mathcal{A}^*$ then verification succeeds. Otherwise, we try the next E_c.
5. If there remain no E_c, we start again with $k = k + 1$.

If there exists a set of equations E_c able to verify the program, this algorithm will find it eventually, or find a counter example. However if there is no set of equations that can verify the program, this algorithm does not terminate.

6 Experiments

The verification technique described above has been integrated in the Timbuk library [16]. We implemented the naive equation generation where all possible equation sets E_c are enumerated. Despite the evident scalability issues of this simple version of the verification algorithm, we have been able to verify a series of properties of several classical higher-order functions: *map*, *filter*, *exists*, *forall*, *foldRight*, *foldLeft* as well as higher-order sorting functions parameterised by an ordering function. Most examples are taken from or inspired by [26,28] and have corresponding TRSs in the $\mathcal{K}$ set defined above. The property φ consists in checking that a finite set of forbidden terms is not reachable (`Patterns` section of Timbuk specifications).

Given $\mathcal{A}$, $\mathcal{R}$ and $\mathcal{A}^*$, the *correctness of the verification*, i.e. the fact that $\mathcal{L}(\mathcal{A}^*) \supseteq \mathcal{R}^*(\mathcal{L}(\mathcal{A}))$, can be checked in a proof assistant embedding a formalisation of rewriting and tree automata. It is enough to prove that (a) $\mathcal{L}(\mathcal{A}^*) \supseteq \mathcal{L}(\mathcal{A})$ and that (b) for all critical pairs $\langle l \to r, \sigma, q\rangle$ of $\mathcal{A}^*$ we have $r\sigma \to_{\mathcal{A}^*}^* q$. Property (a) can be checked using standard algorithms on tree automata. Property (b) can be checked by enumerating all critical pairs of $\mathcal{A}^*$ (there are finitely many) and by proving that all of them satisfy $r\sigma \to_{\mathcal{A}^*}^* q$. Since there exists algorithms for checking properties (a) and (b), the complete proof of correctness can automatically be built in the proof assistant. For instance, the automaton $\mathcal{A}^*$ can be used as a certificate to build the correctness proof in Coq [6] and in Isabelle/HOL [14]. It is also used to build unreachability proofs in

Isabelle/HOL [14]. Besides, since verifying (a) and (b) is automatic, the correctness proof may be run outside of the proof assistant (in a more efficient way) using a formally verified external checker extracted from the formalisation. All our (successful) completion attempts output a `comp.res` file, containing $\mathcal{A}$, $\mathcal{R}$ and $\mathcal{A}^*$, which has been certified automatically using the external certified checker of [6]. Timbuk's site http://people.irisa.fr/Thomas.Genet/timbuk/funExperiments/ lists those verification experiments. Nine of them are automatically proven. Two other examples show that correct counter-examples are generated when the property is not provable. On one example equation generation times out due to our naïve enumeration of equations. For this last case, by providing the right set of equations in `mapTree2NoGen` the verification of the function succeeds.

7 Related Work

When it comes to verifying first-order imperative programs, there exist several successful tools based on abstract interpretation such as ASTREE [3] and SLAM [2]. The use of abstract interpretation for verifying higher-order functional programs has comparatively received less attention. The tree automaton completion technique is one analysis technique able to verify first-order Java programs [4]. Until now, the completion algorithm was guaranteed to terminate only in the case of first-order functional programs [19].

Liquid Types [31], followed by Bounded Refinement Types [33,34], and also Set-Theoretic Types [8,9], are all attempts to enrich the type system of functional languages to prove non-trivial properties on higher-order programs. However, these methods are not automatic. The user has to express the property he wants to prove using the type system, which can be tedious and/or difficult. In some cases, the user even has to specify straightforward intermediate lemmas to help the type checker.

The first attempt in verifying regular properties came with Jones [21] and Jones and Andersen [22]. Their technique computes a grammar over-approximating the set of states reachable by a rewriting systems. However, their approximation is fixed and too rough to prove programs like Example 1 (*filter odd*). Our program and property models are close to those of Jones and Andersen. However, the approximation in our analysis is not fixed and can be automatically adapted to the verification objective.

Ong *et al.* proposes one way of addressing the precision issue of Jones and Andersen's approach using a model checking technique on Pattern Matching Recursion Schemes [28] (PMRS). This technique improves the precision but is still not able to verify functions such as Example 1 (see [32] page 85). As shown in our experiments, our technique handles this example.

Kobayashi *et al.* developed a tree automata-based technique [26] (but not relying on TRS and completion), able to verify regular properties (including safety properties on Example 1). We have verified a selection of examples coming from [26] and observed that we can verify the same regular properties as they can.

Our prototype implementation is inferior in terms of execution time, due to the slow generation of equations. A strength of our approach is that our verification results are certifiable and that they can be used as certificates to build unreachability proofs in proof assistants (see Sect. 6).

Our verification framework is based on regular abstractions and uses a simple abstraction mechanism based on equations. Regular abstractions are less expressive than Higher-Order Recursion Schemes [23,29] or Collapsible Pushdown Automata [7], and equation-based abstractions are a particular case of predicate abstraction [24]. However, the two restrictions imposed in this particular framework result in two strong benefits. First, the precision of the approximation is formally defined and precisely controlled using equations: $\mathcal{L}(\mathcal{A}^*) \subseteq (\mathcal{R}/E)^*(\mathcal{L}(\mathcal{A}))$ [20]. This precision property permits us to prove intricate properties with simple (regular) abstractions. Second, using tree automata-based models facilitates the certification of the verification results in a proof assistant. This significantly increases the confidence in the verification result compared *e.g.*, to verdicts obtained by complex CEGAR-based model-checkers.

8 Conclusion and Future Work

This paper shows that tree automata completion is a simple yet powerful, fully automatic verification technique for higher-order functional programs, expressed as term rewriting systems. We have proved that the completion algorithm terminates on a subset of TRS encompassing common functional programs, and provided experimental evidence of the viability of the approach by verifying properties on fundamental higher-order functions including filtering and sorting.

One remaining question is whether this approach is complete: if there exists a regular approximation of the reachable terms of a functional program, can we build it using equations? We can already answered this question in the positive when $\mathcal{L} = \mathcal{W}(\mathcal{C})$, i.e., all results are first order terms [15]. Extending this result to all kind of results, including higher-order ones, is a promising research topic.

The generation of the approximating equations is automatic but simple-minded, and too simple to turn the prototype into a full verification tool. Further work will look into how sets of contracting equations can be generated in a more efficient manner, notably by taking the structure of the TRS into account and using a CEGAR approach.

The present verification technique is agnostic to the evaluation strategy. An interesting research track would be to experiment completion-based verification techniques with different term rewriting semantics of functional programs such as outlined by Clemente *et al.* [10]. This would permit us to take a particular evaluation strategy into account, and in certain cases, improve the precision of the verification. We already experimented with this in [18]. This is in line with our long-term research goal of providing a light-weight verification tool to assist the working OCaml programmer.

Our work focuses on verifying regular properties represented by tree automata. Dealing with non-regular over-approximations of reachable terms would allow us to verify relational properties like comparing the length of the list before and after *filter*. This is one of the objective of techniques like [24]. Building non-regular over-approximations of reachable terms for TRS, using a form of completion, is possible [5]. However, up to now, adapting automatically the precision of such approximations to a given verification goal is not possible. Extending their approach with equations may provide a powerful verification tool worth pursuing.

References

1. Baader, F., Nipkow, T.: Term Rewriting and All That. Cambridge University Press, New York (1998)
2. Ball, T., Rajamani, S.K.: The SLAM project: debugging system software via static analysis. In: Conference Record of POPL 2002: The 29th SIGPLAN-SIGACT Symposium on Principles of Programming Languages, Portland, OR, USA, 16–18 January 2002, pp. 1–3 (2002)
3. Blanchet, B., Cousot, P., Cousot, R., Feret, J., Mauborgne, L., Miné, A., Monniaux, D., Rival, X.: A static analyzer for large safety-critical software. In: Proceedings of the ACM SIGPLAN 2003 Conference on Programming Language Design and Implementation 2003, San Diego, California, USA, 9–11 June 2003, pp. 196–207 (2003)
4. Boichut, Y., Genet, T., Jensen, T., Le Roux, L.: Rewriting approximations for fast prototyping of static analyzers. In: Baader, F. (ed.) RTA 2007. LNCS, vol. 4533, pp. 48–62. Springer, Heidelberg (2007). https://doi.org/10.1007/978-3-540-73449-9_6
5. Boichut, Y., Chabin, J., Réty, P.: Towards more precise rewriting approximations. In: Dediu, A.-H., Formenti, E., Martín-Vide, C., Truthe, B. (eds.) LATA 2015. LNCS, vol. 8977, pp. 652–663. Springer, Cham (2015). https://doi.org/10.1007/978-3-319-15579-1_51
6. Boyer, B., Genet, T., Jensen, T.: Certifying a tree automata completion checker. In: Armando, A., Baumgartner, P., Dowek, G. (eds.) IJCAR 2008. LNCS (LNAI), vol. 5195, pp. 523–538. Springer, Heidelberg (2008). https://doi.org/10.1007/978-3-540-71070-7_43
7. Broadbent, C.H., Carayol, A., Hague, M., Serre, O.: C-shore: a collapsible approach to higher-order verification. In: ICFP 2013. ACM (2013)
8. Castagna, G., Nguyen, K., Xu, Z., Abate, P.: Polymorphic functions with set-theoretic types: part 2: local type inference and type reconstruction. In: POPL 2015. ACM (2015)
9. Castagna, G., Nguyen, K., Xu, Z., Im, H., Lenglet, S., Padovani, L.: Polymorphic functions with set-theoretic types: part 1: syntax, semantics, and evaluation. In: POPL 2014. ACM (2014)
10. Clemente, L., Parys, P., Salvati, S., Walukiewicz, I.: Ordered tree-pushdown systems. In: FSTTCS 2015. LIPIcs, Schloss Dagstuhl - Leibniz-Zentrum fuer Informatik, vol. 45, pp. 163–177 (2015)
11. Comon, H.: Sequentiality, monadic second-order logic and tree automata. Inf. Comput. **157**(1–2), 25–51 (2000)
12. Comon, H., Dauchet, M., Gilleron, R., Jacquemard, F., Lugiez, D., Löding, C., Tison, S., Tommasi, M.: Tree Automata Techniques and Applications (2008). http://tata.gforge.inria.fr

13. Coq: The Coq proof assistant reference manual: version 8.6 (2016)
14. Felgenhauer, B., Thiemann, R.: Reachability, confluence, and termination analysis with state-compatible automata. Inf. Comput. **253**, 467–483 (2017)
15. Genet, T.: Automata completion and regularity preservation. Technical report, Inria (2017). https://hal.inria.fr/hal-01501744
16. Genet, T., Boichut, Y., Boyer, B., Murat, V., Salmon, Y.: Reachability Analysis and Tree Automata Calculations. IRISA/Université de Rennes 1. http://people.irisa.fr/Thomas.Genet/timbuk/
17. Genet, T., Haudebourg, T., Jensen, T.: Verifying Higher-Order Functional Programs With Tree Automata: Extended Version (2017). https://hal.inria.fr/hal-01614380
18. Genet, T., Salmon, Y.: Reachability analysis of innermost rewriting – extended version. Log. Methods Comput. Sci. **13**(1), 1–35 (2017)
19. Genet, T.: Termination criteria for tree automata completion. J. Log. Algebr. Methods Program. **85**(1), 3–33 (2016)
20. Genet, T., Rusu, V.: Equational approximations for tree automata completion. J. Symb. Comput. **45**(5), 574–597 (2010)
21. Jones, N.D.: Flow analysis of lazy higher-order functional programs. In: Abramsky, S., Hankin, C. (eds.) Abstract Interpretation of Declarative Languages, pp. 103–122. Ellis Horwood, Chichester (1987)
22. Jones, N.D., Andersen, N.: Flow analysis of lazy higher-order functional programs. Theor. Comput. Sci. **375**(1–3), 120–136 (2007)
23. Kobayashi, N.: Types and higher-order recursion schemes for verification of higher-order programs. In: Proceedings of the 36th ACM SIGPLAN-SIGACT Symposium on Principles of Programming Languages, POPL 2009, Savannah, GA, USA, 21–23 January 2009, pp. 416–428 (2009)
24. Kobayashi, N., Sato, R., Unno, H.: Predicate abstraction and CEGAR for higher-order model checking. In: Proceedings of the 32nd ACM SIGPLAN Conference on Programming Language Design and Implementation, PLDI 2011, San Jose, CA, USA, 4–8 June 2011, pp. 222–233 (2011)
25. Kounalis, E.: Testing for the ground (co-)reducibility property in term-rewriting systems. Theor. Comput. Sci. **106**(1), 87–117 (1992)
26. Matsumoto, Y., Kobayashi, N., Unno, H.: Automata-based abstraction for automated verification of higher-order tree-processing programs. In: Feng, X., Park, S. (eds.) APLAS 2015. LNCS, vol. 9458, pp. 295–312. Springer, Cham (2015). https://doi.org/10.1007/978-3-319-26529-2_16
27. Meseguer, J., Palomino, M., Martí-Oliet, N.: Equational abstractions. TCS **403**(2–3), 239–264 (2008)
28. Ong, C.L., Ramsay, S.J.: Verifying higher-order functional programs with pattern-matching algebraic data types. In: Proceedings of the 38th ACM SIGPLAN-SIGACT Symposium on Principles of Programming Languages, POPL 2011, Austin, TX, USA, 26–28 January 2011, pp. 587–598 (2011)
29. Ong, C.H.: On model-checking trees generated by higher-order recursion schemes. In: 21st Annual IEEE Symposium on Logic in Computer Science (LICS 2006), pp. 81–90. IEEE (2006)
30. Paulson, L.C., et al.: The Isabelle reference manual. Technical report, University of Cambridge, Computer Laboratory (1993)
31. Rondon, P.M., Kawaguchi, M., Jhala, R.: Liquid types. In: Proceedings of the ACM SIGPLAN 2008 Conference on Programming Language Design and Implementation, Tucson, AZ, USA, 7–13 June 2008, pp. 159–169 (2008)

32. Salmon, Y.: Analyse d'atteignabilité pour les programmes fonctionnels avec stratégie d'évaluation en profondeur. Ph.D. thesis, University of Rennes 1 (2015)
33. Vazou, N., Bakst, A., Jhala, R.: Bounded refinement types. In: Proceedings of the 20th ACM SIGPLAN International Conference on Functional Programming, ICFP 2015, Vancouver, BC, Canada, 1–3 September 2015, pp. 48–61 (2015)
34. Vazou, N., Rondon, P.M., Jhala, R.: Abstract refinement types. In: Felleisen, M., Gardner, P. (eds.) ESOP 2013. LNCS, vol. 7792, pp. 209–228. Springer, Heidelberg (2013). https://doi.org/10.1007/978-3-642-37036-6_13

Author Index

Aceto, Luca 203
Achilleos, Antonis 203
Ahmed, Amal 146
Altenkirch, Thorsten 293

Baldan, Paolo 165
Bansal, Suguman 420
Bazille, Hugo 403
Bouyer, Patricia 530

Capriotti, Paolo 293
Castellan, Simon 3
Chaudhuri, Swarat 420
Clairambault, Pierre 3
Clouston, Ranald 258

D'Argenio, Pedro R. 384
Dardha, Ornela 91
Demri, Stéphane 476
Devesas Campos, Marco 71
Dijkstra, Gabe 293
Docherty, Simon 441

Edalat, Abbas 459

Fabre, Eric 403
Francalanza, Adrian 203

Gay, Simon J. 91
Genest, Blaise 403
Genet, Thomas 565
Gerhold, Marcus 384
Goncharov, Sergey 313

Hartmanns, Arnd 384
Haudebourg, Timothée 565
Hayman, Jonathan 3
Herbelin, Hugo 276

Ingólfsdóttir, Anna 203

Jaber, Guilhem 20
Jacq, Clément 39
Jensen, Thomas 565

Katsumata, Shin-ya 110
Kesner, Delia 241
Kraus, Nicolai 293

Lasota, Sławomir 548
Le Roux, Stéphane 367
Levy, Paul Blain 71
Liu, Xinxin 221
Lozes, Étienne 476

Maleki, Mehrdad 459
Mansutti, Alessio 476
Melliès, Paul-André 39
Miquey, Étienne 276

New, Max 146
Nordvall Forsberg, Fredrik 293

Padoan, Tommaso 165
Pérez, Guillermo A. 367
Piórkowski, Radosław 548
Pym, David 441

Rabusseau, Guillaume 513
Ríos, Alejandro 241
Rioux, Nick 146

Sabry, Amr 348
Santocanale, Luigi 494
Scherer, Gabriel 146
Schröder, Lutz 313
Sedwards, Sean 384
Sokolova, Ana 331

Toninho, Bernardo 128
Tzevelekos, Nikos 20

Valiron, Benoît 348
van Glabbeek, Rob 183
Vardi, Moshe Y. 420
Viso, Andrés 241
Vizzotto, Juliana Kaizer 348

Winskel, Glynn 3
Woracek, Harald 331

Yoshida, Nobuko 128
Yu, Tingting 221

Zhang, Wenhui 221

Printed in the United States
By Bookmasters